Climate & Oceans

Editor-in-Chief

John H. Steele

Marine Policy Center, Woods Hole Oceanographic Institu~~...~~, ~~woods~~ Hole,
Massachusetts, USA

Editors

Steve A. Thorpe

National Oceanography Centre, University of Southampton,
Southampton, UK
School of Ocean Sciences, Bangor University, Menai Bridge, Anglesey, UK

Karl K. Turekian

Yale University, Department of Geology and Geophysics, New Haven,
Connecticut, USA

Subject Area Volumes from the Second Edition

Climate & Oceans edited by Karl K. Turekian
Elements of Physical Oceanography edited by Steve A. Thorpe
Marine Biology edited by John H. Steele
Marine Chemistry & Geochemistry edited by Karl K. Turekian
Marine Ecological Processes edited by John H. Steele
Marine Geology & Geophysics edited by Karl K. Turekian
Marine Policy & Economics guest edited by Porter Hoagland, Marine Policy Center,
Woods Hole Oceanographic Institution, Woods Hole, Massachusetts
Measurement Techniques, Sensors & Platforms edited by Steve A. Thorpe
Ocean Currents edited by Steve A. Thorpe
The Coastal Ocean edited by Karl K. Turekian
The Upper Ocean edited by Steve A. Thorpe

CLIMATE & OCEANS

A DERIVATIVE OF ENCYCLOPEDIA OF OCEAN SCIENCES, 2ND EDITION

Editor

KARL K. TUREKIAN

Amsterdam • Boston • Heidelberg • London • New York • Oxford
Paris • San Diego • San Francisco • Singapore • Sydney • Tokyo
Academic Press is an imprint of Elsevier

ACADEMIC
PRESS

Academic Press is an imprint of Elsevier
32 Jamestown Road, London NW1 7BY, UK
30 Corporate Drive, Suite 400, Burlington, MA 01803, USA
525 B Street, Suite 1900, San Diego, CA 92101-4495, USA

British Library Cataloguing in Publication Data
A catalogue record for this book is available from the British Library

Library of Congress Control Number: 2009907115

ISBN: 978-0-08-096482-9

For information on all Elsevier publications
visit our website at www.elsevierdirect.com

PRINTED AND BOUND IN ITALY
10 11 12 13 14 10 9 8 7 6 5 4 3 2 1

CONTENTS

Climate and Oceans: Introduction   ix

OCEAN CIRCULATION & CLIMATE

Ocean Circulation   *N C Wells*   3

Science of Ocean Climate Models   *S M Griffies*   14

General Circulation Models   *G R Ierley*   22

Ocean Circulation: Meridional Overturning Circulation   *J R Toggweiler*   27

North Atlantic Oscillation (NAO)   *J W Hurrell*   33

Abrupt Climate Change   *S Rahmstorf*   41

TROPICAL CLIMATE CONTROLS

Pacific Ocean Equatorial Currents   *R Lukas*   49

El Niño Southern Oscillation (ENSO)   *K E Trenberth*   57

El Niño Southern Oscillation (ENSO) Models   *S G Philander*   70

Monsoons, History of   *N Niitsuma, P D Naidu*   76

HEAT & WATER BALANCE

Freshwater Transport and Climate   *S Wijffels*   87

Heat and Momentum Fluxes at the Sea Surface   *P K Taylor*   95

Heat Transport and Climate   *H L Bryden*   103

Sea Surface Exchanges of Momentum, Heat, and Fresh Water Determined by Satellite   *L Yu*   110

Evaporation and Humidity   *K Katsaros*   120

Upper Ocean Heat and Freshwater Budgets   *P J Minnett*   127

CRYOSPHERE

Antarctic Circumpolar Current   *S R Rintoul*   141

Arctic Ocean Circulation   *B Rudels*   154

Sub Ice-shelf Circulation and Processes   *K W Nicholls*   169

Ice–Ocean Interaction   *J H Morison, M G McPhee*   179

Coupled Sea Ice–Ocean Models   *A Beckmann, G Birnbaum*   190

Ice Shelf Stability   *C S M Doake*   201

Noble Gases and the Cryosphere   *M Hood*   210

Sea Ice: Overview   *W F Weeks*   214

Sea Ice Dynamics   *M Leppäranta*   223

Sea Ice   *P Wadhams*   234

Sub-Sea Permafrost   *T E Osterkamp*   252

SATELLITE MEASUREMENTS

Satellite Oceanography, History, and Introductory Concepts   *W S Wilson, E J Lindstrom,*
   J R Apel   265

Satellite Passive-Microwave Measurements of Sea Ice   *C L Parkinson*   280

Satellite Remote Sensing of Sea Surface Temperatures   *P J Minnett*   291

PALEOCLIMATE MARINE RECORD

Millennial-Scale Climate Variability   *J T Andrews*   305

Oxygen Isotopes in the Ocean   *K K Turekian*   314

Cenozoic Climate – Oxygen Isotope Evidence   *J D Wright*   316

Determination of Past Sea Surface Temperatures   *M Kucera*   328

Plio-Pleistocene Glacial Cycles and Milankovitch Variability   *K H Nisancioglu*   344

Paleoceanography   *E Thomas*   354

Paleoceanography, Climate Models in   *W W Hay*   362

Paleoceanography: Orbitally Tuned Timescales   *T D Herbert*   370

Paleoceanography: the Greenhouse World   *M Huber, E Thomas*   378

Holocene Climate Variability   *M Maslin, C Stickley, V Ettwein*   389

Plankton and Climate   *A J Richardson*   397

Coccolithophores   *T Tyrrell, J R Young*   407

Benthic Organisms Overview   *P F Kingston*   416

Benthic Foraminifera   *A J Gooday*   425

Past Climate from Corals   *A G Grottoli*   437

CARBON SYSTEM

Carbon Cycle   *C A Carlson, N R Bates, D A Hansell, D K Steinberg*   449

Carbon Dioxide (CO$_2$) Cycle   *T Takahashi*   459

Ocean Carbon System, Modeling of   *S C Doney, D M Glover*   467

Radiocarbon   *R M Key*   477

Stable Carbon Isotope Variations in the Ocean   *K K Turekian*   493

Cenozoic Oceans – Carbon Cycle Models   *L François, Y Goddéris*   494

EFFECTS & REMEDIATION

Sea Level Change   *J A Church, J M Gregory*   507

Sea Level Variations over Geological Time   *M A Kominz*   513

Glacial Crustal Rebound, Sea Levels, and Shorelines   *K Lambeck*   522

Economics of Sea Level Rise   *R S J Tol*   532

Land–Sea Global Transfers   *F T Mackenzie, L M Ver*   536

Effects of Climate Change on Marine Mammals   *I Boyd, N Hanson*   546

Fisheries and Climate   *K M Brander*   550

Seabird Responses to Climate Change   *David G Ainley, G J Divoky*   558

Carbon Sequestration via Direct Injection into the Ocean   *E E Adams, K Caldeira*   566

Iron Fertilization   *K H Coale*   573

Methane Hydrates and Climatic Effects   *B U Haq*   585

APPENDICES

Appendix 1: SI Units and Some Equivalences   593

Appendix 2: Useful Values   596

Appendix 3: Periodic Table of the Elements   597

Appendix 4: The Geologic Time Scale   598

Appendix 5: Properties of Sea Water   599

Appendix 6: The Beaufort Wind Scale and Seastate   604

INDEX

607

CLIMATE AND OCEANS: INTRODUCTION

The role of the oceans in climate locally has been known for a long time. The warming and cooling of land masses close to the oceans is controlled in part by the ocean temperature since the high heat capacity of water dampens the seasonal changes in heating by sunlight received by adjacent continental areas as a function of the seasons. In lower latitudes the oceans dominate the air temperature of islands giving rise to the description of the climate on these islands as "maritime climates."

These effects globally are more extensive than these two elementary examples. Ocean currents and atmospheric circulation over the oceans cause a variety of interactions such as the ENSO (El Niño Southern Oscillation) process and Gulf Stream transport of heat northward in the Atlantic Ocean–atmosphere interactions are thus an important part of the local as well as the global climate regime as a function of seasons and location.

The role of the cryosphere in influencing climate and in turn being influenced by climate is an important part of the ocean-climate relationship, especially as we seek to understand the role of potentially increasing global warming as a result of the enhancement, by human actions, of radiatively important gases, such as carbon dioxide, and aerosols.

The oceans play an important role in deciphering the history of climate change over the past several million years, during the ice ages, as well as over the past hundreds of million years. The source of our information comes from deep sea cores. There the changing ocean proxies for climate changes are preserved in the sedimentary components. The study of the distribution of the remains of marine tests provide not only evidence of ecological changes with time including temperature, but marine tests also provide the material for detailed studies of oxygen isotope variations, sensitive indicators of temperature and continental ice storage typical of the ice ages. The chemistry of these tests and other recorders of past ocean environments, such as coral reefs, provide many other proxies including element ratios and isotopic indicators of environment. Properly dated cores allow the calibration of patterns of variation as a function of time and associated climate indicators. The role of orbital forcing in controlling climate has been indicated by detailed studies of hundreds of cores raised by standard piston coring techniques and more recently by ocean drilling.

The coupling of the knowledge of contemporary ocean-climate relationships to the ancient record allows for a reconstruction of ancient locations of critical currents. In addition the influence of land mass and elevation on controlling ancient ocean-climate relationships has been inferred from these records.

Karl K. Turekian
Editor

OCEAN CIRCULATION & CLIMATE

OCEAN CIRCULATION

N. C. Wells, Southampton Oceanography Centre, Southampton, UK

Introduction

This article discusses the following aspects of ocean circulation: what is meant by the term ocean circulation; how the ocean circulation is determined by measurements and dynamical processes; the consequences of this circulation on the Earth's climate.

What is Ocean Circulation?

The ocean circulation in its simplest form is the movement of sea water through the ocean, which principally transfers temperature and salinity, from one region to another. Temperature differences between regions cause heat transfers. Similarly, differences in salinity produce transfers of salt. On the time scale of the ocean circulation the inputs and exports of salt into and out of the ocean make a negligible contribution to overall salinity and so variations in salinity occur by the addition and removal of fresh water into and out of the ocean.

Two major processes control the ocean circulation: the action of the wind and the action of small density differences, produced by differences in temperature and salinity, within the ocean. The former process is the wind driven circulation whereas the latter is the thermohaline circulation. Although it is useful to separate these two processes to better understand the ocean circulation, they are not independent from each other.

Ocean circulation is in reality a very complex system, as the flows are not steady in time or space. They are turbulent flows that show variability on scales from the largest scale of the ocean basins (*see* El Niño Southern Oscillation (ENSO), North Atlantic Oscillation (NAO)) to the smallest scales where the energy is finally dissipated as heat. This turbulent structure of the ocean means there are fundamental limitations on the predictability of its behavior.

Because of this inherent complexity oceanographers have approached ocean circulation by using a combination of observational methods including ships, buoys Moorings and satellites, combined with the mathematical methods of dynamical oceanography (*see* General Circulation Models). This integrated approach allows hypotheses to be made that can be tested by comparison with observations. Furthermore, mathematical models of the ocean circulation, based on the dynamical principles, can be constructed and tested against observations (*see* El Niño Southern Oscillation (ENSO) Models).

This article considers how ocean circulation is measured, how the major processes at work are determined and the consequences of the ocean circulation on the climate system.

How is the Ocean Circulation Determined?

The determination of ocean currents involves measurement of the displacement of an element of fluid over a measured time interval. The position of the measurement is defined mathematically in a Cartesian coordinate system (**Figure 1**) where x is positive eastward direction (lines of constant latitude), y is positive northward direction (geographic North Pole), and z is positive upwards; $z = 0$ corresponds to mean sea level. Without ocean currents and tides the sea level would be an equipotential surface, i.e., one of constant potential energy. The z coordinate is perpendicular to the equipotential surface. The origin is the intersection of the Greenwich meridian (Universal meridian) and the equator with mean sea level.

The coordinates of a parcel of water can be determined by the Global Positioning System (GPS). This satellite-based system provides a horizontal position with an accuracy of better than 100 m, which is sufficient for large-scale flows in the ocean. Large-scale flows are at least 10 km in spatial scale and have timescales of at least a day. A pressure device attached to the current meter normally determines the vertical position.

There are two mathematical methods for defining the displacement of the fluid. One is to measure the velocity of the fluid at a fixed point in the ocean, and the other is to follow the element of the fluid and to measure its velocity as it moves through the ocean. The first method is known as a Eulerian description and the second is a Lagrangian description of flow. In principle, the two methods are independent of each other. This means that a Eulerian measurement can not provide Lagrangian currents, and Lagrangian measurements can not provide Eulerian currents.

Eulerian

$$v(x, y, z, t)$$

Measurement of velocity of the fluid at fixed point (x, y, z)

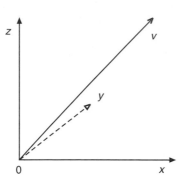

For both diagrams –

x = eastward direction
y = northward direction
(A) z = vertical upward

Lagrangian

$$v(a, t)$$

Measurement of the velocity of a fluid element. a is the position vector from the origin to the fluid element.

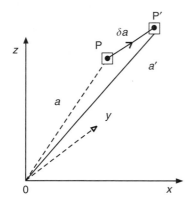

Element of fluid moves a small distance δa in a time δt, from P to P′.

The velocity is the $\dfrac{\delta a}{\delta t}$

In infinitesimal time $\dfrac{\delta a}{\delta t} = v(a, t)$

(B)

Figure 1 Eulerian and Lagrangian specifications of flow.

Having defined the two mathematical methods how the currents are measured in practice is now considered. Initially, these methods will address only the measurement of the horizontal flow. The vertical

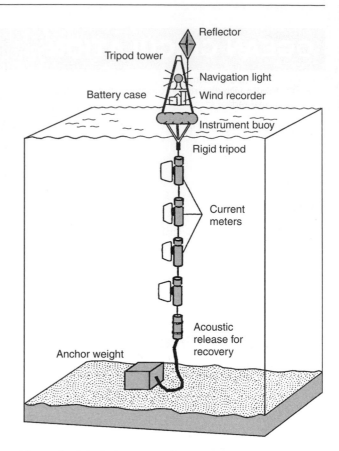

Figure 2 A typical current meter mooring.

flow is difficult to measure directly, and will be discussed later in this article.

First, the Eulerian method is considered. The measurement of the flow at a fixed point in the ocean is only straightforward when a fixed position can be maintained, for instance with a current meter attached to the bottom of the ocean or to a pier on the coast. Most measurements have to be made well away from land. This is achieved by attaching the current meter to a mooring which is attached to weights and then deployed (**Figure 2**). The position of a mooring can be determined by GPS. The current meter may be a rotary device or an acoustic device. The rotary current meter measures the number of revolutions over a fixed period, whereas the acoustic one measures the change in frequency of an emitted sound pulse caused by the ocean current (i.e., it uses the Doppler effect). Moorings may be deployed for periods of up to 2 years. In the analysis of the record it is normal to remove the high frequency variability of less than 1 day caused by tides by filtering the data.

A Lagrangian measurement of current can be determined by following an element of water with a float. The horizontal displacement of the water over

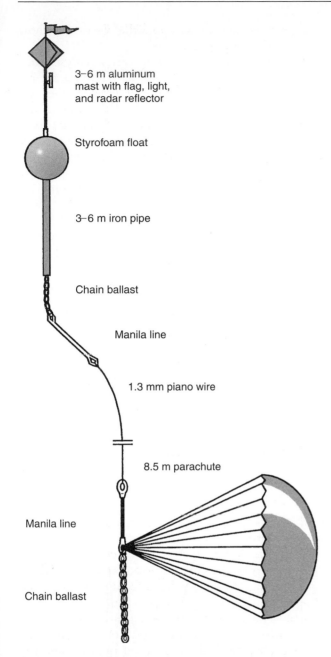

3–6 m aluminum
mast with flag, light,
and radar reflector

Styrofoam float

3–6 m iron pipe

Chain ballast

Manila line

1.3 mm piano wire

8.5 m parachute

Manila line

Chain ballast

Figure 3 A typical drifter with a parachute drogue of a few meters below the surface. It will follow the current at the depth of the parachute.

a small interval of time defines the Lagrangian current. **Figure 3** shows typical float designs that are used. The position of the float can be determined by two methods. A float that has a surface satellite transmitter/receiver can have its position determined by GPS, whereas a subsurface float would use an acoustic navigation system. Some floats can descend to a predetermined depth, maintain that depth for a few weeks and then return to the surface for a position fix. This technique allows the current to be measured down to depths of 1 km below the surface.

Each method gives different information on the flow field. A mooring will give a time series of the horizontal current, whereas a float will give the trajectory of the horizontal displacement of the parcel. It is worth remarking that most of the information on the surface ocean circulation has come from mariners' observations of the ships set, a method which has been used since the nineteenth century. However, these measurements have their limitations since they are neither eulerian nor lagrangian measurements and additional influences (e.g., wind effects on the ship) may cause errors.

This information can be analyzed in many different ways to discern the major current systems. From a set of moorings deployed for a few years across, say, the Gulf Stream, the mean flow (i.e., the average of all the current measurements) and its variability can be determined. The mean flow could be calculated over a particular time period. This time period is limited by the period of deployment, which is of the order of 2 years. This is rather short for a climatological mean, and a much longer period of 10 years is desirable. A few longer time series of currents have been determined for the Gulf Stream in the Florida Straits and for the Antarctic Circumpolar Current in the Drake Passage (*see* Antarctic Circumpolar Current).

Recall ocean currents are turbulent and therefore have variability on a whole range of timescales. Hence the mean flow gives no information on the variability of the flow. However, the statistics of the flow can be calculated, based on the kinetic energy (KE). The kinetic energy/unit volume is defined as:

$$\text{KE} = \frac{1}{2}\rho\left[u^2 + v^2\right]$$

where ρ is the density of the sea water and u and v are the eastward and northward components of the horizontal flow, respectively.

If the time mean current is defined as $\bar{u}$ and u' as the deviation from $\bar{u}$ at any time, the mean kinetic energy (KEM) and eddy kinetic energy (EKE) can be defined by:

$$\text{KEM} = \frac{1}{2}\rho\left[\bar{u}^2 + \bar{v}^2\right]$$

$$\text{EKE} = \frac{1}{2}\rho\left[u'^2 + v'^2\right]$$

These two numbers give quantitative measures on the mean and variability of the flow respectively. The ratio EKE/KEM gives a measure of the relative variability of the flow. If the ratio is very much less than 1 then the flow is steady, whereas if the ratio is approximately equal to 1 then the flow is very variable.

Figure 4 shows the variability of the flow in the Agulhas Current, which is an intense and highly variable current off the coast of South Africa.

Although this ratio gives a measure of the variability of the current, it does not give any idea of the exact time or space scales over which the current is varying. For example, the current may show a slow

change from one season to another or it may show faster variation due to eddies.

To address this variation time series analysis, such as Fourier analysis, can be used to determine the KE of the flow for different time periods. Fourier analysis produces a spectrum of the KE, either in frequency for a time series, or in wave number for a spatial variation in flow. **Figure 5** shows the analysis of a time series into its component frequencies. If the current is varying on all timescales the spectrum would be flat, but if there was only one dominant period, it would peak at that one frequency. This particular analysis shows that the current is varying at the tidal frequency and the inertial frequency both at the high frequency end of the spectrum. The inertial frequency is given by $2\Omega \sin \phi$ where Ω is the rotation rate of the earth and ϕ is the latitude. At the lower frequency end, which corresponds to the ocean circulation frequency, there is a broad band of high kinetic energy. This band is due to eddies which

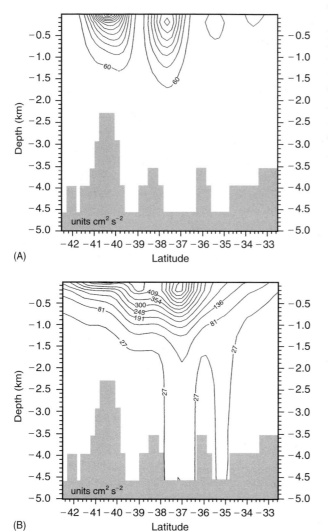

(A)

(B)

Figure 4 The mean kinetic energy (KEM) (A) and the eddy kinetic energy (EKE) (B) in a north–south slice through the Agulhas Current system at 14.4°E. The KEM maximum corresponds to the mean position of the Agulhas Return Current (Eastward flow) between 40° and 41°S, and the Agulhas Current (Westward flow) between 37°S and 38°S. The EKE distribution is much broader than KEM, which shows the large horizontal extent of the flow variability. The ratio of EKE/KEM is typically about a third, which indicates a very variable current system. (Reproduced from Wells NC, Ivchenko V and Best SE (2000) Instabilities in the Agulhas Retroflection Current system: A comparative model study. *Journal of Geophysical Research* 105: 3233–3246.)

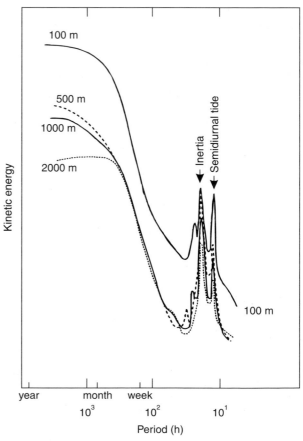

Figure 5 Frequency spectrum of kinetic energy from four depths at site D (39°N, 70°W), north of the Gulf Stream. Note the two high-frequency peaks, coinciding with the inertial period (19 h) and the semidiurnal tide (12.4 h). (Reproduced from Rhines PB (1971) A note on long-period motions at site D. *Deep Sea Research* 18: 21–26.)

cause fluctuations of currents on timescales of weeks to months.

For these mean climatological currents, our knowledge has been augmented by the application of the dynamic method. This method is based on the observation that large-scale ocean currents are in geostrophic balance, over large areas of the ocean. Geostrophic balance means that the Coriolis force balances the horizontal pressure gradient force. The geostrophic flow is a good approximation to the flow in the interior of an ocean outside the equatorial region. The horizontal pressure gradient is dependent on the slope of the ocean surface and the horizontal variation of the density distribution within the ocean. In the future, the former may be determined by satellite measurements of the sea surface height and the geoid[1] but at present we do not have an accurate geoid at sufficiently high resolution to measure the sea surface slope. The latter can be determined from temperature, salinity and pressure measurements that have been made over large ocean areas during the last century. The dynamic method allows the determination of the vertical shear of the horizontal geostrophic current, and therefore to determine the absolute geostrophic current, additional measurements are required. For example if the current has been measured at a particular depth then the dynamic method can be referenced to that depth and the vertical profile of current can be obtained.

The recent World Ocean Circulation Experiment (WOCE) hydrographic program has provided more measurements of the ocean than all previous hydrographic programs and will give the most comprehensive assessment of climatological horizontal ocean flow to date.

Recall that the vertical circulation of the ocean cannot be measured directly because it is technically too difficult. Current indirect methods used to determine the vertical circulation rely on the use of mathematical approaches, such as dynamical models, or the use of chemical tracers.

Observations of temperature and salinity can be inserted into a mathematical ocean general circulation model (see below) which allows the three-dimensional, circulation to be determined, subject to limitations in the accuracy of the model.

Chemical tracers have been inadvertently injected into the ocean from nuclear tests in the 1960s and from industrial processes (e.g., chlorofluorocarbons). Naturally occurring tracers such as ^{14}C also exist.

These tracers can be measured with high accuracy in a few laboratories around the world and from their distributions at different times, the three-dimensional circulation can be estimated. This method reveals the time history of the ocean circulation wherever the tracer is measured. This is very different information from that provided by the methods previously discussed, but nonetheless it can reveal unique aspects of the flow. For example, nuclear fallout deposited in the surface layer of the Nordic seas in the 1960s was located in the deep western boundary current 10 years later.

An Ocean General Circulation Model

An ocean general circulation model is composed of a set of mathematical equations which describe the time-dependent dynamical flows in an ocean basin. The basin is discretized into a set of boxes of regular horizontal dimensions but variable thickness in the vertical dimension. The horizontal flow (northward and eastward components) is predicted by the momentum equation (**Figure 6A**) at the corners of each box (**Figure 7**).

The forcing for the flow may come from the wind stress (the frictional term in the momentum equation) and from the surface buoyancy fluxes, arising from heat and freshwater (precipitation + runoff–evaporation) exchange with the atmosphere and adjacent landmasses. These buoyancy fluxes change the temperature and salinity in the surface layer of the ocean. The surface water masses are then subducted into the interior of the ocean by the vertical and horizontal components of the flow, where they are mixed with other water masses.

The processes of transport and mixing are described by the temperature and salinity equations (**Figure 6B and C**), at the center of each ocean box (**Figure 7**). From these two equations the seawater density, and thence the pressure can be obtained for each box. The horizontal pressure gradient is then determined for the momentum equation, and the vertical velocity is calculated from the horizontal divergence of the flow. This set of time-dependent equations can then be used to describe all the dynamical components of the flow field, provided that suitable initial and boundary conditions are specified.

Wind-driven and Thermohaline Circulation

The wind-driven circulation is considered first. The surface layer of the ocean is directly driven by the surface wind stress and is also subject to the

---

[1] The geoid is an equipotential surface, which would be represented by the sea level of a stationary ocean. Ocean currents cause deviations in sea level from the geoid.

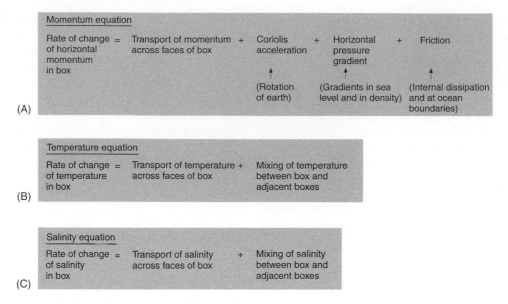

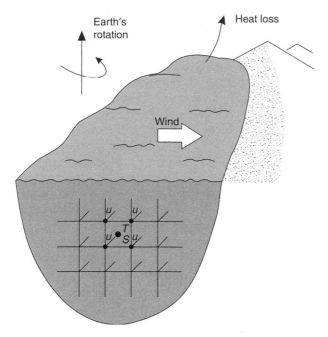

Figure 6   The basic equations for an ocean general circulation model. (A) Momentum equation; (B) temperature equation; (C) salinity equation. (Reproduced from Summerhayes and Thorpe, 1996.)

Figure 7   A schematic of the model boxes in an ocean general circulation model. The equations (**Figure 6**) for momentum are solved at the corners of the boxes (u), while the temperature (T), and salinity (S) equations are solved at the centers of boxes. The model is forced by climatological wind stress, surface heat fluxes, and freshwater fluxes. (Reproduced from Summerhayes and Thorpe, 1996.)

exchange of heat and fresh water between ocean and atmosphere. This layer, which is typically less than 100 m in depth, is referred to as the Ekman layer. That is a steady wind stress causes a transport of the surface water 90° to the right of the wind direction in the Northern Hemisphere and 90° to the left in the Southern Hemisphere. This is due to the combined action of the wind stress on the ocean surface and the Coriolis force. These Ekman flows can converge and produce a downwelling flow into the interior of the ocean. Conversely a divergent Ekman transport will produce an upwelling flow from the interior into the surface layer.

This type of flow is known as Ekman pumping, and is directly related to the Curl of the Wind Stress (*see*). It is of fundamental importance for the driving of the large-scale horizontal circulation, in the upper layer of the ocean. For example, between 30° and 50° latitude the climatological westerly wind, drives an Ekman flow equatorward, whereas between 15° and 30° latitude the trade winds drive an Ekman flow polewards. At about 30° latitude the flows converge and sink into the deeper ocean. Before discussing the influence of Ekman pumping on the interior ocean circulation the role of density is considered.

The density of sea water increases with depth. From hydrographic measurements of density, the horizontal variation of the depth of a chosen density surface can be mapped. These constant density surfaces are known as isopycnals. The flow tends to move along these surfaces and therefore the variations in the depth of these surfaces gives a picture of the horizontal flow in the deep ocean, away from the surface layer and benthic layer. The isopycnal surfaces dip down in the center of the subtropical gyre at about 30°. The formation of this lens of light warm water is related to the climatological distribution of surface winds, which produce a convergence of Ekman transport towards the center of the gyre, and a downwelling of surface waters into the interior of the ocean. At the center of the lens, the sea surface domes upwards reaching a height of 1 m above the sea surface at the rim. Due to hydrostatic forces the main thermocline is depressed downwards to depths of the order of 500–1000 m (**Figure 8**).

The surface horizontal circulation flows anticyclonically around the lens with the strongest currents on the western edge, where the slope of the density surface reaches a maximum. These are geostrophic currents, where there is a balance between the Coriolis force and the horizontal pressure gradient force. Generally, the circulation in the subtropical gyres is clockwise in the Northern Hemisphere and anticlockwise in the Southern Hemisphere. These large-scale horizontal gyres are ultimately caused by the climatological surface wind circulation and are found in all the ocean basins.

The surface layer is also subject to heating and cooling, and the exchange of fresh water between ocean and atmosphere, both of which will change the density of the layer. For example, heat is lost over the Gulf Stream on the rim of the light water lens of the subtropical gyre. Recall that flow tends to follow isopycnal layers and these layers will slope downwards towards the center of the gyre. Cooling of the waters in the Gulf Stream leads to the sinking of surface waters to produce a water mass known as 18°C water. This water, which is removed from the surface layer, will slowly move along the isopycnal layers into the thermocline. As it moves clockwise around the gyre it will be subducted in to the deeper layers of the thermocline, in a spiral-like motion (**Figure 9**). The deepest extent of the main thermocline is located in the subtropical gyre to the west of Bermuda on the eastern edge of the Gulf Stream rather than in the center of the ocean basin.

This asymmetry of the gyre is related to the beta effect, i.e., the change of the Coriolis parameter with latitude.

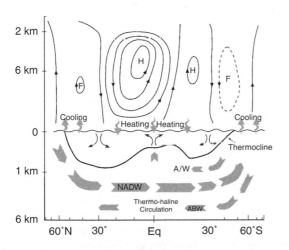

Figure 8 A representation of the meridional average section through the atmosphere for December–February in the Northern Hemisphere. The cells are the Hadley Cell (H) and Ferrel Cell (F). The strength of the cells is represented by the solid contours which are in units of 40 megatonnes/second, whilst the dashed contours are in units of 20 megatonnes/second. Note the predominantly downward motion at ∼30 degree latitude, associated with the subtropical anticyclones, and the strong upward motion at equatorial latitudes which is associated with the Inter-Tropical Convergence Zone. A meridional transect through the Atlantic Ocean, showing the position of the main thermocline. The small arrows represent the wind driven downwelling (Ekman pumping) at ∼30 degree latitude, and the equatorial upwelling, which occurs within and above the main thermocline. The North Atlantic Deep Water (NADW) is produced in the Labrador and Nordic Seas and is the predominant deep water mass by volume. The Antarctic Intermediate Water (AIW) is produced at ∼50°S, and by virtue of its salinity is lighter than the NADW. In contrast Antarctic Bottom Water is the most dense water mass in the worlds ocean and is formed in the Weddell and Ross Seas. These deep flows form part of the thermo-haline circulation. The vertical scales are exaggerated in the lower troposphere and in the upper ocean. The horizontal scale is proportional to the area of the Earth's surface between latitude circles.

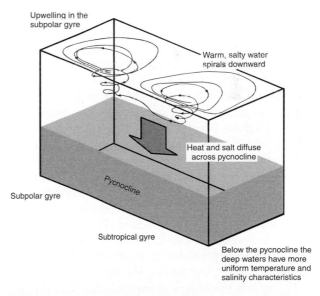

Figure 9 Schematic representation of the wind-driven circulation in the subpolar and subtropical gyre of an ocean basin. The wind circulation causes a convergence of Ekman transport to the center of the subtropical gyre and downwelling into the interior. Conversely in the subpolar gyre there is a divergence of the Ekman transport and upwelling from the interior into the surface layer. This Ekman pumping is responsible for the gyre circulations (see text for details). The western boundary currents are depicted by the closeness of the streamlines. They are caused by the poleward change in the Coriolis force known as the BETA effect. (Reproduced from Bean MS (1997) PhD thesis, University of Southampton.)

The subtropical gyres are one of the most well-studied regions of the ocean, and our understanding is therefore most developed in these regions. These gyres occur in the surface and thermocline regions of the ocean and are primarily controlled by the wind circulation, with modifications due to heating and cooling of the surface. The question now arises of why thermoclines are seen in the ocean. For example, why is the warm water not mixed over the whole depth of the ocean and why is the average ocean temperature about 3°C.

To explain the observed behavior thermohaline circulation, which is generated by small horizontal differences in density, due to temperature and salinity, between low and high latitude is considered. How does it work? Consider an ocean of uniform depth and bounded at the equator and at a polar latitude. We will assume it has initially a uniform temperature and is motionless (for the moment the effect of salinity on density are ignored). This hypothetical ocean is then subject to surface heating at low latitudes and surface cooling at high latitudes. In the lower latitudes the warming will spread downwards by diffusion, whereas in high latitudes the cooling will spread downwards by convection which is a much faster process than diffusion. The heavier colder water will induce a higher hydrostatic pressure at the ocean bottom than will occur at low latitudes. The horizontal pressure gradient at the ocean bottom is directed from the high latitudes to the lower latitudes, and will induce an equatorward abyssal flow of polar water. The flow can not move through the equatorial boundary of our hypothetical ocean and therefore will upwell into the upper layer of the tropical ocean, where it will warm by diffusion. The flow will then return polewards to the high latitudes where it will downwell into deepest layers of the ocean to complete the circuit. It is found that the downwelling occurs in narrow regions of the high latitudes whereas upwelling occurs over a very large area of the tropical ocean. This hypothetical ocean demonstrates the key role of the deep horizontal pressure gradient, caused by horizontal variations in density, for driving the flow.

To explain the observed thermohaline circulation, this hypothetical ocean has to be modified to take into account the Coriolis force, which causes the deep abyssal currents to flow in narrow western boundary currents, the effect of salinity on the density (the haline component of the flow), asymmetries in the buoyancy fluxes between the Northern and Southern Hemispheres and the complex bathemetry of the ocean basins.

There follows a descriptive account of the thermohaline circulation. The deepest water masses in the ocean have their origin in the polar seas. These seas experience strong cooling of the surface, particularly in the winter seasons. In the North Atlantic, there are connections through the Nordic seas to the Arctic Ocean, from which sea ice flows. Heat energy melts the ice in the North Atlantic and the melt water gives rise to further cooling. There are two effects on the density of the water: cooling increases the density whereas surface freshening, due to ice melt, decreases the density of the water. The former process usually dominates and hence denser waters are produced. These dense cold waters flow into the Atlantic through the East Greenland and West Greenland Currents and then into the Labrador Current. These cold waters mix and sink beneath the warm North Atlantic Current.

In addition to surface polar currents there are also deep ocean currents. The cold saline water entering from the Nordic seas mixes as it sinks to the abyssal layers of the ocean and moves southward as a deep current along the western boundary of the Atlantic. This water mass is known as NADW (North Atlantic Deep Water) and it is the most prominent and voluminous of all the deep water masses in the global ocean. It flows into the Antarctic Circumpolar Current from where it flows into the Indian and Pacific Ocean. In addition to NADW, colder denser water, Antarctic Bottom Water (AABW) enters the Southern Ocean from the Antarctic shelf seas. It is not as voluminous as NADW but it flows northwards in the deepest layers into the Atlantic, where it can be distinguished as far north as 30°N. These deep flows upwell into the thermocline and surface waters where they return to the North Atlantic. This global thermohaline circulation has been termed the global conveyor circulation to signify its role in transporting heat and fresh water (**Figures 10** and **11**) around the planet.

How does this circulation explain the thermocline? The rate at which these cold deep abyssal waters are produced can be estimated and it is known for a steady state in the ocean that production has to be balanced by removal. A large-scale upwelling of the abyssal waters into the thermocline produces this removal. Our simple conceptual picture is of warm thermocline water mixing downwards, balanced by a steady upwelling of the cold abyssal layers. Without the upwelling, the warm waters would mix into the deepest layers of the ocean.

The Role of Fresh Water in Ocean Circulation

The present discussion has shown that the wind-driven circulation and the thermohaline circulation

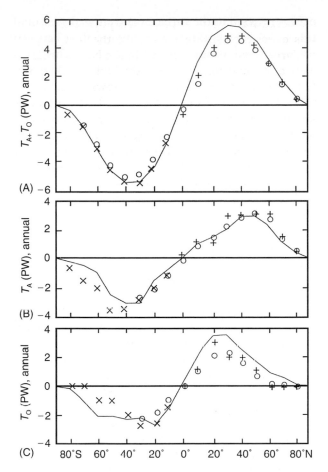

(A)

(B)

(C)   80°S   60°   40°   20°   0°   20°   40°   60°   80°N

Figure 10 Poleward transfer of heat by (A) ocean and atmosphere together ($T_A + T_O$), (B) atmosphere alone (T_A), and (C) ocean alone (T_O). The total heat transfer (A) is derived from satellite measurements at the top of the atmosphere, that of the atmosphere alone (B) is obtained from measurements of the atmosphere, and (C) is calculated as the difference between (A) and (B). (1 Petawatt (PW) = 10^{15} W.) Data compiled from three sources. (Reproduced from Carrissimo BC, Oort AH and Von der Harr TH (1985) Estimating the meridional energy transports in the atmosphere and ocean. *Journal of Physical Oceanography* 15: 52–91.)

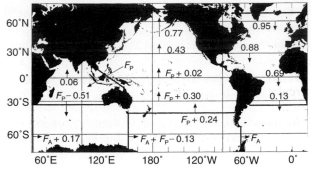

Figure 11 An estimate of the transfer of fresh water ($\times 10^9$ kg s^{-1}) in the world ocean. In polar and equatorial regions precipitation and river run-off exceed evaporation, and hence there is an excess of fresh water, whereas in subtropical regions there is a water deficit. A horizontal transfer of fresh water is therefore required between regions of surplus to regions of deficit. F_P and F_A refer to the freshwater fluxes of the Pacific–Indian throughflow and of the Antarctic Circumpolar Current in the Drake Passage, respectively. (Reproduced from Wijffels SE, Schmitt R, Bryden H and Stigebrandt A (1992) Transport of freshwater by the oceans. *Journal of Physical Oceanography* 22: 155–162.)

and cooling in winter, sink to the deepest layers of the basins. At the Straits of Gibraltar, this dense saline layer flows out beneath the incoming fresher and cooler Atlantic water. This Mediterranean water forms a distinct layer of high salinity water in the eastern Atlantic Ocean. Similar behavior occurs at Bab el Mandeb adjacent to the Gulf of Aden.

The influence of fresh water is more substantial in the polar oceans. A given amount of fresh water will have a greater effect on density at low temperatures than at high temperatures, because the thermal expansion of sea water decreases with decreasing temperature. At higher latitudes there is a net addition of fresh water into the oceans, which arises from the excess of precipitation over evaporation and the melting of sea ice moving towards the equator from the polar regions.

The addition of fresh water adds buoyancy to the surface layer while cooling removes buoyancy, therefore the fresh water will tend to reduce the effect of the cooling. In the Arctic Ocean the surface layer is colder but less dense than the warmer layer at ∼100 m and therefore is in equilibrium. . This stable halocline in the Arctic Ocean reduces the vertical heat flux in to the deep ocean.

In the subpolar oceans, the addition of fresh water reduces the density of the surface layer and can reduce the prevalence of deep convection. This happened in the late 1960s when fresh water, probably from excessive ice in the Arctic Basin, melted in the subpolar gyre. The effect on the thermohaline circulation is unknown, but it is believed from modeling studies that the decrease in the production of

are major components of ocean circulation, which are ultimately driven by the surface wind stress and buoyancy fluxes. Buoyancy fluxes are the net effect of heat exchange and the freshwater exchange with the overlying atmosphere. It has been shown that heat exchange is a major process explaining existence of both the thermocline and the deep abyssal water but what is the role of the fresh water in ocean circulation?

In the subtropics there is net removal of fresh water by evaporation. This increases the salinity of the water which, in turn, increases the density of the thermocline waters. Normally this effect is opposed by heating, which lightens the water. However, in the Mediterranean and the Red Sea evaporation produces salient waters, which by virtue of their salinity deep waters reduced the thermohaline circulation of the ocean.

What are the Consequences of this Circulation on the Climate System?

The effects of the ocean circulation on the climate can be understood in terms of the heat capacity of the ocean. The heat capacity of a column of sea water only 2.6 m deep is equivalent to that of a column of whole atmosphere and therefore the ocean heats and cools on a long timescale compared with the atmosphere.

It is known that there is a poleward gradient of temperature, which is driven by the thermal radiation imbalance between the low and high latitudes. In response to this temperature gradient there is a flow of heat from the warmer to cooler latitudes. Both the atmosphere and ocean circulations transfer this heat from low to high latitudes by a variety of mechanisms. In the low latitudes of the atmosphere there is the Hadley cell which transfers low temperature air in the lowest levels via the trade winds towards the equator and transfers warmer air poleward in the upper troposphere (**Figure 8**). At higher latitudes anticyclones and cyclones and their accompanying upper air jet streams transfer heat polewards. In the ocean, the wind-driven Ekman currents transfer heat as surface waters move across latitude circles. This water is returned deeper in the ocean at a different temperature from that of the surface water. The ocean gyres carry heat towards higher latitudes since the poleward flows of the western part of the gyres are warmer than the equatorward flows in the eastern parts of the gyre. Finally, and not least, is the contribution of the thermohaline circulation, which transports warm surface and thermocline waters to the highest latitudes and returns cold water to lower latitudes. **Figures 10** and **11** show the heat transport and fresh water transport in the ocean.

A major difference between the atmosphere and ocean is the relative speed of their circulation. The atmosphere circulation is a fast system, responding on timescales of days to weeks. For example, weather systems in temperate latitudes grow and decay on timescales of a few days. By contrast, the ocean tends to be slower in its response. The fastest part of the system are the surface Ekman layers which respond to changes in the surface wind circulation on a timescale of one or two days. Changes in wind circulation can cause planetary waves which will change sea level and surface temperature on monthly to seasonal timescales. In particular, the equatorial

oceans respond to the surface wind stress on seasonal timescales, which allows a strong coupling between the ocean and atmosphere to take place. This gives rise to phenomena such as the El Niño Southern Oscillation (*see* El Niño Southern Oscillation (ENSO)). The subtropical gyres respond to changes in the wind circulation on decadal timescales, whereas the deep thermohaline circulation respond on millenial timescales. There is some evidence for rapid changes of local parts of the thermohaline circulation on timescales 50 years.

Observations of the deep ocean are far fewer in number than at the ocean and land surface. The longest continuous data set is a deep station at Bermuda that commenced operations in 1954. Observations from cruises in the earlier part of the century are of unknown quality and therefore it is difficult to know whether differences are due to the use of different instruments or to real changes in the ocean. It is only since the 1950s that such changes have been accurately measured. **Figure 12** shows changes in the temperature for that period of time across the Atlantic. These changes are of the order of a few tenths

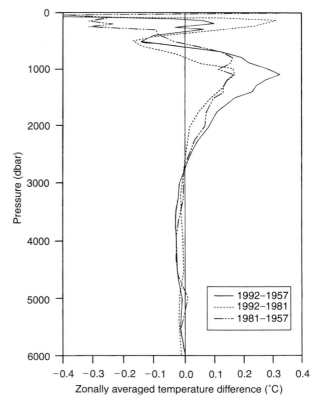

Figure 12 Temperature changes (°C) in the subtropical North Atlantic (24°N), 1957–1992. The measurements have been averaged across 24°N between North Africa and Florida. (Reproduced from Parrilla G, Lavin A, Bryden H, Garcia M and Millard R (1994) Rising temperatures in the subtropical North Atlantic. *Nature* 369: 48–51.)

of a degree over periods of 15 years. As the heat capacity of the oceans is very much larger than that of the atmosphere, these changes in temperature involve very significant changes in the heat content of the ocean. The World Ocean Circulation Experiment from 1990 to 1997 has provided measurements of ocean properties such as temperature, salinity, and chemical tracers as well as current measurements on a global scale. This set of high quality measurements will provide the baseline from which future changes in ocean circulation can be determined.

Despite the brief record of deep ocean observations, sea surface temperature measurements of reasonable quality go back to the late nineteenth century. These measurements can be used to assess changes in the surface layers. Salinity measurements are fewer and not as reliable but, nevertheless, changes can be still detected.

Salinity measurements in the Mediterranean over the last century have shown a warming of the Western Mediterranean Deep Water of 0.1°C and increase of 0.05 in salinity. The reasons for this change are not known, but it has been speculated that the change in salinity may be attributed to a reduction in the freshwater flow due to the damming of the Nile and of rivers flowing into the Black Sea.

An important recently identified question is the stability of the thermohaline circulation. The thermohaline circulation is driven by small density differences and therefore changes in the temperature and salinity arising from global warming may alter the thermohaline circulation. In particular, theoretical modeling of the ocean circulation has shown that the thermohaline circulation may be reduced or turned off completely when significant excess fresh water is added to the subpolar ocean. In the event of thermohaline circulation being significantly reduced or stopped, it may take many centuries before it returns to its present value. (see Abrupt Climate Change, North Atlantic Oscillation (NAO))

In view of the current levels of uncertainty, it is necessary to continue to monitor the ocean circulation, as this will provide the key to the understanding of the present circulation and enhance our ability to predict future changes in circulation.

See also

Abrupt Climate Change. Antarctic Circumpolar Current. El Niño Southern Oscillation (ENSO). Freshwater Transport and Climate. General Circulation Models. Heat Transport and Climate. North Atlantic Oscillation (NAO).

Further Reading

Gill AE (1982) *Atmosphere–Ocean Dynamics*. London: Academic Press.

Siedler G, Church J, and Gould (2001) *Ocean Circulation and Climate*. London: Academic Press.

Summerhayes CP and Thorpe SA (1996) *Oceanography – An Illustrated Guide*. London: Manson Publishing.

Wells NC (1997) *The Atmosphere and Ocean: A Physical Introduction*, 2nd edn. Chichester: John Wiley.

SCIENCE OF OCEAN CLIMATE MODELS

S. M. Griffies, NOAA/GFDL, Princeton, NJ, USA

Published by Elsevier Ltd.

Introduction: Models as Essential Tools for Ocean Science

The rich textures and features of the global ocean circulation and its tracer distributions cannot be fully studied in a controlled laboratory setting using a direct physical analog. Consequently, ocean scientists have increasingly turned to numerical models as a rational experimental tool, along with theoretical methods and a growing array of observational measurements, for rendering a mechanistic understanding of the ocean. Indeed, during the 1990s and early 2000s, global and regional ocean models have become the experimental tool of choice for many oceanographers and climate scientists.

The scientific integrity of computer simulations of the ocean has steadily improved during the past decades due to deepened understanding of the ocean and ocean models, and from enhanced computer power facilitating more realistic representations of the huge range of scales relevant for ocean fluid dynamics. Additionally, ocean models are a key component in earth system models (ESMs), which are used to study interactions between physical, chemical, and biological aspects of the Earth's climate system. ESMs are also used to help anticipate modifications to the Earth's climate arising from humanity's uncontrolled greenhouse experiment. Quite simply, without computer models, our ability to develop a robust and testable scientific basis for ocean and climate dynamics of the past, present, and future would be absent.

There is a tremendous amount of science forming the foundations of ocean models. This science spans a broad interdisciplinary spectrum, including physics, mathematics, chemistry, biology, computer science, climate science, and all aspects of oceanography. This article describes the fundamental principles forming the basis for physical ocean models. We are particularly interested here in global ocean climate models used to study large-scale and long-term phenomena of direct importance to climate.

Scales of Motion and the Subgrid-scale Problem

To appreciate the magnitude of the task required to simulate the ocean circulation, consider the scales of motion involved. We start at the large scale, where a typical ocean basin is on the order of 10^3–10^4 km in horizontal extent, with depths reaching on average to $c.$ 5 km. The ocean's massive horizontal gyre and overturning circulations occupy nearly the full extent of these basins, with typical recirculation times for the horizontal gyres decadal, and overturning timescales millennial.

At the opposite end of the spectrum, the ocean microscale is on the order of 10^{-3} m, which is the scale where molecular viscosity can act on velocity gradients to dissipate mechanical energy into heat. This length scale is known as the Kolmogorov length, and is given by $(v^3/\varepsilon)^{1/4}$, where $v = 10^{-6}$ m^2 s^{-1} is the molecular kinematic viscosity for water, and ε is the energy dissipation rate. In turn, molecular viscosity and the Kolmogorov length imply a timescale $T = L^2/v \approx 1$ s.

When formulating a computational physics problem, it is useful to estimate the number of discrete degrees of freedom required to represent the physical system. Consider a brute force approach, where all the space and timescales described above are explicitly resolved by the ocean simulation. One-second temporal resolution over a millennial timescale climate problem requires more than 3×10^{10} time steps of the model equations. Resolving space into regions of dimension 10^{-3} m for an ocean with volume roughly 1.3×10^{18} m^3 requires 1.3×10^{27} discrete grid cells (roughly 10^4 times larger than Avogadro's number!). These numbers far exceed the capacity of any computer in use today, or for the forseeable future. Consequently, a truncated description of the ocean state is required.

Truncation Methods

Three general approaches to truncation are employed in fluid dynamics. One approach is to coarsen the space time resolution. Doing so introduces a loss of information due to the unresolved small scales. Determining how the resolved scales are affected by the unresolved scales is fundamental to computational fluid dynamics, as well as to a statistical description of fluid turbulence. This is a nontrivial problem in subgrid-scale (SGS) parametrization, a problem intimately related to the turbulence closure problem of fluid dynamics.

The second truncation method filters the continuum equations by truncating the fundamental modes of motion admitted by the equations.

The result is an approximation to the original physical system. The advantage is that reducing the admitted motions also reduces the space timescales required to simulate the system, and/or it simplifies the governing equations thus facilitating computationally cheaper simulations. The hydrostatic approximation is a prime example of filtering used in all present-day global ocean climate models. Here, vertical motions are assumed to possess far less energy than horizontal, thus rendering a simplified vertical momentum balance where the weight of fluid above a point in the ocean determines the pressure at that point. The Boussinesq approximation provides another filtering of the fundamental equations. In this case, the near incompressibility of seawater is exploited to eliminate all acoustic motions by assuming that fluid parcels conserve volume rather than mass. The Boussinesq approximation is commonly used in ocean climate modeling, but it is becoming less so due to its inability to provide a prognostic budget for sea level that includes steric effects. Steric effects are associated with water expansion or contraction arising from density changes, and such changes are a key aspect of the ocean's response to anthropogenic climate change, with a warmer ocean occupying a larger volume which raises the sea level.

A final truncation method considers a much smaller space and time domain, yet maintains the very fine space and time resolution set by either molecular viscosity (direct numerical simulation (DNS)), or somewhat larger viscosity (large eddy simulation (LES)). Both DNS and LES are important for process studies aimed at understanding the mechanisms active in fine-scale features of the ocean. Insights gained via DNS and LES have direct application to the development of rational SGS parametrizations of use for ocean climate models.

Ocean Mesoscale Eddies

The Reynolds number UL/v provides a dimensionless measure of the importance of advective effects relative to viscous or frictional effects. Consider a large-scale ocean current, such as the Gulf Stream, with a velocity scale $U = 1\,\mathrm{m\,s}^{-1}$, length scale $L = 100\,\mathrm{km}$, and molecular viscosity $v = 10^{-6}\,\mathrm{m^2\,s}^{-1}$. In this case, the Reynolds number is $Re \approx 10^{11}$, which is very large and so means that the ocean fluid contains extremely turbulent regimes. At these space and timescales, the effects of rotation and stratification are both critical. The turbulence relevant at this scale is termed geostrophic turbulence. A geostrophically turbulent fluid contains numerous mesoscale eddy features, which result from a conversion of potential energy (imparted to the ocean fluid by the atmospheric forcing) to

kinetic energy through the process of baroclinic instability. Such eddy features are the norm rather than the exception in most of the ocean. They provide a chaotic or turbulent element to the ocean general circulation. Hence, quite generally the ocean flow is not smooth and laminar, unless averaging over many years. Rather, it is turbulent and full of chaotic fine-scale motions which make for an extremely challenging simulation task.

Mesoscale eddies have scales on the order of 100 km in the middle latitudes, 10 km in the higher latitudes, and their timescale for recirculation is on the order of months. The length scale is related to the first baroclinic Rossby radius, which is a scale that arises in the baroclinic instability process. Furthermore, mesoscale eddies are the ocean's analog of atmospheric weather patterns, with atmospheric weather occurring on scales roughly 10 times larger than the ocean eddies. Differences in vertical density stratification account for differences in length scales.

Ocean tracer properties are strongly affected by the mixing and stirring of mesoscale eddies. For example, biological activity is strongly influenced by eddies, especially with the strong upwelling in the cyclonic eddies bringing high nutrient water to the surface. Additionally, eddies are a leading order feature transporting properties such as heat and freshwater poleward across the Antarctic Circumpolar Current. They also mix properties across the time-mean position of the Gulf Stream and Kuroshio Currents, whose jet-like currents continually fluctuate due to baroclinic instability.

Given the smaller scales of mesoscale eddies in the ocean than in the atmosphere, the problem of simulating these ocean features is $10 \times 10 \times 10$ more costly than in the atmosphere. Here, two factors of 10 arise from the horizontal scales, and another from the associated refinement in temporal resolution. It is not rigorously known what grid resolution is required to resolve the ocean's mesoscale eddy spectrum. However, preliminary indications point to 10 km or finer globally, with roughly 50–100 vertical degrees of freedom needed to capture the vertical structure of the eddies, as well as the tight vertical gradients in tracer properties. This resolution is far coarser than that required to resolve the Kolmogorov scale (as required for DNS). Nonetheless, globally resolving the ocean mesoscale eddy spectrum over climate timescales remains beyond our means, with the most powerful computers only just beginning to be applied to such massive simulations.

Ocean climate modelers thus continually seek enhanced computer power in an aim to reduce the level of space time resolution coarsening. The belief, largely reflected in experience with simulations of refined resolution, is that reducing dependence on SGS parametrizations, such as those parametrizing mesoscale eddies, improves the simulation integrity. One reason for this improvement is that many SGS parametrizations are far from relevant for all of the multiple flow regimes of the ocean. Additionally, no SGS parametrization commonly used in ocean climate models includes the stochastic effects of turbulent eddies, and resolving these effects may be important for climate variability and predictability. Finally, refined resolution allows for a better representation of the very complex land–sea boundaries that strongly influence ocean currents and water mass properties. Hence, deducing the impact on ocean climate of explicitly resolved mesoscale eddies, and other fine-scale currents such as occur near boundaries, is the grand challenge of ocean climate modeling in the early twenty-first century.

As grid resolution is refined so that mesoscale eddy features are admitted, the resulting eddy permitting simulations become strongly dependent on the integrity of the numerical methods used for the transport of fluid properties. Although the representation of transport is important at coarse resolution, the fluid is also very viscous due to the huge viscosity required at the noneddying resolutions. Hence, a more energetic simulation, with enhanced energy in the resolved scale and much stronger fronts between ocean properties, generally stresses the ability of transport algorithms to maintain physically appropriate behavior. The integrity of the discrete transport operators is thus arguably the most critical feature of a numerical algorithm that sets the fidelity of eddying simulations.

Although mesoscale eddy permitting simulations are becoming more feasible as computer power increases, there remain important problems where SGS parametrizations must be used to partially capture features of the unresolved mesoscale spectrum, as well as smaller scales. The longer the timescale of the phenomena studied, the greater the need for SGS parametrization. In particular, no mesoscale eddy simulation has yet been run for more than a few decades. Climate modelers must therefore carefully employ truncation methods to make progress. Depending on particulars of the unresolved motions, the resulting SGS parametrizations can, and do, affect in nontrivial manners the simulation's physical integrity. Consequently, a great deal of intellectual energy in ocean model algorithm development relates to establishing numerically robust and physically rational SGS parametrizations. This is an extremely difficult problem. Nonetheless, some progress has been made during the past decades, at least enough to know what not to do in certain cases.

Posing the Problem of Ocean Modeling

The ocean is a forced and dissipative system. Forcing occurs at the upper boundary from interactions with the atmosphere, rivers, and sea ice, and at its lower boundary from the solid Earth. Forcing also occurs from astronomical effects of the Sun and Moon which produce tidal motions.

Important atmospheric forcing occurs over basin scales, with timescales set by the diurnal cycle, synoptic weather variability (days), the seasonal cycle, and interannual fluctuations such as the North Atlantic Oscillation and El Niño. Atmospheric momentum and buoyancy fluxes are predominantly responsible for driving the ocean's large scale horizontal and overturning circulations. Additional influences include forcing at continental boundaries from river inflow and calving glaciers, as well as in polar regions where sea ice dynamics greatly affect the surface buoyancy fluxes. Dissipation in the ocean generally occurs at the microscale, with turbulent processes converting mechanical energy to heat. Energy moves from the large to the small scales through nonlinear advective transport. In short, the ocean circulation emerges from an enormous number of processes spanning a huge space and time range. Furthermore, modes of ocean dynamical motions span space and timescales even greater than the forcing scales, with ocean circulation patterns, especially those in the abyssal regions, maintaining coherence over global scales for millennia.

From a mathematical physics perspective, the problem of modeling the ocean circulation is a problem of establishing and maintaining the kinematic, dynamic, and material properties of the ocean fluid: a fluid which is driven by a multitude of boundary interactions and possesses numerous internal transport and mixing processes. Kinematic balances are set by the geometry of the ocean domain, and by assuming that the fundamental fluid parcel constituents conserve mass. Dynamical balances result from applying Newton's law of motion to the continuum fluid so that the acceleration of a fluid parcel is set by forces acting on the parcel, with the dominant forces being pressure, Coriolis, gravity, and friction. And material balances of tracers, such as salt, heat, and biogeochemical species, are affected by circulation, mixing from turbulence, boundary fluxes, and internal sources and sinks. The ocean modeling problem also involves, at a fundamental level, a rational parametrization of SGS processes, such as fine-scale convective mixing in the upper ocean mixed layer due to strong mechanical and buoyant interactions with the atmosphere, breaking internal waves induced from tides rubbing against the ocean bottom, and mesoscale eddies spawned by baroclinically unstable currents. From these balances emerges the wonderful richness of the ocean circulation and its variety of tracer distributions. Faithfully emulating this system using computer simulations requires a massive effort in scientific and engineering ingenuity and collaboration.

Fundamental Budgets and Methods

We now aim to place the previous discussions onto a more rigorous mathematical basis. Doing so is a necessary first step toward developing a suite of equations amenable to numerical methods for use on computers. Due to limitations in space, the material here will be quite terse and selective, allowing many results to be presented without full derivation and many topics to be omitted.

The basic equations of ocean fluid dynamics can be readily formulated by focusing on the dynamics of a mass conserving parcel of seawater. A fluid parcel is a region that is macroscopically small yet microscopically large. That is, from a macroscopic perspective, the parcel's thermodynamic properties may be assumed uniform, and the methods of continuum mechanics are applicable to describing the mechanics of an infinite number of these effectively infinitesimal parcels. However, from a microscopic perspective, these fluid parcels contain many molecules (on the order of Avogadro's number), and so it is safe to ignore the details of molecular interactions. Regions of a fluid with length scales on the order of 10^{-5} m satisfy these properties of a fluid parcel.

Mass Conservation

The mass of a parcel is written as $dM = \rho \, dV$, where ρ is the mass per volume (i.e., the density), and dV is the parcel's volume. Assuming the parcel mass to be conserved as the parcel moves through the fluid leads to the differential statement of mass conservation

$$\frac{d}{dt} \ln(dM) = 0 \qquad [1]$$

where we assume there to be no internal sources of mass. The time derivative d/dt measures changes occuring with respect to the moving parcel. This Lagrangian or material perspective complements the Eulerian perspective rendered by viewing the fluid from a fixed point in space. Transforming to the Eulerian perspective leads to the relation

$$\frac{d}{dt} = \partial_t + \mathbf{v} \cdot \nabla \qquad [2]$$

where ∂_t measures time changes at a fixed space point, and $\mathbf{v}$ is the parcel's velocity. The transport or advective operator $\mathbf{v} \cdot \nabla$ reveals the fundamentally nonlinear character of fluid dynamics arising from motions of fluid parcels with velocity $\mathbf{v}$. In the Eulerian perspective, mass conservation takes the form

$$\partial_t \rho + \nabla \cdot (\rho \mathbf{v}) = 0 \qquad [3]$$

Tracer Budgets

In addition to freshwater, a seawater parcel contains numerous constituents known as tracers. Heat (a thermodynamical tracer) and salt (a material tracer) are seawater's two active tracers. Heat and salt, along with pressure, determine the mass density through a complex empirical relation known as the equation of state. Density in turn impacts the hydrostatic pressure which then affects currents. Other ocean tracers include radioactive species, such as those introduced in the 1950s and 1960s from nuclear bomb tests; biological tracers such as phytoplankton and zooplankton fundamental to ocean ecosystems; and chemical elements such as carbon, iron, and nitrogen prominent in ocean biogeochemical cycles.

In describing the evolution of tracer within a seawater parcel, it is convenient to consider the tracer concentration C, which represents a mass of tracer per mass of seawater for material tracers. In addition to material tracers, we are concerned with a thermodynamical tracer that measures the heat within a fluid parcel. In this case, C is typically taken to be the potential temperature or potential enthalpy.

The evolution of tracer mass within a Lagrangian parcel of mass conserving fluid is given by

$$\rho \frac{dC}{dt} = -\nabla \cdot \mathbf{J} \qquad [4]$$

where for simplicity we ignore tracer sources. The tracer flux $\mathbf{J}$ arises from SGS transport of tracer in the absence of mass transport. Such transport consists of diffusion and/or unresolved advection. As this flux is not associated with mass transport, it vanishes when the tracer concentration is uniform, in which case the tracer budget reduces to the mass budget of eqn [1]. Use of the material time derivative relation [2] and mass conservation [3] renders the Eulerian conservation law for tracer

$$\partial_t(\rho C) + \nabla \cdot (\rho C \mathbf{v}) = -\nabla \cdot \mathbf{J} \qquad [5]$$

The convergence $-\nabla \cdot (\rho C \mathbf{v})$ is the flux form of advective tracer transport. As stated in the section titled

'Ocean mesoscale eddies', advective transport is critical, especially in eddying simulations where both the tracer and velocity distributions possess non-trivial structure. The flux form of advective tracer transport is employed in ocean climate models rather than the alternative advective form $\rho \mathbf{v} \cdot \nabla C$. The reason is that the flux form is amenable to conservative numerical schemes, as well as to a finite volume interpretation described in the section titled 'Basics of the finite volume method'.

The specification of SGS parametrizations appearing in the flux $\mathbf{J}$ is critical for the tracer equation, especially in models not admitting mesoscale eddies. The original approach, whereby the most common class of ocean climate models employed a diffusion operator oriented according to geopotential surfaces, greatly compromised the simulation's physical integrity for climate purposes. The problem with the horizontally oriented operators is that they introduce unphysically large mixing between simulated water masses. Given the highly ideal fluid dynamics of the ocean interior, most of the tracer transport in the interior arising from eddy effects occurs along a locally referenced potential density direction, otherwise known as a neutral direction. This mixing preserves water mass properties over basin scales for decades. Altering the orientation of the model's diffusion operator from horizontal–vertical to neutral–vertical brought the models more in line with the real ocean. In addition, mesoscale eddies stir tracers in a reversible manner. This stirring corresponds to an antisymmetric component to the SGS tracer transport tensor. The combination of neutral diffusion and skew diffusion have become ubiquitous in all ocean climate models that do not admit mesoscale eddies, even those models not based on the geopotential vertical coordinate.

Linear Momentum Budget

The linear momentum of a fluid parcel is given by $\mathbf{v}\rho \mathrm{d}V$. Through Newton's law of motion, momentum changes in time due to the influence of forces acting on the parcel. The forces acting on the ocean fluid parcel include internal stresses in the fluid arising from friction and pressure; the Coriolis force due to our choice of describing the ocean fluid from a rotating frame of reference; and gravity acting in the local vertical direction. The resulting equation for linear momentum takes the form

$$\rho \frac{\mathrm{d}\mathbf{v}}{\mathrm{d}t} = -\rho g \hat{\mathbf{z}} - f\hat{\mathbf{z}} \wedge \rho \mathbf{v} - \nabla p + \rho \mathbf{F} \qquad [6]$$

In this equation, $g \approx 9.8 \, \mathrm{m\,s^{-2}}$ is the acceleration due to gravity, which is generally assumed constant

for ocean climate modeling. The Coriolis force per mass is written $-f\hat{\mathbf{z}} \wedge \mathbf{v}$, with the Coriolis parameter $f = 2\Omega \sin \phi$, where $\Omega = 7.292 \times 10^{-5}\,\mathrm{s^{-1}}$ is the Earth's angular rotation rate, and ϕ is the latitude. Gradients in pressure p impart an acceleration on the fluid parcel, with the parcel accelerated from regions of high pressure to low.

The friction vector $\mathbf{F}$ arises from the divergence of internal friction stresses. In an ocean model, these stresses must be sufficient to maintain a near-unit-grid Reynolds number. Otherwise, the simulation may go unstable, or at best produce unphysical noise-like features. This constraint on the numerical simulation is unfortunate, since a unit grid Reynolds number requires an effective viscosity many orders of magnitude larger than molecular, since the grid sizes (order 10^4–$10^5\,\mathrm{m}$) are much larger than the Kolmogorov scale ($10^{-3}\,\mathrm{m}$). This level of numerical friction is not based on physics, but arises due to the discrete lattice used for the numerical simulations. Various methods have been engineered to employ the minimal level of friction required to meet this, and other, numerical constraints, so as to reduce the effects of friction on the simulation. Such methods are *ad hoc* at best, and lead to some of the most unsatisfying elements in ocean model practice since the details of the methods can strongly influence the simulation.

The hydrostatic approximation mentioned in the section titled 'Scales of motion and the subgrid-scale problem' exploits the large disparity between horizontal motions, occurring over scales of many tens to hundreds of kilometers, and vertical motions, occurring over scales of tens to hundreds of meters. In this case, it is quite accurate to assume that the moving fluid maintains the hydrostatic balance, whereby the vertical momentum equation takes the form

$$\partial_z p = -\rho g \qquad [7]$$

This approximation is a fundamental feature of the ocean's primitive equations, which are the equations solved by all global ocean climate models in use today. By truncating, or filtering, the vertical momentum budget to the inviscid hydrostatic balance, we are obliged to parametrize strong vertical motions occurring in convective regions, since the primitive equations cannot explicitly represent these motions. This has led to various convective parametrizations in use by ocean climate models. These parametrizations are essential for the models to accurately simulate various deep-water formation processes, especially those occurring in the open ocean due to strong buoyancy fluxes.

General Strategy for Time-stepping Momentum

Acoustic waves are three-dimensional fluctuations in the pressure field. They travel at roughly $1500\,\mathrm{m\,s^{-1}}$. There is no evidence that resolving acoustic waves is essential for the physical integrity of ocean climate models. The hydrostatic approximation filters all acoustic modes except the Lamb wave (an acoustic mode that propogates only in the horizontal direction). As the Lamb wave is close in speed to the external gravity waves, it can generally be subsumed into an algorithm for gravity waves, and so is of little consequence to ocean model algorithms.

External or barotropic gravity waves are roughly 100 times the speed of the next fastest internal wave or advective signal. In their linear form, they travel at speed $\sqrt{gH}$, with H the ocean depth and g the acceleration of gravity. In the deep ocean, they are about 5–10 times slower than acoustic waves.

External gravity waves are nearly two-dimensional in structure, so they are largely represented by dynamics of the vertically integrated fluid column. This property motivates the formulation of primitive equation model algorithms which split the relatively fast vertically integrated dynamics from the slow and more complicated vertically dependent dynamics. Doing so allows for a more efficient time-stepping method to update the ocean's momentum field. Details of this split are often quite complex, and require care to ensure that the overlap between the fast and slow modes is trivial, or else suffer consequences of an unstable simulation. Nonetheless, these algorithms form a fundamental feature in all ocean climate models.

Basics of the Finite Volume Method

The previously derived budgets for infinitesimal fluid parcels represent a starting point for ocean climate model algorithm designs. The next stage is to pose these budgets on a discrete lattice, whereby the continuum fields take on a finite or averaged interpretation. There is actually more than one one way to interpret the relation between the continuum variables and those living on the numerical lattice. We introduce one method here, as it has achieved some recent popularity in the ocean model literature.

The finite volume method takes the flux form continuum equations and integrates them over the finite extent of a discrete model grid cell. The resulting control volume budgets are the basis for establishing an algebraic algorithm amenable to computational methods. For example, vector calculus allows for the parcel budget of a scalar field, such as a tracer concentration described by eqn [5], to be written over an arbitrary finite region as

$$\partial_t \left(\int\!\!\int\!\!\int \rho C\,\mathrm{d}V \right) = - \int\!\!\int \mathrm{d}A_{(\hat{\mathbf{n}})} \hat{\mathbf{n}} \cdot (\mathbf{v}^{\mathrm{rel}} C + \mathbf{J})$$

[8]

The volume integral is taken over an arbitrary fluid region, and the area integral is taken over the bounding surface to that volume, with outward normal $\hat{\mathbf{n}}$ and area element $\mathrm{d}A_{(\hat{\mathbf{n}})}$. The velocity $\mathbf{v}^{\mathrm{rel}}$ is determined by the relative velocity of the fluid parcel, $\mathbf{v}$, and the moving boundary. As the advective and diffusive fluxes penetrate the boundary, they alter the tracer mass in the region. The budget for the vector linear momentum can also be written in this form, with the addition of body forces from gravity and Coriolis which act over the extent of the volume. Once formulated in this manner, the discretization problem shifts from fundamentals to details, with details differing on how one represents the SGS behavior of the continuum fields. This then leads to the multitude of discretization methods available for such processes as transport, time-stepping, etc.

Elements of Vertical Coordinates

The choice of how to discretize the vertical direction is the most important choice in the design of a numerical ocean model. The reason is that much of the model's algorithms and SGS parametrizations are fundamentally influenced by this choice. We briefly outline here some physical considerations which may prejudice a choice for the vertical coordinate. For this purpose, we identify three regimes of the ocean germane to the considerations of a vertical coordinate.

- *Upper ocean mixed layer.* This is a generally turbulent region dominated by transfers of momentum, heat, freshwater, and tracers with the overlying atmosphere, sea ice, rivers, etc. It is a primary region of importance for climate system modeling. It is typically very well mixed in the vertical through three-dimensional convective and turbulent processes. These processes involve non-hydrostatic physics which requires very high horizontal and vertical resolution (i.e., a vertical to horizontal grid aspect ratio near unity) to explicitly represent. A parametrization of these processes is therefore necessary in primitive equation ocean models which exploit the hydrostatic approximation. In this region, it is essential to employ a vertical coordinate that facilitates the representation and parametrization of these highly turbulent processes. Geopotential and pressure coordinates, or their derivatives, are the most commonly used coordinates as they facilitate the use of very refined vertical grid spacing, which can be essential to simulate the strong exchanges between the ocean and atmosphere, rivers, and ice.

- *Ocean interior.* Tracer transport processes in the ocean interior predominantly occur along neutral directions. The transport is largely dominated by large-scale currents and mesoscale eddy fluctuations. Water mass properties in the interior thus tend to be preserved over large space and timescales (e.g., basin and decade scales). This property of the ocean interior is critical to represent in a numerical simulation of ocean climate. A potential density, or isopycnal coordinate, framework is well suited to this task, whereas geopotential, pressure, and terrain-following models have problems associated with numerical truncation errors. The problem becomes more egregious as the model resolution is refined, due to the enhanced levels of eddy activity that pumps tracer variance to the grid scale. Quasi-adiabatic dissipation of this variance is difficult to maintain in nonisopycnal models.

- *Ocean bottom.* The ocean's bottom topography acts as a strong forcing on the overlying currents and so directly influences dynamical balances. In an unstratified ocean, the flow generally follows lines of constant f/H, where f is the Coriolis parameter and H the ocean depth. Additionally, there are several regions where density-driven currents (overflows) and turbulent bottom boundary layer (BBL) processes act as a strong determinant of water mass characteristics. Many such processes are crucial for the formation of deep-water properties in the World Ocean, and for representing coastal processes in regional models. It is for this reason that terrain-following models have been developed over the past few decades, with their dominant application focused on the coastal and estuarine problem.

The commonly used vertical coordinates introduced above can have difficulties accurately capturing all flow regimes. Unfortunately, each regime is important for accurate simulations of the ocean climate system. To resolve this problem, some researchers have proposed the use of hybrid vertical coordinates built from combinations of the traditional choices. The aim is to employ a particular vertical coordinate only in a regime where it is most suitable, with smooth and well-defined transitions to another coordinate when the regime changes. Research into the required hybrid vertical coordinate methods remains an active area in ocean model design.

Ocean Climate Modeling

The use of physical ocean models to simulate the ocean requires understanding and knowledge beyond the fundamentals discussed thus far in this article. Most notably, we require information about boundary fluxes. There are two basic manners that ocean models are generally used for simulations: as components of an ESM, whereby boundary fluxes are computed from atmosphere, sea ice, and river component models, based on interactions with the evolving ocean; or in a stand-alone mode where boundary fluxes are prescribed from a data set, in which case the uncertainties are huge due to sparse measurements over much of the ocean. This uncertainty in observed fluxes greatly handicaps our ability to unambiguously untangle model errors (i.e., errors in numerical methods, parametrizations, and formulations) from flux errors.

Correspondingly, the process of evaluating the fidelity of ocean simulations remains in its infancy relative to the situation in atmospheric modeling, where synoptic weather forecasts provide a stringent test of model fidelity. Nonetheless, ocean observations, including boundary fluxes, are steadily improving, with new observations providing critical benchmarks for evaluating the relevance of ocean simulations. Given the wide-ranging spatial and temporal scales of oceanic phenomena, it is essential that observations be maintained over a wide network in both space and time. In absence of this network, we are unable to provide a mechanistic understanding of the observed ocean climate system.

See also

Coupled Sea Ice-Ocean Models. Heat and Momentum Fluxes at the Sea Surface. Heat Transport and Climate. Ocean Carbon System, Modeling of. Ocean Circulation: Meridional Overturning Circulation.

Further Reading

Bleck R (2002) An oceanic general circulation model frame in hybrid isopycnic–Cartesian coordinates. *Ocean Modelling* 4: 55–88.

Durran DR (1999) *Numerical Methods for Wave Equations in Geophysical Fluid Dynamics*, 470 pp. Berlin: Springer.

Gent PR, Willebrand J, McDougall TJ, and McWilliams JC (1995) Parameterizing eddy-induced tracer transports in ocean circulation models. *Journal of Physical Oceanography* 25: 463–474.

Griffies SM (2004) *Fundamentals of Ocean Climate Models*, 518pp. Princeton, NJ: Princeton University Press.

Griffies SM, Böning C, Bryan FO, *et al.* (2000) Developments in ocean climate modelling. *Ocean Modelling* 2: 123–192.

Hirsch C (1988) *Numerical Computation of Internal and External Flows*. New York: Wiley.

McClean JL, Maltrud ME, and Bryan FO (2006) Quantitative measures of the fidelity of eddy-resolving ocean models. *Oceanography* d192: 104–117.

McDougall TJ (1987) Neutral surfaces. *Journal of Physical Oceanography* 17: 1950–1967.

Müller P (2006) *The Equations of Oceanic Motions*, 302pp. Cambridge, UK: Cambridge University Press.

GENERAL CIRCULATION MODELS

G. R. Ierley, University of California San Diego, La Jolla, CA, USA

Introduction

A general circulation model (GCM) of the ocean is nothing more than that – a numerical model that represents the movement of water in the ocean. Models, and more particularly, numerical models, play an ever-increasing role in all areas of science; in geophysics broadly, and in oceanography specifically. It was perhaps less the early advent of super-computers than the later appearance of powerful personal workstations (tens of megaflops and megabytes) that effected not only a visible revolution in the range of possible computations but also a more subtle, less often appreciated, revolution in the very nature of the questions that scientists ask, and the answers that result.

The range of length scales and timescales in oceanography is considerable. Important dynamics, such as that which creates 'salt fingers' and hence influences the dynamically significant profile of density versus depth, takes place on centimeter scales, while the dominant features in the average circulation cascade all the way to basin scales of several thousand kilometers. Timescales for turbulent events, like waves breaking on shore, are small fractions of seconds, while at the opposite end, scientists have reliably identified patterns in the ocean with characteristic evolution times of order several decades.[1]

Most simply, a 'model' is no more than a mathematical description of a physical system. In the case of physical oceanography, that description includes the following elements:

- The momentum equation ($F = m\dot{v}$, but expressed in terms appropriate to a continuous medium), or often in its place a derivative form, the 'vorticity equation'.

---

[1] It is useful to distinguish extrinsic evolution times, which span geologic time, from intrinsic variation, which characterizes an isolated ocean and atmosphere, thus neglecting such secular influences as orbital variation, change in the earth's rotation rate, variation in the solar constant, etc. Although not all causes of variability have been identified, it is possible that even documented evolution over thousands of years may reflect the latter, intrinsic, variability.

- An equation to express the principle of mass conservation.
- A heat equation, which describes the advection (carrying by the fluid) and diffusion of temperature.
- A similar 'advection–diffusion' equation for salinity.
- An equation of state, which relates the pressure to the density.

Although it may, after the fact, sound obvious, it took scientists many years to appreciate the full role of rotation – which enters Newton's law through a latitude dependent Coriolis force – in generating the observed large scale circulation of ocean.

Elements of these equations can become quite complicated. For example, momentum in the upper ocean is imparted by a complex, not yet fully understood, process of wind–wave interaction. One has to choose whether to represent the action of the wind kinematically, which means that the spatial and temporal variability of the wind must be given beforehand, or dynamically, in which case we must solve not only the set of equations above, but a similar set that describes the simultaneous evolution of the atmosphere. Clearly the dynamical case is the more 'realistic' of the two, but the price is a substantially more involved computation.

It is issues such as these that force one to a choice that often pits understanding and intuition on the one hand against verisimilitude and complete dynamical consistency on the other. As the common aim of most large-scale models is to generate results of maximum realism, it usual that the models are frequently corrected, or 'steered', on the fly by extensive use of data assimilation. It is not merely a matter of slight refinement: no large-scale model (GCM) is yet sufficiently robust that it can be used for forecasting without considerable input of such observationally derived constraints; in the oceanic component, for example, the need to force model agreement at depth by continuously 'relaxing' the solution to the smoothed data of the Levitus atlas (a world-wide compendium of data from many sources, smoothed and interpolated onto a regular grid). How much of this fragility is because of explicit defects (faulty 'subgrid scale modeling' and explicit omissions in the physics e.g., neglect of wave breaking) and how much is due to discrete numerical implementation inconsistent with plausible continuous equations remains an open question.

Their limitations notwithstanding, large numerical models are one vital means by which we grapple with questions about global warming and a host of other environmental issues that affect the way that both we and future generations will live.

Models in Theory and Practice

Historically, computers were initially so limited that the questions posed were often, in effect, slight extensions of preexisting analytic queries. To that extent, such studies continued to conform (at least in principle) to what we may identity as four basic building blocks of most theories, which, in aiming to describe physical reality become subject to constraints. Classical modes to which physical laws are found to conform generally[2] include the following.

- *Expression in quantity, extent, and duration.* The language that offers itself as encompassing all distinct physical conditions and all meaningful physical relations is fundamentally that of mathematics. As others before and since, the great mathematician Eugene Wigner too had a stab at explaining what, in an eponymous essay, he termed 'the unreasonable effectiveness of mathematics in the physical sciences'. It remains a conundrum.

- *Confinement of form.* This attains its purest expression in the Platonic view that mathematical entities are not invented, but discovered, and as such have a prior, if not physical, existence. In extension to physical theories, it corresponds to the belief that there is some true, ultimate, equation association with any natural phenomenon. One important respect in which this idea must be tempered in application to fluids is the mathematical demonstration that a variety of different microscopic laws for interaction may all yield the same generic macroscopic law applicable to behavior at large space scales or timescales. If one's aim is solely to understand the latter, then although for a given problem we might presume that there is indeed a precise, if complicated, microscopic law, one's effort might more profitably be spent understanding the passage to the large-scale limit.

- *Falsifiability.* Implicit in the progress of science is the idea that one makes and then tests hypotheses. But unlike in mathematics, in the physical sciences we cannot show the hypothesis is correct, only that it is wrong. A hypothesis is never vindicated, instead we reach a tentative conclusion of not proven wrong (yet!). While it may stimulate conjecture, and have other worthy ends, the idea of introducing artificial parameters for the express purpose of manufacturing close agreement with reality is nonetheless formally antithetical to the paradigm of testing independent, quantitative predictions.

- *Backward compatibility.* As with confinement of form, the idea of compatibility achieves a purity in mathematics that is not to be expected of the physical sciences. Each new bit of mathematics must fit perfectly into the entire edifice of results already discovered (or invented, as you will). Commonly, though not without exception, in the physical sciences, newer theories are seen to encompass the older theories as special or limiting cases. We speak, for example, of the 'classical limit' as a means of recovering prerelativistic or prequantum results. Indeed, it is only in light of Einstein's theory of special relativity that we can understand the limitations of Newton's law, which we now understand more fully as not a law, but a limiting approximation. Oceanography has families of theories, each nesting one within the next, like a series of oceanographic Matryoshka dolls, the innermost of which is often the theory of 'quasigeostrophy,' which dates from the 1950s.

The unavoidable adoption and resulting sensitivity of GCMs to *ad hoc* parameterizations (e.g., subgrid scale modeling) or necessity set them apart from the traditional pursuit of the scientific method. This distinction was (presciently) appreciated at least by the early 1960s and it changes, or ought to change, one's view of such models as rigorous arbiters of precise truths. And yet while it is true that numerical experiments with GCMs are merely suggestive rather than truly predictive of future evolution of the ocean, the sheer lack of experimental data, to say nothing of the lack of a control, means that theoretical ideas are often assessed on the basis of their success in explaining strictly numerical experiments.

As computers became more powerful it was natural to press for the most realistic model runs possible. And, because the growth in computing power was increasingly realized through distributed[3] as well as mainframe (super), computing, the school of 'kitchen sink' models, which started as a specialized branch off the mainstream of oceanography – largely

---

[2] We speak gently here, since to insist that those four are always either necessary or sufficient would require that we introduce a theory about theories: a metatheory. But we have no notion of how rigorously to evaluate such ideas!

[3] Both virtually, through high speed and increasingly transparent networks, and physically, through desktop workstations of considerable power.

limited in participation to those in close physical proximity to two or three central machines – became a powerful tributary in its own right: an autonomous discipline within oceanography, which naturally began to evolve its own criteria for relevance.

Limits on Numerical Models

We spoke of two sources of error common to large numerical models: difficulties in numerical implementation, and poorly modeled or unrepresented physical processes. In this section we consider specific instances of each, starting with an abstract mathematical point of view. But note that if we could with the wave of hand dispense with these two issues (which one supposes are in principle tractable), the fact that we do not know the exact physical state of the ocean inevitably increases the uncertainty of the results. Moreover, even were we given that exact state at some instant in time, the intrinsic and spontaneous genesis of disorder in such a physical system must forever constrain our predictive power.

There are two key mathematical features of the basic momentum (or vorticity) equation which bear comment: conservation and dissipation. The nonlinear term is fundamental to the initiation and sustenance of turbulence and by itself strictly conserves energy. (Other terms introduce explicit dissipation.) In addition, in two dimensions the term conserves 'enstrophy' (the square of the vorticity) and in three, both 'circulation' and 'helicity' (the dot-product of velocity and vorticity). While one might hope for all such conservation properties to be preserved in numerical implementations, some large models, often those based on curvilinear – as opposed to Cartesian – coördinates – manage only to conserve energy.

Beyond the conserved quantities associated with the nonlinear term, which include the energy and, in general, the so-called 'Noether invariants,' there is a more subtle property associated with the exact (continuous) equation: its associated 'multisymplectic geometry.' Recent mathematical advances make it possible for a discrete numerical model to preserve such structure exactly, though as yet such improvements have not been incorporated into any working GCM. Is it quantitatively important that we do so when basic fluid processes, such as convection, are as yet only crudely modeled? Until the experiment is tried, no one can say. But it is pertinent to note that a similar (that is, 'symplectic') refinement is critical for a numerical solution of sufficient accuracy that one can decide whether planetary orbital motions are chaotic on astronomical timescales.[4]

Finally, strictly speaking, not only ambitious models, but even more confined 'process' models, rest

upon a not yet wholly secure foundation: it is still an open research question whether the basic equation of fluid mechanics (the Navier–Stokes equation) is itself 'globally well-posed' in three dimensions.[5] (It is widely believed to be so, but belief comes cheap. A proof, however, is worth one million dollars(!) – one of seven prizes in a competition recently announced by the Clay Mathematics Institute.)

Even overlooking such foundational matters on which mainly mathematicians would cavil about numerical models, the last three of the four principles above are often violated in more apparent ways in the application and development of present-day models. We illustrate this divergence from traditional norms with a few representative examples to emphasize the sometimes causal (not casual!) link between models and 'reality.' In delineating the borders of the known, the unknown, and the unknowable, it is important to discriminate between deduction and rationalization as competing processes for exploring and explaining those borders.

- Although GCM simulations with a viscosity approaching that of water are at present inconceivable, at least as a thought experiment it is worth bearing in mind that those are the numerical results we would in principle compare against observation to assess a given model.[6] Short of that, GCMs use various formulations that ostensibly mimic the dynamical effects of the unresolved scales of motions. In the simplest instance, this amounts to choosing a numerical viscosity several orders of magnitude larger than that of sea water. But often the value is dictated by purely heuristic numerical considerations: it is set at a threshold value, any decrease below which leads to rapid

---

[4] On dissipation, a deep, though perhaps insufficiently appreciated, mathematical result is that the solution of a parabolic, dissipative system quickly collapses onto a finite-dimensional 'attractor'. This is remarkable. If you think of assigning a point to every one of N molecules of water in the ocean, and tracking the velocity and position of each, the associated 'phase space' – just a record of that evolution – has dimension $6N$. Because the momentum equation is derived on the basis that water is an infinitely divisible continuum, strictly we need to imagine that N approaches infinity. Nonetheless, even in that infinite-dimensional phase space, it remains true that the solution confines itself to only a finite, if quite large, portion.

[5] Hadamard introduced the notion of ill-posedness of partial differential equations. A problem is well-posed when a solution exists, is unique, and depends continuously on the initial data. It is ill-posed when it fails to satisfy one or more of these criteria.

[6] There is a curious division among physical oceanographers as to whether the large-scale flow we observe is, in the end, actually sensitive to the precise value of the viscosity of sea water. Predictably, there are two camps: yes and no.

numerical blowup. It cannot be said to be satis-factory feature that a basic parameter is set not by independent dynamical considerations but for stability reasons, and those pertaining solely to the discrete form of the equations. (The solution to the continuous equation would not blow up!) At times, not only the coefficient, but the actual form of the diffusive operator is adjusted. As above, usually the motivation is intrinsically numerical, so it is not surprising that a catalogue of the various choices shows some, for example, that create artificial sources (or sinks) of vorticity in the flow. Others, subject to the given boundary conditions, do not make mathematical sense in a region where the fluid depth tends to zero (like Atlantic City).

Oceanographers, unfortunately, do not have the luxury of extensive laboratory measurements from which their dissipative parameterizations can be calibrated. A program of direct observation in par-ticular regions where dissipation is thought to be significant is just getting under way. While such measurements will help reveal deficiencies in present formulations, the largest GCMs will probably rely on a solely heuristic approach for some time to come.

- An important, but numerically unresolved, pro-cess in the ocean is that of convection, which typically occurs at small scales in, for example, localized regions of intense surface cooling. The overall thermal structure of the ocean is sensitive to this, a means by which 'bottom water' is formed; a cold, relatively less salty mass that constitutes the deep Atlantic, for example. Be-cause the horizontal resolution is too coarse to encompass the sinking motions, various schemes have been devised to mimic that effect. It has been shown that one of the most common of these leads paradoxically to unacceptable physical (and mathematical) behavior as resolution is improved; it has no verifiable correspondence to a realizable physical process. The temptation with a model that has been extensively tuned to give plausible answers for other observables is to leave well enough alone. Unhappily, a model with one or more such elements whose limits are ill-defined or nonexistent must inevitably produce end results whose errors are typically an opaque mix of ef-fects: some physical, some numerical, some mathematical. In such circumstances, the program of falsifiability of the physical components is apt to be fatally compromised.

The point is not that one should immediately dis-pense with all *ad hoc* parametrizations; excepting those that are simply mathematically ill-posed from the outset, theorists generally do not have better al-ternatives to suggest. But one should always bear in mind the degree to which numerical simulations are sensitive to these components, and seek independent ways in which to constrain their parameters, in isolated settings that test the limits of prediction against known measurements or, failing that, at least against fully converged, adequately resolved simu-lations of a local or regional character. If, within the acceptable parameter range identified, it is found that the original model no longer gives adequate large-scale predictions, then there are more basic problems to be addressed.

Summary

From the numerical side, no computer improvements that can be seen on the horizon seem likely to make reasonably ambitious GCMs accessible to rigorous and extensive parametric and numerical exploration, a prerequisite to their complete understanding. From the mathematical side, it seems to be our funda-mental ignorance about turbulence that most se-verely restricts the range of our grasp, leaving us with an often painfully narrow range of computations to which theoretical remarks can be significantly ad-dressed. For these structural reasons, the gulf be-tween theory and much numerical modeling will probably continue to widen for the foreseeable fu-ture, and thus there may grow to be—indeed some would say it already exists—a division akin to C.P. Snow's 'Two Cultures'.

All the cautions about GCMs notwithstanding, they have become an integral part of the study of physical oceanography. With due regard for the novel capacities and limitations of numerical models, such scientific progress as we do make will more and more often hinge upon judicious computation.

Further Reading

The literature on ocean modeling is not yet productive of definitive treatises, in large measure because the field is yet young and rapidly evolving. Thus in lieu of textbooks or similar references, the reader is directed to the following series of articles.

For some predictions on the perennially intriguing issue of what improvements in large-scale modeling may be driven by plausible increases in computing speed with massively parallel machines see

Semtner A. Ocean and climate modeling *Communica-tions of the ACM* 43 (43): 81–89.

For a look back at the history of one of the single most influential models in physical oceanography, see A.J. Semtner's Introduction to 'A numerical method for the study of the circulation of the World Ocean', which accompanies the reprinting of Kirk Bryan's now classic 1969 article of the title indicated. This pair appears back-to-back, beginning on page 149, in *Journal of Computational Physics,* (1997) 135 (2).

General readers may wish to consult the following succinct review, accessible to a broad audience:

Semtner AJ (1995) Modeling ocean circulation *Science* 269 (5229): 1379–1385.

Finally, for those readers desiring a more in depth appreciation of modeling issues and their implications for specific features of the large scale circulation, consult the careful review.

McWilliams JC (1996) Modeling the oceanic general circulation. In: Lumley JL, Van Dyke M. Read HL (eds) *Annual Review of Fluid Mechanics* Vol. 28, pp. 215–248 Palo Alto, CA: Annual Reviews.

OCEAN CIRCULATION: MERIDIONAL OVERTURNING CIRCULATION

J. R. Toggweiler, NOAA, Princeton, NJ, USA

Introduction

The circulation of the ocean has been traditionally divided into two parts, a wind-driven circulation that dominates in the upper few hundred meters, and a density-driven circulation that dominates below. The latter was once called the 'thermohaline circulation', a designation that emphasized the roles of heating, cooling, freshening, and salinification in its production. Use of 'thermohaline circulation' has all but disappeared among professional oceanographers.

The preferred designation is 'meriodional overturning circulation', hereafter MOC. This usage reflects the sense of the time-averaged flow, which generally consists of a poleward flow of relatively warm water that overlies an equatorward flow of colder water at depth. It also reflects a recognition by oceanographers that most of the work done to drive the circulation, whether near the surface or at depth, comes directly or indirectly from the wind.

Like the thermohaline circulation of old, the MOC is an important factor in the Earth's climate because it transports roughly 10^{15} W of heat poleward into high latitudes, about one-fourth of the total heat transport of the ocean/atmosphere circulation system. Radiocarbon measurements show that it turns over all the deep water in the ocean every 600 years or so. Its upwelling branch is important for the ocean's biota as it brings nutrient-rich deep water up to the surface. It may or may not be vulnerable to the warming and freshening of the Earth's polar regions associated with global warming.

The most distinguishing features of the MOC are observed in the sinking phase, when new deep-water masses are formed and sink into the interior or to the bottom. Large volumes of cold polar water can be observed spilling over sills, mixing violently with warmer ambient water, and otherwise descending to abyssal depths. The main features of the upwelling phase are less obvious. The biggest uncertainty is about where the upwelling occurs and how the upwelled deep water returns to the areas of deep-water formation.

The Cooling Phase – Deep-water Formation

The most vigorous overturning circulation in the ocean today is in the Atlantic Ocean where the upper part of the Atlantic's MOC carries warm, upper ocean water through the Tropics and subtropics toward the north while the deep part carries cold dense polar water southward around the tip of Africa and into the Southern Ocean beyond. The Atlantic's MOC converts roughly $15 \times 106 \, \mathrm{m^3 \, s^{-1}}$ of upper ocean water into deep water. (Oceanographers designate a flow rate of $1 \times 106 \, \mathrm{m^3 \, s^{-1}}$ to be 1 Sv. All the world's rivers combined deliver $c.$ 1 Sv of fresh water to the ocean.)

The MOC in the Atlantic is often characterized as a 'conveyor belt' or more generally as a continuous current or ribbon of flow. It is shown following a path that extends through the Florida Straits and up the east coast of North America as part of the Gulf Stream. The ribbon then cuts to the east across the Atlantic and then northward closer to the coast of Europe. This is somewhat misleading because individual water parcels do not follow this kind of continuous path. The MOC is composed instead of multiple currents that transfer water and water properties along segments of the path. Individual water parcels loop around multiple times and may recirculate all the way around the wind-driven gyres while moving northward in stages.

As the MOC moves northward through the tropical and subtropical North Atlantic, it spans a depth range from the surface down to $\sim 800 \, \mathrm{m}$ and has a mean temperature of some 15–20 °C. During its transit through the Tropics and subtropics, the MOC becomes saltier due to the excess of evaporation over precipitation in this region. It also becomes warmer and saltier by mixing with the salty outflow from the Mediterranean Sea. By the time the MOC has crossed the 50° N parallel into the subpolar North Atlantic the water has cooled to an average temperature of 11–12 °C. Roughly half of the water carried northward by the MOC at this juncture moves into the Norwegian Sea between Iceland and Norway. Part of this flow extends into the Arctic Ocean.

The final stages of this process make the salty North Atlantic water cold and dense enough to sink. Sinking is known to occur in three main places. The densest sinking water in the North Atlantic is formed in the Barents Sea north of Norway where salty water from the Norwegian Current is exposed to the atmosphere on the shallow ice-free Barents shelf. Roughly 2 Sv of water from the Norwegian Current crosses the shelf and sinks into the Arctic basin after being cooled down to 0 °C. This water eventually flows out of the Arctic along the coast of Greenland at a depth of 600–1000 m. The volume of dense water is increased by additional sinking and open ocean convection in the Greenland Sea north of Iceland. The dense water then spills into the North Atlantic over the sill between Greenland and Iceland in Denmark Strait (at about 600-m depth) and through the Faroe Bank Channel between the Faroe Islands and Scotland (at about 800-m depth).

As 0 °C water from the Arctic and Greenland Seas passes over these sills, it mixes with 6 °C Atlantic water beyond the sills and descends the continental slopes down to a depth of 3000 m. The overflows merge and flow southward around the tip of Greenland into the Labrador Sea as part of a deep boundary current that follows the perimeter of the subpolar North Atlantic. A slightly warmer water mass is formed by open ocean convection within the Labrador Sea. The deep water formed in the Labrador Sea increases the volume flow of the boundary current that exits the subpolar North Atlantic beyond the eastern tip of Newfoundland.

Newly formed water masses are easy to track by their distinct temperature and salinity signatures and high concentrations of oxygen. The southward flow of North Atlantic Deep Water (NADW) is a prime example. NADW is identified as a water mass with a narrow spread of temperatures and salinities between 2.0 and 3.5 °C and 34.9 and 35.0 psu, respectively.

Figure 1 shows the distribution of salinity in the western Atlantic which tracks the southward movement of NADW. Newly formed NADW ($S > 34.9$) is easily distinguished from the relatively fresh intermediate water above ($S < 34.6$) and the Antarctic water below ($S < 34.7$). NADW exits the Atlantic south of Africa between 35° and 45° S and joins the eastward flow of the Antarctic Circumpolar Current (ACC). Traces of NADW reenter the Atlantic Ocean through Drake Passage (55–65° S) after flowing all the way around the globe.

The other major site of deep-water formation is the coast of Antarctica. The surface waters around Antarctica, like those over most of the Arctic Ocean, are ice covered during winter and are too fresh to sink. Deep water below 500 m around Antarctica, on the other hand, is relatively warm (1.5 °C) and fairly salty (34.70–34.75 psu). This water mass, known as Circumpolar Deep Water (CDW), penetrates onto the relatively deep continental shelves around

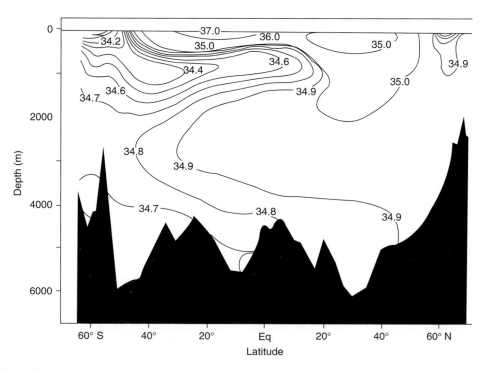

Figure 1 North–south section of salinity down the western Atlantic from Iceland to Drake Passage. The salinity distribution has been contoured every 0.1 psu between 34.0 and 35.0 psu to highlight NADW (34.9–35.0 psu).

Antarctica where it is cooled to the freezing point. Brine rejection from the formation of new sea ice maintains fairly high salinities on the shelf despite the freshening effects of precipitation and the input of glacial meltwater. Very cold shelf water ($-1\,°C$, 34.6–34.7 psu) is then observed descending the continental slope to the bottom in the Weddell Sea. Bottom water is also observed to form off the Adelie coast south of Australia ($\sim 140°\,E$).

The volume of new deep water formed on the Antarctic shelves is not very large in relation to the volume of deep water formed in the North Atlantic, perhaps 3–4 Sv in total. It however entrains a large volume of old Circumpolar Deep Water, as it sinks to the abyss. Even with the entrainment, the deep water formed around Antarctica is denser than the NADW formed in the north. Antarctic Bottom Water (AABW) occupies the deepest parts of the ocean and is observed penetrating northward into the Atlantic, Indian, and Pacific Oceans through deep passages in the mid-ocean ridge system.

Small quantities of deep water are also observed to form in semi-enclosed evaporative seas like the Mediterranean and Red Seas. These water masses are dense owing to their high salinities. They tend to form intermediate- depth water masses after exiting their formation areas in relation to the deep-water and bottom-water masses formed in the North Atlantic and around Antarctica.

The Warming Phase – Upwelling and the Return Flow

The upwelling of deep water back to the ocean's surface was thought at one time to be widely distributed over the ocean. This variety of upwelling was attributed to mixing processes that were hypothesized to be active throughout the interior. The mixing was thought to be slowly transfering heat downward, making the old deep water in the interior progressively less dense so that it could be displaced upward by the colder and saltier deep waters forming near the poles. Since the main areas of deep water formation are located at either end of the Atlantic, and since most of the ocean's area is found in the Indian and Pacific Oceans, it stood to reason that the warming of the return flow should be widely distributed across the Indian and Pacific. Schematic diagrams often depict the closure of the old thermohaline circulation as a flow of warm upper ocean water that passes from the North Pacific through Indonesia, across the Indian Ocean, and then around the tip of Africa into the Atlantic.

Observations made over the last 30 years point instead to turbulent mixing that is intense in some places but is not widespread. Attempts to directly measure the turbulent mixing in the interior have shown that there may only be enough mixing to modify perhaps 10% or 20% of the deep water formed near the poles. There is no indication that any deep water is actually upwelling to the surface in the warm parts of the Indian and Pacific Oceans. Vigorous mixing is found near the bottom, where it is generated by tidal motions interacting with the bottom topography. It is also found in the vicinity of strong wind-driven currents. Thus, the energy that is available to warm the deep waters of the abyss comes mainly from the winds and tides.

It now appears that most of the deep water sinking in the North Atlantic upwells back to the surface in the Southern Ocean. The upwelling occurs within the channel of open water that circles the globe around Antarctica. **Figure 2** shows schematically how this is thought to work. The curved lines in the background are isolines of constant density (also known as isopycnals). Their downward plunge to the north away from Antarctica reflects the flow of the ACC out of the page in the center of the figure. Salty dense water from the North Atlantic is found at the base of the plunging isopycnals. Westerly winds above the ACC (also blowing out of the page) drive the ACC forward and also push the cold fresh water at the surface away from Antarctica to the north. Dense salty water from the North Atlantic is drawn upward in its place. A mixture of the salty dense deep water and the cold fresh surface water is then driven northward out of the channel by the westerly winds and is forced down into the thermocline on the north side of the ACC.

In this way, the winds driving the ACC continually remove water with North Atlantic properties from the interior. Oceanographers are fairly certain that something like this is happening because the winds driving the ACC have become stronger over the last 40 years in response to global warming and the depletion of ozone over Antarctica. The subsurface water around Antarctica has become warmer, saltier, and lower in oxygen as upwelled water from the interior has displaced more of the cold fresh surface water, as shown in **Figure 2**. The thermocline water north of the ACC has also become cooler and fresher.

Numerical experiments with ocean general circulation models show that much of the water forced down into the thermocline around the open channel eventually makes its way into the Atlantic Ocean where it is converted again into deep water in the northern North Atlantic. Thus, the winds over the ACC, in drawing up deep water from the ocean's interior, have been shown to actively enhance the formation of deep water in the North Atlantic.

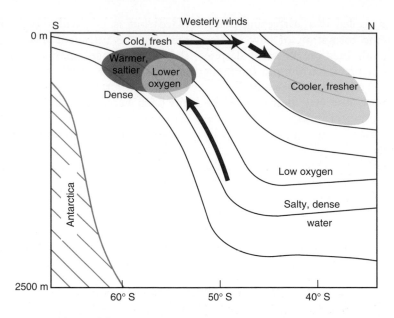

Figure 2   Schematic cross section of the ACC, showing how the winds driving the current in an eastward direction out of the page also drive an overturning circulation. Stronger winds over the last 40 years have drawn more deep water to the surface south of the ACC, which has caused the subsurface waters around Antarctica to become warmer, saltier, and lower in oxygen, and have produced more downwelling to the north, which has made the thermocline waters north of the ACC cooler and fresher. Adapted from Aoki S, Bindoff NL, and Church JA (2005) Interdecadal water mass changes in the Southern Ocean between 30 ° E and 160 ° E. *Geophysical Research Letters* 32 (doi:10.1029/2005GL022220) and reproduced from Toggweiler RR and Russell J (2008) Ocean Circulation in a warming climate. *Nature* 451 (doi:10.1038/nature06590).

The deep water drawn up to the surface around Antarctica is colder than the surface water that is forced down into the thermocline north of the ACC. As this cold water comes into contact with the atmosphere and is carried northward, it takes up solar heat that otherwise would be available to warm the Southern Ocean and Antarctica. The MOC then carries this southern heat across the equator into the high latitudes of the North Atlantic where it is released to the atmosphere when new deep water is formed. The MOC thereby warms the North Atlantic at the expense of a colder Southern Ocean and colder Antarctica. Indeed, sea surface temperatures at 60° N in the North Atlantic are on average *c.* 6 °C warmer than sea surface temperatures at 60° S.

If the warming phase of the MOC involved a downward mixing of heat in low and middle latitudes, as once thought, the MOC would carry tropical heat poleward into high latitudes. The warming phase of the Atlantic's MOC now seems to take place in the south instead. This means that the heat transport by the MOC through the South Atlantic is directed equatorward and is opposed to the heat transport in the atmosphere. The net effect is a weakening of the global heat transport in the Southern Hemisphere and a strengthening of the heat transport in the Northern Hemisphere.

General Theory for the Meridional Overturning Circulation

Figure 3 is a north–south section showing the distribution of potential density through the Atlantic Ocean. Most of the northward flow of the Atlantic's MOC takes place between the 34.0 ($\sim 20\,°C$) and 36.0 ($\sim 8\,°C$) isopycnals, that is, within the main part of the thermocline, but some of the northward flow takes place among the denser isopycnals of the lower thermocline down to $36.6\,g\,kg^{-1}$ ($\sim 1000\,m$). The southward flow of NADW is located, for the most part, below the 37.0 isopycnal.

The isopycnals in the lower thermocline in **Figure 3** are basically flat north of 40° S but rise up to the surface between 40° and 60° S. The rise of the isopycnals marks the eastward flow of the ACC, as shown previously in **Figure 2**. The relatively deep position of these isopycnals north of 40° S reflects the mechanical work done by the winds that drive the ACC. The convergent surface flow forced by the winds north of the ACC pushes the relatively light lower thermocline water down in relation to the cold, dense water that is drawn up by the winds south of the ACC. Numerical experiments carried out in ocean global climate models (GCMs) suggest that all the isopycnals of the lower thermocline would be squeezed up into the main thermocline if the circumpolar channel were closed and the ACC eliminated.

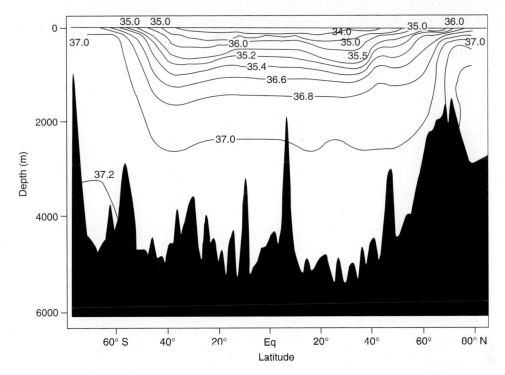

Figure 3 Density structure of the Atlantic thermocline. Seawater density is referenced to a depth of 2000 m and has been zonally averaged across the Atlantic basin (units g kg^{-1}). Contours have been chosen to highlight the lower part of the thermocline. The zonally averaged topography fails to capture the depths of the sills through the Greenland–Iceland–Scotland Ridge (65° N), which are at 600 and 800 m.

The relatively deep lower thermocline puts relatively warm water just to the south of the sills between Greenland, Iceland, and the Faroe Islands in the northern North Atlantic. This warm water is important because it sets up a sharp contrast with the cold water behind the sills, which ultimately drives the flow of dense water over the sills and into the deep Atlantic. In this way, the winds and the ACC in the south contribute as much, if not more, to the overflows in the north as the cooling that takes place at the surface.

The depths of the isopycnals in the lower thermocline lead one to a general theory for the MOC in the Atlantic. Deep-water formation in the North Atlantic converts relatively low-density thermocline water into new deep water. It thereby removes mass from the thermocline and allows the isopycnals of the thermocline to be squeezed upward. The winds in the south have the opposite effect as they draw deep water up to the surface and pump it northward into the thermocline. They convert dense water from the deep ocean into low-density thermocline water and cause the thermocline to thicken downward. In this way, the thermocline thickness reflects a balance between the addition of mass via winds in the south and the loss of mass by deep-water formation in the north. The strength of the MOC should in this sense be proportional to the thermocline thickness and the

transformations in the north and south that convert light water to dense water and vice versa.

A different kind of theory is needed for the overturning of the bottom water formed around Antarctica because the winds cannot play the same role. The upwelling branch of this circulation is also associated with the ACC as in **Figure 2** but the upwelled water in this case flows poleward onto the adjacent Antarctic shelves where it is cooled and sinks back into the interior. The winds in this case help cool the upwelled water by exposing it to the atmosphere and by driving the fresh Antarctic surface waters away but the strength of the bottom water circulation would appear to be limited by the rate at which old bottom waters can be warmed by mixing with the overlying deep water.

Instability of the Meridional Overturning Circulation

Cooling of the ocean in high latitudes makes polar surface waters denser in relation to warmer waters at lower latitudes. Thus cooling contributes to a stronger MOC. The salinity section through the Atlantic Ocean in **Figure 1** gives one a superficial impression that salinification also makes a positive contribution to the MOC. This is actually not true.

The cycling of fresh water between the ocean and atmosphere (the hydrological cycle) results in a net addition of fresh water to the polar oceans which reduces the density of polar surface waters. Thus, the haline part of the old thermohaline circulation is nearly always in opposition to the thermal forcing. The Earth's hydrological cycle is expected to become more vigorous in the future with global warming. This may weaken the MOC in a way which could be fairly abrupt and unpredictable.

NADW is salty because the upper part of the MOC flows through zones of intense evaporation in the Tropics and subtropics. If the rate at which new deep water is forming is relatively high, as it seems to be at the present time, the sinking water removes much of the fresh water added in high latitudes and carries it into the interior. The added fresh water in this case dilutes the salty water being carried into the deep-water formation areas but does not erase the effect of evaporation in low latitudes. Thus, NADW remains fairly salty and is able to export fresh water from the Atlantic basin.

If the rate of deep-water formation is relatively low or the hydrological cycle is fairly strong, the fresh water added in high latitudes can create a low-salinity lid that can reduce or annihilate the MOC. There seems to be a critical input of freshwater input for maintaining the MOC. If the freshwater input is close to this threshold, the overturning becomes unstable and may become prone to wild swings over time.

Coupled (ocean + atmosphere) climate models from the 1990s projected that a fourfold increase in atmospheric CO_2 would increase the hydrological cycle sufficiently that the MOC might collapse. More recent coupled models are predicting that the warming will, in addition, lead to stronger mid-latitude westerly winds that are shifted poleward with respect to their preindustrial position. This change in the westerlies puts stronger Southern Hemisphere westerlies directly over the ACC, which should make the MOC stronger. Thus, the MOC could become stronger or weaker depending on whether the winds or the hydrological cycle dominate in the future.

See also

Antarctic Circumpolar Current. Ocean Circulation.

Further Reading

Aoki S, Bindoff NL, and Church JA (2005) Interdecadal water mass changes in the Southern Ocean between 30°E and 160°E. *Geophysical Research Letters* 32: (doi:10.1029/2005GL022220).

Broecker WS (1991) The great ocean conveyor. *Oceanography* 4: 79–89.

Gnanadesikan A (1999) A simple predictive model for the structure of the oceanic pycnocline. *Science* 283: 2077–2079.

Kuhlbrodt T, Griesel A, Montoya M, Levermann A, Hofmann M, and Rahmstorf S (2007) On the driving processes of the Atlantic meridional overturning circulation. *Reviews of Geophysics* 45: RG2001 (doi:10.1029/2004RG000166).

Manabe S and Stouffer R (1993) Century-scale effects of increased atmospheric CO_2 on the ocean–atmosphere system. *Nature* 364: 215–218.

McCartney MS and Talley LD (1984) Warm-to-cold water conversion in the northern North Atlantic Ocean. *Journal of Physical Oceanography* 14: 922–935.

Munk W and Wunsch C (1998) Abyssal recipes II: Energetics of tidal and wind mixing. *Deep-Sea Research I* 45: 1977–2010.

Orsi A, Johnson G, and Bullister J (1999) Circulation, mixing, and production of Antarctic bottom water. *Progress in Oceanography* 43: 55–109.

Rahmstorf S (1995) Bifurcations of the Atlantic thermohaline circulation in response to changes in the hydrological cycle. *Nature* 378: 145–149.

Rudels B, Jones EP, Anderson LG, and Kattner G (1994) On the intermediate depth waters of the Arctic Ocean. In: Johannessen OM, Muench RD, and Overland JE (eds.) *Geophysical Monograph 85: The Polar Oceans and Their Role in Shaping the Global Environment*, pp. 33–46. Washington, DC: American Geophysical Union.

Russell JL, Dixon KW, Gnanadesikan A, Stouffer RJ, and Toggweiler JR (2006) The Southern Hemisphere westerlies in a warming world: Propping open the door to the deep ocean. *Journal of Climate* 19(24): 6382–6390.

Schmitz WJ and McCartney MS (1993) On the North Atlantic Circulation. *Reviews of Geophysics* 31: 29–49.

Toggweiler JR and Samuels B (1995) Effect of Drake Passage on the global thermohaline circulation. *Deep-Sea Research I* 42: 477–500.

Toggweiler JR and Samuels B (1998) On the ocean's large-scale circulation near the limit of no vertical mixing. *Journal of Physical Oceanography* 28: 1832–1852.

Toggweiler RR and Russell J (2008) Ocean Circulation in a warming climate. *Nature* 451: (doi:10.1038/nature 06590).

Tziperman E (2000) Proximity of the present-day thermohaline circulation to an instability threshold. *Journal of Physical Oceanography* 30: 90–104.

Wunsch C and Ferrari R (2004) Vertical mixing, energy, and the general circulation of the oceans. *Annual Reviews of Fluid Mechanics* 36: 281–314.

NORTH ATLANTIC OSCILLATION (NAO)

J. W. Hurrell, National Center for Atmospheric Research, Boulder, CO, USA

Introduction

Simultaneous variations in weather and climate over widely separated points on Earth have long been noted in the meteorological literature. Such variations are commonly referred to as 'teleconnections'. In the extratropics, teleconnections link neighboring regions mainly through the transient behavior of atmospheric planetary-scale waves. Consequently, some regions may be cooler than average, while thousands of kilometers away warmer conditions prevail. Though the precise nature and shape of these structures vary to some extent according to the statistical methodology and the data set employed in the analysis, consistent regional characteristics that identify the most conspicuous patterns emerge.

Over the middle and high latitudes of the northern hemisphere, a dozen or so distinct teleconnection patterns can be identified during boreal winter. One of the most prominent is the North Atlantic Oscillation (NAO). The NAO dictates climate variability from the eastern seaboard of the USA to Siberia and from the Arctic to the subtropical Atlantic. This widespread influence indicates that the NAO is more than just a North Atlantic phenomenon. In fact, it has been suggested that the NAO is the regional manifestation of a larger scale (hemispheric) mode of variability known as the Arctic Oscillation (see below). Regardless of terminology, meteorologists for more than two centuries have noted the pronounced influence of the NAO on the climate of the Atlantic basin.

Variations in the NAO are important to society and the environment. Through its control over regional temperature and precipitation variability, the NAO directly impacts agricultural yields, water management activities, and fish inventories among other things. The NAO accounts for much of the interannual and longer-term variability evident in northern hemisphere surface temperature, which has exhibited a warming trend over the past several decades to values that are perhaps unprecedented over the past 1000 years.

Understanding the processes that govern variability of the NAO is therefore of high priority, especially in the context of global climate change. This article defines the NAO and describes its relationship to variations in surface temperature and precipitation, as well as its impact on variability in the North Atlantic Ocean and on the regional ecology. It concludes with a discussion of the mechanisms that might influence the amplitude and timescales of the NAO, including the possible roles of the stratosphere and the ocean.

What is the North Atlantic Oscillation?

Like all atmospheric teleconnection patterns, the NAO is most clearly identified when time averaged data (monthly or seasonal) are examined, since time averaging reduces the 'noise' of small-scale and transient meteorological phenomena not related to large-scale climate variability. Its spatial signature and temporal variability are most often defined through the regional sea level pressure field, for which some of the longest instrumental records exist.

The NAO refers to a north–south oscillation in atmospheric mass with centers of action near Iceland and over the subtropical Atlantic from the Azores across the Iberian Peninsula. Although it is the only teleconnection pattern evident throughout the year in the northern hemisphere, its amplitude is largest during boreal winter when the atmosphere is dynamically the most active. During the months December through March, for instance, the NAO accounts for more than one-third of the total variance in sea level pressure over the North Atlantic.

A time series (or index) of more than 100 years of wintertime NAO variability and the spatial signature of the oscillation are shown in **Figures 1** and **2**[1] Differences of >15 hPa occur across the North Atlantic between the two phases of the NAO. In the so-called positive phase, higher than normal surface pressures south of 55°N combine with a broad region of anomalously low pressure throughout the Arctic. Because air flows counterclockwise around low pressure and clockwise around high pressure in the northern hemisphere, this phase of the oscillation is associated with stronger than average westerly winds across the middle latitudes of the Atlantic onto

---

[1] More sophisticated and objective statistical techniques, such as eigenvector analysis, yield time series and spatial patterns of average winter sea level pressure variability very similar to those shown in Figures 1 and 2.

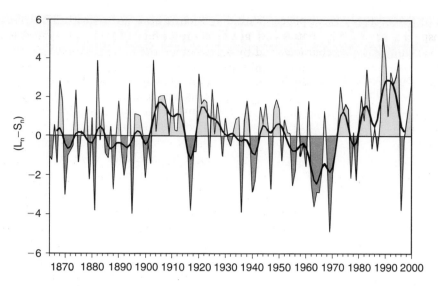

Figure 1 Winter (December–March) index of the NAO based on the difference of normalized sea level pressure between Lisbon, Portugal, and Stykkisholmur/Reykjavik, Iceland from 1864 to 2000. The average winter sea level pressure data at each station were normalized by division of each seasonal pressure by the long-term mean (1864–1983) standard deviation. The heavy solid line represents the index smoothed to remove fluctuations with periods <4 years.

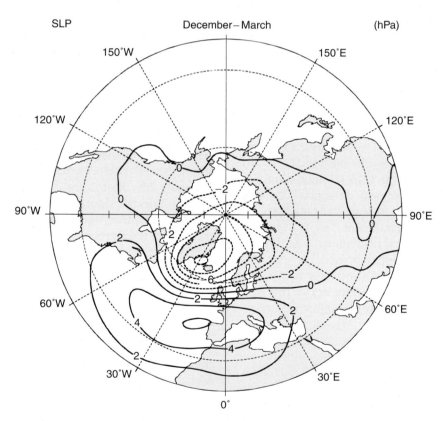

Figure 2 Difference in sea level pressure between years with an NAO index value >1.0 and those with an index value < − 1.0 (high minus low index winters) since 1899. The contour increment is 2 hPa and negative values are dashed.

Europe, with anomalous southerly flow over the eastern USA and anomalous northerly flow across western Greenland, the Canadian Arctic, and the Mediterranean.

The NAO is also readily apparent in meteorological data throughout the depth of the troposphere, and its variability is significantly correlated with changes in the strength of the winter polar vortex in the stratosphere of the northern hemisphere. Within the lower stratosphere, the leading pattern of geopotential height variability is also characterized by a seesaw in mass between the polar cap and the middle latitudes, but with a much more zonally symmetric (or annular) structure than in the troposphere. When heights over the polar region are lower than normal, heights at nearly all longitudes in middle latitudes are higher than normal. In this phase, the stratospheric westerly winds that encircle the pole are enhanced and the polar vortex is 'strong' and anomalously cold. It is this annular mode of variability that has been termed the Arctic Oscillation.

However, the signature of the stratospheric Arctic Oscillation in winter sea level pressure data looks very much like the anomalies associated with the NAO, with centers of action over the Arctic and the Atlantic (**Figure 2**). The 'annular' character of the Arctic Oscillation in the troposphere, therefore, reflects the vertically coherent fluctuations throughout the Arctic more than any coordinated behavior in the middle latitudes outside of the Atlantic basin. That the NAO and Arctic Oscillation reflect essentially the same mode of tropospheric variability is emphasized by the fact that their time series are nearly identical, with differences depending mostly on the details of the analysis procedure.

There is little evidence for the NAO to vary on any preferred timescale (**Figure 1**). Large changes can occur from one winter to the next, and there is also a considerable amount of variability within a given winter season. This is consistent with the notion that much of the atmospheric circulation variability in the form of the NAO arises from processes internal to the atmosphere, in which various scales of motion interact with one another to produce random (and thus unpredictable) variations. On the other hand, there are also periods when anomalous NAO-like circulation patterns persist over many consecutive winters. In the Icelandic region, for instance, sea level pressure tended to be anomalously low during winter from the turn of the century until about 1930 (positive NAO index), while the 1960s were characterized by unusually high surface pressure and severe winters from Greenland across northern Europe (negative NAO index). A sharp reversal has occurred over the past 30 years, with strongly positive NAO index values since 1980 and sea level pressure anomalies across the North Atlantic and Arctic that resemble those in **Figure 2**. In fact, the magnitude of the recent upward trend is unprecedented in the observational record and, based on reconstructions using paleoclimate and model data, perhaps over the past several centuries as well. Whether such low frequency (interdecadal) NAO variability arises from interactions of the North Atlantic atmosphere with other, more slowly varying components of the climate system (such as the ocean), whether the recent upward trend reflects a human influence on climate, or whether the longer timescale variations in the relatively short instrumental record simply reflect finite sampling of a purely random process, are topics of considerable current interest.

Impacts of the North Atlantic Oscillation

Temperature

The NAO exerts a dominant influence on wintertime temperatures across much of the northern hemisphere. Surface air temperature and sea surface temperature (SST) across wide regions of the North Atlantic Ocean, North America, the Arctic, Eurasia, and the Mediterranean are significantly correlated with NAO variability.[2] Such changes in surface temperature (and related changes in rainfall and storminess) can have significant impacts on a wide range of human activities, as well as on marine and terrestrial ecosystems.

When the NAO index is positive, enhanced westerly flow across the North Atlantic during winter moves relatively warm (and moist) maritime air over much of Europe and far downstream across Asia, while stronger northerlies over Greenland and north-eastern Canada carry cold air southward and decrease land temperatures and SST over the northwest Atlantic (**Figure 3**). Temperature variations over North Africa and the Middle East (cooling), as well as North America (warming), associated with the stronger clockwise flow around the subtropical Atlantic high-pressure center are also notable.

The pattern of temperature change associated with the NAO is important. Because the heat storage capacity of the ocean is much greater than that of land, changes in continental surface temperatures are much larger than those over the oceans, so they tend to dominate average northern hemisphere (and global) temperature variability. Especially given the large and coherent NAO signal across the Eurasian continent from the Atlantic to the Pacific (**Figure 3**), it is not surprising that NAO variability explains about one-third of the northern hemisphere interannual surface temperature variance during winter.

The strength of the link between the NAO and northern hemisphere temperature variability has also

---

[2] Sea surface temperatures (SSTs) are used to monitor surface air temperature over the oceans because intermittent sampling is a major problem and SSTs have much greater persistence.

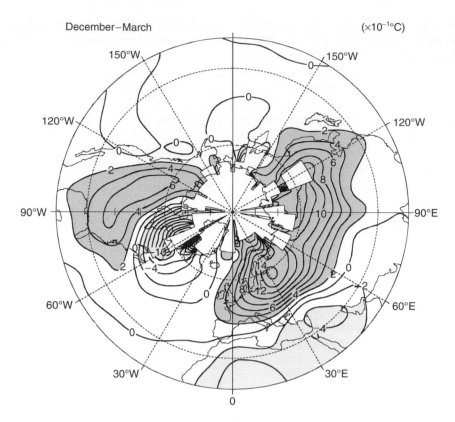

December–March   (×10⁻¹°C)

Figure 3 Changes in land surface and sea surface temperatures (× 10^{-1}°C) corresponding to unit deviations of the NAO index for the winter months (December–March) from 1935 to 1999. The contour increment is 0.2°C. Temperature changes >0.2°C are indicated by dark shading, and those < − 0.2°C are indicated by light shading. Regions with insufficient data are not contoured.

added to the debate over our ability to detect and distinguish between natural and anthropogenic climate change. Since the early 1980s, winter temperatures over much of North America and Eurasia have been considerably warmer than average, while temperatures over the northern oceans have been slightly colder than average. This pattern, which has contributed substantially to the well-documented warming trend in northern hemisphere and global temperatures over recent decades, is quite similar to winter surface temperature changes projected by computer models forced with increasing concentrations of greenhouse gases and aerosols. Yet, it is clear from the above discussion that a significant fraction of this temperature change signal is associated with the recent upward trend in the NAO (**Figures 1** and **3**). How anthropogenic climate change might influence modes of natural climate variability such as the NAO, and the nature of the relationships between increased radiative forcing and interdecadal variability of these modes, remain central research questions.

Precipitation and Storms

Changes in the mean circulation patterns over the North Atlantic are accompanied by changes in the

intensity and number of storms, their paths, and their associated weather. Here, the term 'storms' refers to atmospheric disturbances operating on timescales of about a week or less. During winter, a well-defined storm track connects the North Pacific and North Atlantic basins, with maximum storm activity over the oceans. The details of changes in storminess differ depending on the analysis method and whether the focus is on surface or upper-air features. Generally, however, positive NAO index winters are associated with a northward shift in the Atlantic storm activity, with enhanced activity from southern Greenland across Iceland into northern Europe and a modest decrease in activity to the south. The latter is most noticeable from the Azores across the Iberian Peninsula and the Mediterranean. Positive NAO winters are also typified by more intense and frequent storms in the vicinity of Iceland and the Norwegian Sea.

The ocean integrates the effects of storms in the form of surface waves, so that it exhibits a marked response to long-lasting shifts in the storm climate. The recent upward trend toward more positive NAO index winters has been associated with increased wave heights over the north-east Atlantic and decreased wave heights south of 40°N. Such changes have consequences for the operation and safety of shipping, offshore industries, and coastal development.

Changes in the mean flow and storminess associated with swings in the NAO are also reflected in pronounced changes in the transport and convergence of atmospheric moisture and, thus, the distribution of evaporation and precipitation. Evaporation (E) exceeds precipitation (P) over much of Greenland and the Canadian Arctic during high NAO index winters (**Figure 4**), where changes between high and low NAO index states are on the order of $1\,mm\,d^{-1}$. Drier conditions of the same magnitude also occur over much of central and southern Europe, the Mediterranean, and parts of the Middle East, whereas more precipitation than normal falls from Iceland through Scandinavia.

This pattern, together with the upward trend in the NAO index since the late 1960s (**Figure 1**), is consistent with recent observed changes in precipitation over much of the Atlantic basin. One of the few regions of the world where glaciers have not exhibited a retreat over the past several decades is in Scandinavia, where more than average amounts of precipitation have been typical of many winters since the early 1980s. In contrast, over the Alps, snow depth and duration in recent winters have been among the lowest recorded this century, and the retreat of Alpine glaciers has been widespread. Severe drought has persisted throughout parts of Spain and Portugal as well. As far east as Turkey, river runoff is significantly correlated with NAO variability. There is also some observational and modeling evidence of a declining precipitation rate over much of the

Greenland Ice Sheet over the past two decades, although measurement uncertainties are large.

Ocean Variability

It has long been recognized that fluctuations in SST and the strength of the NAO are related, and there are clear indications that the North Atlantic Ocean varies significantly with the overlying atmosphere. The leading pattern of SST variability during winter consists of a tripolar structure marked, in one phase, by a cold anomaly in the subpolar North Atlantic, a warm anomaly in the middle latitudes centered off Cape Hatteras, and a cold subtropical anomaly between the equator and 30°N. This structure suggests that the SST anomalies are driven by changes in the surface wind and air–sea heat exchanges associated with NAO variations. The relationship is strongest when the NAO index leads an index of the SST variability by several weeks, which highlights the well-known result that extratropical SST responds to atmospheric forcing on monthly and seasonal timescales. Over longer periods, persistent SST anomalies also appear to be related to persistent anomalous patterns of SLP (including the NAO), although the mechanisms which produce SST changes on decadal and longer time-scales remain unclear. Such fluctuations could primarily be the local oceanic response to atmospheric decadal variability. On the other hand, nonlocal dynamical processes in the ocean could also be contributing to the SST variations.

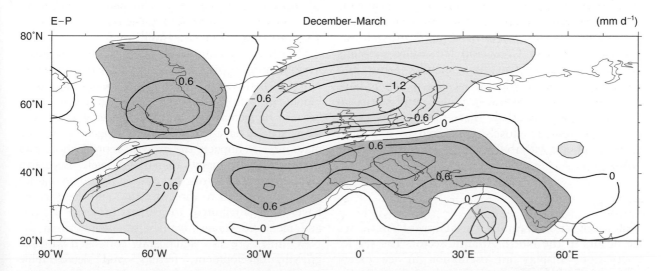

Figure 4 Difference in evaporation (E) minus precipitation (P) between years with an NAO index value >1.0 and those with an index value < − 1.0 (high minus low index winters) since 1958. The contour increment is 0.3 mm d⁻¹. Differences >0.3 mm d⁻¹ are indicated by dark shading, and those <0.3 mm d⁻¹ are indicated by light shading.

Subsurface ocean observations more clearly depict long-term climate variability, because the effect of the annual cycle and month-to-month variability in the atmospheric circulation decays rapidly with depth. These measurements are much more limited than surface observations, but over the North Atlantic they too indicate fluctuations that are coherent with the winter NAO index to depths of 400 m.

The ocean's response to NAO variability also appears to be evident in changes in the distribution and intensity of winter convective activity in the North Atlantic. The convective renewal of intermediate and deep waters in the Labrador Sea and the Greenland/Icelandorwegian Seas contribute significantly to the production and export of North Atlantic Deep Water and, thus, help to drive the global thermohaline circulation. The intensity of winter convection at these sites is not only characterized by large interannual variability, but also interdecadal variations that appear to be synchronized with variations in the NAO (**Figure 1**). Deep convection over the Labrador Sea, for instance, was at its weakest and shallowest in the postwar instrumental record during the late 1960s. Since then, Labrador Sea Water has become progressively colder and fresher, with intense convective activity to unprecedented ocean depths (>2300 m) in the early 1990s. In contrast, warmer and saltier deep waters in recent years are the result of suppressed convection in the Greeland/Icelandorwegian Seas, whereas intense convection was observed there during the late 1960s.

There has also been considerable interest in the occurrence of low salinity anomalies that propagate around the subpolar gyre of the North Atlantic. The most famous example is the Great Salinity Anomaly (GSA). The GSA formed during the extreme negative phase of the NAO in the late 1960s, when clockwise flow around anomalously high pressure over Greenland fed record amounts of fresh water through the Denmark Strait into the subpolar North Atlantic ocean gyre. There have been other similar instances as well, and statistical analyses have revealed that these propagating salinity modes are closely connected to a pattern of atmospheric variability strongly resembling the NAO.

Sea Ice

The strongest interannual variability of Arctic sea ice occurs in the North Atlantic sector. The sea ice fluctuations display an oscillation in ice extent between the Labrador and Greenland Seas. Strong interannual variability is evident in the sea ice changes over the North Atlantic, as are longer-term fluctuations, including a trend over the past 30 years of diminishing (increasing) ice concentration during boreal winter east (west) of Greenland. Associated with the sea ice fluctuations are large-scale changes in sea level pressure that closely resemble the NAO.

When the NAO is in its positive phase, the Labrador Sea ice boundary extends farther south, while the Greenland Sea ice boundary is north of its climatological extent. Given the implied surface wind changes (**Figure 2**), this is qualitatively consistent with the notion that sea ice anomalies are directly forced by the atmosphere, either dynamically via wind-driven ice drift anomalies, or thermodynamically through surface air temperature anomalies (**Figure 3**). The relationship between the NAO index (**Figure 1**) and an index of the North Atlantic ice variations is indeed strong, although it does not hold for all individual winters. This last point illustrates the importance of the regional atmospheric circulation in forcing the extent of sea ice.

Ecology

Changes in the NAO have a wide range of effects on North Atlantic ecosystems. Temperature is one of the primary factors, along with food availability and spawning grounds, in determining the large-scale distribution pattern of fish and shellfish. Changes in SST and winds associated with changes in the NAO have been linked to variations in the production of zooplankton, as well as to fluctuations in several of the most important fish stocks across the North Atlantic. This includes not only longer-term changes associated with interdecadal NAO variability, but also interannual signals as well.

Over land, fluctuations in the strength of the NAO have been linked to variations in plant phenology. In Norway, for example, most plant species have been blooming from 2–4 weeks earlier in recent years because of increasingly warm and wet winters, and many species have been blooming longer. Winter climate directly impacts the growth, reproduction, and demography of many animals, as well. European amphibians and birds have been breeding earlier over the past two to three decades, and these trends have been attributed to earlier growing seasons and increased forage availability. Variations in the NAO index are also significantly correlated with the growth, development, fertility, and demographic trends of large mammals from North America to northern Europe, such as northern ungulates and Canadian lynx.

What are the Mechanisms that Govern North Atlantic Oscillation Variability?

Although the NAO is an internal mode of variability of the atmosphere, surface, stratospheric or anthropogenic processes may influence its phase and amplitude. At present there is no consensus on the process or processes that most influence the NAO, especially on long (interdecadal) timescales. It is quite possible that NAO variability is affected by one or more of the mechanisms described below.

Atmospheric Processes

Atmospheric general circulation model (AGCMs)[3] provide strong evidence that the basic structure of the NAO results from the internal, nonlinear dynamics of the atmosphere. The observed spatial pattern and amplitude of the NAO are well simulated in AGCMs forced with fixed climatological annual cycles of solar insolation and SST, as well as fixed atmospheric trace gas composition. The governing dynamical mechanisms are interactions between the time-mean flow and the departures from that flow. Such intrinsic atmospheric variability exhibits little temporal coherence and, indeed, the timescales of observed NAO variability (**Figure 1**) do not differ significantly from this reference.

A possible exception is the interdecadal NAO variability, especially the strong trend toward the positive index polarity of the oscillation over the past 30 years. This trend exhibits a high degree of statistical significance relative to the background interannual variability in the observed record; moreover, multi-century AGCM experiments like those described above do not reproduce interdecadal changes of comparable magnitude. A possible source of the trend could be through a connection to processes that have affected the strength of the atmospheric circulation in the lower stratosphere.

During winters when the stratospheric westerlies are enhanced, the NAO tends to be in its positive phase. There is a considerable body of evidence to support the notion that variability in the troposphere can drive variability in the stratosphere, but it also appears that some stratospheric control of the troposphere may also occur.

The atmospheric response to strong tropical volcanic eruptions provides some evidence for a stratospheric

influence on the Earth's surface climate. Volcanic aerosols act to enhance north–south temperature gradients in the lower stratosphere by absorbing solar radiation in lower latitudes. In the troposphere, the aerosols exert only a very small direct influence. Yet, the observed response following eruptions is not only lower geopotential heights over the pole with stronger stratospheric westerlies, but also a positive NAO-like signal in the tropospheric circulation.

Reductions in stratospheric ozone and increases in greenhouse gas concentrations also appear to enhance the meridional temperature gradient in the lower stratosphere, leading to a stronger polar vortex. It is possible, therefore, that the upward trend in the NAO index in recent decades (**Figure 1**) is associated with trends in either or both of these quantities. Indeed, a decline in the amount of ozone poleward of 40°N has been observed during the last two decades, and the stratospheric polar vortex has become colder and stronger.

Ocean Forcing of the Atmosphere

In the extratropics, it is clear that the atmospheric circulation is the dominant driver of upper ocean thermal anomalies. A long-standing issue, however, has been the extent to which anomalous extratropical SST feeds back to affect the atmosphere. Most evidence suggests that this effect is quite small compared with internal atmospheric variability. Nevertheless, the interaction between the ocean and atmosphere could be important for understanding the details of the observed amplitude of the NAO, its interdecadal variability, and the prospects for meaningful predictability.

While intrinsic atmospheric variability is random in time, theoretical and modeling evidence suggest that the ocean can respond to it with marked persistence or even oscillatory behavior. On seasonal and interannual timescales, for example, the large heat capacity of the upper ocean can lead to slower changes in SST relative to the faster, stochastic atmospheric forcing. On longer timescales, SST observations display a myriad of variations. Middle and high latitude SSTs over the North Atlantic, for example, were colder than average during the 1970s and 1980s, but warmer than average from the 1930s through the 1950s. Within these periods, shorter-term variations in SST are also apparent, such as a dipole pattern that fluctuates on approximately decadal time scales with anomalies of one sign east of Newfoundland, and anomalies of opposite polarity off the south-east coast of the USA.

A key to whether or not changes in the state of the NAO reflect these variations in the state of the ocean

---

[3] Atmospheric general circulation models consist of a system of equations that describe the large-scale atmospheric balances of momentum, heat, and moisture, with schemes that approximate small-scale processes such as cloud formation, precipitation, and heat exchange with the sea surface and land.

surface is the sensitivity of the atmosphere to middle and high latitude SST and upper ocean heat content anomalies. Most AGCM studies show weak responses to extratropical SST anomalies, with sometimes contradictory results. Yet, some AGCMs, when forced with the time history of observed, global SSTs and sea ice concentrations over the past 50 years or so, show modest skill in reproducing aspects of the observed NAO behavior, especially its interdecadal fluctuations (**Figure 1**).

Such results do not necessarily imply, however, that the extratropical ocean is behaving in anything other than a passive manner. It could be, for instance, that long-term changes in tropical SSTs force a remote atmospheric response over the North Atlantic, which in turn drives changes in extratropical SSTs and sea ice. Some model studies indicate a sensitivity of the North Atlantic atmosphere to tropical SST variations, including variations over the tropical Atlantic which are substantial on both interannual and interdecadal timescales.

The response of the extratropical North Atlantic atmosphere to changes in tropical and extratropical SST distributions, and the role of land processes and sea ice in producing atmospheric variability, are problems which are currently being addressed. Until these are better understood, it is difficult to evaluate the realism of more complicated scenarios that rely on truly coupled interactions between the atmosphere, ocean, land, and sea ice to produce North Atlantic climate variability. It is also difficult to evaluate the extent to which interannual and longer-term variations of the NAO might be predictable.

Glossary

Great salinity anomaly   A widespread freshening of the upper 500–800 m layer of the far northern North Atlantic Ocean, traceable around the subpolar gyre from its origins north of Iceland in the mid-to-late 1960s until it return to the Greenland Sea in the early 1980s.

See also

Evaporation and Humidity. Heat and Momentum Fluxes at the Sea Surface. Sea Ice.

Further Reading

Appenzeller C, Stocker TF, and Anklin M (1998) North Atlantic oscillation dynamics recorded in Greenland ice cores. *Science* 282: 446–449.

Dickson B (1999) All change in the Arctic. *Nature* 397: 389–391.

Hurrell JW (1995) Decadal trends in the North Atlantic Oscillation regional temperatures and precipitation. *Science* 269: 676–679.

Kerr RA (1997) A new driver for the Atlantic's moods and Europe's weather? *Science* 275: 754–755.

National Research Council (1998) Decade-to-century scale climate variability and change. A science strategy. Washington: National Academy Press.

Rodwell MJ, Rowell DP, and Folland CK (1999) Oceanic forcing of the wintertime North Atlantic Oscillation and European climate. *Nature* 398: 320–323.

Shindell DT, Miller RL, Schmidt G, and Pandolfo L (1999) Simulation of recent northern winter climate trends by greenhouse gas forcing. *Nature* 399: 452–455.

Stenseth NC and Co-authors 1999 (1999) Common dynamic structure of Canadian lynx populations within three climatic regions. *Science* 285: 1071–1073.

Sutton R and Allen MR (1997) Decadal predictability of North Atlantic sea surface temperature and climate. *Nature* 388: 563–567.

Thompson DWJ and Wallace JM (1998) The Arctic oscillation signature in the wintertime geopotential height and temperature fields. *Geophysical Research Letters* 25: 1297–1300.

World Climate Research Programme (1998) The North Atlantic Oscillation. *Climate Variability and Predictability (CLIVAR) Initial Implementation Plan.* WCRP no. 103, WMO/TD no. 869, ICPO no. 14, 163–192.

ABRUPT CLIMATE CHANGE

S. Rahmstorf, Potsdam Institute for Climate Impact Research, Potsdam, Germany

Introduction

High-resolution paleoclimatic records from ice and sediment cores and other sources have revealed a number of dramatic climatic changes that occurred over surprisingly short times – a few decades or in some cases a few years. In Greenland, for example, temperature rose by 5–10 °C, snowfall rates doubled, and windblown dust decreased by an order of magnitude within 40 years at the end of the last glacial period. In the Sahara, an abrupt transition occurred around 5500 years ago from a relatively green shrubland supporting significant populations of animals and humans to the dry desert we know today.

One could define an abrupt climate change simply as a large and rapid one – occurring faster than in a given time (say 30 years). The change from winter to summer, a very large change (in many places larger than the glacial–interglacial transition) occurring within 6 months, is, however, not an abrupt change in climate (or weather), it is rather a gradual transition following the solar forcing in its near-sinusoidal path. The term 'abrupt' implies not just rapidity but also reaching a breaking point, a threshold – it implies a change that does not smoothly follow the forcing but is rapid in comparison to it. This physical definition thus equates abrupt climate change with a strongly nonlinear response to the forcing. In this definition, the quaternary transitions from glacial to interglacial conditions and back, taking a few hundred or thousand years, are a prime example of abrupt climate change, as the underlying cause, the Earth's orbital variations (Milankovich cycles), have timescales of tens of thousands of years. On the other hand, anthropogenic global warming occurring within a hundred years is not as such an abrupt climate change as long as it smoothly follows the increase in atmospheric carbon dioxide. Only if global warming triggered a nonlinear response, like a rapid ocean circulation change or decay of the West Antarctic Ice Sheet (WAIS), would one speak of an abrupt climate change.

Paleoclimatic Data

A wealth of paleoclimatic data has been recovered from ice cores, sediment cores, corals, tree rings, and other sources, and there have been significant advances in analysis and dating techniques. These advances allow a description of the characteristics of past climatic changes, including many abrupt ones, in terms of geographical patterns, timing, and affected climatic variables. For example, the ratio of oxygen isotopes in ice cores yields information about the temperature in the cloud from which the snow fell. Another way to determine temperature is to measure the isotopic composition of the nitrogen gas trapped in the ice, and it is also possible to directly measure the temperature in the borehole with a thermometer. Each method has advantages and drawbacks in terms of time resolution and reliability of the temperature calibration. Dust, carbon dioxide, and methane content of the prehistoric atmosphere can also be determined from ice cores.

On long timescales, climatic variability throughout the past 2 My at least has been dominated by the Milankovich cycles in the Earth's orbit around the sun – the cycles of precession, obliquity, and eccentricity with periods of roughly 23 ky, 41 ky, and 100 ky, respectively. Since the middle Pleistocene transition 1.2 Ma, the regular glaciations of our planet follow the 100-ky eccentricity cycle; even though this has only a rather weak direct influence on the solar radiation reaching the Earth, it modulates the much stronger other two cycles. The prevalence of the 100-ky cycle in climate is thus apparently a highly nonlinear response to the forcing that is likely linked to the nonlinear continental ice sheet and/or carbon cycle dynamics. The terminations of glaciations occur rather abruptly (**Figure 1**). Greenland ice cores show that the transition from the last Ice Age to the warm Holocene climate took about 1470 years, with much of the change occurring in only 40 years. The local Greenland response is not typical for the global response; however, since Greenland temperatures can be strongly affected by Atlantic Ocean circulation, which went through rapid changes during deglaciation. Globally, the transition from full Ice Age to Holocene conditions took around 5 ky.

The ice ages were not just generally colder than the present climate but were also punctuated by abrupt climatic transitions. The best evidence for these transitions, known as Dansgaard–Oeschger (D/O) events, comes from the last ice age (**Figure 2**). D/O events typically start with an abrupt warming by up to 12 °C within a few decades or less, followed by gradual cooling over several hundred or thousand

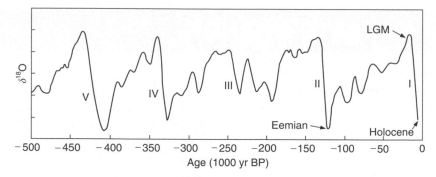

Figure 1 Record of $\delta^{18}O$ from marine sediments (arbitrary units), reflecting mainly the changes in global ice volume during the past 50 ky. Note the rapid terminations (labeled with roman numbers) of glacial periods.

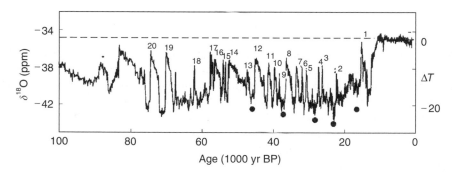

Figure 2 Record $\delta^{18}O$ from the GRIP ice core, a proxy for atmospheric temperature over Greenland (approximate temperature range ΔT (°C) is given on the right). Note the relatively stable Holocene climate during the past 10 ky and before that the much colder glacial climate punctuated by Dansgaard–Oeschger warm events (numbered). The timing of Heinrich events 1 to 6 is marked by black dots.

years. The cooling phase often ends with an abrupt final temperature drop back to cold ('stadial') conditions. Although first seen in the Greenland ice cores, the D/O events are not a local feature of Greenland climate. **Figure 3** shows that subtropical sea surface temperatures in the Atlantic closely mirror the sequence of events in Greenland. Similar records have been found near Santa Barbara, California, in the Cariaco Basin off Venezuela, and off the coast of India. D/O climate change is centered on the North Atlantic and regions with strong atmospheric response to changes in the North Atlantic, with little response in the Southern Ocean or Antarctica. The 'waiting time' between successive D/O events is most often around 1470 years or, with decreasing probability, multiples of this period. This suggests the existence of an as yet unexplained 1470-year cycle that often (but not always) triggers a D/O event. The second major type of abrupt event in glacial times is the Heinrich (H) event. H events involve surging of the Laurentide Ice Sheet through Hudson Strait, occurring in the cold stadial phase of some D/O cycles. They have a variable spacing of several thousand years. The icebergs released to the

North Atlantic during H events leave telltale dropstones in the ocean sediments when they melt, the so-called Heinrich layers. Sediment data suggest that H events shut down or at least drastically reduce the formation of North Atlantic Deep Water (NADW). Records from the South Atlantic and parts of Antarctica show that the cold H events in the North Atlantic were associated with unusual warming there (a fact sometimes referred to as 'bipolar seesaw').

At the end of the last glacial, a particularly interesting abrupt climatic change took place, the so-called Younger Dryas event (12 800–11 500 years ago). Conditions had already warmed to near-interglacial conditions and continental ice sheets were retreating, when within decades the climate in the North Atlantic region switched back to glacial conditions for more than a thousand years. It has been speculated that the cooling resulted from a sudden influx of fresh water into the North Atlantic through St. Lawrence River, when an ice barrier holding back a huge meltwater lake on the North American continent broke. This could have shut down the Atlantic thermohaline circulation (i.e., the circulation driven by temperature and salinity differences), but evidence is controversial.

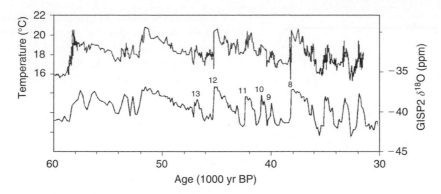

Figure 3 Sea-surface temperatures derived from alkenones in marine sediments from the subtropical Atlantic (Bermuda Rise, upper curve) compared to δ^{18}O values from the GISP2 ice core in Greenland (lower curve).

Alternatively, the Younger Dryas may simply have been the last cold stadial period of the glacial following a temporary D/O warming event.

Does abrupt climate change occur only during glacial times? Early evidence for the last interglacial, the Eemian, suggested abrupt changes there, but has since been refuted. During the present interglacial, the Holocene, climate was much more stable than during the last glacial. However, two abrupt events stand out. One is the 8200-year event that shows up as a cold spike in Arctic ice cores and affected the North Atlantic region. The second major change is the abrupt desertification of the Sahara 5500 years ago. There is much evidence from cave paintings, fire remains, bones, ancient lake sediments, and the like that the Sahara was a partly swampy savannah before this time. The best evidence for the abrupt ending of this benign climate comes from Atlantic sediments off northeastern Africa, which show a sudden and dramatic step-function increase in windblown dust, witnessing a drying of the adjacent continent.

Mechanisms of Abrupt Climate Change

The increased spatial coverage, quality, and time resolution of paleoclimatic data as well as advances in computer modeling have led to a greater understanding of the mechanisms of abrupt climate change, although many aspects are still in dispute and not fully understood.

The simplest concept for a mechanism causing abrupt climatic change is that of a threshold. A gradual change in external forcing (e.g., the change in insolation due to the Milankovich cycles) or in an internal climatic parameter (e.g., the slow buildup or melting of continental ice) continues until a specific threshold value is reached where some qualitative change in climate is triggered. Various such critical thresholds are known to exist in the climate system. Continental ice sheets may have a stability threshold where they start to surge; the thermohaline ocean circulation has thresholds where deep-water formation shuts down or shifts location; methane hydrates in the seafloor have a temperature threshold where they change into the gas phase and bubble up into the atmosphere; and the atmosphere itself may have thresholds where large-scale circulation regimes (such as the monsoon) switch.

For the D/O events, H events, and the Younger Dryas event discussed above, the paleoclimatic data clearly point to a crucial role of Atlantic Ocean circulation changes. Modeling and analytical studies of the Atlantic thermohaline circulation (sometimes called the 'conveyor belt') show that there are two positive feedback mechanisms leading to threshold behavior. The first, called advective feedback, is caused by the large-scale northward transport of salt by the Atlantic currents, which in turn strengthens the circulation by increasing density in the northern latitudes. The second, called convective feedback, is caused by the fact that oceanic convection creates conditions favorable for further convection. These (interconnected) feedbacks make convection and the large-scale thermohaline circulation self-sustaining within certain limits, with well-defined thresholds where the circulation changes to a qualitatively different mode.

Three main circulation modes have been identified both in sediment data and in models (**Figure 4**): (1) a warm or interglacial mode with deep-water formation in the Nordic Seas and large oceanic heat transport to northern high latitudes (**Figure 4(a)**); (2) a cold or stadial mode with deep-water formation south of the shallow sill between Greenland, Iceland, and Scotland and with greatly reduced heat transport

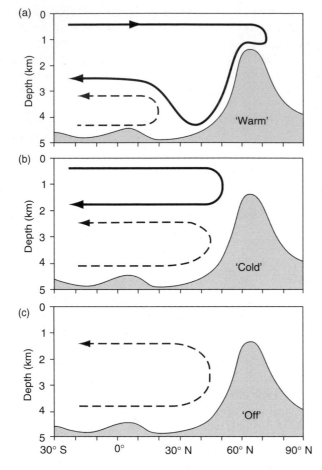

Figure 4 Schematic of three major modes of Atlantic Ocean circulation. (a) 'Warm' or interglacial mode; (b) 'cold' or stadial mode; (c) 'off' or Heinrich mode. In the warm mode the Atlantic thermohaline circulation reaches north over the Greenland–Iceland–Scotland ridge into the Nordic Seas, while in the cold mode it stops south of Iceland. Switches between circulation modes at certain thresholds can pace and amplify climatic changes.

to high latitudes (**Figure 4(b)**); and (3) a 'switched off' or 'Heinrich' mode with practically no deep-water formation in the North Atlantic (**Figure 4(c)**). In the last mode, the Atlantic deep circulation is dominated by inflow of Antarctic Bottom Water (AABW) from the south.

Many features of abrupt glacial climate can be explained by switches between these three circulation modes. Model simulations suggest that the cold stadial mode is the only stable mode in a glacial climate; it prevails during the cold stadial periods of the last glacial. D/O events can be interpreted as temporary incursions of warm Atlantic waters into the Nordic Seas and deep-water formation there, that is, a switch to the warm mode causing abrupt climatic warming in the North Atlantic region. As this mode is not stable in glacial conditions, the

circulation starts to gradually weaken and temperatures start to decline again immediately after the incursion, until the threshold is reached where convection in the Nordic Seas stops and the system reverts to the stable stadial mode. H events can be interpreted as a switch from the stadial mode to the H mode, that is, a shutdown of North Atlantic deep-water formation. As this mode is probably also unstable in glacial conditions, the system spontaneously reverts to the stadial or to the warm mode after a waiting time of centuries, the timescale being determined partly by slow oceanic mixing processes.

This interpretation is consistent with the observed patterns of surface temperature change. The warming during D/O events is centered on the North Atlantic because this is where the change in oceanic heat transport occurs; the warm mode delivers heat to much higher latitudes than does the cold mode. A switch to the H mode, on the other hand, strongly reduces the interhemispheric heat transport from the South Atlantic to the North Atlantic. This cools the Northern Hemisphere while warming the Southern Hemisphere, explaining the 'bipolar seesaw' response in climate. It should also be noted that the initial transient response can differ from the equilibrium response as the oceanic heat storage capacity is large. The patterns of these abrupt changes differ from the longer-timescale (many thousands of years) response to the Milankovich cycles because, for the latter, the slow changes in atmospheric greenhouse gases (e.g., CO_2) and continental ice cover act to globally synchronize and amplify climatic change.

While the threshold behavior of the Atlantic ocean can dramatically shape and amplify climatic change, the question remains what triggers the mode switches. As mentioned above, D/O switches appear to be paced by an underlying 1470-year cyclicity that is as yet unexplained. This could either be an external (astronomical or solar) cycle or an internal oscillation of the climate system, perhaps also involving the Atlantic thermohaline circulation. A superposition of two major shorter solar cycles can, in climate model simulations, trigger events spaced 1470 years apart. The irregularity in D/O event timing is probably the result of the presence of stochastic variability in the climate system as well as the presence of longer-term trends such as the slow buildup of large continental ice sheets.

The ocean circulation change during H events, on the other hand, can be explained by the large amounts of fresh water entering the North Atlantic at these times in the form of icebergs. Simulations show that the observed amounts of fresh water are sufficient to shut down deep-water formation in the North Atlantic. The nonlinear dynamics of ice sheets provide a plausible trigger mechanism. Ice sheets may grow for many thousands of years until their base melts owing to geothermal heating, when the ice sheet becomes unstable and surges.

Thresholds in ocean circulation and continental or sea ice dynamics are not the only mechanisms that can cause abrupt climatic changes. In the desertification of the Sahara in the mid-Holocene, probably neither of these mechanisms were involved. Rather, an unstable positive feedback between vegetation cover (affecting albedo and evapotranspiration) and monsoon circulation in the atmosphere appears to have been responsible.

It is almost certain that there are further nonlinearities in the climate system that could have caused abrupt climatic changes in the past or may do so in the future. We are only beginning to understand abrupt climate change, and the interpretations presented here – while consistent with data and model results – are not the only possible interpretations. Reflecting the state of this science, they are current working hypotheses rather than established and well-tested theory.

Risk of Future Abrupt Changes

The prevalence of abrupt nonlinear (rather than smooth and gradual) climatic change in the past naturally leads to the question whether such changes can be expected in the future, either by natural causes or by human interference. The main outside driving forces of past climatic changes are the Milankovich cycles. Close inspection of these cycles as well as modeling results indicate that we are presently enjoying an unusually quiet period in the climatic effect of these cycles, owing to the present minimum in eccentricity of the Earth's orbit. The next large change in solar radiation that could trigger a new ice age is probably at least 30 ky away. If this is correct, it makes the Holocene an unusually long interglacial, comparable to the Holstein interglacial that occurred around 400 ka when the Earth's orbit went through a similar pattern. This stable orbital situation leaves unpredictable events (such as meteorite impacts or a series of extremely large volcanic eruptions) and anthropogenic interference as possible causes for abrupt climatic changes in the lifetime of the next few generations of humans.

Significant anthropogenic warming of the lower atmosphere and ocean surface will almost certainly occur in this century, raising concerns that nonlinear thresholds in the climate system could be exceeded and abrupt changes could be triggered at some point. Processes that have been (rather speculatively)

mentioned in this context include a collapse of the WAIS, a strongly enhanced greenhouse effect due to melting of permafrost or triggering of methane hydrate deposits at the seafloor, a large-scale wilting of forests when drought-tolerance thresholds are exceeded, nonlinear changes in monsoon regimes, and abrupt changes in ocean circulation.

Of those possibilities, the risk of a change in ocean circulation is probably the best understood and perhaps also the least unlikely. Two factors could weaken the circulation and bring it closer to a threshold: the warming of the surface and a dilution of high-latitude waters with fresh water. The latter could result from an enhanced atmospheric water cycle and precipitation as well as meltwater runoff from Greenland and other glaciers. Both warming and fresh water input reduce surface density and thereby inhibit deep-water formation. Model simulations of global-warming scenarios so far suggest three possible responses: a shutdown of convection in the Labrador Sea, one of the two main NADW formation sites; a complete shutdown of NADW formation (i.e., similar to a switch to the H mode); and a shutdown of AABW formation. A transition to the stadial circulation mode has so far not been simulated, perhaps because convection in the Nordic Seas is strongly wind driven and is more effectively switched off by increased sea ice cover than by warming.

A shutdown of Labrador Sea convection would be a significant qualitative change in the Atlantic Ocean circulation, but would probably affect only the surface climate of a smaller region surrounding the Labrador Sea. Effects on ecosystems and fisheries have not been investigated but could be severe. A complete shutdown of NADW formation would have wider climatic repercussions. Temperatures in northwestern Europe could initially rise several degrees in step with global warming, then abruptly drop back to near present values (the competing effects of raised atmospheric CO_2 and reduced oceanic heat transport almost balancing). If CO_2 levels decline again in future centuries as expected, European temperatures could remain several degrees below present as the Atlantic thermohaline circulation is not expected to recover perhaps for millennia. Further effects of a shutdown of deep-water renewal include reduced oceanic uptake of CO_2 (enhancing the greenhouse effect), shifts in tropical rainfall belts, accelerated global sea level rise (due to a faster warming of the deep oceans), and rapid regional sea level rise in the northern Atlantic.

The probability of major climatic thresholds being crossed in the coming centuries is difficult to establish and largely unknown. Currently, this possibility lies within the (still rather large) uncertainty range for future climate projections, so the risk cannot be ruled out. The IPCC 4th assessment report assigns a probability of up to 10% to a shutdown of the Atlantic overturning circulation within this century.

See also

Ocean Circulation: Meridional Overturning Circulation. Past Climate from Corals.

Further Reading

Abrantes F and Mix A (eds.) (1999) *Reconstructing Ocean History – A Window Into the Future.* New York: Plenum.

Broecker W (1987) Unpleasant surprises in the greenhouse? *Nature* 328: 123.

Clark PU, Webb RS, and Keigwin LD (eds.) (1999) *Mechanisms of Global Climate Change at Millennial Time Scales.* Washington, DC: American Geophysical Union.

Clark PU, Alley RB, and Pollard D (1999) Northern Hemisphere ice sheet influences on global climate change. *Science* 286: 1104–1111.

Houghton JT, Meira Filho LG, and Callander BA (1995) *Climate Change 1995.* Cambridge, UK: Cambridge University Press.

Sachs JP and Lehman SJ (1999) Subtropical North Atlantic temperatures 60,000 to 30,000 years ago. *Science* 286: 756–759.

Stocker T (2000) Past and future reorganisations in the climate system. *Quarterly Science Review (PAGES Special Issue)* 19: 301–319.

Taylor K (1999) Rapid climate change. *American Scientist* 87: 320.

TROPICAL CLIMATE CONTROLS

PACIFIC OCEAN EQUATORIAL CURRENTS

R. Lukas, University of Hawaii at Manoa, Hawaii, USA

Introduction

An essential characteristic of Pacific equatorial ocean currents is that they span the width of the Pacific basin (15 000 km at the Equator), linking to eastern and western boundary flows (**Figures 1** and **2**). While they have long zonal scales, the relatively strong near-equatorial flows have complex vertical and meridional structures. They exhibit energetic variability on timescales from days to years. In particular, the currents of the equatorial Pacific are considerably altered during El Niño/Southern Oscillation (ENSO) events.

These flows are subject to the distinctive physics associated with the equatorward decrease of the vertical component of the earth's rotation vector. The associated vanishing of the horizontal Coriolis force at the Equator results in relatively strong currents for a given wind stress or pressure gradient. Rapid

adjustment of the currents to changing forcing is associated with a special class of internal wave motions termed linear equatorially trapped waves. There are several different types of waves with rich meridional and vertical structure, governed by dispersion relationships that tie zonal wavelength, meridional structure, and wave period together. The fastest waves cross the Pacific in only 2–3 months. With greater distance from the Equator, zonal propagation speeds become slower. The key feature is that these waves can transmit the signals of wind forcing to and from remote locations.

The Pacific equatorial surface currents are primarily wind-driven. Local forcing of the equatorial currents is dominated by surface wind stress (as opposed to heat and/or fresh water fluxes) and its variability. Variable wind forcing results in vertical pumping of the thermocline and subsequent dynamic adjustment, including radiation of equatorially trapped waves. The currents are not forced solely by local winds, however, because equatorially trapped waves carry wind-forcing signals across the entire basin, and similar boundary-trapped waves transmit information about forcing between the equator and higher latitudes. An important portion of the

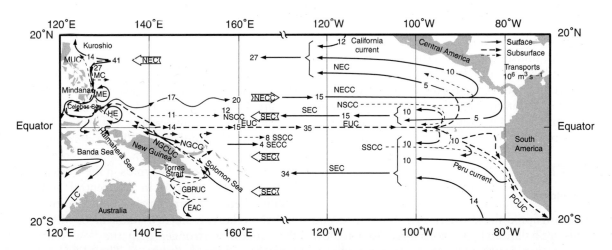

Figure 1 Schematic illustration of the major equatorial currents in the Pacific Ocean and their connections to eastern and western boundary currents. (Note the break in the longitude axis.) Surface currents are indicated by solid lines; subsurface currents are indicated by dashed lines, with deeper currents having lighter weight. Approximate average individual current transports (10^6 m^3 s^{-1}) are provided where known with some confidence. The equatorial surface currents are the North Equatorial Current (NEC), the South Equatorial Current (SEC), the North Equatorial Countercurrent (NECC), and the South Equatorial Countercurrent (SECC). The subsurface equatorial currents discussed here are the Equatorial Undercurrent (EUC), the Northern Subsurface Countercurrent (NSCC), and the Southern Subsurface Countercurrent (SSCC). Eastern boundary currents are the Peru Current and the Peru–Chile Undercurrent (PCUC). Western boundary surface currents are the Kuroshio, the Mindanao Current (MC), the New Guinea Coastal Current (NGCC) and the East Australia Current (EAC). Subsurface flows along the western boundary are the Mindanao Undercurrent (MUC), the New Guinea Coastal Undercurrent (NGCUC), and the Great Barrier Reef Undercurrent (GBRUC). The Mindanao Eddy (ME) and Halmahera Eddy (HE) are indicated.

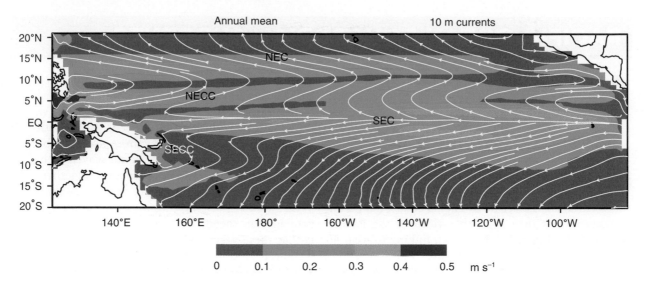

Figure 2 Map of long-term mean surface flow in the tropical Pacific. White lines and arrows indicate the direction of flow, while the colors indicate the speed of flow as given by the color bar. Current names are abbreviated as in **Figure 1**.

equatorial circulation is forced by winds and buoyancy forces (surface heat and fresh water fluxes) far from the equator.

The basic spatial structures and temporal variation of Pacific Ocean equatorial currents are presented here. Because systematic current measurements are available at only a few, widely separated locations, spatial variability is addressed primarily with the ocean assimilation/reanalysis from the US National Oceanic and Atmospheric Administration (NOAA) National Centers for Environmental Prediction, which combines a general circulation model of the ocean with atmospheric and ocean data to estimate the state of the tropical Pacific Ocean every week. Direct current measurements from several sites along the Equator maintained by the NOAA Pacific Marine Environmental Laboratory are used primarily to address temporal variability.

Mean Flow

Interior Flows

Zonal geostrophic flow, where meridional pressure gradients are balanced by Coriolis forces (due to flow on the rotating earth), dominates over meridional flow in the long-term annual-average Pacific equatorial circulation. In the surface layer (upper 50 m or so), currents directly driven by the generally westward Trade Winds typically flow poleward, superimposed on these strong zonal flows (**Figure 2**). The divergence of poleward-flowing surface currents is most pronounced along the Equator, leading to depth-dependent pressure gradients and strong vertical flow (called equatorial upwelling).

Surface currents The time-averaged surface flows (**Figure 2**) are dominated by the westward South Equatorial Current (SEC; between about 3°N and 20°S) and the North Equatorial Current (NEC; between about 10°N and 20°N). A persistent North Equatorial Countercurrent (NECC) flows eastward across the basin in the narrow band between about 5°N and 10°N. A weaker eastward-flowing South Equatorial Countercurrent (SECC) extends eastward from the region of the western boundary, but this flow only intermittently reaches the central and eastern Pacific.

North–south profiles of surface currents at three different longitudes across the Pacific basin clearly show the structure of these major zonal flows (**Figure 3**). The NEC, NECC, and SEC are strongest in the central Pacific where the Trade Winds are strongest. The SEC and NECC are considerably weaker in the west than in the east, reflecting the greater variability of the winds in the western Pacific. Very near the Equator in the central and eastern Pacific, there is a minimum in the speed of the SEC, and the relatively narrow filament of SEC north of the equator is stronger than the flow south of the Equator, except in the west. The SECC occurs between about 3°S and 10°S, but generally dissipates west of the dateline.

The meridional flows are considerably weaker than the zonal flow (**Figure 3**). The average currents have a northward component north of the Equator, and southward to the south of the Equator. The transition occurs rapidly very close to the Equator at all three longitudes, due to east-to-west trade winds and the change in sign of the Coriolis force at the

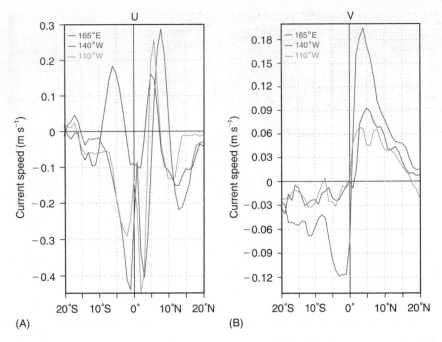

Figure 3 North–south profiles of zonal (A) and meridional (B) current in the western (165°E), central (140°W) and eastern (110°W) Pacific. Positive current is eastward and northward. Note the difference in the velocity scales between the two panels.

Equator. This divergence causes equatorial upwelling, which is responsible for colder surface temperatures along the Equator than just to the north or south.

Subsurface structure The zonal flow of the NECC is unusual among the surface currents in that it has a subsurface maximum near 50 m (**Figure 4**). This is due to its flowing eastward against the prevailing Trade winds. The directly wind-driven flow vanishes below the mixed layer (usually shallower than 100 m), and currents below are zonally-oriented except near the eastern and western boundaries.

The time-averaged subsurface flows are dominated by the eastward Equatorial Undercurrent (EUC; **Figures 4–6**), which is the strongest equatorial Pacific current, reaching speeds of about 1 m s^{-1} between 120°W and 140°W (**Figures 5** and **6**). The EUC is found within the very strong equatorial thermocline just below the westward SEC (**Figures 4** and **5**). The strong mean shear above the core of the EUC gives rise to strong vertical mixing. Note also the strong meridional shear of the zonal flow on either side of the EUC, and between the SEC and NECC.

Much weaker Northern and Southern Subsurface Countercurrents (NSCC and SSCC) flow eastward below the poleward flanks of the EUC (**Figure 4**). The westward Equatorial Intermediate Current (EIC) is found directly below the EUC across the Pacific. Both the EUC and EIC slope upward toward the east

(**Figure 5**), tending to follow shoaling isopycnal surfaces. On the Equator, alternating deep equatorial jets (not shown) are found below the EIC.

The poleward wind-driven meridional flows are mostly confined to the upper 50 m (**Figure 4**). The central Pacific section (**Figure 4E**) shows meridional flow nearly symmetric with respect to the equator, with divergent poleward flow in the near-surface, and convergent equatorward flow near the core of the EUC. This classical picture is not seen in the eastern and western sections, due to a cross-equatorial component of the surface winds, especially in the east.

Boundary Flows

Surface The westward flow of the NEC impinges on the Philippines where it splits near 14°N into a northward-flowing Kuroshio Current and southward Mindanao Current (MC) along the western boundary (**Figures 1** and **2**). The MC has surface speeds exceeding 1 m s^{-1}, and the flow reaches to depths of 300–600 m. The upper 100 m flow splits at the south end of Mindanao Island, with a significant portion entering the Sulawesi Sea, and most of the rest retroflecting into the NECC. A small portion recirculates around the persistent Mindanao Eddy. A portion of the flow in the Sulawesi Sea transits through the Makassar Strait and ultimately through the Indonesian Seas and into the Indian Ocean, while the rest returns to the

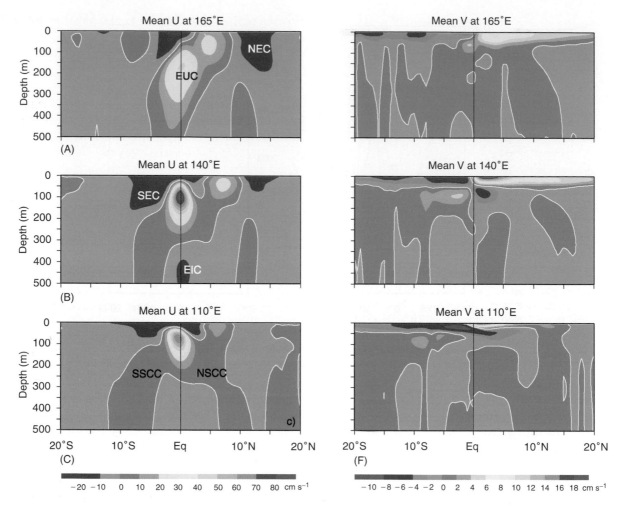

Figure 4 Vertical section showing the zonal (A–C) and meridional (D–F) components of mean flow in the upper 500 m of the western (A, D), central (B, E) and eastern (C, F) Pacific Ocean. Color bars give magnitude of flow (note that zonal and meridional scales are not the same); positive values are eastward and northward. Zero values are indicated by white contours.

Pacific to join the NECC (**Figure 6**). Deeper portions of the MC are also split between the Indonesian Throughflow and the Pacific equatorial circulation, with the latter flowing into the EUC and NSCC.

The SEC impinges on the north-eastern coast of Australia and the complex of islands including New Guinea. Similarly to the NEC, it splits into poleward and equatorward boundary flows near 14°S (**Figure 1**). The poleward branch is the East Australia Current, and the northward branch flows under a shallow southward surface flow as the Great Barrier Reef Undercurrent, eventually becoming the New Guinea Coastal Current (NGCC) and New Guinea Coastal Undercurrent (NGCUC), which follow a convoluted path around topographic features ending up with westward flow along the north coast of New Guinea. In this region, the surface flow of the NGCC reverses seasonally with the Asian winter monsoon westerly winds.

Low-latitude boundary currents in the eastern Pacific are not nearly as strong as along the western boundary, but they play a significant role in closing the circulation of the Pacific Ocean (**Figure 1**). The Peru Current flows northward along the west coast of South America, ultimately turning offshore into the SEC. North of the Equator, the NECC flows into the Gulf of Panama and retroflects around the Costa Rica Dome into the NEC, joining southward flow from the California Current.

Subsurface Along the western Pacific boundary (**Figure 1**), the NGCUC flows westward along the north coast of New Guinea with a maximum near 200 m, but extending to at least 800 m. The upper thermocline waters contribute a small fraction to the Indonesian Throughflow, with the rest retroflecting around the persistent Halmahera Eddy to form (with contributions from the MC) the

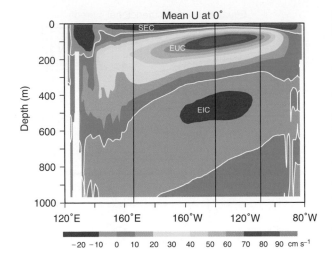

Figure 5 Zonal flow speed in the upper 1000 m in a vertical section along the Equator. Speeds are given in the color bar, with positive values eastward. The vertical lines indicate the longitudes of long-term current meter measurements (see **Figure 7**).

eastward-flowing EUC (**Figure 6**). The deeper portions of the NGCUC flow across the Equator and are traced into the weak Mindanao Undercurrent (MUC) which flows northward below the MC, carrying Antarctic Intermediate Water into the North Pacific.

In the east, the Peru–Chile Undercurrent flows poleward along the west coast of Ecuador, Peru and Chile, basically an extension of the EUC past the Galapagos Islands that then turns southward after converging at the coast of Ecuador. Some of these waters join the westward flows of the Peru Current and SEC through upwelling. The fate of waters

flowing eastward in the NSCC and SSCC is not well known.

Variability

The variability of equatorial currents is complex, spanning a broad range of time and space scales. This variability is largely forced by changing winds; an important fraction of these wind changes are due to sea surface temperature changes in the equatorial zone, these being associated with changes in the currents and winds. These coupled variations include the well-known El Niño phenomenon, but also include the annual cycle.

Annual Cycle

Although the annual cycle of wind forcing near the equator is not as extreme as in mid latitudes, the Pacific Trade Winds vary enough to force significant changes to the currents discussed above. Because of equatorial wave dynamics and coupling with the atmosphere, the relationship of the current variability to the winds is quite complex.

Figure 7 shows the long-term mean plus annual cycle of zonal and meridional current in the near-surface layer and at depth within the EUC for locations in the western, central and eastern Pacific. (Here, the annual cycle is the sum of annual and semiannual harmonics analyzed from direct current measurements.) The range of zonal current variation is about one order of magnitude larger than the meridional variations. The annual harmonic dominates the annual cycle of surface flow, except in the western Pacific where monsoon winds cross the

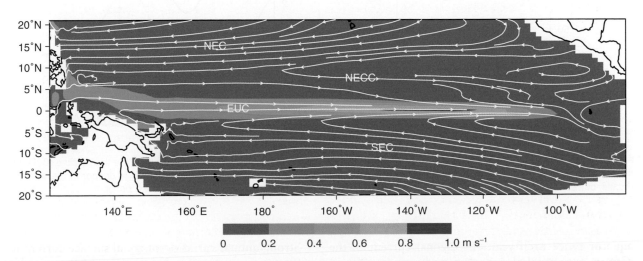

Figure 6 Map of long-term mean flow in the tropical Pacific on the isopycnal surface $\sigma_\Theta = 24.5\,\mathrm{kg m^{-3}}$, which lies within the high-speed core of the Equatorial Undercurrent. White lines and arrows indicate the direction of flow, while the colors indicate the speed of flow as given by the color bar. Current names are abbreviated as in **Figure 1**.

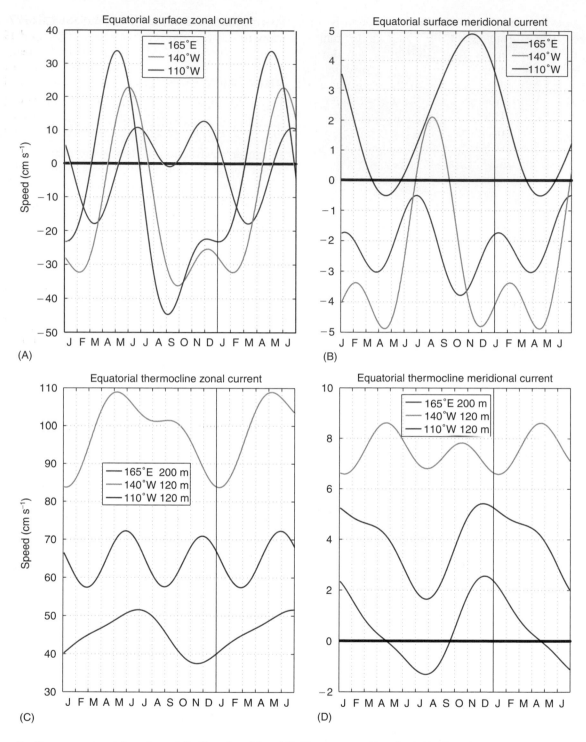

Figure 7   Mean annual variation of zonal (A, C) and meridional (B, D) currents at 10 m depth (A, B) and in the thermocline (C, D) on the Equator at three locations across the Pacific Ocean indicated in **Figure 5**. Because the thermocline is deeper in the western Pacific, currents are presented for a corresponding depth there. Note the different speed scales for each panel.

equator twice each year. Also, zonal current in the eastern equatorial thermocline and meridional current in the central equatorial thermocline show strong semiannual signals.

Strong annual variation of zonal surface current is sufficient to reverse the direction of the flow on the equator, especially during the Northern Hemisphere spring (**Figure 7A**). In the east, this feature occurs nearly every year, and has been erroneously described as a 'surfacing' of the EUC. In the central Pacific, such reversals are mainly observed during strong El Niño events, and its appearance in the annual cycle here may be due to the occurrence of several El Niño events during the record that was analyzed (1984–1998). It is noteworthy that the maximum eastward deviation of the annual cycle appears progressively later toward the west, thought to be coupled with westward propagation of the annual cycle of zonal wind and sea surface temperature.

The annual cycle in the strong eastward flow of the EUC within the thermocline (**Figure 7C**) is not large enough to reverse the current direction. (On interannual timescales, however, the flow may reverse—see below.)

El Niño/Southern Oscillation (ENSO)

The strongest variability of the zonal equatorial currents is associated with El Niño episodes that occurred in 1986–87, 1990–91, 1993, and 1997–98, and with La Niña episodes of 1984, 1988, and 1996. Current meter records that have had their mean and annual cycles removed are presented in **Figure 8**, showing that El Niño variations are large enough to reverse the westward flow of the SEC, especially in the western equatorial Pacific. Strong eastward surface flow in the warm water pool of the western equatorial Pacific (e.g., 1997) has been implicated in the warming of the sea surface in the central and eastern equatorial Pacific. Also, the eastward flow of the EUC is reversed during some of these events (e.g., in **Figure 8B** at 140°W during 1997). This disappearance of the EUC was first observed during the strong 1982–83 El Niño event.

The current systems off the Equator are also affected by ENSO, again through a combination of local wind forcing and remotely-forced baroclinic waves. The NECC strengthens in the early phases of El Niño. The off-equatorial portion of the SEC weakens during El Niño, and strengthens during La Niña. The interannual variations of the NEC show a

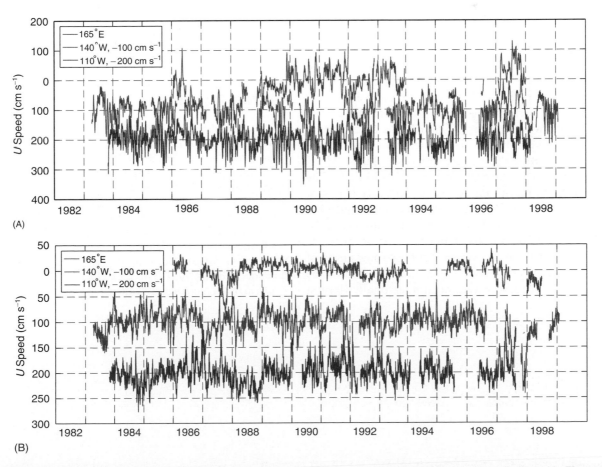

Figure 8 Time series of observed zonal currents, with mean and annual cycle removed, at three locations (indicated in **Figure 5**) along the Equator in the surface layer (A) and in the thermocline (B). Warm El Niño events are indicated by red shading along the time axis; blue shading indicates cold La Niña events. Missing observations are indicated by gaps. Note that time series have been offset from each other for clarity.

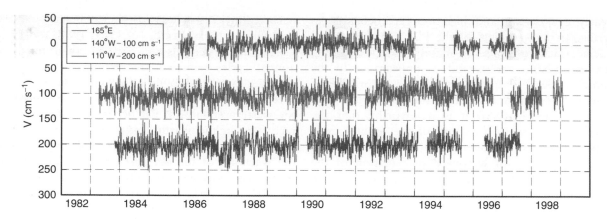

Figure 9 Time series of observed surface layer meridional currents, with mean and annual cycle removed, at three locations along the Equator. Missing observations are indicated by gaps. Note that time series have been offset from each other for clarity.

correlation with the NECC, but there are also strong variations with longer timescales than ENSO.

High-frequency Current Variability

Equatorially trapped waves with periods of a few days to a couple of months are quite energetic and ubiquitous (**Figures 8** and **9**). Zonal current fluctuations are dominated by intraseasonal fluctuations associated with the atmospheric intraseasonal (30–60 days) oscillation; eastward-propagating equatorially trapped Kelvin waves play an important role in transmitting this variability from the western Pacific to the eastern Pacific (e.g., **Figure 8A** during 1997). Meridional current fluctuations tend to have more energy at higher frequencies than the zonal current variations (compare **Figures 8** and **9**). Here, dominant periods are in the range 20–30 days. The amplitude of meridional current variability is largest at the surface in the east, and becomes somewhat smaller in the central Pacific, and much smaller in the west. The dominant mechanism is the tropical instability wave, which arises from the strong shears of the zonal flows discussed earlier. Modulation of the amplitudes of these waves occurs seasonally (largest amplitude in the northern fall) and interannually (small amplitude during El Niño) as the shears are affected by the annual cycle and ENSO.

See also

El Niño Southern Oscillation (ENSO). El Niño Southern Oscillation (ENSO) Models.

Further Reading

Godfrey JS, Johnson GC, McPhaden MJ, Reverdin G, and Wijffels S (2001) The tropical ocean circulation. In: Siedler G and Church J (eds.) *Ocean Circulation and Climate*, pp. 215–246. London: Academic Press.

Hastenrath S (1985) *Climate and Circulation of the Tropics*. Dordrecht: D. Reidel.

Lukas R (1986) The termination of the Equatorial Undercurrent in the Eastern Pacific. *Progress in Oceanography* 16: 63–90.

Neumann G (1968) *Ocean Currents*. Amsterdam: Elsevier.

Philander SGH (1990) *El Niño, La Niña, and the Southern Oscillation*. San Diego: Academic Press.

EL NIÑO SOUTHERN OSCILLATION (ENSO)

K. E. Trenberth, National Center for Atmospheric Research, Boulder, CO, USA

Introduction

A major El Niño began in April of 1997 and continued until May 1998. It has been labeled by some as the 'El Niño of the century' as it was certainly the biggest on record by several measures. It brought with it many weather extremes and unusual weather patterns around the world. Moreover, the event was predicted by climate scientists and received unprecedented news coverage, so that the term 'El Niño' became part of the popular vernacular. Many things were blamed on El Niño, and some of them indeed were influenced by El Niño, although in some instances, the connection was, at best, tenuous. Although El Niño may be relatively new to the public, it has been known to scientists, at least in some respects, for decades and even centuries. El Niño refers to the exceptionally warm sea temperatures in the tropical Pacific, but it is linked to major changes in the atmosphere through the phenomenon known as the Southern Oscillation (SO), in particular, so that the whole phenomenon is called El Niño–Southern Oscillation (ENSO) by scientists. This article outlines the current understanding of ENSO and the physical connections between the tropical Pacific and the rest of the world.

ENSO Events

El Niños are not uncommon. Every three to seven years or so, a pronounced warming occurs of the surface waters of the tropical Pacific Ocean. The warmings take place from the International Dateline to the west coast of South America and result in changes in the local and regional ecology, and are clearly linked with anomalous global climate patterns. In 1997, the warming can be seen by comparing the sea surface temperatures (SSTs) in December 1997 at the peak of the 1997/98 event with those a year earlier (**Figure 1**). As well as the total SST fields, this figure also displays the departures from average.

The warmings have come to be known as 'El Niño events'. Historically, 'El Niño' referred to the appearance of unusually warm water off the coast of Peru, where it was readily observed as an enhancement of the normal warming about Christmas (hence Niño, Spanish for 'the boy Christchild') and only more recently has the term come to be regarded as synonymous with the basinwide phenomenon. The oceanic and atmospheric conditions in the tropical Pacific are seldom close to average, but instead fluctuate somewhat irregularly between the warm El Niño phase of ENSO, and the cold phase of ENSO consisting of basinwide cooling of the tropical Pacific, dubbed 'La Niña events' ('La Niña' is 'the girl' in Spanish). The most intense phase of each event typically lasts about a year.

The SO is principally a global-scale seesaw in atmospheric sea level pressure involving exchanges of air between eastern and western hemispheres (**Figure 2**) centered in tropical and subtropical latitudes with centers of action located over Indonesia and the tropical South Pacific Ocean (near Tahiti). Thus the nature of the SO can be seen from the inverse variations in pressure anomalies (departures from average) at Darwin (12.4°S 130.9°E) in northern Australia and Tahiti (17.5°S 149.6°W) in the South Pacific Ocean (**Figure 3**) whose annual mean pressures are strongly and significantly oppositely correlated. Consequently, the difference in pressure anomalies, Tahiti–Darwin, is often used as a Southern Oscillation Index (SOI). The sequences of swings in the SOI shown in **Figure 3** are discussed below in conjunction with those of SST.

Higher than normal pressures are characteristic of more settled and fine weather, with less rainfall, whereas lower than normal pressures are identified with 'bad' weather, more storminess and rainfall. So it is with the SO. Thus for El Niño conditions, higher than normal pressures over Australia, Indonesia, southeast Asia, and the Philippines signal drier conditions or even droughts. Dry conditions also prevail at Hawaii, parts of Africa, and extend to the northeast part of Brazil and Colombia. On the other end of the seesaw, excessive rains prevail over the central and eastern Pacific, along the west coast of South America, parts of South America near Uruguay, and southern parts of the United States in winter (cf. **Figure 2**) often leading to flooding. When the pressure pattern in **Figure 2** reverses in sign, as for La Niña, the regions favored for drought in El Niño tend to become excessively wet, and vice versa.

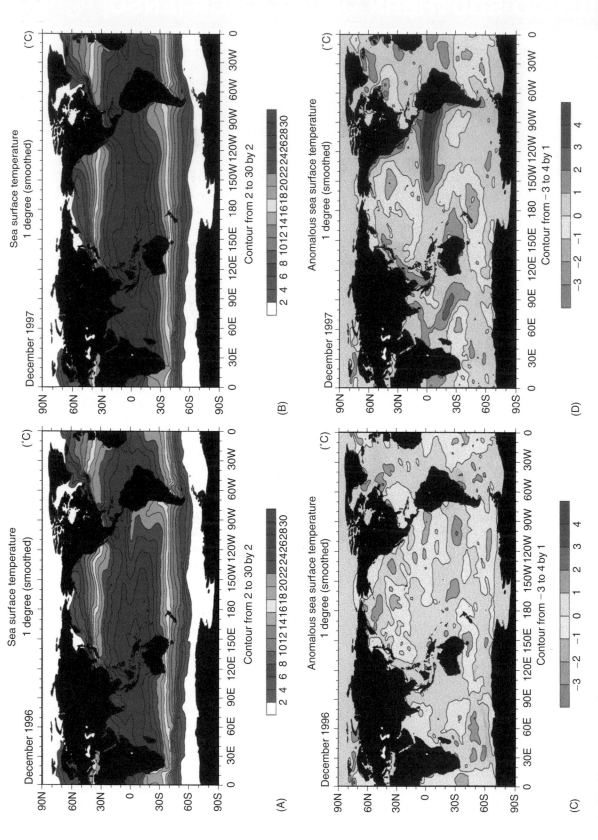

Figure 1 Monthly mean sea surface temperatures in °C for December 1996 (A, C) and 1997 (B, D), before and during the peak of the 1997–98 El Niño event. (A) and (B) show the actual SSTs and (C) and (D) show the anomaly, defined as the departure from the mean for 1950–79 with contour interval 2°C (top) and 1°C (bottom). Based on data from US National Oceanic and Atmospheric Administration.

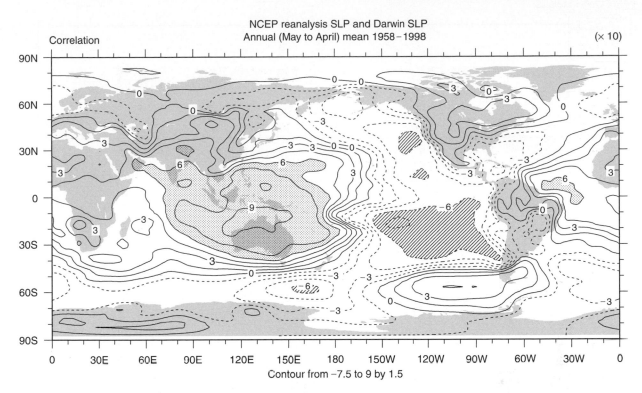

NCEP reanalysis SLP and Darwin SLP
Annual (May to April) mean 1958–1998

(× 10)

Correlation

Contour from −7.5 to 9 by 1.5

Figure 2 Map of correlation coefficients (× 10) of the annual mean sea level pressures with that at Darwin. It shows the Southern Oscillation pattern in the phase corresponding to El Niño events; during La Niña events the signs are reversed. Values exceeding 0.6 in magnitude are stippled or hatched.

The Tropical Pacific Ocean–Atmosphere System

The distinctive pattern of average sea surface temperatures in the Pacific Ocean sets the stage for ENSO events. The pattern in December 1996 (**Figure 1**) is sufficiently close to average to illustrate the main points. One key feature is the 'warm pool' in the tropical western Pacific, where the warmest ocean waters in the world reside and extend to depths of over 150 m with values at the surface >28°C. Other key features include warm waters north of the equator from about 5 to 15°N, much colder waters in the eastern Pacific, and a cold tongue along the equator that is most pronounced about October and weakest in March. The warm pool migrates with the sun back and forth across the equator but the distinctive patterns of SST are brought about mainly by the winds (**Figure 4**).

The existence of the ENSO phenomenon is dependent on the east–west variations in SSTs (**Figure 1**) in the tropical Pacific, and the close links with sea-level pressures (**Figure 3**) and thus surface winds in the tropics (**Figure 4**), which in turn determine the major areas of rainfall (**Figure 5**). The temperature of the surface waters is readily conveyed to the overlying atmosphere and because warm air is less dense

it tends to rise whereas cooler air sinks. As air rises into regions where the air is thinner, the air expands, causing cooling and therefore condensing moisture in the air, which produces rain. Low sea-level pressures are set up over the warmer waters while higher pressures occur over the cooler regions in the tropics and subtropics, and the moisture-laden winds tend to blow toward low pressure so that the air converges, resulting in organized patterns of heavy rainfall. The rain comes from convective cloud systems, often as thunderstorms, and perhaps as tropical storms or even hurricanes, which preferentially occur in the 'convergence zones'. Because the wind is often light or calm right in these zones, they have previously been referred to as the 'doldrums'. Of particular note are the Intertropical Convergence Zone (ITCZ) and the South Pacific Convergence Zone (SPCZ) (**Figure 5**) which are separated by the equatorial dry zone. These atmospheric climatological features play a key role in ENSO as they change in character and move when SSTs change.

The rainfall patterns in the tropics can be illustrated by quantities sensed from satellite (**Figure 5**). There is a strong coincidence between the patterns of SSTs and tropical convection throughout the year, although there is interference from effects of nearby land and monsoonal circulations. The strongest

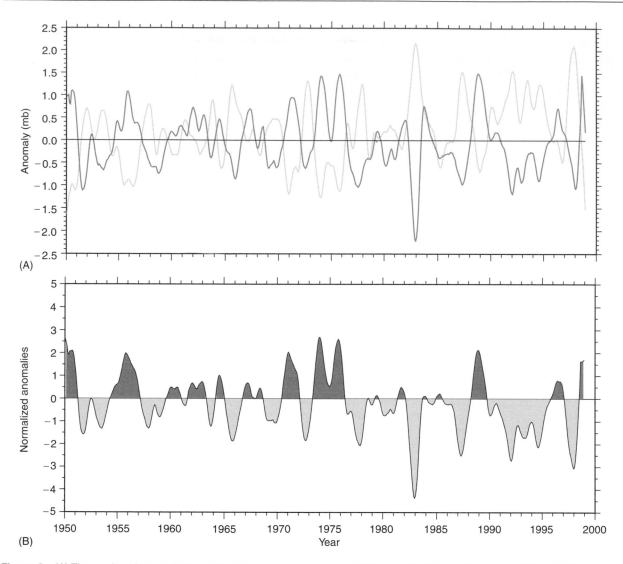

Figure 3   (A) Time series of anomalies in sea level pressures at Darwin (solid line) and Tahiti (light line) from 1950 to 1998 smoothed to remove fluctuations of period less than about a year. (B) Corresponding time series of the Southern Oscillation Index, normalized Tahiti minus Darwin pressures, in standardized units.

seasonal migration of rainfall occurs over the tropical continents, Africa, South America and the Australian–Southeast Asian–Indonesian maritime region. Over most of the Pacific and Atlantic the ITCZ remains in the Northern Hemisphere year round, with convergence of the tradewinds favored by the presence of warmer water. In the subtropical Pacific, the SPCZ also lies over waters warmer than about 27°C. The ITCZ is weakest in January in the Northern Hemisphere when the SPCZ is strongest in the Southern Hemisphere.

The surface winds (**Figure 4**) drive surface ocean currents which determine where the surface waters flow and diverge, and thus where cooler nutrient-rich waters upwell from below. Because of the Earth's rotation, easterly winds along the equator deflect

currents to the right in the Northern Hemisphere and to the left in the Southern Hemisphere and thus away from the equator, creating upwelling along the equator. Thus the winds largely determine the SST distribution, the differential sea levels and the heat content of the upper ocean. The presence of nutrients and sunlight in the cool surface waters along the equator and western coasts of the Americas favors development of tiny plant species (phytoplankton), which in turn are grazed on by microscopic sea animals (zooplankton) which provide food for fish.

Temperatures in the upper ocean are measured by about 70 instrumented buoys moored to the bottom of the ocean throughout the tropical Pacific (**Figure 6**). Thus for December 1996 and 1997 the

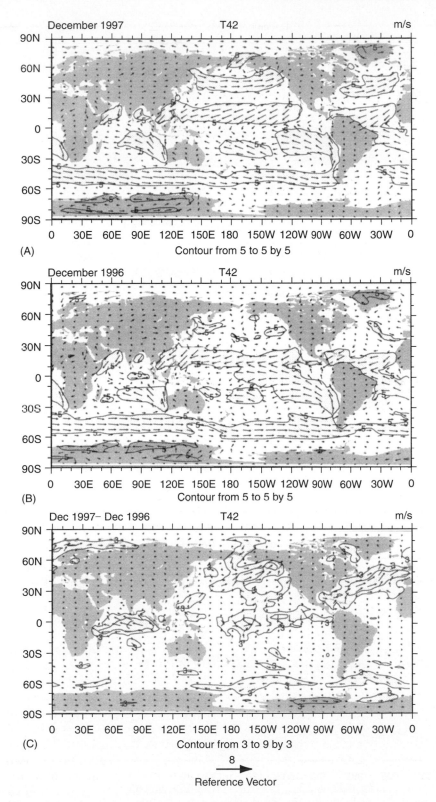

Figure 4 Mean surface winds for December 1996 (A) and 1997 (B) from global analyses by the European Centre for Medium Range Forecasts, and their difference (C). Shading indicates $5\,m\,s^{-1}$ for (A) and (B) and $3\,m\,s^{-1}$ for (C). The reference vector is plotted at bottom.

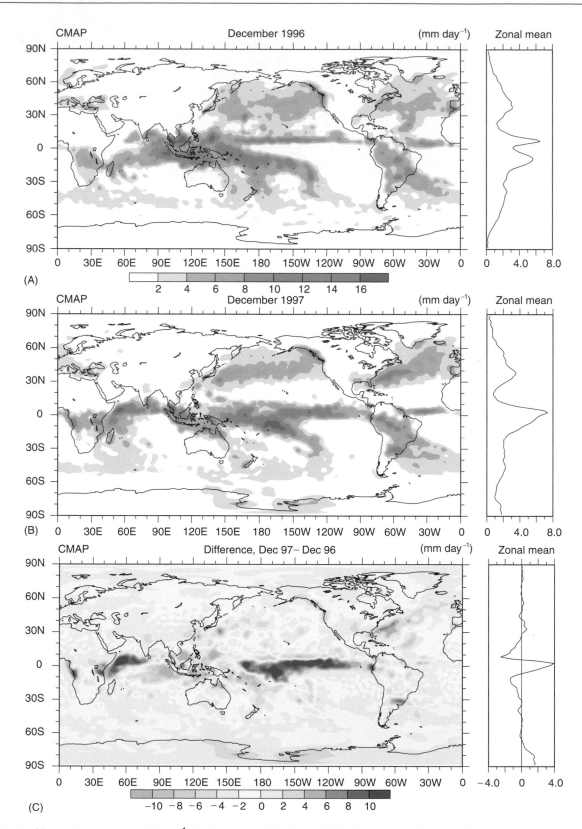

Figure 5   Mean precipitation rate (mm d⁻¹) for December 1996 (A) and 1997 (B), and the difference (C), as constructed from gauge measurements and various estimates from satellite from 1979 to 1995. The zonal mean values are given at right. (Based on data from the Global Precipitation Climatology Project.)

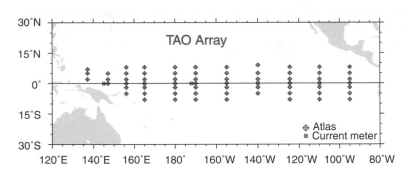

Figure 6 A center piece of the Pacific El Niño observing system is an array of buoys in the tropical Pacific moored to the ocean bottom known as the TAO (Tropical Atmosphere–Ocean) array. The latter is maintained by a multinational group spearheaded in the United States by the National Oceanic and Atmospheric Administration's Pacific Marine Environmental Laboratory (PMEL). Each buoy has a series of temperature measurements on a sensor cable on the upper 500 m of the mooring, and on the buoy itself are sensors for surface wind, sea surface temperature, surface air temperature, humidity, and a transmitter to satellite. Observations are continually made, averaged into hourly values, and transmitted via satellite to centers around the world for prompt processing. Right, an ATLAS (Autonomous Temperature Line Acquisition System) buoy. Courtesy PMEL.

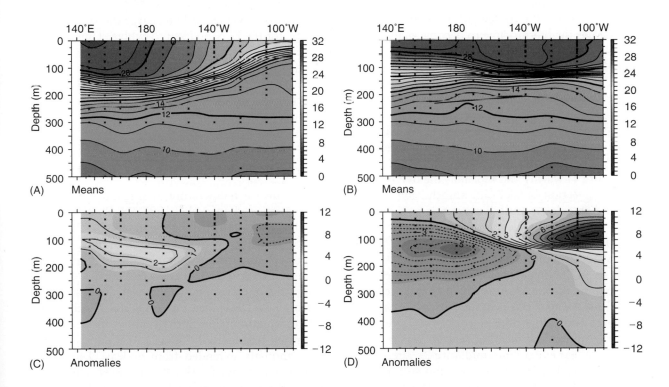

Figure 7 Zonal cross sections along the equator across the Pacific Ocean (2°N to 2°S) from the TAO array temperatures for December 1996 (A, C) and 1997 (B, D). (A, B) total field; (C, D) anomalies. Contours every °C.

temperature structure throughout the equatorial region can be mapped (**Figure 7**). The heat content of the upper ocean depends on the configuration of the thermocline (the region of sharp temperature gradient within the ocean separating the well-mixed surface layers from the cooler abyssal ocean waters). Normally the thermocline is deep in the western tropical Pacific (150 m) and sea level is high as waters driven by the easterly tradewinds pile up. In the eastern Pacific on the equator, the thermocline is very shallow (50 m) and sea level is relatively low. The Pacific sea surface slopes up by about 60 cm from east to west along the equator. The temperatures in December 1996 depict conditions somewhat similar to average, but with signs of the incipient El Niño developing in the western tropical Pacific at 100–150 m depth as a warm anomaly (**Figure 7**) but no such signs at the surface (**Figure 1**).

The tropical Pacific, therefore, is a region where the atmospheric winds are largely responsible for the tropical SST distribution which, in turn, is very much involved in determining the precipitation distribution and the tropical atmospheric circulation. This sets the stage for ENSO to occur.

Mechanisms of ENSO

Most of the interannual variability in the atmosphere in the tropics and a substantial part of the variability over the extratropics is related and tied together through ENSO. ENSO is a natural phenomenon arising from coupled interactions between the atmosphere and the ocean in the tropical Pacific Ocean, and there is good evidence from cores of coral and glacial ice in the Andes that it has been going on for millennia.

During El Niño, the tradewinds weaken (**Figure 4**) which causes the thermocline to become shallower in the west and deeper in the eastern tropical Pacific (**Figure 7**), while sea level falls in the west and rises in the east by as much as 25 cm as warm waters surge eastward along the equator. Equatorial upwelling decreases or ceases and so the cold tongue weakens or disappears (e.g., as in December 1997, **Figure 1**) and the nutrients for the food chain are substantially reduced. The resulting increase in sea temperatures (e.g., **Figure 1**) warms and moistens the overlying air so that convection breaks out, and the convergence zones and associated rainfall move to a new location with a resulting change in the atmospheric circulation (**Figure 5**). A further weakening of the surface trade winds completes the positive feedback cycle leading to an El Niño event. The shifts in the location of the organized rainfall in the tropics and the latent heat released alters the heating patterns of the atmosphere. Somewhat like a rock in a stream of water, the anomalous heating sets up waves in the atmosphere that extend into midlatitudes altering the winds and changing the jet stream and storm tracks (e.g., **Figure 8**). Note especially the strong westerly jets in the Pacific of both hemispheres, and in the Northern (winter) Hemisphere the jet stream in December 1997 extends into California and across the southern United States, carrying with it disturbances that result in extensive rains. Weaker westerlies exist farther north and so the overall storm track shifts towards the equator in the Pacific.

Although the El Niños and La Niñas are often referred to as 'events' which last a year or so, ENSO is oscillatory in nature. The ocean is a source of moisture and its enormous heat capacity acts as the flywheel that drives the system through its memory of the past, resulting in an essentially self-sustained sequence in which the ocean is never in equilibrium with the atmosphere. The amount of warm water in the tropics builds up prior to and is then depleted during El Niño. During the cold phase with relatively clear skies, solar radiation heats up the tropical Pacific Ocean, the heat is redistributed by currents, with most of it being stored in the deep warm pool in the west or off the equator such as at about 10 or 20°N. During El Niño, heat is transported out of the tropics within the ocean toward higher latitudes in response to the changing currents, and increased heat is released into the atmosphere mainly in the form of increased evaporation, thereby cooling the ocean. Added rainfall contributes to a general warming of the global atmosphere that peaks a few months after a strong El Niño event. It has therefore been suggested that the time scale of ENSO is determined by the time required for an accumulation of warm water in the tropics to essentially recharge the system, plus the time for the El Niño itself to evolve. Thus a major part of the onset and evolution of the events is determined by the history of what has occurred one to two years previously. This also means that the future evolution is predictable for several seasons in advance.

Interannual Variations in Climate

The subsurface temperature anomalies which eventually developed into the 1997–98 El Niño were traceable at least from about September 1996 on the equator in the far western Pacific. However, positive subsurface temperature anomalies in the upper 100–200 m in the far western Pacific exceeded 1°C for all the months of 1996, and so this was not a sufficient predictor. By December 1996 (**Figure 7**) subsurface temperature anomalies in the vicinity of the equator had grown to exceed 2.5°C at 150 m depth and the warm anomaly extended from at least 140°E (the westernmost buoy) to 140°W. However, conditions were still below normal in the eastern Pacific. By December 1997, the subsurface warm anomaly had progressed eastward and amplified to produce positive anomalies exceeding 11°C (**Figure 7**) at about 100 m depth, accompanying the surface SST anomalies exceeding 5°C (**Figure 1**). Also note, however, in December 1997 the subsequent cold anomaly of over 5°C in the western equatorial Pacific near 150 m depth, as the warm pool was displaced into the central Pacific. The warm pool in the east continued to diminish with time as the cold anomaly intensified and subsequently spread all the way across the Pacific as part of the signature of the La Niña that

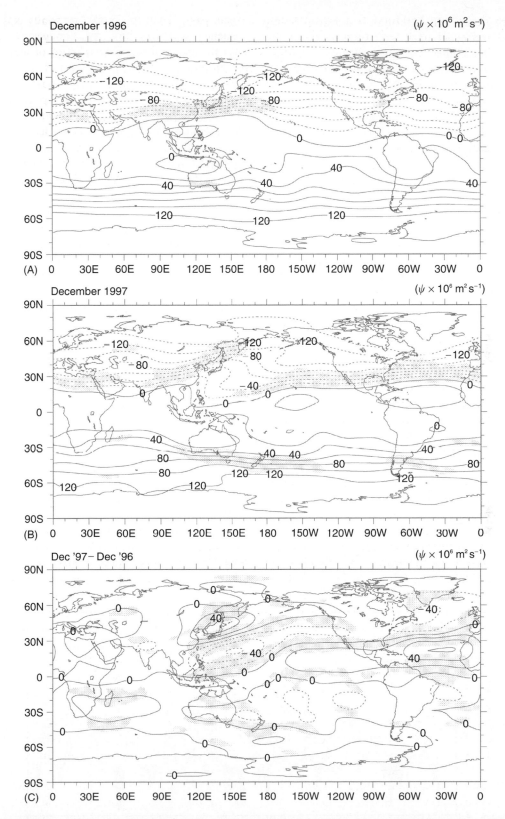

Figure 8 Flow patterns (winds) at the jet stream level, 10 km or so above the surface of the Earth, for December 1996 (A) and 1997 (B), and their differences (C) based on global analyses by the European Centre for Medium Range Prediction. The contours are stream-lines at the 200 mb level and the region where winds exceed $40\,\mathrm{m\,s^{-1}}$ ($20\,\mathrm{m\,s^{-1}}$ for (C)) are stippled to indicate the jet streams. El Niño conditions favor a more vigorous jet stream and storm track from the Pacific to the Gulf of Mexico, bringing heavy rains to the southern United States. Note similar effects in the South Pacific near New Zealand.

began about June 1998, although it did not develop to a full fledged event until later in the year. A key aspect of these changes was the obvious loss of heat content throughout the equatorial Pacific beginning late 1997 and continuing through 1998.

The evolution of SST in several recent ENSO events after 1950 is shown in **Figure 9** for two regions. The region of the Pacific Ocean which is most involved in ENSO is the central equatorial Pacific, especially the area 5°N to 5°S, 170°E to 120°W, whereas the traditional El Niño region is along the coast of South America. The latter is less important for the global changes in weather patterns but is certainly important locally. Variations in both regions are closely related but differ in detail from event to event in relative amplitudes and sequencing. For SSTs, the departure from average required for an El Niño is 0.5°C over the central Pacific region, which is large enough to cause perceptible effects in Pacific rim countries. Larger El Niño events have traceable influences over more extensive regions and even globally.

The ENSO events clearly identifiable in **Figure 9** since 1950 occurred in 1951, 1953, 1957–58, 1963, 1965, 1969, 1972–73, 1976–77, 1982–83, 1986–87, 1990–95 and 1997–98. The 1990–95 event might also be considered three modest events where the conditions in between failed to return to below normal so that they merged together. Worldwide climate anomalies lasting several seasons have been identified with all of these events. The 1997–98 event has the biggest SST departures on record, but for the SOI, the El Niño event of 1982–83 still holds the record (**Figure 3**).

Each El Niño event has its own character. In the El Niño winters of 1992–93, 1994–95, and 1997–98, southern California was battered by storms and experienced flooding and coastal erosion (in part aided by the high sea levels). However, in more modest El Niños (e.g., 1986–87 and 1987–88 winters) California was more at risk from droughts. Because of the enhanced activity in the Pacific and the changes in atmospheric circulation throughout the tropics, there is a decrease in the number of tropical storms and hurricanes in the tropical Atlantic during El Niño. A good example is 1997, one of the quietest Atlantic hurricane seasons on record, whereas the 1990–95 and 1997–98 El Niños terminated before

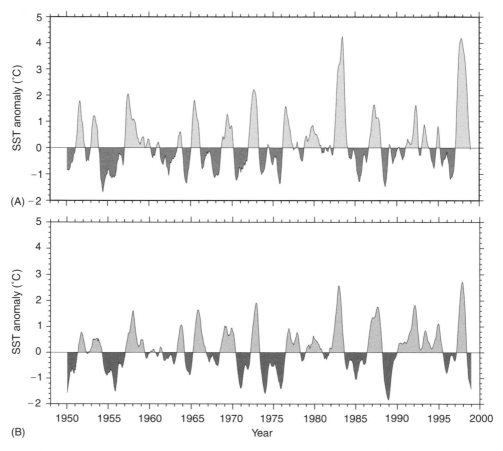

Figure 9 Time series of sea surface temperature (SST) anomalies from 1950 through 1998 relative to the means of 1950–79 for (A) the traditional El Niño area: 0°–10°S 90°W–80°W, and (B) the region most involved in ENSO 5°N–5°S, 170°E–120°W.

the 1995 and 1998 hurricane seasons which unleashed the Atlantic storms and placed those seasons among the most active on record.

The SO has global impacts, however; the connections to higher latitudes (known as teleconnections) tend to be strongest in the winter of each hemisphere and feature alternating sequences of high and low pressures accompanied by distinctive wave patterns in the jet stream (**Figure 8**) and storm tracks in mid-latitudes. Although warming is generally associated with El Niño events in the Pacific and extends, for instance, into western Canada and Alaska, cool conditions typically prevail over the North and South Pacific Oceans. To a first approximation, reverse anomaly patterns occur during the La Niña phase of the phenomenon.

The prominence of the SO has varied throughout the last century (**Figure 10**). Very slow long-term (decadal) variations are present; for instance SOI values are mostly below the long-term mean after 1976. This accompanies the generally above normal SSTs in the western Pacific along the equator (**Figure 9**). The decadal atmospheric and oceanic variations are even more pronounced in the North Pacific and across North America than in the tropics and are also clearly present in the South Pacific, with evidence suggesting that they are at least in part forced from the tropics. Although not yet clear in detail, it is likely that climate change associated with increased greenhouse gases in the atmosphere, which contribute to global warming, are changing ENSO, perhaps by expanding the west Pacific warm pool and making for more frequent and bigger El Niño events.

Impacts

Changes associated with ENSO produce large variations in weather and climate around the world from year to year and often these have a profound impact on humanity because of droughts, floods, heat waves and other changes which can severely disrupt agriculture, fisheries, the environment, health, energy demand, and air quality, and also change the risks of fire. The normal upwelling of cold nutrient-rich and CO_2-rich waters in the tropical Pacific is suppressed during El Niño. The presence of nutrients and sunlight fosters development of phytoplankton and zooplankton to the benefit of many fish species. Therefore El Niño-induced changes in oceanic conditions can have disastrous consequences for fish and seabirds and thus for the fishing and guano industries, for example, along the South American coast. Other marine creatures may benefit so that unexpected harvests of shrimp or scallops occur in some places. Rainfall over Peru and Ecuador can transform barren desert into lush growth and benefit some crops, but can also be accompanied by swarms of grasshoppers, and increases in the populations of toads and insects. Human health is affected by mosquito-borne diseases such as malaria, dengue, and viral encephalitis, and by water-borne diseases such as cholera.

Economic impacts can be large, with losses typically overshadowing gains. One assessment placed losses during the 1997–98 El Niño event at over US$34 billion, although it does not account for the widespread human suffering and loss of life. An estimate of the economic loss of production in

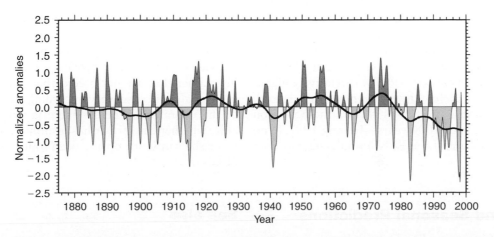

Figure 10 Time series of the Southern Oscillation Index (SOI) based solely on observations at Darwin from 1866 to 1998. Also shown in a curve designed to show the multidecadal fluctuations. The zero corresponds to the mean for the first 100 years 1866 to 1965, highlighting the recent trend for more El Niños.

Queensland, Australia due to drought in the prolonged warm ENSO phase from 1990 to 1995 is $1 billion (Australian) per year. The La Niña in 1998 was linked to the extensive, severe, and persistent 1988 North American spring–summer drought, which brought losses of over US$30 billion, as well as major climate anomalies elsewhere over the globe.

ENSO also plays a prominent role in modulating carbon dioxide exchanges with the atmosphere. The decrease in outgassing of CO_2 during El Niño is enough to reduce the build up of CO_2 in the atmosphere by 50%. El Niño also influences the incidence of fires, which result in more CO_2 emissions, while changing rainfall and temperatures over land through the teleconnections, such that CO_2 uptake by the terrestrial biosphere is enhanced.

In 1997, the strongest drought set in over Indonesia and led to many fires, set as part of activities of farmers and corporations clearing land for agriculture, raging out of control. With the fires came respiratory problems in adjacent areas 1000 km distant and even a commercial plane crash in the area has been linked to the visibility problems. Subsequently in 1998, El Niño-related drought and fires evolved in Brazil, Mexico and Florida. Flooding took place in Peru and Ecuador, as usual with El Niño, and also in Chile, and coastal fisheries were disrupted. The strong winter 1997–98 Northern Hemisphere jet stream created wet conditions from California to Florida. Normally these storms veer to the north toward the Gulf of Alaska or enter North America near British Columbia and Washington, where they could subsequently link up with the cold Arctic and Canadian air masses and bring them down into the United States. Instead, the pattern was persistently favorable for relatively mild conditions over the northern states such that temperatures averaged over 10°C (18°F) above normal in February in the Great Lakes area. Similar changes occurred in the Southern Hemisphere and spread downstream to South America. Globally, it seems that the land temperature for February 1998 was relatively higher than average than for any other month on record. The calendar year 1998 was the warmest year on record – going back 1000 years – beating 1997, in no small part because of the El Niño influences in both those years.

ENSO and Seasonal Predictions

The main features of ENSO have been captured in models, in particular in simplified models which predict the anomalies in SSTs. Lead times for predictions in the tropics of at least six months have been shown to be practicable. For instance, an El Niño was predicted for 1997 in late 1996 and it began in April 1997. However, its full extent was not predicted accurately until about mid-1997, about 6 months in advance. Further improvements in the climate observing system and more realistic and comprehensive models provide prospects for further advances.

It is already apparent that reliable prediction of tropical Pacific SST anomalies can lead to useful skill in forecasting rainfall anomalies in parts of the tropics. Although there are certain common aspects to ENSO events in the tropics, the effects at higher latitudes are more variable. One difficulty is the vigor of weather systems in the extratropics which can override relatively modest ENSO influences from the tropics. Nevertheless, systematic changes in the jet stream and storm tracks do tend to occur on average, thereby allowing useful predictions to be made in some regions, although with some inherent uncertainty, so that the predictions are couched in terms of probabilities.

Skillful seasonal predictions of temperatures and rainfalls have the potential for huge benefits for society, although because the predictability is somewhat limited, a major challenge is to utilize the uncertain forecast information in the best way possible throughout different sectors of society (e.g., crop production, forestry resources, fisheries, ecosystems, water resources, transportation, energy use). The utility of a forecast may vary considerably according to whether the user is an individual versus a group or country. An individual may find great benefits if the rest of the community ignores the information, but if the whole community adjusts (e.g., by growing a different crop), then supply and market prices will change, and the strategy for the best crop yield may differ substantially from the strategy for the best monetary return. On the other hand, the individual may be more prone to small-scale vagaries in weather that are not predictable. Vulnerability of individuals will also vary according to the diversity of the operation: whether there is irrigation available, whether the farmer has insurance, and whether he or she can work off the farm to help out in times of adversity.

See also

Evaporation and Humidity. Heat Transport and Climate.

Further Reading

Glantz MH, Katz RW, and Nicholls N (eds.) (1991) *Teleconnections Linking World-wide Climate Anomalies.* Cambridge: Cambridge University Press.

Glantz MH (1996) *Currents of Change: El Niño's Impact on Climate and Society.* Cambridge: Cambridge University Press.

Suplee C (1999) El Niño/La Niña *National Geographic,* March. 72–95.

National Research Council (1996) *Learning to Predict Climate Variations Associated with El Niño and the Southern Oscillation: Accomplishments and Legacies of the TOGA Program.* Washington: National Academy Press.

Philander SGH (1990) *El Niño, La Niña, and the Southern Oscillation.* London: Academic Press.

Trenberth KE (1997) Short-term climate variations: recent accomplishments and issues for future progress. *Bulletin of the American Meteorological Society* 78: 1081–1096.

Trenberth KE (1997) The definition of El Niño. *Bulletin of the American Meteorological Society* 78: 2771–2777.

Trenberth KE (1999) The extreme weather events of 1997 and 1998. *Consequences* 5(1): 2–15.

EL NIÑO SOUTHERN OSCILLATION (ENSO) MODELS

S. G. Philander, Princeton University, Princeton, NJ, USA

Introduction

The signature of El Niño is the interannual appearance of unusually warm surface waters in the eastern tropical Pacific Ocean. That area is so vast that the effect on the atmosphere is profound. Rainfall patterns are altered throughout the tropics – some regions experience floods, others droughts – and even weather patterns outside the tropics are affected significantly. From an atmospheric perspective, these various phenomena are attributable to the change in the sea surface temperature pattern of the tropical Pacific. Why does the pattern change? Whereas sea surface temperatures depend mainly on the incident solar radiation over most of the globe, the tropics are different. There the winds are of primary importance because of the shallowness of the thermocline, the thin layer of large temperature gradients, at a depth of approximately 100 m, that separates warm surface waters from the cold water at depth. The winds, by causing variations in the depth of the thermocline, literally bring the deep, cold water to the surface in regions where the thermocline shoals. For example, the trade winds that drive warm surface waters westward along the equator expose cold, deep water to the surface in the eastern equatorial Pacific. A relaxation of the winds, such as occurs during El

Niño, permits the warm water to flow back eastward. The changes in the winds are part of the atmospheric response to the altered sea surface temperatures. This circular argument – the winds are both the cause and consequence of sea surface temperature changes – suggests that interactions between the ocean and atmosphere are at the heart of the matter. Those interactions give rise to a broad spectrum of natural modes of oscillation. This result has several important implications. One is that El Niño, even though we tend to regard him as an isolated phenomenon that visits sporadically, is part of a continual fluctuation, known as the Southern Oscillation. El Niño corresponds to one phase of this oscillation, the phase during which sea surface temperatures in the eastern tropical Pacific are unusually high. La Niña is the name for the complementary phase, when temperatures are below normal. (Very seldom are temperatures 'normal', as is evident in **Figure 1**.) To ask why El Niño, or La Niña, occur, is equivalent to asking why a pendulum spontaneously swings back and forth. Far more interesting questions concern the factors that determine the period and other properties of the oscillation, and the degree to which it is self-sustaining or damped. Only strongly damped oscillations, that disappear at times, require a trigger to get going again. Hence a search for the disturbance that triggered a particular El Niño is based on an implicit assumption, which may not be correct at all times, that the Southern Oscillation was damped and had disappeared for a while.

The tools for predicting El Niño are coupled models of the ocean and atmosphere; each provides boundary conditions for the other, sea surface temperatures in the

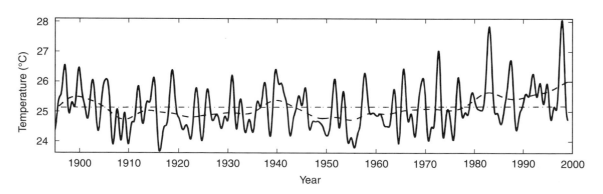

Figure 1 Surface temperature fluctuations at the equator (in °C) in the eastern equatorial Pacific (after removal of the seasonal cycle) over the past 100 years. Temperature maxima correspond to El Niño, minima to La Niña. The smoothly varying dashed line is a 10-year running mean.

one case, the winds in the other case. The following discussion concerns first atmospheric then oceanic models, and finally coupled models.

The Atmosphere

The atmospheric circulation in low latitudes corresponds mainly to direct thermal circulations driven by convection over the regions with the highest surface temperatures. Moisture-bearing trade winds converge onto these regions where the air rises in cumulus towers that provide plentiful rainfall locally. The three main convective zones – they can easily be identified in satellite photographs of the Earth's cloud cover – are over the Congo and Amazon River basins, and the 'maritime continent' of the western equatorial Pacific, south-eastern Asia, and northern Australia. The latter region includes an enormous pool of very warm water that extends to the vicinity of the dateline. Much of the air that rises there subsides over the relatively cold eastern equatorial Pacific where rainfall is minimal; the deserts along the coasts of Peru and California in effect extend far westward over the adjacent ocean. The only region of warm surface waters and heavy rainfall in the eastern tropical Pacific is the doldrums, also known as the Intertropical Convergence Zone, a narrow band between 3° and 10°N approximately, onto which the south-east and north-east trade winds converge.

A warming of the eastern tropical Pacific, such as occurs during El Niño, amounts to an eastward expansion of the pool of warm waters over the western Pacific and causes an eastward migration of the convective zone, thus bringing rains to the east, droughts to the west. The east–west Intertropical Convergence Zone, simultaneously moves equatorward. During La Niña, a westward contraction of the warm waters shifts the convection zone back westward and intensifies the trades. The Southern Oscillation is this interannual back-and-forth movement of air masses across the tropical Pacific. **Figure 2** depicts its complementary El Niño and La Niña states.

A hierarchy of models is available to simulate the atmospheric response to changes in tropical sea surface temperature patterns. The most realistic, the general circulation models used for weather prediction, attempt to incorporate all the important physical processes that determine the atmospheric circulation. Simpler models isolate a few specific processes and, by elucidating their roles, provide physical insight. Particularly useful for capturing the essence of the Southern Oscillation – the departure from the time-averaged state – is a model that treats the atmosphere as a one-layer fluid on a rotating sphere. Motion is driven either by a heat source over the region of unusually high surface temperatures (the moisture carried by the winds that converge onto the source can amplify the magnitude of the heat source), or is driven by sea surface temperature gradients that give rise to pressure gradients in the lower layer of the atmosphere. Such models are widely used as the atmospheric components of relatively simple coupled ocean–atmosphere models.

The complex general circulation models reproduce practically all aspects of the interannual Southern Oscillation over several decades if, in the simulations, the observed sea surface temperature patterns are specified. The models fail to do so if the temperature patterns are allowed only seasonal, not interannual variability. This important result implies that, although weather prediction is limited to a matter of weeks, coupled ocean–atmosphere models should be capable of extended forecasts of certain averaged fields, those associated with the Southern Oscillation, for example. The key difference between weather and climate is that the first is an initial value problem – a forecast of the weather tomorrow requires an accurate description of the atmosphere – whereas climate (from an atmospheric perspective) is a boundary value problem; a change in climate can be induced by altering conditions at the lower boundary of the atmosphere. The crucial conditions are sea surface temperatures in the case of El Niño.

The Oceans

The salient feature of the thermal structure of the tropical oceans is the thermocline, the thin layer of large vertical temperature gradients at a depth of approximately 100 m, that separates warm surface waters from colder water at depth. Sea surface temperature patterns in the tropics tend to reflect variations in the depth of the thermocline and those variations are controlled by the winds. Thus the surface waters are cold where the thermocline is shallow, in the eastern equatorial Pacific for example, and are warm where the thermocline is deep, in the western tropical Pacific. The downward slope of the thermocline, from east to west in the equatorial Pacific, is a consequence of the westward trade winds that, along the equator where the Coriolis force vanishes, drive the surface waters westward. When those winds relax, during El Niño, the warm surface waters in the west return to the east so that the thermocline there deepens and sea surface temperatures rise. To a first approximation, the change from La Niña to El Niño amounts to a horizontal (east–west), adiabatic redistribution of warm surface waters, and is the dynamical response of the ocean to the changes in the winds.

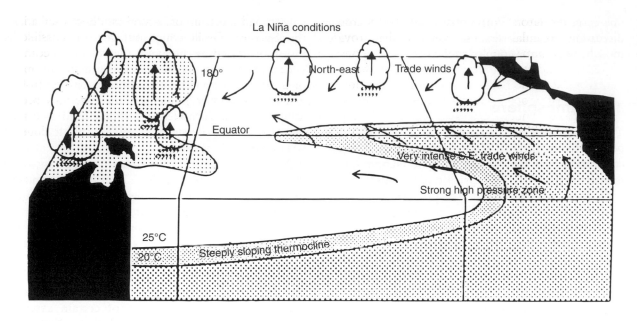

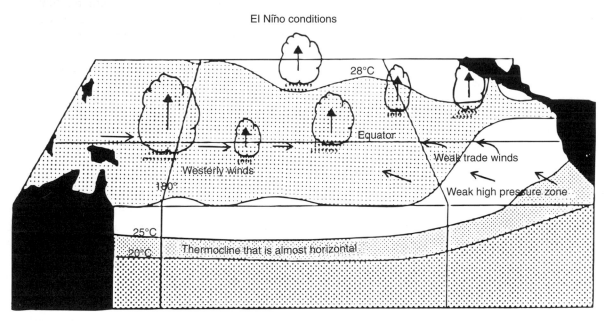

Figure 2 A schematic view of La Niña (top) and El Niño (bottom) conditions. During La Niña, intense trade winds cause the thermocline to have a pronounced slope, down to the west, so that the equatorial Pacific is cold in the east, warm in the west where moist air rises into cumulus towers. The air subsides in the east, a region of little rainfall, except in the doldrums where the south-east and north-east trades converge. During El Niño, the trades along the equator relax, as does the slope of the thermocline when the warm surface waters flow eastward. The change in surface temperatures is associated with an eastward shift of the region of heavy precipitation.

Theoretical studies of the oceanic response to changes in the wind started with studies of the generation of the Gulf Stream. How long, after the sudden onset of the winds, before a Gulf Stream appears and the ocean reaches a state of equilibrium? The answer to this question provides an estimate of the 'memory', or adjustment-time of the ocean, a timescale that turns out to be relevant to the timescale of the Southern Oscillation. In an unbounded ocean,

a Gulf Stream is impossible, so that the generation of such a current depends on the time it takes for information concerning the presence of coasts to propagate east–west across the ocean basin. The speed of the oceanic waves that carry this information, known as Rossby waves, increases with decreasing latitude. It should take on the order of a decade to generate the Gulf Stream from a state of rest, a matter of months to generate the same current near the equator. Fortuitously, the three tropical oceans have similarities and differences that provide a wealth of information about the oceanic response to different wind stress patterns. To explain and simulate tropical phenomena, oceanographers developed a hierarchy of models of which the simplest is the shallow-water model whose free surface is a good analog of the very sharp, shallow tropical thermocline. Studies with that tool showed how a change in the winds over one part of an ocean basin (e.g. the western equatorial Pacific), can influence oceanic conditions in a different and remote part of the basin (e.g. off the coast of Peru). Thus a warming of the surface waters along the coast of Peru during El Niño could be a consequence of a change in the winds over the western equatorial Pacific. The roles of waves and currents in effecting such changes depend on the manner in which the winds vary. If the winds change abruptly, then waves are explicitly present, but if the winds vary slowly, with a timescale that is long compared with the adjustment time of the ocean, then the waves are strictly implicit because the response is an equilibrium one. A detailed description of the response of the Pacific to slowly varying winds first became available for El Niño of 1982 and provided a stringent test for sophisticated general circulation models of the ocean. The success of such models in simulating the measurements bolstered confidence in the models which are capable of reproducing, deterministically, the oceanic aspects of the Southern Oscillation between El Niño and La Niña over an extended period, provided that the surface winds are specified. Today such models serve, on a monthly operational basis, to interpolate measurements from a permanent array of instruments in the Pacific, thus providing a detailed description of current conditions in the Pacific Ocean, a description required as initial conditions for coupled ocean–atmosphere models that predict El Niño.

Interactions between the Ocean and Atmosphere

Suppose that, during La Niña, when intense trades drive the surface waters at the equator westward, a random disturbance causes a slight relaxation of the trades. Some of the warm water in the west then starts flowing eastward, thus decreasing the east–west temperature gradient that maintains the trades. The initial weakening of the winds is therefore reinforced, causing even more warm water to flow eastward. This tit-for-tat (positive feedback) leads to the demise of La Niña, the rise of El Niño. The latter state can similarly be shown to be unstable to random perturbations.

A broad spectrum of natural modes of oscillation – the Southern Oscillation is but one – is possible because of unstable interactions between the ocean and atmosphere. The properties of the different modes depend mainly on the mechanisms that control sea surface temperature, because that is the only oceanic parameter that affects the atmosphere on the timescales of interest here. Those mechanisms include advection by oceanic currents, and vertical movements of the thermocline caused either by local winds, or by remote winds that excite waves that propagate along the thermocline. In one class of coupled ocean–atmosphere modes, sea surface temperature variations depend primarily on advection. Consider an initial perturbation in the form of a confined equatorial region of unusually high sea surface temperatures. That warm patch, superimposed on waters that get progressively warmer from west to east, induces westerly winds to its west, easterly winds to its east. The westerly winds drive convergent currents that advect warm water, in effect extending the patch westward. The easterly winds induce divergent currents that advect cold water, thus lowering sea surface temperatures and contracting the warm patch on its eastern side. The net result is a westward displacement of the original disturbance. A mode of this type is involved in the response of the eastern equatorial Pacific to the seasonal variations in solar radiation.

In certain coupled ocean–atmosphere modes the slow adjustment of the ocean (over a period of months and longer) to a change in winds, in contrast to the short period (weeks) the atmosphere takes to come into equilibrium with altered sea surface temperature, is of great importance. (That is why the ocean, far more than the atmosphere, needs to be monitored to anticipate future developments.) These modes, which are known as 'delayed oscillator' modes because of the lagged response of the oceans, differ from those discussed in the previous paragraph, in being affected by the boundedness of the ocean basin – the presence of north–south coasts. Furthermore, sea surface temperatures now depend mainly on thermocline displacements, not on upwelling induced by local winds. The effect of winds to the west of an, initially, equatorially confined region of unusually warm surface waters can extend far eastward because of certain waves that travel efficiently in that direction along the equator. This type of mode is associated with long timescales, of several years, on which the oceanic response is almost, but not quite, in equilibrium with the gradually changing winds. That distinction is of vital importance because the small departure from equilibrium, the 'memory' of the system, brings about the transition from one phase of the oscillation to the next. If El Niño conditions are in existence then the delayed response of the ocean causes El Niño to start waning after a while, thus setting the stage for La Niña.

To which of the various modes does the observed Southern Oscillation correspond? The answer depends on the period under consideration, because the properties of the dominant mode at a given time (the one likely to be observed) depend on the background state at the time. For example, if the thermocline is too deep then the winds may be unable to bring cold water to the surface so that ocean–atmosphere interactions, and a Southern Oscillation, are impossible. Such a state of affairs can be countered by sufficiently intense easterly winds along the equator because they slope the thermocline down towards the west, shoaling it in the east. These considerations indicate that the time-averaged depth of the thermocline, and the intensity of the easterly winds, are factors that determine the properties of the observed Southern Oscillation. (The temperature difference across the thermocline is another factor.) If the thermocline is shallow, and the winds are intense (this was apparently the case some 20 000 years ago during the last Ice Age), then the Southern Oscillation is likely to have a short period of 1 or 2 years, and to resemble the mode excited by the seasonal variation in solar radiation. In general, the observed mode is likely to be a hybrid one with properties intermediate between those associated with the seasonal cycle, and those known as the delayed oscillator type. During the 1960s and 1970s the thermocline was sufficiently shallow, and the trade winds sufficiently intense, for the dominant mode to have some of the characteristics of the annual mode – a relatively high frequency of approximately 3 years, and westward phase propagation. Since the late 1970s, the background state has changed because of a slight relaxation of the trades, a deepening of the thermocline in the eastern tropical, and a rise in the surface temperatures of that region. This has contributed to a change in the properties of El Niño. During the 1980s and 1990s it was often associated with eastward phase propagation, and with a longer period of recurrence, 5 years. This gradual modulation of the Southern Oscillation cannot explain the differences between one El Niño and the next, why El Niño was very weak in 1992, exceptionally intense in 1997. Those differences are attributable to random disturbances, such as westerly wind bursts that last for a week or two over the far western equatorial Pacific. They can be influential because the background state, for the past few decades, has always been such that interactions between the ocean and atmosphere are marginally unstable at most, or slightly damped so that the continual Southern Oscillation is sustained by sporadic perturbations. A useful analogy is a swinging pendulum that is subject to modest blows at random times. A blow, depending on its timing, can either amplify the oscillation – that apparently happened in 1997 when a burst of westerly winds in March of that year led to a very intense El Niño – or can damp the oscillation. The predictability of El Niño is therefore limited, because the westerly wind bursts cannot be anticipated far in advance. (Several models predicted that El Niño would occur in 1997, but none anticipated its large amplitude.)

These results concerning the stability properties of ocean–atmosphere modes come from analyses of relatively simple coupled models: the ocean is a shallow-water model with a mixed, surface layer in which sea surface temperatures have horizontal variations; the atmosphere is also a single layer of fluid driven by heat sources proportional to sea surface temperature variations. These models deal only with modest departures from a specified background state that can be altered to explore various possible worlds. The results are very helpful in the development of more sophisticated coupled general circulation models of the ocean and atmosphere which have to simulate, not only the interannual Southern Oscillation, but also the background state. At this time the models are capable of simulating with encouraging realism various aspects of the Earth's climate, and of the Southern Oscillation. As yet, the properties of the simulated oscillations do not coincide with those of the Southern Oscillation as observed during the 1980s and 1990s, presumably because the background state has inaccuracies. The models are improving rapidly.

Conclusions

The Southern Oscillation, between complementary El Niño and La Niña states, results from interactions between the tropical oceans and atmosphere. The detailed properties of the oscillation (e.g. its period and spatial structure) depend on long-term averaged background conditions and hence change gradually with time as those conditions change. The ocean, because its inertia exceeds that of the atmosphere by a large factor (oceanic adjustment to a change in forcing is far more gradual than atmospheric adjustment), needs to be monitored in order to anticipate future developments. The development of computer models capable of predicting El Niño is advancing rapidly.

See also

El Niño Southern Oscillation (ENSO). Satellite Remote Sensing of Sea Surface Temperatures.

Further Reading

The *Journal of Geophysical Research* Volume 103 (1998) is devoted to a series of excellent and detailed reviews of various aspects of El Niño and La Niña.

Neelin JD, Latif M, and Jin F-F (1994) Dynamics of coupled ocean–atmosphere models: the tropical problem. *Annual Review of Fluid Mechanics* 26: 617–659.

Philander SGH (1990) *El Niño, La Niña and the Southern Oscillation*. New York: Academic Press.

Zebiak S and Cane M (1987) A model El Niño Southern Oscillation. *Monthly Weather Review* 115: 2262–2278.

Schopf PS and Suarez MJ (1988) Vacillations in a coupled ocean–atmosphere model. *Journal of Atmospheric Science* 45: 549–566.

MONSOONS, HISTORY OF

N. Niitsuma, Shizuoka University, Shizuoka, Japan
P. D. Naidu, National Institute of Oceanography, Dona Paula, India

Introduction

The difference in specific heat capacity between continents and oceans (and specifically between the large Asian continent and the Indian Ocean) induces the monsoons, strong seasonal fluctuations in wind direction, and precipitation over oceans and continents. Over the Indian Ocean, strong winds blow from the southwest during boreal summer, whereas weaker winds from the northeast blow during boreal winter. The monsoons are strongest over the western part of the Indian Ocean and the Arabian Sea (**Figure 1**). The high seasonal variability affects various fluctuations in the environment and its biota which are reflected in marine sediments. Marine sedimentary sequences from the continental margins thus contain a record of the history of monsoonal occurrence and intensity, which may be deciphered to obtain insight in the history of the monsoon system. Such insight is important not only to understand the mechanisms that cause monsoons, but also to understand the influence of monsoons on the global climate system including the Walker circulation, the large-scale, west–east circulation over the tropical ocean associated with convection. The summer monsoons may influence the Southern Oscillation because of interactions between the monsoons and the Pacific trade wind systems.

Geologic Records of the Monsoons

Sedimentary sequences record the effects of the monsoons. One such effect is the upwelling of deeper waters to the surface, induced by the strong southwesterly winds in boreal summer in the Arabian Sea, and the associated high productivity of planktonic organisms. Continents supply sediments to the continental margin through the discharge of rivers, as the result of coastal erosion, and carried by winds (eolian sediments). Monsoon-driven wind direction and strength and precipitation therefore control the sediment supply to the continental margin.

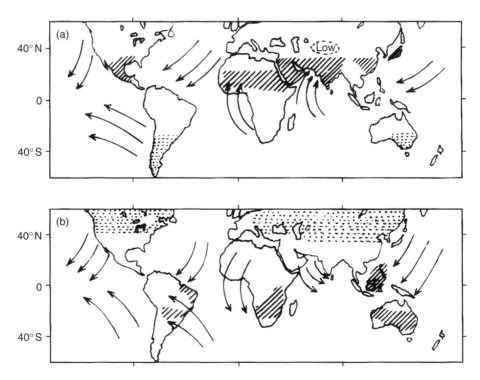

Figure 1 The domains of the monsoon system of the atmosphere during Northern Hemisphere (a) summer and (b) winter. The hatching shows the land areas with maximum surface temperature and stippling indicates the coldest land surface.

Monsoonal precipitation supplies a large volume of fresh water, discharged by rivers from the continents into the oceans, with the flow directed by the topography in the regions of the river mouths. The density of fresh water is less than that of seawater, and the seasonal freshwater flow dilutes surface ocean water which then has a lower density than average seawater. The existence of strong vertical density gradients causes the development of stratification in the water column.

The elevation of the continent governs the circulation of the atmosphere, and influences vegetation patterns. The vegetation cover on the continent directly affects the albedo and heat capacity of the land surface, both of which are important factors in the generation of monsoonal circulation patterns. The high Himalayan mountain range acts as a barrier for air circulation; east–west-oriented mountain ranges particularly affect the course of the jet streams. Continental topography, vegetation coverage, and elevation thus affect the monsoonal circulation and therefore the rate of sediment supply to the continental margin, as well as the composition of continental margin sedimentary sequences.

Sedimentary Indicators of Monsoons

Monsoon-induced seasonal contrasts in wind direction and precipitation are recorded in the sediments by many different proxies. Most of these monsoonal indicators, however, record qualitative and/or quantitative changes of the monsoons in one, but not in both, season. For example, upwelling indicators in the Arabian Sea represent the strength of the southwest (summer) monsoon.

In interpreting the sedimentary record, one must note that bottom-dwelling fauna bioturbates the sediment to depths of c. 10 cm, causing the co-occurrence of sedimentary material deposited at different times within a single sediment sample. Laminated sediments without bioturbation are only deposited in oxygen-minimum zones, where anoxia prevents activity of burrowing metazoa. The time resolution of studies of various monsoonal indicators thus is limited by this depth of bioturbation, but each proxy by itself does record information on the environment in which it formed at the time that it was produced. Single-shell measurements of oxygen and carbon isotopes in foraminiferal tests (see below) thus may show the variability within each sample, thus within the zone of sediment mixing.

Oxygen isotopes. The oxygen isotopic ratio ($\delta^{18}O$) of the calcareous tests of fossil organisms (such as foraminifera) provides information on the oxygen isotopic ratio in seawater and on the temperature of formation of the test. The volume of the polar ice sheets and the local influx of fresh water controls the oxygen isotopic ratio of seawater. The $\delta^{18}O$ record of the volume of the polar ice sheets can be used for global correlation of oxygen isotopic stages, corresponding to glacial and interglacial intervals. Freshwater discharge caused by monsoonal precipitation lowers the $\delta^{18}O$ value of seawater. The temperature record of various species with different depth habits, such as benthic species at the bottom, and planktonic species at various depths below the surface, can provide a temperature profile of the water column. Temperature is the most basic physical parameter, which provides information on the stratification or mixing of the water column, as well as on changes in thermocline depth. Different species of planktonic foraminifera grow in different seasons, and temperatures derived from the oxygen isotopic composition of their tests thus delineates the seasonal monsoonal variation in sea surface temperatures.

Carbon isotopes. The carbon isotopic ratio in the calcareous test of foraminifera provides information on the carbon isotopic ratio of dissolved inorganic carbon (DIC) in seawater and on biofractionation during test formation. Carbon isotopic ratios of various species of foraminifera which live at different depths provide information on the carbon isotopic profile of DIC in the water column. The carbon isotope ratio of DIC in seawater is well correlated with the nutrient concentration in the water, because algae preferentially extract both the lighter carbon isotope (^{12}C) and nutrients to form organic matter by photosynthesis. Decomposition of organic matter releases lighter carbon as well as nutrients, and lowers the carbon isotopic composition of DIC. The carbon isotopic profile with depth thus provides information on the balance of photosynthesis and decomposition.

Rate of sedimentation. Marine sediments are composed of biogenic material produced in the water column (dominantly calcium carbonate and opal), and terrigenous material supplied from the continents. The sediments are transported laterally on the seafloor and down the continental slope, and eventually settle in topographic depressions in the seafloor. Both seafloor topography and the supply of biogenic and lithogenic material control the apparent rate of sedimentation.

Organic carbon content. Organisms produce organic carbon in the euphotic zone, which is strongly recycled by organisms in the upper waters, but a small percentage of the organic material eventually settles on the seafloor. Organisms (including bacteria) decompose the organic carbon on the surface of the seafloor as well as within the sediment, using dissolved oxygen in the process. The flux of organic matter and the availability of oxygen thus control the organic carbon content in the sediments. In high-productivity areas, such as regions where monsoon-induced upwelling occurs, a large amount of organic matter sinks from the sea surface through the water column. The sinking organic matter decomposes and consumes dissolved oxygen, and at mid-water depths an oxygen minimum zone may develop in such high productivity zones. Oxygen concentrations may fall to zero in high-productivity environments, and in these anoxic environments, eukaryotic benthic life becomes impossible, so that there is no bioturbation. The organic carbon content in the sediment deposited below oxygen-minimum zones may become very high, and values of up to 7% organic carbon have been recorded in areas with intense upwelling, such as the Oman Margin.

Calcium carbonate. Three factors control the calcium carbonate content of pelagic sediments: productivity, dilution, and dissolution. Calcium carbonate tests are produced by pelagic organisms, including photosynthesizing calcareous nannoplankton and heterotrophic planktonic foraminifera. At shallow water depths (above the calcite compensation depth, CCD) dissolution is negligible; therefore, the calcium carbonate content in continental margin sediments is regulated by a combination of biotic productivity and dilution by terrigenous sediment.

Magnetic susceptibility. Magnetic susceptibility provides information about the terrigenous material supply and its source. The part of the sediment provided by biotic productivity (calcium carbonate and opaline silica) has no carriers of magnetic material. Magnetic susceptibility is thus used as an indicator of terrigenous supply to the ocean.

Clay mineral composition. Weathering and erosion processes on land lead to the formation of various clay minerals. The clay mineral composition in the sediments is an indicator for the intensity of weathering processes and thus temperature and humidity in the region of origin. The clay mineral composition thus can be used as a proxy for the aridity and vegetation coverage in the continents from which the material derived. In addition, the crystallinity of the clay mineral illite depends on the moisture content of soils in the area of origin. At high moisture, illite decomposes and dehydrates, so that its crystallinity decreases. Therefore, the crystallinity of illite can be used as a proxy for the humidity in its source area.

Eolian dust. Eolian dust consists of fine-grained quartz and clay minerals, such as illite. The content and grain size of eolian dust in marine sediment, therefore, is an indicator of wind strength and direction, as well as of the aridity in the source area.

Fossil abundance and diversity. The abundance and diversity of foraminifera, calcareous nannoplankton, diatoms, and radiolarians depend on the chemical and physical environmental conditions in the oceans. The diversity of nannofossil assemblages decreases with sea surface temperature and is thus generally correlated with latitude. In upwelling areas, low sea surface temperatures caused by the monsoon-driven upwelling disturb the zonal diversity patterns of nannoflora. Therefore, the diversity of nannofossils can be used as an upwelling indicator. In addition, the faunal and floral assemblages can be used to estimate sea surface temperature and salinity as well as productivity.

UK_{37} *ratio.* The biomarker UK_{37} (an alkenone produced by calcareous nannoplankton) is used to estimate sea surface temperatures within the photic zone where the photosynthesizing algae dwell. The records are not always easy to calibrate, especially in the Tropics and at high latitudes, but global calibrations are now available.

Indicators of Present Monsoons

The specific effect of the monsoons on the supply of lithogenic and biogenic material to the seafloor varies in different regions, so that specific tracers cannot be efficiently applied in all oceans.

Arabian Sea

During the boreal summer, strong southwest monsoonal winds produce intense upwelling in the Arabian Sea (**Figure 2**). These upwelling waters are characterized by low temperatures and are highly enriched in nutrients. The process of upwelling fuels the biological productivity in June through August in the Arabian Sea. The weaker, dry, northeast winds which prevail during the boreal winter do not produce upwelling, and productivity is thus lower in the winter months. Thus, the southwest and northeast monsoonal winds produce a strong seasonal contrast in primary productivity in the Arabian Sea. Sediment trap mooring experiments have demonstrated that up to 70% of the biogenic and lithogenic flux to the seafloor occurs during the summer monsoon. Biological and terrestrial particles thus strongly reflect summer conditions, and eventually settle on the seafloor to contribute to a distinct biogeochemical record of monsoonal upwelling. Therefore, regional sediments beneath the areas affected by monsoon-driven upwelling record long-term variations in the

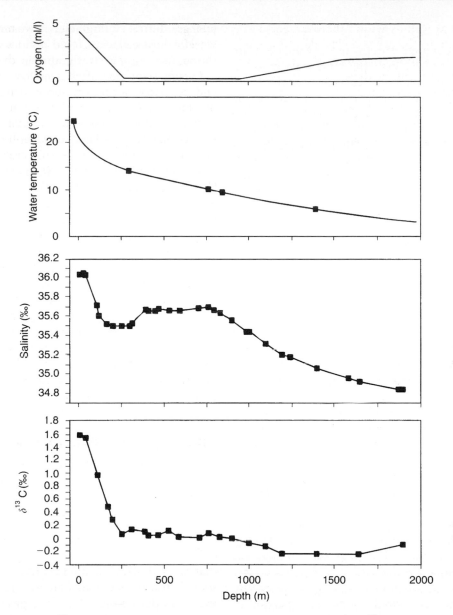

Figure 2   Vertical profiles of $\delta^{13}C$, salinity, temperature, and oxygen values in the Arabian Sea. $\delta^{13}C$ and salinity are from GEOSECS Station 413, while water temperature and oxygen values are from ODP Site 723 in the Oman Margin.

strength and timing of the monsoonal circulation. The following proxies were used to study the upwelling strength in the Arabian Sea.

Globigerina bulloides abundance. Seasonal plankton tows and sedimentary trap data document that the planktonic foraminifer species *G. bulloides* is abundant during the summer upwelling season in the Arabian Sea. Core top data from the upwelling zones of the Arabian Sea show that the dominance of *G. bulloides* in the living assemblage is preserved in the sediment. Changes in the abundance of *G. bulloides* in the sediments thus have been used to infer the history of upwelling intensity in the western Indian Ocean.

Oxygen and carbon isotopes. More recently, it was proposed that oxygen and carbon isotopic difference between the surface and subsurface dwelling planktonic foraminifera can be used as proxy records to reconstruct intensity of upwelling and monsoon.

Lithogenic material. Lithogenic material deposited in the northwest Arabian Sea is dominantly eolian (diameter up to $18.5\,\mu m$), and is transported exclusively during the summer southwest monsoon. The lithogenic grain size in sediment cores thus provides information about the strength of the southwest monsoonal winds and associated upwelling in the Arabian Sea.

China Sea

The surface circulation patterns in the China Sea are also closely associated with the large-scale seasonal reversal of the atmospheric circulation over the Asian continent. High precipitation over Asia during the summer leads to increased input of fresh water into the China Sea, lowering sea surface salinity. Therefore, the $\delta^{18}O$ record of planktonic foraminifera in sediment cores documents the magnitude of freshwater discharge, sea surface salinity, and summer monsoonal rainfall in the past. During the winter, the westerly winds lower the sea surface temperature and deliver a large amount of eolian dust to the South China Sea. The rate of eolian dust supply and sea surface temperature changes in this region thus reflect the strength of the winter monsoon.

Japan Sea

The winter monsoon's westerlies transport eolian dust from the Asian desert regions to the Japan Sea and the Japanese Islands 3–5 days after sandstorms in the source area. The thickness of the dust layer is larger in the western Japan Sea. The concentration and grain size of eolian particles in sediments of Japan Sea thus represent the strength of the winter monsoon.

Continental Records

Lake levels in the monsoon-influenced regions are highly dependent on the monsoon rainfall; therefore researchers have been using lake levels to trace the monsoon history during the late Quaternary period. The layers of stalagmite (carbonate mineral) preserves the oxygen isotopic composition of monsoon rains that were falling when the stalagmite got precipitated. The oxygen isotope composition of stalagmite is proportional to the amount of rainfall: the more the rainfall, the lighter the $\delta^{18}O$ of stalagmite, and vice versa. Thus, the oxygen isotopic ratios of stalagmites from the caves of monsoon-influenced regions have been used to infer the ultra-high-resolution variability of monsoon precipitation.

Variability of Monsoons during Glacial and Interglacials

Arabian Sea

Along the Oman Margin of the Arabian Sea, strong southwest summer monsoon winds induce upwelling. Detailed analyses of various monsoon tracers such as abundance of *G. bulloides* and *Actinomma* spp. (a radiolarian) and pollen reveal that the southwest monsoon winds were more intense during interglacials (warm periods) and weaker during glacials (cold periods), recognized by the oxygen isotopic stratigraphy in the same samples.

Carbon isotope differences between planktonic and benthic foraminifera show lower gradients during interglacials, and higher gradients during glacials. The lower gradients indicate that upwelling was strong (and pelagic productivity high) during interglacials due to a strong summer southwest monsoon. Similarly, the oxygen isotope difference between planktonic and benthic foraminifera along the Oman Margin reflects changes in thermocline depth, associated with summer monsoon-driven upwelling. A larger oxygen isotope difference between planktonic and benthic foraminifera during interglacials reflects the presence of a shallow thermocline, as a result of the strong summer monsoons. Oxygen and carbon isotope records from various planktonic and benthic foraminifera in several cores located within and away from the axis of the Somali Jet along the Oman Margin indicate that sea surface temperatures were lower and varied randomly during interglacials, reflecting strong upwelling induced by a strong summer monsoon (**Figure 3**).

Studies of the oxygen and carbon isotopic composition of individual tests of planktonic foraminifera enable us to understand the seasonal temperature variability induced by monsoons in the Arabian Sea. Such studies show that the seasonality was stronger during glacials and weaker during interglacials, because during glacials the southwest summer monsoon was weaker and the northeast winter monsoon stronger. The variability in seasonal contrast during glacials and interglacials suggests that interannual and interdecadal changes in monsoonal strength were also greater during glacial periods than during interglacials (**Figure 4**).

High-resolution monsoon records from the Arabian Sea show millennial timescale variability of SW monsoon over last 20 ky. Synchronous changes between SW monsoon intensity and variations of temperatures in North Atlantic and Greenland suggest some kind link between monsoons to the high-latitude temperature.

South China Sea

High-resolution studies in the South China Sea document a high rate of delivery of eolian dust during glacial periods, as well as lower sea surface temperatures, documented by the UK_{37} records. These data indicate that during glacial periods the winter monsoon was more intense. During interglacials the sea surface salinity (derived from oxygen isotope records) was much lower than during

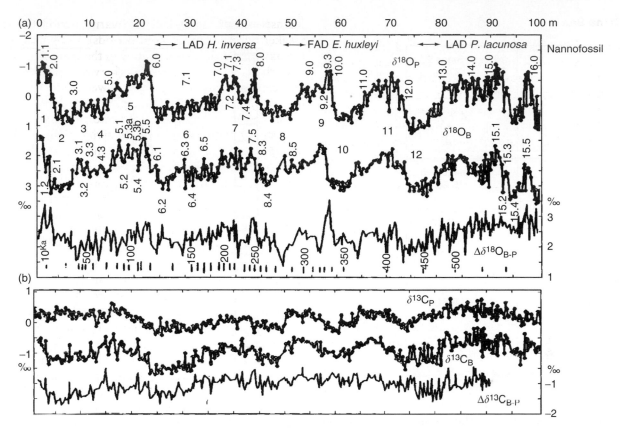

Figure 3 (a) Oxygen isotope profiles of planktonic foraminifera (*Pulleniatina obliquiloculata*, $\delta^{18}O_P$) and benthic foraminifera (*Uvigerina excellens*, $\delta^{18}O_B$), and the difference between oxygen isotopes of planktonic and benthic foraminifera ($\Delta\delta^{18}O_{B-P}$). (b) Carbon isotope profiles of planktonic foraminifera ($\delta^{13}C_P$) and benthic foraminifera ($\delta^{13}C_B$), and the difference between planktonic and benthic records ($\Delta\delta^{13}C_{B-P}$) over last 800 000 years at ODP Site 723 in the Arabian Sea. Large planktonic–benthic differences indicate a more vigorous monsoonal circulation during the summer monsoon.

glacials, probably as a result of increased freshwater discharge from rivers caused by the high summer monsoon precipitation. Over the South China Sea, glacials were thus characterized by an intense winter monsoon, and interglacials by a strong summer monsoon circulation.

Japan Sea

The oceanographic conditions of the Japan Sea are strongly influenced by eustatic sea level changes, because shallow straits connect this sea to the Pacific Ocean. In glacial times, at low sea level, most of the straits were above sea level and the Japan Sea was connected to the ocean only by a narrow channel located on the present continental shelf. River discharge of fresh water into the semi-isolated Japan Sea caused the development of strong stratification, and the development of anoxic conditions, as documented by the occurrence of annually laminated sediments.

The sedimentary sequence in the Japan Sea thus consists of mud, with laminated sections alternating with bioturbated, homogeneous muds. During interglacials the waters at the bottom were oxygenated, although the organic carbon content of the sediment can be high (up to 5%) and diatoms abundant (up to 30% volume). During glacials, conditions on the seafloor alternated between euxinic (anoxic) and noneuxinic. Glacial and interglacial parts of the sediment section can thus be easily recognized.

The sedimentary sequences contain eolian dust transported from the deserts of west China and the Chinese Loess Plateau by the westerly winter monsoon. The eolian dust content of the sediments is thus an indicator of the strength of the winter monsoon. The crystallinity of illite, a main component of the eolian dust, indicates that the source region on the Asian continent was more humid during interglacials. A high content of eolian dust and a high crystallinity of illite during glacial stages indicate that during these intervals the winter monsoon was strong and the summer monsoon weak. Summer monsoons were strong during interglacials, but winter monsoons were strong during glacials.

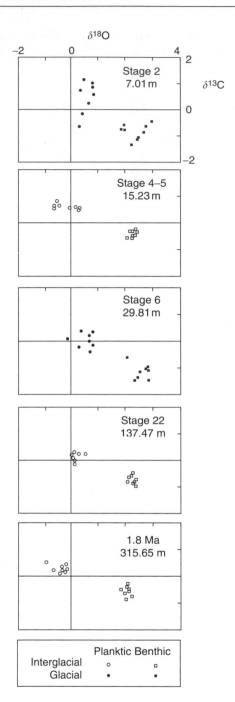

Figure 4 Carbon and oxygen isotope ratios of individual tests of planktonic foraminifera (*Pulleniatina obliquiloculata*) and benthic foraminifera (*Uvigerina excellens*) at ODP Site 723 in the Arabian Sea. The variations in difference between planktonic and benthic values reflect the magnitude of the seasonal changes in surface and bottom waters through time.

Long-term Evolution of the Asian Monsoon

Arabian Sea

Long-term variations of proxies of monsoonal intensity tracers, especially the species diversity of

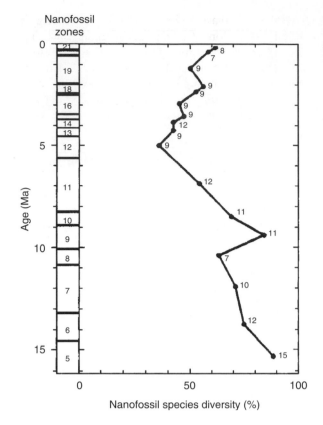

Figure 5 Calcareous nannofossll diversity for the last 15 My in the Indian Ocean. High diversity indicates weak upwelling and low diversity represents strong upwelling. The upwelling intensity is controlled by the strength of the summer monsoon winds over the Indian Ocean. The numbers on the species diversity profile represent number of species.

calcareous nannofossil species, and the abundance of the planktonic foraminifera *G. bulloides* and the radiolarian *Actinomma* spp. (upwelling indicators) show that the evolution of the Asian monsoon started in the late Miocene, at about 9.5 Ma. Between 9.5 and 5 Ma, the monsoon increased noticeably in strength, with smaller fluctuations in monsoonal intensity from 5 to 2 Ma (**Figure 5**).

Upwelling indicators such as the differences in oxygen and carbon isotope between planktonic foraminifera, and the organic carbon content in sediments from the Oman Margin indicate that until about 0.8 Ma the summer monsoon was intense, as during interglacials. Oxygen and carbon isotope ratios in individual tests of planktonic and benthic foraminifera show that from 0.8 Ma onward, the strength of the summer monsoon changed with glacial and interglacial cycles, as described above.

China Sea

In sediments from the South China Sea, magnetic susceptibility and calcium carbonate percentages started to decrease at about 7 Ma, reflecting increased deposition of terrigenous, eolian dust. This increase in dust supply indicates that monsoons started to become noticeable at that time. Both color reflectance and magnetic susceptibility show cyclic fluctuations in monsoonal intensity from 0.8 Ma on, in response to glacial and interglacial cycles (as described above).

Japan Sea

Cyclic changes with large amplitude of magnetic susceptibility and illite crystallinity in sediment cores indicate that the imprint of the monsoon signature in the Japan Sea sediment record started at about 0.8 Ma. The crystallinity of illite was much higher until 0.8 Ma, reflecting that the summer monsoon was weak until that time, as it was afterward during glacials.

Asian Continent

Loess and paleosol sequences on the Chinese Plateau show cyclic fluctuations in magnetic susceptibility and illite crystallinity for the last 0.6 My. Loess and paleosol sequences in the Kathamandu Basin show cyclic fluctuations in magnetic susceptibility and illite crystallinity for the last 1.1 My. The cyclic sequences of low crystallinity, representing a humid climate in the interglacial periods in both areas, is consistent with the marine sediment records of the Japan Sea.

Palearctic elements first became represented in the molluskan faunas in the Siwalik group sediments in the Himalayas after 8 Ma, suggesting the beginning of seasonal migrations of water birds crossing the Himalayas at this time. The diversity of this molluskan fauna increased around 5 Ma, a time when the summer monsoon was strong. The timing of developments in the molluskan faunas in the Himalayas is thus consistent with the long-term evolution of the Asian summer monsoon as derived from marine records.

The Asian Monsoon and the Global Climate System

Uplift of the Himalayas and the Tibetan Plateau occurred coeval with the increase in strength of the Asian monsoon between 9.5 and 5 Ma, as documented by the heavy mineral composition of deposits in the Bengal Fan, derived from the weathering and erosion of the rising Himalayas. Cyclic fluctuations in the strength of the summer monsoon started at about 1.1 Ma in the southern Himalayas and Tibet, whereas in the Arabian Sea, South China Sea, and Japan Sea such changes started at about 0.8 Ma.

Peru Margin

As a result of strong southeasterly trade winds, nutrient-rich water upwells along the Peru coast and reaches the photic zone in a belt that is c. 10 km wide, and parallels the coastline. In this region, upwelling-induced productivity is very high. The organic carbon concentration in the sediments below this high-productivity zone has been used to trace the upwelling strength in the past. Upwelling is absent or less intense during El Niño/Southern Oscillations (ENSO) events, and the Southern Oscillation is linked to the Asian monsoons in the tropical Walker circulation.

Upwelling along the Peru Margin started at around 3.5 Ma, as indicated by an increase in the organic carbon content, and the decrease of sea surface temperatures (derived from UK_{37} records). Upwelling along the Peru Margin thus started after the Asian monsoons reached their full strength at about 5 Ma.

Equatorial Upwelling

The intensity of the southeast Asian monsoon controls the easterly trade winds associated with the north and south equatorial currents, and the strength of the easterly trade winds controls the intensity of equatorial upwelling. In the equatorial Pacific, the intensity of trade winds and equatorial upwelling increased at about 5 Ma (as indicated by a high abundance of siliceous and calcareous pelagic microfossils), at the time that the Asian monsoons developed their full intensity.

Teleconnections

Numerous data sets show synchroneity between changes of monsoon intensity and abrupt climate shifts in Greenland during the deglaciation (between 15 and 16 ky). In addition, monsoon proxy records from the Arabian Sea reveal that the intervals of weak summer monsoon coincide with cold periods in the North Atlantic region and vice versa. All these evidences suggest some kind of teleconnection between monsoon and global climate. However, the exact physical mechanism underlying the link between high-latitude temperature changes and SW monsoon has not been addressed yet.

Cyclicity of Monsoon

Paleomonsoon records show periodicity of SW monsoon at 100, 41, and 23 ky corresponding to Earth's orbital changes of eccentricity, obliquity, and precession, respectively. Subsequently, high-resolution monsoon variability exhibits suborbital periodicities of 2200, 1700, 1500, and 775 years.

Conclusions

The evolution of the Asian monsoon started at around 9.5 Ma, in response to the uplift of the Himalayas. The monsoonal intensity reached its maximum at around 5 Ma, and from that time the associated easterly trade winds caused intense upwelling in the equatorial Pacific. Before 1.1 Ma, the summer monsoon was strong over the Arabian Sea, whereas the winter monsoon was strong over the Japan Sea. The glacial and interglacial cycles in intensity of the monsoons in the Arabian Sea, the South China Sea, and the Japan Sea started around 0.8 Ma, coinciding with the uplift of the Himalayas to their present-day elevation. Therefore, the chronological sequence of monsoonal events and the strength of trade winds and equatorial upwelling suggest that the Asian monsoons (linked to the development of the Himalayan mountains) were an important control on global climate and oceanic productivity.

The Tropics receive by far the most radiative energy from the sun, and the energy received in these regions and the ways in which it is transported to higher latitudes controls the global climate. In the Tropics, the atmospheric circulation over the Asian continent is dominated by the area of highest elevation: the Himalayas and the Tibetan Plateau. The high heat capacity of this region causes the strong seasonality in wind directions, temperature, and rainfall, involving extensive transport of moisture and thus also latent heat from sea to land during summer. The Himalayas–Tibetan Plateau thus influence the transport of sensible and latent heat from low-latitude oceanic areas to mid- and high-latitude land areas. These high mountains act as a mechanical barrier to the air currents, and the north–south contrast across these mountains varied in magnitude with the glacial–interglacial cycles from 0.8 Ma onward, at which time the glacial–interglacial climatic fluctuations reached their largest amplitude. The Asian monsoons thus control the atmospheric heat budget in the Northern Hemisphere, and changes in monsoonal intensity trigger global climate change.

See also

Carbon Cycle. El Niño Southern Oscillation (ENSO). Holocene Climate Variability. Oxygen Isotopes in the Ocean. Stable Carbon Isotope Variations in the Ocean.

Further Reading

Clemens SE, Prell W, Murray D, Shimmield G, and Weedon G (1991) Forcing mechanisms of the Indian Ocean monsoon. *Nature* 353: 720–725.

Fein JS and Stephenes PL (eds.) (1987) *Monsoons.* Chichester, UK: Wiley.

Kutzbach JE and Guetter PJ (1986) The influence of changing orbital parameters and surface boundary conditions on climate simulations for the past 18,000 years. *Journal of Atmospheric Science* 43: 1726–1759.

Naidu PD and Malmgren BA (1995) A 2,200 years periodicity in the Asian Monsoon system. *Geophysical Research Letters* 22: 2361–2364.

Niitsuma N, Oba T, and Okada M (1991) Oxygen and carbon isotope stratigraphy at site 723, Oman Margin. *Proceedings of Ocean Drilling Program Scientific Results* 117: 321–341.

Prell WL, Murray DW, Clemens SC, and Anderson DM (1992) Evolution and variability of the Indian Ocean summer monsoon: Evidence from western Arabian Sea Drilling Program. *Geophysical Monographs* 70: 447–469.

Prell WL and Niitsuma N, et al. (eds.) (1991) *Proceedings of the Ocean Drilling Program Scientific Results, Vol. 117.* College Station, TX: Ocean Drilling Program.

Street FA and Grove AT (1979) Global maps of lake-level fluctuations since 30,000 years BP. *Quaternary Research* 12: 83–118.

Takahashi K and Okada H (1997) Monsoon and quaternary paleoceanography in the Indian Ocean. *Journal of the Geological Society of Japan* 103: 304–312.

Wang L, Sarnthein M, and Erlenkeuser H (1999) East Asian monsoon climate during the late Pleistocene: High-resolution sediment records from the South China Sea. *Marine Geology* 156: 245–284.

Wang P, Prell WL, and Blum MP (2000) Leg 184 summary: Exploring the Asian monsoon through drilling in the South China Sea. *Proceedings of the Ocean Drilling Program, Initial Reports* 184: 1–77.

HEAT & WATER BALANCE

FRESHWATER TRANSPORT AND CLIMATE

S. Wijffels, CSIRO Marine Research, TAS, Australia

Introduction

The ocean is the largest reservoir of water on the planet, consisting of 96% of the total available surface water (**Figure 1**) and it covers 75% of the earth's surface. It is no surprise then that the majority of water cycling through the atmosphere derives from the ocean: about 12.2 Sv (1 Sverdrup = $1 \times 10^6 \, \mathrm{m}^3 \, \mathrm{s}^{-1}$) precipitates over the oceans compared to only 3.5 Sv over land (**Figure 1**), while 13.5 Sv evaporates from the oceans compared to 2.2 Sv of evapotranspiration over land. Total runoff into the global oceans is around 1.3–1.5 Sv, which must be balanced in the long term by a slightly higher total evaporation than precipitation over the oceans.

In the following the term transport is used to refer to processes within the ocean and atmosphere, while the term flux is used for processes between these media.

At any point on the ocean's surface, the total fresh water flux is often a small residual between the two nearly equal and opposite fluxes of precipitation and evaporation. These fluxes between the ocean and atmosphere have quite different spatial patterns, and together create a rich structure in the mean annual fresh water flux at the ocean surface (**Figure 2**). The main sources of atmospheric water vapour are the subtropical oceans under the atmospheric high-pressure belts; the main atmospheric water sinks are the tropical convergence zones (particularly the eastern Indian Ocean/western Pacific) and the polar oceans. Since runoff is a relatively small component of the global fluxes, the transport of moisture within the atmosphere from surface sources to sinks is compensated by an equal and opposite ocean transport of fresh water. The atmosphere's poleward transport of moisture also carries latent heat, and this heat transport comprises as much as 1.5 PW (1.5×10^{15} W) of the total of 4 PW of poleward atmospheric energy transport. The compensating ocean fresh water transport is therefore a fundamental parameter in the planetary energy budget.

Surface fresh water fluxes impact on the oceans in several ways. They change the salinity of surface waters, imprinting them with properties characteristic of their formation regions. For example, intermediate waters formed in the subpolar regions where excess precipitation occurs are traceable far from their source regions because of their low salinity. Fresh water inputs from runoff or excess precipitation can also profoundly influence local air–sea interaction: the formation of fresh light surface layers can suppress convective mixing and thus isolate warmer, saltier deep waters from atmospheric cooling. Such mechanisms are observed in ocean models that display strong sensitivity to fresh water forcing, especially at high latitudes where deep convective mixing occurs that renews the near-bottom waters in the ocean. Changes in high-latitude fresh water forcing of the Atlantic are suspected to have changed the global thermohaline circulation in the past. The sensitivity of ocean models to changes in fresh water forcing is also manifest in the slowing down of the modeled global ocean thermohaline circulation when high-latitude precipitation increases under greenhouse-gas forcing.

Primarily because of the difficulty of measuring rainfall rates over the oceans, estimates of surface fresh water fluxes have been, to date, too uncertain to constrain model behavior well. In the realm of atmospheric modeling, the lack of a reliable benchmark against which to compare the models' moisture transport is a difficulty – the differences among models are often smaller than those among observations. Atmospheric models often overestimate the poleward transport of moisture compared to estimates based on direct observations. However, these observations are mostly made over the land, with

Global Water Reservoirs and Fluxes

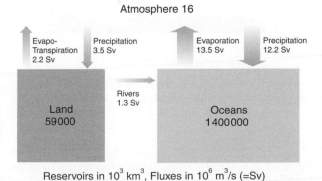

Figure 1 Global water reservoirs and fluxes. (Adapted with permission from Schmitt, 1995.)

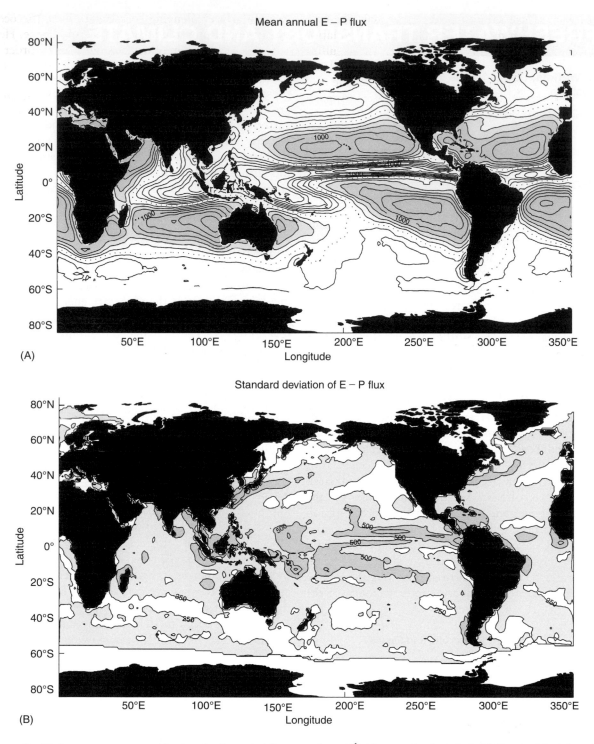

Figure 2   (A) Average mean annual fresh water flux out of the ocean (mm y^{-1}). The zero line is dotted. (B) Standard deviation of 10 estimates of the above showing where the uncertainties are largest. The contour interval is 250 mm y^{-1} in both (A) and (B). (Reproduced with permission from Wijffels, 2001.)

very few made over the oceans that cover the majority of the surface area of the planet.

When forced with observed air–sea fresh water fluxes, ocean models can drift off to unrealistic states, and it has been difficult to distinguish the cause: either inaccurate model physics or errors in the forcing fields. To avoid this problem (and because fresh water forcing has been considered less important compared to thermal forcing) the practice of forcing the model's surface salinity field back to an observed surface salinity field has become the convention in ocean modeling. Such a flux formulation for salinity is physically unjustifiable. The resulting surface flux and salinity fields in the model are unrealistic. This prevents the use of salinity (the next most commonly observed quantity after temperature), to identify and correct physical errors in ocean models.

Methods of Fresh Water Flux and Transport Estimation

In the past, most estimates of ocean fresh water transport derive from surface observations (ship and island) of rainfall rates and parameters (such as wind speed and temperature) used to estimate evaporation. The resulting surface fresh water flux fields can be integrated over the surface area of ocean basins, runoff from the continents added, and the result compared with estimates of ocean fresh water transport based on ocean measurements at specific locations.

Surface Fluxes

In recent years, many new estimates of both atmospheric and surface moisture fluxes have appeared. They are based on the development of data-assimilating atmospheric general circulation models and the availability of new satellite datasets that can be used to deduce evaporation, precipitation, and the net moisture content of the atmospheric column.

Estimates based on data-assimilating atmospheric models have several difficulties, as described by Trenberth and Guillemont in 1999. First, the model output often does not obey total mass conservation making budget calculations difficult. Second, there is a lack of atmospheric profile data over the oceans; the assimilation of scant island station data into these models produces 'bulls eyes' in the surface flux fields, indicating differences between the models and observations.

Estimates of evaporation at the surface rely on empirical relations between the flux and parameters based on either radiometric data measured from satellites or marine meteorological measurements such as wind speed, relative humidity, and sea surface temperature. Though constantly improving, these flux formulas, which are required to apply under all conditions, suffer small biases. When accumulated over large areas such as ocean basins, these flux biases can dominate the totals. The accuracy of the ship-based measurements can also be poor and vary between vessels. Precipitation is

particularly challenging to estimate over the ocean because it is sporadic in both time and space. Here, satellite estimates may be the only way to progress, but these also rely on empirical algorithms that require 'tuning.'

The range (as measured by the standard deviation) of current estimates of the mean annual surface fresh water flux (**Figure 2B**) is globally about $250 \, \text{mm} \, \text{y}^{-1}$, which if integrated over the surface area of the Pacific Ocean north of 30°S adds up to 1 Sv of fresh water transport, which is as large as the natural transport. The largest uncertainty occurs over the tropics and the area affected by mid-latitude storm tracks, as well as a region in the Southeast Pacific off Chile. These are all regions where precipitation is high, thus confirming that the main uncertainty in the total water flux derives from precipitation estimates.

Direct Estimates of Ocean Fresh Water Transport

Ocean fresh water transports can be directly estimated in the same way as those of heat: by examining the flux budgets of volumes of ocean enclosed by long hydrographic lines. The technique is reliant on being able to determine the steady-state portion of the velocity and the salinity field.

For a volume of ocean enclosed by a hydrographic section, salt conservation applies in the steady state, as the transport of salt through the atmosphere and in runoff is negligible:

$$\int \int \rho \, S v \, \mathrm{d}x \, \mathrm{d}z = T_I^S \qquad [1]$$

where ρ is the *in situ* density, S the salinity (e.g., 0.035), v the cross-track velocity (into the volume) and x the along-track distance. T_I^S represents the total salt transport associated with the interbasin exchange, such as the flow through Bering Strait or the Indonesian Throughflow.

Mass conservation is written as

$$\int \int \rho v \, \mathrm{d}x \, \mathrm{d}z + [P - E + R] = T_I^M \qquad [2]$$

where E, P and R are the net fluxes into the surface of the ocean volume of, respectively, evaporation, precipitation, and runoff, and T_I^M the interbasin mass transport. The fresh water part of the above total mass transport is just

$$\int \int \rho v (1 - S) \, \mathrm{d}x \, \mathrm{d}z + [P - E + R] = T_I^M - T_I^S \qquad [3]$$

The $P - E + R$ eqn [3] is not a useful approach because the mass transport across an ocean section has uncertainties that are larger than the fresh water transport. However, the errors in the total salt and mass transports across a section are strongly correlated, and these errors can be largely canceled through defining an areal average salinity and its deviation for the section:

$$\bar{S} = \frac{\int\int S \, dx \, dz}{\int\int dx \, dz}; \quad S' = S - \bar{S} \qquad [4]$$

For simplicity, we also assume that the interbasin transport of salt occurs at a known salinity, S_I, so that the associated salt transport is just $T_I^S = S_I \times T_T^M$. Combining eqns [1], [2] and [4], the surface fresh water flux can now be written as a simple product of the salinity deviation and the velocity field:

$$[P - E + R] = \frac{T_I^M S_I' - \int\int \rho \upsilon \, S' dx \, dz}{\bar{S}} \qquad [5]$$

Here the first term on the right is referred to as the 'leakage' term associated with the total cross-section transport (the interbasin exchange) and the second term is due to correlations of salinity and velocity across the section, which effect a fresh water transport. In practice, υ is found from density and wind-stress measurements using the geostrophic and Ekman assumptions, and often inverse techniques, while S is directly measured along ocean sections.

Deriving error estimates for the product of eqn [5] is challenging, since the statistics of the $\upsilon S'$ term are not well known. Simple scaling arguments were used by Wijffels to show that the expected uncertainty in the direct transport estimates based on eqn [5] might be 0.17 Sv outside of the tropics, but as large as 0.3 Sv in the tropics because of uncertainties in the near-surface wind-driven component. Going beyond the simple scaling argument requires simultaneous time-series of both velocity and salinity over basin scales, measurements that are not likely to be available in the short term.

Comparison of Direct and Indirect Transport Estimates

Runoff from the continents must be added to the surface $E - P$ fluxes integrated over the ocean basins in order to predict the ocean transport of fresh water. Despite attempts to catalog the runoff of major rivers, there are few global estimates of runoff. Here Baumgartner and Riechel's 1975 compilation is utilized, which roughly agrees with the runoff deduced from recent atmospheric analyses. The lack of global runoff datasets makes assessing the errors in the runoff fluxes difficult and adds uncertainty to estimates of ocean fresh water transport.

All three major ocean basins exchange large amounts of sea water through linking passages: the Southern Ocean, the Indonesian Archipelago, and Bering Strait. As these sea water exchanges are much larger than the fresh water exchanges through the atmosphere, it is simpler to present only the divergent part of the ocean fresh water transport (in contrast to the full transports reported by Wijffels and colleagues in 1992). This is equivalent to removing an unknown constant equal to the Pacific–Indian Throughflow for the Indian and South Pacific Oceans, and the Bering Strait flow in the North Pacific and Atlantic. Only the divergence part of the fresh water transport relative to the entrances of the Bering and Throughflow straits (South of Mindinao in the Philippines) will be presented below.

While the size and salinity of the Bering Strait flow are relatively well known, those of the Pacific–Indian Throughflow are not. Hence, investigators have had to make assumptions about them in order to generate an estimate of the fresh water divergence over the South Pacific and Indian Oceans – that is, to calculate the 'leakage' term in eqn [5]. As direct estimates for the long-term average Throughflow range between 5 and 10 Sv, it remains a large source of uncertainty in the freshwater budgets of the South Pacific and Indian Oceans.

Basin Balances

Most direct transport estimates derive from single-section or regional analyses of long hydrographic lines, many of which were completed during the World Ocean Circulation Experiment during the 1990s. To date, few truly global syntheses have been made and so we report Wijffels' year 2000 compilation. The divergent part of the ocean fresh water transport in the three major ocean basins is shown in **Figure 3**.

According to the surface flux estimates, the Indian Ocean north of 30°S undergoes net evaporation; that is, the ocean circulation must import fresh water to the Indian basin, from which the atmosphere exports it. Only two latitudes are currently constrained by direct ocean transport estimates in this basin: 32°S and 18°S, shown by the location of the vertical bars in **Figure 3A**. Over the large evaporative zone between latitudes 15° and 40°S, the indirect transport estimates are fairly consistent (they have similar slopes) and are also in reasonable agreement with the direct estimates. It is in the regions of high precipitation north of 10°S and south of 40°S that the transport curves diverge, confirming again the large differences between estimates of precipitation over

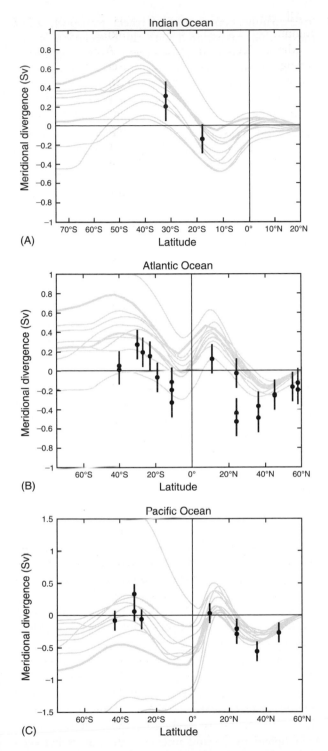

(A)

(B)

(C)

Figure 3 The divergent part of the ocean fresh water transport (Sv) in each ocean basin. Indirect estimates based on surface flux climatologies and atmospheric analyses are shown as gray continuous lines while direct ocean estimates at discrete latitudes are shown in black with error bars. The 'classical' climatology of Baumgartner and Reichel used by Wijffels et al. in 1992 is shown as the thicker gray line. (A) Indian Ocean relative to 30°N; (B) Atlantic/Arctic relative to Bering Strait; and (C) Pacific transport relative to Bering Strait and the Throughflow channels off Mindanao. (Adapted with permission from Wijffels, 2001.)

the ocean. The reader may note in particular the 0.5 Sv variability in the net fresh water divergence north of 18°S, where the monsoons are active.

The Atlantic Ocean is the best-covered by direct transport estimates (**Figure 3B**), which are remarkably consistent, except for those at 24°N, which are from three occupations of a trans-ocean section spanning 30 years. Nearly all of the major transport maxima are delineated by the direct estimates. Again, the indirect estimates diverge most strongly over regions of high precipitation in the tropics and polar regions. When integrated over the Atlantic between 40°S and Bering Strait (we have included the entire Arctic Ocean in the Atlantic), the indirect transport divergences range between 1.0 and 0.0 Sv, while, surprisingly, the direct estimates indicate very little net fresh water divergence over the basin. This difference could be due to an underestimate of runoff to the Arctic/Atlantic as well as an underestimate of $P - E$ to the basin.

Problems with biases in the indirect transport estimates are even more pronounced in the Pacific Ocean owing to its huge size (**Figure 3C**). Indirect transport estimates vary wildly over the South Pacific, where in situ atmospheric data and marine observations are very scarce. Here again, despite the different assumptions made to close the ocean mass balance, Throughflow sizes and different data sets, the direct ocean fresh water transport estimates are quite consistent, and show much less scatter than the indirect estimates. Remarkably, despite a 25 years' difference between section occupations, two direct estimates of the fresh water divergence made near 30°S are indistinguishable.

Interbasin Exchange

One of the first attempts to deduce the exchange of fresh water between ocean basins was made by Baumgartner and Reichel. Lacking 'control' points for the nondivergent part, they assumed zero fresh water transport across the Atlantic equator, and could thus integrate runoff and surface fluxes to deduce the ocean transport. Using new estimates of transport through Bering Strait, Wijffels et al. in 1992 also used the Baumgartner and Reichel climatology to predict the ocean fresh water transport. They deduced that the Pacific received an excess of precipitation and runoff over evaporation of 0.5 Sv, which was then redistributed through the Indonesian Throughflow, Bering Strait, and Southern Ocean to the more evaporative Atlantic and Indian Oceans.

The new direct ocean estimates indicate a quite different interocean fresh water exchange. **Figure 3C** shows that the fresh water divergence over the Pacific between Bering Strait and 30°S is near zero: there is a net balance of evaporation, precipitation, and runoff over that basin. Direct transport estimates for the Atlantic/Arctic also suggest a net divergence of fresh water that is much smaller than previously thought. The new direct estimate of a 0.24 Sv convergence between Bering Strait and 30°S in the Atlantic is roughly half that predicted by Baumgartner and Reichel (**Figure 3B**), while the direct estimates at 40°S indicate almost no net divergence over the Atlantic/Arctic. The Indian Ocean direct transport estimates, however, remain consonant with net excess evaporation over precipitation over that basin (**Figure 3A**).

Since the Pacific Ocean and Atlantic/Arctic Oceans cannot be the source of the excess ocean fresh water required to supply the Indian deficit, only one possibility remains: excess precipitation and ice melt over the Southern Ocean. The newly available direct estimates imply a fresh water source of about 0.5 Sv south of 30°S, highlighting the importance of the Southern Ocean in the global ocean fresh water balance.

Mechanisms of Ocean Fresh Water Transport

In 1981 Stommel and Csanady pointed out that ocean heat and fresh water transport is related to the rates of conversion of water from one temperature–salinity class to another. They went further and attempted to model this process with salinity as a simple function of temperature. Recent analyses of fresh water transports across ocean sections and in general circulation models showed this assumption to be wrong, though the underlying idea remains powerful, as it links surface fluxes to water mass inventories and exchanges in temperature–salinity space. Stommel and Csanady's approach has also been recast in terms of density classes of water, which expresses the competition between surface fluxes and interior ocean mixing in controlling exchange between density classes. The challenge in analysing ocean sections will be in distinguishing the water mass conversion at the surface from that due to internal mixing. Use of the fresh water fluxes will be critical.

How and which elements of the circulation achieve the ocean fresh water transport is also of great interest. The definition of a tracer transport mechanism across an ocean section is still, however, somewhat *ad hoc*. In their pioneering work in 1982, Hall and Bryden chose to form zonal averages (and deviations) of velocity and properties on pressure surfaces. They termed the resulting products the 'overturning' component of tracer transport, while the residual (associated with the correlation of velocity and tracer at a pressure level) was termed the 'horizontal' or gyre component. A similar decomposition can be carried out within density layers and, as density is largely determined by temperature, more closely relates back to Stommel and Csanady's suggestion. Fresh water divergence is also achieved by the interbasin flows, the 'leakage' term in eqn [5]. Unfortunately, few detailed decompositions of the ocean fresh water transport across hydrographic lines are available; those that are, however, reveal interesting mechanisms and cases where different circulation components can provide canceling fresh water transports.

For example, at 10°N in the Pacific, the small net fresh water divergence over the Pacific relative to Bering Strait is due to a balance between three major mechanisms: (1) net export of very fresh water through Bering Strait to the Arctic; (2) a northward fresh water transport by a shallow meridional circulation where the northward Ekman transport in the upper 100 m is fresh and there is a compensating salty southward thermocline flow (100–300 m); (3) southward fresh water transport is achieved by a 300–450 m deep horizontal gyre where salty South Pacific waters flow north in the eastern Pacific and fresh intermediate water flows south in the western Pacific. The deep and bottom water circulations in the low-latitude Pacific achieve little net fresh water transport.

In the Indian Ocean a large net evaporation of 0.31 Sv is estimated to occur north of 32°S. Three mechanisms act to import this fresh water to the ocean basin: (1) a leakage term associated with the inflow of fresh Indonesian Archipelago waters that are evaporated and leave across 32°S as salty thermocline waters; (2) upwelling of deep and intermediate waters and their export as saltier thermocline waters; and (3) horizontal inflow of fresh Antarctic intermediate water in the east that leaves as saltier intermediate water in the west. This latter transport mechanism was also found to be an important fresh water transport mechanism at 32°S in the Pacific. It is likely that the recirculation of subtropical mode and intermediate waters between the Southern Ocean and the Southern Hemisphere subtropical gyres may be the single most important mechanism for balancing the large net flux (0.57 Sv) received by the ocean south of 30°S from the atmosphere and ice flows.

The ability of ocean general circulation models to reproduce the estimated fresh water transports and their mechanisms will be a stringent test of the models' realism. In ocean-only models, surface-flux forcing will determine the net equilibrium transports (unless the problematic relaxation boundary conditions are used), but internal model physics will determine how this transport is achieved. It is also noteworthy that the fresh water transports effected by the subtropical gyres through the vS' term in eqn [5] is not accounted for in simple box models of the global thermohaline circulation, which allow only a single salinity and temperature to represent the major water mass pools. Such models must fold this upper-ocean gyre transport into the deep-water component of the global thermohaline circulation, confusing the role of fresh water forcing as a control on the circulation. More detailed studies of how the ocean transports fresh water are required to isolate the relative roles of the shallow wind-driven gyres and the deep circulation in balancing the surface forcing.

Global Budgets

Direct ocean transport estimates are available in all ocean basins at five latitude bands, which allows the total global meridional fresh water transport to be examined. Since few of the major rivers flow meridionally, the zonally integrated meridional ocean transport of fresh water is largely equal and opposite to that in the atmosphere. Therefore, these estimates can be compared with direct estimates of atmospheric moisture transport or those produced by atmospheric general circulation models (**Figure 4**). Based on a comparison with Oort and Piexoto's 1983 global atmospheric estimates, international model intercomparison studies concluded that most atmospheric models overestimate the poleward transport of moisture. Direct ocean measurements are still too sparse to shed conclusive light on this issue. In the northern hemisphere high latitudes, the direct estimates agree better with Oort and Piexoto's than those from Atmospheric Model Intercomparison Project (AMIP), while in the tropics and southern hemisphere the opposite is true. To more usefully constrain the total meridional moisture transport in the atmosphere, more direct ocean fresh water transport estimates are needed as well as a better estimate of their errors – those shown in **Figure 4** are based on simple scale arguments and so are rather conservative.

Future Directions

Despite many new estimates of surface fresh water fluxes over the oceans having been made, their use in assessing atmospheric models and forcing ocean

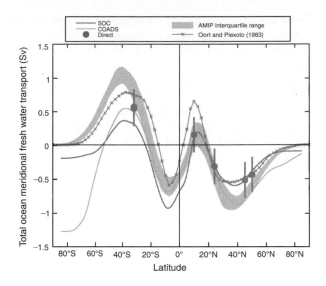

Figure 4 Various estimates of the total ocean meridional fresh water transport (Sv). Estimates based on ocean data are shown as black circles; thin lines are indirect estimates based on two recent surface flux climatologies with the Baumgartner and Reichel's 1975 continental runoff added. Gray shaded is the interquartile range for the atmospheric models participating in AMIP. Oort and Piexoto's 1983 direct atmospheric estimate is marked as x-x. The two surface flux climatologies are as follows. SOC: Josey SA, Kent EC, Oakley D and Taylor PK (1996) A new global air-sea heat and momentum flux climatology. *International WOCE Newsletter* 24: 3–5. COADS: da Silva AM, Young C and Levitus S (1994) *Atlas of Surface Marine Data*, vol. 1: *Algorithms and procedures*. NOAA Atlas NESDIS 6. (Reproduced with permission from Wijffels, 2001.)

models will depend on the accuracy of their basin-wide integrals. These can only be assessed by fresh water transport estimates from ocean data. Direct ocean fresh water transports are not as well reported or analyzed as their companion heat transports. While most estimates are fairly consonant with each other and with error estimates based on simple scaling arguments, others are quite anomalous. Without a detailed breakdown of the mechanisms making up these transports, tracking down the source of these differences is next to impossible.

There may be enormous potential in the idea of 'tuning' surface flux products by using direct ocean estimates to remove flux biases. This might in turn lead to products that are accurate enough to directly force ocean climate models with confidence, and thus allow meaningful use of salinity as an ocean model diagnostic.

Monitoring for changes in ocean fresh water storage may also now be feasible with the availability of salinity sensors that are stable over long deployments on floats and buoys.

Estimates of ocean fresh water transport will remain reliant on transport-resolving temperature–salinity sections, until such time as data-assimilating ocean models are sufficiently accurate to capture the essential ocean fresh water transport mechanisms.

See also

Abrupt Climate Change. Heat and Momentum Fluxes at the Sea Surface. Heat Transport and Climate. Upper Ocean Heat and Freshwater Budgets.

Further Reading

Baumgartner A and Reichel E (1975) *The World Water Balance*. New York: Elsevier.

Hall MM and Bryden HL (1982) Direct estimates and mechanisms of ocean heat transport. *Deep-Sea Research* 29: 339–359.

Oort AH and Peixoto's JP (1983) Global angular momentum and energy balance requirements from observations. *Advances in Geophysics* 25: 355–490.

Schmitt R (1995) The ocean component of the global water cycle. *Review of Geophysics* supplement: 1395–1409.

Schmitt R (1999) The ocean's response to the freshwater cycle. In: Browning KA and Gurney RJ (eds.) *Global Energy and Water Cycles*. Cambridge: Cambridge University Press.

Stommel HM and Csanady GT (1980) A relation between the TS curve and global heat and atmospheric water transports. *Journal of Geophysical Research* 85: 495–501.

Trenberth KE and Guillemot C (1999) Estimating evaporation-minus-precipitation as a residual of the atmospheric water budget. In: Browning KA and Gurney RJ (eds.) *Global Energy and Water Cycles*. Cambridge: Cambridge University Press.

Wijffels SE, Schmitt RW, Bryden HL, and Stigebrandt A (1992) Transport of freshwater by the oceans. *Journal of Physical Oceanography* 22: 155–162.

Wijffels SE (2001) Ocean transport of freshwater. In: Church J, Gould J, and Siedler G (eds.) *Ocean Circulation and Climate*. London: Academic Press.

HEAT AND MOMENTUM FLUXES AT THE SEA SURFACE

P. K. Taylor, Southampton Oceanography Centre, Southampton, UK

Introduction

The maintenance of the earth's climate depends on a balance between the absorption of heat from the sun and the loss of heat through radiative cooling to space. For each 100 W of the sun's radiative energy entering the atmosphere nearly 40 W is absorbed by the ocean – about twice that adsorbed in the atmosphere and three times that falling on land surfaces. Much of this oceanic heat is transferred back to the atmosphere by the local sea to air heat flux. The geographical variation of this atmospheric heating drives the weather systems and their associated winds. The wind transfers momentum to the sea causing waves and the wind-driven currents. Major ocean currents transport heat polewards and at higher latitudes the sea to air heat flux significantly ameliorates the climate. Thus the heat and momentum fluxes through the ocean surface form a crucial component of the earth's climate system.

The total heat transfer through the ocean surface, the net heat flux, is a combination of several components. The heat from the sun is the short-wave radiative flux (wavelength 0.3–3 μm). Around noon on a sunny day this flux may reach about 1000 W m^{-2} but, when averaged over 24 h, a typical value is 100–300 W m^{-2} varying with latitude and season. Part of this flux is reflected from the sea surface – about 6% depending on the solar elevation and the sea state. Most of the remaining short-wave flux is absorbed in the upper few meters of the ocean. In calm weather, with winds less than about 3 m s^{-1}, a shallow layer may be formed during the day in which the sea is warmed by a few degrees Celsius (a 'diurnal thermocline'). However, under stronger winds or at night the absorbed heat becomes mixed down through several tens of metres. Thus, in contrast to land areas, the typical day to night variation in sea surface sea and air temperatures is small, $<1°C$. Both the sea and the sky emit and absorb long-wave radiative energy (wavelength 3–50 μm). Because, under most circumstances, the radiative temperature of the sky is colder than that of the sea, the downward long-wave flux is usually smaller than the upward flux. Hence the net long-wave flux acts to cool the surface, typically by 30–80 W m^{-2} depending on cloud cover.

The turbulent fluxes of sensible and latent heat also typically transfer heat from sea to air. The sensible heat flux is the transfer of heat caused by difference in temperature between the sea and the air. Over much of the ocean this flux cools the sea by perhaps 10–20 W m^{-2}. However, where cold wintertime continental air flows over warm ocean currents, for example the Gulf Stream region off the eastern seaboard of North America, the sensible heat flux may reach 100 W m^{-2}. Conversely warm winds blowing over a colder ocean region may result in a small sensible heat flux into the ocean – a frequent occurrence over the summertime North Pacific Ocean. The evaporation of water vapor from the sea surface causes the latent heat flux. This is the latent heat of vaporization which is carried by the water vapor and only released to warm the atmosphere when the vapor condenses to form clouds. Usually this flux is significantly greater than the sensible heat flux, being on average 100 W m^{-2} or more over large areas of the ocean. Over regions such as the Gulf Stream latent heat fluxes of several hundred W m^{-2} are observed. In foggy conditions with the air warmer than the sea, the latent heat flux can transfer heat from air to sea. In summertime over the infamous fog-shrouded Grand Banks off Newfoundland the mean monthly latent heat transfer is directed into the ocean, but this is an exceptional case.

Measuring the Fluxes

The standard instruments for determining the radiative fluxes measure the voltage generated by a thermopile which is exposed to the incident radiation. Typically the incoming short-wave radiation is measured by a pyranometer which is mounted in gimbals for use on a ship or buoy (**Figure 1**). For better accuracy the direct and scattered components should be determined separately but, apart from at the Baseline Surface Radiation Network stations which are predominantly situated on land, at present this is rarely done. The reflected short-wave radiation is normally determined from the sun's elevation

Figure 1 A pyranometer used for measuring short-wave radiation. The thermopile is covered by two transparent domes. (Photograph courtesy of Southampton Oceanography Centre.)

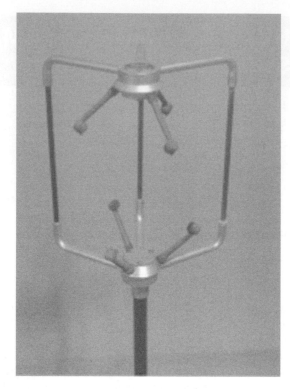

Figure 2 The sensing head of a three-component ultrasonic anemometer. The wind components are determined from the different times taken for sound pulses to travel in either direction between the six ceramic transducers. (Photograph courtesy of Southampton Oceanography Centre.)

and lookup tables based on the results of previous experiments. The pyrgeometer used to determine the long-wave radiation is similar to the pyranometer but uses a coated dome to filter out, as far as possible, the effects of the short-wave heating. Because the air close to the sea surface is normally near to the sea temperature, the use of gimbals is less important. However, a clear sky view is required and a number of correction terms have to be calculated for the temperature of the dome and any short-wave leakage. Again, only the downward component is normally measured; the upwards component is calculated from knowledge of the sea temperature and emissivity of the sea surface.

The turbulent fluxes may be measured in the near-surface atmosphere using the eddy correlation method. If upward moving air in an eddy is on average warmer and moister than the downward moving air, then there is an upwards flux of sensible heat and water vapor and hence also an upward latent heat flux. Similarly the momentum flux, or wind stress, may be determined from the correlation between the horizontal and vertical wind fluctuations. Since a large range of eddy sizes may contribute to the flux, fast response sensors capable of sampling at 10 Hz or more must be exposed for periods of the order of 30 min for each flux determination. Three-component ultrasonic anemometers (**Figure 2**) are relatively robust and, by also determining the speed of sound, can provide an estimate of the sonic temperature flux, a function of the heat and moisture fluxes. The sensors used for determining the fluctuations in temperature and humidity have previously tended to be fragile and prone to contamination by salt particles which are ever-present in the marine atmosphere. However, improved sonic thermometry, and new techniques for water vapor measurement, such as microwave refractometry or differential infrared absorption instruments, are now becoming available.

Despite these improvements in instrumentation, obtaining accurate eddy correlation measurements over the sea remains very difficult. If the instrumentation is mounted on a buoy or ship the six components of the wave-induced motion of the measurement platform must be measured and removed from the signal. The distortion both of the turbulence and the mean wind by ship, buoy or fixed tower must be minimized and, as far as possible, corrected for. Thus eddy correlation measurements are not routinely obtained over the ocean, rather they are used in special air–sea interaction experiments to calibrate other less direct methods of flux estimation. For example, in the inertial dissipation method, fluctuations of the wind, temperature, or humidity at a few Hertz are measured and related (through turbulence theory) to the fluxes. This method is less sensitive to flow distortion or platform motion, but relies on various assumptions about the formation and dissipation of turbulent quantities, which may not be valid under some conditions. It has been implemented on a semi-routine basis on some research ships to increase the range of available flux data.

The most commonly used method of flux estimation is variously referred to as the bulk (aerodynamic) formulae. These formulae relate the difference between the value of temperature, humidity or wind ('x' in [1]) at some measurement height, z, and the value assumed to exist at the sea surface – respectively the sea surface temperature, 98% saturation humidity (to allow for salinity effects), and zero wind (or any nonwind-induced water current). Thus the flux F_x of some quantity x is:

$$F_x = \rho U_z C_{xz}(x_z - x_0) \qquad [1]$$

where ρ is the air density, and U_z the wind speed at the measurement height. While appearing intuitively correct (for example, blowing over a hot drink will cool it faster) these formulae can also be derived from turbulence theory. The value for the transfer coefficient, C_{xz}, characterizes both the surface roughness applicable to x and the relationship between F_x and the vertical profile of x. This varies with the atmospheric stability, which itself depends on the momentum, sensible heat, and water vapor fluxes, as well as the measurement height. Thus, although it may appear simple, Eqn [1] must be solved by iteration, initialized using the equivalent neutral value of C_{xz} at some standard height (normally 10 m), C_{x10n}. Typical neutral values (determined using eddy correlation or inertial dissipation data) are shown in **Table 1**. Many research problems remain. For example: C_{D10n} is expected to depend on the state of development of the wave field, but can this be successfully characterized by the ratio of the predominant wave speed to the wind speed (the wave age), or by the wave height and steepness, or is a spectral representation of the wave field required? What are the effects of waves propagating from other regions (i.e., swell waves)? What is the behavior of C_{D10n} in low wind speed conditions? Furthermore C_{E10n} and C_{H10n} are relatively poorly defined by the available experimental data, and recent bulk algorithms have used theoretical models of the ocean surface (known as surface renewal theory)

to predict these quantities from the momentum roughness length.

Sources of Flux Data

Until recent years the only source of data for flux calculation routinely available from widespread regions of the world's oceans was the weather reports from merchant ships. Organized as part of the World Weather Watch system of the World Meteorological Organisation, these 'Voluntary Observing Ships (VOS)' are asked to return coded weather messages at 00 00, 06 00, 12 00, and 18 00 h GMT daily, also recording the observation (with further details) in the ship's weather logbook. The very basic set of instruments provided will normally include a barometer and a means of measuring air temperature and humidity – typically wet and dry bulb thermometers mounted in a hand swung sling psychrometer or a fixed, louvered 'Stevenson' screen. Sea temperature is obtained using a thermometer and an insulated bucket, or by reading the temperature gauge for the engine cooling water intake. Depending on which country recruited the VOS an anemometer and wind vane might be provided, or the ship's officers might be asked to estimate the wind velocity from observations of the sea state using a tabulated 'Beaufort scale'. Because of the problems of adequately siting an anemometer and maintaining its calibration, these visual estimates are not necessarily inferior to anemometer-based values.

Thus the VOS weather reports include all the variables needed for calculating the turbulent fluxes using the bulk formulae. However, in many cases the accuracy of the data is limited both by the instrumentation and its siting. In particular, a large ship can induce significant changes in the local temperature and wind flow, since the VOS are not equipped with radiometers. The short-wave and long-wave fluxes must be estimated from the observer's estimate of the cloud amount plus (as appropriate) the solar elevation, or the sea and air temperature and

Table 1 Typical values (with estimated uncertainties) for the transfer coefficients[a]

Flux	Transfer coefficients	Typical values
Momentum	Drag coefficient $C_{D10n}(\times 1000)$	$= 0.61\ (\pm 0.05) + 0.063\ (\pm 0.005)\ U_{10n}$ $(U_{10n} > 3\,m\,s^{-1}) = 0.61 + 0.57/U_{10n} < 3\,m\,s^{-1}$
Sensible heat	Stanton no., U_{H10n}	$1.1\ (\pm 0.2) \times 10^{-3}$
Latent heat	Dalton no., U_{E10n}	$1.2\ (\pm 0.1) \times 10^{-3}$

[a]Neither the low wind speed formula for C_{D10n}, nor the wind speed below which it should be applied, are well defined by the available, very scattered, experimental data. It should be taken simply as an indication that, at low wind speeds, the surface roughness increases as the wind speed decreases due to the dominance of viscous effects.

humidity. The unavoidable observational errors and the crude form of the radiative flux formulae imply that large numbers of reports are needed, and correction schemes must be applied, before satisfactory flux estimates can be obtained. While there are presently nearly 7000 VOS, the ships tend to be concentrated in the main shipping lanes. Thus whilst coverage in most of the North Atlantic and North Pacific is adequate to provide monthly mean flux values, elsewhere data is mainly restricted to relatively narrow, major trade routes. For most of the southern hemisphere the VOS data is only capable of providing useful values if averaged over several years, and reports from the Southern Ocean are very few indeed. These shortcomings of VOS-derived fluxes must be borne in mind when studying the flux distribution maps presented below.

Satellite-borne sensors offer the potential to overcome these sampling problems. They are of two types, passive sensors which measure the radiation emitted from the sea surface and the intervening atmosphere at visible, infrared, or microwave frequencies, and active sensors which transmit microwave radiation and measure the returned signal. Unfortunately these remotely sensed data do not allow all of the flux components to be adequately estimated. Sea surface temperature has been routinely determined using visible and infrared radiometers since about 1980. Potential errors due, for example, to changes in atmospheric aerosols following volcanic eruptions, mean that these data must be continually checked against ship and buoy data. Algorithms have been developed to estimate the net surface short-wave radiation from top of the atmosphere values; those for estimating the net surface long wave are less successful. The surface wind velocity can be determined to good accuracy by active scatterometer sensors by measuring the microwave radiation backscattered from the sea surface. Unfortunately scatterometers are relatively costly to operate, since they demand significant power from the spacecraft and, to date, few have been flown. The determination of near-surface air temperature and humidity from satellite is hindered by the relatively coarse vertical resolution of the retrieved data. A problem is that the radiation emitted by the near-surface air is dominated by that originating from the sea surface. Statistically based algorithms for determining the near-surface humidity have been successfully demonstrated. More recently neural network techniques have been applied to retrieving both air temperature and humidity; however, at present there is no routinely available product. Thus the satellite flux products for which useful accuracy has been demonstrated are presently limited to momentum, short-wave radiation, and latent heat flux.

Numerical weather prediction (NWP) models (as used in weather forecasting centers) estimate values of the air–sea fluxes as a necessary part of their calculations. Since these models assimilate most of the available data from the World Weather Watch system, including satellite data, radiosonde profiles, and surface observations, it might be expected that NWP models represent the best source of flux data. However, there are other problems. The vertical resolution of these models is relatively poor and many of the near-surface processes which affect the fluxes have to be represented in terms of larger-scale parameters. Improvements to these models are normally judged on the resulting quality of the weather forecasts, not on the accuracy of the surface fluxes; sometimes these may become worse. Indeed, the continual introduction of model changes results in time discontinuities in the output variables. This makes the determination of interannual variations difficult. Because of this, NWP centres such as the European Centre for Medium Range Weather Forecasting (ECMWF) and the US National Centers for Environmental Prediction (NCEP) have reanalyzed the past weather and have gone back several decades. The surface fluxes from these reanalyses are receiving much study. Those presently available appear less accurate than fluxes derived from VOS data in regions where there are many VOS reports; in sparsely sampled regions the model fluxes may be more accurate. There are particular weaknesses in the short-wave radiation and latent heat fluxes. New reanalyses are planned and efforts are being made to improve the flux estimates; eventually these reanalyses will provide the best source of flux data for many purposes.

Regional and Seasonal Variation of the Momentum Flux

The main features of the wind regimes over the global oceans have long been recognized and descriptions are available in many books on marine meteorology (see Further Reading). The major features of the wind stress variability derived from ship observations from the period 1980–93 will be summarized here, using plots for January and July to illustrate the seasonal variation. The distribution of the heat fluxes will be discussed in the next section.

In northern hemisphere winter (**Figure 3A**) large wind stresses due to the strong midlatitude westerly winds are obvious in the North Atlantic and the North Pacific west of Japan. To the south of these regions the extratropical high pressure zones result in low wind stress values, south of these is the north-

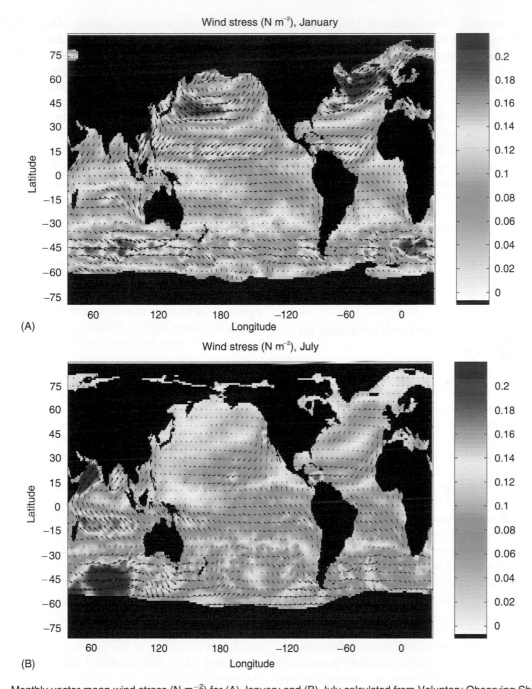

Figure 3 Monthly vector mean wind stress (N m^{-2}) for (A) January and (B) July calculated from Voluntary Observing Ship weather reports for the period 1980–93. (Adapted with permission from Josey SA, Kent EC and Taylor PK (1998) *The Southampton Oceanography Centre (SOC) Ocean–Atmosphere Heat, Momentum and Freshwater Flux Atlas*. SOC Report no. 6.)

east trade wind belt. The Inter-Tropical Convergence Zone (ITCZ) with very light winds is close to the equator in both oceans. In the summertime southern hemisphere the south-east trade wind belt is less well marked. The extratropical high pressure regions are extensive but, despite it being summer, high winds and significant wind stress exist in the midlatitude southern ocean. The north-east monsoon dominates the wind patterns in the Indian Ocean and the South China Sea (where it is particularly strong). The ITCZ is a diffuse region south of the equator with relatively strong south-east trade winds in the eastern Indian Ocean.

In northern hemisphere summer (**Figure 3B**) the wind stresses in the midlatitude westerlies are very much decreased. Both the north-east and the south-east trade wind zones are evident respectively to the north and south of the ITCZ. This is predominantly north of the equator. The south-east trades are particularly strong in the Indian Ocean and feed into a very strong south-westerly monsoon flow in the Arabian Sea. The ship data indicate very strong winds in the Southern Ocean south west of Australia. These are also evident in satellite scatterometer data, which suggest that the winds in the Pacific sector of the Southern Ocean, while still strong, are somewhat less than those in the Indian Ocean sector. In contrast the ship data appear to show very light winds. The reason is that in wintertime there are practically no VOS observations in the far south Pacific. The analysis technique used to fill in the data gaps has, for want of other information, spread the light winds of the extratropical high pressure region farther south than is realistic; a good example of the care needed in interpreting the flux maps available in many atlases.

Regional and Seasonal Variation of the Heat Fluxes

The global distribution of the mean annual net heat flux is shown in **Figure 4A**. The accuracy and method of determination of such flux distributions will be discussed further below, here they will be used to give a qualitative description. Averaged over the year the ocean is heated in equatorial regions and loses heat in higher latitudes, particularly in the North Atlantic. However, this mean distribution is somewhat misleading, as the plots for January (**Figure 4B**) and July (**Figure 4C**) illustrate. The ocean loses heat over most of the extratropical winter hemisphere and gains heat in the extratropical summer hemisphere and in the tropics throughout the year. The relative magnitude of the individual flux components is illustrated in **Figure 5** for three representative sites in the North Atlantic Ocean. At the Gulf Stream site (**Figure 5A**) the large cooling in winter dominates the incoming solar radiation in the annual mean. However, even at this site the mean monthly short-wave flux in summer is greater than the cooling. Indeed the effect of the longer daylight periods increases the mean short-wave radiation to values similar to or larger than those observed in equatorial regions (**Figure 5C**). The midlatitude site (**Figure 5B**) is typical of large areas of the ocean. The ocean cools in winter and warms in summer, in each case by around $100 \, \mathrm{W \, m^{-2}}$. The annual mean flux is small – around $10 \, \mathrm{W \, m^{-2}}$ – but cannot be neglected because of the very large ocean areas involved. At this site, and generally over the ocean, this annual balance is between the sum of the latent heat flux and net long-wave flux which cool the ocean, and the net short-wave heating. Only in very cold air flows, as over the Gulf Stream in winter, is the sensible heat flux significant.

As regards the interannual variation of the surface fluxes, the major large-scale feature over the global ocean is the El Niño-Southern Oscillation system in the equatorial Pacific Ocean. The changes in the net heat flux under El Niño conditions are around $40 \, \mathrm{W \, m^{-2}}$ in the eastern equatorial Pacific. For extratropical and midlatitude regions the interannual variability of the summertime net heat flux is typically about 20–$30 \, \mathrm{W \, m^{-2}}$, being dominated by the variations in latent heat flux. In winter the typical variability increases to about 30–$40 \, \mathrm{W \, m^{-2}}$, although in particular areas (such as over the Gulf Stream) variations of up to $100 \, \mathrm{W \, m^{-2}}$ can occur. The major spatial pattern of interannual variability in the North Atlantic is known as the North Atlantic Oscillation (NAO). This represents a measure of the degree to which mobile depressions, or alternatively near stationary high pressure systems, occur in the midlatitude westerly zone.

Accuracy of Flux Estimates

It has been shown that, although the individual flux components are of the order of hundreds of $\mathrm{W \, m^{-2}}$, the net heat flux and its interannual variability over much of the world ocean is around tens of $\mathrm{W \, m^{-2}}$. Furthermore it can be shown that a flux of $10 \, \mathrm{W \, m^{-2}}$ over 1 year would, if stored in the top $500 \, \mathrm{m}$ of the ocean, heat that entire layer by about $0.15°\mathrm{C}$. Temperature changes on a decadal time scale are at most a few tenths of a degree, so the global mean budget must balance to better than a few $\mathrm{W \, m^{-2}}$. For these various reasons there is a need to measure the flux components, which vary on many time and space scales, to an accuracy of a few $\mathrm{W \, m^{-2}}$. Given the available data sources and methods of determining the fluxes described in the previous sections, it is not surprising that this level of accuracy cannot be achieved at present.

To take an example, in calculating the flux maps shown in **Figure 4** from VOS data many corrections were applied to the VOS observations to attempt to remove biases caused by the methods of observation. For example, air temperature measurements were corrected for the heat island caused by the ship heating up in sunny, low wind conditions. The wind speeds were adjusted depending on the anemometer heights on different ships. Corrections were applied to sea temperatures calculated from engine room intake data. Despite these and other corrections, the global annual mean flux showed about $30 \, \mathrm{W \, m^{-2}}$ excess heating of the ocean. Previous climatologies

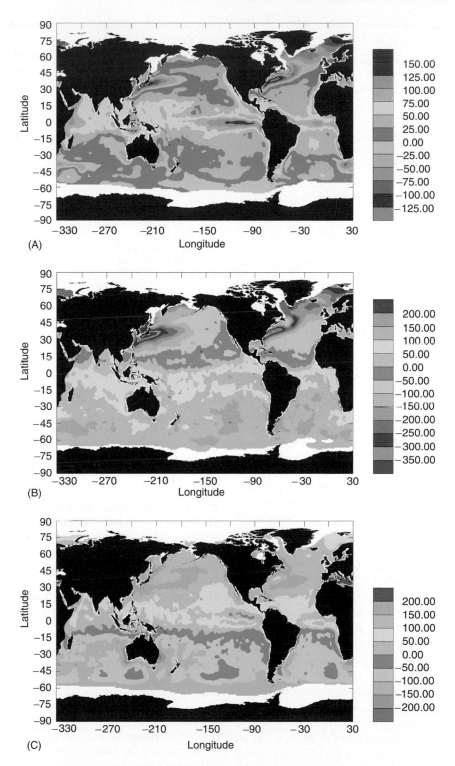

Figure 4 Variation of the net heat flux over the ocean, positive values indicate heat entering the ocean: (A) annual mean, (B) January monthly mean, (C) July monthly mean. (Adapted with permission from Josey SA, Kent EC and Taylor PK (1998) *The Southampton Oceanography Centre (SOC) Ocean–Atmosphere Heat, Momentum and Freshwater Flux Atlas*. SOC Report no. 6.)

calculated from ship data had shown similar biases and the fluxes had been adjusted to remove the bias, or to make the fluxes compatible with estimates of the meridional heat transport in the ocean. However, comparison of the unadjusted flux data with accurate data from air–sea interaction buoys showed good agreement between the two. This suggests that adjusting the fluxes globally is not correct and that

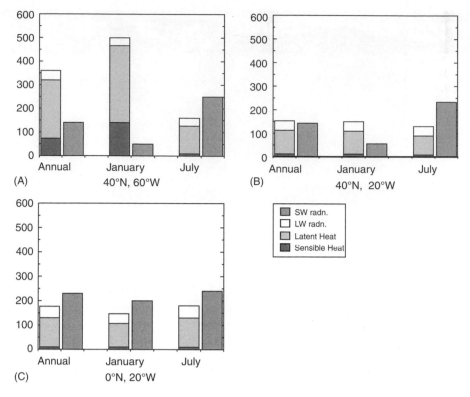

Figure 5 Mean heat fluxes at three typical sites in the North Atlantic for the annual mean, and the January and July monthly means. In each case the left-hand column shows the fluxes which act to cool the ocean while the right-hand column shows the solar heating. (A) Gulf Stream site (40°N, 60°W), (B) midlatitude site (40°N, 20°W), (C) equatorial site (0°N, 20°W).

regional flux adjustments are required; however, the exact form of these corrections is presently not shown.

In the future, computer models are expected to provide a major advance in flux estimation. Recently coupled numerical models of the ocean and of the atmosphere have been run for many simulated years during which the modeled climate has not drifted. This suggests that the air–sea fluxes calculated by the models are in balance with the simulated oceanic and atmospheric heat transports. However, it does not imply that the presently estimated flux values are realistic. Errors in the short-wave and latent heat fluxes may compensate one another; indeed in a typical simulation the sea surface temperature stabilized to a value which was, over large regions of the ocean, a few degrees different from that which is observed. Nevertheless the estimation of flux values using climate or NWP models is a rapidly developing field and improvements will doubtless occur in the next few years. There will be a continued need for *in situ* and satellite data for assimilation into the models and for model development and verification. However, it seems very likely that in future the most accurate routine source of the air–sea flux data will be from numerical models of the coupled ocean–atmosphere system.

See also

El Niño Southern Oscillation (ENSO). El Niño Southern Oscillation (ENSO) Models. Evaporation and Humidity. Freshwater Transport and Climate. Heat Transport and Climate. North Atlantic Oscillation (NAO). Satellite Passive-Microwave Measurements of Sea Ice. Satellite Remote Sensing of Sea Surface Temperatures. Upper Ocean Heat and Freshwater Budgets.

Further Reading

Browning KA and Gurney RJ (eds.) (1999) *Global Energy and Water Cycles*. Cambridge: Cambridge University Press.

Dobson F, Hasse L, and Davis R (eds.) (1980) *Air–Sea Interaction, Instruments and Methods*. New York: Plenum Press.

Kraus EB and Businger JA (1994) *Atmosphere–Ocean Interaction*, 2nd edn. New York: Oxford University Press.

Meteorological Office (1978) *Meteorology for Mariners*, 3rd edn. London: HMSO.

Stull RB (1988) *An Introduction to Boundary Layer Meteorology*. Dordrecht: Kluwer Academic.

Wells N (1997) *The Atmosphere and Ocean: A Physical Introduction*, 2nd edn. London: Taylor and Francis.

HEAT TRANSPORT AND CLIMATE

H. L. Bryden, University of Southampton, Southampton, UK

Introduction: The Global Heat Budget

The Earth receives energy from the sun (**Figure 1**) principally in the form of short-wave energy (sunlight). The amount of solar radiation is quantified by the 'solar constant' which satellite radiometers have measured since about 1985 to have a mean value of $1366\,\mathrm{W\,m^{-2}}$, an 11-year sunspot cycle of amplitude $1.5\,\mathrm{W\,m^{-2}}$, and a maximum at maximum sunspot activity. A fraction of the sunlight is reflected directly back into space and this fraction is termed the albedo. Brighter areas like snow in polar regions have high albedo (0.8) reflecting most of the short-wave radiation back to space, while darker areas like the ocean have small albedo (0.05) and small reflection. Averaged over the Earth's surface, the albedo is about 0.3. Overall, the net incoming short-wave radiation (incoming minus reflected) peaks in equatorial regions and decreases to small values in polar latitudes (**Figure 2**).

The Earth radiates energy back to space in the form of long-wave, black-body radiation proportional to the fourth power of the absolute temperature at the top of the atmosphere. Because the temperature at the top of the atmosphere is relatively uniform with latitude varying only from 200 to 230 K, there is only a small latitudinal variation in outgoing radiation (**Figure 1**). Over a year, the net incoming radiation equals the net outgoing radiation within our ability to measure the radiation, thus maintaining the overall heat balance of the Earth. For the radiation budget as a function of latitude, however, there is more incoming short-wave radiation at equatorial and tropical latitudes and more outgoing long-wave radiation at polar latitudes. To maintain this heating–cooling distribution, the atmosphere and ocean must transport energy poleward from the Tropics toward the Pole and the maximum poleward energy transport in each hemisphere occurs at a latitude of *c.* 35°, where there is a change from net incoming to net outgoing radiation, and the maximum has a magnitude of about 5.8 PW (1 petawatt (PW) $= 10^{15}$ W).

As recently as the mid-1990s, it was controversial whether the ocean or the atmosphere was responsible for the majority of the energy transport. Oceanographers found a maximum ocean heat transport of about 2 PW at 25–30° N, while meteorologists reported a maximum atmospheric transport of about 2.5 PW from the analysis of global radiosonde network observations. Thus, there was a missing petawatt of energy transport in the Northern Hemisphere for the combined ocean–atmosphere system. Recent analyses combining observations and models suggest that the atmosphere does carry the additional petawatt that observational analyses alone could not find

Figure 1 Schematic of the Earth's radiation budget. Q is incoming, short-wave solar radiation; αQ is reflected solar radiation where α is albedo, and E is outgoing, long-wave black-body radiation.

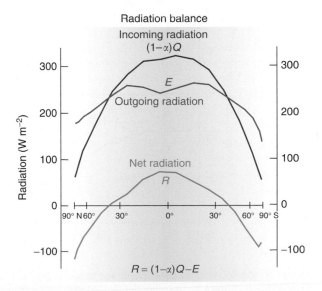

Figure 2 Latitudinal profiles of net incoming short-wave radiation, outgoing long-wave radiation, and the net radiative heating of the Earth. Note the latitudinal scale is stretched so that it is proportional to the surface area of the Earth.

due to a sparsity of radiosonde observations over the ocean. Now it is generally accepted that the atmosphere carries the majority of energy transport at 35° N, though some oceanographers point out that half of the maximum atmospheric transport is due to latent heat (water vapor) transport that can be considered to be a joint ocean–atmosphere process (**Figure 3**).

Here we will concentrate on ocean heat transport, especially on the mechanisms of ocean heat transport in the Northern Hemisphere. Conservation of energy in the ocean is effectively expressed as conservation of heat, where heat is defined to be $\rho C_p \Theta$, where ρ is seawater density, C_p is specific heat of seawater at constant pressure, and Θ is potential temperature, the temperature of a water parcel brought adiabatically (without heat exchange) to the sea surface from depth. Because ρC_p is nearly constant at about $4.08 \times 10^6 \, \mathrm{J \, m^{-3} \, {}^\circ C^{-1}}$, heat conservation is essentially expressed as conservation of potential temperature. There are many subtleties to the definitions that are described in entries for density, potential temperature, and heat, but here we use traditional definitions of potential temperature, density, and specific heat based on the internationally recognized equation of state for seawater.

Ocean heat transport is then the flow of heat through the ocean, $\rho C_p \Theta v$, where v is the water velocity. Such definition depends on the temperature scale and has little meaning until it is considered for a given volume of the ocean. Because of mass conservation, there is no net mass transport into or out of a fixed ocean volume over long timescales (neglecting the relatively miniscule contributions from evaporation minus precipitation), so it is the ocean heat transport convergence that is meaningful, that is the amount of heat transport into the volume minus the heat transport out of the volume. Since mass is conserved, the heat transport convergence is proportional to the mass transport times the difference in temperature between the inflow and the outflow across the boundaries of the volume.

For a complete latitude band like 25° N, where the Atlantic Ocean and Pacific Ocean volume north of 25° N can be considered to be an enclosed ocean, the heat transport convergence is commonly referred to as the ocean heat transport at 25° N. Individually the Atlantic and Pacific Oceans are nearly enclosed with only a small throughflow connecting them in Bering Straits, so the Atlantic heat transport at 25° N is commonly referred to, even though it is formally the heat transport convergence between 25° N and Bering Straits and similarly for Pacific heat transport at 25° N. Such definitions of heat transport are generally used throughout the Atlantic north of 30° S and the Pacific north of about 10° N, where each ocean

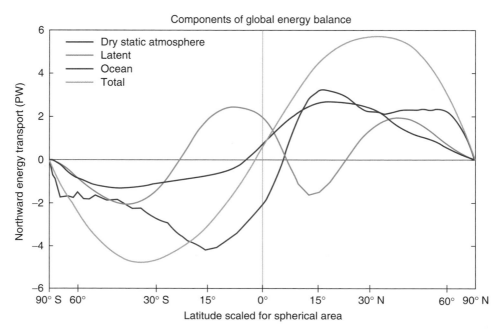

Figure 3 Components of atmosphere and ocean energy transports required to balance the net radiational heating/cooling of the Earth following **Figure 2**. The standard atmospheric energy transport is here divided into the dry static atmospheric energy transport and the latent heat transport. Latent heat transport is fundamentally a joint atmosphere–ocean process since the atmospheric water vapor transport is balanced by an opposing oceanic freshwater transport. The ocean heat transport is determined by integrating over the oceans the spatial distribution of atmosphere–surface heat exchange calculated by subtracting the atmospheric energy transport divergence from the radiative heating at the top of the atmosphere.

basin is closed except for the small Bering Straits transport.

Air–Sea Heat Exchange

Conservation of heat means that any convergence of ocean heat transport is balanced by heat loss to the atmosphere (the amount of heat gain or loss through the ocean bottom is small in comparison to exchanges with the atmosphere). Thus charts of air–sea energy exchange are primary sources for our understanding of ocean heat transport, where it occurs and how big it is. Estimates of air–sea energy exchange have long been made based on measurements of cloud cover, surface air and water temperatures, wind speed, humidity, and bulk formula exchange coefficients to calculate the size of the radiational heating and latent heat cooling of the ocean surface and the sensible heat exchange between the ocean and atmosphere. Combining such shipboard observations on a global scale produces air–sea flux climatologies giving air–sea exchange by month and region. From such climatologies, the global distribution of annual averaged air–sea energy exchange (**Figure 4**) shows that the ocean gains heat over much of the equatorial and tropical regions and gives up large amounts of heat over the warm poleward flowing western boundary currents like the Gulf Stream, Kuroshio or Agulhas Current, and over open-water subpolar and polar regions. Ocean heat

transport convergence for any arbitrary ocean volume can technically be estimated by summing up the air–sea energy exchange over the surface area of the ocean volume. There is a problem, however, in that the air–sea energy exchange estimates have an uncertainty of about $30\,\mathrm{W\,m^{-2}}$. One way to determine this uncertainty is to sum the air–sea exchanges globally and to find that there is on average a heat gain by the ocean of $30\,\mathrm{W\,m^{-2}}$, in each of the two state-of-the-art air–sea exchange climatologies. It is of course possible to remove this bias, either uniformly, by region or by component (radiative, latent, or sensible heat exchange); despite careful comparison with buoy measurements with bulk formula estimates at several oceanic locations, there is no consensus on how to remove the $30\,\mathrm{W\,m^{-2}}$ uncertainty in air–sea energy exchange.

Distribution of Ocean Heat Transport

Estimating ocean heat transport convergence from *in situ* oceanographic measurements is a second method to quantify the role of the ocean in the global heat balance. The advantage of this direct approach is that the mechanisms of ocean heat transport are examined rather than just their overall effect in terms of the air–sea exchange. It was first reliably applied at 25° N in the Atlantic, a latitude where the warm northward Gulf Stream flow of about 30 Sv (1 Sv = $1 \times 10^6\,\mathrm{m^3\,s^{-1}}$) through Florida Straits is regularly

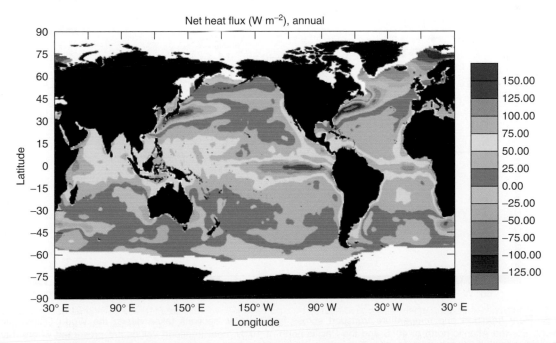

Figure 4 Global distribution of the annual mean net heat gain by the ocean as determined from bulk formula calculations. Positive values indicate a gain of heat by the oceans. SOC climatology – Josey SA, Kent EC, and Taylor PK (1999) New insights into the ocean heat budget closure problem from analysis of the SOC air–sea flux climatology. *Journal of Climate* 12: 2856–2880.

monitored by submarine telephone cable voltage (the varying flow of conducting seawater through a magnetic field produces varying voltage in the cable which is continuously measured). This is the latitude of the easterly trade winds whose westward wind stress generates a northward surface water transport of about 4 Sv in the Ekman layer. Because the Atlantic is closed to the north except for a small (1 Sv) flow from the Pacific through Bering Straits, a nearly equal amount of southward flow of 35 Sv must cross the mid-ocean section between the Bahamas and Africa. The vertical distribution of this southward return flow (and its temperature) has been estimated from trans-atlantic hydrographic sections where temperature and salinity profiles are used to derive geostrophic velocity profiles and the reference-level velocity for the geostrophic velocity profiles is set to make the southward geostrophic transport equal to the northward Gulf Stream plus Ekman transport.

Analysis of three hydrographic sections along 25° N made in 1957, 1981, and 1992 suggests that the ocean heat transport is 1.3 PW. Estimates of the error in heat transport using this direct method are about 0.3 PW, implying that direct ocean heat transport estimates are more accurate than air–sea flux estimates for areas larger than 30° latitude × 30° longitude (approximately, $30\,W\,m^{-2} \times 3100\,km \times 3100\,km = 0.3\,PW$). In terms of mechanisms, the Gulf Stream carries warm water (transport-weighted temperature of 19 °C) northward and the northward surface wind-driven Ekman transport is also warm (25 °C). The compensating southward mid-ocean flow is about equally divided between a recirculating thermocline flow above 800 m depth with an average temperature close to the 19 °C Gulf Stream flow and a cold deep-water flow with an average temperature less than 3 °C. Overall then, the northward heat transport of 1.3 PW across 25° N is due to a net northward transport of about 18 Sv of warm upper layer waters balanced by a net southward transport of cold deep waters. Deep-water formation in the Nordic and Labrador Seas of the northern Atlantic

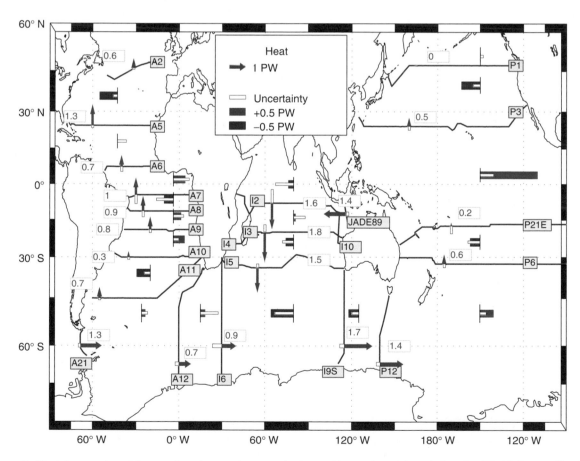

Figure 5 Heat transport and temperature transport across hydrographic sections taken during the World Ocean Circulation Experiment. For the Atlantic north of 40° S and the Pacific north of 10° N, heat transport values are presented where there is no net mass transport across the sections. For the remaining sections, temperature transports are presented where there is a net mass transport across the section and the temperature transport includes the net transport multiplied by temperature in degrees Celsius.

connects the northward flowing warm waters and southward flowing cold deep waters. This overall vertical circulation is commonly called the Atlantic meridional overturning circulation.

During the World Ocean Circulation Experiment 1990–99, transoceanic hydrographic sections were made and ocean heat transport estimated in each ocean basin. The Atlantic heat transport is northward from $40°$ S to $55°$ N; it is of the order 0.5 PW at the southern boundary of the Atlantic, increases in the tropical regions as heat is gained at the sea surface, reaches a maximum near $25°$ N, and then decreases northward as the ocean gives up heat to the atmosphere (**Figure 5**). At all latitudes, the northward heat transport is due to the meridional overturning circulation where the cold deep-water transport of c. 15–20 Sv persists through the Atlantic and the compensating northward upper water flow changes temperature, warming in tropical regions and then cooling in northern subtropical and subpolar latitudes.

The North Pacific, north of $10°$ N where the basin is essentially closed, exhibits a similar pattern of northward heat transport, but the heat transport achieves a maximum of only 0.8 PW at $25°$ N, 50% less than the North Atlantic despite the Pacific being more than twice as wide as the Atlantic. The Pacific is different from the Atlantic in that there is no substantial deep meridional overturning circulation, for there is no deep-water formation in the North Pacific. The Pacific heat transport is due to a horizontal circulation where warm waters flow northward in the Kuroshio western boundary current and then recirculate southward over the vast zonal extent of the Pacific, all at depths shallower than 1000 m. The Kuroshio has about the same size and temperature structure as the Gulf Stream, but the zonal temperature distribution in the Pacific exhibits much colder upper water temperatures in mid-ocean and particularly along the eastern boundary of western North America than does the Atlantic along Europe and Africa. Thus, the northward heat transport in the North Pacific is due to the horizontal upper water circulation where warm water flows northward in the Kuroshio, loses heat to the atmosphere at latitudes south of $50°$ N, and then recirculates southward at colder temperatures in the mid- and eastern Pacific.

South of $10°$ N in the Pacific and throughout the Indian Ocean, ocean heat transports are somewhat ambiguous because of the throughflow from the Pacific to the Indian Ocean through the Indonesian archipelago. The throughflow transport is not yet well defined but there is a substantial observational effort now underway to quantify its mass transport and associated temperature structure. Because mass is not conserved for individual Pacific or Indian zonal sections south of $10°$ N, heat transports across such zonal sections in the literature generally depend on the temperature scale used as well as on the size of the throughflow assumed. These fluxes should properly be called temperature transports and to be interpreted properly should include both a net mass transport across each section and a transport-weighted temperature. Reported northward temperature transports in the South Pacific are mainly due to a net northward transport (to balance the throughflow) times temperature in degrees Celsius. Similarly, southward temperature transports in the south Indian Ocean tend to be large because they include a substantial net southward mass transport (to balance the throughflow) times an average temperature. Careful ocean heat transport divergence estimates for the Indian Ocean taking into account the throughflow transport and temperature generally suggest that the Indian Ocean gains heat from the atmosphere between 0.4 and 1.2 PW depending on the size and temperature of the throughflow. For the South Pacific, the ocean heat transport convergence or divergence is ambiguous for a normal range of throughflow transport, so it is uncertain whether the South Pacific as a whole gains heat or loses heat to the atmosphere.

It is possible to combine the South Pacific and Indian sections to enclose a confined ocean volume north of say $32°$ S. For such combinations where mass is conserved, the Indo-Pacific Ocean heat transport at $32°$ S is meaningful and estimates are that it is southward with a magnitude of 0.4–1.2 PW. For the complete latitude band at about $30°$ S combining Indo-Pacific and Atlantic Ocean heat transports, the ocean heat transport is southward or poleward but with only a maximum value of about 0.5 PW since the northward Atlantic heat transport partially cancels the southward Indo-Pacific heat transport. Thus, it appears that the ocean contributes much less than the atmosphere to the poleward heat transport in the Southern Hemisphere required to balance the Earth's radiation budget.

Eddy Heat Transport

In addition to heat transport by the steady ocean circulation, temporally and spatially varying currents with scales of 10–100 days and 10–100 km, which are called mesoscale features or eddies, can also transport heat. Correlations between time-varying velocity and temperature fluctuations can transport heat, $\rho C_p \langle \Theta' v' \rangle$ (where primes denote fluctuations and angular brackets indicate time averages), even when there is no net velocity or mass transport, that is, $\langle v' \rangle$ is zero. Such eddy heat transport can be substantial in regions where there are strong lateral temperature gradients like the Gulf Stream or Kuroshio extensions separating the subtropical from the subpolar gyre. In the zonally unbounded Southern Ocean where there are several thermal fronts associated with the Antarctic Circumpolar Current, eddy heat transport is the dominant mechanism for transporting heat poleward. Here eddy motions are observed to have colder temperature when they are flowing northward and warmer temperature when following southward though there is no average flow over the eddy scales. The resulting eddy heat transport, typically of the order $4 \times 10^3 \, \mathrm{W \, m^{-2}}$ for $\langle \Theta' v' \rangle = 0.1 \, {}^\circ \mathrm{C \, cm \, s^{-1}}$, is southward, downgradient from high temperature on the northern side toward cold temperature on the southern side of the front, and this downgradient heat flux is a signature of the baroclinic instability process by which eddies form and grow on the potential energy stored in the large-scale lateral temperature distribution. For the 3500 m depth and 20 000 km zonal extent of the Southern Ocean, this poleward eddy heat flux amounts to 0.3 PW across a latitude of 60° S.

Individual eddies or rings of isolated water mass properties may also transport heat. For example, Agulhas rings formed with a core of Indian Ocean water properties in the retroflection area south of Africa are observed to transit across the South Atlantic. These eddies have relatively warm water cores and their heat transport is often estimated by multiplying their heat content anomaly by an estimated number of how many such rings are formed each year. Such calculation is somewhat ambiguous because it is not clear how the mass is returned and what its temperature is. Similar estimates have been made with Gulf Stream and Kuroshio rings, both warm core and cold core, and with meddies formed from the outflow of Mediterranean water. While individual rings are impressive, it is not yet clear whether they carry a significant amount of heat compared with the annual averaged air–sea exchange in any region.

Future Developments

There is a third method, the residual method, for estimating ocean heat transport that takes the difference between the energy transport required to maintain the Earth's radiation budget and the atmospheric energy transport to define the ocean heat transport as a residual. In its original implementation, the residual method could only be applied to estimate zonally averaged ocean heat transport into the polar cap north of any given latitude because atmosphere energy transport could only be determined for a complete zonal integral. In a recent development, the atmospheric energy transport divergence is estimated on a grid point basis from a globally consistent model analysis that assimilates atmospheric observations and radiation variables. From such analysis, the surface energy flux can be estimated at each grid point as the difference between radiation input at the top of the atmosphere and atmospheric transport divergence. Presently, the radiation input can only be accurately estimated for the intensive period of Earth Radiation Budget Experiment from 1985 to 1989. Imposing constraints that the annual averaged surface flux over land should be zero and that the net global air–sea flux should be zero leads to realistic charts of air–sea heat exchange from which ocean heat transport divergence can be estimated. Careful comparison of such ocean heat transport convergence with existing air–sea flux climatologies and with direct estimates of ocean heat transport convergence has not yet been done.

There are many outstanding questions on how ocean heat transport will change under changing climate conditions. As atmospheric CO_2 has increased, the ocean has warmed up by $14 \times 10^{22} \, \mathrm{J}$ over the past 40 years. Such warming represents a heat flux of only 0.1 PW or $0.3 \, \mathrm{W \, m^{-2}}$ averaged over the ocean surface area, so it is unlikely to be detectable in local estimates of air–sea exchange that have uncertainties of $30 \, \mathrm{W \, m^{-2}}$ or in direct estimates of ocean heat transport convergence with uncertainties of 0.3 PW. Instead, local estimates of ocean heat content change over decadal timescales provide a sensitive estimate of how the difference between ocean heat transport convergence and air–sea exchange is changing in a changing climate.

There may be changes in ocean circulation that will lead to measurable changes in air–sea heat exchange and ocean heat transport. For example, most coupled climate models predict that the Atlantic meridional overturning circulation will slow down by order of 50% over the next century as atmospheric CO_2 increases. Because Atlantic heat transport is presently closely related to the strength of the meridional overturning circulation, Atlantic ocean heat transport could reduce measurably. In addition, the absence of a meridional overturning circulation in coupled climate models leads to much colder (10 °C lower) temperatures in the northern Atlantic that greatly reduce the amount of heat given up by the ocean to the atmosphere in northern latitudes. In fact, there has been a recent suggestion that the Atlantic meridional circulation decreased by 30% since 1992. The heat transport decreased by only about 15% from 1.3 to 1.1 PW, however, as the horizontal gyre circulation increased to transport more heat northward. Thus, the Atlantic heat transport may not reduce proportionately with a decreased meridional overturning circulation, because in the absence of a meridional overturning circulation it is possible that the Atlantic will become more like the Pacific with a horizontal gyre circulation that still transports a substantial amount of heat northward.

If Atlantic ocean heat transport reduces under changing climate, will the atmospheric energy transport act to compensate with larger northward energy transport? Presently, the Atlantic Ocean circulation transports more than 20% of the maximum energy transport required to balance the Earth's radiation budget. The hypothesized Bjerkenes compensation mechanism suggests that a reduction in ocean heat transport will be compensated by increased atmospheric energy transport. For extratropical latitudes, atmospheric transport is primarily effected by eddies, cyclones, and anticyclones. Will a reduction in ocean heat transport then be accompanied by increased mid-latitude storminess and increased atmospheric energy transport? Or will the overall radiation budget for the Earth system be fundamentally altered?

Clearly, it is of interest to monitor the changes in ocean circulation and heat transport, most importantly in the Atlantic where there are concerns that increasing atmospheric CO_2 may lead relatively quickly to substantial changes in ocean circulation and heat transport. A program to monitor the Atlantic meridional overturning circulation and associated heat transport started in 2004 and such monitoring may provide the first evidence for changes in Atlantic ocean heat transport.

Further Reading

Bryden HL (1993) Ocean heat transport across 24°N latitude. In: McBean GA and Hantel M (eds.) *Geophysical Monograph Series, Vol. 75 Interactions between Global Climate Subsystems: The Legacy of Hann*, pp. 65–75. Washington, DC: American Geophysical Union.

Bryden HL and Beal LM (2001) Role of the Agulhas Current in Indian Ocean circulation and associated heat and freshwater fluxes. *Deep-Sea Research I* 48: 1821–1845.

Bryden HL and Imawaki S (2001) Ocean heat transport. In: Siedler G, Church J, and Gould J (eds.) *Ocean Circulation and Climate*, pp. 455–474. New York: Academic Press.

Bryden HL, Longworth HR, and Cunningham SA (2005) Slowing of the Atlantic meridional overturning circulation at 25°N. *Nature* 438: 655–657.

Cunningham SA, Kanzow T, Rayner D, et al. (2007) Temporal variability of the Atlantic meridional overturning circulation at 26.5°N. *Science* 317: 935–938.

Ganachaud A and Wunsch C (2000) Improved estimates of global ocean circulation, heat transport and mixing from hydrographic data. *Nature* 408: 453–457.

Josey SA, Kent EC, and Taylor PK (1999) New insights into the ocean heat budget closure problem from analysis of the SOC air–sea flux climatology. *Journal of Climate* 12: 2856–2880.

Lavín A, Bryden HL, and Parrilla G (1998) Meridional transport and heat flux variations in the subtropical North Atlantic. *Global Atmosphere and Ocean System* 6: 269–293.

Levitus S, Antonov J, and Boyer T (2005) Warming of the World Ocean, 1955–2003. *Geophysical Research Letters* 32 (doi:10.1029/2004GL021592).

Shaffrey L and Sutton R (2006) Bjerknes compensation and the decadal variability of the energy transports in a coupled climate model. *Journal of Climate* 19(7): 1167–1181.

Trenberth KE and Caron JM (2001) Estimates of meridional atmosphere and ocean heat transports. *Journal of Climate* 14: 3433–3443.

Trenberth KE, Caron JM, and Stepanaik DP (2001) The atmospheric energy budget and implications for surface fluxes and ocean heat transports. *Climate Dynamics* 17: 259–276.

Vellinga M and Wood RA (2002) Global climatic impacts of a collapse of the Atlantic thermohaline circulation. *Climatic Change* 54: 251–267.

SEA SURFACE EXCHANGES OF MOMENTUM, HEAT, AND FRESH WATER DETERMINED BY SATELLITE

L. Yu, Woods Hole Oceanographic Institution, Woods Hole, MA, USA

Introduction

The ocean and the atmosphere communicate through the interfacial exchanges of heat, fresh water, and momentum. While the transfer of the momentum from the atmosphere to the ocean by wind stress is the most important forcing of the ocean circulation, the heat and water exchanges affect the horizontal and vertical temperature gradients of the lower atmosphere and the upper ocean, which, in turn, modify wind and ocean currents and maintain the equilibrium of the climate system. The sea surface exchanges are the fundamental processes of the coupled atmosphere–ocean system. An accurate knowledge of the flux variability is critical to our understanding and prediction of the changes of global weather and climate.

The heat exchanges include four processes: the short-wave radiation (Q_{SW}) from the sun, the outgoing long-wave radiation (Q_{LW}) from the sea surface, the sensible heat transfer (Q_{SH}) resulting from air–sea temperature differences, and the latent heat transfer (Q_{LH}) carried by evaporation of sea surface water. Evaporation releases both energy and water vapor to the atmosphere, and thus links the global energy cycle to the global water cycle. The oceans are the key element of the water cycle, because the oceans contain 96% of the Earth's water, experience 86% of planetary evaporation, and receive 78% of planetary precipitation.

The amount of air–sea exchange is called sea surface (or air–sea) flux. Direct flux measurements by ships and buoys are very limited. Our present knowledge of the global sea surface flux distribution stems primarily from bulk parametrizations of the fluxes as functions of surface meteorological variables that can be more easily measured (e.g., wind speed, temperature, humidity, cloud cover, precipitation, etc.). Before the advent of satellite remote sensing, marine surface weather reports collected from voluntary observing ships (VOSs) were the backbone for constructing the climatological state of the global flux fields. Over the past two decades, satellite remote sensing has become a mature technology for remotely sensing key air–sea variables. With continuous global spatial coverage, consistent quality, and high temporal sampling, satellite measurements not only allow the construction of air–sea fluxes at near-real time with unprecedented quality but most importantly, also offer the unique opportunity to view the global ocean synoptically as an entity.

Flux Estimation Using Satellite Observations

Sea Surface Wind Stress

The *Seasat-A* satellite scatterometer, launched in June 1978, was the first mission to demonstrate that ocean surface wind vectors (both speed and direction) could be remotely sensed by active radar backscatter from measuring surface roughness. Scatterometer detects the loss of intensity of transmitted microwave energy from that returned by the ocean surface. Microwaves are scattered by wind-driven capillary waves on the ocean surface, and the fraction of energy returned to the satellite (backscatter) depends on both the magnitude of the wind stress and the wind direction relative to the direction of the radar beam (azimuth angle). By using a transfer function or an empirical algorithm, the backscatter measurements are converted to wind vectors. It is true that scatterometers measure the effects of small-scale roughness caused by surface stress, but the retrieval algorithms produce surface wind, not wind stress, because there are no adequate surface-stress 'ground truths' to calibrate the algorithms. The wind retrievals are calibrated to the equivalent neutral-stability wind at a reference height of 10 m above the local-mean sea surface. This is the 10-m wind that would be associated with the observed surface stress if the atmospheric boundary layer were neutrally stratified. The 10-m equivalent neutral wind speeds differ from the 10-m wind speeds measured by anemometers, and these differences are a function of atmospheric stratification and are normally in the order of $0.2 \, \text{m s}^{-1}$. To compute the surface wind stress, τ, the conventional bulk formulation is then employed:

$$\tau = (\tau_x, \tau_y) = \rho c_d W(u, v) \qquad [1]$$

where τ_x and τ_y are the zonal and meridional components of the wind stress; W, u, and v are the scatterometer-estimated wind speed at 10 m and its zonal component (eastward) and meridional component (northward), respectively. The density of surface air is given by ρ and is approximately equal to $1.225 \, \text{kg m}^{-3}$, and c_d is a neutral 10-m drag coefficient.

Scatterometer instruments are typically deployed on sun-synchronous near-polar-orbiting satellites that pass over the equator at approximately the same local times each day. These satellites orbit at an altitude of approximately 800 km and are commonly known as Polar Orbiting Environmental Satellites (POES). There have been six scatterometer sensors aboard POES since the early 1990s. The major characteristics of all scatterometers are summarized in **Table 1**. The first European Remote Sensing (ERS-1) satellite was launched by the European Space Agency (ESA) in August 1991. An identical instrument aboard the successor ERS-2 became operational in 1995, but failed in 2001. In August 1996, the National Aeronautics and Space Administration (NASA) began a joint mission with the National Space Development Agency (NASDA) of Japan to maintain continuous scatterometer missions beyond ERS satellites. The joint effort led to the launch of the NASA scatterometer (NSCAT) aboard the first Japanese Advanced Earth Observing Satellite (ADEOS-I). The ERA scatterometers differ from the NASA scatterometers in that the former operate on the C band ($\sim 5 \, \text{GHz}$), while the latter use the Ku band ($\sim 14 \, \text{GHz}$). For radio frequency band, rain attenuation increases as the signal frequency increases. Compared to C-band satellites, the higher frequencies of Ku band are more vulnerable to signal quality problems caused by rainfall. However, Ku-band satellites have the advantage of being more sensitive to wind variation at low winds and of covering more area.

Rain has three effects on backscatter measurements. It attenuates the radar signal, introduces volume scattering, and changes the properties of the sea surface and consequently the properties of microwave signals scattered from the sea surface. When the backscatter from the sea surface is low, the additional volume scattering from rain will lead to an overestimation of the low wind speed actually present. Conversely, when the backscatter is high,

attenuation by rain will reduce the signal causing an underestimation of the wind speed.

Under rain-free conditions, scatterometer-derived wind estimates are accurate within $1 \, \text{m s}^{-1}$ for speed and $20°$ for direction. For low (less than $3 \, \text{m s}^{-1}$) and high winds (greater than $20 \, \text{m s}^{-1}$), the uncertainties are generally larger. Most problems with low wind retrievals are due to the weak backscatter signal that is easily confounded by noise. The low signal/noise ratio complicates the ambiguity removal processing in selecting the best wind vector from the set of ambiguous wind vectors. Ambiguity removal is over 99% effective for wind speed of $8 \, \text{m s}^{-1}$ and higher. Extreme high winds are mostly associated with storm events. Scatterometer-derived high winds are found to be underestimated due largely to deficiencies of the empirical scatterometer algorithms. These algorithms are calibrated against a subset of ocean buoys – although the buoy winds are accurate and serve as surface wind truth, few of them have high-wind observations.

NSCAT worked flawlessly, but the spacecraft (ADEOS-I) that hosted it demised prematurely in June 1997 after only 9 months of operation. A replacement mission called QuikSCAT was rapidly developed and launched in July 1999. To date, QuikSCAT remains in operation, far outlasting the expected 2–3-year mission life expectancy. QuikSCAT carries a Ku-band scatterometer named SeaWinds, which has accuracy characteristics similar to NSCAT but with improved coverage. The instrument measures vector winds over a swath of 1800 km with a nominal spatial resolution of 25 km. The improved sampling size allows approximately 93% of the ocean surface to be sampled on a daily basis as opposed to 2 days by NSCAT and 4 days by the ERS instruments. A second similar-version SeaWinds instrument was placed on the ADEOS-II mission in December 2002. However, after only a few months of operation, it followed the unfortunate path of NSCAT and failed in October 2003 due – once again – to power loss.

The Advanced Scatterometer (ASCAT) launched by ESA/EUMETSAT in March 2007 is the most recent satellite designed primarily for the global measurement of sea surface wind vectors. ASCAT is flown on the first of three METOP satellites. Each METOP has a design lifetime of 5 years and thus, with overlap, the series has a planned duration of 14 years. ASCAT is similar to ERS-1/2 in configuration except that it has increased coverage, with two 500-km swaths (one on each side of the spacecraft nadir track).

The data collected by scatterometers on various missions have constituted a record of ocean vector winds for more than a decade, starting in August 1992. These satellite winds provide synoptic global

Table 1 Major characteristics of the spaceborne scatterometers

Characteristics	Scatterometer						
	SeaSat-A	ERS-1	ERS-2	NSCAT	SeaWinds on QuikSCAT	SeaWinds on ADEOS II	ASCAT
Operational frequency	Ku band	C band	C band	Ku band	Ku band	Ku band	C band
	14.6 GHz	5.255 GHz	5.255 GHz	13.995 GHz	13.402 GHz	13.402 GHz	5.255 GHz
Spatial resolution	50 km × 50 km with 100-km spacing	50 km × 50 km	50 km × 50 km	25 km × 25 km	25 km × 25 km	25 km × 6 km	25 km × 25 km
Scan characteristics	Two-sided, double 500 km swaths separated by a 450 km nadir gap	One-sided, single 500-km swath	One-sided, single 500-km swath	Two-sided, double 600-km swaths separated by a 329-km nadir gap	Conical scan, one wide swath of 1800 km	Conical scan, one wide swath of 1800 km	Two-sided, double 500-km swaths separated by a 700-km nadir gap
Daily coverage	Variable	41%	41%	77%	93%	93%	60%
Period in service	Jul. 1978–Oct. 1978	Aug. 1991–May. 1997	May. 1995–Jan. 2001	Sep. 1996–Jun. 1997	Jun. 1999–current	Dec. 2002–Oct. 2003	Mar. 2007–current

view from the vantage point of space, and provide excellent coverage in regions, such as the southern oceans, that are poorly sampled by the conventional observing network. Scatterometers have been shown to be the only means of delivering observations at adequate ranges of temporal and spatial scales and at adequate accuracy for understanding ocean–atmosphere interactions and global climate changes, and for improving climate predictions on synoptic, seasonal, and interannual timescales.

Surface Radiative Fluxes

Direct estimates of surface short-wave (SW) and long-wave (LW) fluxes that resolve synoptic to regional variability over the globe have only become possible with the advent of satellite in the past two decades. The surface radiation is a strong function of clouds. Low, thick clouds reflect large amounts of solar radiation and tend to cool the surface of the Earth. High, thin clouds transmit incoming solar radiation, but at the same time, they absorb the outgoing LW radiation emitted by the Earth and radiate it back downward. The portion of radiation, acting as an effective 'greenhouse gas', adds to the SW energy from the sun and causes an additional warming of the surface of the Earth. For a given cloud, its effect on the surface radiation depends on several factors, including the cloud's altitude, size, and the particles that form the cloud. At present, the radiative heat fluxes at the Earth's surface are estimated from top-of-the-atmosphere (TOA) SW and LW radiance measurements in conjunction with radiative transfer models.

Satellite radiance measurements are provided by two types of radiometers: scanning radiometers and nonscanning wide-field-of-view radiometers. Scanning radiometers view radiance from a single direction and must estimate the hemispheric emission or reflection. Nonscanning radiometers view the entire hemisphere of radiation with a roughly 1000-km field of view. The first flight of an Earth Radiation Budget Experiment (ERBE) instrument in 1984 included both a scanning radiometer and a set of nonscanning radiometers. These instruments obtain good measurements of TOA radiative variables including insolation, albedo, and absorbed radiation. To estimate surface radiation fluxes, however, more accurate information on clouds is needed.

To determine the physical properties of clouds from satellite measurements, the International Satellite Cloud Climatology Project (ISCCP) was established in 1983. ISCCP pioneered the cross-calibration, analysis, and merger of measurements from the international constellation of operational weather satellites. Using geostationary satellite measurements with polar orbiter measurements as supplemental when there are no geostationary measurements, the ISCCP cloud-retrieval algorithm includes the conversion of radiance measurements to cloud scenes and the inference of cloud properties from the radiance values. Radiance thresholds are applied to obtain cloud fractions for low, middle, and high clouds based on radiance computed from models using observed temperature and climatological lapse rates.

In addition to the global cloud analysis, ISCCP also produces radiative fluxes (up, down, and net) at the Earth's surface that parallels the effort undertaken by the Global Energy and Water Cycle Experiment – Surface Radiation Budget (GEWEX-SRB) project. The two projects use the same ISCCP cloud information but different ancillary data sources and different radiative transfer codes. They both compute the radiation fluxes for clear and cloudy skies to estimate the cloud effect on radiative energy transfer. Both have a 3-h resolution, but ISCCP fluxes are produced on a 280-km equal-area (EQ) global grid while GEWEX-SRB fluxes are on a $1° \times 1°$ global grid. The two sets of fluxes have reasonable agreement with each other on the long-term mean basis, as suggested by the comparison of the global annual surface radiation budget in **Table 2**. The total net radiation differs by about $5 \, \mathrm{W \, m^{-2}}$, due mostly to the SW component. However, when compared with ground-based observations, the uncertainty of these fluxes is about $10–15 \, \mathrm{W \, m^{-2}}$. The main cause is the uncertainties in surface and near-surface atmospheric properties such as surface skin temperature, surface air and near-surface-layer temperatures and humidity, aerosols, etc. Further improvement requires improved retrievals of these properties.

In the late 1990s, the Clouds and the Earth's Radiant Energy System (CERES) experiment was

Table 2 Annual surface radiation budget (in $\mathrm{W \, m^{-2}}$) over global oceans. Uncertainty estimates are based on the standard error of monthly anomalies

Data 21-year mean 1984–2004	Parameter		
	SW Net downward	LW Net downward	SW+LW Net downward
ISCCP (Zhang et al., 2004)	173.2 ± 9.2	-46.9 ± 9.2	126.3 ± 11.0
GEWEX-SRB (Gupta et al., 2006)	167.2 ± 13.9	-46.3 ± 5.5	120.9 ± 11.9

equipped with the first spaceborne precipitation radar (PR) along with a microwave radiometer (TMI) and a visible/infrared radiometer (VIRS). Coincident measurements from the three sensors are complementary. PR provides detailed vertical rain profiles across a 215-km-wide strip. TMI (a five-frequency conical scanning radiometer) though has less vertical and horizontal fidelity in rain-resolving capability, and it features a swath width of 760 km. The VIRS on TRMM adds cloud-top temperatures and structures to complement the description of the two microwave sensors. While direct precipitation information from VIRS is less reliable than that obtained by the microwave sensors, VIRS serves an important role as a bridge between the high-quality but infrequent observations from TMI and PR and the more available data and longer time series data available from the geostationary visible/infrared satellite platforms.

The TRMM satellite focuses on the rain variability over the tropical and subtropical regions due to the low inclination. An improved instrument, AMSR, has extended TRMM rainfall measurements to higher latitudes. AMSR is currently aboard the *Aqua* satellite and is planned by the Global Precipitation Measurement (GPM) mission to be launched in 2009. Combining rainfall estimates from visible/infrared with microwave measurements is being undertaken by the Global Climatology Project (GPCP) to produce global precipitation analyses from 1979 and continuing.

Summary and Applications

The satellite sensor systems developed in the past two decades have provided unprecedented observations of geophysical parameters in the lower atmosphere and upper oceans. The combination of measurements from multiple satellite platforms has demonstrated the capability of estimating sea surface heat, fresh water, and momentum fluxes with sufficient accuracy and resolution. These air–sea flux data sets, together with satellite retrievals of ocean surface topography, temperature, and salinity (**Figure 1**), establish a complete satellite-based observational infrastructure for fully monitoring the ocean's response to the changes in air–sea physical forcing.

Atmosphere and the ocean are nonlinear turbulent fluids, and their interactions are nonlinear scale-dependent, with processes at one scale affecting processes at other scales. The synergy of various satellite-based products makes it especially advantageous to study the complex scale interactions between the atmosphere and the ocean. One clear example is the satellite monitoring of the development of the El Niño–Southern Oscillation (ENSO) in 1997–98. ENSO is the largest source of interannual variability in the global climate system. The phenomenon is characterized by the appearance of extensive warm surface water over the central and eastern tropical Pacific Ocean at a frequency of *c.* 3–7 years. The 1997–98 El Niño was one of the most severe events experienced during the twentieth century. During the

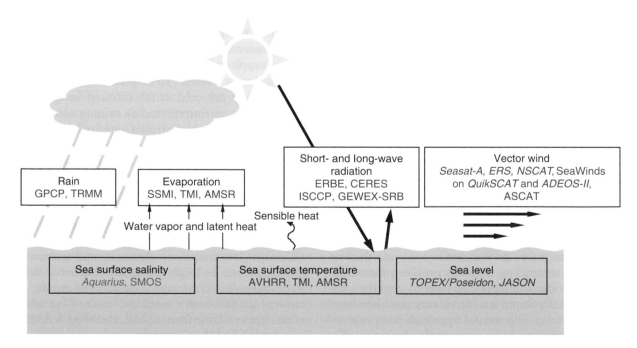

Figure 1 Schematic diagram of the physical exchange processes at the air–sea interface and the upper ocean responses, with corresponding sensor names shown in red.

peak of the event in December 1997 (**Figure 2**), the SST in the eastern equatorial Pacific was more than 5 °C above normal, and the warming was accompanied by excessive precipitation and large net heat transfer from the ocean to the atmosphere.

The 1997–98 event was also the best observed thanks largely to the expanded satellite-observing capability. One of the major observational findings was the role of synoptic westerly wind bursts (WWBs) in the onset of El Niño. **Figure 3** presents the evolution of zonal wind from NSCAT scatterometer combined with SSM/I-derived wind product, sea surface height (SSH) from TOPEX altimetry, and SST from AVHRR imagery in 1996–98. The appearance of the anomalous warming in the eastern basin in February 1997 coincided with the arrival of the downwelling Kelvin waves generated by the WWB of December 1996 in the western Pacific. A series of subsequent WWB-induced Kelvin waves further enhanced the eastern warming, and fueled

the El Niño development. The positive feedback between synoptic WWB and the interannual SST warming in making an El Niño is clearly indicated by satellite observations. On the other hand, the synoptic WWB events were the result of the development of equatorial twin cyclones under the influence of northerly cold surges from East Asia/western North Pacific. NSCAT made the first complete recording of the compelling connection between near-equatorial wind events and mid-latitude atmospheric transient forcing.

Clearly, the synergy of various satellite products offers consistent global patterns that facilitate the mapping of the correlations between various processes and the construction of the teleconnection pattern between weather and climate anomalies in one region and those in another. The satellite observing system will complement the *in situ* ground observations and play an increasingly important role in understanding the cause of global climate changes

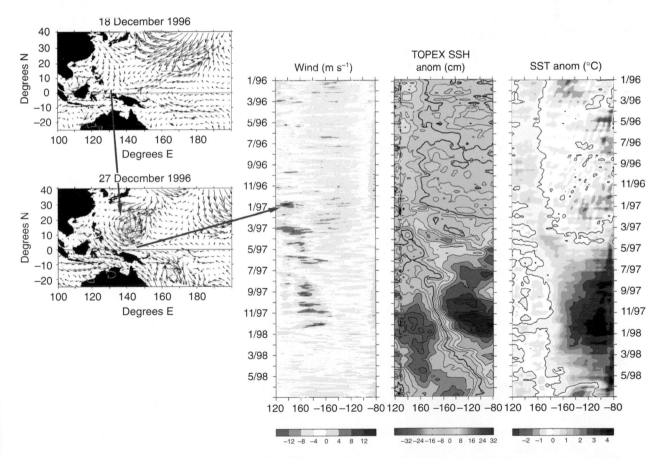

Figure 2 (First column) An example of the scatterometer observations of the generation of the tropical cyclones in the western tropical Pacific under the influence of northerly cold surges from East Asia/western North Pacific. The effect of westerly wind bursts on the development of El Niño is illustrated in the evolution of the equatorial sea level observed from *TOPEX/Poseidon* altimetry and SST from AVHRR. The second to fourth columns show longitude (horizontally) and time (vertically, increasing downwards). The series of westerly wind bursts (second column, SSM/I wind analysis by Atlas *et al.* (1996)) excited a series of downwelling Kelvin waves that propagated eastward along the equator (third column), suppressed the thermocline, and led to the sea surface warming in the eastern equatorial Pacific (fourth column).

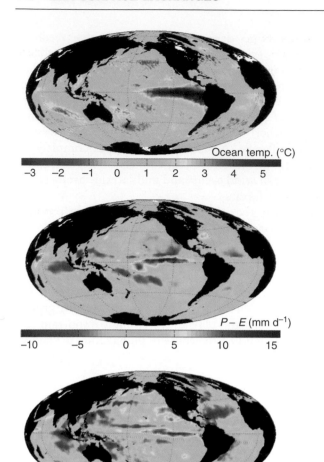

Ocean temp. (°C)

-3 -2 -1 0 1 2 3 4 5

P – E (mm d⁻¹)

-10   -5   0   5   10   15

Net heat loss (W m⁻²)

-80 -60 -40 -20 0 20 40 60 80 100 120

Figure 3 Satellite-derived global ocean temperature from AVHRR (top), precipitation minus evaporation from GPCP and WHOI OAFlux, respectively (middle), and net heat loss $(Q_{LH} + Q_{SH} + Q_{LW} - Q_{SW})$ from the ocean (bottom) during the El Niño in Dec. 1997. The latent and sensible heat fluxes $Q_{LH} + Q_{SH}$ are provided by WHOI OAFlux, and the short- and long-wave radiative fluxes are by ISCCP.

and in improving the model skills on predicting weather and climate variability.

Nomenclature

c_d	drag coefficient
c_e	turbulent exchange coefficient for latent heat
c_h	turbulent exchange coefficient for sensible heat
c_p	specific heat capacity of air at constant pressure
E	evaporation
L_e	latent heat of vaporization
q_a	specific humidity at a reference height above the sea surface
q_s	specific humidity at the sea surface
Q_{LH}	latent heat flux
Q_{SH}	sensible heat flux
T_a	temperature at a reference height above the sea surface
T_s	temperature at the sea surface
u	zonal component of the wind speed
v	meridional component of the wind speed
W	wind speed
ρ	density of surface air
ρ_w	density of sea water
τ	wind stress
τ_x	zonal component of the wind stress
τ_y	meridional component of the wind stress

See also

El Niño Southern Oscillation (ENSO). Evaporation and Humidity. Heat and Momentum Fluxes at the Sea Surface. Heat Transport and Climate. Satellite Oceanography, History and Introductory Concepts. Satellite Remote Sensing of Sea Surface Temperatures. Upper Ocean Heat and Freshwater Budgets.

Further Reading

Adler RF, Huffman GJ, Chang A, *et al.* (2003) The Version 2 Global Precipitation Climatology Project (GPCP) monthly precipitation analysis (1979–present). *Journal of Hydrometeorology* 4: 1147–1167.

Atlas R, Hoffman RN, Bloom SC, Jusem JC, and Ardizzone J (1996) A multiyear global surface wind velocity dataset using SSM/I wind observations. *Bulletin of the American Meteorological Society* 77: 869–882.

Bentamy A, Katsaros KB, Mestas-Nuñez AM, *et al.* (2003) Satellite estimates of wind speed and latent heat flux over the global oceans. *Journal of Climate* 16: 637–656.

Chou S-H, Nelkin E, Ardizzone J, Atlas RM, and Shie C-L (2003) Surface turbulent heat and momentum fluxes over global oceans based on the Goddard satellite retrievals, version 2 (GSSTF2). *Journal of Climate* 16: 3256–3273.

Gupta SK, Ritchey NA, Wilber AC, Whitlock CH, Gibson GG, and Stackhouse RW, Jr. (1999) A climatology of surface radiation budget derived from satellite data. *Journal of Climate* 12: 2691–2710.

Liu WT and Katsaros KB (2001) Air–sea flux from satellite data. In: Siedler G, Church J, and Gould J (eds.) *Ocean Circulation and Climate*, pp. 173–179. New York: Academic Press.

Kubota M, Iwasaka N, Kizu S, Konda M, and Kutsuwada K (2002) Japanese Ocean Flux Data Sets with Use of Remote Sensing Observations (J-OFURO). *Journal of Oceanography* 58: 213–225.

Wentz FJ, Gentemann C, Smith D, and Chelton D (2000) Satellite measurements of sea surface temperature through clouds. *Science* 288: 847–850.

Yu L and Weller RA (2007) Objectively analyzed air–sea heat fluxes (OAFlux) for the global oceans. *Bulletin of the American Meteorological Society* 88: 527–539.

Zhang Y-C, Rossow WB, Lacis AA, Oinas V, and Mishchenko MI (2004) Calculation of radiative fluxes from the surface to top of atmosphere based on ISCCP and other global data sets: Refinements of the radiative transfer model and the input data. *Journal of Geophysical Research* 109: D19105 (doi:10.1029/2003JD004457).

Relevant Websites

http://winds.jpl.nasa.gov
  – Measuring Ocean Winds from Space.

http://www.gewex.org
  – The Global Energy and Water Cycle Experiment (GEWEX).

http://precip.gsfc.nasa.gov
  – The Global Precipitation Climatology Project.

http://isccp.giss.nasa.gov
  – The International Cloud Climatology Project.

http://oaflux.whoi.edu
  – The Objectively Analyzed air–sea Fluxes project.

http://www.ssmi.com
  – The Remote Sensing Systems Research Company

http://www.gfdi.fsu.edu
  – The SEAFLUX Project, Geophysical Fluid Dynamics Institute.

http://eosweb.larc.nasa.gov
  – The Surface Radiation Budget Data, Atmospheric Science Data Center.

EVAPORATION AND HUMIDITY

K. Katsaros, Atlantic Oceanographic and
Meteorological Laboratory, NOAA, Miami, FL, USA

Introduction

Evaporation from the sea and humidity in the air above
the surface are two important and related aspects of the
phenomena of air–sea interaction. In fact, most sub-
sections of the subject of air–sea interaction are related
to evaporation. The processes that control the flux of
water vapor from sea to air are similar to those for
momentum and sensible heat; in many contexts, the
energy transfer associated with evaporation, the latent
heat flux, is of greatest interest. The latter is simply the
internal energy carried from the sea to the air during
evaporation by water molecules. The profile of water
vapor content is logarithmic in the outer layer, from a
few centimeters to approximately 30 m above the sea,
as it is for wind speed and air temperature under
neutrally stratified conditions. The molecular transfer
rate of water vapor in air is slow and controls the flux
only in the lowest millimeter. Turbulent eddies dominate the vertical exchange beyond this laminar layer.
Modifications to the efficiency of the turbulent transfer
occur due to positive and negative buoyancy forces.
The relative importance of mechanical shear-generated
turbulence and density-driven (buoyancy) fluxes was
formulated in the 1940s, the Monin-Obukhov theory,
and the field developed rapidly into the 1960s. New
technologies, such as the sonic anemometer and
Lyman-alpha hygrometer, were developed, which
allowed direct measurements of turbulent fluxes. Furthermore, several collaborative international field experiments were undertaken. A famous one is the
'Kansas' experiment, whose data were used to formulate modern versions of the 'flux profile' relations, i.e.,
the relationship between the profile in the atmosphere
of a variable such as humidity, and the associated turbulent flux of water vapor and its dependence on atmospheric stratification.

The density of air depends both on its temperature
and on the concentration of water vapor. Recent
improvements in measurement techniques and the
ability to measure and correct for the motion of a
ship or aircraft in three dimensions have allowed
more direct measurements of evaporation over the
ocean. The fundamentals of turbulent transfer in the
atmosphere will not be discussed here, only the
special situations that are of interest for evaporation
and humidity. As the water molecules leave the sea,
they remove heat and leave behind an increase in the
concentration of sea salts. Evaporation, therefore,
changes the density of salt water, which has consequences for water mass formation and general
oceanic circulation.

This article will focus on how humidity varies in
the atmosphere, on the processes of evaporation, and
how it is modified by the other phenomena discussed
under the heading of air–sea interaction. All processes occurring at the air–sea interface interact and
modify each other, so that none are simple and linear
and most result in feedback on the phenomenon itself. The role of wind, temperature, humidity, wave
breaking, spray, and bubbles will be broached and
some fundamental concepts and equations presented.
Methods of direct measurements and estimation
using *in situ* mean measurements and satellite
measurements will be discussed. Subjects requiring
further research are also explored.

History/Definitions and Nomenclature

Many ways of measuring and defining the quantity
of the invisible gas, water vapor, in the air have developed over the years. The common ones have been
gathered together in **Table 1**, which gives their name,
definition, SI units, and some further explanations.
These quantitative definitions are all convertable one
into another. The web-bulb temperature may seem
rather anachronistic and is completely dependent on
a rather crude measurement technique, but it is still a
fundamental and dependable measure of the quantity
of water vapor present in the air.

Evaporation or turbulent transfer of water vapor
in the air was first modeled in analogy with down–
gradient transfer by molecular conduction in solids.
The conductivity was replaced by an 'Austaush' coefficient, A_e, or eddy diffusion coefficient, leading to
the expression:

$$E = -A_e \rho \frac{\partial \bar{q}}{\partial z} \qquad [1]$$

where E is the evaporation rate, ρ the air density, $\bar{q}$ is
mean atmospheric humidity, and z represents the
vertical coordinate. Assuming no advection, steady
state, and no accumulation of water vapor in the
surface layer of the atmosphere (referred to as 'the

Table 1 Measures of humidity

Nomenclature	Units (SI)	Definition
Absolute humidity	kg m^{-3}	Amount of water vapor in the volume of associated moist air
Specific humidity	g kg^{-1}	The mass of water per unit mass of moist air (or equivalently in the same volume)
Mixing ratio	g kg^{-1}	The ratio of the mass of water as vapor to the mass of dry air in the same volume
Saturation humidity	Any of the above units	Can be given in terms of all three units and refers to the maximum amount the air can hold at its current temperature in terms of absolute or specific humidity, corresponds to 100% relative humidity
Relative humidity (RH)	%	Percent of saturation humidity that is actually in the air
Vapor pressure	hPa (or mb)	The partial pressure of the water vapor in the air
Dew point temperature	°K, °C	The temperature at which dew would form based on the actual amount of water vapor in the air. Dew point depression compared to actual temperature is a measure of the 'dryness' of the air
Wet-bulb temperature	°K, °C	This is a temperature obtained by the wetted thermometer of the pair of thermometers used in a psychrometer[a] (see Measurements chapter)

[a] A psychrometer is a measuring device consisting of two thermometers (mercury in glass or electronic), where one thermometer is covered with a wick wetted with distilled water. The device is aspirated with environmental air (at an air speed of at least 3 m s^{-1}). The evaporation of the distilled water cools the air passing over the wet wick, causing a lowering of the wet thermometer's temperature, which is dependent on the humidity in the air.

constant flux layer'), the A_e is a function of z as the turbulence scales increase away from the air–sea interface and the gradient is a decreasing function of height, z, as the distance from the source of water vapor, the sea surface, increases.

Determining E by measuring the gradient of q has not proved to be a good method because of the difficulties of obtaining differences of q accurately enough and in knowing the exact heights of the measurements well enough (say from a ship or a buoy on the ocean). The A_e must also be determined, which would require measurements of the intensity of the turbulent exchange in some fashion. The so-called direct method for evaluating the vapor flux in the atmosphere requires high frequency measurements. This method has been refined during the past 35 years or so, and has produced very good results for the turbulent flux of momentum (the wind stress). Fewer projects have been successful in measuring vapor flux over the ocean, because the humidity sensors are easily corrupted by the presence of spray or miniscule salt particles on the devices, which being hygroscopic, modify the local humidity.

Evaporation, E, can be measured directly today by obtaining the integration over all scales of the turbulent flux, namely, the correlation between the deviations from the mean of vertical velocity (w') and humidity (q') at height (z) within the constant flux layer. This correlation, resulting from the averaging of the vapor conservation equation (in analogy to the Reynolds stress term in the Navier–Stokes equation) can be measured directly, if sensors are available that resolve all relevant scales of fluctuations.

The correlation equation is

$$\overline{\rho w \cdot q} = \bar{\rho}\bar{w}\bar{q} + \bar{\rho}\overline{w'q'}, \qquad [2]$$

where w and q are the instantaneous values and the overbar indicates the time-averaged means. The product of the averages is zero since $\bar{w} = 0$. Much discussion and experimentation has gone into determining the time required to obtain a stable mean value of the eddy flux $\bar{\rho}\overline{q'w'}$. For the correlation term $\bar{\rho}\overline{w'q'}$ to represent the total vertical flux, there has to be a spectral gap between high and low frequencies of fluctuations, and the assumption of steady state and horizontal homogeneity must hold. The required averaging time is of the order of 20 min to 1 h.

Another commonly used method, the indirect or inertial dissipation method, also requires high frequency sensing devices, but relies on the balance between production and destruction of turbulence to be in steady state. The dissipation is related to the spectral amplitude of turbulent fluctuations in the inertial subrange, where the fluctuations are broken down from large-scale eddies to smaller and smaller scales, which happens in a similar fashion regardless of scale of the eddies responsible for the production of turbulence in the atmospheric boundary layer. The magnitude of the spectrum in the inertial subrange is, therefore, a measure of the total energy of the turbulence and can be interpreted in terms of the turbulent flux of water vapor. The advantage of this method over the eddy correlation method is that it is less dependent on the corrections for flow distortion and motion of the ship or the buoy platform, but it requires corrections for atmospheric stratification and other predetermined coefficients. It would not give the true flux if the production of turbulence was changing, as it does in changing sea states. Most of the time, the direct flux is not measured by either the direct or the indirect method; we resort to a parameterization of the flux in terms of so-called 'bulk' quantities.

The bulk formula has been found from field experiments where the total evaporation E has been measured directly together with mean values of q and wind speed, U, at one height, $z = a$ (usually referred to as 10 m by adjusting for the logarithmic vertical gradient), and the known sea surface temperature.

$$E = \overline{\rho w' q'} = \bar{\rho} \cdot C_{E_a} \overline{U_a} (\overline{q_s} - \overline{q_a}) \qquad [3]$$

where $\overline{q_s}$ is the saturation specific humidity at the air–sea interface, a function of sea surface temperature (SST). Air in contact with a water surface is assumed to be saturated. Above sca water the saturated air has 98% of the value of water vapor density at saturation over a freshwater surface, due to the effects of the dissolved salts in the sea. C_{E_a} is the exchange coefficient for water vapor evaluated for the height a. Experiments have shown C_{E_a} to be almost constant at $1.1–1.2 \times 10^{-3}$ for $U < 18\,\mathrm{m\ s^{-1}}$, for neutral stratification, *i.e.* no positive or negative buoyancy forces acting and at a height of 10 m, written as $C_{E_{10N}}$. However, measurements show large variability in $C_{E_{10N}}$ which may be due to the effects of sea state, such as sheltering in the wave troughs for large waves and increased evaporation due to spray droplets formed in highly forced seas with breaking waves. Results from a field experiment, the Humidity Exchange Over the Sea (HEXOS) experiment in the North Sea, are shown in **Figure 1**. Its purpose was to address the question of what happens to evaporation or water (vapor) flux at high wind speeds. However, the wind only reached $18\,\mathrm{m\ s^{-1}}$ and the measurements showed only weak, if any, effects of the spray. Theories suggest that the effects will be stronger above $25\,\mathrm{m\ s^{-1}}$. More direct measurements are still required before these issues can be settled, especially for wind speeds $> 20\,\mathrm{m\ s^{-1}}$ (*see* Further Reading and the section on meteorological sensors for mean measurements for a discussion of the difficulties of making measurements over the sea at high wind speeds).

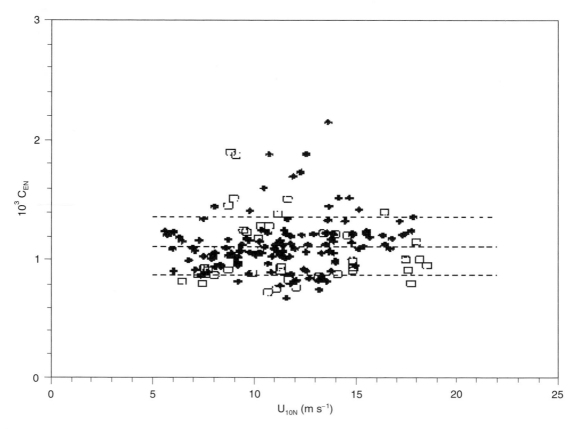

Figure 1 Vapor flux exchange coefficients from two simultaneous measurement sets: the University of Washington (crosses) and Bedford Institute of Oceanography (squares) data. Thick dashed line is the average value, 1.12×10^{-3}, for 170 data points. Thin dashed lines indicate standard deviations (from DeCosmo et al., 1996).

Clausius–Clapeyron Equation

The Clausius–Clapeyron equation relates the latent heat of evaporation to the work required to expand a unit mass of liquid water into a unit mass of water as vapor. The latent heat of evaporation is a function of absolute temperature. The Clausius–Clapeyron equation expresses the dependence of atmospheric saturation vapor pressure on temperature. It is a fundamental concept for understanding the role of evaporation in air–sea interaction on the large scale, as well as for gaining insight into the process of evaporation from the sea (or Earth's) surface on the small scale. Note first of all that the Clausius–Clapeyron equation is highly non-linear, viz:

$$\frac{d \ln p_v}{dT} = \frac{\Delta H_{vap}}{RT^2} \qquad [4]$$

where p_v is the vapor pressure, T is absolute temperature (°K), and ΔH_{vap} is the value of the latent heat of evaporation, R is the gas constant for water vapor $= 461.53 \, \text{J} \, \text{kg}^{-1} \, \text{°K}^{-1}$. The dependence of vapor pressure on temperature is presented in a simplified form as:

$$e_s = 610.8 \exp\left[19.85\left(1 - \frac{T_0}{T}\right)\right](P_a) \qquad [5]$$

where e_s is vapor pressure in pascals, T_0 is a reference temperature set to $0°\text{C} = 273.16 \, °\text{K}$, and T is the actual temperature in °K which is accurate to 2% below 30°C (**Figure 2**).

Figure 2 displays the saturation vapor pressure and the pressure of atmospheric water vapor for 60% relative humidity. On the right-hand side of the figure, the ordinate gives the equivalent specific humidity values (for a near surface total atmospheric pressure of 1000 hpa). This figure illustrates that the atmosphere can hold vastly larger amounts of water as vapor at temperatures above 20°C than at temperatures below 10°C. For constant relative humidity, say 60%, the difference in specific humidity or vapor pressure in the air compared with the amount at the air–sea interface, if the sea is at the same temperature as the air, is about three times at 30°C what it would be at 10°C. Therefore, evaporation is driven much more strongly at tropical latitudes compared with high latitudes (cold sea and air) for the same mean wind and relative humidity as illustrated by eqn [4] and **Figure 2**.

Tropical Conditions of Humidity

By far, most of the water leaving the Earth's surface evaporates from the tropical oceans and jungles, providing the accompanying latent heat as the fuel that drives the atmospheric 'heat engines,' namely, thunderstorms and tropical cyclones. Such extreme and violent storms depend for their generation on the enormous release of latent heat in clouds to create the vertical motion and compensating horizontal accelerated inflows. Tropical cyclones do not form over oceanic regions with temperatures <26°C, and

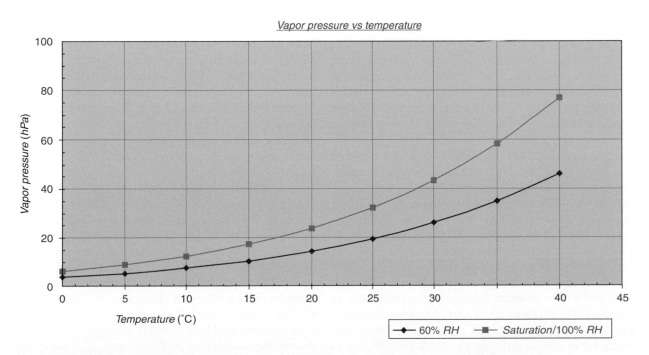

Figure 2 Vapor pressure (hPa) as a function of temperature for two values of relative humidity, 60% and 100%.

temperature increases of only 1° or 2°C sharply enhance the possibility of formation.

Latitudinal and Regional Variations

The Clausius–Clapeyron equation holds the secrets to the role of water vapor for both weather and climate. Warm moist air flowing north holds large quantities of water. As the air cools by vertical motion, contact with cold currents, and loss of heat by infrared radiation, the air reaches saturation and either clouds, storms and rain form, or fog (over cold surfaces) and stratus clouds. The warmer and moister the original air, the larger the possible rainfall and the larger the release of latent heat. Latitudinal, regional, and seasonal variations in evaporation and atmospheric humidity are all related to the source of heat for evaporation (upper ocean heat content) and the capacity of the air to hold water at its actual temperature. Many other processes such as the dynamics behind convergence patterns and the development of atmospheric frontal zones contribute to the variability of the associated weather.

Vertical Structure of Humidity

The fact that the source of moisture is the ocean, lakes, and moist ground explains the vertical structure of the moisture field. Lenses of moist air can form aloft. However, when clouds evaporate at high elevations where atmospheric temperature is low, the absolute amounts of water vapor are also low for that reason.

Thus, when the surface air is continually mixed in the atmospheric boundary layer with drier air, being entrained from the free atmosphere across the boundary layer inversion, it usually has a relative humidity less than 100% of what it could hold at its actual temperature. The exceptions are fog, clouds, or heavy rain, where the air has close to 100% relative humidity. The process of exchange between the moist boundary layer air and the upper atmosphere allows evaporation to continue. Deep convection in the inter-tropical convergence zone brings moist air up throughout the whole of the troposphere, even over-shooting into the stratosphere. Moisture that does not rain out locally is available for transport poleward. The heat released in these clouds modifies the temperature of the air. Similarly, over the warm western boundary currents, such as the Gulf Stream, Kuroshio, and Arghulas Currents, substantial evaporation and warming of the atmosphere takes place. Without the modifying effects of the hydrologic cycle of evaporation and precipitation

on the atmosphere, the continents would have more extreme climates and be less habitable.

Sublimation–Deposition

The processes of water molecules leaving solid ice and condensing on it are called sublimation and deposition, respectively. These processes occur over the ice-covered polar regions of the ocean. In the cold regions, this flux is much less than that from open leads in the sea ice due to the warm liquid water, even at 0°C.

At an ice surface, water vapor saturation is less than over a water surface at the same temperature. This simple fact has consequences for the hydrologic cycle, because in a cloud consisting of a mixture of ice and liquid water particles, the vapor condenses on the ice crystals and the droplets evaporate. This process is important in the initial growth of ice particles in clouds until they become large enough to fall and grow by coalescence of droplets or other ice crystals encountered in their fall. Similar differences in water vapor occur for salty drops, and the vapor pressure over a droplet also depends on the curvature (radius) of the drop. Thus, particle size distribution in clouds and in spray over the ocean are always changing due to exchange of water vapor. For drops to become large enough to rain out, a coalescence-type growth process must typically be at work, since growth by condensation is rather slow.

Sources of Data

Very few direct measurements of the flux of water vapor are available over the ocean at any one time. The mean quantities ($\overline{U}$, $\overline{q_a}$, SST) needed to evaluate the bulk formula are reported regularly from voluntary observing ships (VOS) and from a few moored buoys. However, most of such buoys do not measure surface humidity, only a small number in the North Atlantic and tropical Pacific Oceans do so. The VOS observations are confined to shipping lanes, which leaves a huge void in the information available from the Southern Hemisphere. Alternative estimates of surface humidity and the water vapor flux include satellite methods and the surface fluxes produced in global numerical models, in particular, the re-analysis projects of the US Weather Service's National Center for Environmental Prediction (NCEP) and the European Center for Medium Range Weather Forecasts (ECMWF). The satellite method has large statistical uncertainty and, thus, requires weekly to monthly averages for obtaining reasonable accuracy ($\pm 30\,\mathrm{W\,m^{-2}}$ and $\pm 15\,\mathrm{W\,m^{-2}}$ for the weekly and monthly latent heat flux). Therefore, these data are most useful for climatological estimates and for checking the numerical models' results.

Estimation of Evaporation by Satellite Data

The estimation of evaporation/latent heat flux from the ocean using satellite data also relies on the bulk formula. The computation of latent heat flux by the bulk aerodynamic method requires SST, wind speed ($\overline{U}_{10_N}$), and humidity at a level within the surface layer $\overline{q_a}$, as seen in eqn [3]. Therefore, evaluation of the three variables from space is required. Over the ocean, $\overline{U}_{10_N}$ and SST have been directly retrieved from satellite data, but $\overline{q_a}$ has not. A method of estimating $\overline{q_a}$ and latent heat flux from the ocean using microwave radiometer data from satellites was proposed in the 1980s. It is based on an empirical relation between the integrated water vapor W (measured by spaceborne microwave radiometers) and $\overline{q_a}$ on a monthly timescale. The physical rationale is that the vertical distribution of water vapor through the whole depth of the atmosphere is coherent for periods longer than a week. The relation does not work well at synoptic and shorter timescales and also fails in some regions during summer. Modification of this method by including additional geophysical parameters has been proposed with some overall improvement, but the inherent limitation is the lack of information about the vertical distribution of q near the surface.

Two possible improvements in E retrieval include obtaining information on the vertical structures of humidity distribution and deriving a direct relation between E and the brightness temperatures (T_B) measured by a radiometer. Recent developments provide an algorithm for direct retrieval of boundary layer water vapor from radiances observed by the Special Sensor Microwave/Imager (SSM/I) on operational satellites in the Defense Meteorological Satellite Program since 1987. This sensor has four frequencies, 19.35, 22, 37, and 85.5 GHz, all except the 22 GHz operated at both horizontal and vertical polarizations. The 22 GHz channel at vertical polarization is in the center of a weak water vapor absorption line without saturation, even at high atmospheric humidity. The measurements are only possible over the oceans, because the oceans act as a relatively uniform reflecting background. Over land, the signals from the ground overwhelm the water vapor information.

Because all the three geophysical parameters, $\overline{U}_{10_N}$, W, and SST, can be retrieved from the radiances at the frequencies measured by the older microwave radiometer, launched in 1978 and operating to 1985 – the Scanning Multichannel Microwave Radiometer (SMMR) on Nimbus-7 (similar to SSM/I, but with 10.6 and 6.6 GHz channels as well, and no 85 GHz channels) – the feasibility of retrieving E directly from the measured radiances was also demonstrated. SMMR measures at 10 channels, but only six channels were identified as significantly useful in estimating E. SSM/I, the operational microwave radiometer that followed SMMR, lacks the low-frequency channels which are sensitive to SST, making direct retrieval of E from T_B unfeasible. The microwave imager (TMI) on the Tropical Rainfall Measuring Mission (TRMM), launched in 1998, includes low-frequency measurements sensitive to SST and could, therefore, allow direct estimates of evaporation rates. **Figure 3** gives an example of global monthly mean values of humidity obtained solely with satellite data from SSM/I.

To calculate q_s, gridded data of sea surface temperature can also be used, such as those provided operationally by the US National Weather Service based on infrared observations from the Advanced Very High Resolution Radiometer (AVHRR) on operational polar-orbiting satellites. The exact coincident timing is not so important for SST, since SST varies slowly due to the large heat capacity of water, and this method can only provide useful accuracies when averages are taken over 5 days to a week. Wind speed is best obtained from scatterometers, rather than from the microwave radiometer, in regions of heavy cloud or rain, since scatterometers (which are active radars) penetrate clouds more effectively. Scatterometers have been launched in recent times by the European Space Agency (ESA) and the US National Aeronautic and Space Administration (NASA) (the European Remote Sensing Satellites 1 and 2 in 1991 and 1995, the NASA scatterometer, NSCAT, on a Japanese short-lived satellite in 1996, and the QuikSCAT satellite in 1999).

Future Directions and Conclusions

Evaporation has been measured only up to wind speeds of $18 \, \mathrm{m \, s^{-1}}$. The models appear to converge on the importance of the role of sea spray in evaporation, indicating that its significance grows beyond about $20 \, \mathrm{m \, s^{-1}}$. However, the source function of spray droplets as a function of wind speed or wave breaking has not been measured, nor are techniques for measuring evaporation in the presence of droplets well–developed, whether for rain or sea spray. Since evaporation and the latent heat play such important roles in tropical cyclones and many other weather phenomena, as well as in oceanic circulation, there is great motivation for getting this important energy and mass flux term right. The bulk model is likely to

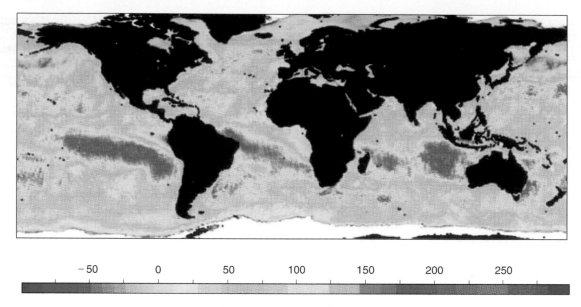

Figure 3 Global distribution of monthly mean latent heat flux in W m^{-2} for September 1987. (Reproduced with permission from Schulz *et al.*, 1997.)

be the main method used for estimating evaporation for some time to come. Development of more direct satellite methods and validating them should be an objective for climatological purposes. Progress in the past 30 years has brought the estimate of evaporation on a global scale to useful accuracy.

See also

Satellite Remote Sensing of Sea Surface Temperatures.

Further Reading

Bentamy A, Queffeulou P, Quilfen Y, and Katsaros KB (1999) Ocean surface wind fields estimated from satellite active and passive microwave instruments. *Institute of Electrical and Electronic Engineers, Transactions, Geoscience Remote Sensing* 37: 2469–2486.

Businger JA, Wyngaard JC, lzumi Y, and Bradley EF (1971) Flux-profile relationships in the atmospheric surface layer. *Journal of Atmospheric Science* 28: 181–189.

DeCosmo J, Katsaros KB, Smith SD, *et al.* (1996) Air–sea exchange of water vapor and sensible heat: The Humidity Exchange Over the Sea (HEXOS) results. *Journal of Geophysical Research* 101: 12001–12016.

Dobson F, Hasse L, and Davies R (eds.) (1980) *Instruments and Methods in Air–sea Interaction.* New York: Plenum Publishing.

Donelan MA (1990) Air–sea Interaction. In: LeMehaute B and Hanes DM (eds.) *The Sea*, Vol. 9, pp. 239–292. New York: John Wiley.

Esbensen SK, Chelton DB, Vickers D, and Sun J (1993) An analysis of errors in Special Sensor Microwave Imager

evaporation estimates over the global oceans. *Journal of Geophysical Research* 98: 7081–7101.

Geernaert GL (ed.) (1999) *Air–sea Exchange Physics, Chemistry and Dynamics.* Dordrecht: Kluwer Academic Publishers.

Geernaert GL and Plant WJ (eds.) (1990) *Surface Waves and Fluxes*, Vol. 2. Dordrecht: Kluwer Academic Publishers.

Katsaros KB, Smith SD, and Oost WA (1987) HEXOS – Humidity Exchange Over the Sea: A program for research on water vapor and droplet fluxes from sea to air at moderate to high wind speeds. *Bulletin of the American Meteoroloical Society* 68: 466–476.

Kraus EB and Businger JA (eds.) (1994) *Atmosphere–Ocean Interaction* 2nd ed. New York: Oxford University Press.

Liu WT and Katsaros KB (2001) Air–sea fluxes from satellite data. In: Siedler G, Church J, and Gould J (eds.) *Ocean Circulation and Climate.* Academic Press

Liu WT, Tang W, and Wentz FJ (1992) Precipitable water and surface humidity over global oceans from SSM/I and ECMWF. *Journal of Geophysical Research* 97: 2251–2264.

Makin VK (1998) Air–sea exchange of heat in the presence of wind waves and spray. *Journal of Geophysical Research* 103: 1137–1152.

Schneider SH (ed.) (1996) *Encyclopedia of Climate and Weather.* New York: Oxford University Press.

Schulz J, Meywerk J, Ewald S, and Schlüssel P (1997) Evaluation of satellite-derived latent heat fluxes. *Journal of Climate* 10: 2782–2795.

Smith SD (1988) Coefficients for sea surface wind stress, heat flux, and wind profiles as a function of wind speed and temperature. *Journal of Geophysical Research* 93: 15467–15472.

UPPER OCEAN HEAT AND FRESHWATER BUDGETS

P. J. Minnett, University of Miami, Miami, FL, USA

Introduction

Most of the solar energy reaching the surface of the Earth is absorbed by the upper ocean. Some of this is released locally, often within the course of the following night, but some heat is retained for longer periods and is moved around the planet by the oceanic surface currents. Subsequent heat release to the atmosphere helps determine the patterns of weather and climate around the globe.

While maps of sea surface temperature measured from satellites are now commonplace, it is the underlying reservoir of heat stored in the upper ocean that has the impact on the atmospheric circulation and weather, not only over the oceans but also over the continents downstream. Because the specific heat of water is much greater than that for air, the thermal capacity of a layer of the ocean about 3-m thick is the same as that of the entire atmosphere above. The upper ocean heat content, however, is not so accessible to measurements by satellite-borne instruments and is therefore less well described, and its properties less well understood.

The density of seawater is determined in a nonlinear fashion by temperature and salinity and, to a much lesser degree, by pressure. Warmer, fresher seawater is less dense than cooler, saltier water. The viscosity of seawater is very low and so the fluid is very sensitive to flow generation by density differences. However, as a result of the rotation of the Earth, oceanic flow is not simply a redistribution of mass so that the surfaces of constant density coincide with surfaces of constant gravitational force; deviations are supported by balancing the horizontal pressure forces, caused by the variable distribution of density, with the Coriolis force (geostrophy). Vertical exchanges between the upper ocean and the deeper layers are inhibited by layers of density gradients, called pycnoclines, some of which are permanent features of the ocean, and others, generally close to the surface, are transient, existing for a day or less. Upper ocean salinity, through its contribution to controlling the ocean density, is therefore an important variable in determining the density distribution of the upper ocean and the availability of oceanic heat to drive atmospheric processes.

The range of sea surface temperatures, and, by extension, the mixed layer temperature, extends from $-1.8\,°C$, the freezing point of seawater, to above $30\,°C$ in the equatorial regions, especially in the western Pacific Ocean and eastern Indian Ocean. In particularly favorable situations, surface temperatures in excess of $35\,°C$ may be found, such as in the southern Red Sea.

The lowest upper ocean salinities are found in the vicinity of large river outflows and are close to zero. For most of the open ocean, upper ocean salinities lie in the range of 34–37. (Ocean salinity is measured as a dimensionless ratio with a multiplier of 10^{-3}. A salinity of 35 means that $1\,kg$ of seawater contains $35\,g$ of dissolved salts.) Unlike elevated surface temperatures that result in a lowering of the surface density and a stable near-surface water column, increasing surface salinities by evaporation lead to increasing density and an unstable situation where the denser surface waters sink.

Governing Processes

The upper ocean heat and salt (or freshwater) distributions are determined by the fluxes of heat and moisture through the ocean surface, the horizontal divergence of heat and salinity by advection, and by fluxes through the pycnocline at the base of the upper ocean 'mixed layer'. This can be expressed for heat content per unit area, H, by

$$\frac{\Delta H}{\Delta t} = Q_{surf} + Q_{horiz} + Q_{base}$$

where Q_{surf} represents the heat fluxes through the ocean surface, Q_{horiz} the divergence of advective heat flux in the column extending from the surface to the depth of the mixed layer, and Q_{base} is the vertical heat flux through the pycnocline at the base of the mixed layer, often presumed to be small in comparison with the surface exchanges. The surface heat flux has three components: the radiative fluxes, the turbulent fluxes, and the heat transport by precipitation. The radiative fluxes are the sum of the shortwave contribution from the sun, and the net infrared flux, which is in turn the difference between the incident atmospheric emission and the emission from the sea surface. The turbulent fluxes comprise those of sensible and latent heat. A similar expression can be used for the upper ocean freshwater budget, where the fluxes are simply those of water. The surface exchanges are the difference between the mass fluxes due to precipitation and evaporation, and the horizontal advective fluxes can be best framed in terms of the divergence of salinity.

The depth of the mixed layer is often not easy to determine and there are several approaches used in the literature, including the depth at which the temperature is cooler than the surface temperature, and values of 0.1, 0.2, or 0.5 K are commonly used. Another is based on an increase in density, and a value of $0.125 \, \text{kg m}^{-3}$ is often used. These are both proxies for the parameter that is really desired, which is the depth to which turbulent mixing occurs, thereby connecting the atmosphere to the heat stored in the upper ocean. In situations of low wind speed and high insolation, a significant shallow pycnocline can develop through temperature stratification, and this decouples the 'mixed' layer beneath from the atmosphere above. Nevertheless, in most discussions of the surface heat and salt budget, these diurnal effects are discounted and the depth of integration is to the top of the seasonal pycnocline, or in the absence of the seasonal pycnocline, to the depth of the top of the permanent pycnocline.

Surface Heat Exchanges

The heat input at the surface is primarily through the absorption of insolation. Of course this heating occurs only during daytime and is very variable in the course of a day because of the changing solar zenith angle, and by modulation of the atmospheric transparency by clouds, aerosols, and variations in water vapor. At a given location, there is also a seasonal modulation. In the Tropics, with the sun overhead on a very clear day, the instantaneous insolation can exceed $1000 \, \text{W m}^{-2}$. The global average of insolation is about $170 \, \text{W m}^{-2}$. The reflectivity of the sea surface in the visible part of the spectrum is low and depends on the solar zenith angle and the surface roughness, and thereby on surface wind speed. For a calm surface with the sun high in the sky, the integrated reflectance, the surface albedo, is about 0.02, with an increase to ~ 0.06 for a solar zenith angle of $60°$. Having passed through the sea surface the solar irradiance, L_λ, is absorbed along the propagation path, z, according to Beer's law:

$$\text{d}L_\lambda/\text{d}z = -\kappa_\lambda L_\lambda$$

where the absorption coefficient, κ_λ, is dependent on the wavelength of the light (red being absorbed more

quickly than blue) and on the concentration of suspended and dissolved material in the surface layer, such as phytoplankton. When the wind is low, the near-surface density stratification that results from the absorption of heat causes the temperature increase to be confined to the near-surface layers, causing the growth of a diurnal thermocline. This is usually eroded by heat loss back to the atmosphere during the following night. If the wind speed during the day is sufficiently high, greater than a few meters per second, the subsurface turbulence spreads the heat throughout the mixed layer. There are a few locations where the insolation is high, the water is very clear, and the mixed layer depth sufficiently shallow that a small fraction of the solar radiation penetrates the entire mixed layer and is absorbed in the underlying pycnocline.

Although the absorption and emission of thermal infrared radiation are confined to the ocean surface skin layer of a millimeter or less, the net infrared budget is a component of the surface heat flux that indirectly contributes to the upper ocean heat budget. The infrared budget is the difference between the emission, given by $\varepsilon\sigma T^4$, where ε is the broadband infrared surface emissivity, σ is Stefan–Boltzmann constant, and T is the absolute temperature of the sea surface. For $T = 20 \,°\text{C}$, the surface emission is $\sim 410 \, \text{W m}^{-2}$. The incident infrared radiation is the emission from greenhouse gases (such as CO_2 and H_2O), aerosols, and clouds, and as such is very variable. For a dry, cloud-free polar atmosphere, the incident atmospheric radiation can be $< 200 \, \text{W m}^{-2}$, whereas for a cloudy tropical atmosphere, $400 \, \text{W m}^{-2}$ can be exceeded. The net infrared flux at the surface is generally in the range of $0–100 \, \text{W m}^{-2}$, with an average of about $50 \, \text{W m}^{-2}$.

The turbulent heat fluxes at the ocean surface are so called because the vertical transport is accomplished by turbulence in the lower atmosphere. They can be considered as having two components: the sensible heat flux that results from a temperature difference between the sea surface and the overlying atmospheric boundary layer, and the latent heat flux that results from evaporation at the sea surface. The sensible heat flux depends on the air–sea temperature difference and the latent heat flux on the atmospheric humidity near the sea surface. Both have a strong wind speed dependence. Since the ocean is usually warmer than the atmosphere in contact with the sea surface, and since the atmosphere is rarely saturated at the surface, both components usually lead to heat being lost by the ocean. The global average of latent heat loss is about $90 \, \text{W m}^{-2}$ but sensible heat loss is only about $10 \, \text{W m}^{-2}$. Extreme events, such as cold air outbreaks from the eastern coasts of continents over warm western boundary currents, can lead to much higher turbulent heat fluxes, even exceeding $1 \, kW \, m^{-2}$.

The final component of the surface heat budget is the sensible heat flux associated with precipitation. Rain is nearly always cooler than the sea surface and so precipitation causes a reduction of heat content in the upper ocean. Typical values of this heat loss are about $2-3 \, W \, m^{-2}$ in the Tropics, but in cases of intense rainfall values of up to $200 \, W \, m^{-2}$ can be attained.

Surface Freshwater Exchanges

Over most of the world's ocean, the flux of fresh water through the ocean surface is the difference between evaporation and precipitation. The loss of fresh water at the sea surface through evaporation is linked to the latent heat flux through the latent heat of evaporation. Clearly, precipitation exhibits very large spatial and temporal variability, especially in the Tropics where torrential downpours associated with individual cumulonimbus clouds can be very localized and short-lived. Estimates of annual, globally averaged rainfall over the oceans is about 1 m of fresh water per year, but there are very large regional variations with higher values in areas of heavy persistent rain, such as the Intertropical Convergence Zone (ITCZ) which migrates latitudinally with the seasons. Over much of the mid-latitude oceans, drizzle is the most frequent type of precipitation according to ship weather reports.

The global distribution of evaporation exceeds that of oceanic precipitation, with the difference being made up by the freshwater influx from rivers and melting glaciers.

Advective Fluxes

The determination of the amount of heat and fresh water moved around the upper oceans is not straightforward as the currents are not steady, exhibiting much temporal and spatial variation. The upper ocean currents are driven both by the surface wind stress, including the large-scale wind patterns such as the trade winds and westerlies, and by the large-scale density differences that give rise to the thermohaline circulation that links all oceans at all depths. The strong western boundary surface currents, such as the Gulf Stream, carry much heat poleward, but have large meanders and shed eddies into the center of the ocean basins. Indeed, the ocean appears to be filled with eddies. Thus the measurements of current speed and direction, and temperature and salinity taken at one place at one time could be quite different when repeated at a later date.

Measurements

Much of what we know about the upper ocean heat and salt distribution has been gained from analysis of measurements from ships. Large databases of shipboard measurements have been compiled to produce a 'climatological' description of upper ocean heat and salt content. In some ocean areas, such as along major shipping lanes, the sampling density of the temperature measurements is sufficient to provide descriptions of seasonal signals, and the length of measurements sufficiently long to indicate long-term climate fluctuations and trends, but these interpretations are somewhat contentious. In other ocean areas, the data are barely adequate to confidently provide an estimate of the mean state of the upper ocean.

Temperature is a much simpler measurement than salinity and so there is far more information on the distribution of upper ocean heat than of salt. Historically temperatures were measured by mercury-in-glass thermometers which recorded temperatures at individual depths. Water samples could also be taken for subsequent chemical analysis for salinity. The introduction of continuously recording thermometers, such as platinum resistance thermometers and later thermistors, resulted in measurements of temperature profiles, and the use of expendable bathythermographs (XBTs) meant that temperature profiles could be taken from moving ships or aircraft. The continuous measurement of salinity was a harder problem to solve and is now accomplished by calculating salinity from measurements of the ocean electrical conductivity. The standard instruments for the combined measurements of temperature and conductivity are referred to as CTDs (conductivity–temperature–depth) and are usually deployed on a cable from a stationary research ship, although some have been installed in towed vehicles for measurements behind a moving ship in the fashion of a yo-yo. In recent years, CTDs have been mounted in autonomous underwater vehicles (AUVs) that record profiles along inclined saw-tooth paths through the upper ocean (say to 600 m) and which periodically break surface to transmit data by satellite telemetry. Similarly, autonomous measurements from deep-water (to 2000 m) floats are transmitted via satellite when they surface. In the ARGO project, begun in 2000, over 3000 floats have been deployed throughout the global ocean. The floats remain at depth for about 10 days, drifting with the currents, and then make CTD measurements as they come to surface where their positions are fixed by the Global Positioning System (GPS), and the profile data transmitted to shore. Where time series of profiles are required at a particular location, internally recording

CTDs can be programmed to run up and down wires moored to the seafloor. These have been used effectively in the Arctic Ocean, but the instruments have to be recovered to retrieve the data as the presence of ice prevents the use of a surface float for data telemetry. Additional sensors, such as transmissometers to measure turbidity, often augment the CTD measurements. Further information is supplied by a network of moored buoys that now span the tropical Pacific and Atlantic Oceans and which support sensors at fixed depths.

The spatial distribution of upper ocean temperatures can be derived from satellite measurements of the sea surface temperatures which can now be made with global accuracies of 0.4 K or better using infrared and microwave radiometers. Such data sets now extend back a couple of decades. In 2009 and 2010, two new low-frequency microwave radiometers capable of measuring open ocean salinity are planned for launch (Aquarius is a NASA instrument, and SMOS – Soil Moisture, Ocean Salinity – is an ESA mission). To convert these satellite measurements of surface temperature and surface salinity into upper ocean heat and salt contents requires knowledge of the mixed layer depth, and while this is not directly accessible from satellite measurements, it can be inferred from measurements of ocean surface topography, derived from satellite altimetry, through the use of a simple upper ocean model. Such upper ocean heat content estimates are now being routinely derived and used in an experimental mode to assist in hurricane forecasting and research.

Several lines across ocean basins have been sampled from a fleet of research ships in the framework of the World Ocean Circulation Experiment (WOCE) which took place between 1990 and 2002. Many of these sections are currently being reoccupied in the Repeat Hydrography Program to determine changes on decadal scales.

Distributions

Heat

The quantitative specification of the upper ocean heat content remains rather uncertain, not so much because of our ability to measure the sea surface temperature (**Figure 1**), which is generally a good estimate of the mixed layer temperature, especially at night, but in determining the depth of the mixed layer. If the objective is to estimate the heat potentially available to the atmosphere, then the depth of the mixed layer based on density stratification in the pycnocline is more appropriate than the depth based on a temperature gradient in the thermocline,

although this is often used because of the availability of more data. On an annual basis, this can lead to significant differences in the estimates of the depth of the oceanic layer that can supply heat to the atmosphere (**Figure 2**).

The difference between the tops of the thermocline and pycnocline results from density stratification caused by vertical salinity gradients (a halocline). In the low-latitude oceans, this is sometimes called a 'barrier layer' and can be as thick as the overlying isothermal layer, that is, halving the thickness of the upper ocean layer in contact with the atmosphere compared to that which would be estimated using temperature profiles alone. The barrier layers are probably caused by the subduction of more saline waters underneath water freshened by rainfall or river runoff.

At high latitudes, the nonlinear relationship between seawater density and temperature and salinity means that density is nearly independent of temperature. The depth of the surface layer is therefore determined by the vertical salinity profile. During ice formation, brine is released from the freezing water and this destabilizes the surface layer, causing convective mixing. During ice melt, the release of fresh water stabilizes the upper ocean. In the Arctic Ocean, the depth of the mixed layer is determined by the depth of the halocline.

The annual means of course do not reveal the details of the seasonal cycle in heat content, which in turn reflect the seasonal patterns of the surface fluxes and advective transports. **Figure 3** shows the global distributions of the surface fluxes for January and July. The patterns in the insolation (short-wave heat flux) reflect the changes in the solar zenith angle, and the seasonal changes in cloud cover and properties. The seasonal patterns of the surface winds are apparent in the turbulent fluxes.

The seasonal changes in the sea surface temperatures and upper ocean heat content are the summations of small daily residuals of local heating and cooling. Under clear skies and low winds, the absorption of insolation in the upper ocean leads to a stabilization of the surface layer through the formation of a near-surface thermocline. While the surface heat budget remains positive (i.e., the insolation exceeds heat loss through turbulent heat loss and the net infrared radiation), the diurnal thermocline grows with an attendant increase in the sea surface temperature. As the insolation decreases and the surface heat budget changes sign, the surface heat loss results in a fall in surface temperature and the destabilization of the near-surface layer. The resultant convective instability erodes the thermal stratification, returning the upper layer to a state close to

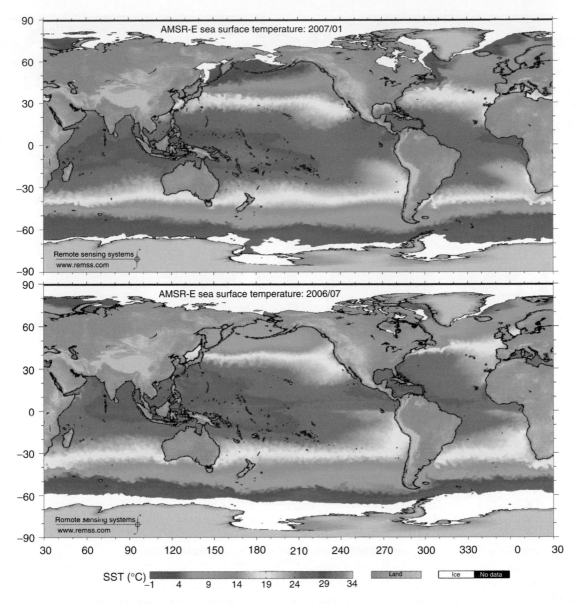

SST (°C) ‑1 4 9 14 19 24 29 34   Land   Ice   No data

Figure 1 Global distributions of the sea surface temperature measured by the advanced microwave scanning radiometer for the Earth Observing System on the NASA satellite *Aqua*. Monthly averaged fields for Jan. 2007 and Jul. 2006 are shown. AMSR-E data are produced by Remote Sensing Systems and sponsored by the NASA Earth Science REASoN DISCOVER Project and the AMSR-E Science Team. Data are available at http://www.remss.com.

that before diurnal heating began. In the heating season, on average, there will be more heat in the upper ocean at the end of the diurnal cycle, and in the cooling season there will be less. On days when the wind speed is greater than a few meters per second, the wind-induced turbulent mixing prevents the growth of the diurnal thermocline and the heat input during the day, and removed at night, is distributed throughout the mixed layer. **Figure 4** shows measurements of the diurnal heating, expressed as a difference between the 'skin' temperature and a bulk temperature at the depth of a few meters, as a function of wind speed and time of day.

The magnitude of the surface temperature signal of diurnal warming is very strongly dependent on wind speed, and can be eroded very quickly if winds increase in the course of a day.

The upper ocean of course exhibits variability on timescales longer than a year, often with profound consequences around the globe. The best known is the El Niño–Southern Oscillation that results in a marked change in sea surface temperature, depth of the mixed layer, and consequently upper ocean heat content in the equatorial Pacific Ocean. The perturbations to the atmospheric circulation have effects on weather patterns, including rainfall, around the

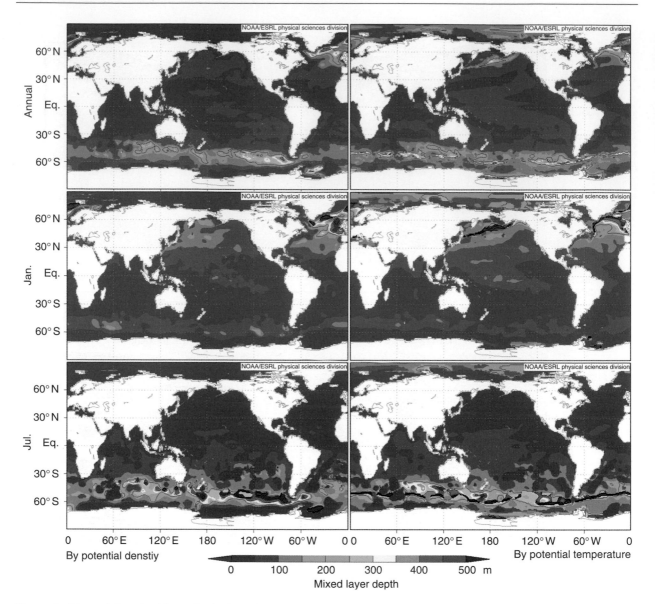

Figure 2 Maps of the mixed layer depth. Annual averages are shown along with monthly means for January and July. The left column shows mixed layer depths based on a potential density difference criterion, and the right column on a potential temperature difference. The deepest values are found at high latitudes in the winter hemisphere. A discrepancy in the estimates by a factor of 2 is seen in some regions. This translates into an equivalent uncertainty in the estimate of the heat content of the upper ocean. The figure is based on images produced at the NOAA-CIRES Climate Diagnostics Center, Boulder, Colorado, from their website at http://www.cdc.noaa.gov.

globe. Other multiyear features include the North Atlantic Oscillation, Arctic Oscillation, and Pacific Decadal Oscillation, in all of which there is a shift in both atmospheric circulation and oceanic response. In the case of the Arctic Oscillation, determined from the strength of the polar vortex relative to mid-latitude surface pressure, a negative phase results in high surface pressures in the Arctic, and a more uniform distribution of sea ice. This is considered the normal situation. The positive phase results in lower surface pressure fields over the Arctic Ocean, a

thinning of the ice cover, and intrusion of relatively warm Atlantic water into the Arctic Basin. These have consequences on the upper ocean salinity and density stratification, and on interactions with the atmosphere, although the complexities of these feedbacks are poorly understood.

The heat and salinity content advected through imaginary boundaries extending across oceans from coast to coast and from the surface to depth are a measure of the transport of heat and salt. For example, to maintain the Earth's radiative equilibrium

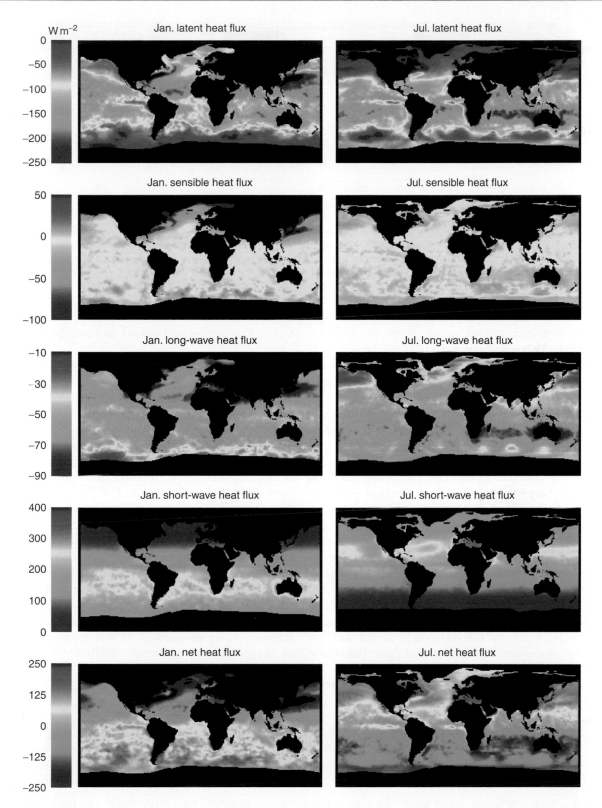

Figure 3 Distributions of the components of the surface heat fluxes for Jan. and Jul. The warm colors indicate warming or less cooling of the ocean, and the cool colors indicate cooling or less warming of the ocean. The data are from the UK National Oceanography Center surface flux climatology (v1.1) and were obtained from http://www.noc.soton.ac.uk.

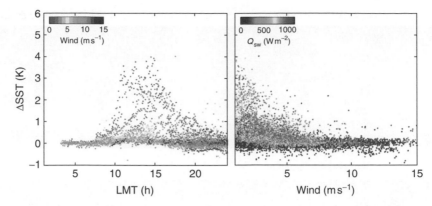

Figure 4 Signatures of diurnal heating revealed in the temperature difference between a radiometrically measured ocean skin temperature and a bulk temperature at a depth of 2 m. The measurements were taken in the Caribbean Sea from the Royal Caribbean International cruise liner *Explorer of the Seas*, which has been equipped as a research vessel. The temperature differences are plotted as a function of local mean time (LMT) and colored by wind speed (left), and as function of wind speed, colored by LMT (left). The largest temperature differences occur in early afternoon on days when the winds are low. Figure provided by Dr. C. L. Gentemann.

with the sun and space the combined heat transport of the atmosphere and ocean from the Tropics toward the Poles is about 5.5×10^{15} W. How this is partitioned between the atmosphere and ocean is the subject of much research. Within the Atlantic Ocean, the northward transport of heat across $24°$N is about 1.3×10^{15} W. Interestingly the average heat transport in the Atlantic is northward, even south of the Equator: 0.3×10^{15} W northward at $30°$S and 0.6×10^{15} W at $11°$S, although these estimates include transport at depth. The differences in heat transport between such lines at different latitudes provide estimates of the net heat absorbed by the upper ocean or given up to the atmosphere within the surface area of the oceans enclosed by the sections. Thus 77 ± 57 W m^{-2} are estimated to be released by the Atlantic Ocean to the atmosphere between $36°$ and $48°$N, but only 8 ± 33 W m^{-2} between $22°$ and $36°$N. In the North Pacific, 39 ± 19 W m^{-2} flow to the atmosphere between $24°$ and $48°$N. Similarly the differences in the salt (or freshwater) content advected across these imaginary boundaries indicate the imbalance between precipitation plus continental runoff and evaporation.

Fresh Water

The large-scale patterns of upper ocean salinity (**Figure 5**) mirror the distribution of the annual freshwater flux at the sea surface (**Figure 6**), which is determined by the difference between rainfall and evaporation. The patterns of the components of the freshwater flux are quite zonal in character, with a band of heavy rainfall in the ITCZ and the maxima in evaporation occurring in the regions of the trade winds. There is very little known variability in the seasonal distribution of surface salinity, with the exceptions being in coastal regions where river run off often has a seasonal modulation, especially in the Bay of Bengal where the rainfall influencing the river discharge is dominated by the monsoons.

The precipitation over the Bay of Bengal also shows a strong monsoonal influence, but over much of the oceans the seasonal variability is relatively

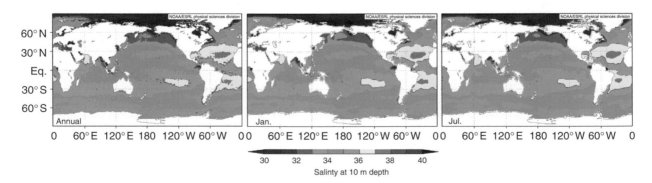

Figure 5 Global distribution of salinity at a depth of 10 m as a global average (left) and monthly averages for Jan. (center) and Jul. (right). The figure is based on images produced at the NOAA-CIRES Climate Diagnostics Center, Boulder, Colorado (http://www.cdc.noaa.gov).

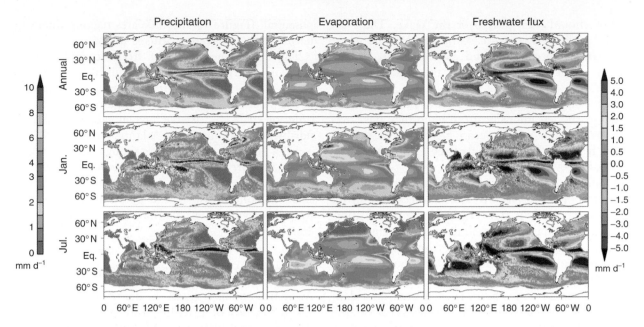

Figure 6 Global distributions of precipitation, evaporation, and freshwater flux at the ocean surface. Annual means are shown in the top row besides monthly averages for Jan. (middle row) and Jul. (bottom row). The color scale is at left for precipitation (positive mass flux into the ocean) and evaporation (positive mass flux into the atmosphere). The color scale for the freshwater flux is at right (positive mass flux into the atmosphere). The figures were generated from the Hamburg Ocean Atmosphere Parameters and Fluxes from Satellite Data Set (HOAPS) (http://www.hoaps.zmaw.de).

muted. Similarly for evaporation, although variations in the Northern Hemisphere signal are greater than those in the Southern Hemisphere. Pronounced maxima in the wintertime evaporation occur over the Gulf Stream and Kuroshio, and represent enhanced moisture fluxes from the sea surface driven by cold, dry air flowing off the continents over the warm surface waters of the north-flowing currents (**Figure 6**).

On shorter timescales, there is pronounced variability in rainfall associated with the passage of weather fronts at mid-latitudes and with individual clouds in the Tropics. These small-scale, short-duration rainfall events hinder the accurate determination of the freshwater flux into the sea surface. In the Tropics, the rainfall associated with individual cumulonimbus clouds has a diurnal signature, especially in the vicinity of islands, even small atolls, where the diurnal sea breeze can trigger convection that results in rainfall, either directly into the ocean, or as runoff from land.

The Arctic Ocean is a particularly interesting area regarding the local freshwater budget as the vertical stability is constrained by the salinity gradients in the halocline. Freshwater volumes in the Arctic Ocean are often calculated relative to a seawater salinity of 34.8. The fresher surface waters are sustained by riverine inflow, primarily from the great Siberian rivers and the Mackenzie River in Canada, that between them annually contribute about 3200 km³.

The inflow from the Pacific Ocean through the Bering Strait is about 2500 km³. The freshwater outflow is mainly through the Canadian Archipelago as liquid (~ 3200 km³) and through Fram Strait (~ 2400 km³ as liquid and ~ 2300 km³ as ice). The contribution of precipitation minus evaporation is c. 2000 km³. The fresh water generated by brine rejection during ice formation would be $\sim 10\,000$ km³, which is a relatively small proportion of the riverine and Bering Strait input. The residence time of fresh water in the Arctic Ocean is about 10 years.

Severe Storms

An important consequence of variations in the upper ocean heat content is severe storm generation and intensification. The prediction of the strength and trajectory of land-falling hurricanes and cyclones benefits from knowledge of the upper ocean heat content in the path of the storm. A surface temperature of 26 °C is generally accepted as being necessary for hurricane development, but the rate of development depends on the heat in the upper ocean available to drive the storm's intensification. The passage of a severe storm leaves a wake that is identifiable as a depression of the surface temperature of several degrees and a deficit in the upper ocean heat content. These may survive for several days and can influence the development of subsequent storms should they pass over the wake. There are several well-documented cases in the Atlantic where hurricanes approaching land have suddenly lost intensity as they follow, or cross, the path of a prior storm.

The converse is also true and hurricanes can undergo sudden intensification when they pass over regions of high upper ocean heat content, as can result from the meandering of the Loop Current in the Gulf of Mexico, for example.

Monitoring the upper ocean heat content has become important for severe storm forecasting in the Tropics, especially in terms of sudden intensification. Using a combination of satellite measurements of sea surface temperature, sea surface topography, and a simple ocean model, the spatial distribution of the heat content between the surface and the estimated depth of the 26 °C isotherm is calculated on a daily basis. This is referred to as the 'tropical cyclone heat

potential' (**Figure 7**) and indicates regions where intensification of severe storms is likely.

The rate of heat transfer from ocean to atmosphere in a hurricane is very difficult to measure, and varies greatly with the size, intensity, and stage of development of the storm. Estimates range in the order of 10^{13}–10^{14} W. We have already seen that the northward heat flux in the Atlantic Ocean at 24° N is $\sim 1.3 \times 10^{15}$ W. Thus, even though severe storms grow and are sustained by large heat fluxes, the magnitudes of the associated flow are relatively small in comparison to the poleward oceanic heat transport which is ultimately released to the atmosphere.

Reactions to Climate Change

Away from polar regions, the density of seawater is a strong function of its temperature, and a consequence

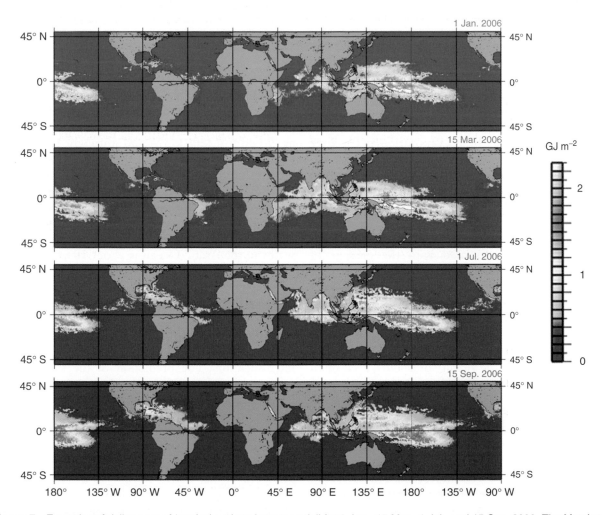

Figure 7   Examples of daily maps of 'tropical cyclone heat potential' for 1 Jan., 15 Mar., 1 Jul., and 15 Sep. 2006. The March and September dates correspond roughly to the peak of cyclone activity in each hemisphere. The maps were derived from satellite measurements of sea surface temperature, sea surface topography, and a simple ocean model. The figure was derived from images generated by the NOAA Atlantic Oceanographic and Meteorological Laboratory (AOML), Miami, Florida (http://www.aoml.noaa.gov).

of increasing temperatures as a result of global change is the expansion of the upper ocean, which will contribute to sea level rise. In fact, about half of the observed rise in global sea level during the twentieth century of 1–2 mm yr^{-1} can be attributable to expansion of the warming upper ocean.

In addition to thermal expansion, another major impact of climate change is the increase in the upper ocean freshwater budget (reduction in salinity) as the land ice (glaciers and ice caps of Greenland and Antarctica) melt and the runoff enter the high-latitude oceans. This will result in an increase in the stability of the upper ocean and a consequent likely reduction in the mixed layer depths, especially in winter (**Figure 2**). Here the atmosphere is coupled to the heat available in a very deep ocean layer. Mixing heat and fresh water from the upper ocean to depth is also a driver of the global thermohaline circulation and disruption to this will also have significant consequences on the near-surface components of the circulation and on the details of the poleward transfer of heat. This will impact global weather patterns, including rainfall over the ocean and land, in ways that are difficult to predict.

Future Developments

Improvements in our ability to determine the upper ocean heat and freshwater budgets, and monitor their changes with time, will occur in the near future with new satellite missions that will both continue the existing time series of sea surface temperature, topography and rainfall, and also introduce new variables: notably sea surface salinity.

Additional information on the subsurface distributions of heat and fresh water will be provided by the autonomous profiling floats of the ARGO project that measure temperature and salinity from about 2000-m depth to a few meters below the surface. These will be augmented by AUVs, or 'gliders', roaming the oceans taking measurements along undulating paths, transmitting the data via satellite communications when they break the surface.

The interpretation of the measurements, from both *in situ* and space-borne sensors, will be aided by increasingly complex, high-resolution models of the ocean state and the coupled ocean–atmosphere system.

See also

Evaporation and Humidity. Heat and Momentum Fluxes at the Sea Surface. Heat Transport and Climate. Ocean Circulation.

Further Reading

Chen SS and Houze RA (1997) Diurnal variation and lifecycle of deep convective systems over the tropical Pacific warm pool. *Quarterly Journal of the Royal Meteorological Society* 123: 357–388.

Foltz GR, Grodsky SA, Carton JA, and McPhaden MJ (2003) Seasonal mixed layer heat budget of the tropical Atlantic Ocean. *Journal of Geophysical Research* 108: 3146 (doi:10.1029/2002JC001584).

Gill AE (1982) *Atmosphere–Ocean Dynamics.* San Diego, CA: Academic Press.

Hasegawa T and Hanawa K (2003) Decadal-scale variability of upper ocean heat content in the tropical Pacific. *Geophysical Research Letters* 30: 1272 (doi:10.1029/2002GL016843).

Josey SA, Kent EC, and Taylor PK (1999) New insights into the ocean heat budget closure problem from analysis of the SOC air–sea flux climatology. *Journal of Climate* 12: 2856–2880.

Levitus S, Antonov J, and Boyer T (2005) Warming of the world ocean, 1955–2003. *Geophysical Research Letters* 32: L02604 (doi:10.1029/2004GL021592).

Macdonald AM (1998) The global ocean circulation: A hydrographic estimate and regional analysis. *Progress in Oceanography* 41: 281–382.

Peixoto JO and Oort AH (1992) *Physics of Climate.* New York: American Institute of Physics.

Serreze MC, Barrett AP, Slater AG, *et al.* (2006) The large-scale freshwater cycle of the Arctic. *Journal of Geophysical Research* 111: C11010 (doi:10.1029/2005JC003424).

Shay LK, Goni GJ, and Black PG (2000) Effects of a warm oceanic feature on hurricane Opal. *Monthly Weather Review* 128: 1366–1383.

Siedler G, Church J, and Gould J (eds.) (2001) *Ocean Circulation and Climate: Observing and Modelling the Global Ocean.* San Diego, CA: Academic Press.

Willis JK, Roemmich D, and Cornuelle B (2004) Interannual variability in upper ocean heat content, temperature, and thermosteric expansion on global scales. *Journal of Geophysical Research* 109: C12036 (doi:10.1029/2003JC002260).

Relevant Websites

http://www.remss.com
– AMSR Data, Remote Sensing Systems.
http://aquarius.gsfc.nasa.gov
– Aquarius Mission Website, NASA.
http://www.esr.org
– Aquarius/SAC-D Satellite Mission, ESR.
http://www.hoaps.zmaw.de
– Hamburg Ocean Atmosphere Parameters and Fluxes from Satellite Data.
http://www.ghrsst-pp.org
– High-Resolution SSTs from Satellites, GHRSST-PP.
http://www.noc.soton.ac.uk
– NOC Flux Climatology, at Ocean Observing and Climate pages of the National Oceanography Centre

(NOC), and The World Ocean Circulation Experiment (WOCE) 1990–2002, NOC, Southhampton.

http://ushydro.ucsd.edu
– Repeat Hydrography Project.

http://www.cdc.noaa.gov
– Search for Gridded Climate Data at PSD, ESRL Physical Sciences Division, NOAA.

http://www.esa.int
– SMOS, The Living Planet Programme, ESA.

http://www.aoml.noaa.gov
– Tropical Cyclone Heat Potential, Atlantic Oceanographic and Meteorological Laboratory (AOML).

CRYOSPHERE

ANTARCTIC CIRCUMPOLAR CURRENT

S. R. Rintoul, CSIRO Antarctic Climate and
Ecosystems Cooperative Research Centre,
Hobart, TAS, Australia

Introduction

The Drake Passage between the South American and
Antarctic continents is the only band of latitudes
where the ocean circles the Earth, unblocked by land.
The existence of this oceanic channel has profound
implications for the global ocean circulation and
climate. The Drake Passage gap permits the Antarctic Circumpolar Current (ACC) to exist, a system
of ocean currents which flows from west to east
along a roughly 25 000-km-long path circling Antarctica. In terms of transport, the ACC is the largest
current in the world ocean, carrying about
$137 \pm 8 \times 10^6 \, \mathrm{m}^3 \, \mathrm{s}^{-1}$ through the Drake Passage.
The wind-driven ocean circulation theories that ex-
plain much of the ocean current patterns observed
at other latitudes do not apply in an unbounded
channel and the unique dynamics of the ACC have
long been a puzzle for oceanographers. The three-
dimensional circulation in the ACC belt is now
understood to reflect the interplay of wind and
buoyancy exchange with the atmosphere, water mass
modification, eddy fluxes of heat and momentum, and
strong interactions between the flow and bathymetry.

The strong eastward flow of the ACC has several
important implications for the global ocean circulation and its influence on regional and global climate.
By transporting water between the major ocean
basins, the ACC tends to smooth out differences in
water properties between the basins. The interbasin
connection allows a global-scale pattern of ocean
currents to be established, known as the thermohaline
circulation (*see* Ocean Circulation: Meridional Overturning Circulation), which transports heat, moisture,
and carbon dioxide around the globe and strongly
influences the Earth's climate. The strong flow of the
ACC is associated with steeply sloping density surfaces, which shoal to the south across the current and
bring dense waters to the surface in the high-latitude
Southern Ocean. Where the dense waters are exposed
at the sea surface, they exchange heat, moisture,
and gases like oxygen and carbon dioxide with the
atmosphere. In this sense, the Southern Ocean provides a window to the deep sea. The ecology and

biogeography of the Southern Ocean are influenced
strongly by the ACC and the overturning circulation
plays an important role in the marine carbon and
nutrient cycles. The west-to-east flow of the ACC inhibits north–south exchanges across the current and
isolates Antarctica from the warm waters to the north;
the present glacial climate of Antarctica was not established until the South American and Antarctic
continents began to drift apart about 30 Ma, opening
a circumpolar channel.

Structure of the Antarctic Circumpolar Current

A schematic view of the major currents of the
Southern Hemisphere oceans south of 20° S is shown
in **Figure 1**. The flow of the ACC is focused in several
jets, associated with sharp cross-stream gradients (or
fronts) in temperature, salinity, and other properties.
The three main fronts of the ACC – the Subantarctic
Front, Polar Front, and southern ACC front – are
indicated by the arrows circling Antarctica. To
the south of the circumpolar flow of the ACC,
clockwise gyres are found in the Weddell Sea, Ross
Sea, and the Australian–Antarctic Basin. A westward
flow associated with the Antarctic Slope Front and
Antarctic Coastal Current is found near the continental shelf break around much of Antarctica. To the
north of the ACC, water flows to the east in the
southern limb of the large anticlockwise subtropical
gyres in each basin. Exchanges of water masses between the ACC and the gyre circulations to the north
and south are important components of the global
circulation.

The distribution of water properties on a transect
crossing the ACC is illustrated in **Figure 2**. The
temperature and salinity of water masses are largely
set at the sea surface, where there is active exchange
with the atmosphere; nutrient and oxygen concentrations are also influenced by biological processes.
Water masses carry these characteristics with them as
they sink from the surface into the ocean interior.
The major water masses of the Southern Ocean are
associated with various property extrema that reflect
their circulation and formation history. For example,
the Subantarctic Mode Water is formed by deep
convection in winter on the northern flank of the
ACC, producing deep well-mixed layers that are rich
in oxygen. The Antarctic Intermediate Water is the
name given to the prominent salinity minimum layer

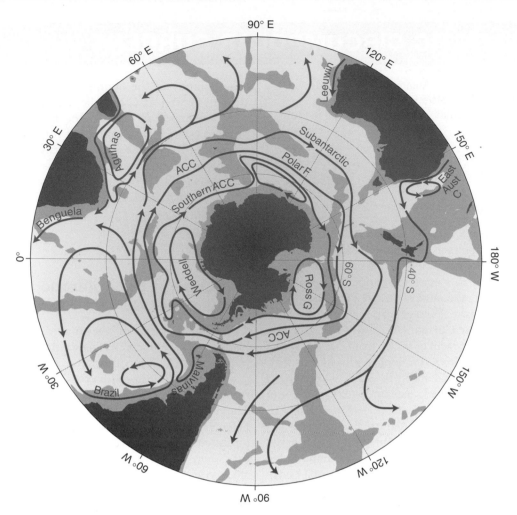

Figure 1 A schematic view of the major ocean currents of the Southern Hemisphere oceans south of 20° S. Depths shallower than 3500 m are shaded. C, current; G, gyre; F, front; ACC, Antarctic Circumpolar Current.

north of the ACC. The Circumpolar Deep Water (CDW) is often divided into two layers: Upper CDW corresponds to the oxygen minimum layer, and Lower CDW corresponds to the deep salinity maximum layer. The relatively fresh layer near the sea-floor is Antarctic Bottom Water, which sinks near Antarctica and carries water rich in oxygen and chlorofluorocarbons into the deep ocean.

Water properties at a given depth change dramatically as the Southern Ocean is crossed from north to south (**Figure 2**). Surfaces of constant temperature, salinity, density, and other properties slope upward to the south. As a result, density layers found at 3000-m depth in subtropical latitudes approach the sea surface near Antarctica. The shoaling of density surfaces to the south is associated with the strong eastward flow of the ACC. Tilted density surfaces create pressure forces which drive ocean currents and an accompanying Coriolis force to balance the pressure force. (This balance of forces,

known as geostrophy, describes the dynamics of all large-scale ocean currents (*see* Ocean Circulation).) In the Southern Hemisphere, an increase in density to the south supports an eastward flow (relative to the seafloor), as observed in the ACC.

The rise of temperature, salinity, and density surfaces to the south occurs in a series of steps, or rapid transitions, rather than as a uniform slope across the Southern Ocean (**Figure 2**). These rapid transitions are known as fronts. Because the strength of an ocean current is proportional to the magnitude of the horizontal density gradient, each of the fronts is associated with a maximum in velocity. Most of the flow of the ACC is concentrated in the fronts, with smaller transports observed between the fronts (**Figure 2(f)**). The zones between the fronts also coincide with regions of relatively uniform water properties at each depth. Unlike many fronts at lower latitudes, the ACC fronts extend from the sea surface to the seafloor. The current jets associated

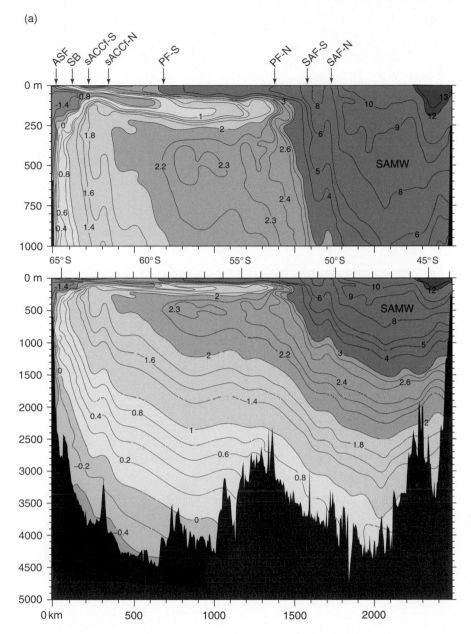

Figure 2 Property distributions along a roughly north–south section across the Southern Ocean at 140° E south of Australia (World Ocean Circulation Experiment section SR3): (a) potential temperature (°C), (b) salinity (PSS78), (c) neutral density (kg m^{-3}), (d) oxygen (μmol kg^{-1}), (e) chlorofluorocarbon 11 (CFC-11) (pM kg^{-1}), and (f) transport at each station pair (solid line, left axis) and cumulative transport from south to north (dashed line, right axis). Transport is in units of sverdrups (1 Sv = 10^{6} m^{3} s^{-1}). Contours slope upward from north to south; regions where the slope of the contours is steep correspond to the ACC fronts and to transport maxima. SAMW, Subantarctic Mode Water; LCDW, Lower Circumpolar Deep Water; AAIW, Antarctic Intermediate Water; UCDW, Upper Circumpolar Deep Water; CDW, Circumpolar Deep Water CDW; AABW, Antarctic Bottom Water. SAF, Subantarctic Front; PF, Polar Front; sACCf, southern ACC front; SB, southern boundary; ASF, Antarctic Slope Front; N and S indicate northern and southern branches of the primary fronts. The positive and negative peaks in transport labeled SAZ indicate a strong recirculation in the Subantarctic Zone north of the ACC at the time this section was occupied. Sections are adapted from Orsi AH and Whitworth T, III (2005) In: Sparrow M, Chapman P, and Gould J (eds.) *Hydrographic Atlas of the World Ocean Circulation Experiment (WOCE), Vol. 1: Southern Ocean*. Southampton: International WOCE Project Office (ISBN 0-904175-49-90), with permission (http://www.soc.soton.ac.uk).

with the fronts therefore also extend throughout the water column. The deep-reaching nature of the ACC fronts reflects the weak stratification of the Southern Ocean, where the change in density from surface waters to deep waters is small compared to lower latitudes. The mean current speeds of the ACC jets are relatively modest, typically less than 0.5 m s^{-1} (about 1 knot), with much weaker flow between the jets. However, because the current is broad and deep, it carries a large transport.

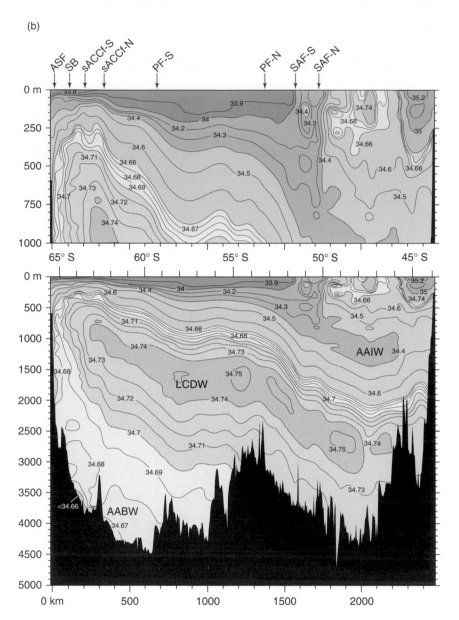

Figure 2 *Continued.*

Most studies of the ACC now recognize three main fronts. From north to south, these fronts are the Subantarctic Front, the Polar Front, and the southern ACC front. The southern limit of the ACC domain is defined by the 'southern boundary of the ACC', the southernmost streamline to pass through Drake Passage, which is also often associated with a weak front and current core. These fronts of the ACC are circumpolar in extent and can be found on any north–south transect across the Southern Ocean. Because the fronts extend to the seafloor and the stratification in the Southern Ocean is relatively weak, the position of the fronts is strongly influenced by bathymetry.

Between the fronts lie zones with more or less uniform physical and chemical characteristics: the Subantarctic Zone between the Subantarctic and Subtropical Fronts, the Polar Frontal Zone between the Subantarctic and the Polar Fronts, and the Antarctic Zone south of the Polar Front. Within each zone, the water properties tend to be similar at each depth and follow a similar seasonal cycle. For example, the Subantarctic Zone is characterized by very deep surface mixed layers in winter and low nutrient concentrations (close to zero for silicic acid and somewhat higher concentrations of nitrate and phosphate). The Antarctic Zone is characterized by fresh surface waters, shallow summer mixed layers,

(c)

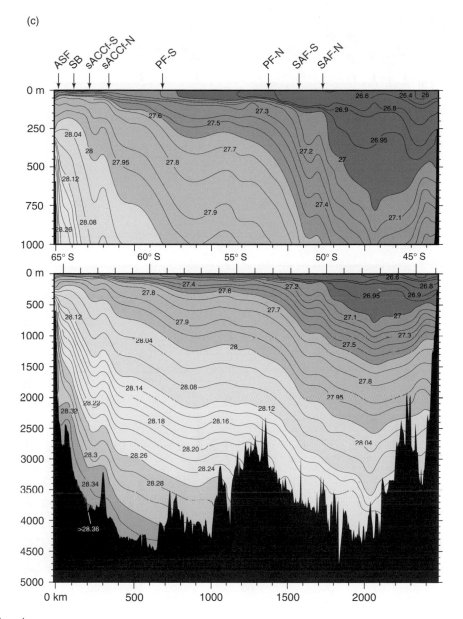

Figure 2 *Continued.*

and high concentrations of major nutrients like nitrate and silicic acid, but low concentrations of micronutrients like iron.

The zones delimited by the fronts of the ACC also define biogeographic zones populated by distinct species assemblages. For example, waters south of the Polar Front tend to be dominated by large phytoplankton such as diatoms (who need silicic acid) and large zooplankton, while coccolithophores and small zooplankton dominate north of the Subantarctic Front. The distribution and foraging patterns of larger animals (e.g., fish, seabirds, and marine mammals) are also influenced by the frontal structure of the ACC. In some cases, the fronts themselves tend to be associated with higher primary productivity. The higher productivity near fronts can be caused by advection of micronutrients by the current or by upwelling caused by eddies or by topographic interactions. The currents of the ACC can also play a direct role in ecosystem dynamics, for example, by carrying krill and larvae from the Antarctic Peninsula to South Georgia.

Circulation of the Antarctic Circumpolar Current

A map of the elevation of the sea surface (dynamic height) shows how the position and intensity of the current varies along its circumpolar path

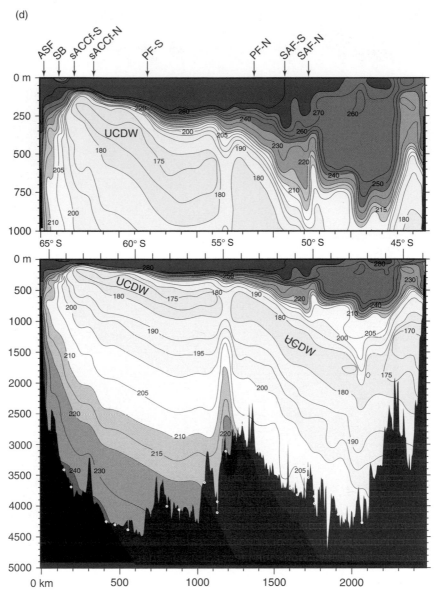

Figure 2 *Continued.*

(**Figure 3**). Contours of dynamic height are approximate streamlines for the flow, so the current flow is rapid in regions where the contours are closely spaced and weak in regions where the contours are widely separated. The steering of the fronts by large bathymetric features is clearly illustrated in **Figure 3**.

Recent studies using high-resolution sampling from ships and satellites have revealed a more complex structure to the ACC than previously appreciated. The ACC fronts consist of multiple branches, which merge and diverge in different regions and at different times along the circumpolar path of the current system (**Figures 3** and **4**). The multiple jets in the ACC reflect the tendency for geophysical flows on a sphere to self-organize into narrow, elongated, persistent zonal flows; similar circulation patterns form in the Earth's atmosphere and on other planets. The position of the fronts varies with time, but generally over a relatively small latitude range at any given longitude (typically $\pm 1°$ of latitude). The variability is larger downstream of major bathymetric features and in regions where the ACC fronts interact with the strong boundary currents of the subtropical gyres to the north (e.g., south of Africa).

Transport of the Antarctic Circumpolar Current

During the World Ocean Circulation Experiment (WOCE) in the 1990s, the transport through Drake

(e)

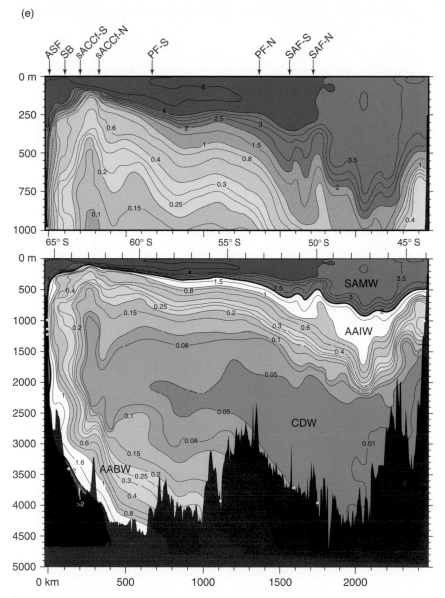

Figure 2 *Continued.*

Passage was estimated to be $137 \pm 8 \times 10^6\,m^3\,s^{-1}$, similar to the estimates made in the late 1970s. Because the Atlantic basin is nearly closed to the north of the Southern Ocean, the net transport between Africa and Antarctica must be very close to the Drake Passage transport (to within *c.* $1 \times 10^6\,m^3\,s^{-1}$). The transport between Australia and Antarctica must be somewhat greater, to compensate for the flow from the Pacific to the Indian Oceans through the Indonesian archipelago. Repeat transects during WOCE showed that the baroclinic transport south of Australia is $147 \pm 10 \times 10^6\,m^3\,s^{-1}$, consistent with estimates that about $10\text{--}15 \times 10^6\,m^3\,s^{-1}$ flows through the Indonesian passages.

The fronts, in particular the Subantarctic and Polar Fronts, carry most of the ACC transport. The relative contribution of these two fronts to the total transport varies around the circumpolar path. For example, south of Australia the Subantarctic Front carries 4 times more water to the east than the Polar Front, while in Drake Passage the transport carried by the two fronts is roughly equal in magnitude.

The transport of water masses by the ACC also changes with longitude. For example, the ACC carries an excess of intermediate density water into the Atlantic through Drake Passage, which is compensated by an excess of deep water leaving the basin south of Africa. These changes in water mass transports by the ACC reflect water mass transformations in the Atlantic basin, where relatively light Antarctic Intermediate Water is converted to denser

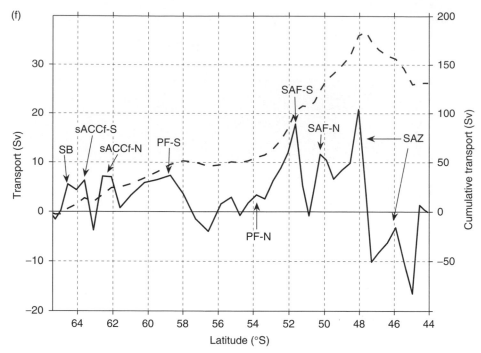

Figure 2 *Continued.*

North Atlantic Deep Water. The ACC also transports vast amounts of heat, fresh water, nutrients, carbon, and other properties between the ocean basins.

The transport of the ACC varies over a range of timescales. Multiyear deployments of bottom pressure recorders in Drake Passage during the late 1970s and 1990s suggest a standard deviation in net transport of about $8-10 \times 10^6 \, \mathrm{m^3 \, s^{-1}}$. For periods shorter than about 6 months, most of the variability is due to changes in sea level (i.e., changes in the barotropic, or depth-independent, flow). Models and sea level measurements suggest these barotropic motions are highly correlated with changes in wind stress and tend to follow bathymetric contours (more precisely, the flow is along lines of constant planetary vorticity, where planetary vorticity is given by the Coriolis parameter (equal to twice the rotation rate of the Earth multiplied by the sine of the latitude) divided by the ocean depth). For longer periods, variations in the density field (and hence the baroclinic, or depth-varying, flow) also become important.

Dynamics of the Antarctic Circumpolar Current

The absence of continental barriers in the latitude band of Drake Passage makes the dynamics of the ACC distinctly different in character from those of currents at other latitudes. Simple wind-driven ocean circulation theory (the Sverdrup balance), which generally does a good job of describing the circulation of the upper ocean in basins bounded by continents, cannot be applied in the usual way in a continuous ocean channel. The dynamical balance of the ACC has therefore been a topic of great interest for many years.

The strong westerly winds over the Southern Ocean have long been recognized to help drive the ACC. The winds drive surface waters to the left of the wind, causing upwelling to the south of the wind stress maximum and downwelling to the north. This pattern of upwelling and downwelling helps to establish the tilt of density surfaces across the ACC and therefore its geostrophic flow.

Despite the absence of continental barriers, the Sverdrup theory of wind-driven currents has been applied in the Southern Ocean by assuming that relatively shallow bathymetric features act as 'effective continents'. However, such calculations assume no interaction between the current and the bathymetry, which we know to be a poor assumption for the deep-reaching ACC. In addition, wind is not the only factor driving the ACC. The atmosphere also drives ocean currents by exchanging heat and fresh water with the ocean, causing the density of seawater to change. Exchange of fresh water can result from

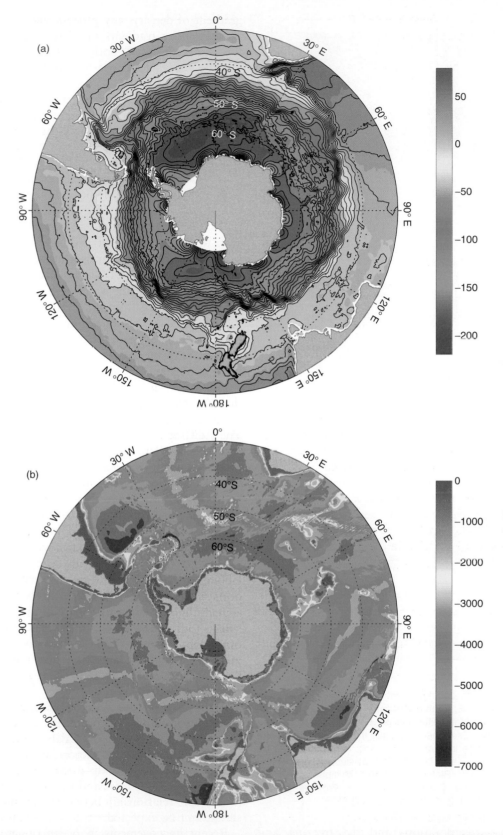

Figure 3 (a) Absolute sea level of the Southern Ocean derived from drifter data, satellite altimetry, and surface winds. Contours of sea level are approximate streamlines for the flow; where the lines are close together, the flow velocity is large. The small-scale features near the Antarctic coast are spurious and likely result from lack of sampling by floats and satellites in the sea ice zone. (b) Sea floor bathymetry (contour interval 500 m). (a) Courtesy Serguei Sokolov; data are from Niiler PP, Maximenko NA, and McWilliams JC (2003) Dynamically balanced absolute sea level of the global ocean derived from near-surface velocity observations. *Geophysical Research Letters* 30: 2164. (b) Courtesy Serguei Sokolov; bathymetry data are from ETOPO5, NOAA, National Geophysical Data Centre.

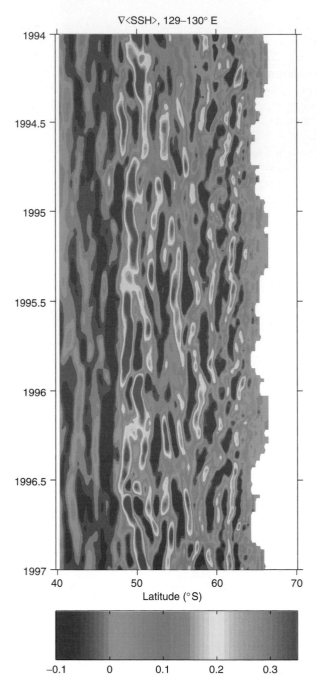

∇<SSH>, 129–130° E

Latitude (°S)

Figure 4 The north–south gradient of sea surface height (in m per 100 km) at 130° E, south of Australia. Large height gradients indicate strong currents. The multiple bands of large gradient correspond to the jets of the ACC. Note that the jets merge and diverge, and change in intensity with time, and also persist for many months at the same latitude. Reproduced with permission from Sokolov S and Rintoul SR (2007) Multiple jets of the Antarctic Circumpolar Current. *Journal of Physical Oceanography* 37: 1394–1412. © Copyright [2007] AMS.

gradient of density, any process that produces horizontal density gradients will drive ocean currents. In the case of the ACC, both the strong westerly winds and the air–sea exchange of buoyancy play a part in driving the current.

The momentum supplied to the Southern Ocean by the wind needs to be compensated in some way. The question of what balances the wind forcing has been a topic of debate for many decades. Recent studies have confirmed the early hypothesis by W. Munk and E. Palmén that interaction of the ACC with seafloor topography provides a force to balance the wind. This force, known as the bottom form stress, results when the ocean currents are organized such that there is higher pressure on one side of a ridge on the seafloor than is found on the other side. In the case of the ACC, higher pressure is generally found on the west side of topographic ridges or hills, providing a force from the solid Earth to the ocean that balances the wind stress at the sea surface. While these pressure differences are too small to observe directly, realistic numerical simulations clearly show this force balance in action.

Eddies produced by dynamical instabilities of the ACC fronts play a crucial role in establishing the momentum and heat balance of the Southern Ocean. The ACC has some of the most vigorous eddy activity observed in the ocean (**Figure 5**). Eddies are produced when dynamical processes release some of the energy stored in the sloping of density surfaces across the ACC, converting some of the energy in the mean flow into motions that vary with time, or eddies. This process is called baroclinic instability. The eddies transfer momentum vertically from the sea surface to the deep ocean, helping to set up the system of deep currents that interact with bathymetry to provide the bottom form stress. Both transient eddies (motions that vary with time) and standing eddies (deviations of the flow from the east–west average) contribute to the momentum transport. At the same time, the eddies carry heat poleward across the ACC, to compensate the heat lost to the cold atmosphere near Antarctica.

Attempts to relate ACC transport variability to variations in wind have been inconclusive. It is now understood that the dynamical balance of the ACC depends on a number of factors, including wind and buoyancy forcing, eddy–mean flow interaction, topographic form stress and the ocean stratification, so a simple relationship between transport and wind should not be expected.

Recent improvements in ocean observing systems have allowed changes in the ACC to be assessed for the first time. Comparison of temperature profiles collected since the 1950s suggests that much of the

precipitation, evaporation, and the freezing and melting of ice, both sea ice and glacial ice in the form of icebergs and ice shelves. Because the speed of ocean currents is proportional to the horizontal

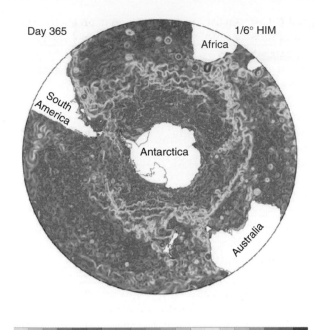

Day 365   1/6° HIM

-5 -3 -2 -1.8 -1.6 -1.4 -1.2 -1 -0.9 -0.8 -0.7 -0.6 -0.5 -0.4 -0.3 -0.2 -0.1 0 0.2 0.4 0.6
log$_{10}$ of magnitude of velocity averaged over top 100 m in m s^{-1}

Figure 5 A snapshot of surface speed from a high-resolution numerical simulation of the ACC. The filamented, eddy-rich structure of the ACC stands out clearly. Adapted with permission from Hallberg R and Gnanadesikan A (2006) The role of eddies in determining the structure and response of the Southern Ocean overturning: Results from the Modelling Eddies in the Southern Ocean project. *Journal of Physical Oceanography* 36: 2232–2252. © Copyright [2007] AMS.

Southern Ocean has warmed. The warming is largest in the ACC belt and is consistent with a southward shift of the current, allowing warm water to move south into areas previously occupied by cooler water. The southward movement of the ACC has been linked to a southward shift and strengthening of the westerly winds. The change in the winds, in turn, has been linked both to loss of ozone in the polar stratosphere and to enhanced greenhouse warming. The transport of the ACC has apparently not changed much over this time period, despite the change in wind forcing. Recent studies suggest the ACC may be in a regime in which the transport is insensitive to changes in wind: as the wind forcing increases, the ACC starts to speed up, but this causes more vigorous eddy fluxes that dissipate the extra energy in the mean flow.

The Antarctic Circumpolar Current and the Overturning Circulation

The eastward flow of the ACC is dynamically linked to a weaker circulation in the north–south plane. The distribution of water properties on transects across the Southern Ocean clearly reveals water masses spreading across the ACC. For example, the salinity maximum of the Lower CDW and oxygen minimum of the Upper CDW can be traced as they shoal from depths of 2000–3500 m north of the ACC to approach the sea surface south of the Polar Front (**Figure 2**). The high-oxygen, low-salinity waters formed in the Southern Ocean (Antarctic Intermediate Water and Antarctic Bottom Water) can be followed as they cross the ACC and enter the basins to the north.

These distributions reflect an ocean circulation pattern known as the overturning circulation (**Figure 6**). Deep water spreads to the south across the ACC and upwells at the sea surface. Some of the upwelled deep water is driven north beneath the westerly winds, gains heat and fresh water from the atmosphere, and therefore becomes less dense, and ultimately sinks to form Antarctic Intermediate Water and Subantarctic Mode Water. Deep water that upwells further south and closer to Antarctica is converted to denser Antarctic Bottom Water and returns to the north. The result of these water mass transformations is a circulation in the north–south plane that consists of two counter-rotating cells. According to the residual mean theory, the strength of the net overturning circulation (mean flow plus the eddy contribution) is determined by the surface buoyancy forcing.

The Southern Ocean overturning cells play an important part in the global-scale overturning circulation. The Southern Ocean imports deep water from the basins to the north, and exports bottom water and intermediate water. Recent studies suggest the conversion of deep water to intermediate water in the Southern Ocean is a key link in the global overturning circulation. For decades it has been assumed that the sinking of dense water in the polar regions was balanced by widespread upwelling at lower latitudes. However, measurements of mixing rates and large-scale tracer budgets suggest that mixing in the interior of the ocean is an order of magnitude too weak to support the upwelling required. The transformation of deep water to intermediate water by air–sea buoyancy exchange in the Southern Ocean provides an alternative means of connecting the upper and lower limbs of the global overturning circulation. Mixing likely also makes a contribution in regions of rough bathymetry, including the Southern Ocean, where elevated mixing rates have been measured.

Eddies spawned by the ACC make an important contribution to the overturning circulation. The eddies transfer mass across the Drake Passage gap,

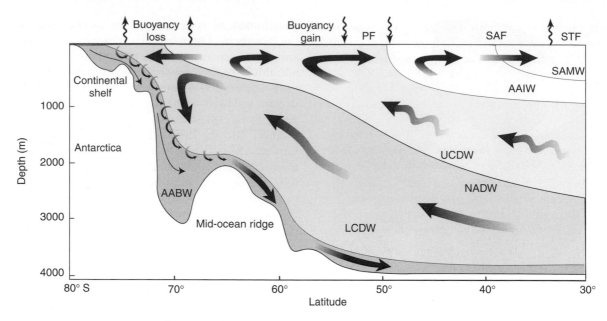

Figure 6 A schematic view of the Southern Ocean overturning circulation. Upper Circumpolar Deep Water (UCDW) upwells at high latitude and gains buoyancy (from heating, precipitation, and ice melt) as it is driven north, to ultimately sink as Antarctic Intermediate Water (AAIW) or Subantarctic Mode Water (SAMW). Lower Circumpolar Deep Water (LCDW) and North Atlantic Deep Water (NADW) upwell closer to the Antarctic continent, are made more dense (by cooling and salt rejected during sea ice formation), and sink to form Antarctic Bottom Water (AABW). The major southern ocean fronts are indicated. STF, Subtropical Front; SAF, Subantarctic Front; PF, Polar Front. Reproduced with permission from Speer K, Rintoul SR, and Sloyan B (2000) The diabatic Deacon cell. *Journal of Physical Oceanography* 30: 3212–3222. © Copyright [2007] AMS.

where the absence of land barriers means that there can be no net east–west pressure gradient and therefore no net north–south flow. Furthermore, the same forces that drive the overturning circulation (wind and buoyancy exchange) also drive the ACC. The west-to-east flow of the ACC and the overturning circulations cannot be understood in isolation. The two are intimately linked, and eddy fluxes, topographic interactions, and wind and buoyancy forcing are all important ingredients of the dynamical coupling between them.

The Southern Ocean overturning also has a large influence on global biogeochemical cycles. Upwelling of nutrient-rich deep water south of the ACC returns nutrients to the surface layer. The nutrients are exported from the Southern Ocean to lower latitudes by the overturning circulation, ultimately supporting a large fraction of global primary production. Water masses at the surface of the Southern Ocean exchange oxygen and carbon dioxide with the atmosphere and carry 'ventilated' water to the interior of the ocean when the water masses sink. Where deep water upwells south of the ACC, carbon dioxide is released to the atmosphere; where water sinks from the sea surface, carbon dioxide is carried from the atmosphere into the ocean. As a result of the overturning circulation, more of the carbon released by human activities is accumulating just north

of the ACC than in any other latitude band of the ocean.

Summary

The ACC is the largest current in the world ocean, carrying about $137 \pm 8 \times 10^6 \, \text{m}^3 \, \text{s}^{-1}$ from west to east around Antarctica. By connecting the ocean basins, the ACC allows water masses and climate anomalies to propagate between the basins. The current flow is concentrated in a number of circumpolar fronts, which extend from the sea surface to the seafloor. The fronts also mark the boundaries between zones with distinct physical, chemical, and ecological characteristics. Eddies produced by dynamical instabilities of the fronts play an important part in the dynamics of the ACC by transporting momentum vertically and heat and mass poleward. Both wind and buoyancy forcing contribute to driving the ACC. Interaction between the deep-reaching flow and the bottom topography establish bottom form stresses to balance the wind forcing. The strong eastward flow of the ACC is intimately connected to an overturning circulation made up of two counter-rotating cells. Water mass transformations driven by exchange of heat and moisture with the atmosphere connect the upper and lower limbs of the thermohaline circulation. The transport and storage of heat, fresh water, and carbon dioxide by the ACC have a significant influence on global and regional climate.

See also

Carbon Cycle. Heat Transport and Climate. Ocean Circulation. Ocean Circulation: Meridional Overturning Circulation.

Further Reading

Cunningham S, Alderson SG, King BA, and Brandon MA (2003) Transport and variability of the Antarctic Circumpolar Current in Drake Passage. *Journal of Geophysical Research* 108: 8084 (doi:10.1029/2001 JC001147).

Deacon G (1984) *The Antarctic Circumpolar Ocean.* London: Cambridge University Press.

Gille ST (2002) Warming of the Southern Ocean since the 1950s. *Science* 295: 1275–1277.

Gordon AL and Molinelli E (1986) *Southern Ocean Atlas.* Washington, DC and New Delhi: National Science Foundation and Amerind Publishing.

Hallberg R and Gnanadesikan A (2006) The role of eddies in determining the structure and response of the Southern Ocean overturning: Results from the Modelling Eddies in the Southern Ocean project. *Journal of Physical Oceanography* 36: 2232–2252.

Munk WH and Palmén E (1951) Note on the dynamics of the Antarctic Circumpolar Current. *Tellus* 3: 53–55.

Niiler PP, Maximenko NA, and McWilliams JC (2003) Dynamically balanced absolute sea level of the global ocean derived from near-surface velocity observations. *Geophysical Research Letters* 30: 2164.

Nowlin WD Jr. and Klinck JM (1986) The physics of the Antarctic Circumpolar Current. *Review of Geophysics and Space Physics* 24: 469–491.

Olbers D, Borowski D, Volker C, and Wolff J-O (2004) The dynamical balance, transport and circulation of the Antarctic Circumpolar Current. *Antarctic Science* 16: 439–470.

Orsi AH and Whitworth T, III (2005) In: Sparrow M, Chapman P, and Gould J (eds.) *Hydrographic Atlas of the World Ocean Circulation Experiment (WOCE), Vol. 1: Southern Ocean.* Southampton: International WOCE Project Office (ISBN 0-904175-49-9).

Orsi AH, Whitworth T, III, and Nowlin WD (1995) On the meridional extent and fronts of the Antarctic Circumpolar Current. *Deep-Sea Research I* 42: 641–673.

Rintoul SR, Hughes C, and Olbers D (2001) The Antarctic Circumpolar Current system. In: Siedler G, Church J, and Gould J (eds.) *Ocean Circulation and Climate,* pp. 271–302. London: Academic Press.

Sokolov S and Rintoul SR (2007) Multiple jets of the Antarctic Circumpolar Current. *Journal of Physical Oceanography* 37: 1394–1412.

Speer K, Rintoul SR, and Sloyan B (2000) The diabatic Deacon cell. *Journal of Physical Oceanography* 30: 3212–3222.

Whitworth T, III (1980) Zonation and geostrophic flow of the Antarctic Circumpolar Current at Drake Passage. *Deep-Sea Research* 27: 497–507.

Whitworth T, III and Nowlin WD (1987) Water masses and currents of the Southern Ocean at the Greenwich Meridian. *Journal of Geophysical Research* 92: 6462–6476.

Relevant Website

http://www.soc.soton.ac.uk
– Electronic Atlas of WOCE Data.

ARCTIC OCEAN CIRCULATION

B. Rudels, Finnish Institute of Marine Research, Helsinki, Finland

Introduction

The Arctic Ocean is the northernmost part of the Arctic Mediterranean Sea, which also comprises the Greenland Sea, the Iceland Sea, and the Norwegian Sea (the Nordic Seas) and is separated from the North Atlantic by the 500–850-m-deep Greenland–Scotland Ridge. The Arctic Ocean is a small, $9.4 \times 10^6 \text{km}^2$, enclosed ocean. Its boundaries to the south are the Eurasian continent, Bering Strait, North America, Greenland, Fram Strait, and Svalbard. The shelf break from Svalbard southward to Norway closes the boundary. The Arctic Ocean lies almost entirely north of and occupies most of the region north of the polar circle. More than half of its area, 53%, consists of large, shallow shelves, the broad Eurasian marginal seas: the Barents Sea (200–300 m), the Kara Sea (50–100 m), the Laptev Sea (<50 m), the East Siberian Sea (<50 m) and the Chukchi Sea (50–100 m), and the narrower shelves north of North America and Greenland. The deep Arctic Ocean comprises two major basins, the Eurasian Basin and the Canadian (also called Amerasian) Basin, separated by the approximately 1600-m-deep Lomonosov Ridge. The Eurasian Basin is further divided into the Nansen and Amundsen basins by a mid-ocean ridge (the Gakkel Ridge), while the Canadian Basin is separated by the Alpha Ridge and the Mendeleyev Ridge into the Makarov and the Canada Basins. The Amundsen Basin is the deepest (~4500 m), while the maximum depths of the Makarov and the Nansen Basins are ~4000 m. The Canada Basin is slightly shallower (~3800 m) but by far the largest (**Figure 1**).

The Arctic Ocean water masses are primarily of Atlantic origin. Atlantic waters (AWs) enter the Arctic Ocean from the Nordic Seas through the 2600-m-deep Fram Strait and over the ~200-m-deep sills in the Barents Sea. The Arctic Ocean also receives low-salinity Pacific water through the shallow (45 m) and narrow (50 km) Bering Strait. The outflows occur through Fram Strait and through the shallow (150–230 m) and narrow channels in the Canadian Arctic Archipelago.

The physical oceanography of the Arctic Ocean is shaped by the severe high-latitude climate and the large freshwater input from runoff, ~ 0.1 Sv ($1\,\text{Sv} = 1 \times 10^6 \text{m}^3\,\text{s}^{-1}$), and net precipitation, ~ 0.07 Sv. The Arctic Ocean is a strongly salinity stratified ocean that allows the surface water to cool to freezing temperature and ice to form in winter and to remain throughout the year in the central, deep part of the ocean.

The water masses in the Arctic Ocean can, because of the stratification, be identified by different vertical layers. Here five separate layers will be distinguished:

1. The ~ 50-m-thick upper, low-salinity polar mixed layer (PML) is homogenized in winter by freezing and haline convection, while in summer the upper 10–20 m become diluted by sea ice meltwater. The salinity, S, ranges from 30 to 32.5 in the Canadian Basin and between 32 and 34 in the Eurasian Basin.
2. The 100–250-m-thick halocline, with salinity increasing with depth, while the temperature remains close to freezing, $32.5 < S < 34.5$ (**Figure 2**).
3. The 400–700-m-thick Atlantic layer historically defined as subsurface water with potential temperature, θ, above $0\,^\circ\text{C}$, $34.5 < S < 35$.
4. The intermediate water below the Atlantic layer that communicates freely across the Lomonosov Ridge, $-0.5\,^\circ\text{C} < \theta < 0\,^\circ\text{C}$, $34.87 < S < 34.92$.
5. The deep and bottom waters in the different basins, $-0.55\,^\circ\text{C} < \theta < -0.5\,^\circ\text{C}$, $34.92 < S < 34.96$ (Canadian Basin), $-0.97\,^\circ\text{C} < \theta < -0.5\,^\circ\text{C}$, $34.92 < S < 34.945$ (Eurasian Basin).

There are large lateral variations of the characteristics in these layers that depend upon the circulation and upon the mixing processes in the Arctic Ocean (**Figures 2** and **3**). A more detailed classification, especially for the deeper water masses, is given in **Table 1**. It should be kept in mind that this classification is not unique and that several others exist in the literature.

Circulation

The circulation of the uppermost layers of the Arctic Ocean has mainly been inferred from the ice drift, determined from satellites and from drifting buoys, while at deeper levels primarily the distributions of temperature and salinity and more recently of other tracers have been used to deduce the movements of the different water masses, often with

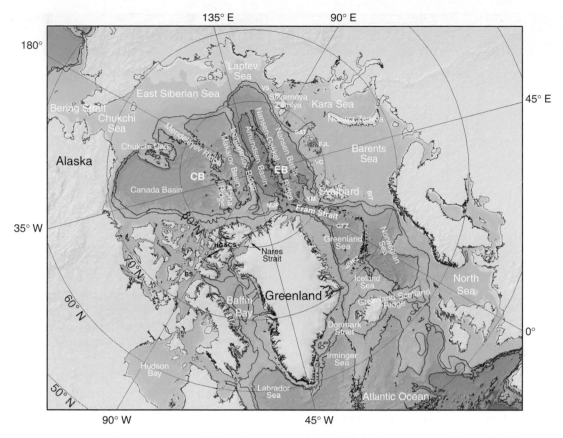

Figure 1 Map of the Arctic Mediterranean Sea showing geographical and bathymetric features. The bathymetry is from IBCAO (the International Bathymetric Chart of the Arctic Ocean; Jakobsson *et al.*, 2000; 2008) and the projection is Lambert equal area. The 500 and 2000 m isobaths are shown. All maps used here are made by Martin Jakobsson (personal communication). BIT, Bear Island Trough; CB, Canadian Basin; EB, Eurasian Basin; GFZ, Greenland Fracture Zone; MJP, Morris Jessup Plateau; JMFZ, Jan Mayen Fracture Zone; SAT, St. Anna Trough; YM, Yermak Plateau; VC, Victoria Channel; VS, Vilkiltskij Strait; FJL, Franz Josef Land; BS, Barrow Strait; HG & CS, Hell Gate and Cardigan Sound.

the assumption that the circulation is largely controlled by the bathymetry. Direct, moored current measurements have been scarce and mostly confined to the continental slope and to the Lomonosov Ridge. These measurements have confirmed the importance of the bathymetry for the circulation. In the deep basins current measurements have been made from drifting ice camps and more recently also from autonomous ice-tethered platforms, relaying the observations via satellite to shore. Subsurface drifters are just beginning to be used and observational efforts during the International Polar Year (IPY) 2007–09 are likely to significantly increase the knowledge of the circulation in the Arctic Ocean.

The motions of the ice cover and the surface water are predominantly forced by the wind, and the atmospheric high-pressure cell over the Arctic creates the anticyclonic Beaufort gyre in the Canada Basin. Ice leaks from the offshore side of the gyre and joins the Transpolar Drift (TPD) that brings ice from the Canada Basin toward Fram Strait. A second branch originating from the Siberian shelves, mainly from

the Laptev Sea, carries ice across the Eurasian Basin. About 90% of the ice export (0.09 Sv) passes through Fram Strait (**Figure 4**).

The PML and the halocline are maintained by river runoff, ice melt, and the inflow of low-salinity water through Bering Strait. The lowest surface salinities and the thickest halocline are therefore observed in the Canada Basin. The inflow through Bering Strait, although affected by local winds, is in the last instance driven by a higher sea level in the North Pacific as compared to the Arctic Ocean. This creates a pressure gradient that forces the Pacific water northward into the Arctic Ocean. The Bering Strait inflow continues across the Chukchi Sea in four branches. One branch enters the East Siberian Sea, while one of the central inflow branches enters the Arctic Ocean along the Herald Canyon west of the Chukchi Plateau, and the other passes via the Central Gap east of the Chukchi Plateau into the Canada Basin. The easternmost branch reaches the Canada Basin along the Barrow Canyon close to Alaska. River runoff, mainly from the Mackenzie

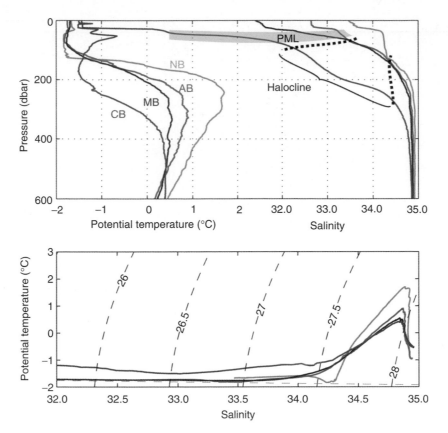

Figure 2 Potential temperature and salinity profiles and θS curves from the upper layers of the Nansen Basin (NB, dark yellow), Amundsen Basin (AB, green), Makarov Basin (MB, magenta), and Canada Basin (CB, blue). The PML and the halocline are indicated in the salinity profiles. Above the PML, the low-salinity layer due to seasonal ice melt is seen. The temperature maximum in the Canada Basin is due to the presence of Bering Strait Summer Water (BSSW) and the temperature minimum below indicates the upper halocline with S ~ 33.1, deriving from the colder, more saline Bering Strait Winter Water (BSWW) and from brine release and haline convection in the Chukchi Sea. No halocline is present in the Nansen Basin, only a deep winter mixed layer between the thermocline and the seasonal ice melt layer with temperature close to freezing. The curved shape of the Nansen Basin thermocline as seen in the θS diagram suggests wintertime haline convection with dense, saline parcels penetrating into the thermocline. Adapted from Rudels B, Jones EP, Schauer U, and Eriksson P (2004) Atlantic sources of the Arctic Ocean surface and halocline waters. *Polar Research* 23: 181–208.

River, adds freshwater to the Canada basin. The runoff peaks in early summer (June). The river runoff as well as most of the Pacific water becomes trapped in the anticyclonic Beaufort gyre, forming an oceanic high-pressure cell in the southern Canada Basin. The Pacific water leaves the Beaufort gyre and the Arctic Ocean mainly through the Canadian Arctic Archipelago, but a smaller fraction also exits, at least intermittently, through Fram Strait.

Warm AW crosses the Greenland–Scotland Ridge and continues toward the Arctic Ocean in the Norwegian Atlantic Current. The Norwegian Atlantic Current splits as it reaches the Bear Island Trough. One part enters the Barents Sea together with the Norwegian Coastal Current, which carries low-salinity water from the Baltic Sea and runoff from the Norwegian coast. The remaining part continues as the West Spitsbergen Current to Fram

Strait. There the current again splits. Some AW recirculates westward in the strait to join the southward-flowing East Greenland Current, and the rest enters the Arctic Ocean in two streams. One stream flows over the Svalbard shelf and slope, the other passes west and north around the Yermak Plateau and then continues eastward, eventually joining the inner stream at the continental slope east of Svalbard. The deeper outer stream also transports intermediate and deep waters from the Nordic Seas into the Arctic Ocean.

As the AW enters the Arctic Ocean, it encounters, and melts, sea ice, and the upper part becomes colder and less saline. In winter this upper layer is homogenized by convection and mechanical stirring and cooled to freezing temperature. The salinity of the upper layer is 34.2–34.4, and it is advected with the warm AW core eastward along the Eurasian

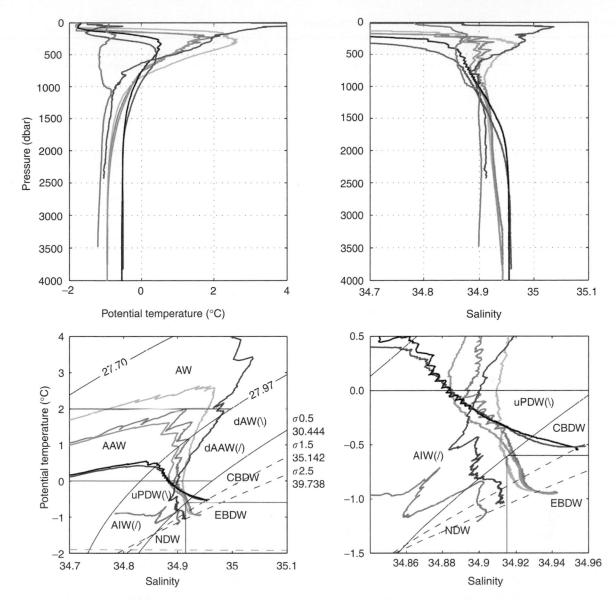

Figure 3 Characteristics of the water columns in different parts of the Arctic Mediterranean. Upper row: potential temperature and salinity profiles; lower row: θS curves. Green: The Greenland Sea, the ultimate source of the Arctic Intermediate Water (AIW) and the Nordic Seas Deep Water (NDW) entering the Arctic Ocean. Red: The West Spitsbergen Current in Fram Strait carrying warm Atlantic Water (AW), AIW, and NDW into the Arctic Ocean. Dark yellow: The Fram Strait branch at the continental slope of the Nansen Basin. Magenta: The interior Nansen Basin. Cyan: The Amundsen Basin. Black: The Makarov Basin. Blue: The Canada Basin. The shift in depth of the temperature maximum between the Makarov Basin and the Canada Basin is due to the stronger presence of Pacific water (PW) in the Canada Basin, displacing the deeper part of the water column. Note that the Canadian Basin deep waters becomes warmer than the Eurasian Basin deep waters below 1000 m and more saline below 1500 m, above the sill depth of the Lomonosov Ridge. In the θS diagrams, (/) indicates a stratification unstable in temperature or salinity, (\\) indicates a stratification stable in both components. AAW, Arctic Atlantic Water; dAW, dense Atlantic Water; dAAW, dense Arctic Atlantic Water; uPDW, upper Polar Deep Water; CBDW, Canadian Basin Deep Water; EBDW, Eurasian Basin Deep Water. The $\sigma_{1.5}$ isopycnal shows the density at the sill depth of the Lomonosov Ridge and the $\sigma_{2.5}$ isopycnal the density at sill depth in Fram Strait.

continental slope. In summer, seasonal ice melt creates a low-salinity upper layer, which is removed in winter by freezing and the upper layer is homogenized down to the thermocline. No cold halocline is present between the mixed layer and the AW in the Nansen Basin (see **Figure 2**). The Arctic Intermediate Water (AIW) and the Nordic Seas Deep Water (NDW) that enter the Arctic Ocean in the West Spitsbergen Current can be identified west and north of the Yermak Plateau as less saline and colder anomalies. Farther to the east these signals in temperature and salinity gradually disappear, but signs

Table 1 Simplified water mass classification for the Arctic Ocean

Water masses	Abbreviation	Definition	Main source or origin
Upper waters ($\sigma_\theta < 27.70$)			
Polar mixed layer	PML	$32 < S < 34$	Arctic Ocean
(Upper) halocline		$32.5 < S < 33.5$	Chukchi Sea
(Lower) halocline		$33.5 < S < 34.5$	Nansen Basin, Barents Sea
Intermediate waters I ($27.70 < \sigma_\theta < 27.97$)			
Atlantic Water	AW	$2 < \theta$	West Spitsbergen Current
Arctic Atlantic Water	AAW	$0 < \theta < 2$	Arctic Ocean (transformed)
Intermediate waters II ($27.97 < \sigma_\theta, \sigma_{0.5} < 30.444$)			
Dense Atlantic Water	DAW	$0 < \theta$, unstable in S (/)	West Spitsbergen Current
Dense Arctic Atlantic Water	DAAW	$0 < \theta$, stable in θ and S (\\)	Arctic Ocean (transformed)
Arctic Intermediate Water	AIW	$\theta < 0$, unstable in S or θ (/)	Greenland Sea
Upper Polar Deep Water	UPDW	$\theta < 0$, stable in S and θ (\\)	Arctic Ocean
Deep waters ($30.444 < \sigma_{0.5}$)			
Nordic Seas Deep Water	NDW	$S < 34.915$	Greenland Sea
Canadian Basin Deep Water	CBDW	$-0.6 < \theta$, $34.915 < S$	Canadian Basin
Eurasian Basin Deep Water	EBDW	$\theta < -0.6$, $34.915 < S$	Eurasian Basin

of the Nordic Seas waters can still be detected by other tracers, for example, CFCs.

In the Barents Sea, the AW remains in contact with the atmosphere during most of its transit. It is cooled significantly and becomes freshened by net precipitation and by the melting of sea ice. Some of the water entering the Barents Sea returns as colder, denser water to the Norwegian Sea in the Bear Island Channel, but the major part reaches the eastern Barents Sea. In the Barents Sea, the AW becomes separated into three different water masses. (1) The bulk of the inflow is cooled and freshened by air–sea interaction and becomes denser. (2) Some of the AW reaches the shallow areas west of Novaya Zemlya and becomes transformed into saline, dense bottom water by the ice formation and brine rejection. (3) The upper part of the AW interacts with sea ice and a less saline, upper layer is formed by ice melt, which in winter becomes homogenized down to the thermocline. These waters all enter the Arctic Ocean, mainly by passing between Novaya Zemlya and Franz Josef Land into the Kara Sea and then continuing in the St. Anna Trough to the Arctic Ocean. However, some water reaches the Arctic Ocean directly from the Barents Sea along the Victoria Channel.

The Barents Sea branch follows the eastern side of the St. Anna Trough and then continues along the continental slope as an almost 1000-m-thick, cold, and weakly stratified water column, displacing the warmer, more saline Fram Strait inflow branch from the slope and deflecting the denser basin waters toward larger depths. Strong isopycnal mixing between the two branches takes place and inversions and irregular intrusive layers are formed (**Figure 5**, right column).

North of the Laptev Sea the two inflow branches have largely merged and bottom temperatures above 0 °C indicate that Fram Strait branch water again is present at the slope. The boundary current splits at the Lomonosov Ridge with one part continuing into the Canadian Basin, the rest returning along the ridge and the Amundsen Basin toward Fram Strait. The AW temperature in the boundary current is significantly reduced already north of the Laptev Sea. This could partly be due to mixing between the two branches and partly be the result of heat loss to the upper layers and the ice. However, it may also be an indication that some of the Fram Strait branch recirculates already in the Nansen Basin. The returning Fram Strait branch then becomes more prominent in the northern Nansen Basin, while the Barents Sea branch dominates in the Amundsen Basin and along the Lomonosov Ridge (**Figure 5**, left column).

The bulk of the Barents Sea branch in the St. Anna Trough is denser, colder, and less saline than the Atlantic layer and forms a distinct salinity minimum beneath the AW in the Amundsen Basin and above the Gakkel Ridge. This minimum identifies the upper Polar Deep Water (uPDW) in the Eurasian Basin and is located higher in the water column and is more distinct than the AIW minimum in the Nansen Basin deriving from the Nordic Seas. This is about the only area of the Arctic Ocean where a part of the water column is unstably stratified in salinity (**Figures 3** and **5**).

One part of the Barents Sea inflow, mostly comprising water from the Norwegian Coastal Current,

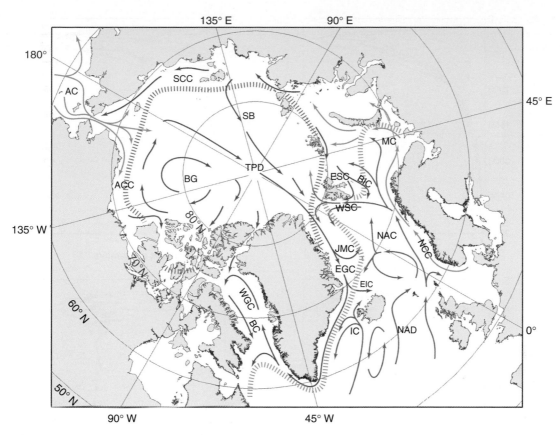

Figure 4 The circulation of the upper layers of the Arctic Mediterranean Sea. Warm Atlantic currents are indicated by red arrows, cold, less-saline polar and arctic currents by blue arrows. Low-salinity transformed currents are shown by green arrows. The maximum ice extent is shown by a blue and the minimum ice extent by red striped line. AC, Anadyr Current; ACC, Alaskan Coastal Current; BC, Baffin Current; BIC, Bear Island Current; BG, Beaufort gyre; EGS, East Greenland Current; EIC, East Iceland Current; ESC, East Spitsbergen Current; IC, Irminger Current; JMC, Jan Mayen Current; MC, Murman Current; NAD, North Atlantic Drift; NAC, Norwegian Atlantic Current; NCC, Norwegian Coastal Current; SB, Siberian branch (of the Transpolar Drift); SCC, Siberian Coastal Current; TPD, Transpolar Drift; WGC, West Greenland Current; WSC, West Spitsbergen Current.

enters the Kara Sea through the Kara Strait south of Novaya Zemlya. It mixes with the runoff from Ob and Yenisey, forming the low-salinity water present on the Kara Sea shelf. Most of this shelf water flows through the Vilkiltskij Strait into the Laptev Sea, where it receives additional freshwater, mostly from the Lena River. The main export of shelf water from the Eurasian shelves to the deep basins occurs across the Laptev Sea shelf break into the Amundsen Basin, and the low-salinity shelf water overruns the boundary current and the two inflow branches. The upper parts of the Fram Strait and the Barents Sea branches become isolated from the ice, the sea surface, and the atmosphere and evolve into halocline waters.

The Fram Strait branch supplies the halocline in the Amundsen Basin and the Makarov Basin and the lower halocline, beneath the Pacific water, in the northern part of the Canada Basin. The Barents Sea branch halocline, which remains at the continental slope, is affected by the stronger mixing occurring over the steep topography and becomes warmer by mixing with AW from below. It eventually supplies the lower halocline in the southern Canada Basin. The shelf water outflow supplies the PML in the Amundsen and Makarov Basins but it has high enough salinity also to contribute to the halocline waters in the Canada Basin, beneath most of the Pacific waters (**Figure 6**).

The properties of the Atlantic and intermediate waters in the Canada Basin are distinctly different from those in the Eurasian Basin, indicating that the waters of the boundary current become transformed along their paths in the Canadian Basin. The temperature maximum in the Atlantic layer becomes colder and less saline, while in the intermediate water range the temperature and the salinity increase and the intermediate uPDW salinity minimum disappears. The θS curves below the temperature maximum approach a straight line toward higher salinities and lower temperatures with increasing depth (**Figure 3**).

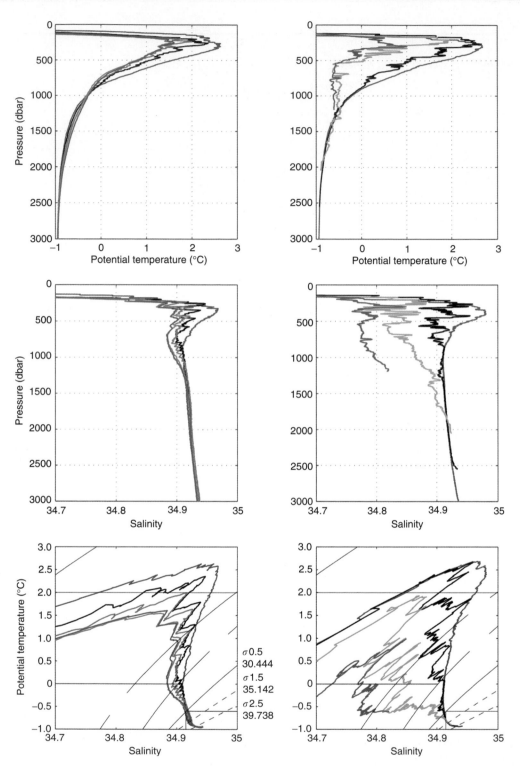

Figure 5 Potential temperature and salinity profiles and θS curves showing the interaction and interleaving between the Fram Strait branch and the Barents Sea branch north of Severnaya Zemlya (right column) and the water properties of the Nansen and Amundsen basins offshore of the Fram Strait branch (left column). Red stations: Fram Strait branch, blue station: the Barents Sea branch, black and cyan stations: active mixing between the branches. Interleaving is present not only in the AW but also in the deeper layers. Offshore of the Fram Strait branch, the warm, saline intrusions in the Nansen Basin (black and magenta stations) suggest a close recirculation of the Fram Strait branch in the Nansen Basin, while the colder, less saline intrusions in the Amundsen Basin (green and blue stations) indicate that the intermediate part of the water column here is dominated by Barents Sea branch waters.

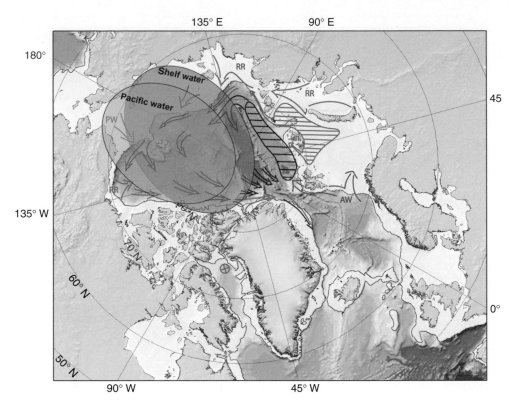

Figure 6 Circulation of the Atlantic-derived halocline waters and the distribution of Pacific water (orange) and Eurasian river runoff (green). The proposed source areas for the Fram Strait branch lower halocline water (black) and the Barents Sea branch lower halocline water (blue), and the circulation of these waters in the Arctic Ocean are indicated. RR, river runoff; PW, Pacific water; AW, Atlantic water. The cross indicates possible contribution of Barents Sea branch lower halocline water to the Baffin Bay bottom water. The isoline shown is 500 m. Based on Rudels B, Jones EP, Schauer U, and Eriksson P (2004) Atlantic sources of the Arctic Ocean surface and halocline waters. *Polar Research* 23: 181–208.

The boundary current circulates cyclonically around the Canada Basin and splits at the different ridges (**Figure 7**). One loop enters the Makarov Basin along the Mendeleyev Ridge, and one loop penetrates into the northern Canada Basin at the Chukchi Plateau. The boundary current also enters the Canada Basin at the Alpha Ridge and the Makarov Basin at the North American side of the Lomonosov Ridge. These circulation loops and their interactions with the boundary current induce a time lag that makes the Atlantic and intermediate waters of the different basins distinct, even if the waters originate from the same inflow across the Lomonosov Ridge.

Processes

The climatic forcing on the Arctic Ocean is strong; the large variation of incoming solar radiation, the severe cooling during the polar night, and the intense weather systems, either locally formed polar lows or cyclones advected from lower latitudes, all contribute in forming the Arctic Ocean characteristics. The strong stability of the deep Arctic Ocean basins confines the effects of these forcing fields to the ice-ocean surface and to the PML, but the shallow Arctic shelves, which experience the largest impact of the forcing, the strong runoff in summer and excessive ice formation in winter, do not only add low-salinity upper water and ice to the central basins but also create waters dense enough to break through the stratification, transforming the deeper advected layers.

Ice formation releases brine, and in lee polynyas, areas of open water where the ice is removed by offshore winds, sufficient ice may form to allow the released brine to overcome the initial low salinity on the shelves and create saline dense bottom waters at freezing temperature. These bottom waters eventually cross the shelf break and descend into the deep basins until their densities match those of the surrounding. They then merge with the ambient water. Less-dense plumes add colder water to the upper layers, supplying water to the halocline.

This occurs in the Chukchi Sea, where the water of the Canada Basin upper halocline with salinity 33.1 is formed. In the Nansen Basin no halocline is

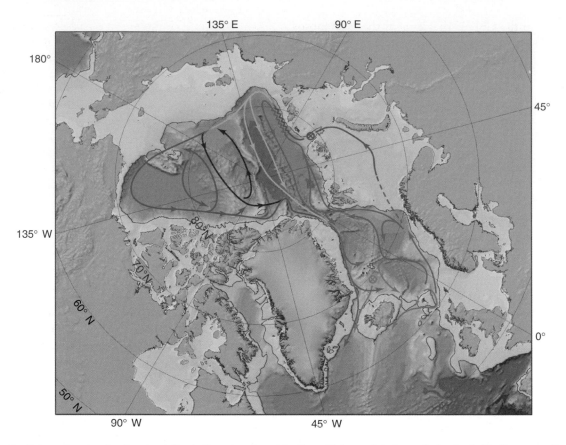

Figure 7 Schematics showing the circulation in the subsurface Atlantic and intermediate layers in the Arctic Mediterranean Sea. The interactions between the Barents Sea and Fram Strait inflow branches north of the Kara Sea as well as the recirculation and different inflow streams in Fram Strait and the overflows across the Greenland–Scotland Ridge are indicated. The isoline shown is 500 m. Based on Rudels B, Jones EP, Anderson LG, and Kattner G (1994) On the intermediate depth waters of the Arctic Ocean. In: Johannesen OM, Muench RD, and Overland JE (eds.) *AGU Geophysical Monographs 85: The Polar Oceans and Their Role in Shaping the Global Environment*, pp. 33–46. Washington, DC: American Geophysical Union.

present and the density range between the upper mixed layer and the AW is small. The shelf outflows then either enter the mixed layer or sink into and cool the thermocline and the Atlantic layer. The formation of halocline waters in the Eurasian Basin occurs, when the boundary current is overrun by the less saline shelf water, and the initial mixed layers become transformed into halocline waters.

Comparison between the intermediate and deep waters in the Nordic Seas and in the Arctic Ocean shows that the shelf-slope convection not only brings salt but also heat into the deeper layers (**Figures 3** and **5**). This implies that the denser, more saline plumes are entraining warmer water, as they pass through the intermediate Atlantic layer. Observations of outflowing, cold and dense water from Storfjorden in southern Svalbard have shown that as the outflow reaches deeper than 1500 m, it has become warmer than the ambient water, supporting the idea of entraining boundary plumes. No plumes have yet been followed from their sources down the slope in the Arctic Ocean, but saline, warmer, and denser

bottom layers have been observed deeper than 2000 m at the continental slope north of Severnaya Zemlya.

The deep waters of the Canadian and Eurasian Basins differ significantly, the Canadian Basin Deep Water (CBDW) being warmer and more saline than the Eurasian Basin Deep Water (EBDW). The difference in temperature was first believed due to the presence of the Lomonosov Ridge, blocking the cold deep water from the Nordic Seas, but the temperature difference starts well above the sill depth of the Lomonosov Ridge, showing that heat as well as salt is added to the uPDW and the CBDW in the Canadian Basin.

The differences between the Amundsen and Nansen Basins are small but clear. The Nansen Basin is slightly less saline than the Amundsen Basin between 1600 and 2600 m due to a stronger presence of NDW and AIW. In the Amundsen Basin a mid-depth (1700 m) salinity maximum or at least a sharp bend in the θS curves is observed. It is strongest closer to Greenland but is present over most of the basin. This maximum derives from the CBDW that crosses the Lomonosov Ridge and then penetrates into the central Amundsen Basin from the Greenland continental slope.

In the Nansen, Amundsen, and Canada Basins, the salinity of the deep water increases toward the bottom, while the temperature goes through a minimum and then increases slightly before an isothermal and isohaline bottom layer is reached (**Figure 3**). Shelf-slope convection cannot explain the temperature minimum in the deep waters, but it could be due to advection between the basins. The deep water in the Makarov Basin has at the sill depth (2400 m) of the Alpha Ridge similar θS characteristics as the temperature minimum in the Canada Basin and could supply the minimum. In the Eurasian Basin this explanation does not work. A minimum is present in both the Amundsen and Nansen Basins and these minima lie too deep to derive from the inflow of colder NDW. One possibility is that the St. Anna Trough intermittently conduits colder, denser bottom water, formed in the Barents Sea, into the Nansen Basin. This cold water would enter below the warm layers of the Fram Strait branch and by entraining less heat it could contribute colder water to the deeper part of the water column than the slope convection occurring farther east around Severnaya Zemlya.

Geothermal heat flux has been proposed as an energy source for heating and stirring the bottom layers, keeping them homogenous and gradually increasing their temperature, thereby creating the overlying temperature minimum. The high bottom salinity could then be a remnant from convection events that took place several hundred years ago and brought dense, saline, and cold water to the bottom, where it now is isolated and gradually becomes warmer, thicker, and less dense. Whether the bottom layers are kept homogenous by geothermal heating and convection from below, or if the stirring is due to mechanical mixing as the boundary plumes enter the bottom layer is an open question. However, geothermal heating and ventilation by shelf-slope convection are not two mutually exclusive processes and could operate simultaneously.

The Makarov Basin is different. The salinity becomes constant with depth, while the temperature still decreases, creating a temperature-stratified layer above the isothermal and isohaline bottom water (**Figure 3**). This rules out the gradual heating of an old, fossil bottom layer but requires that colder, rather than more saline, water renews the bottom layer. This then excludes shelf-slope convection as a dominant process. One possibility is that colder water from the Amundsen Basin is brought across the Lomonosov Ridge, either intermittently, forced by, for example, topographically trapped waves, or

flowing through yet undetected passages in the ridge. The central part of the Lomonosov Ridge, where such an exchange was believed to occur, has recently been surveyed. A gap was found but not as deep as expected, barely 1900 m, and more critically, the water mass properties observed during the survey indicated that the flow of the densest water was from the Makarov Basin to the Amundsen Basin – in the wrong direction. This inflow of Makarov Basin water followed the Lomonosov Ridge toward Greenland and is likely to be one source, perhaps the most important one, for the mid-depth salinity maximum in the Amundsen Basin.

A flow onto the shelves is needed to compensate for the supply to the PML in summer and the outflow of dense water to the deeper layers and the export of ice in winter. This flow could come from the Arctic Ocean, along the bottom in summer and at the surface in winter, or along horizontally separated inflow and outflow paths across the shelf break. It could also be supplied from behind, like the Bering Strait inflow and the inflow to the Barents Sea. These inflow shelves differ from the inner shelves of the Kara, Laptev, and East Siberian Seas, where a two-way exchange across the shelf break might be more likely, even necessary. However, the difference is not as large as it might appear. Most of the Barents Sea inflow actually passes the Kara Sea before it enters the deep basins and much of the runoff from Ob and Yenisey, together with a substantial fraction of the Barents Sea inflow, continues into the Laptev Sea and then enters either the Amundsen Basin or the East Siberian Sea.

Such eastward flow is the usual fate of river plumes, which tend to follow the coastline eastward. This also holds for the East Siberian Coastal Current, which carries the runoff from perhaps the most 'inner' shelf of the Arctic Ocean, that of the East Siberian Sea, into the Chukchi Sea. The East Siberian Coastal Current occasionally, but rarely, passes south through Bering Strait into the Bering Sea.

The circulation pattern in the Arctic Ocean outlined above gives the impression that water masses are advected, with little mixing, in loops through the different basin, replacing one vintage water mass with another. This is not entirely the case. The structure of the salinity and temperature profiles shows sign of intrusive mixing and the presence of lenses or eddies of anomalous water masses in the water column. Eddies were first reported in the Canadian Basin halocline and were there considered related to the outflow of dense water from the Chukchi Sea. These eddies are mostly anticyclonic, 10–20 km in diameter and highly energetic with maximum velocities above $0.3 \, \mathrm{m \, s^{-1}}$. Eddies have subsequently been observed in all water masses of the Arctic Ocean, although the velocities associated with the deeper-lying eddies have so far not been determined. Their water characteristics imply that the eddies have traveled considerable distances as coherent water bodies without their anomalous properties being removed by smaller-scale mixing, for example double-diffusive convection, merging the eddies with the Arctic Ocean water columns in different basins (**Figure 8**).

The intrusive layers are particularly intriguing (**Figures 5** and **8**). The largest property amplitudes are found at the frontal zones but the layers appear to reach over entire basins, making their extent perhaps the largest one observed in the World Ocean. The inversions allow for the release of the potential energy stored in the unstably stratified component by double-diffusive convection, which can drive the interleaving layers across the basins, and this has been suggested as a mechanism for transfer of heat from the boundary current into the deep basins. For the layers to expand, along-layer gradients in heat and salt are required. In the interior of the basins such gradients are often absent, and an alternative explanation for the wide extent of the intrusive layers is that they are formed in the frontal zones, expand initially, driven by double-diffusive fluxes, until the potential energy, available in the stratification, is removed. After this the layers remain as fossil structures that are advected around the main circulation gyres. The intrusions between the Fram Strait and the Barents Sea branch water observed in the Amundsen Basin and over the Gakkel Ridge have been interpreted as fossil structures, initially formed by the interactions between the Fram Strait and the Barents Sea branches north of the Kara Sea and now being advected toward Fram Strait (**Figure 5**).

Interleaving layers are observed in all background stratifications: saltfinger unstable, diffusively unstable, as well as when both components are stably stratified. In the stable–stable situation, disturbances extending across the front are necessary to create the initial inversions. Differential diffusion, taking into account the more rapid diffusion of heat in weakly turbulent surroundings, has been proposed as a mechanism for creating interleaving in a stable–stable stratification. The time required to create layers would then be in the order of years. This appears long, since interleaving layers are found very close to the area where the parent water masses first meet also when both components are stably stratified.

Variability

Until 1990 an underlying assumption has been that the Arctic Ocean, at least in its deeper parts, is reasonably quiet and unchanging and that observations made during a longer period, 10–20 years, could be merged and used to describe the basic hydrographic conditions. The observations in the 1990s proved the Arctic Ocean to be as variable and changing as any other ocean.

An inflow of anomalously warm AW was reported in 1990 and has been observed propagating in the boundary current and into the different basins, changing the characteristics of the Atlantic layer in the Arctic Ocean. This warm inflow event persisted for almost a decade. Colder water then entered the Arctic Ocean for a short period after which the inflowing AW again became warmer. These inflow events can be traced upstream and originate from the input of warmer AW across the Greenland–Scotland Ridge. The pulses have also been followed in the Arctic Ocean, tracing several of the suggested circulation loops in the basins, giving timescales for the movements along the different loops. Model work has reproduced many of these events and their pathways around the Arctic Ocean.

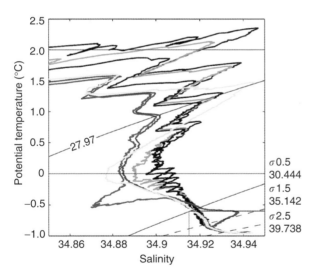

Figure 8 θS diagram showing eddies present in the intermediate and deep layers on a section taken in 1996 across the Eurasian Basin. The red station shows the warm and saline Canadian Basin Deep Water, the blue station indicates an isolated lens of cold, low-salinity Barents Sea branch water. The cyan and black stations show an eddy of Barents Sea branch water (cyan station) with warmer and more saline water in both the slope and the basin directions (black stations) and surrounded by interleaving structures like a 'meddy'. This is in contrast to the slope to basin decrease in salinity and temperature seen in the interleaving in the Atlantic layer (although here also an eddy (not shown) was detected). Finally an eddy of warm, saline Fram Strait branch water, yellow station, was present in the colder, Barents Sea branch dominated Atlantic layer in the Amundsen Basin.

The first pulse has now left the Eurasian Basin and partly exited the Arctic Ocean through Fram Strait. In the Amundsen Basin it has been replaced by colder water, while the part that entered the Makarov Basin at the Mendeleev Ridge has circulated around the Makarov Basin and is now found at the Makarov Basin side of the Lomonosov Ridge, practically removing the temperature front previously present along the ridge. The pulse has also entered the northern Canada Basin at the Chukchi Plateau. Its movements around and south of the Chukchi Plateau have taken comparably long, and it has been proposed that the AW here enters not mainly in the boundary current but directly into the basin as intrusive, double-diffusively driven layers. Older water, previously found in the southern Canada Basin, has shifted northward along the slope and is seen penetrating along the Alpha Ridge into the northern Canada Basin.

The 1990 inflow event coincided with a strong, positive state of the North Atlantic Oscillation (NAO) and of the Arctic Oscillation (AO), which also affected the distribution of the runoff from the Siberian rivers. Instead of entering the Amundsen Basin from the Laptev Sea it continued eastward to the Makarov Basin and the northern Canada Basin. The upper Pacific water lens, which in the 1970s extended over the entire Canadian Basin to the Lomonosov Ridge, then contracted to the Canada Basin. Similar scenarios have been reproduced in model studies.

The shifting of the Pacific/Atlantic surface front as well as the river water (shelf water) front counterclockwise toward the Canada Basin elevated the effects of the warmer AW. In the Amundsen Basin, and partly in the Makarov Basin, it approached closer to the sea surface, into the levels previously occupied by the halocline, thus magnifying the increase of both temperature and salinity at these levels. The shifting of the river water front also caused an increase in salinity of the surface layers of the Amundsen and Makarov Basins. The area with a deep winter mixed layer, previously confined to the Nansen Basin, expanded into the Amundsen basin, almost reaching the Lomonosov Ridge. The winter convection reached down to 120–130 m and actually caused a temperature decrease immediately above the Atlantic layer. The river water front has during the 2000s moved back into the Amundsen Basin, almost as far as the Gakkel Ridge, indicating that the shelf water outflow from the Laptev Sea again primarily enters the Amundsen Basin, recreating the PML–halocline structure in that basin.

The inflow in the Barents Sea branch also appears to vary. The intermediate salinity minimum in the Eurasian Basin has become more pronounced and less saline uPDW has crossed the Lomonosov Ridge in the boundary current along the continental slope and entered the Canadian Basin. This has made the uPDW characteristics in the Makarov Basin more similar to those in the Eurasian Basin, with a curved rather than a straight θS curve below the temperature maximum. A question is, if these anomalies in the Canadian Basin will stay long enough to be removed by shelf-slope convection and interior mixing processes, recreating the older, smooth, θS characteristics, or if this inflow will create entirely new θS structures? If so, does this mean that the gyre circulation has changed and the communication between the Eurasian and Canadian Basins has increased, leading to a more rapid ventilation of the Canadian Basin? Historical data from the Russian archives indicate that sudden changes have occurred before, and the situation now observed may not be unique. The use of CTDs instead of Nansen bottles also reveals structures in the water column previously not resolved.

The circulation pattern in the Arctic Ocean responds to long periodic (decadal) variations in the atmospheric circulation, the NAO and the AO. The positive AO state increases the inflow of AW and weakens the Beaufort gyre, while the negative state leads to a well-developed Beaufort gyre and a smaller inflow of AW. The negative state then retains the fresh water and the ice, while the positive state acts to reduce the storage of ice and low-salinity upper waters, which, together with a larger inflow of AW, increases the mean salinity of the Arctic Ocean waters.

Perhaps the most prominent change observed in the Arctic Ocean has been the retreat of the ice cover. A reduction of the minimum ice extent of $1.5-2 \times 10^6 \, km^2$ ($>20\%$) between 1979 and 2005 has been observed. Comparison between submarine observations of ice thicknesses 20 years apart indicates a thinning of the ice, from 3.1 to 1.8 m. However, the magnitude of this change has been contested. Thickness observations at the same position years apart might not be relevant, since the distribution of the ice will depend upon the forcing of the ice field, which will vary between the different years. Nevertheless, even disregarding changes in ice thickness, the ice cover has become significantly reduced.

A thinner and less extensive ice cover indicates a loss of ice storage, which could either be due to reduced formation of ice, or to an increased export of ice out of the Arctic Ocean. The amount of freshwater released by the reduction of the ice cover could have contributed to perhaps the largest fraction of the freshening of the Nordic Seas and to the subpolar gyre reported in recent years, but this does not give any information about where the phase change occurred, in the Arctic Ocean or south of Fram Strait. The time series of the ice export are not long enough to provide an answer and the knowledge of the liquid freshwater export is uncertain, if not completely absent.

The presence of the halocline between the PML and the Atlantic layer makes the stirring in the PML entrain cold water from below, and the ice cover is isolated from the heat stored in the Atlantic layer. The situation in the mid-1990s with no river (shelf) water present in the Amundsen Basin could lead to the entrainment of warm AW into the uppermost layer and thus bring heat to the sea/ice surface. This sensible heat rather than latent heat of ice formation could then be supplied to the atmosphere and less ice would form in winter. The heat might also directly melt the ice, but if the stirring of the mixed layer, driving the entrainment of heat from below, is due to haline convection, a change from freezing to melting would stop the convection, and the entrainment as well as the heat transport from below would cease.

Transports

The last 10 years have seen large programs focusing on measuring the transports through the key passages of the Arctic Ocean. Much of this activity has been coordinated by the ASOF (Arctic and Subarctic Ocean Fluxes) program but other projects have also been involved.

The observations of the Bering Strait inflow have largely confirmed the transport estimates proposed by Russian researchers 50 years ago. A mean inflow of 0.8 Sv of low-salinity ($S \leq 32$) water takes place. The seasonal variations are large; the inflow is 1.2 Sv in summer and 0.4 Sv in winter. The recent observations have shown that the freshwater transport through Bering Strait might be substantially (20%) higher than previously estimated, making it two-thirds of the river runoff. This reevaluation is due to the inclusion of the transport in the low-salinity Alaskan Coastal Current.

The transports through the Canadian Arctic Archipelago, notoriously difficult to measure due to the remoteness, the severe climate, and the proximity to the magnetic north pole that makes direction determinations extremely difficult, have in recent years been measured in the Hell Gate and Cardigan Strait (the Jones Sound) and in Barrow Strait (Lancaster Sound). This gives observed transports through two of the three main passages through the Archipelago. Transport measurements have also been made in Nares Strait, but year-long transport estimates from Nares Strait are not yet available.

The fluxes through the narrow Hell Gate are directed out of the Arctic Ocean, barotropic, and almost constant, while in the neighboring Cardigan Strait the flow is weaker and reversals are observed. The combined average transport is estimated to 0.3 Sv. The transports in the wider Barrow Strait are largely barotropic on the southern side of the channel and directed eastward, toward Lancaster Sound and Baffin Bay. A weak, baroclinic westward flow is observed on the northern side, indicating transport of runoff and penetration of water from Baffin Bay and Lancaster Sound. The flow is highly variable with an estimated mean transport of 0.75 Sv from the Arctic Ocean to Baffin Bay.

Models indicate that the Barrow Strait (Lancaster Sound) might contribute one-third to one-half of the total outflow through the Canadian Arctic Archipelago. Should this be correct the total transport would be 1.5–2 Sv. It is low-salinity, primarily Pacific, water that passes through the archipelago, but the bottom water in Baffin Bay, $-0.5\,°C$ and 34.5, likely derives from the Arctic Ocean through Nares Strait and would then be supplied by lower halocline waters.

The transport of the AW to the Barents Sea between Bjørnøya and Norway has been measured continuously for 10 years. The mean net transport is into the Barents Sea, but there is a return flow of transformed, colder, and denser water. The mean net transport of AW to the Barents Sea has been estimated to 1.5 Sv. However, short periodic variations are large and in spring, due to changing wind conditions, a whole month of small net inflow, occasionally even outflow, has been observed. There are also indications of variability on longer timescales, 3–4 years, but no trend has been detected. The AW is warm and saline at the entrance, but it loses much of its heat during transit and does not contribute heat to the central Arctic Ocean. To the inflow of AW should be added a transport of 0.7 Sv of less saline (34.4) water from the Norwegian Coastal Current, increasing the net inflow to ~ 2.2 Sv.

Fram Strait, which also has been monitored regularly since 1997, has a two-directional flow, and not only polar surface water (PSW), comprising the PML and halocline waters, and AW but also intermediate and deep water masses pass through the strait. The flow is largely barotropic and highly variable in space and time. Eddies are present, both barotropic and baroclinic, which complicates the transport estimates. Most of the steady flow occurs in the northward-flowing West Spitsbergen Current to the east and the southward-flowing East Greenland Current to the west. The inflow comprises warm AW and colder, less-saline AIW and NDW, while the outflow carries sea ice, low-salinity PSW, cool Arctic Atlantic Water (AAW), uPDW with temperature between 0 and $-0.5\,°C$ and CBDW, seen as a salinity maximum at 1700 m, and the colder, but also saline EBDW close to the bottom.

The observed total northward and southward transports are large, 10–15 Sv. The mean net transport is much smaller and lies between 1.5 and 2.5 Sv southward, out of the Arctic Ocean. A large recirculation is present in the strait. This appears partly associated with the barotropic eddies that drift westward along the sill of the strait and carry water from the West Spitsbergen Current to the East Greenland Current.

The transports through Fram Strait of largest climate importance are the export of ice and the heat carried by the AW. There is a large interannual variability but no trend has been found. The estimated mean ice export is ~ 0.9 Sv, while the outflow of PSW is ~ 1 Sv. The inflow of AW and heat also varies strongly from year to year but larger transports and higher temperatures have been observed in the last years, giving a flux ~ 50 TW. The average heat transport since 1997 is 35–40 TW. This is estimated relative to an assumed mean temperature of $-0.1\,^{\circ}$C in the Arctic Ocean. However, since the mass budget is not closed, the heat transport will depend upon the choice of reference temperature. The long-term mean inflow of AW, $>2\,^{\circ}$C, is $c.\ 3$ Sv.

Whether an increased heat transport through Fram Strait, connected with warmer AW and perhaps a stronger flow, also leads to more heat being available for the Arctic, for ice melt and for the atmosphere, is not yet clear. The fact that pulses of warm AW can be traced around the Arctic Ocean could imply that most of the heat is not lost but only stored and will eventually leave the Arctic Ocean through Fram Strait. The heat advected into the Arctic Ocean through Bering Strait is located closer to the sea surface and could have a larger impact on the heat flux to the ice cover from below and on the thickness of the ice cover. The large retreat in ice extent reported in the last years has mostly occurred in the southern Canada Basin, close to the Chukchi Sea.

Significance for Climate

The influence of the Arctic Ocean on the circulation at lower latitudes is mainly through the export of freshwater as ice and as low-salinity PSW, and through the export of dense intermediate and deep waters that contribute to the Greenland–Scotland overflow and to the North Atlantic Deep Water, enforcing the thermohaline circulation and the Atlantic meridional overturning circulation (AMOC). Of the around 6-Sv overflow water supplied to the North

Atlantic Deep Water by the Arctic Mediterranean about 3 Sv have passed through the Arctic Ocean.

The outflow of ice and less-dense surface water could increase the stability of the water column in the convection areas to the south, in the Nordic Seas, and in the Labrador Sea. However, the fresh water largely remains in the East Greenland Current and mostly bypasses the Greenland Sea and Iceland Sea gyres. In recent years, ice has not been formed in the central Greenland Sea and the Greenland Sea has been dominated by thermal convection.

The AAW and the uPDW contribute to the Denmark Strait overflow, while the CBDW and the EBDW mainly enter the Greenland Sea, where they supply the mid-depth (1800 m) temperature maximum and the deep salinity maximum. In recent years, the local convection in the Greenland Sea has not penetrated through the temperature maximum and only less-dense AIW has been formed. The production of AIW has lead to a more direct contribution of the Greenland Sea to the AMOC, especially to the East Greenland Current and the Denmark Strait overflow. The denser Arctic Ocean deep water masses, previously continuing in the East Greenland Current to Denmark Strait are now entering the Greenland Sea and gradually replacing the old, colder, and less saline Greenland Sea deep and bottom waters, making the Greenland Sea water column more 'Arctic' in character.

See also

Heat Transport and Climate. North Atlantic Oscillation (NAO). Ocean Circulation. Ocean Circulation: Meridional Overturning Circulation. Sea Ice: Overview. Upper Ocean Heat and Freshwater Budgets.

Further Reading

Björk G, Jakobsson M, Rudels B, et al. (2007) Bathymetry and deep-water exchange across the central Lomonosov Ridge at 88–89° N. Deep-Sea Research I 54: 1197–1208 (doi:10.1016/j.dsr.2007.05.010).

Björk G, Söderqvist J, Winsor P, Nikolopoulos A, and Steele M (2002) Return of the cold halocline to the Amundsen Basin of the Arctic Ocean: Implications for the sea ice mass balance. Geophysical Research Letters 29(11): 1513 (doi:10.1029/2001GL014157).

Coachman LK and Aagaard K (1974) Physical oceanography of the Arctic and sub-Arctic seas. In: Herman Y (ed.) Marine Geology and Oceanography of the Arctic Ocean, pp. 1–72. New York: Springer.

Dickson B, Meincke J, and Rhines P (eds.) (2008) *Arctic–Subarctic Ocean Fluxes: Defining the Role of the Northern Seas in Climate*. New York: Springer.

Dickson RR, Rudels B, Dye S, Karcher M, Meincke J, and Yashayaev I (2007) Current estimates of freshwater Arctic and subarctic seas. *Progress in Oceanography* 73: 210–230 (doi:10.1016/j.pocean.2006.12.003).

Fahrbach E, Meincke J, Østerhus S, *et al.* (2001) Direct measurements of volume transports through Fram Strait. *Polar Research* 20: 217–224.

Jakobsson M, Cherkis NZ, Woodward J, Macnab R, and Coakley B (2000) New grid of Arctic bathymetry aids scientists and mapmakers. *EOS, Transactions of American Geophysical Union* 81(9): 89–96.

Jakobsson M, Macnab R, Mayer L, *et al.* (2008) An improved bathymetric portrayal of the Arctic Ocean: Implications for ocean modelling and geological, geophysical and oceanographic analyses. *Geophysical Research Letters* 35: L07602 (doi:10.1029/2008 GL0335220).

Johannesen OM, Muench RD, and Overland JE (eds.) (1994) *AGU Geophysical Monographs 85: The Polar Oceans and Their Role in Shaping the Global Environment*. Washington, DC: American Geophysical Union.

Jones EP, Rudels B, and Anderson LG (1995) Deep water of the Arctic Ocean: Origin and circulation. *Deep-Sea Research* 42: 737–760.

Leppäranta M (ed.) (1998) *Physics of Ice-Covered Seas*. Helsinki: Helsinki University Press.

Lewis EL, Jones EP, Lemke P, Prowse T, and Wadhams P (eds.) (2000) *The Freshwater Budget of the Arctic Ocean*. Dordrecht: Kluwer Academic Publishers.

Merryfield WJ (2002) Intrusions in double-diffusively stable Arctic waters: Evidence for differential mixing? *Journal of Physical Oceanography* 32: 1452–1459.

Nansen F (1902) Oceanography of the North Polar Basin. *Scientific Results III(9): The Norwegian North Polar Expedition 1893–96*. Oslo: Jacob Dybwad.

Peterson BJ, McClelland J, Curry R, Holmes RM, Walsh JE, and Aagaard K (2006) Trajectory shifts in the Arctic and subarctic freshwater cycle. *Science* 313: 1061–1066.

Prinsenberg SJ and Hamilton J (2005) Monitoring the volume, freshwater and heat fluxes passing through Lancaster Sound in the Canadian Arctic Archipelago. *Atmosphere-Ocean* 43: 1–22.

Quadfasel D, Sy A, Wells D, and Tunik A (1991) Warming of the Arctic. *Nature* 350: 385.

Rudels B, Jones EP, Anderson LG, and Kattner G (1994) On the intermediate depth waters of the Arctic Ocean. In: Johannesen OM, Muench RD, and Overland JE (eds.) *AGU Geophysical Monographs 85: The Polar Oceans and Their Role in Shaping the Global Environment*, pp. 33–46. Washington, DC: American Geophysical Union.

Rudels B, Jones EP, Schauer U, and Eriksson P (2004) Atlantic sources of the Arctic Ocean surface and halocline waters. *Polar Research* 23: 181–208.

Rudels B, Muench RD, Gunn J, Schauer U, and Friedrich HJ (2000) Evolution of the Arctic Ocean boundary current north of the Siberian shelves. *Journal of Marine Systems* 25: 77–99.

Schauer U, Fahrbach E, Østerhus S, and Rohardt G (2004) Arctic warming through Fram Strait: Oceanic heat transports from 3 years of measurements. *Journal of Geophysical Research* 109: C06026 (doi:10.1029/2003 JC001823).

Serreze MC, Barrett A, Slater AJ, *et al.* (2006) The large-scale freshwater cycle in the Arctic. *Journal of Geophysical Research* 111: C11010 (doi:10.1029/2005 JC003424).

Smith WO, Jr. (ed.) (1990) *Polar Oceanography, Part A: Physical Sciences*. San Diego: Academic Press.

Smith WO, Jr. and Grebmeier JM (eds.) (1995) *Coastal and Estuarine Studies: Arctic Oceanography, Marginal Ice Zones and Continental Shelves*. Washington, DC: American Geophysical Union.

Steele M and Boyd T (1998) Retreat of the cold halocline layer in the Arctic Ocean. *Journal of Geophysical Research* 100: 881–994.

Stein R and MacDonald RW (eds.) (2004) *The Organic Carbon Cycle in the Arctic Ocean*, 362pp. Berlin: Springer.

Swift JH, Aagaard K, Timokhov L, and Nikiforov EG (2005) Long-term variability of Arctic Ocean waters: Evidence from a reanalysis of the EWG data set. *Journal of Geophysical Research* 110: C03012 (doi:10.1029/ 2004JC002312).

Timmermanns M-L, Garrett C, and Carmack E (2003) The thermohaline structure and evolution of the deep water in the Canada Basin, Arctic Ocean. *Deep-Sea Research I* 50: 1305–1321 (doi:10.1016/S0967-0637(03)00125-0).

Untersteiner N (ed.) (1986) *The Geophysics of Sea Ice*. New York: Plenum.

Wadhams P, Gascard J-C, and Miller L (1990) Topical studies in oceanography: The European Subpolar Ocean Programme: ESOP. *Deep-Sea Research II* 46: 1011–1530.

Walsh D and Carmack EC (2002) A note on evanescent behavior of Arctic thermohaline intrusions. *Journal of Marine Research* 60: 281–310.

Wassmann P (ed.) (2006) Special Issue: Structure and Function of Contemporary Food Webs on Arctic Shelves: A Pan-Arctic Comparison. *Progress in Oceanography* 71(2–4): 123–477.

Wheeler PA (1997) Topical studies in oceanography: 1994 Arctic Ocean section. *Deep-Sea Research II* 44: 1483–1758.

Woodgate RA and Aagaard K (2005) Revising the Bering Strait freshwater flux into the Arctic Ocean. *Geophysical Research Letters* 32: L02602 (doi:1029/2004 GL021747).

Woodgate RA, Aagaard K, Muench RD, *et al.* (2001) The Arctic boundary current along the Eurasian slope and the adjacent Lomonosov Ridge. *Deep-Sea Research I* 48: 1757–1792.

SUB ICE-SHELF CIRCULATION AND PROCESSES

K. W. Nicholls, British Antarctic Survey,
Cambridge, UK

Introduction

Ice shelves are the floating extension of ice sheets. They extend from the grounding line, where the ice sheet first goes afloat, to the ice front, which usually takes the form of an ice cliff dropping down to the sea. Although there are several examples on the north coast of Greenland, the largest ice shelves are found in the Antarctic where they cover 40% of the continental shelf. Ice shelves can be up to 2 km thick and have horizontal extents of several hundreds of kilometers. The base of an ice shelf provides an intimate link between ocean and cryosphere. Three factors control the oceanographic regime beneath ice shelves: the geometry of the sub-ice shelf cavity, the oceanographic conditions beyond the ice front, and tidal activity. These factors combine with the thermodynamics of the interaction between sea water and the ice shelf base to yield various glaciological and oceanographic phenomena: intense basal melting near deep grounding lines and near ice fronts; deposition of ice crystals at the base of some ice shelves, resulting in the accretion of hundreds of meters of marine ice; production of sea water at temperatures below the surface freezing point, which may then contribute to the formation of Antarctic Bottom Water; and the upwelling of relatively warm Circumpolar Deep Water.

Although the presence of the ice shelf itself makes measurement of the sub-ice shelf environment difficult, various field techniques have been used to study the processes and circulation within sub-ice shelf cavities. Rates of basal melting and freezing affect the flow of the ice and the nature of the ice–ocean interface, and so glaciological measurements can be used to infer the ice shelf's basal mass balance. Another indirect approach is to make ship-based oceanographic measurements along ice fronts. The properties of in-flowing and out-flowing water masses give clues as to the processes needed to transform the water masses. Direct measurements of oceanographic conditions beneath ice shelves have been made through natural access holes such as rifts, and via access holes created using thermal (mainly hot-water) drills. Numerical models of the sub-ice shelf regime have been developed to complement the field measurements. These range from simple one-dimensional models following a plume of water from the grounding line along the ice shelf base, to full three-dimensional models coupled with sea ice models, extending out to the continental shelf-break and beyond.

The close relationship between the geometry of the sub-ice shelf cavity and the interaction between the ice shelf and the ocean implies a strong dependence of the ice shelf/ocean system on the state of the ice sheet. During glacial climatic periods the geometry of ice shelves would have been radically different to their geometry today, and ice shelves probably played a different role in the climate system.

Geographical Setting

By far the majority of the world's ice shelves are found fringing the Antarctic coastline (**Figure 1**). Horizontal extents vary from a few tens to several hundreds of kilometers, and maximum thickness at the grounding line varies from a few tens of meters to 2 km. By area, the Ross Ice Shelf is the largest at around 500 000 km^2. The most massive, however, is the very much thicker Filchner-Ronne Ice Shelf in the southern Weddell Sea. Ice from the Antarctic Ice Sheet flows into ice shelves via fast-moving ice streams (**Figure 2**). As the ice moves seaward, further nourishment comes from snowfall, and, in some cases, from accretion of ice crystals at the ice shelf base. Ice is lost by melting at the ice shelf base and by calving of icebergs at the ice front. Current estimates suggest that basal melting is responsible for around 25% of the ice loss from Antarctic ice shelves; most of the remainder calves from the ice fronts as icebergs.

Over central Antarctica the weight of the ice sheet depresses the lithosphere such that the seafloor beneath many ice shelves deepens towards the grounding line. The effect of the lithospheric depression has probably been augmented during glacial periods by the scouring action of ice on the seafloor: at the glacial maxima the grounding line would have been much closer to the continental shelf-break. Since ice shelves become thinner towards the ice front and float freely in the ocean, a typical sub-ice shelf cavity has the shape of a cavern that dips downwards towards the grounding line (**Figure 2**).

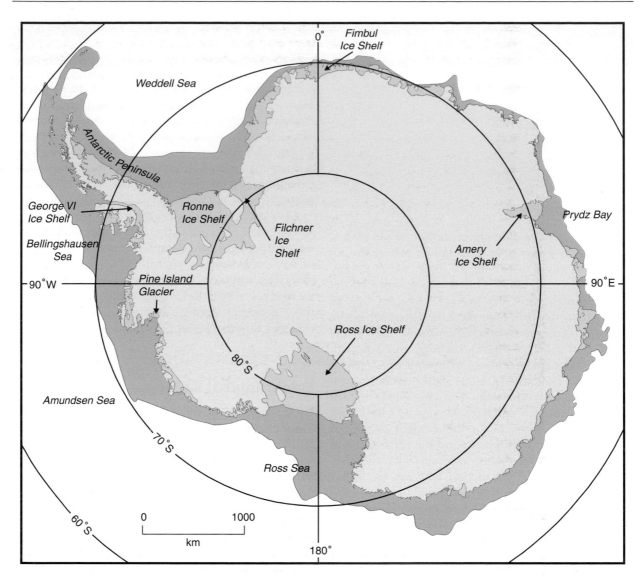

Figure 1   Map showing ice shelves (blue) covering about 40% of the continental shelf (dark gray) of Antarctica.

This geometry has important consequences for the ocean circulation within the cavity.

Oceanographic Setting

The oceanographic conditions over the Antarctic continental shelf depend on whether relatively warm, off-shelf water masses are able to cross the continental shelf-break.

For much of Antarctica a dynamic barrier at the shelf-break prevents advection of circumpolar deep water (CDW) onto the continental shelf itself. In these regions the principal process determining the oceanographic conditions is production of sea ice in coastal polynyas and leads, and the water column is largely dominated by high salinity shelf water (HSSW). Long residence times over some of the broader continental shelves, for example in the Ross and southern Weddell seas, enable HSSW to attain salinities of over 34.8 PSU. HSSW has a temperature at or near the surface freezing point (about $-1.9°C$), and is the densest water mass in Antarctic waters. Conditions over the continental shelves of the Bellingshausen and Amundsen seas (**Figure 1**) represent the other extreme. There, the barrier at the shelf-break appears to be either weak or absent. At a temperature of about $1°C$, CDW floods the continental shelf.

Between these two extremes there are regions of continental shelf where tongues of modified warm deep water (MWDW) are able to penetrate the shelf-break barrier (**Figure 3**), in some cases reaching as far as ice fronts. MWDW comes from above the warm core of CDW: the continental shelf effectively skims

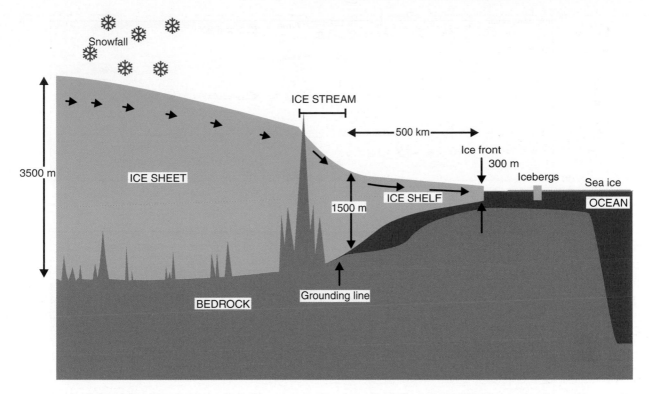

Figure 2 Schematic cross-section of the Antarctic ice sheet showing the transition from ice sheet to ice stream to ice shelf. Also shown is the depression of the lithosphere that results in the deepening of the seabed towards the continental interior.

off the shallower and cooler part of the water column.

What Happens When Ice Shelves Melt into Sea Water?

The freezing point of fresh water is 0°C at atmospheric pressure. When the water contains dissolved salts, the freezing point is depressed: at a salinity of around 34.7 PSU the freezing point is −1.9°C. Sea water at a temperature above −1.9°C is therefore capable of melting ice. The freezing point of water is also pressure dependent. Unlike most materials, the pressure dependence for water is negative: increasing the pressure decreases the freezing point. The freezing point T_f of sea water is approximated by:

$$T_f = aS + bS^{3/2} - cS^2 - dp$$

where $a = -5.75 \times 10^{-2}°C\,PSU^{-1}$, $b = 1.710523 \times 10^{-3}°C\,PSU^{-3/2}$, $c = -2.154996 \times 10^{-4}°C\,PSU^{-2}$ and $d = -7.53 \times 10^{-4}°C\,dbar^{-1}$. S is the salinity in PSU, and p is the pressure in dbar. Every decibar increase in pressure therefore depresses the freezing point by 0.75 m°C. The depression of the freezing point with pressure has important consequences for the interaction between ice shelves and the ocean. Even though HSSW is already at the surface freezing point, if it can

be brought into contact with an ice shelf base, melting will take place. As the freezing point at the base of deep ice shelves can be as much as 1.5°C lower than the surface freezing point, the melt rates can be high.

When ice melts into sea water the effect is to cool and freshen. Consider unit mass of water at temperature T_0, and salinity S_0 coming into contact with the base of an ice shelf where the *in situ* freezing point is T_f. The water first warms m kg of ice to the freezing point, and then supplies the latent heat necessary for melting. The resulting mixture of melt and sea water has temperature T and salinity S. If the initial temperature of the ice is T_i, the latent heat of melting is L, the specific heat capacity of sea water and ice, c_w and c_i, then heat and salt conservation requires that:

$$(T - T_f)(1 + m)c_w + m(c_i(T_f - T_i) + L)$$
$$= (T_o - T_f)c_w$$
$$S(1 + m) = S_o$$

Eliminating m, and then expressing T as a function of S reveals the trajectory of the mixture in T–S space as a straight line passing through (S_0, T_0), with a gradient given by:

$$\frac{dT}{dS} = \frac{L}{S_o c_w} + \frac{(T_f - T_i)c_i}{S_o c_w} + \frac{(T_o - T_f)}{S_o}$$

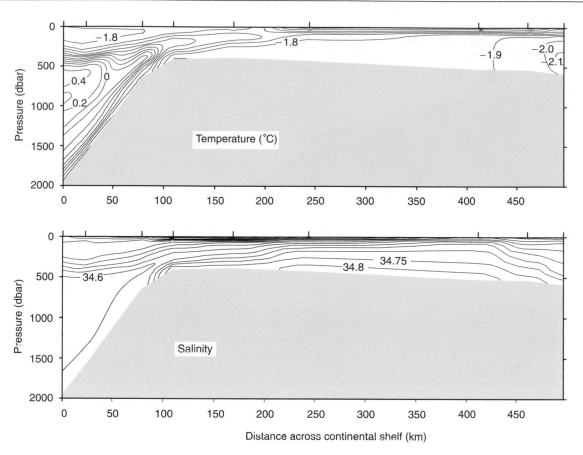

Figure 3 Hydrographic section over the continental slope and across the open continental shelf in the southern Weddell Sea, as far as the Ronne Ice Front. Water below the surface freezing point (−1.9°C) can be seen emerging from beneath the ice shelf. The majority of the continental shelf is dominated by HSSW, although in this location a tongue of warmer MWDW penetrates across the shelf-break. The station locations are shown by the heavy tick marks along the upper axes.

The gradient is dominated by the first term, which evaluates to about 2.4°C PSU^{-1}. In polar waters the third term is two orders of magnitude lower than the first; the second term results from the heat needed to warm the ice, and, at about a tenth the size of the first term, makes a measurable contribution to the gradient. This relationship allows the source water for sub-ice shelf processes to be found by inspection of the T–S properties of the resultant water masses. Examples of T–S plots from beneath ice shelves in warm and cold regimes are shown in **Figure 4**.

Two important passive tracers are introduced into sea water when glacial ice melts. When water evaporates from the ocean, molecules containing the lighter isotope of oxygen, ^{16}O, evaporate preferentially. Compared with sea water the snow that makes up the ice shelves is therefore low in ^{18}O. By comparing the $^{18}O/^{16}O$ ratios of the outflowing and inflowing water it is possible to calculate the concentration of melt water, provided the ratio is known for the glacial ice. Helium contained in the air bubbles in the ice is also introduced into the sea water when the ice melts. As helium's solubility in

water increases with increasing water pressure, the concentration of dissolved helium in the melt water can be an order of magnitude greater than in ambient sea water, which has equilibrated with the atmosphere at surface pressure.

Modes of Sub-ice Shelf Circulation

Various distinguishable modes of circulation appear to be possible within a sub-ice shelf cavity. Which mode is active depends primarily on the oceanographic forcing from seaward of the ice front, but also on the geometry of the sub-ice shelf cavity. Thermohaline forcing drives three modes of circulation, although the tidal activity is thought to play an important role by supplying energy for vertical mixing. Another mode results from tidal residual currents.

Thermohaline Modes

Cold regime external ventilation Over the parts of the Antarctic continental shelf dominated by the

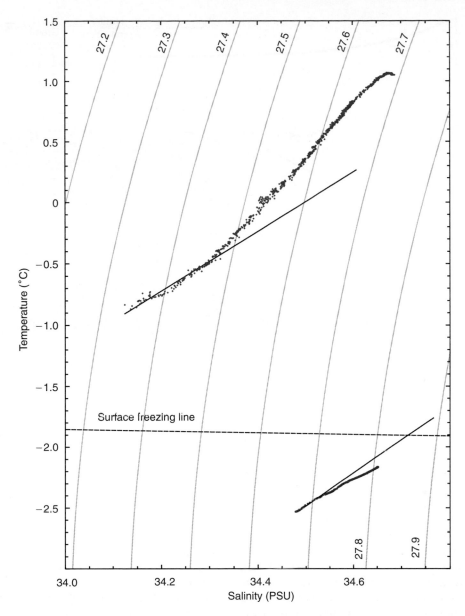

Figure 4 Temperature and salinity trajectories from CTD stations through the George VI Ice Shelf (red) and Ronne Ice Shelf (blue). The cold end of each trajectory corresponds to the base of the ice shelf. The straight lines are at the characteristic gradient for ice melting into sea water. For the Ronne data, as the source water will be HSSW at the surface freezing point, the intersection of the characteristic with the broken line gives the temperature and salinity of the source water. The isopycnals (gray lines) are referenced to sea level.

production of HSSW, such as in the southern Weddell Sea, the Ross Sea, and Prydz Bay, the circulation beneath large ice shelves is driven by the drainage of HSSW into the sub-ice shelf cavities. The schematic in **Figure 5** illustrates the circulation mode. HSSW drains down to the grounding line where tidal mixing brings it into contact with ice at depths of up to 2000 m. At such depths HSSW is up to 1.5°C warmer than the freezing point, and relatively rapid melting ensues (up to several meters of ice per year). The HSSW is cooled and diluted,

converting it into ice shelf water (ISW), which is defined as water with a temperature below the surface freezing point.

ISW is relatively buoyant and rises up the inclined base of the ice shelf. As it loses depth the *in situ* freezing point rises also. If the ISW is not entraining sufficient HSSW, which is comparatively warm, the reduction in pressure will result in the water becoming *in situ* supercooled. Ice crystals are then able to form in the water column and possibly rise up and accrete at the base of the ice shelf. This 'snowfall' at

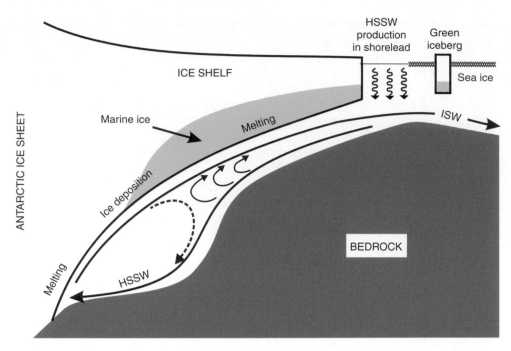

Figure 5 Schematic of the two thermohaline modes of sub-ice shelf circulation for a cold regime ice shelf.

the ice shelf base can build up hundreds of meters of what is termed 'marine ice'. Entrainment of HSSW, and the possible production of ice crystals, often result in the density of the ISW finally matching the ambient water density before the plume has reached the ice front. The plume then detaches from the ice shelf base, finally emerging at the ice front at mid-water depths.

The internal Rossby radius beneath ice shelves is typically only a few kilometers, and so rotational effects must be taken into account when considering the flow in three dimensions. HSSW flows beneath the ice shelf as a gravity current and is therefore gathered to the left (in the Southern Hemisphere) by the Coriolis force. As an organized flow, it then follows bathymetric contours. Once converted into ISW, the flow is again gathered to the left, following either the coast, or topography in the ice base. If the ISW plume fills the cavity, conservation of potential vorticity would demand that it follow contours of constant water column thickness. The step in water column thickness caused by the ice front then presents a topographic obstacle for the outflow of the ISW. However, the discontinuity can be reduced by the presence of trenches in the seafloor running across the ice front. This has been proposed as the mechanism that allows ISW to flow out from beneath the Filchner Ice Shelf, in the southern Weddell Sea (**Figure 1**).

Initial evidence for this mode of circulation came from ship-based oceanographic observations along the ice front of several of the larger ice shelves. Water

with temperatures up to $0.3°C$ below the surface freezing point indicated interaction with ice at a depth of at least 400 m, and the $^{18}O/^{16}O$ ratio confirmed the presence of glacial melt water at a concentration of several parts per thousand. Nets cast near ice fronts for biological specimens occasionally recovered masses of ice platelets, again from depths of several hundred meters. The ISW flowing from beneath the Filchner Ice Shelf has been traced overflowing the shelf-break and descending the continental slope, ultimately to mix with deep waters and form bottom water.

Evidence from the ice shelf itself comes in the form of glaciological measurements. By assuming a steady state (the ice shelf neither thickening nor thinning with time at any given point) conservation arguments can be used to derive the basal mass balance at individual locations. The calculation needs measurements of the local ice thickness variation, the horizontal spreading rate of the ice as it flows under its own weight, the horizontal speed of the ice, and the surface accumulation rate. This technique has been applied to several ice shelves, but is time-consuming, and has rarely been used to provide a good areal coverage of basal mass balance. However, it has demonstrated that high basal melt rates do indeed exist near deep grounding lines; that the melt rates reduce away from the grounding line; that further still from the grounding line, melting frequently switches to freezing; and that the balance usually returns to melting as the ice front is approached.

One-dimensional models have been to study the development of ISW plumes from the grounding line to where they detach from the ice shelf base. The most sophisticated includes frazil ice dynamics, and suggests that the deposition of ice at the base depends not only on its formation in the water column, but also on the flow regime being quiet enough to allow the ice to settle at the ice base. As the flow regime usually depends on the basal topography, the deposition is often highly localized. For example, a reduction in basal slope reduces the forcing on the buoyant plume, thereby slowing it down and possibly allowing any ice platelets to be deposited.

Deposits of marine ice become part of the ice shelf itself, flowing with the overlying meteoric ice. This means that, although the marine ice is deposited in well-defined locations, it moves towards the ice front with the flow of the ice and may or may not all be melted off by the time it reaches the ice front. Icebergs that have calved from Amery Ice Front frequently roll over and reveal a thick layer of marine ice. Impurities in marine ice result in different optical properties, and these bergs are often termed 'green icebergs'.

Ice cores obtained from the central parts of the Amery and Ronne ice shelves have provided other direct evidence of the production of marine ice. The interface between the meteoric and marine ice is clearly visible – the ice changes from being white and bubbly, to clear and bubble-free. Unlike normal sea ice, which typically has a salinity of a few PSU, the salinity of marine ice was found to be below 0.1 PSU. The salinity in the cores is highest at the interface itself, decreasing with increasing depth. A different type of marine ice was found at the base of the Ross Ice Shelf. There, a core from near the base showed 6 m of congelation ice with a salinity of between 2 and 4 PSU. Congelation ice differs from marine ice in its formation mechanism, growing at the interface directly rather than being created as an accumulation of ice crystals that were originally formed in the water column.

Airborne downward-looking radar campaigns have mapped regions of ice shelf that are underlain by marine ice. The meteoric (freshwater) ice/marine ice interface returns a characteristically weak echo, but the return from marine ice/ocean boundary is generally not visible. By comparing the thickness of meteoric ice found using the radar with the surface elevation of the freely floating ice shelf, it is possible to calculate the thickness of marine ice accreted at the base. In some parts of the Ronne Ice Shelf basal accumulation rates of around $1 \, \mathrm{m\,a^{-1}}$ result in a marine ice layer over 300 m thick, out of a total ice column depth of 500 m. Accumulation rates of that magnitude would be expected to be associated with high ISW fluxes. However, cruises along the Ronne Ice Front have been unsuccessful in finding commensurate ISW outflows.

Internal recirculation Three-dimensional models of the circulation beneath the Ronne Ice Shelf have revealed the possibility of an internal recirculation of ISW. This mode of circulation is driven by the difference in melting point between the deep ice at the grounding line, and the shallower ice in the central region of the ice shelf. The possibility of such a recirculation is indicated in **Figure 5** by the broken line. Intense deposition of ice in the freezing region salinifies the water column sufficiently to allow it to drain back towards the grounding line. In three dimensions, the recirculation consists of a gyre occupying a basin in the topography of water column thickness. The model predicts a gyre strength of around one Sverdrup ($10^6 \, \mathrm{m^3 \, s^{-1}}$).

This mode of circulation is effectively an 'ice pump' transporting ice from the deep grounding line regions to the central Ronne Ice Shelf. The mechanism does not result in a loss or gain of ice overall. The heat used to melt the ice at the grounding line is later recovered in the freezing region. The external heat needed to maintain the recirculation is therefore only the heat to warm the ice to the freezing point before it is melted. Ice leaves the continent at a temperature of around $-30 \, ^\circ \mathrm{C}$, and has a specific heat capacity of around $2010 \, \mathrm{J\,kg^{-1}{}^\circ C^{-1}}$. As the latent heat of ice is $335 \, \mathrm{kJ\,kg^{-1}}$, the heat required for warming is less than 20% of that required for melting. To support an internal redistribution of ice therefore requires a small fraction of the external heat that would be needed to melt and remove the ice from the system entirely. A corollary is that a recirculation of ISW effectively decouples much of the ice shelf base from external forcings that might be imposed, for example, by climate change.

Apart from the lack of a sizable ISW outflow from beneath the Ronne Ice Front, evidence in support of an ISW recirculation deep beneath the ice shelf is scarce, as it would require observations beneath the ice. Direct measurements of conditions beneath ice shelves are limited to a small number of sites. Fissures through George VI and Fimbul ice shelves (**Figure 1**) have allowed instruments to be deployed with varying degrees of success. The more important ice shelves, such as the Ross, Amery and Filchner-Ronne system have no naturally occurring access points. Instead, access holes have to be created using hot water, or other thermal-type drills. In the late 1970s researchers used various drilling techniques to gain access to the cavity at one location beneath the

Ross Ice Shelf before deploying various instruments. During the 1990s several access holes were made through the Ronne Ice Shelf, and data from these have lent support both to the external mode of circulation, and most recently, to the internal recirculation mode first predicted by numerical models.

Warm regime external ventilation   The flooding of the Bellingshausen and Amundsen seas' continental shelf by barely modified CDW results in very high basal melt rates for the ice shelves in that sector. The floating portion of Pine Island Glacier (**Figure 1**) has a mean basal melt rate estimated to be around $12 \, \mathrm{m \, a}^{-1}$, compared with estimates of a few tens of centimeters per year for the Ross and Filchner-Ronne ice shelves. Basal melt rates for Pine Island Glacier are high even compared with other ice shelves in the region. George VI Ice Shelf on the west coast of the Antarctic Peninsula, for example, has an estimated mean basal melt rate of $2 \, \mathrm{m \, a}^{-1}$. The explanation for the intense melting beneath Pine Island Glacier can be found in the great depth at the grounding line. At over $1100 \, \mathrm{m}$, the ice shelf is $700 \, \mathrm{m}$ thicker than George VI Ice Shelf, and this results in not only a lower freezing point, but also steeper basal slopes. The steep slope provides a stronger buoyancy forcing, and therefore greater turbulent heat transfer between the water and the ice.

The pattern of circulation in the cavities beneath warm regime ice shelves is significantly different to its cold regime counterpart. Measurements from ice front cruises show an inflow of warm CDW ($+1.0°C$), and an outflow of CDW mixed with glacial melt water. **Figure 6** shows a two-dimensional schematic of this mode of circulation. Over the open continental shelf the ambient water column consists of CDW overlain by colder, fresher water left over from sea ice production during the previous winter. Although the melt water-laden outflow is colder, fresher, and of lower density than the inflow, it is typically warmer and saltier than the overlying water, but of similar density. Somewhat counterintuitively, therefore, the products of sub-glacial melt are often detected over the open continental shelf as relatively warm and salty intrusions in the upper layers. Again, measurements of oxygen isotope ratio, and also helium, provide the necessary confirmation that the upwelled CDW contains melt water from the base of ice shelves. In the case of warm regime ice shelves, melt water concentrations can be as high as a few percent.

Tidal Forcing

Except for within a few ice thicknesses of grounding lines, ice shelves float freely in the ocean, rising and falling with the tides. Tidal waves therefore propagate through the ice shelf-covered region, but are modified by three effects of the ice cover: the ice shelf base provides a second frictional surface, the draft of the ice shelf effectively reduces the water column thickness, and the step change in water column thickness at the ice front presents a topographic feature that has significant consequences for the generation of residual tidal currents and the propagation of topographic waves along the ice front.

Conversely, tides modify the oceanographic regime of sub-ice shelf cavities. Tidal motion helps transfer heat and salt beneath the ice front. This is a result

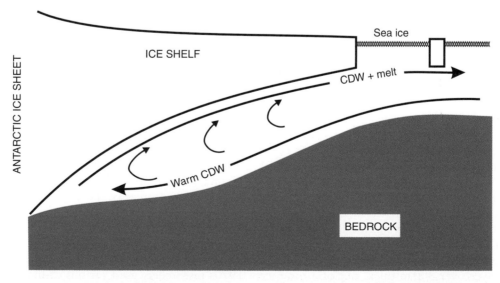

Figure 6   Schematic of the thermohaline mode of sub-ice shelf circulation for a warm regime ice shelf.

both of the regular tidal excursions, which take water a few kilometers into the cavity, and of residual tidal currents which, in the case of the Filchner-Ronne Ice Shelf, help ventilate the cavity far from the ice front. The effect of the regular advection of potentially seasonally warmed water from seaward of the ice shelf is to cause a dramatic increase in basal melt rates in the vicinity of the ice front. Deep beneath the ice shelf, tides and buoyancy provide the only forcing on the regime. Tidal activity contributes energy for vertical mixing, which brings the warmer, deeper waters into contact with the base of the ice shelf. **Figure 7A** shows modeled tidal ellipses for the M_2 semidiurnal tidal constituent for the southern Weddell Sea, including the sub-ice shelf domain. A map of the modeled residual currents for the area of the ice shelf is shown in **Figure 7B**. Apart from the activity near the ice front itself, a residual flow runs along the west coast of Berkner Island, deep under the ice shelf. However, this flow probably makes only a minor contribution to the ventilation of the cavity.

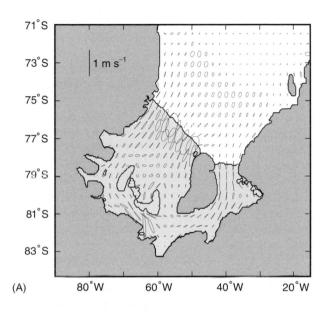

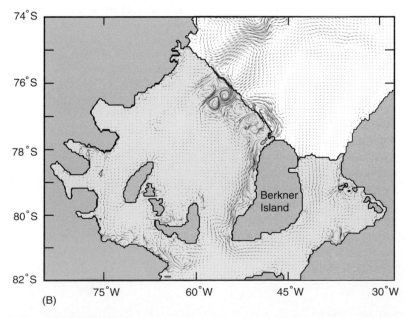

Figure 7 Results from a tidal model of the southern Weddell Sea, in the vicinity of the Ronne Ice Shelf. (A) The tidal ellipses for the dominant M_2 species. (B) Tidally induced residual currents.

How Does the Interaction between Ice Shelves and the Ocean Depend on Climate?

The response to climatic changes of sub-ice shelf circulation depends on the response of the oceanographic conditions over the open continental shelf. In the case of cold regime continental shelves, a reduction in sea ice would lead to a reduction in HSSW production. Model results, together with the implications of seasonality observed in the circulation beneath the Ronne Ice Shelf, suggest that drainage of HSSW beneath local ice shelves would then reduce, and that the net melting beneath those ice shelves would decrease as a consequence. Some general circulation models predict that global climatic warming would lead to a reduction in sea ice production in the southern Weddell Sea. Reduced melting beneath the Filchner-Ronne Ice Shelf would then lead to a thickening of the ice shelf. Recirculation beneath ice shelves is highly insensitive to climatic change. The thermohaline driving is dependent only on the difference in depths between the grounding lines and the freezing areas. A relatively small flux of HSSW is required to warm the ice in order to allow this mode to operate.

The largest ice shelves are in a cold continental shelf regime. If intrusions of warmer off-shelf water were to become more dominant in these areas, or if the shelf-break barrier were to collapse entirely and the regime switch from cold to warm, then the response of the ice shelves would be a dramatic increase in their basal melt rates. There is some evidence from sediment cores that such a change might have occurred at some point in the last few thousand years in what is now the warm regime Bellingshausen Sea. Evidence also points to the possibility that one ice shelf in that sector, the floating extension of Pine Island Glacier (**Figure 1**), might be a remnant of a much larger ice shelf.

During glacial maxima the Antarctic ice sheet thickens and the ice shelves become grounded. In many cases they ground as far as the shelf-break. There are two effects. The continental shelf becomes very limited in extent, and so there is little possibility for the production of HSSW; and where the ice shelves overhang the continental shelf-break, the only possible mode of circulation will be the warm regime mode. Substantial production of ISW during glacial conditions is therefore unlikely.

See also

Holocene Climate Variability. Ice–Ocean Interaction. Ice-shelf Stability. Sea Ice: Overview. Sea Ice: Overview.

Further Reading

Jenkins A and Doake CSM (1991) Ice–ocean interactions on Ronne Ice Shelf, Antarctica. *Journal of Geophysical Research* 96: 791–813.

Jacobs SS, Hellmer HH, Doake CSM, Jenkins A, and Frolich RM (1992) Melting of ice shelves and the mass balance of Antarctica. *Journal of Glaciology* 38: 375–387.

Nicholls KW (1997) Predicted reduction in basal melt rates of an Antarctic ice shelf in a warmer climate. *Nature* 388: 460–462.

Oerter H, Kipfstuhl J, Determann J, *et al.* (1992) Ice-core evidence for basal marine shelf ice in the Filchner-Ronne Ice Shelf. *Nature* 358: 399–401.

Williams MJM, Jenkins A, and Determann J (1998) Physical controls on ocean circulation beneath ice shelves revealed by numerical models. In: Jacobs SS and Weiss RF (eds.) *Ocean, Ice, and Atmosphere: Interactions at the Antarctic Continental Margin, Antarctic Research Series 75*, pp. 285–299. Washington DC: American Geophysical Union.

ICE–OCEAN INTERACTION

J. H. Morison, University of Washington, Seattle, WA, USA

M. G. McPhee, McPhee Research Company, Naches, WA, USA

Introduction

The character of the sea ice cover greatly affects the upper ocean and vice versa. In many ways ice-covered seas provide ideal examples of the planetary boundary layer. The under-ice surface may be uniform over large areas relative to the vertical scale of the boundary layer. The absence of surface waves simplifies the boundary layer processes. However, thermodynamic and mechanical characteristics of ice–ocean interaction complicate the picture in unique ways. We discuss a few of those unique characteristics.

We deal first with how momentum is transferred to the water and introduce the structure of the boundary layer. This will lead to a discussion of the processes that determine the fluxes of heat and salt. Finally, we discuss some of the unique characteristics imposed on the upper ocean by the larger-scale features of a sea ice cover.

Drag and Characteristic Regions of the Under-ice Boundary Layer

To understand the interaction of the ice and water, it is useful to consider three zones of the boundary layer: the molecular sublayer, surface layer, and outer layer (**Figure 1**).Under a reasonably smooth and uniform ice boundary, these can be described on the basis of the influence of depth on the terms of the equation for a steady, horizontally homogeneous boundary layer (eqn [1]).

$$ifV = \frac{\partial}{\partial z}\left(v\frac{\partial V}{\partial z}\right) + \frac{\partial}{\partial z}\left(K\frac{\partial V}{\partial z}\right) - \rho^{-1}\nabla_h p \qquad [1]$$

The coordinate system is right-handed with z positive upward and the origin at the ice under-surface. V is the horizontal velocity vector in complex notation ($V = u + iv$), ρ is water density, and p is pressure. An eddy diffusivity representation is used for turbulent shear stress, $K(\partial V/\partial z) = \overline{V'w'}$, where K is the eddy

diffusivity. The term $v(\partial V/\partial z)$ is the viscous shear stress, where v is the kinematic molecular viscosity. The pressure gradient term, $\rho^{-1}\nabla_h p$ is equal to $\rho^{-1}(\partial p/\partial x + i\partial p/\partial y)$.

The stress gradient term due to molecular viscosity is of highest inverse order in z. It varies as z^{-2}, and therefore dominates the stress balance in the molecular sublayer (**Figure 1**) where z is vanishingly small. As a result the viscous stress, $v(\partial V/\partial z)$, is effectively constant in the molecular sublayer, and the velocity profile is linear.

The next layer away from the boundary is the surface layer. Here the relation between stress and velocity depends on the eddy viscosity, which is proportional to the length scale and velocity scale of turbulent eddies. The length scale of the turbulent eddies is proportional to the distance from the boundary, $|z|$. Therefore, the turbulent stress term varies as z^{-1} and becomes larger than the viscous term beyond z greater than $(1/k)(v/u_{*0})$, typically a fraction of a millimeter. The velocity scale in the surface layer is u_{*0}, where ρu_{*0}^2 is equal to τ_0, the average shear stress at the top of the boundary layer. Thus, K is equal to $ku_{*0}|z|$, where Von Kármän's constant, k, is equal to 0.4. Because the turbulent stress term dominates the equations of motion, the stress is roughly constant with depth in the surface layer. This and the linear z dependence of the eddy coefficient result in the log-layer solution or 'law of the wall' (eqn [2]).

$$\frac{u}{u_*} = \frac{1}{k}\ln z + C = \frac{1}{k}\ln\frac{z}{z_0} \qquad [2]$$

$C = -(\ln z_0)/k$ is a constant of integration. Under sea ice the surface layer is commonly 1–3 m thick.

The surface layer is where the influence of the boundary roughness is imposed on the planetary boundary layer. In the presence of under-ice roughness, the average stress the ice exerts on the ocean, τ_0, is composed partly of skin friction due to shear and partly of form drag associated with pressure disturbances around pressure ridge keels and other roughness elements. Observations under very rough ice have shown a decrease in turbulent stress toward the surface, presumably because more of the momentum transfer is taken up by pressure forces on the rough surface. The details of this drag partition are not known. Drag partition is complicated further for cases in which stratification exists at depths shallow compared to the depth of roughness elements.

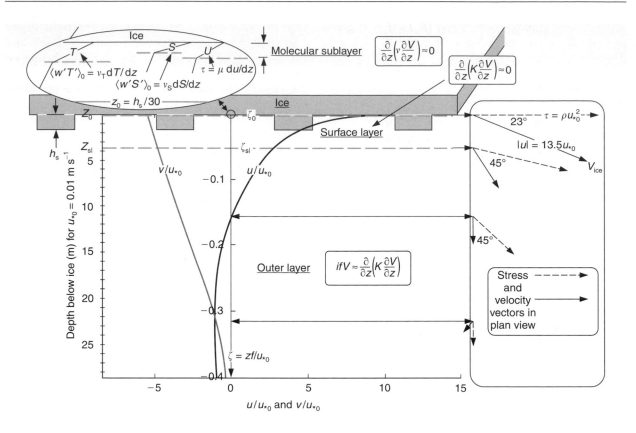

Figure 1 Illustration of three regions of the planetary boundary layer under sea ice: molecular sublayer, surface layer, and outer layer. The velocity profiles are from the Rossby similarity solution (eqns [8], [9], [10] and [11]) for $u_{*0} = 0.01$ m s^{-1}, $z_0 = 0.06$ m, $\eta_* = 1$. The stress and velocity vector comparisons are from the same solution.

Then it also becomes possible to transfer momentum by internal wave generation. However, for many purposes τ_0 is taken as the turbulent stress evaluated at z_0. Laboratory studies of turbulent flow over rough surfaces suggest that z_0 may be taken equal to $h_s/30$, where h_s is the characteristic height of the roughness elements.

In rare situations the ice surface may be so smooth that bottom roughness and form drag are not factors in the drag partition. In such a hydrodynamically smooth situation, the turbulence is generated by shear induced instability in the flow. The surface length scale, z_0, is determined by the level of turbulent stress and is proportional to the molecular sublayer thickness according to the empirically derived relation $z_0 = 0.13(\nu/u_{*0})$.

In the outer layer farthest from the boundary, the Coriolis and pressure gradient terms in eqn [1], which have no explicit z dependence, are comparable to the turbulent stress terms. The presence of the Coriolis term gives rise to a length scale, h, for the outer boundary layer equal to u_{*0}/f under neutral stratification. This region is far enough from the boundary so that the turbulent length scale becomes

independent of depth and in neutral conditions has been found empirically to be $\lambda = \xi_n u_{*0}/f$, where ξ_n is 0.05. For neutral stratification, u_* and h are the independent parameters that define the velocity profile over most of the boundary layer. The ratio of the outer length scale to the surface region length scale, z_0, is the surface friction Rossby number, $R_0 = u_{*0}/(z_0 f)$.

Solutions for the velocity in the outer layer can be derived for a wide range of conditions if we non-dimensionalize the equations with these Rossby similarity parameters, u_{*0}/f and u_{*0}. However, the growth and melt of the ice produce buoyancy flux that strongly affects mixing. Melting produces a stabilizing buoyancy flux that inhibits turbulence and contracts the boundary layer. Freezing causes a destabilizing buoyancy flux that enhances turbulence and thickens the boundary layer. We can account for the buoyancy flux effect by adjusting the Rossby parameters dealing with length scale. We define the scale of the outer boundary layer as $h_m = u_{*0}\eta_*/f$. If the mixing length of the turbulence in the outer layer is $\lambda_m = \xi_n u_{*0}\eta_*^2/f$, it interpolates in a reasonable way between known values of λ_m for neutral stratification

$(\xi_n u_{*0}/f)$ and stable stratification $(R_c L)$ if η_* is given as eqn [3].

$$\eta_* = \left(1 + \frac{\xi_n u_*}{f} \frac{1}{R_c L}\right)^{1/2} \qquad [3]$$

R_c is the critical Richardson number; the Obukhov length, L, is the ratio of shear and buoyant production of turbulent energy, $\rho u_{*0}^3/kg\langle \rho'w'\rangle$; and $-\langle \rho'w'\rangle g/\rho$ is the turbulent buoyancy flux. With this Rossby similarity normalization of the equations of motion, we can derive analytical expressions for the under-ice boundary layer profile that are applicable to a range of stratification.

For large $|z|$, V will approach the free stream geostrophic velocity, $\bar{V}_g = U_g + iV_g = f^{-1}\rho^{-1}\nabla_h p$. Here we will assume this is zero. However, surface stress-driven absolute velocity solutions can be superimposed on any geostrophic current. We also ignore the time variation and viscous terms and define a normalized stress equal to $\Sigma = (K\partial V/\partial z)/u_{*0}^2$. The velocity is nondimensionalized by the friction velocity and the boundary layer thickness, $U = Vfh_m/u_{*0}^2$, and depth is nondimensionalized by the boundary layer thickness scale, $\zeta = z/h_m$. With these changes eqn [1] becomes eqn [4].

$$iU = \partial\Sigma/\partial\zeta \qquad [4]$$

In terms of nondimensional variables the constitutive law is given by eqn [5].

$$\Sigma = K_*\partial U/\partial\zeta \qquad [5]$$

The nondimensional eddy coefficient is given by eqn [6].

$$K_* = ku_{*0}\lambda_m/fh_m^2 = k\xi_n \qquad [6]$$

Eqn [6] is the Rossby similarity relation that is the key to providing similarity solutions for stable and neutral conditions. It even provides workable results for slightly unstable conditions.

Eqns [4] and [5] can be combined in an equation for nondimensionalized stress (eqn [7]).

$$(i/K_*)\Sigma = d\Sigma d/\zeta \qquad [7]$$

This has the solution eqn [8].

$$\Sigma = e^{\hat{\delta}\zeta} \qquad [8]$$

$$\hat{\delta} = (i/K_*)^{1/2} \qquad [9]$$

Eqn [8] attenuates and rotates (to the right in the Northern Hemisphere) with depth. It duplicates the salient features found in data and sophisticated numerical models.

In the outer layer, eqns [5] and [8] are satisfied for nondimensional velocity given by eqn [10].

$$U = -i\hat{\delta}e^{\hat{\delta}\zeta} \quad \text{for } \zeta \leq \zeta_0 \qquad [10]$$

Thus the velocity is proportional to stress but rotated $45°$ to the right.

As we see in the derivation of the law of the wall [2], the surface layer variation of the eddy viscosity with depth is critical to the strong shear present there. Thus eqn [10] will not give a realistic profile in the surface layer. We define the nondimensional surface layer thickness, ζ_{sl}, as the depth where the surface layer mixing length, $|z|$, becomes equal to the outer layer mixing length, $\lambda_m = \xi_n u_{*0}\eta_*^2/f$. We find ζ_{sl} is equal to $-\eta_*\xi_n$ and applying the definition [6] gives K_* as $K_{*sl} = -k\zeta/\eta_*$ in the surface layer. If we approximate the stress profile [8] by a Taylor series, we can integrate [5] with K_{*sl} substituted for K_* to obtain the velocity profile in the surface layer.

$$U(\zeta) - U(\zeta_{sl}) = \frac{\eta_*}{k}\left[\ln\left(\frac{\zeta_{sl}}{\zeta_0}\right) + \hat{\delta}(\zeta_{sl} - \zeta)\right] \quad \text{for } \zeta \geq \zeta_0$$
$$[11]$$

Eqn [11] is analogous to [2] except for the introduction of the $\hat{\delta}(\zeta_{sl} - \zeta)$ term. This is the direct result of accounting for the stress gradient in the surface layer. This term is small compared to the logarithmic gradient.

Figure 1 illustrates the stress and velocity vectors at various points in the boundary layer as modeled by eqns [8] through [11]. For neutral conditions the nondimensional boundary layer thickness is typically 0.4 (dimensional thickness is $0.4u_*/f$). Through the outer layer, the velocity vector is $45°$ to the right of the stress vector as a consequence of the $-i\hat{\delta} \propto e^{-i(45°)}$ multiplier in [10]. As the ice surface is approached through the surface layer, the stress vector rotates $10–20°$ to the left to reach the surface direction. However, in the surface layer the velocity shear in the direction of the surface stress is great because of the logarithmic profile. Thus, as the surface is approached, the velocity veers to the left twice as much as stress. At the surface the velocity is about $23°$ to the right of the surface stress.

It is commonly useful to relate the stress on under-ice surface to the relative velocity between ice and water a neutral-stratification drag coefficient,

$\rho u_{*0}^2 = \rho C_z V_{(z)}^2$ where C_z is the drag coefficient for depth z. If z is in the log-layer, eqn [2] can be used to derive the relation between ice roughness and the drag coefficient. We find that $C_z = k^2[\ln(z/z_0)]^{-2}$. Clearly values of the drag coefficient can vary widely depending on the under-ice roughness. Typical values of z_0 range from 1 to 10 cm under pack ice. A commonly referenced value for the Arctic is 6 cm, which produces a drag coefficient at the outer edge of the log layer of 9.4×10^{-3} (**Figure 1**).

If the reference depth is outside the log layer, the drag coefficient formulation is poorly posed because of the turning in the boundary layer. For neutral conditions, eqns [10] and [11] can be used to obtain a Rossby similarity drag law that yields the non-dimensional surface drift relative to the geostrophic current for unit nondimensional surface stress (eqn [12]).

$$U_0 = \frac{V_0}{u_{*0}} = \frac{1}{k}([\ln(R_0) - A] - iB) \qquad [12]$$

Here

$$A = \left(1 - \ln \xi_n - \sqrt{\frac{k}{2\xi_n}} + \sqrt{\frac{\xi_n}{2k}}\right) \cong 2.2$$

$$B = \sqrt{\frac{k}{2\xi_n}} + \sqrt{\frac{\xi_n}{2k}} \cong 2.3 \qquad [13]$$

This Rossby similarity drag law for outside the surface layer results in a surface stress that is proportional to $V^{1.8}$ rather than V^2, a result that is supported by observational evidence, and can be significant at high velocities.

Heat and Mass Balance at the Ice–Ocean Interface: Wintertime Convection

The energy balance at the ice–ocean interface not only exerts major influence over the ice mass balance but also dictates the seasonal evolution of upper ocean salinity and temperature structure. At low temperature, water density is controlled mainly by salinity. Salt is rejected during freezing, so that buoyancy flux from basal growth (or ablation), combined with turbulent mixing during storms, determines the depth of the well-mixed layer.

Vertical motion of the ice–ocean interface depends on isostatic adjustment as the ice melts or freezes. The interface velocity is $w_0 + w_i$ where $w_0 = -(\rho_{ice}/\rho)\dot{h}_b$, $\dot{h}_b$ is the basal growth rate, and w_i represents isostatic adjustment to runoff of surface melt and percolation of water through the ice cover. In an infinitesimal control volume following the ice–ocean interface, conservation of heat and salt may be expressed in kinematic form as eqns [14] and [15].

$$\dot{q} = \langle w'T'\rangle_0 - w_0 Q_L \text{(with units K m s}^{-1}) \qquad [14]$$

$$(w_0 + w_i)(S_0 - S_{ice}) = \langle w'S'\rangle_0 \text{(with units psu m s}^{-1}) \qquad [15]$$

where $\dot{q} = H_{ice}/(\rho c_p)$ is flux (H_{ice}) conducted away from the interface in the ice; ρ is water density; c_p is specific heat of seawater; $\langle w'T'\rangle_0$ is the kinematic turbulent heat flux from the ocean; Q_L is the latent heat of fusion (adjusted for brine volume) divided by c_p; S_0 is salinity in the control volume, S_{ice} is ice salinity, and $\langle w'S'\rangle_0$ is turbulent salinity flux. Fluid in the control volume is assumed to be at its freezing temperature, approximated by the freezing line (eqn [16]).

$$T_0 = -mS_0 \qquad [16]$$

By standard closure, turbulent fluxes are expressed in terms of mean flow properties (eqns [17] and [18]).

$$\langle w'T'\rangle_0 = c_h u_{*0}\delta T \qquad [17]$$

$$\langle w'S'\rangle_0 = c_S u_{*0}\delta S \qquad [18]$$

u_{*0} is the square root of kinematic turbulent stress at the interface (friction velocity); $\delta T = T - T_0$ and $\delta S = S - S_0$ are differences between far-field and interface temperature and salinity; and c_h and c_S are turbulent exchange coefficients termed Stanton numbers.

The isostatic basal melt rate, w_0 is the key factor in interface thermodynamics, and in combination with w_i it determines the salinity flux. A first-order approach to calculating w_0 that is often sufficiently accurate (relative to uncertainties in forcing parameters) when melting or freezing is slow, is to assume that $S_0 = S$, the far-field salinity, and that c_h is constant. Combining [14], [16], and [17] gives eqn [19].

$$w_0 = \frac{c_h u_{*0}(T + mS) - \dot{q}}{Q_L} \qquad [19]$$

Salinity flux is determined from [15]. Note the c_S is not used, and that this technique fixes (unrealistically) the temperature at the interface to be the mixed layer freezing temperature.

A more sophisticated approach is required when melting or freezing is intense. Manipulation of [14] through [18] produces a quadratic equation for w_0 (eqn [20]).

$$\frac{S_L}{u_{*0}}w_0^2 + (S_T + S_L c_S - S_{ice})w_0 + (u_{*0}c_S + w_i)S_T$$
$$+ u_{*0}c_S S - w_i S_{ice} = 0 \qquad [20]$$

$$S_T = \left(\frac{\dot{q}}{c_h u_{*0}} - T\right)/m \quad \text{and} \quad S_L = Q_L/(m c_h) \qquad [21]$$

Here c_h and c_S (turbulent Stanton numbers for heat and salt) are both important and not necessarily the same. Melting or freezing will decrease or increase S_0 relative to far-field salinity, with corresponding changes in T_0.

The Marginal Ice Zone Experiments (MIZEX) in the 1980s showed that existing ice–ocean turbulent transfer models overestimated melt rates by a wide factor. It became clear that the rates of heat and mass transfer were less than momentum transfer (by an order of magnitude or more), and were being controlled by molecular effects in thin sublayers adjacent to the interface. If it is assumed that the extent of the sublayers is proportional to the bottom roughness scale, z_0, then dimensional analysis suggests that the Stanton numbers (nondimensional heat and salinity flux) should depend mainly on two other dimensionless groups, the turbulent Reynolds number, $Re_* = u_{*0}z_0/v$, where v is molecular viscosity, and the Prandtl (Schmidt) numbers, $v/v_{T(S)}$, where v_T and v_S are molecular diffusivities for heat and salt. Laboratory studies of heat and mass transfer over hydraulically rough surfaces suggested approximate expressions for the Stanton numbers of the form shown in eqn [22].

$$c_{h(S)} = \frac{\langle w'T(S)'\rangle_0}{u_{*0}\delta T(S)} \propto (Re_*)^{-1/2}\left(\frac{v}{v_{T(S)}}\right)^{-2/3} \qquad [22]$$

The Stanton number, c_h, has been determined in several turbulent heat flux studies since the original MIZEX experiment, under differing ice types with z_0 values ranging from less than a millimeter (eastern Weddell Sea) to several centimeters (Greenland Sea MIZ). According to [22], c_h should vary by almost a factor of 10. Instead, it is surprisingly constant, ranging from about 0.005 to 0.006, implying that the Reynolds number dependence from laboratory results cannot be extrapolated directly to sea ice.

If the Prandtl number dependence of [14] holds, the ratio $c_h/c_S = (v_h/v_S)^{2/3}$ is approximately 30. Under conditions of rapid freezing, the solution of [20] with this ratio leads to significant supercooling of the water column, because heat extraction far outpaces salt injection in what is called double diffusion. This result has caused some concern. Because the amount of heat represented by this supercooling is substantial, it has been hypothesized that ice may spontaneously form in the supercooled layer and drift upward in the form of frazil ice crystals. This explanation has not been supported by ice core sampling, which shows no evidence of widespread frazil ice formation beyond that at the surface of open water.

The physics of the freezing process suggest that the seeming paradox of the supercooled boundary layer may be realistic without spontaneous frazil formation. When a parcel of water starts to solidify into an ice crystal, energy is released in proportion to the volume of the parcel. At large scales this manifests itself as the latent heat of fusion. However, as the parcel solidifies, energy is also required to form the surface of the solid. This surface energy penalty is proportional to the surface area of the parcel and depends on other factors including the physical character of any nucleating material. In any event, if the parcel is very small the ratio of parcel volume to surface area will be so small that the energy released as the volume solidifies is less than the energy needed to create the new solid surface. For this reason, ice crystals cannot form even in supercooled water without a nucleating site of sufficient size and suitable character. In the clean waters of the polar regions, the nearest suitable site may only be at the underside of the ice cover where the new ice can form with no nucleation barrier. Therefore, it is possible to maintain supercooled conditions in the boundary layer without frazil ice formation.

Furthermore, recent results suggest that supercooling in the uppermost part of the boundary layer may be intrinsic to the ice formation process. Sea ice is a porous mixture of pure ice and high-salinity liquid water (i.e., brine). The bottom surface of a growing ice floe consists of vertically oriented pure ice platelets separated by vertical layers of concentrated brine. This platelet–brine sandwich (on edge) structure is on the scale of a fraction of a millimeter, and its formation is controlled by molecular diffusion of heat and salt. The low solid solubility of the salt in the ice lattice results in an increase of the salinity of water in the layer above the advancing freezing interface. Because heat diffuses more rapidly than salt at these scales, the cold brine tends to supercool the water below the ice–water interface. With this local supercooling, any disturbance of the ice bottom will tend to grow spontaneously. The conditions of sea ice growth are such that this instability is always present. Continued growth results in additional rejection of salt, some fraction of which is trapped in the brine layers, and consequently the interfacial region of the ice sheet continues to experience constitutional supercooling. Also, anisotropy in the molecular attachment efficiency intrinsic to the crystal structure of the ice platelets creates an additional supercooling in the interfacial region. The net result is that heat is extracted from the top of the water column at the rate needed to maintain its temperature near but slightly below the equilibrium freezing temperature as salt is added. This and the convective processes in the growing ice may imply that $c_h/c_S = 1$ during freezing. The situation with melting may be quite different, since the physical properties of the interface change dramatically.

Observations to date suggest that c_h remains relatively unchanged with variable ice type and mixed layer temperature elevation above freezing. A value of 5.5×10^{-3} is representative. c_S is not so well known, since direct measurements of $\langle w'S' \rangle$ are relatively rare. The dependence of the exchange coefficients on Prandtl and Schmidt numbers is not clear, and will only be resolved with more research.

Effects of Horizontal Inhomogeneity: Wintertime Buoyancy Flux

Although the under-ice surface may be homogeneous over ice floes hundreds of meters in extent, the key fluxes of heat and salt are characteristically nonuniform. As ice drifts under the action of wind stress, the ice cover is deformed. Some areas are forced together, producing ridging and thick ice, and some areas open in long, thin cracks called leads. In special circumstances the ice may form large, unit-aspect-ratio openings called polynyas. In winter the openings in the ice expose the sea water directly to cold air without an intervening layer of insulating sea ice. This results in rapid freezing. As the ice forms, it rejects salt and results in unstable stratification of the boundary layer beneath open water or thin ice. These effects are so important that, even though such areas may account for less than 10% of the ice cover, they may account for over half the total ice growth and salt flux to the ocean. Thus the dominant buoyancy flux is not homogeneous but is concentrated in narrow bands or patches. Similarly, in the summer solar radiation is reflected from the ice but is nearly completely absorbed by open water. Fresh water from summertime surface melt tends to drain into leads, making them sources of fresh water flux as well.

The effect of wintertime convection in leads is illustrated in **Figure 2**. It shows two extremes in the upper ocean response. **Figure 2A** shows what we might expect in the case of a stationary lead. As the surface freezes, salt is rejected and forms more dense water that sinks under the lead. This sets up a circulation with fresh water flowing in from the sides near the surface and dense water flowing away from the lead at the base of the mixed layer.

Figure 2B illustrates the case in which the lead is embedded in ice moving at a velocity great enough to produce a well-developed turbulent boundary layer (e.g. $0.2 \, \text{m s}^{-1}$). If the mixed layer is fully turbulent, the cellular convection pattern may not occur; rather, the salt rejected at the surface may simply mix into the surface boundary layer.

The impact of nonhomogeneous surface buoyancy flux on the boundary layer can also be characterized by the equations of motion. The viscous terms in eqn [1] can be neglected at the scales we discuss here, but the possibility of vertical motion associated with large-scale convection requires that we include the vertical component of velocity. For steady state we have eqn [23].

$$\bar{V} \cdot \nabla \bar{V} + \bar{f} \times \bar{V} = \frac{\partial}{\partial z}\left(K \frac{\partial \bar{V}}{\partial} \right) - \rho^{-1} \nabla p \qquad [23]$$

$\bar{V}$ is the velocity vector including the mean vertical velocity w; $\bar{f}$ is the Coriolis parameter times the vertical unit vector. The advective acceleration term, $\bar{V} \cdot \nabla \bar{V}$, and pressure gradient term are necessary to account for the horizontal inhomogeneity that is caused by the salinity flux at the lead surface.

The condition that separates the free convection regime of **Figure 2A** and the forced convection regime of **Figure 2B** is expressed by the relative magnitude of the pressure gradient, $\rho^{-1}\nabla_b p$, and turbulent stress, $\partial/\partial z(K\partial V/\partial z)$, terms in [23]. This ratio can be derived with addition of mass conservation and salt conservation equations, and if we assume the vertical equation is hydrostatic, $\partial p/\partial z = -g\rho = -gMS$, where M is the sensitivity of density to salinity. If we nondimensionalize the equations by the ice velocity U_i, mixed-layer depth, d, average salt flux at the lead surface, F_S, and friction velocity, u_{*0}, the ratio of the pressure gradient term to the turbulent stress term scales as eqn [24].

$$L_0 = \frac{gMF_S d}{\rho_0 U_i u_{*0}^2} \qquad [24]$$

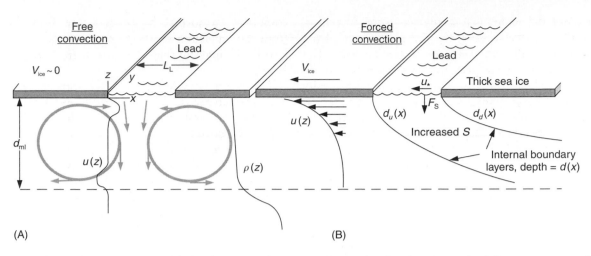

Figure 2 Modes of lead convection. (A) The free convection pattern that results when freezing and salt flux are strong, and the relative velocity of the ice is low. Cellular patterns of convective overturning are driven by pressure gradients that arise from the salinity disturbance due to ice formation. (B) The forced convection regime that exists when ice motion is strong. The salinity flux and change in surface stress in the lead cause a change in the character of the boundary layer that grows deeper downstream. The balance of forces is primarily Coriolis and turbulent diffusion of momentum. (From Morison JH, MCPhee MG, Curtin T and Paulson CA (1992) The oceanography of winter leads. *Journal of Geophysical Research* 97: 11199–11218.)

If this lead number is small because the ice is moving rapidly or the salt flux is small, the pressure gradient term is not significant in [23]. In this forced convection case, illustrated in **Figure 2B**, the boundary layer behaves as in the horizontally homogeneous case except that salt is advected and diffused away from the lead in the turbulent boundary layer.

If the lead number is large because the ice is moving slowly or the salt flux is large, the pressure gradient term is significant. In this free convection case the salinity disturbance is not advected away, but builds up under the lead. This creates pressure imbalances that can drive the type of cellular motion shown in **Figure 2A**.

Figure 3 shows conditions for which the lead number is unity for a range of ice thickness. Here the salt flux has been parametrized in terms of the air–sea temperature difference, and stress has been parametrized in terms of U_i. The figure shows the locus of points where L_0 is equal to unity. For typical winter and spring conditions, L_0 is close to 1, indicating that a mix of free and forced convection is common. Conditions where lead convection features have been observed are also shown in **Figure 3**. Most of these are in the free convection regime, probably because they are more obvious during quiet conditions.

There have been several dedicated efforts to study the effects of wintertime lead convection. The most recent example was the 1992 Lead Experiment (LeadEx) in the Beaufort Sea. **Figure 4** illustrates the average salinity profile at 9 m under a nearly stationary lead. The data was gathered with an autonomous underwater vehicle. Using the vehicle vertical motion as a proxy for vertical water velocity, it is also possible to estimate the salt flux $w'S'$. The lead was moving at 0.04 m s^{-1}, and estimates of salt flux put L_0 between 4 and 11 (free convection in **Figure 3**). Salinity increased in the downstream direction across the lead and reached a sharp maximum

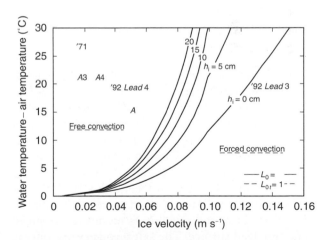

Figure 3 Air–water temperature difference versus U_i for L_0 equal to 1 for various ice thicknesses, h_i. Also shown are the temperature difference and ice velocity values for several observations of lead convection features such as underice plumes. Most of these are in the free convection regime: '71 denotes the AIDJEX pilot study; A3 denotes the 1974 AIDJEX Lead Experiment – lead 3 (ALEX3); A4 denotes ALEX4; A denotes the 1976 Arctic Mixed Layer Experiment; and '92 Lead 4 denotes the 1992 LeadEx lead 4. LeadEx lead 3 ('92 Lead 3) was close to $L_0 = 1$. (From Morison JH, McPhee MG, Curtin T and Paulson CA (1992) The oceanography of winter leads. *Journal of Geophysical Research* 97: 11199–11218.)

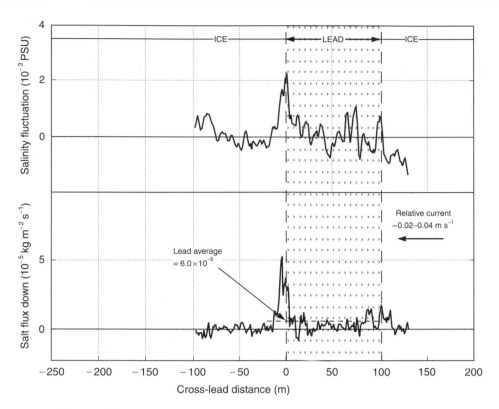

Figure 4 Composites of S' and $w'S'$ at 9 m depth measured with an autonomous underwater vehicle during four runs under lead 4 at the 1992 Lead Experiment. The horizontal profile data have been collected in 1 m bins. (From Morison JH, McPhee MG (1998) Lead convection measured with an autonomous underwater vechicle, *Journal of Geophysical Research* 103: 3257–3281.)

at the downstream edge. The salt flux was highest near the lead edges, but particularly at the downstream edge. With even a slight current, the downstream edge plume is enhanced by several factors. The vorticity in the boundary layer reinforces the horizontal density gradient at the downstream edge and counters the gradient at the upstream edge. The salt excess is greatest at the downstream edge by virtue of the salt that is advected from the upstream lead surface. The downstream edge plume is also enhanced by the vertical motion of water at the surface due to water the horizontal flow being forced downward under the ice edge.

Figure 5 shows the salt flux beneath a 1000 m wide lead moving at 0.14 m s^{-1} with L_0 equal to about 1 (**Figure 3**). Here the salt flux is more evenly spread under the lead surface. The salt flux derived from the direct $w'S'$ correlation method does show some enhancement at the lead edge. This may be partly due to the influence of pressure gradient forces and the reasons cited for the free convection case described above. The other factor that influences the convective pattern is the lead width. In the case of the 100 m lead in even a weak current, the convection may not be fully developed until the downstream edge is reached. For the 1000 m lead of the second case, the convection under the downstream portion of the

lead was a fully developed unstable boundary layer. The energy-containing eddies filled the mixed layer and their dominant horizontal wavelength was equal to about twice the mixed layer depth.

Effects of Horizontal Inhomogeneity: Summertime Buoyancy Flux

The behavior of the boundary layer under summer leads is relatively unknown compared to the winter lead process. Because of the important climate consequences, it is a subject of increasing interest. Summertime leads are thought to exhibit a critical climate-related feature of air–sea–ice interaction, ice-albedo feedback. This is because leads are windows that allow solar radiation to enter the ocean. The proportion of radiation that is reflected (albedo) from sea ice and snow is high (0.6–0.9) while that from open water is low (0.1). The fate of the heat that enters summer leads is important. If it penetrates below the draft of the ice, it warms the boundary layer and is available to melt the bottom of the ice over a large area. If most of the heat is trapped in the lead above the draft of the ice, it will be available to melt small pieces of ice and the ice

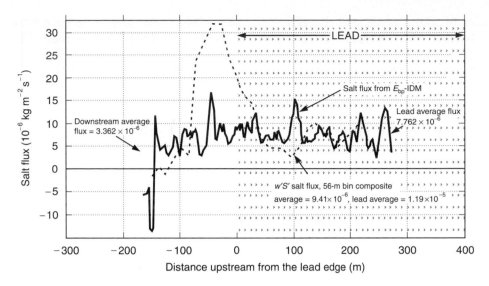

Figure 5 Composite average for autonomous underwater vehicle runs 1 to 5 at lead 3 of the 1992 Lead Experiment. The salinity is band-passed at 1 rad m^{-1} and is indicative of the turbulence level and is used to estimate the salt flux by the E_{bp} – IDM method of Morison and McPee (1998). The salt flux is elevated in the lead and decreases beyond about 72 m downstream of the lead edge. The composite average of $w'S'$ in 56 m bins for the same runs is also shown in the center panel. The average flux and the decrease downstream are about the same as given by the E_{bp} – IDM method, but $w'S'$ suggests elevated fluxes near the lead edge. (From Morison JH, McPhee MG (1998) Lead convection measured with an autonomous underwater vechicle, *Journal of Geophysical Research* 103: 3257–3281.)

floe edges. In the latter case the area of ice will be reduced and the area of open water increased. This allows even more solar radiation to enter the upper ocean, resulting in a positive feedback. This process may greatly affect the energy balance of an ice-covered sea. The critical unknown is the partition of heating between lateral melt of the floe edges and bottom melt.

There are fundamental similarities between the summertime and wintertime lead problems. The equations of motion ([15]–[24]) are virtually identical. Only the sign of the buoyancy flux is opposite. The heat flux is important to summer leads and tends to decrease the density of the surface waters. However, as with winter leads, the buoyancy flux is controlled mainly by salt. As the top surface of the ice melts, much of the water that does not collect in melt ponds on the ice surface instead runs into the leads. If the ambient ice velocity is low, ice melt from the bottom surface will tend to flow upward and collect in the leads as well. Thus leads are the site of a concentrated flux of fresh water accumulated over large areas of ice. If this flux, F_S, into the lead is negative enough relative to the momentum flux represented by u_{*0}, the lead number, L_0, will be a large negative number and shear production of turbulent energy will not be able to overcome the stabilizing buoyant production. This means turbulent mixing will be weak beneath the lead surface and a layer of fresh water will accumulate near the surface of the lead. The stratification

at the bottom of this fresh water layer may be strong enough to prevent mixing until a storm produces a substantial stress. This will be made even more difficult than in the winter situation because of the effect of stabilizing buoyancy flux on the boundary layer generally. The only way the fresh water will be mixed downward is by forced convection; there is no analogue to the wintertime free convection regime.

When there is sufficient stress to mix out a summertime lead, the pattern must resemble that of the forced convection regime in **Figure 2A**. At the upstream edge of the lead, fresh warm water will be mixed downward in an internal boundary layer that increases in thickness downstream until it reaches the steady-state boundary layer thickness appropriate for that buoyancy flux or the ambient mixed layer depth. The rate of growth should scale with the local value of u_{*0} (or perhaps $u_{*0}\eta_*$). At the downstream edge, another boundary layer conforming to the under-ice buoyancy flux and surface stress will begin to grow at a rate roughly scaling with the local u_{*0}. In spite of the generally stabilizing buoyancy flux, this process has the effect of placing colder, more saline water from under the ice on top of fresher and warmer (consequently lighter) water drawn from the lead. Thus, even embedded in the stable summer boundary layer, the horizontal inhomogeneity due to leads may create pockets of instability and more rapid mixing than might be expected on the basis of average conditions.

Recent studies of summertime lead convection at the 1997–98 Surface Heat Budget of the Arctic experiment saw the salinity decrease in the upper 1 m of leads to near zero and temperatures increase to more than 0°C. Only when ice velocities were driven by the wind to speeds of nearly 0.2 m s^{-1} were these layers broken down and the fresh, warm water mixed into the upper ocean. At these times the heat flux measured at 5 m depth reached values over 100 W m^{-2}. The criteria for the onset of mixing are being studied along with the net effect of the growing internal boundary layers. Even with an understanding of the mixing process, it will be a challenge to apply this information to larger-scale models, because the mixing is nonlinearly dependent on the history of calm periods and strong radiation.

Internal Waves and their Interaction with the Ice Cover

One of the first studies of internal waves originated with observations made by Nansen during his 1883 expedition. It did not actually involve interaction with the ice cover, but with his ship the *Fram*. He found that while cruising areas of the Siberian shelf covered with a thin layer of brackish water, the *Fram* had great difficulty making any headway. It was hypothesized by V. Bjerknes and proved by Ekman that this 'dead water' phenomenon was caused by the drag of the internal wave wake produced by the ship's hull as it passed through the shallow surface layer. This suggests that internal wave generation by deep keels may cause drag on moving ice. Evidence of internal wave generation by keels has been observed by several authors, but estimates of the amount of drag vary widely. This is due mainly to wide differences in the separation of the stratified pycnocline and the keels.

The drag produced by under-ice roughness of amplitude h_0 with horizontal wavenumber β moving at velocity V_i (magnitude v_i) over a pycnocline with stratification given by Brunt–Vaisala frequency, N, a depth d below the ice–ocean interface, can be expressed as an effective internal wave stress (eqn [25]), where C_{wd} (eqn [26]) accounts for the drag that would exist if there were no mixed layer between the ice and the pycnocline.

$$\Sigma_{iw} = -\Gamma C_{wd} V_i \qquad [25]$$

$$C_{wd} = \tfrac{1}{2}\beta_x^2 h_0 [(\beta_c^2/\beta_x^2) - 1]^{1/2} \qquad [26]$$

The wavenumber in the direction of the relative ice velocity, V_i, is β_x, and β_c is the critical wave number

above which the waves are evanescent ($\beta_c = N/v_i$). Γ is an attenuation factor that accounts for the separation of the pycnocline from the ice by the mixed layer of depth d (eqn [27]).

$$\Gamma = \left(\sinh^2(\beta d) \left\{ \left[\coth(\beta d) - \frac{\beta \Delta b}{v_i^2 \beta_x^2} \right]^2 + \frac{N^2}{v_i^2 \beta_x^2} - 1 \right\} \right)^{-1} \qquad [27]$$

Δb is the strength of the buoyancy jump at the base of the mixed layer. For wavenumbers of interest and d much bigger than about 10 m, Γ becomes small and internal wave drag is negligible. Thus it is not a factor in the central Arctic over most of the year. However, in the summer pack ice, and many times in the marginal ice zone, stratification will extend to or close to the surface. Then internal wave drag can be at least as important as form drag.

The ice cover also uniquely affects the ambient internal wave field. In most of the world ocean the internal wave energy level, when normalized for stratification, is remarkably uniform. It has been established by numerous studies that the internal wave energy in the Arctic Ocean is typically several times lower. In part this may be due to the absence of surface gravity waves. The other likely reason is that friction on the underside of the ice damps internal waves. Decomposing the internal wave field into vertical modes, one finds the mode shapes for horizontal velocity are a maximum at the surface. This is perfectly acceptable in the open water situation. However, at the horizontal scales of most internal waves, an ice cover imposes a surface boundary condition of zero horizontal velocity. The effect of this can be estimated by assuming that a time-varying boundary layer is associated with each spectral component of the internal wave field. This is not rigorously correct because all the modes interact in the same nonlinear boundary layer, and are thereby coupled. However, in the presence of a dominant, steady current due to ice motion, the effect on the internal wave modes can be linearized and considered separately. The near-surface internal wave velocity can be approximated as a sum of rotary components (eqn [28]).

$$V(z) = \sum_{n=0}^{M} D_n(z) e^{i\omega_n t} = \sum_{n=0}^{M} [A_n(z) + iB_n(z)] e^{i\omega_n t} \qquad [28]$$

The internal wave motion away from the boundary $D_{\infty n}$ can be subtracted from the linear time-varying boundary layer equation (eqn [1] with the addition of the time variation acceleration, $\partial V/\partial t$). This yields an equation for each rotary component of velocity in the boundary layer (eqn [29]).

$$i(\omega_n + f)(D_n - D_{\infty n}) = \frac{\partial}{\partial z} K \frac{\partial D_n}{\partial z} \qquad [29]$$

$$D_n = 0 \quad \text{at } z = z_0$$
$$D_n = D_{\infty n} \quad \text{at } z = d$$

This oscillating boundary layer equation can be solved for K of the form $K = ku_{*0}z\exp(-6|\omega + f|z/u_{*0})$. When we do this for representative internal wave conditions in the Arctic and compute the energy dissipation, we find the timescale required to dissipate the internal wave energy through under-ice friction is 32 days. This is a factor of 3 smaller than is typical for open ocean conditions. Assuming a steady state with internal wave forcing and other dissipation mechanisms in place, the under-ice boundary layer will result in a 75% reduction in steady-state internal wave energy. This suggests the effect of the under-ice boundary layer is critical to the unique character of internal waves in ice-covered seas.

Outstanding Issues

The outstanding issue of ice–ocean interaction is how the small-scale processes in the ice and at the interface affect the exchange between the ice and water. This is arguably most urgent in the case of heat and salt exchange during ice growth. When we apply laboratory-derived concepts for the diffusion of heat and salt to the ice–ocean interface, we get results that are not supported by observation, such as spontaneous frazil ice formation and large ocean heat flux under thin ice.

These results are causing significant errors in large-scale models. They stem from a molecular sublayer model of the ice–ocean interface (**Figure 1**) and the difference between the molecular diffusivities of heat and salt. What seems to be wrong is the molecular sublayer model. Recent results in the microphysics of ice growth reveal that the structure and thermodynamics of the growing ice produce instabilities and convection within the ice and extending into the water. The ice surface is thus not a passive, smooth surface covered with a thin molecular layer. Rather it is field of jets emitting plumes of supercooled, high-salinity water at a very small scale. This type of

unstable convection likely tends to equalize the diffusion of heat and salt relative to the apparently unrealistic parameterizations we are using now.

Similarly, we do not really understand how the turbulent stress we might measure in the surface layer is converted to drag on the ice. Certainly a portion of this is through viscous friction in the molecular sublayer. However, in most cases the underside of the ice is not hydrodynamically smooth, which suggests that pressure force acting on the bottom roughness elements are ultimately transferring a large share of the momentum. Understanding this will require perceptual breakthroughs in our view of how turbulence and the mean flow interact with a rough surface buried in a boundary layer. Achieving this understanding is complicated greatly by a lack of contemporaneous measurements of turbulence and under-ice topography at the appropriate scales. This drag partition problem is general and not limited to the under-ice boundary layer. However, the marvelous laboratory that the under-ice boundary layer provides may be the place to solve it.

See also

Coupled Sea Ice–Ocean Models. Sea Ice. Sea Ice: Overview. Sub Ice-Shelf Circulation and Processes.

Further Reading

Johannessen OM, Muench RD, and Overland JE (eds.) (1994) *The Polar Oceans and Their Role in Shaping the Global Environment: The Nansen Centennial Volume.* Washington, DC: American Geophysical Union.

McPhee MG (1994) On the turbulent mixing length in the oceanic boundary layer. *Journal of Physical Oceanography* 24: 2014–2031.

Morison JH, McPhee MG, and Maykutt GA (1987) Boundary layer, upper ocean and ice observations in the Greenland Sea marginal ice zone. *Journal of Geophysical Research* 92(C7): 6987–7011.

Morison JH and McPhee MG (1998) Lead convection measured with an autonomous underwater vehicle. *Journal of Geophysical Research* 103(C2): 3257–3281.

Smith WO (ed.) (1990) *Polar Oceanography.* San Diego, CA: Academic Press.

Wettlaofer JS (1999) Ice surfaces: macroscopic effects of microscopic structure. *Philosphical Transactions of the Royal Society of London A* 357: 3403–3425.

COUPLED SEA ICE–OCEAN MODELS

A. Beckmann and G. Birnbaum, Alfred-Wegener-Institut für Polar und Meeresforschung, Bremerhaven, Germany

Introduction

Oceans and marginal seas in high latitudes are seasonally or permanently covered by sea ice. The understanding of its growth, movement, and decay is of utmost importance scientifically and logistically, because it affects the physical conditions for air–sea interaction, the large-scale circulation of atmosphere and oceans and ultimately the global climate (e.g., the deep and bottom water formation) as well as human activities in these areas (e.g., ship traffic, offshore technology).

Coupled sea ice–ocean models have become valuable tools in the study of individual processes and the consequences of ice–ocean interaction on regional to global scales. The sea ice component predicts the temporal evolution of the ice cover, thus interactively providing the boundary conditions for the ocean circulation model which computes the resulting water mass distribution and circulation.

A number of important feedback processes between the components of the coupled system can be identified which need to be adequately represented (either resolved or parameterized) in coupled sea ice–ocean models (see **Figure 1**):

- ice growth through freezing of sea water, the related brine release and water mass modification;
- polynya maintenance by continuous oceanic upwelling;
- lead generation by lateral surface current shear and divergence;
- surface buoyancy loss causing oceanic convection; and
- pycnocline stabilization in melting regions.

Not all of these are equally important everywhere, and it is not surprising that numerous variants of coupled sea ice–ocean models exist, which differ in physical detail, parameterizational sophistication, and numerical formulation. Models for studies with higher resolution usually require a higher level of complexity.

The main regions for applying coupled sea ice–ocean models are the Arctic Ocean, the waters surrounding Antarctica, and marginal seas of the Northern Hemisphere (e.g., Baltic Sea, Hudson Bay). A universally applicable model needs to include (either explicitly or by adequate parameterization) the

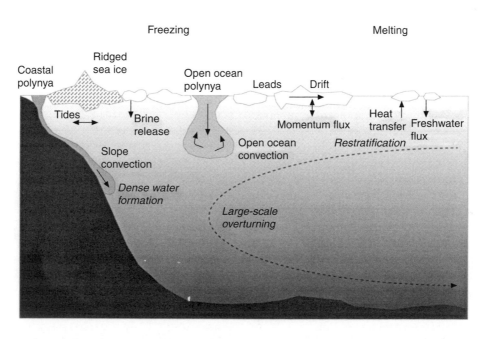

Figure 1 Cartoon of coupled sea ice–ocean processes. Effects in the ocean are in *italics*.

specific mechanisms of each region, e.g., mainly seasonal variations of the ice cover or the presence of thick, ridged multiyear ice.

This article describes the philosophy and design of large- and mesoscale prognostic dynamic–thermodynamic sea ice models which are coupled to primitive equation ocean circulation models (*see* General Circulation Models). The conservation principles, the most widely used parameterizations, several numerical and coupling aspects, and model evaluation issues are addressed.

Basics

Sea water and sea ice as geophysical media are quite different; whereas the liquid phase is continuous, three-dimensional, and largely incompressible, the solid phase can be best characterized as granular, two-dimensional and compressible. Both share a high degree of nonlinearity, and many direct feedbacks between oceanic and sea ice processes exist.

Today's coupled sea ice–ocean models are Eulerian; granular Lagrangian models, which consider the floe–floe interaction explicitly, exist but have so far not been fully coupled to ocean circulation models. Originally designed for use with large-scale coarse resolution ocean models, sea ice in state-of-the-art models is treated as a continuous medium. Following the continuum hypothesis only the average effect of a large number of ice floes is considered, assuming that averaged ice volume and velocities are continuous and differentiable functions of space and time.

Thus, very similar numerical methods are being applied to both sea water and sea ice, which greatly facilitates the coupling of models of these two components of the climate system. Conceptually, the continuum approach limits the applicability of sea ice models to grid spacings that are much larger than typical floes, i.e., several to several tens of kilometers. Typical values are 100–300 km for global climate studies, 10–100 km for regional climate simulations and about 2 km for process studies and quasi-operational forecasts. The latter cases are stretching the continuum concept for sea ice quite a bit, but still give reasonable results.

State-of-the-art coupled sea ice–ocean models are based on two principles for the description of sea ice, the conservation of mass and momentum, covering its thermodynamics and dynamics. Mass conservation for snow is also taken into account. A snow layer modifies the thermal properties of the ice cover through an increased albedo and reduced conductivity. This leads to delayed surface melting and lower basal freezing rates. In the following description sea ice is frozen sea water, and ice is the sum of sea ice and snow.

The temporal change of sea ice and snow volume due to local sources/sinks and drift is described by the mass conservation equation

$$\frac{\partial h_{(i,s)}}{\partial t} + \nabla \cdot \left(\vec{v}_i h_{(i,s)} \right) = S_{h_{(i,s)}}^{thdyn} \quad [1]$$

where $h_{(i,\,s)}$ is the sea ice and snow volume per unit area (with $h = h_i + h_s$), respectively, v_i is the two-dimensional ice velocity vector and $S_{h_{(i,s)}}^{thdyn}$ denotes the thermodynamic sources and sinks for sea ice and snow. The corresponding ice velocities are obtained from the momentum equation

$$(\rho_i h_i + \rho_s h_s) \left[\frac{\partial \vec{v}_i}{\partial t} + \vec{v}_i \cdot \nabla \vec{v}_i + f \vec{k} \times \vec{v}_i + g \nabla \mathscr{H} \right]$$
$$= \vec{\tau}_{ai} - \vec{\tau}_{iw} + \vec{\mathscr{F}}_i \quad [2]$$

where the local time rate of change, advection of momentum and the accelerations due to Coriolis are included. The wind stress $\vec{\tau}_{ai}$ is external to the coupled sea ice–ocean model; the ocean surface current stress $\vec{\tau}_{iw}$ and the sea surface height $\mathscr{H}$ is part of the coupling to the ocean. The so-called ice stress term $\vec{\mathscr{F}}_i$ summarizes all internal forces generated by floe–floe interactions.

Subgridscale Parameterizations

The granular nature of the medium, combined with the strong sensitivity of both thermodynamics and dynamics on the number, size, and thickness of individual ice floes requires the inclusion of a subgridscale structure of the modeled ice.

Ice Classes

An obvious assumption is that of a subgridscale ice thickness distribution. In this widely used approach the predicted ice volume h is thought to be the average of several compartments, the thermodynamic ice classes, which represent both thinner and thicker ice and possibly include open water. The relative contribution of an ice class is fixed (e.g., uniform between 0 and twice the average ice thickness). For each ice class, a separate thermodynamic balance is computed; the resulting fluxes are then averaged according to their relative areal coverage.

Ice Categories

As the subgridscale distribution of ice thickness will change with time and location, an even more sophisticated approach considers time-evolution of the volume of each compartment, which is then called an ice category. The form of these prognostic equations follows the conservation eqn [1]. Different subgridscale ice velocities are not taken into account; the advection of all compartments takes place with the resolved velocity field.

Models with several ice categories are in use. A minimum requirement, however, has been identified in the discrimination between ice-covered areas and open water. Then the prognostic equation for ice volume [1] is accompanied by a formally similar equation for ice concentration, i.e., the percentage of ice-covered area per unit area A,

$$\frac{\partial A}{\partial t} + \nabla \cdot (\vec{v}_i A) = S_A^{thdyn} + S_A^{thdyn} \quad [3]$$

Thermodynamic sources and sinks S_A^{thdyn} for ice concentration are chosen empirically and involve parameterizations of subgridscale thermodynamic melting and freezing. The conceptual *ansatz*:

$$S_A^{thdyn} = \frac{1-A}{h_{cls}} \max\left[0, \frac{\partial h}{\partial t}\right] + \frac{A}{h_{opn}} \min\left[\frac{\partial h}{\partial t}, 0\right] \quad [4]$$

describes the formation of new ice between the ice floes with the first term on the right hand side; here h_{cls} is the so-called lead closing parameter. The second term parameterizes basal melting of sea ice with a similar approach involving h_{opn}. The coefficients are often derived from the assumed internal structure of the ice-covered portion of the grid cell, the ice classes. The dynamic source/sink term S_A^{thdyn} follows from the constitutive law (see section on dynamics below). Conceptually, the ice concentration (or compactness) A has to lie between 0 and 1, which has to be enforced separately.

With the introduction of an ice concentration, the ice volume h has to be replaced by the actual ice thickness

$$h^* = \frac{h}{A} \quad [5]$$

i.e., the mean value of individual floe height, such that the total ice volume per grid box is not affected by this approach.

An approach using two categories, open water and ice-covered areas with an internal structure (ice classes), has been proven highly adequate for a large number of situations.

Thermodynamics

Mass conservation for ice is closely tied to the heat balance at its surfaces. Sea ice forms, if the freezing temperature of sea water is reached. The surface freezing point T_f (in K) is a function of salinity, estimated by the polynomial approximation

$$T_f = 273.15 - 0.0575 S_w + 1.710523 \times 10^{-3} S_w^{3/2} - 2.154996 \times 10^{-4} S_w^2 \quad [6]$$

where S_w is the sea surface salinity.

The majority of today's sea ice thermodynamics models is based on a one-dimensional (vertical) heat diffusion equation, which for sea ice (without snow cover) reads

$$\rho_i c_{pi} \frac{\partial T_i}{\partial t} = \frac{\partial}{\partial z}\left(k_i \frac{\partial T_i}{\partial z}\right) + K_i I_{oi} \exp[-K_i z] \quad [7]$$

Here, T_i, ρ_i, c_{pi} and k_i are the sea ice temperature, density, specific heat, and thermal conductivity, respectively. The net short-wave radiation at the sea ice surface is I_{oi} and K_i is the bulk extinction coefficient.

If a snow cover is present, the penetrating short wave radiation is neglected and a second prognostic equation for the snow is solved:

$$\rho_s c_{ps} \frac{\partial T_s}{\partial t} = \frac{\partial}{\partial z}\left(k_s \frac{\partial T_s}{\partial z}\right) \quad [8]$$

At the snow–sea ice interface, the temperatures and fluxes have to match.

Assuming that ice exists, the local time rate of change of ice thickness due to freezing of sea water or melting of ice is the result of the energy fluxes at the surface and the base of the ice. At the surface, the ice temperature and thickness change is determined from the energy balance equation:

$$Q_{a(i,s)} = (1 - \alpha_{(i,s)})(1 - I_{oi})\mathcal{R}_{SW}^{\downarrow} + \mathcal{R}_{LW}^{\downarrow} - \varepsilon_{(i,s)}\sigma_0 T_{0(i,s)}^4 + Q_l + Q_s + Q_c$$

$$= \begin{cases} 0 & \text{if } T_{o(i,s)} < T_{m(i,s)} \\ -\rho_{(i,s)} L_{(i,s)} \frac{\partial h^*}{\partial t} & \text{if } T_{o(i,s)} = T_{m(i,s)} \end{cases} \quad [9]$$

where $T_{o(i,s)}$ and $T_{m(i,s)}$ are the surface and melting temperatures, respectively, $\rho_{(i,s)}$ and $L_{(i,s)}$ are the density and heat of fusion for sea ice and snow. Besides the conductive heat flux in the ice $Q_c = k_{(i,s)} \partial T_{(i,s)}/\partial t$ the following atmospheric fluxes are considered: downward short-wave radiation $\mathcal{R}_{SW}^{\downarrow}(\phi, \lambda, A_{cl})$, net long-wave radiation $\mathcal{R}_{LW}^{\downarrow}(T_a, A_{cl}) - \varepsilon_{(i,s)}\sigma_o T_{o(i,s)}^4$, as well as sensible $Q_s(\vec{v}_a, T_a, T_{o(i,s)})$ and latent $Q_l(\vec{v}_a, q_a, q_{o(i,s)})$ heat fluxes. The albedos $\alpha_{(i,s)}$ and emissivities $\in_{(i,s)}$ are dependent on the surface structure of the medium (sea ice, snow). The atmospheric forcing data are:

- near-surface wind velocity $\vec{v}_a$;
- near-surface atmospheric temperature T_a;
- near-surface atmospheric dew point temperature T_d, or specific humidity q_a;
- cloudiness A_{cl};
- precipitation P and evaporation E (needed for the sea ice and snow mass balance).

Note that the wind velocity is also needed for the atmospheric forcing in the momentum [2].

At the base of the ice (the sea surface), an imbalance of the conductive heat flux in the sea ice (Q_c) and the turbulent heat flux from the ocean

$$Q_{sw} = \rho_w c_{pw} c_h u_\star \left(T_f - T_{ml} \right) \qquad [10]$$

leads to a change in ice thickness:

$$Q_{iw} = -Q_{ow} - Q_c = -\rho_i L_i \frac{\partial h_i^*}{\partial t} \qquad [11]$$

Here, T_{ml} is the ocean surface and mixed layer temperature, c_h is the heat transfer coefficient and $u_\star$ is the friction velocity. In general, the main sink for ice volume is basal melting due to above-freezing temperatures in the oceanic mixed layer. The source for snow is a positive rate of P–E, if the air temperature is below the freezing point of fresh water. The main source for sea ice is basal freezing. However, the formation of additional sea ice on the upper ice surface is possible through a process called flooding. This conversion of snow into sea ice takes place when the weight of the snow exceeds the buoyancy of the ice and sea water intrudes laterally.

Often, the vertical structure of temperature is approximated by simple zero-, one- or two-layer formulations, with the resulting internal temperature profile being piecewise linear. The most simple approach, the zero-layer model, eliminates the capacity of the ice to store heat. However, it has been used successfully in areas where sea ice is mostly seasonal and thus relatively thin (<1 m).

The specifics of the brine-related processes in the sea ice are difficult to implement in models. As a consequence, sea ice models usually assume constant sea ice salinity S_i of about 5 PSU (practical salinity units) to calculate the heat of fusion and the vertical heat transfer coefficient. The errors arising from this assumption are largest during the early freezing processes, when salt concentrations are considerably higher.

The open water part of each grid cell, where the atmosphere is in direct contact with the ocean, is treated like any other air–sea interface. The thermodynamic eqns [9] and [11] are modified to the radiative and heat fluxes between ocean and atmosphere. In the case of heat loss resulting in an ocean temperature below the freezing point T_f, new ice is formed:

$$
\begin{aligned}
Q_{aw} &= (1 - \alpha)(1 - I_{ow})\mathscr{R}_{SW}^{\downarrow} + \mathscr{R}_{LW}^{\downarrow} - \varepsilon_w \sigma_0 T_{0w}^4 \\
&\quad + Q_l + Q_s + Q_{sw} \\
&= \begin{cases} 0 & \text{if } T_{ow} > T_f \\ -\rho_i L_i \frac{\partial b^*}{\partial t} & \text{if } T_{ow} = T_f \end{cases}
\end{aligned} \qquad [12]
$$

In the case of above-freezing ocean surface temperatures, Q_{sw} follows from [10] with T_f replaced by T_{ow}. An illustration summarizing the fluxes is given in **Figure 2**.

The solution of eqns [7], [8], [9] and [11] is conceptually straightforward but algebraically complicated in that it involves iterative solution of the energy balance equation to obtain the surface temperature.

Dynamics

Driven by wind and surface ocean currents, sea ice grown locally is advected horizontally. Free drift (the absence of internal ice stresses) is a good approximation for individual ice floes. In a compact ice cover, however, internal stresses will resist further compression and react to shearing stresses. These internal ice forces are expressed as the divergence of the isotropic two-dimensional internal stress tensor

$$\vec{\mathscr{F}}_i = \nabla \cdot \sigma \qquad [13]$$

which depends on the stress–strain relationship, where the deformation rates are proportional to the spatial derivatives of ice velocities. A general form of the constitutive law is

$$\sigma = \begin{pmatrix} \eta\left(\frac{\partial u_i}{\partial x} - \frac{\partial v_i}{\partial y}\right) + \zeta\left(\frac{\partial u_i}{\partial x} + \frac{\partial v_i}{\partial y}\right) - \frac{1}{2}P_i & \eta\left(\frac{\partial u_i}{\partial y} + \frac{\partial v_i}{\partial x}\right) \\ \eta\left(\frac{\partial u_i}{\partial y} - \frac{\partial v_i}{\partial x}\right) & \eta\left(\frac{\partial v_i}{\partial y} - \frac{\partial u_i}{\partial x}\right) + \zeta\left(\frac{\partial u_i}{\partial x} + \frac{\partial v_i}{\partial y}\right) - \frac{1}{2}P_i \end{pmatrix} \qquad [14]$$

here, ζ and η are nonlinear viscosities for compression and shear. P_i is the ice strength.

The most widely used sea ice rheology is based on the viscous–plastic approach. Introduced as the result of the ice dynamics experiment AIDJEX, it has proven to be a universally applicable rheology. It treats the ice as a linear viscous fluid for small deformation rates and as a rigid plastic medium for larger deformation rates. Simpler rheologies have been tested but could not reproduce the observed ice distributions and thicknesses nearly as well as the viscous–plastic approach. The viscosities are then

$$\zeta = e^2 \eta = \frac{P_i}{2\Delta} \qquad [15]$$

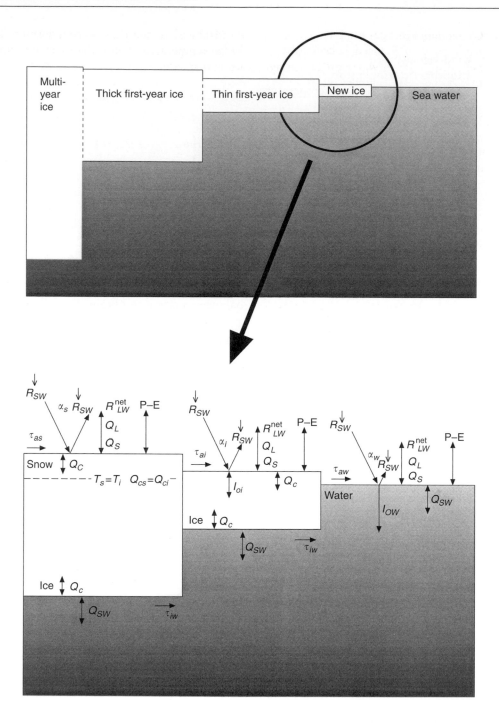

Figure 2 Schematic of the concept of ice classes/categories, surface energy balance and the ice–ocean flux coupling. The upper panel shows a grid cell covered with several classes/categories of ice, including open water. The lower panel is a detailed view at the fluxes between atmosphere, ice, and ocean, in the three cases: snow-covered sea ice, pure sea ice, and open water.

where an elliptic yield curve of ellipticity e is assumed, with the deformation rate given by

$$\Delta = \left[\left\{ \left(\frac{\partial u_i}{\partial x} \right)^2 + \left(\frac{\partial v_i}{\partial y} \right)^2 \right\} (1 + e^{-2}) + \left(\frac{\partial u_i}{\partial y} + \frac{\partial v_i}{\partial x} \right)^2 e^{-2} \right.$$
$$\left. + 2 \frac{\partial u_i}{\partial x} \frac{\partial v_i}{\partial y} (1 - e^{-2}) \right]^{1/2}$$

[16]

In particular, internal forces are only important for densely packed ice floe fields, i.e., for ice concentrations exceeding 0.8. Most sea ice models take this into account by assuming that the ice strength is

$$P_i = P^* h \exp[-C^*(1 - A)]$$

[17]

where P^* and C^* are empirical parameters. The same functional dependence is also used successfully to describe the generation of open water areas through shear deformation, which is parameterized by

$$S_A^{dyn} = -0.5\left(\Delta - \left|\nabla \cdot \vec{v}_i\right|\right)\exp\left[- C^*(1 - A)\right] \quad [18]$$

Subgridscale processes like ridging and rafting can be successfully parameterized this way.

Coupling

Numerical ocean circulation models are described in article General Circulation Models. A schematic illustration of the interactions in a coupled sea ice–ocean model is given in **Figure 3**. The coupling between the sea ice and ocean components is done via fluxes of heat, salt, and momentum. They enter the ocean model through the surface boundary conditions to the vertically diffusive/viscous terms. Given the relative (to the depth of the ocean) small draught of the ice, it is assumed not to deform the sea surface, all boundary conditions are applied at the air–sea interface.

Due to the presence of subgridscale ice categories, the fluxes have to be weighed with the areal coverage of open water, and ice of different thickness. The resulting boundary conditions for the simplest two-category (ice and open water) case are:

$$A_v^M \frac{\partial \vec{v}_w}{\partial z}\bigg|_{z=0} = A\vec{\tau}_{iw} + (1 - A)\vec{\tau}_{aw} \quad [19]$$

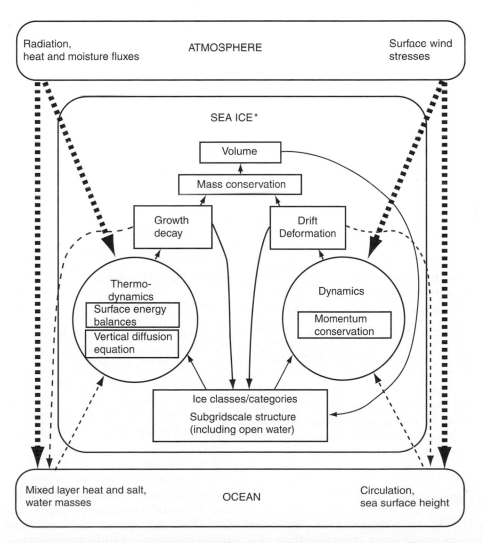

Figure 3 Concept of a coupled sea ice–ocean numerical model with prescribed atmospheric forcing. Thick broken arrows represent the atmospheric forcing, thin dashed arrows represent the coupling pathways between ice and ocean, and thin arrows indicate how the ice volume is computed from thermodynamical and dynamical principles (mass and momentum conservation), with the assumption of a subgridscale ice distribution. *Includes both sea ice and snow.

$$A_v^T \frac{\partial T_w}{\partial z}\bigg|_{z=0} = -\frac{1}{\rho_w c_p}(AQ_{iw} + (1-A)Q_{aw}) \quad [20]$$

$$A_v^s \frac{\partial S_w}{\partial z}\bigg|_{z=0} = (S_w - S_i)\frac{\rho_i}{\rho_w}\frac{\partial h_i}{\partial t}$$

$$+ S_w \begin{cases} P - E & \text{if } h = 0 \\ P - E & \text{if } h > 0, T_a > 0 \\ (1-A)(P-E) & \text{if } h > 0, T_a < 0 \\ \frac{\rho_s}{\rho_w}\frac{\partial h^s}{\partial t} & \text{if } h_s > 0, T_a > 0 \end{cases} \quad [21]$$

where the freshwater flux is converted into a salt flux, and the momentum exchange is parameterized (like the atmospheric wind forcing $\vec{\tau}_{aw}$ and $\vec{\tau}_{ai}$) in the form of the usual quadratic drag law:

$$\vec{\tau}_{iw} = \rho_w C_w |\vec{v}_i - \vec{v}_w|$$
$$\left[(\vec{v}_i - \vec{v}_w)\cos\theta_{iw} + \vec{k} \times (\vec{v}_i - \vec{v}_w)\sin\theta_{iw}\right] \quad [22]$$

Here, C_w is the drag coefficient and θ_{iw} the rotation angle. The vertical viscosities and diffusivities (A_v^M, A_v^T, A_v^S) are ocean model parameters.

The sea surface height required to compute the ice momentum balance is either taken directly from the ocean model (for free sea surface models) or computed diagnostically from the upper ocean velocities using the geostrophic relationship.

The described coupling approach can be used between any sea ice and ocean model, irrespective of vertical resolution, or the use of a special surface mixed layer model. However, the results may suffer from a grid spacing that does not resolve the boundary layer sufficiently well.

A technical complication results from the different timescales in ocean and ice dynamics. With the implicit solution of the ice momentum equations, time steps of several hours are possible for the evolution of the ice. General ocean circulation models, on the other hand, require much smaller time steps, and an asynchronous time-stepping scheme may be most efficient. In that case, the fluxes to the ocean model remain constant over the long ice model step, whereas the velocities of the ocean model enter the fluxes only in a time-averaged form, mainly to avoid aliasing of inertial waves.

Numerical Aspects

Equations [1]–[3] are integrated as an initial boundary problem, usually on the same finite difference grid as the ocean model. A curvilinear coordinate system may be used to conform the ice model grid to an irregular coastline or to locally increase the resolution. The horizontal grid is usually staggered, either of the 'B' or 'C' type. Both have advantages and disadvantages: the 'B' grid has been favored because of the more convenient formulation of the stress terms and the better representation of the Coriolis term; the 'C' grid avoids averaging for the advection and pressure gradient terms. The treatment of coastal boundaries and the representation of flow through passages is also different.

Due to the large nonlinear viscosities in the viscous plastic approach, an explicit integration of the momentum equations would require time steps of the order of seconds, whereas the thickness equations can be integrated with time steps of the order of hours. Therefore, the momentum equations are usually solved implicitly. This leads to a nonlinear elliptic problem, which is solved iteratively. An explicit alternative has been developed for elastic–viscous–plastic rheology.

Other general requirements for numerical fluid dynamics models also apply: a positive definite and monotonic advection scheme is desired to avoid negative ice volume and concentration with the numerical implementation and algorithms depending on the computer architecture (serial, vector, parallel). Finally, the implementation needs to observe the singularities of the system of ice equations, which occur when h, A and Δ approach zero. Minimum values have to be specified to avoid vanishing ice volumes, concentrations, and deformation rates.

Model Evaluation

Modeling systems need to be validated against either analytical solutions or observational data. The various simplifications and parameterizations, as well as the specifics of the numerical implementation of both components' thermodynamics and dynamics and their interplay make this quite an extensive task. Analytical solutions of the fully coupled sea ice–ocean system are not known, and so model validation and optimization has to rely on geophysical observations.

Since *in situ* measurements in high latitude ice-covered regions are sparse, remote sensing products are being used increasingly to improve the spatial and temporal coverage of the observational database. Data sets of sea ice concentration, thickness and drift, ocean sea surface temperature, salinity, and height, are currently available, as well as hydrography and transport estimates.

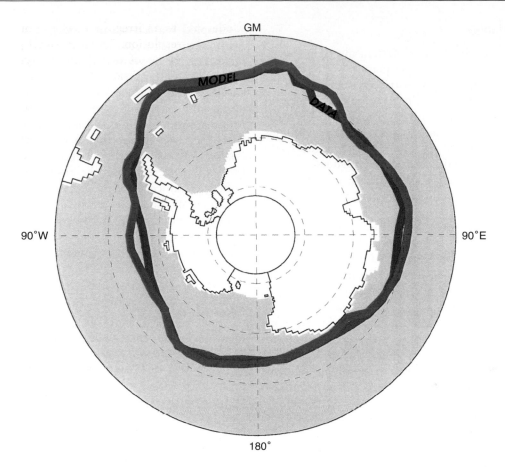

Figure 4 September 1987 simulated (blue) and remotely observed (red) sea ice edge around Antarctica. After Timmermann *et al.*, 2001.

Ice Variables

The most widely used validation variable for sea ice models is the ice concentration, i.e., the percentage of ice-covered area per unit area, which can be obtained from satellite observations. From these observations, maps of monthly mean sea ice extent are constructed, and compared to model output (see **Figure 4**). It should be noted though that the modeled ice concentrations represent a subgridscale parameterization with an empirically determined source/sink relationship such that an optimization of a model with respect to this quantity may be misleading.

A more rigorous model evaluation focuses on sea ice thickness or drift, which is a conserved quantity and more representative of model performance. Unfortunately, 'ground truth' values of these variables are available in few locations and over relatively short periods only (e.g., upward looking sonar, ice buoys) and comparison of point measurements with coarse resolution model output is always problematic. A successful example is shown in **Figure 5**. The routine derivation of ice thickness estimates from satellite observations will be a major step in the validation

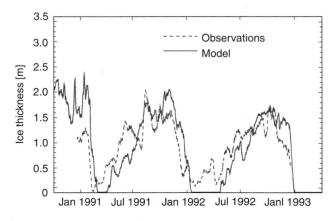

Figure 5 Time series of simulated (blue) and ULS (upward looking sonar) measured (red) sea ice thickness at 15°W, 70°S in the Weddell Sea. After Timmermann *et al.*, 2001.

attempts. The evaluation of ice motion is done through comparison between satellite-tracked and modeled ice buoys. Thickness and motion of first-year ice in free drift is usually represented well in the models, as long as atmospheric fields resolving synoptic weather systems are used to drive the system.

Ocean Variables

The success of the coupled system also depends on the representation of oceanic quantities. The model's temperature and salinity distribution as well as the corresponding circulation need to be consistent with prior knowledge from observations. The representation of water masses (characteristics, volumes, formation locations) can be validated against the existing hydrographic observational database, but a rigorous quantification of water mass formation products has been done only in few cases.

Parameter Sensitivities

Systematic evaluations of coupled model results have shown that a few parameter and conceptual choices are most crucial for model performance. For the sea ice component, these are the empirical source/sink terms for ice concentration, as well as the rheology. The most important oceanic processes are the vertical mixing (parameterizations), especially in the case of convection. In general, the formulation of the heat transfer between the oceanic mixed layer and the ice is central to the coupled system.

Finally, the performance of a coupled sea ice–ocean model will depend on the quality of the atmospheric forcing data; products from the weather centers (European Centre for Medium Range Weather Forecasts, National Centers for Environmental Prediction/National Center for Atmospheric Research) provide consistent, but still rather coarsely resolved atmospheric fields, which have their lowest overall quality in high latitudes, especially in areas of highly irregular terrain and for the P–E and cloudiness fields. Some errors, even systematic ones, are presently unavoidable.

All these data products are available with different temporal resolution. Unlike for stand-alone ocean models, which can be successfully run with climatological monthly mean forcing data, winds, sampled daily or 6-hourly, have been found necessary to produce the observed amount of ridging and lead formation in sea ice models.

Conclusions

A large amount of empirical information is needed for coupled sea ice–ocean models and the often strong sensitivity to variations of these makes the optimization of such modeling systems a difficult task. Yet, several examples of successful simulation of fully coupled ice–ocean interaction exist, which qualitatively and quantitatively compare well with the available observations.

Coupled sea ice–ocean modeling is an evolving field, and much needs to be done to improve parameterizations of vertical (and lateral) fluxes at the ice–ocean interfaces. For climate studies, water mass variability on seasonal and interannual timescales needs to be captured by the model. For operational forecast purposes, the ice thickness distribution in itself is most important; here atmospheric data quality and assimilation methods become crucial. Obvious next steps may be the inclusion of tides, icebergs, and ice shelves.

Ultimately, however, a fully coupled atmosphere–ice–ocean model is required for the simulation of phenomena that depend on feedback between the three climate system components.

See also

Arctic Ocean Circulation. General Circulation Models. Ice–Ocean Interaction. Sea Ice: Overview. Upper Ocean Heat and Freshwater Budgets.

Nomenclature

c_h	heat transfer coefficient
$c_{p(i,s,w)}$	specific heat of sea ice/snow/water $(\text{J kg}^{-1}\text{ K}^{-1})$
e	ellipticity
f	Coriolis parameter (s^{-1})
g	gravitational acceleration (m s^{-2})
h	ice (sea ice plus snow) volume per unit area (m)
$h_{(i,s)}$	sea ice/snow volume per unit area (m)
h^*	actual ice thickness (m)
$h^*_{(i,s)}$	actual sea ice/snow thickness (m)
h_{cls}, h_{opn}	lead closing/opening parameter (m)
$k_{(i,s)}$	thermal conductivity of sea ice/snow $(\text{W m}^{-1}\text{ K}^{-1})$
$\vec{k}$	vertical unit vector
q_a	atmospheric specific humidity
$q_{o(i,s)}$	surface specific humidity of sea ice/snow
t	time (s)
$\vec{v_i} = (u_i, v_i)$	horizontal ice velocity (m s^{-1})
$u_\star$	friction velocity (m s^{-1})
$\vec{v_a} = (u_a, v_a)$	wind velocity (m s^{-1})
$\vec{v}_w = (u_{w_1}, v_w)$	ocean surface velocity (m s^{-1})
x, y, z	spatial directions (m)
A	ice concentration
A_{cl}	cloudiness
A_v^M, A_v^T, A_v^S	oceanic vertical mixing coefficients $(\text{m}^2\text{ s}^{-1})$

C^*	empirical parameter
C_w	oceanic drag coefficient
$I_{o(i,w)}$	short wave radiation penetrating sea ice/water (W m^{-2})
E	evaporation (m s^{-1})
K_i	bulk extinction coefficient (m^{-1})
$L_{(i,s)}$	heat of fusion (J kg^{-1})
P	precipitation (m s^{-1})
P_i	ice strength (N m^{-1})
P^*	ice strength parameter (N m^{-2})
$Q_{a(i,s,w)}$	net energy flux between atmosphere and sea ice/snow/water (W m^{-2})
Q_c	conductive heat flux in the ice (W m^{-2})
Q_l, Q_s	atmospheric latent/sensible heat flux (W m^{-2})
Q_{iw}	turbulent heat flux at the ocean surface (W m^{-2})
Q_{sw}	oceanic sensible heat flux (W m^{-2})
$\mathcal{R}_{SW}^{\downarrow}, \mathcal{R}_{LW}^{\downarrow}$	downward short/long wave radiation (W m^{-2})
$S_{(i,w)}$	sea ice/sea water salinity (PSU)
T_a, T_i, T_w	air/ice/water temperature (K)
T_d	dew point temperature (K)
T_f	freezing temperature of sea water (K)
T_{ml}	oceanic mixed layer temperature (K)
$T_{m(i,s)}$	melting temperature of sea ice/snow at the surface (K)
$T_{o(i,s,w)}$	sea ice/snow/water surface temperature (K)
$\mathscr{F} \rightarrow$	internal ice forces (N m^{-2})
$\mathscr{H}$	sea surface elevation (m)
$S_A^{(thdyn,dyn)}$	source/sink terms for ice concentration (s^{-1})
$S_{h(i,s)}^{thdyn}$	source/sink terms for sea ice/snow volume per unit area (m s^{-1})
$\alpha_{(i,sw)}$	sea ice/snow/sea water albedo
Δ	ice deformation rate (s^{-1})
$\varepsilon_{(i,sw)}$	sea ice/snow/sea water emissivity
η, ζ	nonlinear viscosities (kg s^{-1})
$\nabla = (\frac{\partial}{\partial x}, \frac{\partial}{\partial y})$	horizontal gradient operator
λ, ϕ	geographical longitude/latitude (deg)
$\rho_{(a,i,s,w)}$	densities of air/sea ice/snow/water (kg m^{-3})
σ	two-dimensional stress tensor (N m^{-1})
σ_o	Stefan-Boltzmann constant (W m^{-2} K^{-4})
θ_{iw}	turning angle (deg)
$\overrightarrow{\tau_{ai}}, \overrightarrow{\tau_{iw}}, \overrightarrow{\tau_{aw}}$	air-ice/ice-water/air-water stress (N m^{-2})

Appendix

A typical parameter set for simulations with coupled dynamic–thermodynamic ice–ocean models (e.g., Timmermann et al., 2001), as shown in **Figures 4 and 5**, are

$\rho_a = 1.3$ kg m-3
$\rho_i = 910$ kg m-3
$\rho_s = 290$ kg m-3
$\rho_w = 1027$ kg m-3
$e = 2$
$C^* = 20$
$P^* = 2000$ N m^{-2}
$h_{cls} = 1$ m
$h_{opn} = 2$ h*
$C_w = 3 \times 10^{-3}$
$c_h = 1.2 \times 10^{-3}$
$\alpha_w = 0.1$
$\alpha_i = 0.75$
$\alpha_i = 0.65$ (melting)
$\alpha_s = 0.8$
$\alpha_w = 0.7$ (melting)
$K_i = 0.04$ m^{-1}
$S_i - 5$ PSU
$\theta = 10$ degrees
$c_{pi} = 2000$ J K^{-1} kg^{-1}
$c_{pw} - 4000$ J K^{-1} kg^{-1}
$c_{pa} = 1004$ J K^{-1} kg^{-1}
$L_i = 3.34 \times 10^5$ J kg^{-1}
$L_s = 1.06 \times 10^5$ J kg^{-1}
$k_i = 2.1656$ W m^{-1} K^{-1}
$k_s = 0.31$ W m^{-1} K^{-1}

Further Reading

Curry JA and Webster PJ (1999) *Thermodynamics of Atmospheres and Oceans*. London: Academic Press. International Geophysics Series..

Fichefet T, Goosse H, and Morales Maqueda M (1998) On the large-scale modeling of sea ice and sea ice–ocean interaction. In: Chassignet EP and Verron J (eds.) *Ocean Modeling and Parameterization*, pp. 399–422. Dordrecht: Kluwer Academic.

Haidvogel DB and Beckmann A (1999) *Numerical Ocean Circulation Modeling*. London: Imperial College Press.

Hibler WD III (1979) A dynamic-thermodynamic sea ice model. *Journal of Physical Oceanography* 9: 815–846.

Kantha LH and Clayson CA (2000) *Numerical Models of Oceans and Oceanic Processes*. San Diego: Academic Press.

Leppäranta M (1998) The dynamics of sea ice. In: Leppäranta M (ed.) *Physics of Ice-Covered Seas*, vol. 1, 305–342.

Maykut GA and Untersteiner N (1971) Some results from a time-dependent thermodynamic model of sea ice. *Journal of Geophysical Research* 76: 1550–1575.

Mellor GL and Häkkinen S (1994) A review of coupled ice–ocean models. In: Johannessen OM, Muench RD and Overland JE (eds) *The Polar Oceans and Their Role in Shaping the Global Environment*. AGU Geophysical Monograph, 85, 21–31.

Parkinson CL and Washington WM (1979) A large-scale numerical model of sea ice. *Journal of Geophysical Research* 84: 311–337.

Timmermann R, Beckmann A and Hellmer HH (2001) Simulation of ice–ocean dynamics in the Weddell Sea. Part I: Model description and validation. *Journal of Geophysical Research* (in press).

ICE SHELF STABILITY

C. S. M. Doake, British Antarctic Survey, Cambridge, UK

Introduction

Ice shelves are floating ice sheets and are found mainly around the Antarctic continent (**Figure 1**). They can range in size up to 500 000 km² and in thickness up to 2000 m. Most are fed by ice streams and outlet glaciers, but some are formed by icebergs welded together by sea ice and surface accumulation. They exist in embayments where the shape of the bay and the presence of ice rises, or pinning points, plays an important role in their stability. Ice shelves lose mass by basal melting, which modifies the properties of the underlying water mass and eventually influences the circulation of the global ocean, and by calving. Icebergs, sometimes more than 100 km in length, break off intermittently from the continent and drift to lower latitudes, usually breaking up and melting by the time they reach the Antarctic Convergence. The lowest latitude that ice shelves can exist is determined by the mean annual air temperature. In the Antarctic Peninsula a critical isotherm of about $-5°C$ seems to represent the limit of viability. In the last 40 years or so, several ice shelves have disintegrated in response to a measured atmospheric warming trend in the western and northern part of the Antarctic Peninsula. However, this warming trend is not expected to affect the stability of the larger ice shelves further south such as Filchner–Ronne or Ross, in the near future.

What are Ice Shelves?

Physical and Geographical Setting

An ice shelf is a floating ice sheet, attached to land where ice is grounded along the coastline. Nourished mainly by glaciers and ice streams flowing off the land, ice shelves are distinct from sea ice, which is formed by freezing of sea water. Most ice shelves are found in Antarctica where the largest can cover areas of 500 000 km² (e.g. Ross Ice Shelf and Filchner–Ronne Ice Shelf). Thicknesses vary from nearly 2000 m, around for example parts of the grounding line of Ronne Ice Shelf, to about 100 m at the seaward edge known as the ice front.

Typically, ice shelves exist in embayments, constrained by side walls until they diverge too much for the ice to remain in contact. Thus the geometry of the coastline is important for determining both where an ice shelf will exist and the position of the ice front. There are often localized grounding points on sea bed shoals, forming ice rises and ice rumples, both in the interior of the ice shelf and along the ice front. These pinning points provide restraint and cause the ice shelf to be thicker than if it were not pinned. Ice flows from the land to the ice front. Input velocities range from near zero at a shear margin (e.g., with a land boundary), to several hundred meters per year at the grounding line where ice streams and outlet glaciers enter. At the ice front velocities can reach up to several kilometers per year. A characteristic of ice shelves is that the (horizontal) velocity is almost the same at all depths, whereas in glaciers and grounded ice sheets the velocity decreases with depth (**Figure 2**).

In the Antarctic, most ice shelves have net surface accumulation although there may be intensive summer melt which floods the surface. The basal regime is controlled by the subice circulation. Basal melting is often high near both the grounding line and the ice front. Marine ice can accumulate in the intermediate areas, where water at its *in situ* freezing point upwells and produces frazil ice crystals. Surface temperatures will be near the mean annual air temperature whereas the basal temperature will be at the freezing point of the water. Therefore the coldest ice is normally in the upper layers.

Ice shelves normally form where ice flows smoothly off the land as ice streams or outlet glaciers. In some areas, however, ice breaks off at the coastline and reforms as icebergs welded together by frozen sea ice and surface accumulation. The processes of formation and decay are likely to operate under very different conditions. An ice shelf is unlikely to reform under the same climatic conditions which caused it to decay. Complete collapse is a catastrophic process and rebuilding requires a major change in the controlling parameters. However, there is evidence of cyclicity on periods of a few hundred years in some areas.

Collapse of the northernmost section of Larsen Ice Shelf within a few days in January 1995 indicates that, after retreat beyond a critical limit, ice shelves can disintegrate rapidly. The breakup history of two northern sections of Larsen Ice Shelf (Larsen A and Larsen B) between 1986 and 1997 has been used to determine a stability criterion for ice shelves.

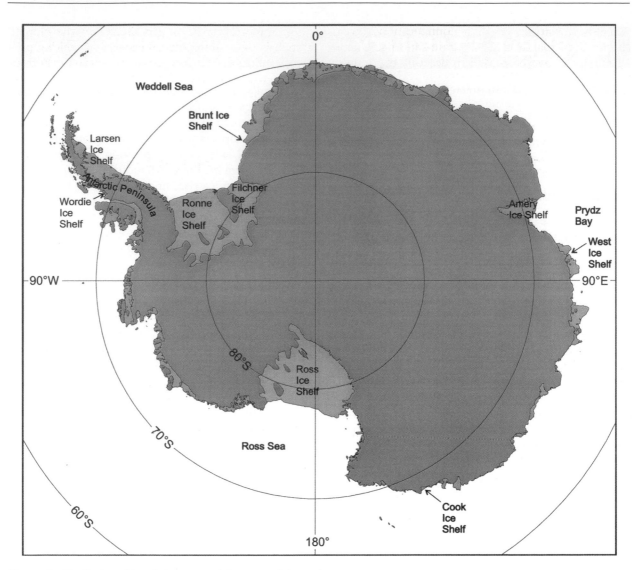

Figure 1   Distribution of ice shelves around the coast of Antarctica.

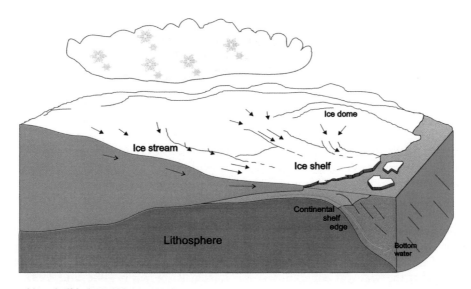

Figure 2   Cartoon of ice shelf in bay with ice streams.

Analysis of various ice-shelf configurations reveals characteristic patterns in the strain-rates near the ice front which have been used to describe the stability of the ice shelf.

Why are Ice Shelves Important?

Ice shelves are one of the most active parts of the ice sheet system. They interact with the inland ice sheet, glaciers and ice streams which flow into them, and with the sea into which they eventually melt, either directly or as icebergs. A typical Antarctic ice shelf will steadily advance until its ice front undergoes periodic calving, generating icebergs. Calving can occur over a wide range of time and spatial scales. Most ice fronts will experience a quasi-continuous 'nibbling' away of the order of tens to hundreds of meters per year, where ice cliffs collapse to form bergy bits and brash ice. The largest icebergs may be more than 100 km in size, but will calve only at intervals of 50 years or more. Thus when charting the size and behavior of an ice shelf it can be difficult to separate and identify stable changes from unstable ones.

Mass balance calculations show that calving of icebergs is the largest factor in the attrition of the Antarctic Ice Sheet. Estimates based on both ship and satellite data suggest that iceberg calving is only slightly less than the total annual accumulation. Ice shelf melting at the base is the other principal element in the attrition of the ice sheet, with approximately 80% of all ice shelf melting occurring at distances greater than 100 km from the ice front. Although most of the mass lost in Antarctica is through ice shelves, there is still uncertainty about the role ice shelves play in regulating flow off the land, which is the important component for sea-level changes. Because ice shelves are already floating, they do not affect sea level when they breakup.

Ice shelves are sensitive indicators of climate change. Those around the Antarctic Peninsula have shown a pattern of gradual retreat since about 1950, associated with a regional atmospheric warming and increased summer surface melt. The effects of climate change on the mass balance of the Antarctic Ice Sheet and hence global sea level are unclear.

There have been no noticeable changes in the inland ice sheet from the collapse of the Antarctic Peninsula ice shelves. This is because most of the margins with the inland ice sheet form a sharp transition zone, making the ice sheet dynamics independent of the state of the ice shelf. This is not the case with a grounding line on an ice stream where there is a smooth transition zone, but generally the importance of the local stresses there diminishes rapidly upstream.

An important role of ice shelves in the climate system is that subice-shelf freezing and melting processes influence the formation of Antarctic Bottom Water, which in part helps to drive the oceanic thermohaline circulation (see Sub Ice-Shelf Circulation and Processes).

How have Ice Shelves Changed in the Past?

As the polar ice sheets have waxed and waned during ice age cycles, ice shelves have also grown and retreated. During the last 20–25 million years, the Antarctic Ice Sheet has probably been grounded out to the edge of the continental shelf many times. It is unlikely that any substantial ice shelves could have existed then, because of unsuitable coastline geometry and lack of pinning points, although the extent of the sea ice may have been double the present day area. At the last glacial maximum, about 20 000 years ago, grounded ice extended out to the edge of the continental shelf in places, for example in Prydz Bay, but not in others such as the Ross Sea. When the ice sheets began to retreat, some of the large ice shelves seen today probably formed by the thinning and eventual flotation of the formerly grounded ice sheets.

Large ice shelves also existed in the Arctic during the Pleistocene. Abundant geologic evidence shows that marine Northern Hemisphere ice sheets disappeared catastrophically during the climatic transition to the current interglacial (warm period). Only a few small ice shelves and tidewater glaciers exist there today, in places like Greenland, Svalbard, Ellesmere Island and Alaska.

There have been periodic changes in Antarctic ice-shelf grounding lines and ice fronts during the Holocene. The main deglaciation on the western side of the Antarctic Peninsula which started more than 11 000 years ago was initially very rapid across a wide continental shelf. By 6000 years ago, the ice sheet had cleared the inner shelf, whereas in the Ross Sea retreat of ice shelves ceased about the same time. Retreat after the last glacial maximum was followed by a readvance of the grounding line during the climate warming between 7000 and 4000 years ago. Open marine deposition on the continental shelf is restricted to the last 4000 years or so. Short-term cycles (every few hundred years) and longer-term events (approximately 2500 year cycles) have been detected in marine sediment cores that are likely related to global climate fluctuations.

More recently, ice-shelf disintegration around the Antarctic Peninsula has been associated with a regional atmospheric warming which has been occurring since about 1950. Ice front retreat of marginal ice shelves elsewhere in Antarctica (e.g., Cook Ice Shelf, West Ice Shelf) is probably also related to atmospheric warming. There is little sign of any significant impact on the grounded ice.

Where are Ice Shelves Disintegrating Now? Two Case Studies

The detailed history of the breakup of two ice shelves, Wordie and Larsen, illustrate some of the critical features of ice-shelf decay.

Wordie Ice Shelf

Retreat of Wordie Ice Shelf on the west coast of the Antarctic Peninsula has been documented using high resolution visible satellite images taken since 1974 and the position of the ice front in 1966 mapped from aerial photography. First seen in 1936, the ice front has fluctuated in position but with a sustained retreat starting around 1966 (**Figure 3**). The ice shelf area has decreased from about 2000 km^2 in 1966 to about 700 km^2 in 1989. However, defining the position of the ice front can be very uncertain, with large blocks calving off to form icebergs being difficult to classify as being either attached to, or separated from, the ice shelf.

Ice front retreat until about 1979 occurred mainly by transverse rifting along the ice front, creating icebergs up to 10 km by 1 km. The western area was the first to be lost, between 1966 and 1974. Results from airborne radio-echo sounding between 1966 and 1970 suggested that the ice shelf could be divided into a crevassed eastern part and a rifted western part where brine could well-up and infiltrate the ice at sea level. By 1974 there were many transverse rifts south of Napier Ice Rise and by 1979 longitudinal rifting was predominant. Many longitudinal rifts had formed upstream of several ice rises by 1979, some following preexisting flowline features. A critical factor in the break-up was the decoupling of Buffer Ice Rise; by 1986 ice was streaming past it apparently unhindered, in contrast to the compressive upstream folding seen in earlier images. Between 1988 and 1989 the central part of the ice shelf, consisting of broken ice in the lee of Mount Balfour, was lost, exposing the coastline and effectively dividing the ice shelf in two. By 1989, longitudinal rifting has penetrated to the grounding line north of Mount Balfour. Further north, a growing shear zone marked the boundary between fast-flowing ice from the north Forster Ice Piedmont and slower moving

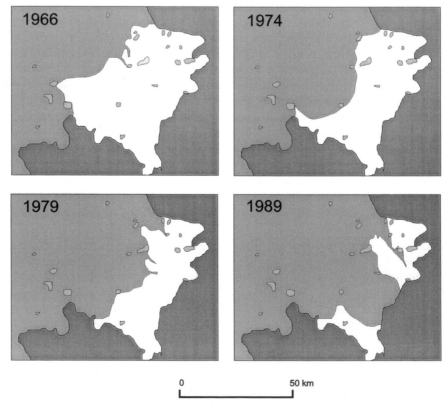

Figure 3   Cartoon showing retreat of Wordie ice front between 1966 and 1989.

ice from Hariot Glacier. There has been no discernible change in the grounding line position.

The major ice rises have played several roles in controlling ice-shelf behavior. When embedded in the ice shelf, they created broken wakes downstream, and zones of compression upstream which helped to stabilize the ice shelf. During ice front retreat, they temporarily pinned the local ice front position and also acted as nucleating points for rifting which quickly stretched upstream, suggesting that a critical fracture criterion had been exceeded. At this stage, an ice rise, instead of protecting the ice shelf against decay, aided its destruction by acting as an indenting wedge.

Breakup was probably triggered by a climatic warming which increased ablation and the amount of melt water. Laboratory experiments show that the fracture toughness of ice is reduced at higher temperatures and possibly by the presence of water. Instead of refreezing in the upper layers of firn, free water could percolate down into crevasses and, by increasing the pressure at the bottom, allow them to grow into rifts or possibly to join up with basal crevasses. Processes like these would increase the production rate of blocks above that required for a 'steady state' ice front position. The blocks will drift away as icebergs if conditions, such as bay geometry and lack of sea ice, are favorable. Thus, ice front retreat would be, *inter alia*, a sensitive function of mean annual air temperature.

Some ice shelves, such as Brunt, are formed from blocks that break off at the coast line and, unable to float away, are 'glued' together by sea ice and snowfall. These heterogeneous ice shelves contrast with those, such as Ronne, where glaciers or ice streams flow unbroken across the grounding line to form a more homogeneous type of ice shelf. Before 1989, Wordie Ice Shelf consisted of a mixture of both types, the main tongues being derived from glacier inputs, while the central portion was formed from blocks breaking off at the grounding line and at the sides of the main tongues. The western rifted area that broke away between 1966 and 1974 was described in early 1967 as 'snowed-under icebergs' and it was the heterogeneous central part that broke back to the coastline around Mount Balfour in 1988/89. Increased ablation would not only enhance rifting along lines of weakness but would also loosen the 'glue' that held the blocks together.

Larsen Ice Shelf

The most northerly ice shelf in the Antarctic, Larsen Ice Shelf, extends in a ribbon down the east coast of the Antarctic Peninsula from James Ross Island to the Ronne Ice Shelf. It consists of several distinct ice shelves, separated by headlands. The ice shelf in Prince Gustav Channel, between James Ross Island and the mainland, separated from the main part of Larsen Ice Shelf in the late 1950s and finally disappeared by 1995.

The section between Sobral Peninsula and Robertson Island, known as Larsen A, underwent a catastrophic collapse at the end of January 1995 when it disintegrated into a tongue of small icebergs, bergy bits and brash ice. Previously the ice front had been retreating for a number of years, but in a more controlled fashion by iceberg calving. Before the final breakup, the surface that had once been flat and smooth had become undulating, suggesting that rifting completely through the ice had occurred. Collapse during a period of intense north-westerly winds and high temperatures was probably aided by a lack of sea ice, allowing ocean swell to penetrate and add its power to increasing the disintegration processes. The speed of the collapse and the small size of the fragments of ice (the largest icebergs were less than 1 km in size) implicate fracture as the dominant process in the disintegration (**Figure 4**).

The ice front of the ice shelf known as Larsen B, between Robertson Island and Jason Peninsula, steadily advanced for about 6 km from 1975 until 1992. Small icebergs broke away from the heavily rifted zone south of Robertson Island after July 1992 and a major rift about 25 km in length had opened up by then. This rift formed the calving front when an iceberg covering 1720 km^2 and smaller pieces corresponding to a former ice shelf area of 550 km^2 broke away between 25 and 30 January 1995, coincident with the disintegration of Larsen A. The ice front has retreated continuously by a few kilometers per year since then (to 1999) and the ice shelf is considered to be under threat of disappearing completely.

Why do Ice Shelves Break-up?

Calving and Fracture

Calving is one of the most obvious processes involved in ice shelf breakup. Although widespread, occurring on grounded and floating glaciers, including ice shelves, as well as glaciers ending in freshwater lakes, it is not well understood. The basic physics, tensile propagation of fractures, may be the same in all cases but the predictability of calving may be quite different, depending on the stress field, the basal boundary conditions, the amount of surface water, etc. Fracture of ice has been studied mainly in laboratories and applying these results to the

Figure 4   Photo of the Larsen breakup. Picture taken approximately 10 km north of Robertson Island, looking west north-west. In the far background the mountains of the Antarctic Peninsula can be seen.

conditions experienced in naturally occurring ice masses requires extrapolations which may not be valid. Questions such as how is crevasse propagation influenced by the ice shelf geometry and by the physical properties of the ice such as inhomogeneity, crystal anisotropy, temperature and presence of water, need to be answered before realistic models of the calving process can be developed. The engineering concepts of fracture mechanics and fracture toughness promise a way forward.

Fracture of ice is a critical process in ice shelf dynamics. Mathematical models of ice shelf behavior based on continuum mechanics, which treat ice as a nonlinear viscous fluid will have to incorporate fracture mechanics to describe iceberg calving and how ice rises can initiate fracture both upstream and downstream. High stresses generated at shear margins around ice rises, or at the 'corner points' of the ice shelf where the two ends of the ice front meet land, can exceed a critical stress and initiate crevassing. These crevasses will propagate under a favorable stress regime, and if there is sufficient surface water may rift through the complete thickness. Transverse crevasses parallel to the ice front act as sites for the initiation of rifting and form lines where calving may eventually occur.

Calving is a very efficient method of getting rid of ice from an ice shelf – once an iceberg has formed, it can drift away within a few days. Sometimes,

however, an iceberg will ground on a sea bed shoal, perhaps for several years. Once an Antarctic iceberg has drifted north of the Antarctic Convergence, it usually breaks-up very quickly in the warmer waters.

Climate Warming

Circumstantial evidence links the retreat of ice shelves around the Antarctic Peninsula to a warming trend in atmospheric temperatures. Observations at meteorological stations in the region show a rise of about 2.5°C in 50 years. Ice shelves appear to exist up to a climatic limit, taken to be the mean annual −5°C isotherm, which represents the thermal limit of ice-shelf viability. The steady southward migration of this isotherm has coincided with the pattern of ice-shelf disintegration.

There are too few sea temperature measurements to show if the observed atmospheric warming has resulted in warmer waters, and thus increased basal melting. Although basal melting may have increased, the indications are that it is surface processes that have played the dominant role in causing breakup. The mechanism for connecting climate warming with ice-shelf retreat is not fully understood, but increased summer melting producing substantial amounts of water obviously plays a part in enhancing fracture processes, leading to calving.

The retreat of ice shelves around the Antarctic Peninsula that has occurred in the last half of the twentieth century has raised questions about whether or not this reflects global warming or whether it is only a regional phenomenon. The warming trend seems to be localized to the Antarctic Peninsula region and there is no significant correlation with temperature changes in the rest of Antarctica.

Stress Patterns

Numerical models of ice shelves have been used to examine their behavior and stability criteria. The strain-rate field can be specified by the principal values ($\dot{\varepsilon}_1$ and $\dot{\varepsilon}_2$ where $\dot{\varepsilon}_1 > \dot{\varepsilon}_2$) and the direction of the principal axes. A simple representation is given by the trajectories, which are a set of orthogonal curves whose directions at any point are the directions of the principal axes. Analyses of the strain-rate trajectories for Filchner Ronne Ice Shelf and for different ice shelf configurations of Larsen Ice Shelf show characteristic patterns of a 'compressive arch' and of isotropic points (**Figure 5**). The 'compressive arch' is seen in the pattern formed by the smallest principal component ($\dot{\varepsilon}_2$) of the strain-rate trajectories. Seaward of the arch both principal strain-rate components are extensive, whereas inland of the arch the $\dot{\varepsilon}_2$ component is compressive. The arch extends from the two ends of the ice front across the whole width of the ice shelf. It is a generic feature of ice shelves studied so far and structurally stable to small perturbations. It is probably related to the geometry of the ice shelf bay. A critical arch, consisting entirely of compressive trajectories, appears to correspond to a criterion for stability. If the ice front breaks back through the arch then an irreversible retreat occurs, possibly catastrophically, to another stable configuration. The exact location of the critical arch cannot yet be determined *a priori*, but it is probably close to the compressive arch delineated by the transition from extension to compression for $\dot{\varepsilon}_2$.

Another pattern seen in the strain-rate trajectories is that of isotropic points. They are indicators of generic features in the surface flow field which are stable to small perturbations in the flow. This means that their existence should not be sensitive either to reasonable errors in data used in the model or to simplifications in the model itself. They act as (permanent) markers in a complicated (tensor) strain-rate field and are often located close to points where the two principal strain-rates are equal or where the velocity field is stationary (usually either a maximum or a saddle point). Isotropic points are classified by a number of properties, but only two categories can be reliably distinguished observationally, by the way the trajectory of either of the principal strain-rates varies around a path enclosing the isotropic point. In one case the trajectory varies in a prograde sense and the isotropic point is called a 'monstar', whereas in the other case the trajectory varies in a retrograde sense and the isotropic point is called a 'star' (**Figure 5**).

The two categories of isotropic points can be identified in the model strain-rate trajectories, 'stars' occurring near input glaciers and 'monstars' near ice fronts if they are in a stable configuration. Icebergs calving off Filchner Ice Shelf in 1986 moved the position of the 'monstar' up to the newly formed ice front, whereas on Ronne Ice Shelf the 'monstar' is about 50 km inland of the ice front. This suggests that the existence of a monstar can be used as a 'weak' indicator of a stable ice front. Calving can remove a monstar even if the subsequent ice front position is not necessarily an unstable one and may readvance.

How Vulnerable are the Large Ice Shelves and how Stable are Grounding Lines?

The two largest ice shelves (Ross and Filchner–Ronne) are too far south to be attacked by the atmospheric warming that is predicted for the twenty-first century. Basal melting will increase if warmer water intrudes onto the continental shelf, but the effects of a warming trend could be counterintuitive. If the warming was sufficient to reduce the rate of sea ice formation in the Weddell Sea, then the production of High Salinity Shelf Water (HSSW) would reduce as well. This would affect the subice shelf circulation, replacing the relatively warm HSSW under the Filchner–Ronne Ice Shelf with a colder Ice Shelf Water which would reduce the melting and thus thicken the ice shelf. Further climate warming would restore warmer waters and eventually thin the ice shelf by increasing the melting rates, possibly to values of around 15 m year^{-1} as seen under Pine Island Glacier. The likelihood of this is discussed in Sub Ice-Shelf Circulation and Processes.

The supplementary question is what effect, if any, would the collapse of the ice shelves have on the ice sheet, especially the West Antarctic Ice Sheet (WAIS). It has been a tenet of the latter part of the twentieth century that the WAIS, known as a marine ice sheet because much of its bed is below sea level, is potentially unstable and may undergo disintegration if the fringing ice shelves were to disappear. However, the theoretical foundations of this belief are shaky and more careful consideration of the relevant

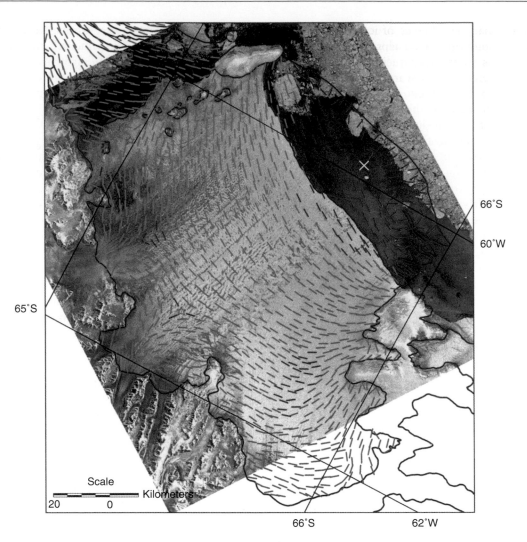

Figure 5 Modeled strain-rate trajectories on Larsen Ice Shelf superimposed on a Radarsat image taken on 21 March 1997. The trajectory of the smallest principal strain-rate ($\dot{\varepsilon}_2$) is negative (compressing flow, blue) over nearly all the Larsen Ice Shelf in its post-iceberg calving configuration (January 1995) but is positive (red) in the region where the iceberg calved. The pattern shows isotropic points ('stars,' green) near where glaciers enter the ice shelf and, for the 'iceberg' area a 'monstar' (yellow cross) near the ice front. 'Stars' are seen for all the different ice front configurations, indicating that no fundamental change is occurring near the grounding line due to retreat of the ice front. A 'monstar' is only seen for Larsen B before the iceberg calved in 1995. The disappearance of the isotropic point is perhaps a 'weak' indication that the iceberg calving event was greater than expected for normal calving from a stable ice shelf and suggests further retreat may be expected.

dynamics suggests that the WAIS is no more vulnerable than any other part of the ice sheet.

The problem lies in how to model the transition zone between ice sheet and ice shelf. If the transition is sharp, where the basal traction varies over lengths of the order of the ice thickness, then it can be considered as a passive boundary layer and does not affect the mechanics of the sheet or shelf to first order. Mass conservation must be respected but there is no need to impose further constraints. Requiring the ice sheet to have the same thickness as the ice shelf at the grounding line permits only two stable grounding line positions for a bed geometry which deepens inland. Depending on the depth of the bed

below sea level, the ice sheet will either shrink in size until it disappears (or, if the bed shallows again, the grounding line is fixed on a bed near sea level), or grow to the edge of the continental shelf. The lack of intermediate stable grounding line positions in this kind of model has been used to support the idea of instability of marine ice sheets. However, permitting a jump in ice thickness at the transition zone means the ice sheet system can be in neutral equilibrium, with an infinite number of steady-state profiles.

Another kind of transition zone is a smooth one, where the basal traction varies gradually, for example, along an ice stream. The equilibrium dynamics have not been worked out for a full three-dimensional flow, but it seems that the presence of ice streams destroys the neutral equilibrium and helps to stabilize marine ice sheets. This emphasizes the importance of understanding the dynamics of ice streams and their role in the marine ice sheet system.

Attempts to include moving grounding lines in whole ice sheet models suffer from incomplete specification of the problem. Assumptions built into the models predispose the results to be either too stable or too unstable. Thus there are no reliable models that can analyze the glaciological history of the Antarctic Ice Sheet. Predictive models rely on linearizations to provide acceptable accuracy for the near future but become progressively less accurate the longer the timescale.

See also

Ice–Ocean Interaction. Sub Ice-Shelf Circulation and Processes.

Further Reading

Doake CSM and Vaughan DG (1991) Rapid disintegration of Wordie Ice Shelf in response to atmospheric warming. *Nature* 350: 328–330.

Doake CSM, Corr HFJ, Rott H, Skvarca P, and Young N (1998) Breakup and conditions for stability of the northern Larsen Ice Shelf, Antarctica. *Nature* 391: 778–780.

Hindmarsh RCA (1993) Qualitative dynamics of marine ice sheets. In: Peltier WR (ed.) *Ice in the Climate System*, NATO ASI Series, vol. 12, pp. 67–99. Berlin: Springer-Verlag.

Kellogg TB and Kellogg DE (1987) Recent glacial history and rapid ice stream retreat in the Amundsen Sea. *Journal of Geophysical Research* 92: 8859–8864.

Robin G and de Q (1979) Formation, flow and disintegration of ice shelves. *Journal of Glaciology* 24(90): 259–271.

Scambos TA, Hulbe C, Fahnestock M, and Bohlander J (2000) The link between climate warming and break-up of ice shelves in the Antarctic Peninsula. *Journal of Glaciology* 46(154): 516–530.

van der Veen CJ (ed.) (1997) *Calving Glaciers: Report of a Workshop 28 February–2 March 1997*. BPRC Report No. 15. Columbus, Ohio: Byrd Polar Research Center, The Ohio State University.

van der Veen CJ (1999) *Fundamentals of Glacier Dynamics*. Rotterdam: A.A. Balkema.

Vaughan DG and Doake CSM (1996) Recent atmospheric warming and retreat of ice shelves on the Antarctic Peninsula. *Nature* 379: 328–331.

NOBLE GASES AND THE CRYOSPHERE

M. Hood, Intergovernmental Oceanographic
Commission, Paris, France

Introduction

Ice formation and melting strongly influence a wide
range of water properties and processes, such as
dissolved gas concentrations, exchange of gases between the atmosphere and the ocean, and dense
water formation. As water freezes, salt and gases
dissolved in the water are expelled from the growing
ice lattice and become concentrated in the residual
water. As a result of the increased salt content, this
residual water becomes more dense than underlying
waters and sinks to a level of neutral buoyancy,
carrying with it the dissolved gas load. Dense water
formation is one of the primary mechanisms by
which atmospheric and surface water properties are
transported into the interior and deep ocean, and
observation of the effects of this process can answer
fundamental questions about ocean circulation and
the ocean–atmosphere cycling of biogeochemically
important gases such as oxygen and carbon dioxide.
Because it is not possible to determine exactly when
and where dense water formation will occur, it is not
an easy process to observe directly, and thus information about the rates of dense water formation and
circulation is obtained largely through the observation of tracers. However, when dense water formation is triggered by ice formation, interaction of
surface water properties with the ice and the lack of
full equilibration between the atmosphere and the
water beneath the growing ice can significantly
modify the concentrations of the tracers in ways that
are not yet fully understood. Consequently, the information provided by tracers in these ice formation
areas is often ambiguous.

A suite of three noble gases, helium, neon, and
argon, have the potential to be excellent tracers in
the marine cryosphere, providing new information
about the interactions of dissolved gases and ice, the
cycling of gases between the atmosphere and ocean,
and mixing and circulation pathways in high latitude
regions of the world's oceans and marginal seas. The
physical chemistry properties of these three gases
span a wide range of values, and these differences
cause them to respond to varying degrees to physical
processes such as ice formation and melting or the

transfer of gas between the water and air. By observing the changes of the three tracers as they respond to these processes, it is possible to quantify the
effect the process has on the gases as a function of the
physical chemistry of the gases. Subsequently, this
'template' of behavior can be used to determine the
physical response of any gas to the process, using
known information about the physical chemistry of
the gas. Although this tracer technique is still being
developed, results from laboratory experiments and
field programs have demonstrated the exciting potential of the nobel gases to provide unique, quantitative information on a range of processes that it is
not possible to obtain using conventional tracers.

Noble Gases in the Marine Environment

The noble gases are naturally occurring gases
found in the atmosphere. **Table 1** shows the abundance of the noble gases in the atmosphere as a
percentage of the total air composition, and the
concentrations of the gases in surface sea water when
in equilibrium with the atmosphere.

Other sources of these gases in sea water include
the radioactive decay of uranium and thorium to
helium-4 (^{4}He), and the radioactive decay of potassium (^{40}K) to argon (^{40}Ar). For most areas of the
surface ocean, these radiogenic sources of the noble
gases are negligible, and thus the only significant
source for these gases is the atmosphere.

The noble gases are biogeochemically inert and are
not altered through chemical or biological reactions,
making them considerably easier to trace and quantify as they move through a system than other gases
whose concentrations are modified through reactions. The behavior of the noble gases is largely
determined by the size of the molecule of each gas
and the natural affinity of each gas to reside in a
gaseous or liquid state. The main physical chemistry

Table 1 Noble gases in the atmosphere and sea water

Gas	Abundance in the atmosphere (%)	Concentration in seawater ($cm^3 g^{-1}$)
Helium	0.0005	3.75×10^{-8}
Neon	0.002	1.53×10^{-7}
Argon	0.9	2.49×10^{-4}

parameters of interest are the solubility of the gas in liquid, the temperature dependence of this solubility, and the molecular diffusivity of the gas. The suite of noble gases have a broad range of these properties, and the behavior of the noble gases determined by these properties, can serve as a model for the behavior of most other gases.

One unique characteristic of the noble gases that makes them ideally suited as tracers of the interactions between gases and ice is that helium and neon are soluble in ice as well as in liquids. It has been recognized since the mid-1960s that helium and neon, and possibly hydrogen, should be soluble in ice because of the small size of the molecules, whereas gases having larger atomic radii are unable to reside in the ice lattice. These findings, however, were based on theoretical treatises and carefully controlled laboratory studies in idealized conditions. It was not until the mid-1980s that this process was shown to occur on observable scales in nature, when anomalies in the concentrations of helium and neon were observed in the Arctic.

The solubility of gases in ice can be described by the same principles governing solubility of gases in liquids. Solubility of gases in liquids or ice occurs to establish equilibrium, where the affinities of the gas to reside in the gaseous, liquid, and solid state are balanced. The solubility process can be described by two principle mechanisms:

1. creation of a cavity in the solvent large enough to accommodate a solute molecule;
2. introduction of the solute molecule into the liquid or solid surface through the cavity.

In applying this approach to the solubility of gases in ice, it follows that if the atomic radius of the solute gas molecule is smaller than the cavities naturally present in the lattice structure of ice, then the energy required to make a cavity in the solvent is zero, and the energy required for the solubility process is then only a function of the energy required to introduce the solute molecule into the cavity. For this reason, the solubility of a gas molecule capable of fitting in the ice lattice is greater than its solubility in a liquid. The solubilities of helium and neon in ice have been determined in two separate laboratory studies, and although the values agree for the solubility of helium in ice, the values for neon disagree. The size of neon is very similar to the size of a cavity in the ice lattice, and the discrepancies between the two reported values for the solubility of neon in ice may result from small differences in the experimental procedure.

During ice formation, most gases partition between the water and air phases to try to establish equilibrium under the changing conditions, whereas helium and neon additionally partition into the ice phase. As water freezes, salt and gases are rejected from the growing ice lattice, increasing the concentrations of salt and gas in the residual water. Helium and neon partition between the water and ice reservoirs according to their solubility in water and ice. The concentrations of the gases in the residual water that have been expelled from the ice lattice, predominantly oxygen and nitrogen, can become so elevated through this process that the pressure of the dissolved gases in the water exceeds the *in situ* hydrostatic pressure and gas bubbles form. The gases then partition between the water, the gas bubble, and the ice according to the solubilities of the gases in each phase. This three-phase partitioning process can occur either at the edge of the growing ice sheet at the ice–water interface, or in small liquid water pockets, called 'brine pockets' in salt water systems, entrained in the ice during rapid ice formation.

Table 2 quantitatively describes how the noble gases partition between the three phases when a system containing these three phases is in equilibrium in fresh water at 0°C. The numbers represent the amount of the gas found in one phase relative to the other. For example, the first row describes the amount of each gas that would reside in the gaseous bubble phase relative to the liquid phase; thus for helium, there would be 106.8 times more helium present in the bubble than in the water. This illustrates the small solubility of helium in water and its strong affinity for the gas phase. Because helium is 1.9 times more soluble in ice than in water, helium partitions less strongly between the bubble and ice phases compared to the partition between the bubble and water phases. The two numbers shown for neon represent the two different estimates for the solubility of neon in the ice phase. One estimate suggests that neon is less soluble in ice than in water, whereas the other suggests that it is more soluble in ice.

Table 2 Noble gas partitioning in three phases

Partition phases	Helium	Neon	Argon
Bubble to water	106.8	81.0	18.7
Bubble to ice	56.9	90.0, 56.3	∞
Ice to water	1.9	0.9, 1.4	0

Application of the Noble Gases as Tracers

The noble gases have been used as tracers of air–sea gas exchange processes for more than 20 years. Typically, the noble gases are observed over time at a single location in the ocean along with other meteorological and hydrodynamic parameters to characterize and quantify the behavior of each of the gases in response to the driving forces of gas exchange such as water temperature, wind speed, wave characteristics, and bubbles injected from breaking waves. Because both the amount and rate of a gas transferred between the atmosphere and ocean depend on the solubility and diffusivities of the gas, the noble gases have long been recognized as ideal tracers for these processes. In addition, argon and oxygen have very similar molecular diffusivities and solubilities, making argon an excellent tracer of the physical behavior of oxygen. By comparing the relative concentration changes of argon and oxygen over time, it is possible to account for the relative contributions of physical and biological processes (such as photosynthesis by phytoplankton in the surface ocean) to the overall concentrations, thus constraining the biological signal and allowing for estimates of the biological productivity of the surface ocean.

The observations of anomalous helium and neon concentrations in ice formation areas and the suggestion that these anomalies could be the result of solubility of these gases in the ice were made in 1983, and since that time, a number of laboratory and field studies have been conducted to characterize and quantify these interactions. The partitioning of the noble gases among the three phases of gas, water, and ice creates a very distinctive 'signature' of the noble gas concentrations left behind in the residual water. Noble gas concentrations are typically expressed in terms of 'saturation', which is the concentration of a gas dissolved in the water relative to its equilibrium with the atmosphere at a given temperature. For example, a parcel of water at standard temperature and pressure containing the concentrations of noble gases shown in column 2 of **Table 1** would be said to have a saturation of 100%. Saturations that deviate from this 100% can arise when equilibration with the atmosphere is incomplete, either because the equilibration process is slow relative to some other dynamic process acting on the system (for example, rapid heating or cooling, or injection of bubbles from breaking waves), or because full equilibration between the water and atmosphere is prevented, as in the case of ice formation.

Typical saturations for the noble gases in the surface ocean range from 100 to 110% of atmospheric equilibrium, due mostly to the influx of gas from bubbles. Ice formation, however, can lead to quite striking saturations of -70 to -60% for helium and neon and $+230\%$ for argon in the relatively undiluted residual water. Ice melting can also lead to large anomalous saturations of the noble gases, showing the reverse of the freezing pattern for the gas saturations, where helium and neon are supersaturated while argon is undersaturated with respect to the atmosphere.

The interactions of noble gases and ice have been well-documented and quantified in relatively simple freshwater systems. Observations of large noble gas anomalies in a permanently ice-covered antarctic lake were quantitatively explained using the current understanding of the solubility of helium and neon in ice and the partitioning of the gases in a three-phase system. Characteristics of ice formed from salt water are more complex than ice formed from fresh water, and the modeling of the system more complex. Using a set of equations developed in 1983 and measurements of the ice temperature, salinity, and density, it is possible to calculate the volume of the brine pockets in the ice and the volume of bubbles in the ice. With this type of information, a model of the ice and the dissolved gas balance in the various phases in the ice and residual water can be constructed. Such an ice model was developed during a field study of gas–ice interactions in a seasonally ice-covered lagoon, and the model predicted the amount of argon, nitrogen, and oxygen measured in bubbles in similar types of sea ice. No measurements are available for the amount of helium and neon in the bubbles of sea ice to verify the results for these gases. It is also possible to predict the relative saturations of the noble gases in the undiluted residual water at the ice–water interface, and this unique fingerprint of the noble gases can then serve as a tracer of the mixing and circulation of this water parcel as it leaves the surface and enters the interior and deep ocean. In this manner, the supersaturations of helium from meltwater have been successfully used as a tracer of water mass mixing and circulation in the Antarctic, and the estimated sensitivity of helium as a tracer for these processes is similar to the use of the conventional tracer, salinity, for these processes.

As an illustration of the ways in which the noble gases can be used to distinguish between the effects of ice formation, melting, injection of air bubbles from breaking waves, or temperature changes on dissolved gases, **Figure 1** shows a vector diagram of the characteristic changes of helium compared to argon resulting from each of these processes.

From a starting point of equilibrium with the atmosphere (100% saturation), both helium and argon saturations increase as a result of bubbles injected from breaking waves. Ice formation increases the saturation of argon and decreases the saturation of helium, whereas ice melting has the opposite effect. Changes in temperature with no gas exchange with

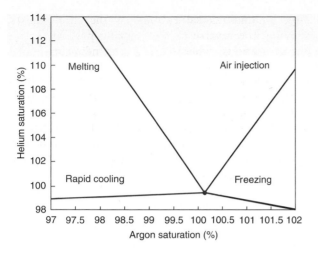

Figure 1 Vector diagram of helium and argon saturation changes in response to upper ocean processes.

the atmosphere to balance this change can lead to modest changes in the saturations of the gases, where the gas saturations decrease with decreasing temperature and increase with increasing temperature. The trends presented here are largely qualitative indicators, since quantitative assessment of the changes depend on the exact nature of the system being studied. However, this diagram does illustrate the general magnitude of the changes that these processes have on the noble gases and conversely, the ability of the noble gases to differentiate between these effects.

Conclusions

The use of the noble gases as tracers in the marine cryosphere is in its infancy. Our understanding of the interactions of the noble gases and ice have progressed from controlled, idealized laboratory conditions to natural freshwater systems and simple salt water systems, and the initial results from these studies are extremely encouraging. This technique is currently being developed more fully to provide quantitative information about the interactions of dissolved gases and ice, and to utilize the resulting effects of these interactions to trace water mass mixing and circulation in the range of dynamic ice formation environments. Water masses in the interior and deep ocean originating in ice formation and melting areas have been shown to have distinct noble gas ratios, which are largely imparted to the water mass at the time of its formation in the surface ocean. By understanding and quantifying the processes responsible for these distinct ratios, we will be

able to learn much about where and how the water mass was formed and the transformations it has experience since leaving the surface ocean. These issues are important for our understanding of the global cycling of gases between the atmosphere and the ocean and for revealing the circulation pathways of water in the Arctic, Antartic, and high latitude marginal seas. The noble gases could represent a significant addition to the set of tracers typically used to study these processes.

See also

Arctic Ocean Circulation. Ice–Ocean Interaction. Oxygen Isotopes in the Ocean. Sea Ice: Overview. Stable Carbon Isotope Variations in the Ocean. Sub Ice-Shelf Circulation and Processes.

Further Reading

Bieri RH (1971) Dissolved noble gases in marine waters. *Earth and Planetary Science Letters* 10: 329–333.

Cox GFN and Weeks WF (1982) Equations for determining the gas and brine volumes in sea ice samples, *USA Cold Regions Research and Engineering Laboratory Report* 82-30, Hanover, New Hampshire.

Craig H and Hayward T (1987) Oxygen supersaturations in the ocean: biological vs. physical contributions. *Science* 235: 199–202.

Hood EM, Howes BL, and Jenkins WJ (1998) Dissolved gas dynamics in perennially ice-covered Lake Fryxell, Antarctica. *Limnology and Oceanography* 43(2): 265–272.

Hood EM (1998) *Characterization of Air–sea Gas Exchange Processes and Dissolved Gas/ice Interactions Using Noble Gases*. PhD thesis, MIT/WHOI, 98–101.

Kahane A, Klinger J, and Philippe M (1969) Dopage selectif de la glace monocristalline avec de l'helium et du neon. *Solid State Communications* 7: 1055–1056.

Namoit A and Bukhgalter EB (1965) Clathrates formed by gases in ice. *Journal of Structural Chemistry* 6: 911–912.

Schlosser P (1986) Helium: a new tracer in Antarctic oceanography. *Nature* 321: 233–235.

Schlosser P, Bayer R, Flodvik A, *et al.* (1990) Oxygen-18 and helium as tracers of ice shelf water and water/ice interaction in the Weddell Sea. *Journal of Geophysical Research* 95: 3253–3263.

Top Z, Martin S, and Becker P (1988) A laboratory study of dissolved noble gas anomaly due to ice formation. *Geophysical Research Letters* 15: 796–799.

Top Z, Clarke WB, and Moore RM (1983) Anomalous neon–helium ratios in the Arctic Ocean. *Geophysical Research Letters* 10: 1168–1171.

SEA ICE: OVERVIEW

W. F. Weeks, Portland, OR, USA

Introduction

Sea ice, any form of ice found at sea that originated from the freezing of sea water, has historically been among the least-studied of all the phenomena that have a significant effect on the surface heat balance of the Earth. Fortunately, this neglect has recently lessened as the result of improvements in observational and operational capabilities in the polar ocean areas. As a result, considerable information is now available on the nature and behavior of sea ice as well as on its role in affecting the weather, the climate, and the oceanography of the polar regions and possibly of the planet as a whole.

Extent

Although the majority of Earth's population has never seen sea ice, in area it is extremely extensive: 7% of the surface of the Earth is covered by this material during some time of the year. In the northern hemisphere the area covered by sea ice varies between 8×10^6 and 15×10^6 km^2, with the smaller number representing the area of multiyear (MY) ice remaining at the end of summer. In summer this corresponds roughly to the contiguous area of the United States and to twice that area in winter, or to between 5% and 10% of the surface of the northern hemisphere ocean. At maximum extent, the ice extends down the western side of the major ocean basins, following the pattern of cold currents and reaching the Gulf of St. Lawrence (Atlantic) and the Okhotsk Sea off the north coast of Japan (Pacific). The most southerly site in the northern hemisphere where an extensive sea ice cover forms is the Gulf of Bo Hai, located off the east coast of China at 40°N. At the end of the summer the perennial MY ice pack of the Arctic is primarily confined to the central Arctic Ocean with minor extensions into the Canadian Arctic Archipelago and along the east coast of Greenland.

In the southern hemisphere the sea ice area varies between 3×10^6 and 20×10^6 km^2, covering between 1.5% and 10% of the ocean surface. The amount of MY ice in the Antarctic is appreciably less than in the Arctic, even though the total area affected by sea ice in the Antarctic is approximately a third larger than in the Arctic. These differences are largely caused by differences in the spatial distributions of land and ocean. The Arctic Ocean is effectively landlocked to the south, with only one major exit located between Greenland and Svalbard. The Southern Ocean, on the other hand, is essentially completely unbounded to the north, allowing unrestricted drift of the ice in that direction, resulting in the melting of nearly all of the previous season's growth.

Geophysical Importance

In addition to its considerable extent, there are good reasons to be concerned with the health and behavior of the world's sea ice covers. Sea ice serves as an insulative lid on the surface of the polar oceans. This suppresses the exchange of heat between the cold polar air above the ice and the relatively warm sea water below the ice. Not only is the ice itself a good insulator, but it provides a surface that supports a snow cover that is also an excellent insulator. In addition, when the sea ice forms with its attendant snow cover, it changes the surface albedo, α (i.e., the reflection coefficient for visible radiation) of the sea from that of open water ($\alpha = 0.15$) to that of newly formed snow ($\alpha = 0.85$), leading to a 70% decrease in the amount of incoming short-wave solar radiation that is absorbed. As a result, there are inherent positive feedbacks associated with the existence of a sea ice cover. For instance, a climatic warming will presumably reduce both the extent and the thickness of the sea ice. Both of these changes will, in turn, result in increases in the temperature of the atmosphere and of the sea, which will further reduce ice thickness and extent. It is this positive feedback that is a major factor in producing the unusually large increases in arctic temperatures that are forecast by numerical models simulating the effect of the accumulation of greenhouse gases.

The presence of an ice cover limits not only the flux of heat into the atmosphere but also the flux of moisture. This effect is revealed by the common presence of linear, local clouds associated with individual leads (cracks in the sea ice that are covered with either open water or thinner ice). In fact, sea ice exerts a significant influence on the radiative energy balance of the complete atmosphere–sea ice–ocean system. For instance, as the ice thickness increases in the range between 0 and 70 cm, there is an increase in the radiation absorption in the ice and a decrease in the ocean. There is also a decrease in the radiation adsorption by the total atmosphere–ice–ocean system. It is also known that the upper 10 cm of the ice can absorb over 50% of the total solar radiation, and that decreases in ice extent produce increases in atmospheric moisture or cloudiness, in turn altering the surface radiation budget and increasing the amount of precipitation. Furthermore, all the ultraviolet and infrared radiation is absorbed in the upper 50 cm of the ice; only visible radiation penetrates into the lower portions of thicker ice and into the upper ocean beneath the ice. Significant changes in the extent and/or thickness of sea ice would result in major changes in the climatology of the polar regions. For instance, recent computer simulations in which the ice extent in the southern hemisphere was held constant and the amount of open water (leads) within the pack was varied showed significant changes in storm frequencies, intensities and tracks, precipitation, cloudiness, and air temperature.

However, there are even less obvious but perhaps equally important air–ice and ice–ocean interactions. Sea ice drastically reduces wave-induced mixing in the upper ocean, thereby favoring the existence of a 25–50 m thick, low-salinity surface layer in the Arctic Ocean that forms as the result of desalination processes associated with ice formation and the influx of fresh water from the great rivers of northern Siberia. This stable, low-density surface layer prevents the heat contained in the comparatively warm (temperatures of up to $+3°C$) but more saline denser water beneath the surface layer from affecting the ice cover. As sea ice rejects roughly two-thirds of the salt initially present in the sea water from which the ice forms, the freezing process is equivalent to distillation, producing both a low-salinity component (the ice layer itself) and a high-salinity component (the rejected brine). Both of these components play important geophysical roles. Over shallow shelf seas, the rejected brine, which is dense, cold, and rich in CO_2, sinks to the bottom, ultimately feeding the deep-water and the bottom-water layers of the world ocean. Such processes are particularly effective in regions where large polynyas exist (semipermanent open water and thin-ice areas at sites where climatically much thicker ice would be anticipated).

The 'fresh' sea ice layer also has an important geophysical role to play in that its exodus from the Arctic Basin via the East Greenland Drift Stream represents a fresh water transport of 2366 km^3 y^{-1} (c. 0.075 Sv). This is a discharge equivalent to roughly twice that of North America's four largest rivers combined (the Mississippi, St. Lawrence, Columbia, and Mackenzie) and in the world is second only to the Amazon. This fresh surface water layer is transported with little dispersion at least as far as the Denmark Strait and in all probability can be followed completely around the subpolar gyre of the North Atlantic. Even more interesting is the speculation that during the last few decades this fresh water flux has been sufficient to alter or even stop the convective regimes of the Greenland, Iceland and Norwegian Seas and perhaps also of the Labrador Sea. This is a sea ice-driven, small-scale analogue of the so-called halocline catastrophe that has been proposed for past deglaciations, when it has been argued that large fresh water runoff from melting glaciers severely limited convective regimes in portions of the world ocean. The difference is that, in the present instance, the increase in the fresh water flux that is required is not dramatic because at near-freezing temperatures the salinity of the sea water is appreciably more important than the water temperature in controlling its density. It has been proposed that this process has contributed to the low near-surface salinities and heavy winter ice conditions observed north of Iceland between 1965 and 1971, to the decrease in convection described for the Labrador Sea during 1968–1971, and perhaps to the so-called 'great salinity anomaly' that freshened much of the upper North Atlantic during the last 25 years of the twentieth century. In the Antarctic, comparable phenomena may be associated with freezing in the southern Weddell Sea and ice transport northward along the Antarctic Peninsula.

Sea ice also has important biological effects at both ends of the marine food chain. It provides a substrate for a special category of marine life, the ice biota, consisting primarily of diatoms. These form a significant portion of the total primary production and, in turn, support specialized grazers and species at higher trophic levels, including amphipods, copepods, worms, fish, and birds. At the upper end of the food chain, seals and walruses use ice extensively as a platform on which to haul out and give birth to young. Polar bears use the ice as a platform while hunting. Also important is the fact that in shelf seas such as the Bering and Chukchi, which are well mixed in the winter, the melting of the ice cover in the spring lowers the surface salinity, increasing the stability of the water column. The reduced mixing concentrates phytoplankton in the near-surface photic zone, thereby enhancing the overall intensity of the spring bloom. Finally, there are the direct effects of sea ice on human activities. The most important of these are its barrier action in limiting the use of otherwise highly advantageous ocean routes between the northern Pacific regions and Europe and its contribution to the numerous operational difficulties that must be overcome to achieve the safe extraction of the presumed oil and gas resources of the polar shelf seas.

Properties

Because ice is a thermal insulator, the thicker the ice, the slower it grows, other conditions being equal. As sea ice either ablates or stops growing during the summer, there is a maximum thickness of first-year (FY) ice that can form during a specific year. The exact value is, of course, dependent upon the local climate and oceanographic conditions, reaching values of slightly over 2 m in the Arctic and as much as almost 3 m at certain Antarctic sites. It is also clear that during the winter the heat flux from areas of open water into the polar atmosphere is significantly greater than the flux through even thin ice and is as much as 200 times greater than the flux through MY ice. This means that, even if open water and thin ice areas comprise less than 1–2% of the winter ice pack, lead areas must still be considered in order to obtain realistic estimates of ocean–atmosphere thermal interactions.

If an ice floe survives a summer, during the second winter the thickness of the additional ice that is added is less than the thickness of nearby FY ice for two reasons: it starts to freeze later and it grows slower. Nevertheless, by the end of the winter, the second-year ice will be thicker than the nearby FY ice. Assuming that the above process is repeated in subsequent years, an amount of ice is ablated away each summer (largely from the upper ice surface) and an amount is added each winter (largely on the lower ice surface). As the year pass, the ice melted on top each summer remains the same (assuming no change in the climate over the ice), while the ice forming on the bottom becomes less and less as a result of the increased insulating effect of the thickening overlying ice. Ultimately, a rough equilibrium is reached, with the thickness of the ice added in the winter becoming equal to the ice ablated in the summer. Such steady-state MY ice floes can be layer cakes of ten or more annual layers with total thicknesses in the range 3.5–4.5 m. Much of the uncertainty in estimating the equilibrium thickness of such floes is the result of uncertainties in the oceanic heat flux. However, in sheltered fiord sites in the Arctic where the oceanic heat flux is presumed to be near zero, MY fast ice with thicknesses up to roughly 15–20 m is known to occur. Another important factor affecting MY ice thickness is the formation of melt ponds on the upper ice surface during the summer in that the thicknesses

and areal extent of these shallow-water bodies is important in controlling the total amount of short-wave radiation that is absorbed. For instance, a melt pond with a depth of only 5 cm can absorb nearly half the total energy absorbed by the whole system. The problem here is that good regional descriptive characterizations of these features are lacking as the result of the characteristic low clouds and fog that occur over the Arctic ice packs in the summer. Particularly lacking are field observations on melt pond depths as a function of environmental variables. Also needed are assessments of how much of the meltwater remains ponded on the surface of the ice as contrasted with draining into the underlying sea water. Thermodynamically these are very different situations.

Conditions in the Antarctic are, surprisingly, rather different. There, surface melt rates within the pack are small compared to the rates at the northern boundary of the pack. The stronger winds and lower humidities encountered over the pack also favor evaporation and minimize surface melting. The limited ablation that occurs appears to be controlled by heat transfer processes at the ice–water interface. As a result, the ice remains relatively cold throughout the summer. In any case, as most of the Antarctic pack is advected rapidly to the north, where it encounters warmer water at the Antarctic convergence and melts rapidly, only small amounts of MY ice remain at the end of summer.

Sea ice properties are very different from those of lake or river ice. The reason for the difference is that when sea water freezes, roughly one-third of the salt in the sea water is initially entrapped within the ice in the form of brine inclusions. As a result, initial ice salinities are typically in the range 10–12‰. At low temperatures (below $-8.7°C$), solid hydrated salts also form within the ice. The composition of the brine in sea ice is a unique function of the temperature, with the brine composition becoming more saline as the temperature decreases. Therefore, the brine volume (the volumetric amount of liquid brine in the ice) is determined by the ice temperature and the bulk ice salinity. Not only is the temperature of the ice different at different levels in the ice sheet but the salinity of the ice decreases further as the ice ages ultimately reaching a value of $\sim 3‰$ in MY ice. Brine volumes are usually lower in the colder upper portions of the ice and higher in the warmer, lower portions. They are particularly low in the above-sea-level part of MY ice as the result of the salt having drained almost completely from this ice. In fact, the upper layers of thick MY ice and of aged pressure ridges produce excellent drinking water when melted. As brine volume is the single most important parameter controlling the thermal, electrical, and mechanical properties of sea ice, these properties show associated large changes both vertically in the same ice sheet and between ice sheets of differing ages and histories. To add complexity to this situation, exactly how the brine is distributed within the sea ice also affects ice properties.

There are several different structural types of sea ice, each with characteristic crystal sizes and preferred crystal orientations and property variations. The two most common structural types are called congelation and frazil. In congelation ice, large elongated crystals extend completely through the ice sheet, producing a structure that is similar to that found in directionally solidified metals. In the Arctic, large areas of congelation ice show crystal orientations that are so similar as to cause the ice to have directionally dependent properties in the horizontal plane as if the ice were a giant single crystal. Frazil, on the other hand, is composed of small, randomly oriented equiaxed crystals that are not vertically elongated. Congelation is more common in the Arctic, while frazil is more common in the Antarctic, reflecting the more turbulent conditions characteristically found in the Southern Ocean.

Two of the more unusual sea ice types are both subsets of so-called 'underwater ice.' The first of these is referred to as platelet ice and is particularly common around margins of the Antarctic continent at locations where ice shelves exist. Such shelves not only comprise 30% of the coastline of Antarctica, they also can be up to 250 m thick. Platelet ice is composed of a loose open mesh of large platelets that are roughly triangular in shape with dimensions of 4–5 cm. In the few locations that have been studied, platelet ice does not start to develop until the fast ice has reached a thickness of several tens of centimeters. Then the platelets develop beneath the fast ice, forming a layer that can be several meters thick. The fast ice appears to serve as a superstrate that facilitates the initial nucleation of the platelets. Ultimately, as the fast ice thickens, it incorporates some of the upper platelets. In the McMurdo Sound region, platelets have been observed forming on fish traps at a depth of 70 m. At locations near the Filchner Ice Shelf, platelets have been found in trawls taken at 250 m. This ice type appears to be the result of crystal growth into water that has been supercooled a fraction of a degree. The mechanism appears to be as follows. There is evidence that melting is occurring on the bottom of some of the deeper portions of the Antarctic ice shelves. This results in a water layer at the ice–water interface that is not only less saline and therefore less dense than the underlying seawater, but also is exactly at its freezing point at that depth because it is in direct contact with the shelf ice. When this water starts to flow outward and upward along the base of the shelf, supercooling develops as a result of adiabatic decompression. This in turn drives the formation of the platelet ice.

The second unusual ice type is a special type of frazil that results from what has been termed suspension freezing. The conditions necessary for its formation include strong winds, intense turbulence in an open water area of a shallow sea and extreme sub-freezing temperatures. Such conditions are characteristically found either during the initial formation of an ice cover in the fall or in regions where polynya formation is occurring, typically by newly formed ice being blown off of a coast or a fast ice area leaving in its wake an area of open water. When such conditions occur, the water column can become supercooled, allowing large quantities of frazil crystals to form and be swept downward by turbulence throughout the whole water column. Studies of benthic microfossils included in sea ice during such events suggest that supercooling commonly reaches depths of 20–25 m and occasionally to as much as 50 m. The frazil ice crystals that form occur in the form of 1–3 mm diameter discoids that are extremely sticky. As a result, they are not only effective in scavenging particulate matter from the water column but they also adhere to material on the bottom, where they continue to grow fed by the supercooled water. Such so-called anchor ice appears to form selectively on coarser material. The resulting spongy ice masses that develop can be quite large and, when the turbulence subsides, are quite buoyant and capable of floating appreciable quantities of attached sediment to the surface. There it commonly becomes incorporated in the overlying sea ice. In rivers, rocks weighing as much as 30 kg have been observed to be incorporated into an ice cover by this mechanism. Recent interest in this subject has been the result of the possibility that this mechanism has been effective in incorporating hazardous material into sea ice sheets, which can then serve as a long-distance transport mechanism.

Drift and Deformation

If sea ice were motionless, ice thickness would be controlled completely by the thermal characteristics of the lower atmosphere and the upper ocean. Such ice sheets would presumably have thicknesses and physical properties that change slowly and continuously from region to region. However, even a casual examination of an area of pack ice reveals striking local lateral changes in ice thicknesses and characteristics. These changes are invariably caused by ice movements produced by the forces exerted on the ice by winds and currents. Such motions are rarely uniform and lead to the build-up of stresses within ice sheets. If these stresses become large enough, cracks may form and widen, resulting in the formation of leads. Such features can vary in width from a few meters to several kilometers and in length from a few hundred meters to several hundred kilometers. As mentioned earlier, during much of the year in the polar regions, once a lead forms it is immediately covered with a thin skim of ice that thickens with time. This is an ever-changing process associated with the movement of weather systems as one lead system becomes inactive and is replaced by another system oriented in a different direction. As lead formation occurs at varied intervals throughout the ice growth season, the end result is an ice cover composed of a variety of thicknesses of uniform sheet ice.

However, when real pack ice thickness distributions are examined (**Figure 1**), one finds that there is a significant amount of ice thicker than the 4.5–5.0 m maximum that might be expected for steady-state MY ice floes. This thicker ice forms by the closing of leads, a process that commonly results in the piling of broken ice fragments into long, irregular features referred to as pressure ridges. There are many small ridges and large ridges are rare.

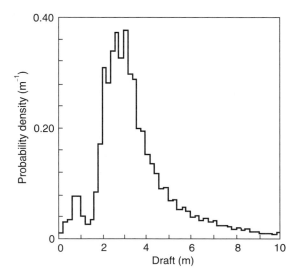

Figure 1 The distribution of sea ice drafts expressed as probability density as determined via the use of upward-looking sonar along a 1400 km track taken in April 1976 in the Beaufort Sea. All ice thicker than ∼4 m is believed to be the result of deformation. The peak probabilities that occur in the range between 2.4 and 3.8 m represent the thicknesses of undeformed MY ice, while the values less than 1.2 m come from ice that ice that recently formed in newly formed leads.

Nevertheless, the large ridges are very impressive, the largest free-floating sail height and keel depth reported to date in the Arctic being 13 and 47 m, respectively (values not from the same ridge). Particularly heavily deformed ice commonly occurs in a band of ∼150 km running between the north coast of Greenland and the Canadian Arctic Islands and the south coast of the Beaufort Sea. The limited data available on Antarctic ridges suggest that they are generally smaller and less frequent than ridges in the Arctic Ocean. The general pattern of the ridging is also different in that the long sinuous ridges characteristic of the Arctic Ocean are not observed. Instead, the deformation can be better described as irregular hummocking accompanied by the extensive rafting of one floe over another. Floe sizes are also smaller as the result of the passage of large-amplitude swells through the ice. These swells, which are generated by the intense Southern Ocean storms that move to the north of the ice edge, result in the fracturing of the larger floes while the large vertical motions facilitate the rafting process.

Pressure ridges are of considerable importance for a variety of reasons. First, they change the surface roughness at the air–ice and water–ice interfaces, thereby altering the effective surface tractions exerted by winds and currents. Second, they act as plows, forming gouges in the sea floor up to 8 m deep when they ground and are pushed along by the ungrounded pack as it drifts over the shallower (<60 m) regions of the polar continental shelves. Third, as the thickest sea ice masses, they are a major hazard that must be considered in the design of offshore structures. Finally, and most importantly, the ridging process provides a mechanical procedure for transferring the thinner ice in the leads directly and rapidly into the thickest ice categories.

Considerable information on the drift and deformation of sea ice has recently become available through the combined use of data buoy and satellite observations. This information shows that, on the average, there are commonly two primary ice motion features in the Arctic Basin. These are the Beaufort Gyre, a large clockwise circulation located in the Beaufort Sea, and the Trans-Polar Drift Stream, which transports ice formed on the Siberian Shelf over the Pole to Fram Strait between Greenland and Svalbard. The time required for the ice to complete one circuit of the gyre averages 5 years, while the transit time for the Drift Stream is roughly 3 years, with about 9% of the sea ice of the Arctic Basin (919 000 km²) moving south through Fram Strait and out of the basin each year. There are many interesting features of the ice drift that exist over shorter time intervals. For instance, recent observations show that the Beaufort Gyre may run backward (counter-clockwise) over appreciable periods of time, particularly in the summer and fall. There have even been suggestions that such reversals can occur on decadal timescales. Typical pack ice velocities range from 0 to $20\,\text{cm s}^{-1}$, although extreme velocities of up to $220\,\text{cm s}^{-1}$ (4.3 knots) have been recorded during storms. During winter, periods of zero ice motion are not rare. During summers, when considerable open water is present in the pack, the ice appears to be in continuous motion. The highest drift velocities are invariably observed near the edge of the pack. Not only are such locations commonly windy, but the floes are able to move toward the free edge with minimal inter-floe interference. Ice drift near the Antarctic continent is generally westerly, becoming easterly further to the north, but in all cases showing a consistent northerly diverging drift toward the free ice edge.

Trends

Considering the anticipated geophysical consequences of changes in the extent of sea ice, it is not surprising that there is considerable scientific interest in the subject. Is sea ice expanding and thickening, heralding a new glacial age, or retreating and thinning before the onslaught of a greenhouse-gas-induced heatwave? One thing that should be clear from the preceding discussion is that the ice is surprisingly thin and variable. Small changes in meteorological and oceanographic forcing should result in significant changes in the extent and state of the ice cover. They could also produce feedbacks that might have significant and complex climatic consequences.

Before we examine what is known about sea ice variations, let us first examine other related observations that have a direct bearing on the question of sea ice trends. Land station records for 1966–1996 show that the air temperatures have increased, with the largest increases occurring winter and spring over both north-west North America and Eurasia, a conclusion that is supported by increasing permafrost temperatures. In addition, meteorological observations collected on Russian NP drifting stations deployed in the Arctic Basin show significant warming trends for the spring and summer periods. It has also recently been suggested that when proxy temperature sources are considered, they indicate that the late twentieth-century Arctic temperatures were the highest in the past 400 years.

Recent oceanographic observations also relate to the above questions. In the late 1980s the balance between the Atlantic water entering the Arctic Basin and the Pacific water appears to have changed, resulting in an increase in the areal extent of the more saline, warmer Atlantic water. In addition, the Atlantic water is shallower than in the past, resulting in temperature increases of as much as $2°C$ and salinity increases of up to 2.5‰ at depths of $200\,\text{m}$. The halocline, which isolates the cold near-surface layer and the overlying sea ice cover from the underlying warmer water, also appears to be thinning, a fact that could profoundly affect the state of the sea ice cover and the surface energy budget in the Arctic. Changes revealed by the motions of data buoys placed on the ice show that there has been a weakening of the Beaufort Sea Gyre and an associated increased divergence of the ice peak. There are also indications that the MY ice in the center of the Beaufort Gyre is less prevalent and thinner than in the past and that the amount of surface melt increased from $\sim 0.8\,\text{m}$ in the mid-1970s to ~ 2 m in 1997. This conclusion is supported by the operational difficulties encountered by recent field programs such as SHEBA that attempted to maintain on-ice measurements. The increased melt is also in agreement with observed decreases in the salinity of the near-surface water layer.

It is currently believed that these changes appear to be related to atmospheric changes in the Polar Basin where the mean atmospheric surface pressure is decreasing and has been below the 1979–95 mean every year since 1988. Before about 1988–99 the Beaufort High was usually centered over 180° longitude. After this time the high was both weaker and typically confined to more western longitudes, a fact that may account for lighter ice conditions in the western Arctic. There also has been a recent pronounced increase in the frequency of cyclonic storms in the Arctic Basin.

So are there also direct measurements indicating decreases in ice extent and thickness? Historical data based on direct observations of sea ice extent are rare, although significant long-term records do exist for a few regions such as Iceland where sea ice has an important effect on both fishing and transportation. In monitoring the health of the world's sea ice covers the use of satellite remote sensing is essential because of the vast remote areas that must be surveyed. Unfortunately, the satellite record is very short. If data from only microwave remote sensing systems are considered, because of their all-weather capabilities, the record is even shorter, starting in 1973. As there was a 2-year data gap between 1976 and 1978, only 25 years of data are available to date. The imagery shows that there are definitely large seasonal, interannual and regional variations in ice extent. For instance, a decrease in ice extent in the Kara and Barents Seas contrasts with an increase in the Baffin Bay/Davis Strait region and out-of-phase fluctuations occur between the Bering and the Okhotsk Seas. The most recent study, which examined passive microwave data up to December 1996, concludes that the areal extent of Arctic sea ice has decreased by 2.9% ± 0.4% per decade. In addition, record or near-record minimum areas of Arctic sea ice have been observed in 1990, 1991, 1993, 1995, and 1997. A particularly extreme recession of the ice along the Beauford coast was also noted in the fall of 1998. Russian ice reconnaissance maps also show that a significant reduction in ice extent and concentration has occurred over much of the Russian Arctic Shelf since 1987.

Has a systematic variation also been observed in ice thickness? Unfortunately there is, at present, no satellite-borne remote sensing technique that can measure sea ice thicknesses effectively from above. There is also little optimism about the possibilities of developing such techniques because the extremely lossy nature of sea ice limits penetration of electromagnetic signals. Current ice thickness information comes from two very different techniques: *in situ* drilling and upward-looking, submarine-mounted sonar. Although drilling is an impractical technique for regional studies, upward-looking sonar is an extremely effective procedure. The submarine passes under the ice at a known depth and the sonar determines the distance to the underside of the ice by measuring the travel times of the sound waves. The result is an accurate, well-resolved under-ice profile from which ice draft distributions can be determined and ice thickness distributions can be estimated based on the assumption of isostacy. Although there have been a large number of under-ice cruises starting with the USS *Nautilus* in 1958, to date only a few studies have been published that examine temporal variations in ice thickness in the Arctic. The first compared the results of two nearly identical cruises: that of the USS *Nautilus* in 1958 with that of the USS *Queenfish* in 1970. Decreases in mean ice thickness were observed in the Canadian Basin (3.08–2.39 m) and in the Eurasian Basin (4.06–3.57 m). The second study has compared the results of two Royal Navy cruises made in 1976 and 1987, and obtained a 15% decrease in mean ice thickness for a 300 000 km² area north of Greenland. Although these studies showed similar trends, the fact that they each only utilized two years' data caused many scientists to feel that a conclusive trend had not been established. However, a recent study has been able to examine this problem in more detail by comparing data from three submarine cruises made in the 1990s (1993,

1996, 1997) with the results of similar cruises made between 1958 and 1976. The area examined was the deep Arctic Basin and the comparisons used only data from the late summer and fall periods. It was found that the mean ice draft decreased by about 1.3 m from 3.1 m in 1958–76 to 1.8 m in the 1990s, with a larger decrease occurring in the central and eastern Arctic than in the Beaufort and Chukchi Seas. This is a very large difference, indicating that the volume of ice in the region surveyed is down by some 40%. Furthermore, an examination of the data from the 1990s suggests that the decrease in thickness is continuing at a rate of about 0.1 m y^{-1}.

Off the Antarctic the situation is not as clear. One study has suggested a major retreat in maximum sea ice extent over the last century based on comparisons of current satellite data with the earlier positions of whaling ships reportedly operating along the ice edge. As it is very difficult to access exactly where the ice edge is located on the basis of only ship-board observations, this claim has met with some skepticism. An examination of the satellite observations indicates a very slight increase in areal extent since 1973. As there are no upward-looking sonar data for the Antarctic Seas, the thickness database there is far smaller than in the Arctic. However, limited drilling and airborne laser profiles of the upper surface of the ice indicate that in many areas the undeformed ice is very thin (60–80 cm) and that the amount of deformed ice is not only significantly less than in the Arctic but adds roughly only 10 cm to the mean ice thickness (**Figure 2**).

What is one to make of all of this? It is obvious that, at least in the Arctic, a change appears to be under way that extends from the top of the atmosphere to depths below 100 m in the ocean. In the middle of this is the sea ice cover, which, as has been shown, is extremely sensitive to environmental changes. What is not known is whether these changes are part of some cycle or represent a climatic regime change in which the positive feedbacks associated with the presence of a sea ice cover play an important role. Also not understood are the interconnections between what is happening in the Arctic and other changes both inside and outside the Arctic. For instance, could changes in the Arctic system drive significant lower-latitude atmospheric and oceanographic changes or are the Arctic changes driven by more dynamic lower-latitude processes? In the Antarctic the picture is even less clear, although changes are known to be underway, as evidenced by the recent breakup of ice shelves along the eastern coast of the Antarctic Peninsula. Not surprisingly, the scientific community is currently devoting considerable energy to attempting to answer these

Figure 2   (A) Ice gouging along the coast of the Beaufort Sea. (B) Aerial photograph of an area of pack ice in the Arctic Ocean showing a recently refrozen large lead that has developed in the first year. The thinner newly formed ice is probably less than 10 cm thick. (C) A representative pressure ridge in the Arctic Ocean. (D) A rubble field of highly deformed first-year sea ice developed along the Alaskan coast of the Beaufort Sea. The tower in the far distance is located at a small research station on one of the numerous off-shore islands located along this coast. (E) Deformed sea ice along the NW Passage, Canada. (F) Aerial photograph of pack ice in the Arctic Ocean.

questions. One could say that a cold subject is heating up.

See also

Antarctic Circumpolar Current. Arctic Ocean Circulation. Sea Ice.

Further Reading

Cavelieri DJ, Gloersen P, Parkinson CL, Comiso JC, and Zwally HJ (1997) Observed hemispheric asymmetry in global sea ice changes. *Science* 278 5340: 1104–1106.

Dyer I and Chryssostomidis C (eds.) (1993) *Arctic Technology and Policy.* New York: Hemisphere.

Leppäranta M (1998) *Physics of Ice-covered Seas, 2 vols.* Helsinki: Helsinki University Printing House.

McLaren AS (1989) The underice thickness distribution of the Arctic basin as recorded in 1958 and 1970. *Journal of Geophysical Research* 94(C4): 4971–4983.

Morison JH, Aagaard K and Steele M (1998) *Study of the Arctic Change Workshop.* (Report on the Study of the Arctic Change Workshop held 10–12 November 1997, University of Washington, Seattle, WA). Arctic System Science Ocean–Atmosphere–Ice Interactions Report No. 8 (August 1998).

Rothrock DA, Yu Y and Maykut G (1999) Thinning of the Arctic sea ice cover. *Geophysical Research Letters* 26.

Untersteiner N (ed.) (1986) *The Geophysics of Sea Ice. NATO Advanced Science Institutes Series B, Physics 146.* New York: Plenum Press.

SEA ICE DYNAMICS

M. Leppäranta, University of Helsinki, Helsinki, Finland

Introduction

Sea ice grows, drifts, and melts under the influence of solar, atmospheric, oceanic, and tidal forcing. Most of it lies in the polar seas but seasonally freezing, smaller basins exist in lower latitudes. In sizeable basins, solid sea ice lids are statically unstable and break into fields of ice floes, undergoing transport as well as opening and ridging which altogether create the exciting sea ice landscape as it appears to the human eye.

The history of sea ice dynamics science initiates from the drift of Nansen's ship *Fram* in 1893–96 in the Arctic Ocean. The average drift speed was 2% of the wind speed and the drift direction deviated 30° to the right from the wind. Much data were collected in Soviet Union North Pole Drifting Stations program commenced in 1937, where science camps drifted from the North Pole to the Greenland Sea. The closure of the sea ice dynamics problem was completed in the 1950s as rheology and conservation laws were established. In the 1970s, two major findings came out in the Arctic Ice Dynamics Joint Experiment (AIDJEX) program: plastic rheology and thickness distribution. Recent progress contains granular flow models, ice kinematics mapping with microwave satellite imagery, anisotropic rheologies, and scaling laws. Methods for remote sensing of sea ice thickness have been slowly progressing, which is the most critical point for the further development of the theory and models of sea ice dynamics.

Drift Ice Medium

Ice State

Drift ice is a peculiar geophysical medium (**Figure 1**). It is granular (ice floes are the basic elements, grains) and the motion takes place on the sea surface plane as a two-dimensional system. The compactness of floe fields may easily change, that is, the medium is compressible, the rheology shows highly nonlinear properties, and by freezing and melting an ice source/sink term exists. The full ice drift problem includes the following unknowns: ice state (a set of relevant material properties), ice velocity, and ice stress. The system is closed by the equations for ice conservation, motion, and rheology.

A sea ice landscape consists of ice floes with ridges and other morphological features, and leads and polynyas. It can be divided into zones of different dynamic character. The central pack is free from immediate influence from the boundaries, and the length scale is the size of the basin. Land fast or fast ice is the immobile coastal sea ice zone extending from the shore to about 10–20-m depths (in Antarctic, grounded icebergs may act as tie points and extend the fast ice zone to deeper waters). Next to fast ice is the shear zone (width 10–200 km), where the mobility of the ice is restricted by the geometry of the boundary and strong deformation takes place. Marginal ice zone (MIZ) lies along the boundary to open sea. It is loosely characterized as the zone, which 'feels the presence of the open ocean' and extends to a distance of 100 km from the ice edge. Well-developed MIZs are found along the oceanic ice edge of the polar oceans. They influence the mesoscale ocean dynamics resulting in ice edge eddies, jets, and upwelling/downwelling.

The horizontal structure of a sea ice cover is well revealed by optical satellite images (**Figure 1(a)**). Ice floes are described by their thickness h and diameter d, and we may examine the drift of an individual floe or a field of floes. For continuum models to be valid for a floe field, the size of continuum material particles D must satisfy $d \ll D \ll \Lambda$, where Λ is the gradient length scale. The ranges are in nature $d \sim 10^1$–10^4 m, $D \sim 10^3$–10^5 m, and $\Lambda \sim 10^4$–10^6 m. As $D \to \Lambda$, discontinuities build up, and as $D \to d$ we have a set of n individual floes.

In the continuum approach, an ice state J is defined for the material description, $\dim(J)$ being the number of levels. The first attempt was $J = \{A, H\}$, where A is ice compactness and H is mean thickness. Three-level ice states $J = \{A, H_u, H_d\}$ decomposing the ice into undeformed ice thickness H_u and deformed ice thickness H_d have been used, and the fine-resolution approach is to take the thickness distribution $p(h)$ for the ice state. The thickness classes are fixed, arbitrarily spaced, and their histogram contains the state as the class probabilities π_k:

$$J = \{\pi_0, \pi_1, \pi_2, \dots\}, \pi_k$$
$$= \text{Prob}\{h = h_k\}, \quad \sum_k \pi_k = 1 \quad [1]$$

(a)

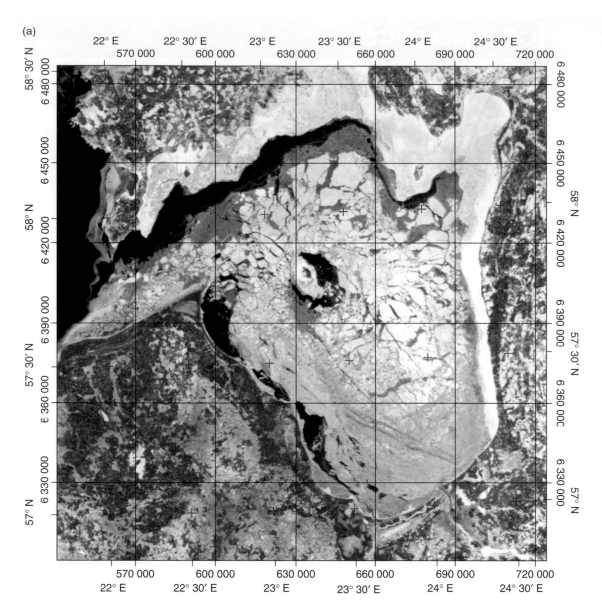

Figure 1   (a) A moderate-resolution imaging spectroradiometer (MODIS) image (NASA's *Terra/Aqua* satellite) of the sea ice cover in the Gulf of Riga, 3 Mar. 2003. The width of the basin is 120 km. (b) Sea ice landscape of the heavy pack ice in the Arctic Ocean, north of Svalbard. (c) Sea ice landscape from the Weddell Sea, Antarctica, showing a first-year ice floe field.

Kinematics

Sea ice kinematics has been mapped using drifters and sequential remote-sensing imagery. In the Arctic Ocean, the long-term drift pattern consists of the Transpolar Drift Stream on the Eurasian side and the Beaufort Sea Gyre on the American side, as illustrated by the historical data in **Figure 2**. In the Antarctica, the governing feature is two annuluses rotating in opposite directions forced by the East Wind and West Wind zones, with meridional movements in places to interchange ice floes between the annuluses, in particular, northward along the Antarctic Peninsula in the Weddell Sea.

Figure 3 shows a typical 1-week time series of sea ice velocity together with wind data. The ice followed the wind with essentially no time lag but sometimes the ice made 'unexpected' steps due to its internal friction. In general, frequency spectra of ice velocity reach highest levels at the synoptic timescales, and a secondary peak appears at the inertial period. Exceptionally very-high ice velocities of more than $1\,\mathrm{m\,s^{-1}}$ have been observed in transient currents in straits and along coastlines. An extreme case

(b)

(c)

Figure 1 Continued.

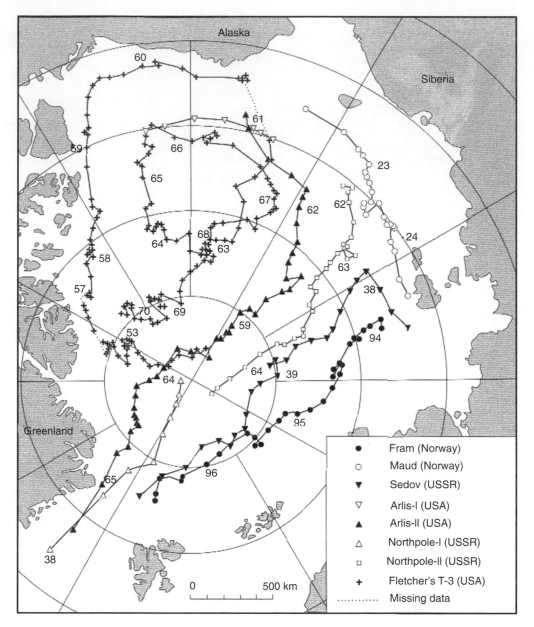

Figure 2   Paths of drifting stations in the Arctic Ocean. The numbers show the year (1894–1970) and marks between are at monthly intervals. Reproduced from Hibler WD, III (1980) Sea ice growth, drift and decay. In: Colbeck S (ed.) *Dynamics of Snow and Ice Masses*, pp. 141–209. New York: Academic Press.

is an 'ice river' phenomenon where a narrow (≈ 0.5 km) band in close ice moves at the speed of up to $3 \, \mathrm{m \, s^{-1}}$. Long-term data of sea ice deformation was obtained in AIDJEX in 1975. The magnitude of strain rate and rotation was $0.01 \, \mathrm{d^{-1}}$ and their levels were higher by 50–100% in summer. In the MIZ, the level is up to $0.05 \, \mathrm{h^{-1}}$ in short, intensive periods. Leads open in divergent directions while ridges form in convergent directions, and both these processes may occur in pure two-dimensional shear deformation. Also, drift ice has a particular asymmetry in the deformation: leads open and close, pressure

(ridged) ice forms, but cannot 'unform' since there is no restoring force.

The ice conservation law tells how ice-state components are changed by advection, mechanical deformation, and thermodynamics. The conservation of ice volume is always required:

$$\frac{\partial H}{\partial t} + \boldsymbol{u} \cdot \nabla H = -H \nabla \cdot \boldsymbol{u} + \phi(H) \qquad [2]$$

where $\boldsymbol{u}$ is ice velocity and $\phi(H)$ is the thermal growth and melt rate. If $H \sim 1 \, \mathrm{m}$ and $\nabla \cdot \boldsymbol{u} \sim -0.1 \, \mathrm{d^{-1}}$, the

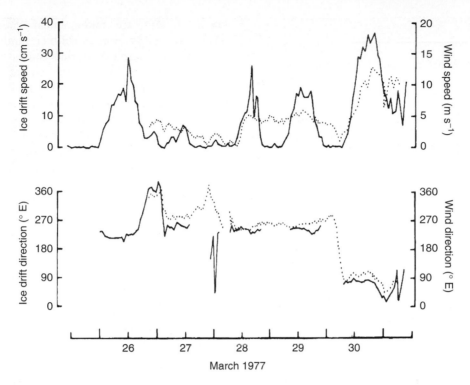

Figure 3 Sea ice (solid lines) and wind velocity (dotted lines) time series, Baltic Sea, Mar. 1977. Reproduced from Leppäranta M (1981) An ice drift model for the Baltic Sea. *Telbus* 33(6): 583–596.

mechanical growth rate is $\sim 10\,\mathrm{cm}\,\mathrm{d}^{-1}$, while apart from thin ice thermal growth rates are usually less than $1\,\mathrm{cm}\,\mathrm{d}^{-1}$. Mechanical growth events are short, so that in the long run thermal production usually overcomes mechanical production. But in regions of intensive ridging, such as the northern coast of Greenland, ice thickness is more than twice the thermal equilibrium thickness of multiyear ice. The conservation law of ice compactness is similar, except that the compactness cannot be more than unity.

Formally, the ice conservation law is expressed for the ice state as

$$\frac{\partial J}{\partial t} + \boldsymbol{u} \cdot \nabla J = \Psi + \Phi \qquad [3]$$

where Ψ and Φ are the change of ice state due to mechanics and thermodynamics, respectively. In multilevel cases, the question is that how deformed ice is produced. In the three-level case $J = \{A, H_\mathrm{u}, H_\mathrm{d}\}$, all ridging adds on the thickness of deformed ice, but when using the thickness distribution additional assumptions are needed for the redistribution. Formally, one may take the limit $\Delta h_k \to 0$ to obtain the conservation law for the spatial density of thickness:

$$\frac{\partial \pi}{\partial t} + \boldsymbol{u} \cdot \nabla \pi = \psi - \pi \nabla \cdot \boldsymbol{u} + \pi \frac{\partial \phi(h)}{\partial h} \qquad [4]$$

where ψ is the mechanical thickness redistributor. This operator closes and opens leads by changing the height of the peak at zero, and under convergence of the compact ice it takes, thin ice is transformed into deformed ice with multiple thickness; for example, 1-m-thick ice of area A_1 is changed into k m ice with area A_1/k, $k \sim 10$ (or over a range of thicknesses), and the probabilities are changed accordingly. The thermodynamic change is straightforward as ice growth and melting advect the distribution in the thickness space as determined by the energy budget. The thermodynamic term is not further discussed below.

Rheology

The mechanisms behind the rheology are floe collisions, floe breakage, shear friction between floes, and friction between ice blocks and potential energy production in ridging. By observational evidence, it is known that (1) stress level ≈ 0 for $A < 0.8$, (2) yield strength > 0 for $A \approx 1$, and (3) tensile strength $\approx 0 \ll$ shear strength $<$ compressive strength. The rheological law is given in its general form as

$$\boldsymbol{\sigma} = \boldsymbol{\sigma}(J, \boldsymbol{\varepsilon}, \dot{\boldsymbol{\varepsilon}}) \qquad [5]$$

where $\boldsymbol{\sigma}$ is stress, $\boldsymbol{\varepsilon}$ is strain, and $\dot{\boldsymbol{\varepsilon}}$ is strain rate (**Figure 4**). The stress is very low for compactness less

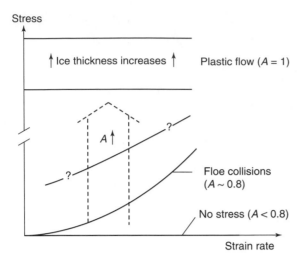

Figure 4 Schematic presentation of the sea ice rheology as a function of ice compactness A and thickness h. The cut in the ordinate axis tells of a jump of several orders of magnitude. Reproduced from Leppäranta M (2005) *The Drift of Sea Ice*, 266pp. Heidelberg: Springer-Praxis.

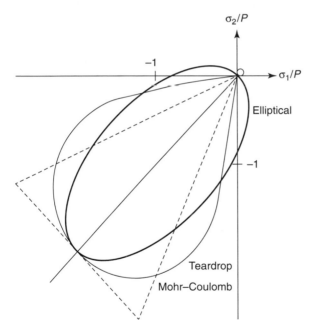

Figure 5 Plastic yield curves for drift ice. Reproduced from Leppäranta M (2005) *The Drift of Sea Ice*, 266pp. Heidelberg: Springer-Praxis.

than 0.8. At higher ice concentrations, the significance of floe collisions and shear friction between ice floes increases, and finally with ridge formation a plastic flow results. How the rheology changes from the superlinear collision rheology to a plastic law is not known. In the plastic regime, the yield strength increases with increasing ice thickness.

Two-dimensional yielding is specified using a yield curve $F(\sigma_1, \sigma_2) = 0$ (Figure 5). In numerical modeling,

lower stresses are taken linear elastic (AIDJEX model) or linear viscous (Hibler model). Drucker's postulate for stable materials states that the yield curve serves as the plastic potential, and consequently, the failure strain is directed perpendicular to the yield curve. Drift ice is strain hardening in compression, and therefore, ridging may proceed only to a certain limit. Hibler model is an excellent tool for numerical modeling since it allows an explicit solution for the stress as a function of the strain rate:

$$\boldsymbol{\sigma} = \frac{P}{2}\left(\frac{\dot{\varepsilon}_\mathrm{I}}{\max(\Delta, \Delta_\mathrm{o})} - 1\right)\mathbf{I} + \frac{P}{2e^2\max(\Delta, \Delta_\mathrm{o})}\dot{\boldsymbol{\varepsilon}}' \quad [6]$$

where $P = P^* h \ \exp[-C(1-A)]$ is the compressive strength, P^* is the strength level constant and C is the strength reduction for lead opening, $\Delta = \sqrt{\dot{\varepsilon}_\mathrm{I}^2 + (\dot{\varepsilon}_\mathrm{II}/e)^2}$, $\dot{\varepsilon}_\mathrm{I}$ and $\dot{\varepsilon}_\mathrm{II}$ are strain-rate invariants equal to divergence and twice the maximum rate of shear, Δ_o is the maximum viscous creep rate, e is the aspect ratio of the yield ellipse, and $\dot{\boldsymbol{\varepsilon}}'$ is the deviatoric strain rate. The normal parameter values are $P^* = 25 \text{ kPa}$, $C = 20$, $e = 2$, and $\Delta_\mathrm{o} = 10^{-9} \text{ s}^{-1}$.

Equation of Motion

The equation of motion is first integrated through the thickness of the ice for the two-dimensional system (Figure 6). This is straightforward. Integration of the divergence of internal ice stress brings the air and ocean surface stresses (τ_a and τ_w) and internal friction, and the other terms are just multiplied by the mean thickness of ice. The result is

$$\rho H \left[\frac{\partial \boldsymbol{u}}{\partial t} + \boldsymbol{u} \cdot \nabla \boldsymbol{u} + f \mathbf{k} \times \boldsymbol{u}\right]$$
$$= \nabla \cdot \boldsymbol{\sigma} + \boldsymbol{\tau}_\mathrm{a} + \boldsymbol{\tau}_\mathrm{w} - \rho H g \boldsymbol{\beta} \quad [7]$$

where ρ is ice density, f is the Coriolis parameter, g is acceleration due to gravity, and $\boldsymbol{\beta}$ is the sea surface slope. The air and water stresses are written as

$$\boldsymbol{\tau}_\mathrm{a} = \rho_\mathrm{a} C_\mathrm{a} U_\mathrm{ag}(\cos \theta_\mathrm{a} + \sin \theta_\mathrm{a} \mathbf{k} \times) U_\mathrm{ag} \quad [8a]$$

$$\boldsymbol{\tau}_\mathrm{w} = \rho_\mathrm{w} C_\mathrm{w} |U_\mathrm{wg} - \boldsymbol{u}|(\cos \theta_\mathrm{w} + \sin \theta_\mathrm{w} \mathbf{k} \times)$$
$$(U_\mathrm{wg} - \boldsymbol{u}) \quad [8b]$$

where ρ_a and ρ_w are air and water densities, C_a and C_w are air and water drag coefficients, θ_a and θ_w are the air- and water-boundary layer angles, and U_ag and U_wg are the geostrophic wind and current velocities. Representative Arctic – the best-known region – parameters are $C_\mathrm{a} = 1.2 \times 10^{-3}$, $\theta_\mathrm{a} = 25°$,

$C_w = 5 \times 10^{-3}$, and $\theta_w = 25°$. In a stratified fluid, the drag parameters also depend on the stability of the stratification; in very stable conditions, $C_a \approx 10^{-4}$ and $\theta_a \approx 35°$ while in unstable conditions $C_a \approx 1.5 \times 10^{-3}$ and $\theta_a \to 0$.

The sea ice dynamics problem can be divided into three categories: (1) stationary ice, (2) free drift, and (3) drift in the presence of internal friction. There are three timescales: local acceleration $T_I = H/(C_w U)$, Coriolis period f^{-1}, and adjustment of the thickness field $T_D = L/U$. These are well separated, $T_I \ll f^{-1} \ll T_D$. **Table 1** shows the result of the magnitude analysis of the equation of motion based on the typical scales.

Stationary Ice

In a stationary ice field, we have, by definition, $\boldsymbol{u} \equiv 0$. Then we have

$$\nabla \cdot \boldsymbol{\sigma} + \boldsymbol{\tau}_a + \boldsymbol{\tau}_w - \rho g h \boldsymbol{\beta} = 0 \qquad [9]$$

The ice is forced by $F = \tau_a + \tau_w - \rho g h \boldsymbol{\beta}$, and the stationarity is satisfied as long as the internal ice stress is beneath the yield level. A natural dimensionless number for this situation is $X_o = PH/FL$: for $X_o > 1$ the forcing is below the strength of the ice.

Free Drift

In free drift, by definition $\nabla \cdot \boldsymbol{\sigma} \equiv 0$. Since the momentum advection is very small and $T_I \sim 1\,\mathrm{h}$, an algebraic steady-state equation results as a very good approximation. Expressing the tilt term via the surface geostrophic current $(g\boldsymbol{\beta} = -f\mathbf{k} \times U_{wg})$, the solution is

$$\boldsymbol{u} = \boldsymbol{u}_a + U_{wg} \qquad [10]$$

where $\boldsymbol{u}_a$ is the wind-driven ice drift, $u_a/U_a = \alpha$, and the direction of ice motion deviates the angle of θ from the wind direction (**Figure 7**). The solution is described by drag ratio ('Na', from Nansen who first documented the wind drift factor) and ice drift Rossby number:

$$Na = \sqrt{\frac{\rho_a C_a}{\rho_w C_w}}, \quad Ro = \frac{\rho}{\rho_w C_w} \frac{fH}{U} \qquad [11a]$$

With $Ro \to 0$, $\alpha \to Na$, and $\theta \to \theta_a - \theta_w$, for the geostrophic reference velocities, in neutral conditions, $Na \approx 2\%$ and $\theta \approx 0$. The latter property is known as

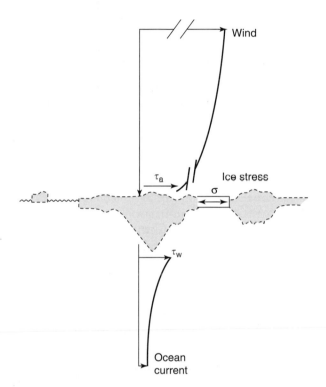

Figure 6 The ice drift problem. Reproduced from Leppäranta M (2005) *The Drift of Sea Ice*, 266pp. Heidelberg: Springer-Praxis.

Table 1 Scaling of the equation of motion of drift ice

Term	Scale	Value	Comments
Local acceleration	$\rho HU/T$	-3	-1 for rapid changes ($T = 10^3\,\mathrm{s}$)
Advective acceleration	$\rho HU^2/L$	-4	Long-term effects may be significant
Coriolis term	ρHfU	-2	Mostly less than -1
Internal friction	PH/L	$-1\,(-\infty)$	Compact ice, $A > 0.9$ ($A < 0.8$)
Air stress	$\rho_a C_a U_{ag}^2$	-1	Mostly significant
Water stress	$\rho_w C_w |U - U_{wg}|^2$	-1	Mostly significant
Sea surface tilt	ρHfU_{wg}	-2	Mostly less than or equal to -2

The representative elementary scales are: $H = 1\,\mathrm{m}$, $U = 10\,\mathrm{cm\,s}^{-1}$, $P = 10\,\mathrm{kPa}$, $U_a = 10\,\mathrm{m\,s}^{-1}$, $U_{wg} = 5\,\mathrm{cm\,s}^{-1}$, $T = 1$ day, and $L = 100\,\mathrm{km}$. The 'Value' column gives the ten-based logarithm of the scale in pascals.
Reproduced from Leppäranta M (2005) *The Drift of Sea Ice*, 266pp. Heidelberg: Springer-Praxis.

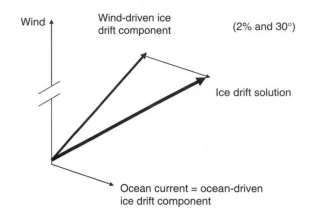

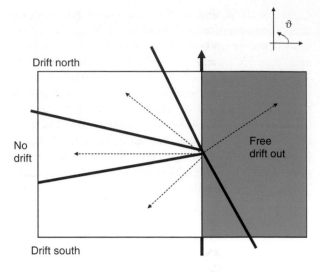

Figure 7 The free drift solution as the vector sum of wind-driven ice drift and geostrophic ocean current.

Figure 8 Steady-state solution of wind-driven zonal flow, Northern Hemisphere. Wind vector is drawn from the crosspoint of the thick lines as indicated by the dashed vectors and the resulting ice velocity is free drift, northward or southward boundary flow, or fast ice, depending on the ice strength and wind direction.

the Zubov's isobaric drift rule. With increasing Rossby number, the wind factor decreases and deviation angle turns more to the right (left) in the Northern (Southern) Hemisphere.

Ice Drift in the Presence of Internal Friction

In compact ice the forcing may become large and break the yield criterion. The ice starts motion, and importance of the internal friction is then described by the friction number:

$$X = \frac{1}{\rho_w C_w U^2} \frac{PH}{L} \qquad [11b]$$

Sea ice rheology has a distinct asymmetry in that during closing the stress may be very high but during opening it is nearly zero. In consequence, the forcing frequencies show up unchanged in ice velocity spectra.

In a channel with a closed end, the full steady-state solution is a stationary ice field. The functional form of the compressive strength, $P = P^*h \exp[-C(1-A)]$, results in a very sharp ice edge, and beyond the edge zone the thickness increases by $dh/dx = \tau_a/P^*$. In the spin-down the ice flows as long as the internal stress overcomes the yield level. Longitudinal boundary zone flow offers a more general frame, representing coastal shear zone or marginal ice zone. The y-axis is aligned along the longitudinal direction, and the system of equations is

$$\rho h \left(\frac{\partial u}{\partial t} + u \frac{\partial u}{\partial x} - fv \right) = \frac{\partial \sigma_{xx}}{\partial x}$$
$$+ \tau_{ax} + \tau_{wx} - \rho hg\beta_x \qquad [12a]$$

$$\rho h \left(\frac{\partial v}{\partial t} + u \frac{\partial v}{\partial x} + fu \right) = \frac{\partial \sigma_{xy}}{\partial x}$$
$$+ \tau_{ay} + \tau_{wy} - \rho hg\beta_y \qquad [12b]$$

$$\boldsymbol{\sigma} = \boldsymbol{\sigma}(h, A, \dot{\boldsymbol{\varepsilon}}), \dot{\varepsilon}_{xx} = \frac{\partial u}{\partial x}, \dot{\varepsilon}_{yx}$$
$$= \dot{\varepsilon}_{yx} = \frac{1}{2} \frac{\partial v}{\partial x}, \dot{\varepsilon}_{yy} = 0 \qquad [12c]$$

$$\frac{\partial \{A, h\}}{\partial t} + \frac{\partial u\{A, h\}}{\partial x} = 0 \quad (0 \leq A \leq 1) \quad [12d]$$

Consider the full steady state in the Northern Hemisphere (**Figure 8**). The free drift solution results when the direction of the wind stress (ϑ) satisfies $-90° + \theta < \vartheta < 90° + \theta$. Since $\theta \sim 30°$, the angle ϑ must be between $-60°$ and $120°$. Otherwise the ice will stay in contact with the coast influenced by the coastal friction, and the land boundary condition implies $u \equiv 0$. Also, in quite general conditions, $|\sigma_{xy}/\sigma_{xx}| = \gamma = \text{constant} \approx 2$, and an algebraic equation can be obtained for the longitudinal velocity v. The ice may drift north if $\tau_{ay} + \gamma\tau_{ax} > 0$, which means that $90° + \theta < \vartheta < 180° - \arctan(\gamma) \approx 150°$, and then $\sigma_{xy} > 0$. Equations [12] give

$$v = \sqrt{\frac{\tau_{ay} + \gamma\tau_{ax}}{C_N} + \left(\frac{\gamma\rho hf}{2C_N} \right)^2} - \frac{\gamma\rho hf}{2C_N} \qquad [13]$$

where $C_N = \rho_w C_w(\cos\theta_w - \gamma\sin\theta_w)$. The southward flow is analyzed in a similar way. The velocity solution is independent of the exact form of the rheology and even of the absolute magnitude of the stresses as long as the proportionality $|\sigma_{xy}| = \gamma|\sigma_{xx}|$ holds. The general solution is illustrated in **Figure 8** for the Northern Hemisphere case, and the Southern Hemisphere case is symmetric. The resulting ice compactness increases to almost 1 in a very narrow ice edge zone, and further in the ice, thickness increases due to ridge formation.

Numerical Modeling

In mesoscale and large-scale sea ice dynamics, all workable numerical models are based on the continuum theory. A full model consists of four basic elements: (1) ice state J, (2) rheology, (3) equation of motion, and (4) conservation of ice. The elements (1) and (2) constitute the heart of the model and are up to the choice of the modeler: one speaks of a three-level ($\dim(J) = 3$) viscous-plastic sea ice model, etc. The unknowns are ice state, ice velocity, and ice stress, and the number of independent variables is $\dim(J) + 2 + 3$. Any proper ice state has at least two levels.

The model parameters can be grouped into those for (a) atmospheric and oceanic drag, (b) rheology, (c) ice redistribution, and (d) numerical design. The primary geophysical parameters are the drag coefficients and compressive strength of ice. The drag coefficients together with the Ekman angles tune the free drift velocity, while the compressive strength tunes the length scale in the presence of internal friction. The secondary geophysical parameters come from the rheology (other than the compressive strength) and the ice state redistribution scheme. The redistribution parameters would be probably very important but the distribution physics lacks good data. The numerical design parameters include the choice of the grid; also since the system is highly nonlinear, the stability of the solution may require smoothing techniques. Since the continuum particle size D is fairly large, the grid size can be taken as $\Delta x \sim D$.

Because the inertial timescale of sea ice is quite small, the initial ice velocity can be taken as zero. At solid boundary, the no-slip condition is employed, while in open boundary the normal stress is zero (a practical way is to define open water as ice with zero thickness and avoid an explicit open boundary).

In short-term modeling, the timescale is 1 h–10 days. The objectives are basic research of the dynamics of drift ice and coupled ice–ocean system, ice forecasting, and applications for marine technology.

In particular, the basic research has involved rheology and thickness redistribution. Leads up to 20 km wide may open and close and heavy-pressure ridges may build up in a 1-day timescale, which has a strong influence on shipping, oil drilling, oil spills, and other marine operations. Also, changes in ice conditions, such as the location of the ice edge, are important for weather forecasting over a few days.

In long-term modeling, the timescale is 1 month–100 years. The objectives are basic research, ice climatology, and global climate. The role of ice dynamics is to transport ice with latent heat and freshwater and consequently modify the ice boundary and air–sea interaction. Differential ice drift opens and closes leads which means major changes to the air–sea heat fluxes. Mechanical accumulation of ice blocks, like ridging, adds large amount to the total volume of ice.

An example of long-term simulations in the Weddell Sea is shown in **Figure 9**. The drag ratio Na and Ekman angle were 2.4% and 10°, both a bit low, for the Antarctica, but it can be explained that the upper layer water current was the reference in the water stress and not the geostrophic flow. The compressive strength constant was taken as $P^* = 20$ kPa. The grid size was about 165 km, and the model was calibrated with drift buoy data for the ice velocity and upward-looking sonar data for the ice thickness. There is a strong convergence region in the southwest part of the basin, and advection of the ice shows up in larger ice thickness northward along the Antarctic peninsula. The width of the compressive region east of the peninsula is around 500 km.

The key areas of modeling research are now ice thickness distribution and its evolution, and use of satellite synthetic aperture radars (SARs) for ice kinematics. The scaling problem and, in particular, the downscaling of the stress from geophysical to local (engineering) scale is examined for combining scientific and engineering knowledge and developing ice load calculation and forecasting methods. The physics of drift ice is quite well represented in short-term ice forecasting models, in the sense that other questions are more critical for their further development, and the user interface is still not very good. Data assimilation methods are coming into sea ice models which give promises for the improvement of both the theoretical understanding and applications.

Concluding Words

The sea ice dynamics problem contains interesting basic research questions in geophysical fluid dynamics. But perhaps, the principal science motivation is connected to the role of sea ice as a dynamic

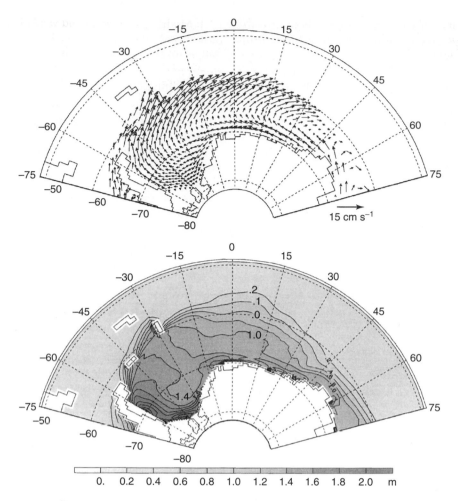

Figure 9   Climatological (a) sea ice velocity and (b) sea ice thickness in the Weddell Sea according to model simulations. Reproduced from Timmermann R, Beckmann A, and Hellmer HH (2000) Simulation of ice–ocean dynamics in the Weddell Sea I: Model configuration and validation. *Journal of Geophysical Research* 107(C3): 10, with permission from the Americal Geophysical Union.

air–ocean interface. The transport of ice takes sea ice (with latent heat and fresh water) to regions, where it would not be formed by thermodynamic processes, and due to differential drift leads open and close and hummocks and ridges form. Sea ice has an important role in environmental research. Impurities are captured into the ice sheet from the seawater, sea bottom, and atmospheric fallout, and they are transported with the ice and later released into the water column. The location of the ice edge is a fundamental boundary condition for the marine biology in polar seas. A recent research line for sea ice dynamics is in paleoclimatology and paleoceanography. Data archive of drift ice and icebergs exists in marine sediments, and via its influence on ocean circulation, the drift ice has been an active agent in the global climate history.

In the practical world, three major questions are connected with sea ice dynamics. Sea ice models have been applied for tactical navigation to provide short-term forecasts of the ice conditions. Ice forcing

on ships and fixed structures are affected by the dynamical behavior of the ice. Sea ice information service is an operational routine system to support shipping and other marine operations such as oil drilling in ice-covered seas. In risk assessment for oil spills and oil combating, proper oil transport and dispersion models for ice-covered seas are needed.

See also

Coupled Sea Ice-Ocean Models. Ice–Ocean Interaction. Ice-shelf Stability. Sea Ice. Sea Ice: Overview.

Further Reading

Coon MD, Knoke GS, Echert DC, and Pritchard RS (1998) The architecture of an anisotropic elastic-plastic sea ice mechanics constitutive law. *Journal of Geophysical Research* 103(C10): 21915–21925.

Dempsey JP and Shen HH (eds.) (2001) *IUTAM Symposium on Scaling Laws in Ice Mechanics*, 484pp. Dordrecht: Kluwer.

Doronin YuP and Kheysin DYe (1975) *Morskoi Led (Sea Ice)*, (English trans. (1977), 323pp. New Delhi: Amerind). Leningrad: Gidrometeoizdat.

Hibler WD, III (1979) A dynamic-thermodynamic sea ice model. *Journal of Physical Oceanography* 9: 815–846.

Hibler WD, III (1980) Sea ice growth, drift and decay. In: Colbeck S (ed.) *Dynamics of Snow and Ice Masses*, pp. 141–209. New York: Acamemic Press.

Hibler WD, III (2004) Modelling sea ice dynamics. In: Bamber JL and Payne AJ (eds.) *Mass Balance of the Cryosphere: Observations and Modelling of Contemporary and Future Changes*, 662pp. Cambridge, UK: Cambridge University Press.

Leppäranta M (1981) An ice drift model for the Baltic Sea. *Telbus* 33(6): 583–596.

Leppäranta M (ed.) (1998) *Physics of Ice-Covered Seas*, vols. 1 and 2, 823pp. Helsinki: Helsinki University Press.

Leppäranta M (2005) *The Drift of Sea Ice*, 266pp. Heidelberg: Springer-Praxis.

Pritchard RS (ed.) (1980) *Proceedings of the ICSI/AIDJEX Symposium on Sea Ice Processes and Models*, 474pp. Seattle, WA: University of Washington Press.

Richter-Menge JA and Elder BC (1998) Characteristics of ice stress in the Alaskan Beaufort Sea. *Journal of Geophysical Research* 103(C10): 21817–21829.

Rothrock DA (1975) The mechanical behavior of pack ice. *Annual Review of Earth and Planetary Sciences* 3: 317–342.

Timmermann R, Beckmann A, and Hellmer HH (2000) Simulation of ice–ocean dynamics in the Weddell Sea I: Model configuration and validation. *Journal of Geophysical Research* 107(C3): 10.

Timokhov LA and Kheysin DYe (1987) *Dynamika Morskikh L'dov*, 272pp. Leningrad: Gidrometeoizdat.

Untersteiner N (ed.) (1986) *Geophysics of Sea Ice*, 1196pp. New York: Plenum.

Wadhams P (2000) *Ice in the Ocean*, 351pp. Amsterdam: Gordon & Breach Science Publishers.

Relevant Websites

http://psc.apl.washington.edu
– AIDJEX Electronic Library, Polar Science Center (PSC).

http://www.aari.nw.ru
– Arctic and Antarctic Research Institute (AARI).

http://ice-glaces.ec.gc.ca
– Canadian Ice Service.

http://www.fimr.fi
– Finnish Ice Service, Finnish Institute of Marine Research.

http://www.hokudai.ac.jp
– Ice Chart Off the Okhotsk Sea Coast of Hokkaido, Sea Ice Research Laboratory, Hokkaido University.

http://IABP.apl.washington.edu
– Index of Animations, International Arctic Buoy Programme (IABP).

http://nsidc.org
– National Snow and Ice Data Center (NSIDC).

http://www.awi.de
– Sea Ice Physics, The Alfred Wegener Institute for Polar and Marine Research (AWI).

SEA ICE

P. Wadhams, University of Cambridge,
Cambridge, UK

Introduction

This article considers the seasonal and interannual variability of sea ice extent and thickness in the Arctic and Antarctic, and the downward trends which have recently been shown to exist in Arctic thickness and extent. There is no evidence at present for thinning or retreat of the Antarctic sea ice cover.

Sea Ice Extent

Arctic

The seasonal cycle The best way of surveying sea ice extent and its variability is by the use of satellite imagery, and the most useful imagery on the large scale is passive microwave, which identifies types of surface through their natural microwave emissions, a function of surface temperature and emissivity. **Figure 1** shows ice extent and concentration maps for the Arctic for each month, averaged over the period 1979–87, derived from the multifrequency scanning multichannel microwave radiometer (SMMR) sensor aboard the *Nimbus-7* satellite. This instrument gives ice concentration and, through comparison of emissions at different frequencies, that percentage of the ice cover which is multiyear ice (i.e., ice which has survived at least one summer of melt). The ice concentrations are estimated to be accurate to $\pm 7\%$. This is an excellent basis for considering the seasonal cycle, although ice extent, particularly in summer, is now significantly less than these figures show.

At the time of maximum advance, in February and March (**Figure 1(a)**), the ice cover fills the Arctic Basin. The Siberian shelf seas are also ice-covered to the coast, although the warm inflow from the Norwegian Atlantic Current keeps the western part of the Barents Sea open. There is also a bight of open water to the west of Svalbard, kept open by the warm West Spitsbergen Current and formerly known as Whalers' Bay because it allowed sailing whalers to reach high latitudes. It is here that the open sea is found closest to the Pole in winter – beyond 81° in some years. The east coast of Greenland has a sea ice cover along its entire length (although in mild winters the ice fails to reach Cape Farewell); this is transported out of Fram Strait by the Transpolar Drift Stream and advected southward in the East Greenland Current, the strongest part of the current (and so the fastest ice drift) being concentrated at the shelf break. At 72–75° N these averaged maps show a distinct bulge in the ice edge, visible from January until April with an ice concentration of 20–50%. During any particular year, this bulge will often appear as a tongue, called Odden, composed mainly of locally formed pancake ice, which covers the region influenced by the Jan Mayen Current (a cold eastward offshoot of the East Greenland Current).

Moving round Cape Farewell there is a thin band of ice off West Greenland (called the 'Storis'), the limit of ice transported out of the Arctic Basin, which often merges with the dense locally formed ice cover of Baffin Bay and Davis Strait. The whole of the Canadian Arctic Archipelago, Hudson Bay, and Hudson Strait are ice-covered, and on the western side of Davis Strait the ice stream of the Labrador Current carries ice out of Baffin Bay southward toward Newfoundland. The southernmost ice limit of this drift stream is usually the north coast of Newfoundland, where the ice is separated by the bulk of the island from an independently formed ice cover filling the Gulf of St. Lawrence, with the ice-filled St. Lawrence River and Great Lakes behind. Further to the west, a complete ice cover extends across the Arctic coasts of NW Canada and Alaska and fills the Bering Sea, at somewhat lower concentration, as far as the shelf break. Sea ice also fills the Sea of Okhotsk and the northern end of the Sea of Japan, with the north coast of Hokkaido experiencing the lowest-latitude sea ice (44°) in the Northern Hemisphere.

In April, the ice begins to retreat from its low-latitude extremes. By May, the Gulf of St. Lawrence is clear, as is most of the Sea of Okhotsk and some of the Bering Sea. The Odden ice tongue has disappeared and the ice edge is retreating up the east coast of Greenland. By June, the Pacific south of Bering Strait is ice-free, with the ice concentration reducing in Hudson Bay and several Arctic coastal locations. August and September (**Figure 1(b)**) are the months of greatest retreat, constituting the brief Arctic summer. During these months the Barents and Kara Seas are ice-free as far as the shelf break, with the Arctic pack retreating to, or beyond, northern Svalbard and Franz Josef Land. The Laptev and East

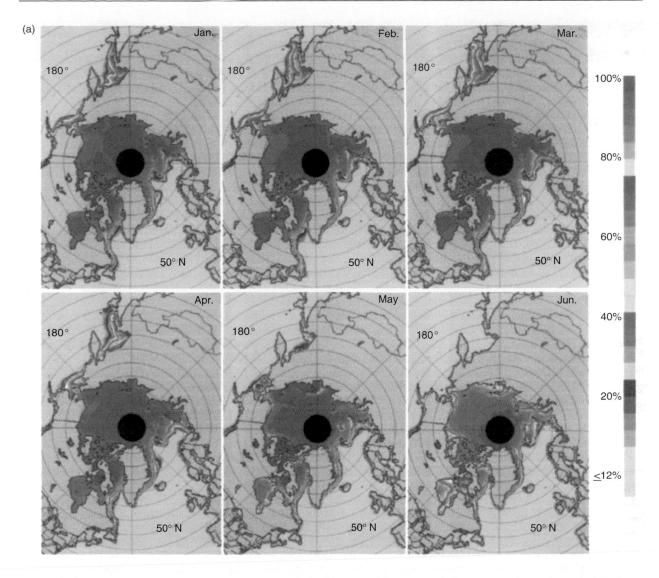

Figure 1 Ice extent and concentration maps for the Arctic for (a) winter, (b) summer months, averaged over the period 1979–87, derived from the multifrequency SMMR sensor (scanning multichannel microwave radiometer) aboard the *Nimbus-7* satellite. From Gloersen P, Campbell WJ, Cavalieri DJ, Comiso JC, Parkinson CL, and Zwally HJ (1992) *Arctic and Antarctic Sea Ice, 1978–1987: Satellite Passive-Microwave Observations and Analysis*, Rept. NASA SP-511, 290pp. Washington, DC: National Aeronautics and Space Administration.

Siberian seas are generally ice-free, with ice in bad years remaining to block choke points such as the Vilkitsky Strait south of Severnaya Zemlya. This allows marine transport through a Northern Sea Route across the top of Russia, but with a need for icebreaker escort through the central ice-choked region. In East Greenland, the ice has retreated northward to about 72–73° (a latitude which varies greatly from year to year), while the whole system of Baffin Bay, Hudson Bay, and Labrador is ice-free. Occasionally a small mass of ice, called the 'Middle Ice', remains at the northern end of Baffin Bay. In the Canadian Arctic Archipelago, the winter fast ice which filled the channels usually breaks up and

partly melts or moves out, but in some years ice remains to clog vital channels, and the Northwest Passage is not such a dependably navigable seaway as the Northern Sea Route. There is usually a slot of open water across the north of Alaska, but again in some years the main Arctic ice edge moves south to touch the Alaskan coast, making navigation very difficult for anything but a full icebreaker.

By October, new ice has formed in many of the areas which were open in summer, especially around the Arctic Ocean coasts, and in November–January there is steady advance everywhere toward the winter peak. The Sea of Okhotsk acquires its first ice cover in December, and the Odden starts to appear;

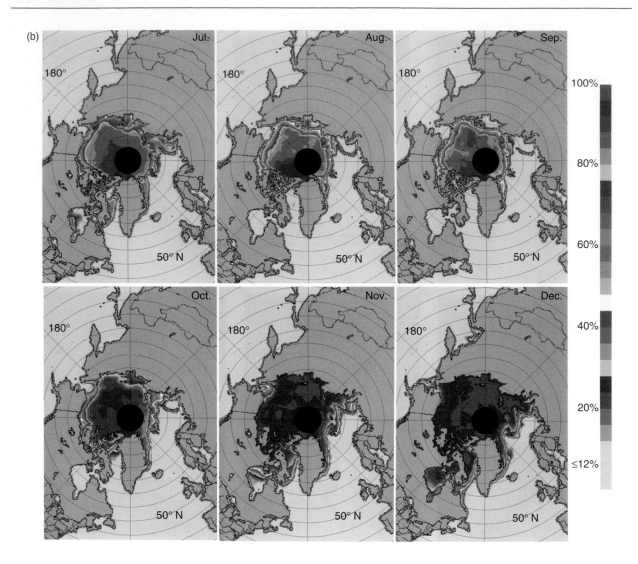

Figure 1 Continued

Baffin Bay and Hudson Bay are already fully ice-covered.

The averaged seasonal cycle for the 1979–87 era (**Figure 2(b)**) gives a maximum extent – 'extent' here is defined as the total area of sea within the 15% ice concentration contour – of $15.7 \times 10^6 \, \text{km}^2$ in late March, and a minimum of $9.3 \times 10^6 \, \text{km}^2$ in early September. For sea ice area, derived as extent multiplied by concentration, the figures are 13.9×10^6 and $6.2 \times 10^6 \, \text{km}^2$ in winter and summer. It is noteworthy that in 2005 the March extent was 14.8 million km^2, only a little lower than the 1980s average, while the September extent was significantly lower at 5.6 million km^2 and a new record low extent of 4.1 million km^2 was reached in September 2007.

Results from SMMR multiyear ice retrievals show that in the Arctic, multiyear ice is found in the highest concentrations within the central Arctic Ocean, in the area controlled by the Beaufort Gyre. This is not surprising, since the area is permanently ice-covered and floes circulate on closed paths which take 7–10 years for a complete circuit. It was found that multiyear fractions of 50–60% are typical for the gyre region, rising to 80% in the very center. Multiyear fractions of 30–40% are found in the part of the Transpolar Drift Stream fringing the Beaufort Gyre, while in the rest of this current and in peripheral areas of the Arctic the multiyear fraction is 20% or less.

Sea ice retreat   The seasonal cycle described above varies in detail from year to year, and there is evidence from an extension of the record to the present day that a steady decrease of the overall ice extent in the Arctic has been taking place. Analyses

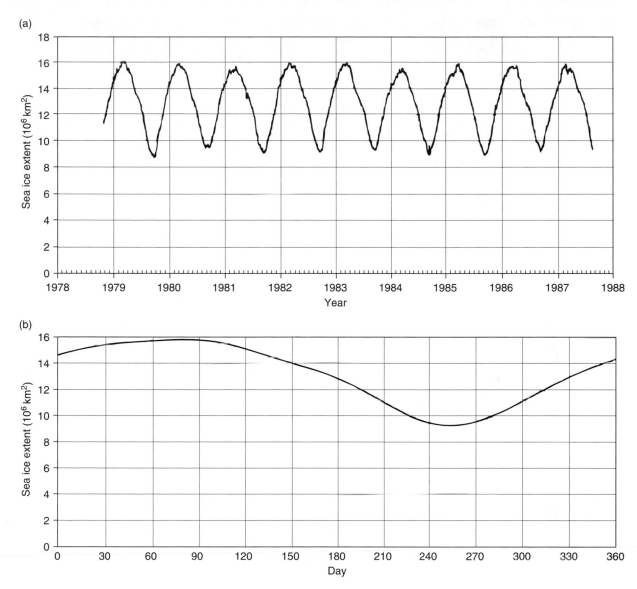

Figure 2 (a) The cycle of Arctic sea ice extent for the 1979–87 period; (b) the averaged seasonal cycle. From Gloersen P, Campbell WJ, Cavalieri DJ, Comiso JC, Parkinson CL, and Zwally HJ (1992) *Arctic and Antarctic Sea Ice, 1978–1987: Satellite Passive-Microwave Observations and Analysis*, Rept. NASA SP-511, 290pp. Washington, DC: National Aeronautics and Space Administration.

where data sets from the SMMR and newer special sensor microwave/imager (SSM/I) sensors have been combined and reconciled show that the sea ice extent in the Arctic has declined at a decadal rate of some 2.8–3% since 1978, with a more rapid recent decline of 4.3% between 1987 and 1994. **Figure 3** shows how an apparently fairly stable annual cycle of large amplitude (**Figure 3(a)**) reveals a distinct downward trend of area (**Figure 3(b)**) when anomalies from interannual monthly means are considered. **Figure 3(c)** shows that the downward trend occurs for every season of the year; the estimated mean annual loss of ice area is $34\,300 \pm 3700\,\mathrm{km}^2$. The rate of decline is greater in

summer (about 7% per decade for September) than in winter (about 2% per decade).

Current analyses suggest that the rate of decline of area is increasing, and the projections of models that have contributed to the *Intergovernmental Panel on Climate Change Fourth Assessment Report* indicate that the Arctic Basin could be essentially ice-free in summer (September) before 2040. This will convert the Arctic into a seasonal sea ice cover, matching the present state of the Antarctic.

The steady hemispheric decline in sea ice extent masks more violent regional changes. In the Bering Sea, there was a sudden downward shift of sea ice area in 1976, indicating a regime shift in the wind

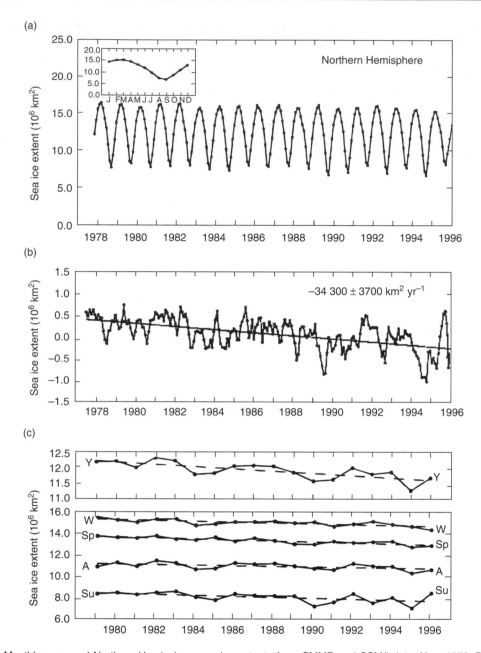

Figure 3 (a) Monthly averaged Northern Hemisphere sea ice extents from SMMR and SSM/I data, Nov. 1978–Dec. 1996. Inset shows average seasonal cycle. (b) Monthly deviations of the extents from the 18-year average, with linear trend shown. (c) Yearly and seasonally averaged ice extents: W = Jan.–Mar., Sp = Apr.–Jun., Su = Jul.–Sep., A = Oct.–Dec. From Parkinson C, Cavalieri D, Gloersen P, Zwally H, and Comiso J (1999) Arctic sea ice extents, areas, and trends, 1978–1996. *Journal of Geophysical Research* 104: 20837–20856.

stress field as the Aleutian Low moved its position. In the Arctic Basin, a passive microwave analysis of the length of the ice-covered season during 1979–86 showed a seesaw effect, with amelioration in the Russian Arctic, Greenland, Barents, and Okhotsk Seas and a worsening in the Labrador Sea, Hudson Bay, and Beaufort Sea. A further analysis extended the coverage to 1978–96 and confirmed these results: the Kara/Barents Sea region had the highest rate of decline in area, of 10.5% per decade, followed by the seas of Okhotsk and Japan and the central Arctic Basin at 9.7% and 8.7%, respectively. Lesser declines were experienced by the Greenland Sea (4.5%), Hudson Bay (1.4%), and the Canadian Arctic Archipelago (0.6%). Increases were registered in the Bering Sea (1%, the starting date being later than the 1976 collapse), Gulf of St. Lawrence (2%), and Baffin Bay/Labrador Sea (3.1%). Taking the more modern data into account it is clear that the seesaw effect discovered over 1979–86 has been largely subsumed into a general retreat.

A particularly important area of sea ice retreat has been the central part of the Greenland Sea gyre, in the vicinity of 75° N 0–5° W. This was normally the site of strong wintertime convection, driven by salt fluxes from local ice growth over the cold Jan Mayen Current, which produces the tongue-like Odden feature. Cold off-ice winds moved newly formed ice eastward within the tongue so that the net salt flux in the western part of the feature was strongly positive, inducing convection. Tracer experiments have shown that deep convection has failed to reach the bottom since about 1971 and in recent years has been greatly reduced in volume and confined to the uppermost 1000 m, while ice production has also been reduced, with no Odden forming at all in 1994, 1995, and 1998–2005. The salt flux produces the causal link between ice retreat and convection shut-off.

The overall hemispheric retreat of Arctic sea ice is clearly related to human-induced global warming. Climatic simulations by GCMs predict that the effects of increased greenhouse gas forcing should be amplified in the polar regions, particularly the Arctic, mainly through the ice-albedo feedback effect. However, an additional cause of some of the recent changes can be identified as a changed pattern of atmospheric circulation in high latitudes. In the North Atlantic sector of the Northern Hemisphere, this can be represented by the North Atlantic Oscillation (NAO) index, the wintertime difference in pressure between Iceland and Portugal, which was low or negative through most of the 1950s to 1970s but which has been rising since the 1980s and which was highly positive throughout the 1990s. A high positive NAO index is associated with an anomalous low-pressure center over Iceland which involves enhanced W and NW winds over the Labrador Sea (cold winds which cause increased cooling, hence increased convection), enhanced E winds over the Greenland Sea in the 72–75° latitude range (causing a reduction in local ice growth in the Odden ice tongue, and a reduced separation between growth and decay regions, hence a reduced rate of convection); enhanced NE winds in the Fram Strait area, causing an increased area flux of ice through the strait (although the ice may have a reduced thickness, so the volume flux is not necessarily increased); and an enhanced wind-driven flow of the North Atlantic Current, allowing more warm water to enter the Atlantic layer of the Arctic Ocean.

Within the Arctic basin, this pattern is incorporated into a large-scale wintertime pattern (with associated index) called the Arctic Oscillation (AO),

which involves a seesaw of sea level pressure (SLP) between the Arctic basin and the surrounding zonal ring. The anomaly appears to extend into the upper atmosphere, and so represents an oscillation in the strength of the whole polar vortex. **Figure 4** shows the results of an analysis of the differences in SLP over the Arctic Ocean between high- and low-NAO years (also corresponding to different phases of the AO index). What had been thought of as the Arctic 'norm', that is, a high over the Beaufort Sea leading to the familiar Beaufort Gyre and Transpolar Drift Stream as the resulting free drift (along the isobars) ice circulation pattern, is actually the result of a low NAO (**Figure 4(b)**). With a high NAO (**Figure 4(a)**) the Beaufort High is suppressed and squeezed toward the Alaskan coast, causing a reduction in the area and strength of the Beaufort Gyre, and a tendency for ice produced on the Siberian shelves to turn east and perform a longer circuit within the basin before emerging from the Fram Strait (this also applies to the trajectory of fresh water from Siberian rivers). The weakening of the Beaufort Gyre may well explain anomalously low summer sea ice extents observed in the Beaufort Sea in 1996–98 (**Figure 5**), since locally melting ice is not replaced by new inputs of ice from the NE. The difference field (**Figure 4(c)**) aptly demonstrates these changes if one considers the differential ice drift vectors as occurring along the isobars shown. **Figures 4(a)** and **4(b)** represent two distinct patterns of Arctic Ocean circulation, which may be called 'cyclonic' and 'anticyclonic', respectively.

Antarctic

The seasonal cycle The sea ice cover in the Antarctic is one of the most climatically important features of the Southern Hemisphere. Its enormous seasonal variation in extent greatly outstrips that of Arctic sea ice, and makes it second only to Northern Hemisphere snow extent as a varying cryospheric feature on our planet's surface. **Figure 6** shows monthly averaged sea ice extent and concentration maps for the Antarctic, derived in the same way as **Figure 1** from SMMR passive microwave data, and covering the same period 1979–87.

With the seasons reversed, the maximum ice extent occurs in August and September. At its maximum (**Figure 6(a)**), the ice cover is circumpolar in extent. Moving clockwise, the ice limit reaches 55° S in the Indian Ocean sector at about 15° E, but lies at c. 60° S around most of the rest of East Antarctica, then slips even further south to 65° S off the Ross Sea. The edge moves slightly north again to 62° S at 150° W, then again shifts southward to 66° S off the Amundsen Sea before moving north again to engulf

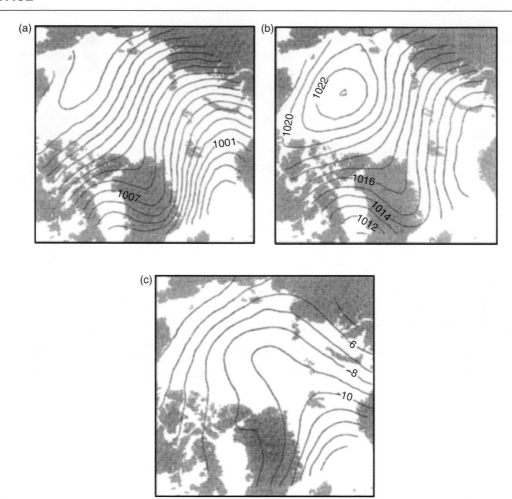

Figure 4 Pressure fields over the Arctic Ocean corresponding to (a) high NAO index, (b) low NAO index, (c) the difference field. From Kwok R and Rothrock D (1999) Variability of Fram Strait ice flux and North Atlantic Oscillation. *Journal of Geophysical Research* 104: 5177–5189.

the South Shetland and South Orkney Islands off the Antarctic Peninsula and complete the circle. The zonal variation in latitude of this winter maximum therefore amounts to some 11°. It has been found that the winter advance of the ice edge follows closely the advance of the 271.2° C isotherm in surface air temperature (freezing point of seawater) and almost coincides with this isotherm at the time of maximum advance. The ice limit is therefore mainly determined thermodynamically, with the gross zonal variations in the winter ice limit matching zonal variations in the freezing isotherm (due to the distribution of continents in the Southern Hemisphere). Smaller-scale variations in the maximum ice limit may be related to deflections in the Antarctic Circumpolar Current as it crosses submarine ridges. We note that within the ice limit the ice concentration is generally less than the almost 100% concentration found in the Arctic Ocean in winter. Even in the areas of greatest concentration, the central Weddell

and Ross seas, it is only in the range 92–96%, while there is a broad marginal ice zone facing the open Southern Ocean over which the concentration steadily diminishes over an outer band of width 200–300 km. This has been found to be a zone over which the advancing winter ice edge is composed of pancake ice, maintained as small cakes by the turbulent effect of the strong wave field.

Ice retreat begins in October and is rapid in November and December. Again the retreat is circumpolar but has interesting regional features. In the sector off Enderby Land at 0–20° E, a large gulf opens up in December to join a coastal region of reduced ice concentration which opens in November. This is a much attenuated version of a winter polynya which was detected in the middle of the pack ice in this sector during 1974–76, but which has only recurred very occasionally as an open water feature since that date, for example, in 1994. It was known as the Weddell Polynya and lay over the Maud Rise, a

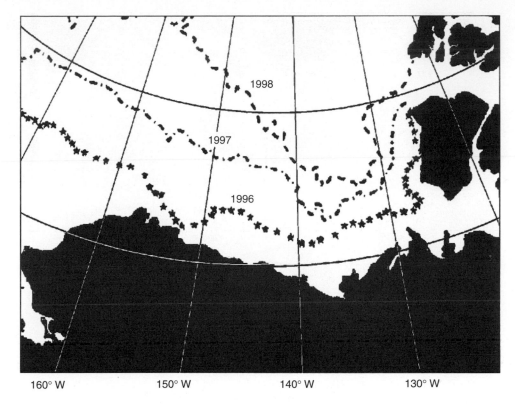

Figure 5 Beaufort Sea ice edge retreat in recent summers.

plateau of reduced water depth. The area was investigated in winter 1986 by the Winter Weddell Sea Project (WWSP) cruise of PFS *Polarstern*, and it was found that the region is already part of the Antarctic Divergence, where upwelling of warmer water can occur, and that additional circulating currents and the doming of isopycnals over the rise could allow enough heat to reach the surface to keep the region ice-free in winter. Since the occurrence is irregular, the region is presumably balanced on the edge of instability. The 1986 winter cover was of high concentration but was very thin. The December distribution also shows an open water region appearing in the Ross Sea, the so-called Ross Sea Polynya, with ice still present to the north. In November and December, a series of small coastal polynyas can be seen to be actively opening along the East Antarctica coast.

By January (**Figure 6(b)**), further retreat has occurred. The Ross Sea is now completely open, East Antarctica has only a narrow fringe of ice around it, and large ice expanses are confined to the eastern Ross Sea, the Amundsen–Bellingshausen Sea sector (60–140° W), and the western half of the Weddell Sea. The month of furthest retreat is February. Ice remains in these three regions, but most of the East Antarctic coastline is almost ice-free as is the tip of the Antarctic Peninsula. This is the season when supply ships can reach Antarctic bases, when tourist ships visit Antarctica, and when most oceanographic research cruises are carried out. It can be seen that the ice concentration in the center of the western Weddell Sea massif is still 92–96%. This is the region which bears the most resemblance to the central Arctic Ocean; it is the only part of the Antarctic to contain significant amounts of multiyear ice, it is very difficult to navigate, and consequently even its bathymetry is not as well known as that of other parts of the Antarctic Ocean.

By March, the very short Antarctic summer is over and ice advance begins. The first advances take place within the Ross and Weddell Seas, then circumpolar advance begins in April. During May and June, the Weddell Sea ice swells out to the northeast, while around the whole of Antarctica the ice edge continues to advance until the August peak.

Figure 7 is the Antarctic equivalent of **Figure 2**. The annual cycle of ice extent can be seen to have a much higher amplitude than in the Arctic, and the year-to-year variability of the peaks and troughs is also somewhat greater. During the 8.8-year record, the February average extent varies from 3.4×10^6 to $4.3 \times 10^6 \, \text{km}^2$, while the September average extent varies from 15.5×10^6 to $19.1 \times 10^6 \, \text{km}^2$, both covering ranges of ± 10–12%. The overall average cycle (**Figure 7(b)**) shows a retreat which is steeper than the advance. The mean minimum extent, at the

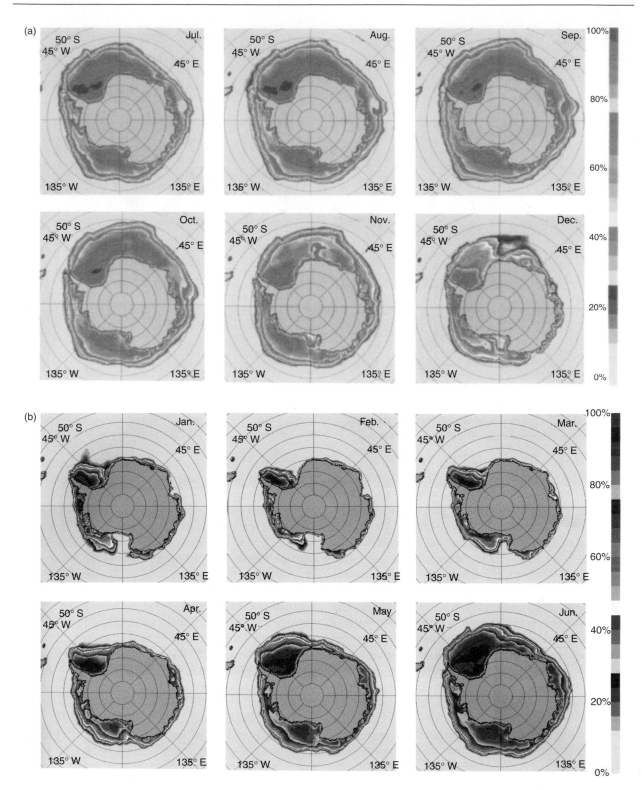

Figure 6 Sea ice extent and concentration maps for the Antarctic for (a) winter, (b) summer months, from SMMR passive microwave data averaged over the period 1979–87. From Gloersen P, Campbell WJ, Cavalieri DJ, Comiso JC, Parkinson CL, and Zwally HJ (1992) *Arctic and Antarctic Sea Ice, 1978–1987: Satellite Passive-Microwave Observations and Analysis*, Rept. NASA SP-511, 290pp. Washington, DC: National Aeronautics and Space Administration.

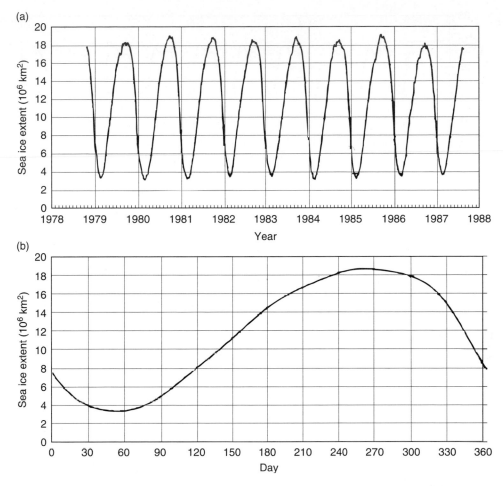

Figure 7 (a) The cycle of Antarctic sea ice extent for the 1979–87 period; (b) the averaged seasonal cycle. From Gloersen P, Campbell WJ, Cavalieri DJ, Comiso JC, Parkinson CL, and Zwally HJ (1992) *Arctic and Antarctic Sea Ice, 1978–1987: Satellite Passive-Microwave Observations and Analysis*, Rept. NASA SP-511, 290pp. Washington, DC: National Aeronautics and Space Administration.

end of February, is $3.6 \times 10^6 \, \text{km}^2$, while the mean maximum extent, in the middle of September, is $18.8 \times 10^6 \, \text{km}^2$. Because of low average ice concentrations within the pack, the corresponding minimum and maximum ice areas are 2.1×10^6 and $15.0 \times 10^6 \, \text{km}^2$.

The winter ice extent in the Antarctic exceeds that of the Arctic winter, while the summer minima are very much lower. This implies that the combined Arctic and Antarctic sea ice extent should be greatest during the Arctic summer. **Figure 8** shows that in fact the peak occurs in October after a plateau during the summer and that the global minimum occurs in late February. The range is *c.* $19-29 \times 10^6 \, \text{km}^2$, with a high interannual variability for both maxima and minima.

Interannual variability Observational data on Antarctic sea ice extent show no significant trend, even relative to the first systematic ice edge maps made in the 1930s. A 1997 study based on whaling records, which suggested that a sudden retreat occurred in the 1960s, has not been confirmed. Passive microwave data for the 1988–94 period show no evidence of an overall trend in extent, but some evidence suggesting that anomaly patterns propagate eastward, offering support for the idea of an Antarctic circumpolar wave in surface pressure, wind, temperature, and sea ice extent. During 7 years of the study, ice seasons shortened in the E Ross Sea, Amundsen Sea, W Weddell Sea, offshore eastern Weddell Sea, and East Antarctica between 40° and 80° E. Ice seasons lengthened in the W Ross Sea, Bellingshausen Sea, central Weddell Sea, and the region 80–135° E. Earlier evidence of a major ice retreat in the Bellingshausen Sea in the summer of 1988–91 was shown to be a short-lived phenomenon. A longer-term statistical analysis of passive microwave data from 1978 onward gave a small and not statistically significant upward trend in overall Antarctic ice extent, of some 1.3% per

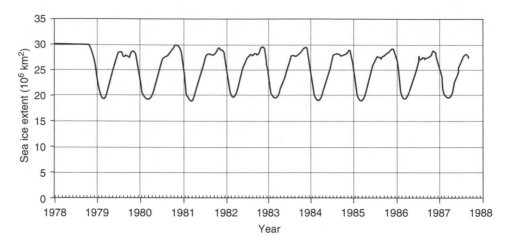

Figure 8   Combined global cycle of sea ice extent, 1979–87. From Gloersen P, Campbell WJ, Cavalieri DJ, Comiso JC, Parkinson CL, and Zwally HJ (1992) *Arctic and Antarctic Sea Ice, 1978–1987: Satellite Passive-Microwave Observations and Analysis*, Rept. NASA SP-511, 290pp. Washington, DC: National Aeronautics and Space Administration.

decade. The conclusion is that there is currently no strong trend in ice extent, in direct contrast to the Arctic.

Sea Ice Thickness

Arctic

With satellite-based altimetry techniques only recently introduced and still not fully validated, our knowledge of the regional and temporal variability of ice thickness in the Arctic comes mainly from upward sonar profiling by submarines. Therefore our level of knowledge depends on whether submarines have been able to operate in the area concerned. Until now, data have been obtained mainly from British submarines operating in the Greenland Sea and Eurasian Basin since 1971, and from US submarines operating in the Canada and Eurasian Basins since 1958.

Results from marginal seas show that the ice in Baffin Bay is largely thin first-year ice with a modal thickness of 0.5–1.5 m. In the southern Greenland Sea, too, the ice, although composed largely of partly melted multiyear ice, also has a modal thickness of about 1 m, with the decline in mean thickness from Fram Strait giving a measure of melt rate and thus freshwater input to the Greenland Sea at different latitudes. Over the Arctic Basin itself, there is a gradation in mean ice thickness from the Soviet Arctic, across the Pole and toward the coasts of north Greenland and the Canadian Arctic Archipelago, where the highest mean thicknesses of *c.* 7–8 m are observed. These overall variations are in accord with the predictions of numerical models which take account of ice dynamics and deformation as well as ice thermodynamics. The overall basin mean is *c.* 5 m in winter and 4 m in summer.

In order to assess whether significant changes are occurring in a region of the Arctic it is necessary to obtain area-averaged observations of mean ice thickness over the same region using the same equipment at different seasons or in different years. Ideally the region should be as large as possible, to allow us to assess whether changes are basin-wide or simply regional. Also the measurements should be repeated annually in order to distinguish between a fluctuation and a trend. Because of the unsystematic nature of Arctic submarine deployments this goal has not yet been achieved, but a number of comparisons have been carried out which strongly suggest that a significant thinning has been occurring. Some of these have been made possible by very large new data sets which have been obtained from the US SCICEX (Scientific Ice Expeditions) civilian submarine program during 1993–99.

SCICEX data obtained in September–October of 1993, 1996, and 1997 have been compared with data obtained during six summer cruises during the period 1958–76. Twenty-nine crossing places were identified, where a submarine track from the recent period crossed one from the early period, and the corresponding tracks (of average length 160 km) were compared in thickness. In each case, the mean thicknesses obtained were adjusted to a standard date of 15 September using an ice–ocean model to account for seasonal variability. The 29 matched data sets were divided into six geographical regions (**Figure 9**). The decline in mean ice draft was significant for every region and increased across the Arctic from the Canada Basin toward Europe – it was 0.9 m in the Chukchi Cap and Beaufort Sea, 1.3 m in the Canada Basin, 1.4 m near the North Pole, 1.7 m in the Nansen Basin, and 1.8 m in the

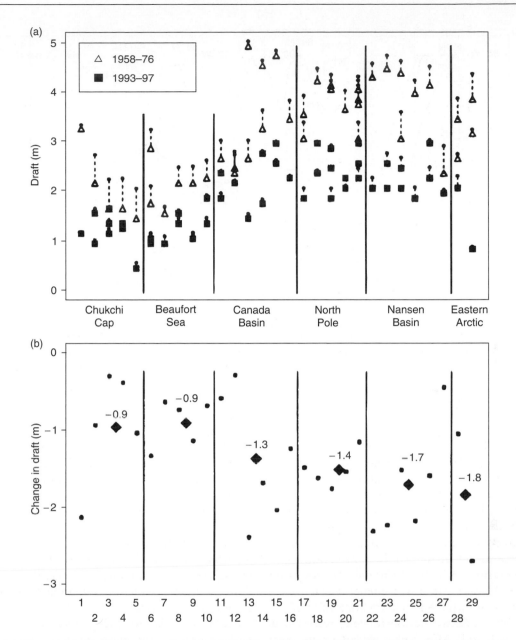

Figure 9 (a) Mean ice drafts at crossings of early cruises with cruises in the 1990s. Early data (1958–76) are shown by open triangles and those from the 1990s by solid squares, both seasonally adjusted to 15 September. The small dots show the original data before the seasonal adjustment. The crossings are grouped into six regions separated by the solid lines and named appropriately. (b) Changes in mean ice draft at cruise crossings (dots) from the early data to the 1990s. The change in the mean ice draft for all crossings in each region is shown by a large diamond. From Rothrock DA, Yu Y, and Maykut GA (1999) Thinning of the Arctic sea-ice cover. *Geophysical Research Letters* 26(23): 3469–3472.

Eastern Arctic. Overall, the mean change in draft was from 3.1 m in the early period to 1.8 m in the recent period, a decline of 42%.

The authors of the study commented that the decline in mean draft could arise thermodynamically from any of the following flux increases:

1. a $4 \, W \, m^{-2}$ increase in ocean heat flux;
2. a $13 \, W \, m^{-2}$ increase in poleward atmospheric heat transport; or

3. a $23 \, W \, m^{-2}$ increase in downwelling short-wave radiation during summer.

Clearly a change in ice dynamics can also produce a change in mean ice draft, although it is not known what change in wind forcing would be needed to account for the magnitude and distribution of the observed draft decrease.

This is the most extensive comparison so far, but it should be noted that all data sets involved are from

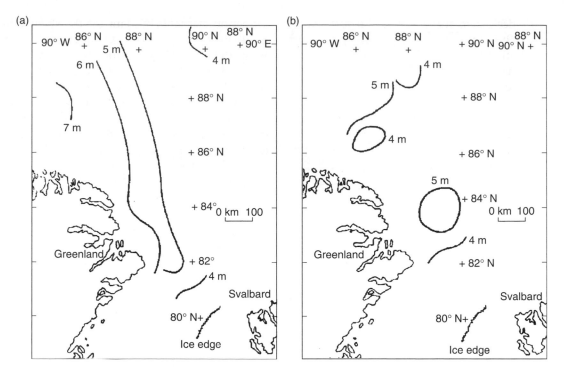

Figure 10 Contour maps of mean ice drafts from Eurasian basin measured from British submarines, October 1976 and May 1987. From Wadhams P (1990) Evidence for thinnig of the Arctic ice cover north of Greenland. *Nature* 345: 795–797.

summer, mostly late summer, so that the reported decline refers to only one season of the year, and that most track comparisons occur over the North Pole region and Canada Basin, with few in the Eurasian Arctic and none south of 84° in the Eurasian Basin.

Complementary to this study are comparisons from the Eurasian Basin and Greenland Sea made using data from British submarine cruises. One comparison (**Figure 10**) involved data sets from a triangular region extending from Fram Strait and the north of Greenland to the North Pole, recorded in October 1976 and May 1987. Mean drafts were computed over 50-km sections, and each value was positioned at the centroid of the section concerned; the results were contoured to give the maps shown in **Figure** 10. There was a decrease of 15% in mean draft averaged over the whole area (300 000 km²), from 5.34 m in 1976 to 4.55 m in 1987. Profiles along individual matching track lines showed that the decrease was concentrated in the region south of 88° N and between 30° and 50° W. By comparison of the entire shape of the probability density functions of ice draft, the conclusion was that the main contribution to the loss of volume was the replacement of multiyear and ridged ice by young and first-year ice. For instance, taking ice of 2–5-m thickness as an indicator of undeformed multiyear ice fraction, this declined from 47.6% in 1976 to 39.1% in 1987, a

relative decline of 18%. This is in agreement with a recent finding that multiyear ice fraction in the Arctic (estimated from passive-microwave data) suffered a 14% decrease during the period 1978–98.

The British study did not correct for seasonal variability between the 1976 measurements, made in October, and those of 1987, made in April–May. If this is done using a model, the decrease in mean ice draft (standardized to 15 September) becomes much greater at 42%, since April–May is the time of greatest ice thickness. This is in excellent agreement with results for the entire overall US data set, yet occurred within a period of only 11 years. This indicates either that thinning occurs faster in the Eurasian Basin than elsewhere in the Arctic or that it is invalid to compare data sets from different times of year simply by standardizing to 'summer' through use of a model.

The latter problem is largely overcome in an analysis of a more recent British data set, obtained in September 1996 by HMS *Trafalgar*. These data can be compared directly with results from October 1976. The two submarines followed similar courses between 81° and 90° N on about the 0° meridian, and it was found that about 2100 km of track from each submarine, when divided into 100-km sections, was close enough in correspondence to the other so as to enable them to be counted as 'crossing tracks'.

The overall decline in mean ice thickness between 1976 and 1996 was 43%, in remarkably close agreement with US results. The mean drafts in 1° bins of latitude are shown in **Table 1**.

It can be seen that there was a significant decrease of mean draft at every latitude, but that the decline is largest just north of Fram Strait and near the Pole itself. A characteristic of the ice cover observed from below was the large amount of completely open water present at all latitudes. A seasonality

correction to the 1976 data for the slight difference in mean draft between October and September brings the ratio to 59.0% for September, a decline of 41%. Thus the British and the US data are in remarkably good agreement in describing a very significant thickness decrease in Arctic Ocean sea ice.

A cautionary note must be sounded in that these significant thickness decreases from spatially averaged data conceal large random variabilities at given locations. Time series of ice draft at fixed locations have been obtained from moored upward sonar systems, of which the most comprehensive set spans Fram Strait. An analysis of data from 1991–98 showed that interseasonal and interannual variability in thickness far exceed any trend, although of course the length of the data set is only 7 years.

A possible direct cause of the observed thinning is the recent discovery that the Atlantic sublayer in the Arctic Ocean, which lies beneath the polar surface water and which derives from the North Atlantic Current, has warmed substantially (by 1–2°C at 200 m depth) and increased its range of influence relative to water of Pacific origin. The front separating the two water types has now shifted from the Lomonosov to the Alpha-Mendeleyev Ridge. This warmer and shallower sublayer should increase the ocean heat flux into the bottom of the ice. This is

Table 1 Mean drafts in 1° bins of latitude in 1976 and 1996

Latitude range	Mean draft (m)		1996 as % of 1976
	1996	1976	
81–82	1.57	5.84	26.9
82–83	2.15	5.87	36.6
83–84	2.88	4.90	58.7
84–85	3.09	4.64	66.6
85–86	3.54	4.57	77.4
86–87	3.64	4.64	78.5
87–88	2.36	4.60	51.2
88–89	3.24	4.41	73.4
89–90	2.19	3.94	55.5
Overall	2.74	4.82	56.8

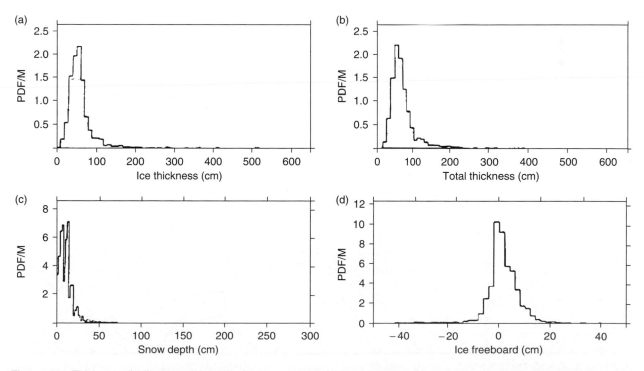

Figure 11 Thickness distributions of Antarctic first-year sea ice (a), ice plus snow (b), and snow alone (c). The distribution of ice freeboards is shown in (d); a negative value permits water infiltration. From Wadhams P, Lange MA, and Ackley SF (1987) The ice thickness distribution across the Atlantic sector of the Antarctic Ocean in midwinter. *Journal of Geophysical Research* 92: 14535–14552.

enhanced by the fact that the structure of the polar surface layer has itself changed. In the Eurasian basin, there was formerly a cold halocline layer in the 100–200-m-depth range, where temperature stayed cold with increasing depth despite salinity rising. Its existence was due to riverine input from Siberia, which has recently diverted eastward due to a changed atmospheric circulation, causing a retreat of the cold halocline and possibly associated with the recently observed summertime retreat of sea ice in the Beaufort Sea sector (**Figure 5**).

Antarctic

In the Antarctic our knowledge of ice thickness is much less extensive than in the Arctic, since systematic data have been obtained mainly by repetitive drilling except for the use of moored upward sonar at certain sites in the Weddell Sea. The winter pack ice in the Antarctic is of global importance because of its vast extent and large seasonal cycle, so the determination of winter ice thickness remains a high research priority.

In 1986, the first deep penetration into the circumpolar Antarctic pack during early winter, the time of ice edge advance, was accomplished by the WWSP cruise of PFS *Polarstern*, during which systematic ice thickness measurements (at 1-m intervals along lines of about 100 holes) were made throughout the eastern part of the Weddell–Enderby Basin, from the ice edge to the coast, covering Maud Rise and representing a typical cross section of the first-year circumpolar Antarctic pack during the season of advance. After a spring cruise in 1988, a second winter cruise was carried out in 1989: the Winter Weddell Gyre Study (WWGS) involved a crossing of the Weddell Sea from the tip of the Antarctic Peninsula to Kap Norvegia in the east during September–October, and thus allowed the multiyear ice regime of the western Weddell Sea to be studied in midwinter.

In the advancing Antarctic pack, composed of first-year consolidated pancake ice, the ice thickness distribution was as shown in **Figure 11**. Note the peak at the very low value of 50–60 cm, with a peak in snow cover thickness at 14–16 cm. The snow cover was sufficient to push the ice surface below

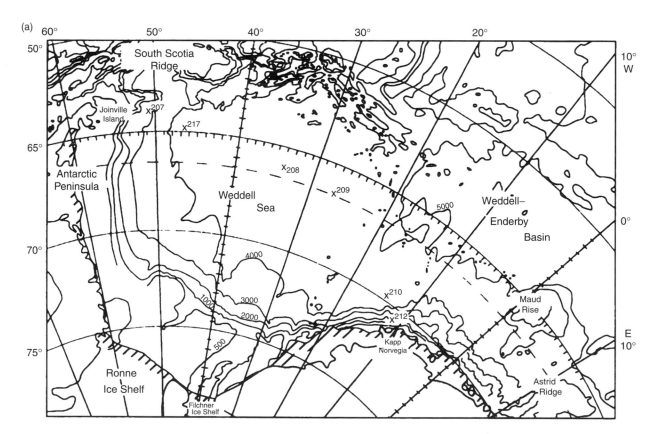

Figure 12 (a) Locations of six moored upward-looking echo sounders in the Weddell Sea. (b) Mean ice drafts (December 1990–92) from these sounders. From Strass VH and Fahrbach E (1998) Temporal and regional variation of sea ice draft and coverage in the Weddell Sea obtained from upward looking sonars. In: Jeffries MO (ed.) *Antarctic Sea Ice: Physical Processes, Interactions and Variability*, Antarctic Research Series, vol. 74, pp. 123–140. Washington, DC: American Geophysical Union.

(b)

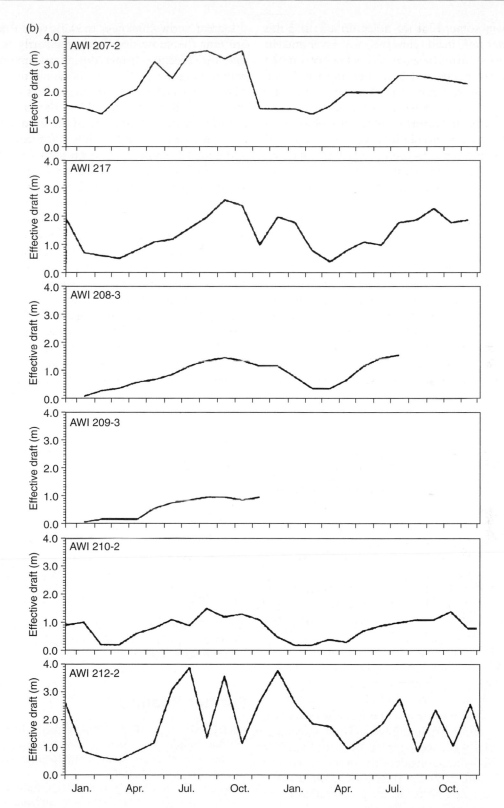

Figure 12 Continued

water level in some 17% of holes drilled, and this leads to water infiltration into the snow layer and the formation of a new type of ice, snow-ice, at the boundary between ice and overlying snow. It is reasonable to suppose that the pancake ice-forming mechanism is typical of the entire circumpolar advancing ice edge in winter (neglecting embayments such as the Ross and Weddell Seas).

Multiyear ice was measured in 1989 in the western Weddell Sea; the Weddell Gyre carries ice from the eastern Weddell Sea deep into high southern latitudes in the southern Weddell Sea off the Filchner–Ronne Ice Shelf, and then northward up the eastern side of the peninsula to be encountered in the NW Weddell Sea. This journey takes about 18 months, and so permits much of the ice to mature into multiyear (strictly, second-year) ice. Multiyear ice can be identified by its structure in cores, and by the very thick snow cover which it acquires, which is almost always sufficient to depress the ice surface below the waterline. Ice drilling showed that the mean thickness of undeformed multiyear ice (1.17 m) was about double that of first-year ice (0.60 m). The presence of ridging roughly doubles the mean draft of the 100-m floe sections in which it occurs (0.60–1.03 m in first-year; 1.17–2.51 m in multiyear). In addition, snow is very much deeper on multiyear ice (0.63–0.79 m) than on first-year (0.16–0.23 m).

Drilling from ships has the three advantages that data can be obtained from many locations, that first-year ice and multiyear ice can be clearly discriminated, and that the detailed structure of ridges can be measured. In other respects, however, moored upward-looking echo sounders (ULESs) give far more information. Data were collected from six such ULES systems moored across the Weddell Sea (**Figure 12(a)**) during a 2-year period from December 1990 to December 1992. The results for mean drafts (**Figure 12(b)**) are in good agreement with drilling data for winter, but also reveal the annual cycle ('effective draft' in this figure is true mean draft, that is, including the open water component). It can be seen that the westernmost ULES, in the Weddell Sea outflow (207), has a cycle ranging from just over 1 m in summer to about 3 m in winter, in good agreement with ridged multiyear ice sampled from drilling. The thickness diminishes considerably over the central Weddell Sea (208, 209) but then rises again near the Enderby Land coast (212) to a very variable mean value. This last ULES, very close to the coast, is in a shear zone where much ridging can occur as well as deformation around grounded icebergs.

In summary, Antarctic sea ice of a given age is much thinner on average than Arctic sea ice, but the overlying snow cover can be thicker. The reasons for the great snow thickness in multiyear Antarctic ice are that the snow does not necessarily melt during the first summer, while during its second year it enters the inner part of the Weddell Sea where precipitation is greater. In the central Arctic, snow depth may reach 40 cm by the end of the first winter, but the snow melts in summer so that snow thickness on multiyear ice is a function only of time of year. In the Antarctic the snow thickness is sufficient to push the ice–snow interface below sea level in 17% of cases sampled for first-year ice and up to 53% for second- and multiyear ice. It has been estimated that the resulting snow-ice (or so-called 'meteoric ice') makes up 16% of the ice mass in the Weddell Sea. Finally, in the Antarctic, much of the ice has a fine-grained structure of randomly oriented crystals, formed from the freezing of a frazil ice suspension to form pancakes, then the freezing together of pancakes to form consolidated pancake ice, the typical first-year ice type forming in the advancing winter ice edge region. In the Arctic, most ice has formed by congelation growth and so shows a crystal fabric of columnar-grained ice with horizontal c-axes. A mechanical difference is that in the Antarctic most ridges appear to be formed by buckling and crushing of the material of the floes themselves, and so are composed of a small number of fairly thick blocks extending to modest depths – typically 6 m or less. In the Arctic, ridges tend to be formed by the crushing of thin ice in refrozen leads between floes, and so are composed of a large mass of small blocks, extending to greater depths – typically 10–20 m, with significant numbers extending to 30 m or more and even to 40–50 m in extreme cases.

See also

Arctic Ocean Circulation. Coupled Sea Ice-Ocean Models. Ice–Ocean Interaction. Sea Ice Dynamics. Sea Ice: Overview.

Further Reading

Ackley SF and Weeks WF (eds.) (1990) *CRREL Monograph 90-1: Sea Ice Properties and Processes.* Hanover, NH: US Army Cold Regions Research and Engineering Laboratory.

Gloersen P, Campbell WJ, Cavalieri DJ, Comiso JC, Parkinson CL, and Zwally HJ (1992) *Arctic and Antarctic Sea Ice, 1978–1987: Satellite Passive-Microwave Observations and Analysis*, Rept. NASA SP-511, 290pp, Rept. NASA SP-511, 290pp. Washington, DC: National Aeronautics and Space Administration.

Jeffries MO (ed.) (1998) *Antarctic Research Series 74: Antarctic Sea Ice: Physical Processes, Interactions and*

Variability. Washington, DC: American Geophysical Union.

Kwok R and Rothrock D (1999) Variability of Fram Strait ice flux and North Atlantic Oscillation. *Journal of Geophysical Research* 104: 5177–5189.

Leppäranta M (ed.) (1998) *Physics of Ice-Covered Seas*, 2 vols. Helsinki: University of Helsinki Press.

Leppäranta M (2005) *The Drift of Sea Ice*. New York: Springer.

Parkinson C, Cavalieri D, Gloersen P, Zwally H, and Comiso J (1999) Arctic sea ice extents, areas, and trends, 1978–1996. *Journal of Geophysical Research* 104: 20837–20856.

Richter-Menge J, Overland J, Proshutinsky A, *et al.* (2006) *State of the Arctic Report. NOAA OAR Special Report*, 36pp. Seattle, WA: NOAA/OAR/PMEL.

Rothrock DA, Yu Y, and Maykut GA (1999) Thinning of the Arctic sea-ice cover. *Geophysical Research Letters* 26(23): 3469–3472.

Strass VH and Fahrbach E (1998) Temporal and regional variation of sea ice draft and coverage in the Weddell Sea obtained from upward looking sonars. In: Jeffries MO (ed.) *Antarctic Sea Ice: Physical Processes, Interactions and Variability*, Antarctic Research Series, vol. 74, pp. 123–140. Washington, DC: American Geophysical Union.

Wadhams P (1990) Evidence for thinning of the Arctic ice cover north of Greenland. *Nature* 345: 795–797.

Wadhams P (2000) *Ice in the Ocean*, 335pp. London: Gordon and Breach.

Wadhams P, Dowdeswell JA, and Schofield AN (eds.) (1996) *The Arctic and Environmental Change*, 193pp. London: Gordon and Breach.

Wadhams P, Gascard J-C, and Miller L (eds.) (1999) *Special Issue: The European Subpolar Ocean Programme: ESOP. Deep-Sea Research II* 46(6–7): 1011–1530.

Wadhams P, Lange MA, and Ackley SF (1987) The ice thickness distribution across the Atlantic sector of the Antarctic Ocean in midwinter. *Journal of Geophysical Research* 92: 14535–14552.

Wheeler PA (ed.) (1997) *Special Issue: 1994 Arctic Ocean Section. Deep-Sea Research II* 44(8).

Zwally HJ, Comiso JC, Parkinson CL, Campbell WJ, Carsey FD, and Gloersen P (1983) *Antarctic Sea Ice 1973–1976: Satellite Passive Microwave Observations*, Rept. SP-459. Washington, DC: NASA.

SUB-SEA PERMAFROST

T. E. Osterkamp, University of Alaska,
Alaska, AK, USA

Introduction

Sub-sea permafrost, alternatively known as sub-marine permafrost and offshore permafrost, is defined as permafrost occurring beneath the seabed. It exists in continental shelves in the polar regions (**Figure 1**). When sea levels are low, permafrost aggrades in the exposed shelves under cold subaerial conditions. When sea levels are high, permafrost degrades in the submerged shelves under relatively warm and salty boundary conditions. Sub-sea permafrost differs from other permafrost in that it is relic, warm, and generally degrading. Methods used to investigate it include probing, drilling, sampling, drill hole log analyses, temperature and salt measurements, geological and geophysical methods (primarily seismic and electrical), and geological and geophysical models. Field studies are conducted from boats or, when the ocean surface is frozen, from the ice cover. The focus of this article is to review our understanding of sub-sea permafrost, of processes ocurring within it, and of its occurrence, distribution, and characteristics.

Sub-sea permafrost derives its economic importance from current interests in the development of offshore petroleum and other natural resources in the continental shelves of polar regions. The presence and characteristics of sub-sea permafrost must be considered in the design, construction, and operation of coastal facilities, structures founded on the seabed, artificial islands, sub-sea pipelines, and wells drilled for exploration and production.

Scientific problems related to sub-sea permafrost include the need to understand the factors that control its occurrence and distribution, properties of warm permafrost containing salt, and movement of heat and salt in degrading permafrost. Gas hydrates that can occur within and under the permafrost are a potential abundant source of energy. As the sub-sea permafrost warms and thaws, the hydrates destabilize, producing gases that may be a significant source of global carbon.

Nomenclature

'Permafrost' is ground that remains below 0°C for at least two years. It may or may not contain ice. 'Ice-bearing' describes permafrost or seasonally frozen soil that contains ice. 'Ice-bonded' describes ice-bearing material in which the soil particles are mechanically cemented by ice. Ice-bearing and ice-bonded material may contain unfrozen pore fluid in addition to the ice. 'Frozen' implies ice-bearing or ice-bonded or both, and 'thawed' implies non-ice-bearing. The 'active layer' is the surface layer of sediments subject to annual freezing and thawing in areas underlain by permafrost. Where seabed temperatures are negative, a thawed layer ('talik') exists near the seabed. This talik is permafrost but does not contain ice because soil particle effects, pressure, and the presence of salts in the pore fluid can depress the freezing point 2°C or more. The boundary between a thawed region and ice-bearing permafrost is a phase

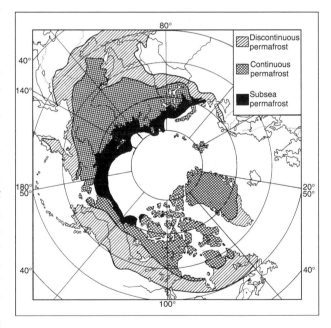

Figure 1 Map showing the approximate distribution of sub-sea permafrost in the continental shelves of the Arctic Ocean. The scarcity of direct data (probing, drilling, sampling, temperature measurements) makes the map highly speculative, with most of the distribution inferred from indirect measurements, primarily water temperature, salinity, and depth (100 m depth contour). Sub-sea permafrost also exists near the eroding coasts of arctic islands, mainlands, and where seabed temperatures remain negative. (Adapted from Pewe TL (1983). *Arctic and Alpine Research* 15(2):145–156 with the permission of the Regents of the University of Coloroado.)

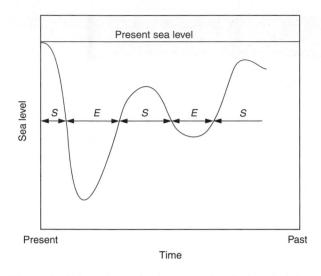

Figure 2 Schematic sea level curve during the last glaciation. The history of emergence (E) and submergence (S) can be combined with paleotemperature data on sub-aerial and sub-sea conditions to construct an approximate thermal boundary condition for sub-sea permafrost at any water depth.

boundary. 'Ice-rich' permafrost contains ice in excess of the soil pore spaces and is subject to settling on thawing.

Formation and Thawing

Repeated glaciations over the last million years or so have caused sea level changes of 100 m or more (**Figure 2**). When sea levels were low, the shallow continental shelves in polar regions that were not covered by ice sheets were exposed to low mean annual air temperatures (typically -10 to $-25°C$). Permafrost aggraded in these shelves from the exposed ground surface downwards. A simple conduction model yields the approximate depth (X) to the bottom of ice-bonded permafrost at time t, (eqn [1]).

$$X(t) = \sqrt{\frac{2K(T_e - T_g)t}{h}} \qquad [1]$$

K is the thermal conductivity of the ice-bonded permafrost, T_e is the phase boundary temperature at the bottom of the ice-bonded permafrost, T_g is the long-term mean ground surface temperature during emergence, and h is the volumetric latent heat of the sediments, which depends on the ice content. In eqn [1], K, h, and T_e depend on sediment properties. A rough estimate of T_g can be obtained from information on paleoclimate and an approximate value for t can be obtained from the sea level history (**Figure 2**). Eqn [1] overestimates X because it neglects geothermal heat flow except when a layer of ice-bearing permafrost from the previous transgression remains at depth.

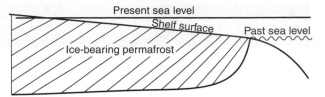

Figure 3 A schematic illustration of ice-bearing sub-sea permafrost in a continental shelf near the time of minimum sea level. Typical thicknesses at the position of the present shoreline would have been about 400–1000 m with shelf widths that are now typically 100–600 km.

Timescales for permafrost growth are such that hundreds of meters of permafrost could have aggraded in the shelves while they were emergent (**Figure 3**).

Cold onshore permafrost, upon submergence during a transgression, absorbs heat from the seabed above and from the geothermal heat flux rising from below. It gradually warms (**Figures 4 and 5**), becoming nearly isothermal over timescales up to a few millennia (**Figure 4**, time t_3). Substantially longer times are required when unfrozen pore fluids are present in equilibrium with ice because some ice must thaw throughout the permafrost thickness for it to warm.

A thawed layer develops below the seabed and thawing can proceed from the seabed downward, even in the presence of negative mean seabed temperatures, by the influx of salt and heat associated with the new boundary conditions. Ignoring seabed erosion and sedimentation processes, the thawing rate at the top of ice-bonded permafrost during submergence is given by eqn [2].

$$\dot{X}_{top} = \frac{J_t}{h} - \frac{J_f}{h} \qquad [2]$$

J_t is the heat flux into the phase boundary from above and J_t is the heat flux from the phase boundary into the ice-bonded permafrost below. J_t depends on the difference between the long-term mean temperature at the seabed, T_s, and phase boundary temperature, T_p, at the top of the ice-bonded permafrost. For $J_t = J_t$, the phase boundary is stable. For $J_t < J_t$, refreezing of the thawed layer can occur from the phase boundary upward. For thawing to occur, T_s must be sufficiently warmer than T_p to make $J_t > J_t$. T_s is determined by oceanographic conditions (currents, ice cover, water salinity, bathymetry, and presence of nearby rivers). T_p is determined by hydrostatic pressure, soil particle effects, and salt concentration at the phase boundary (the combined effect of *in situ* pore fluid salinity, salt transport from the seabed through the thawed layer, and changes in concentration as a result of freezing or thawing).

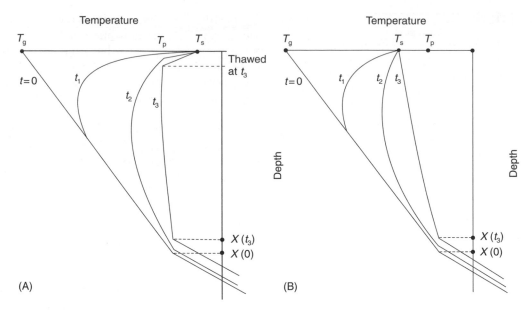

Figure 4 Schematic sub-sea permafrost temperature profiles showing the thermal evolution at successive times (t_1, t_2, t_3) after submergence when thawing occurs at the seabed (A) and when it does not (B). T_g and T_s are the long-term mean surface temperatures of the ground during emergence and of the seabed after submergence. T_p is the phase boundary temperature at the top of the ice-bonded permafrost. $X(0)$ and $X(t_3)$ are the depths to the bottom of ice-bounded permafrost at times $t = 0$ and t_3. (Adapted with permission from Lachenbruch AH and Marshall BV (1977) Open File Report 77-395. US Geological Survey, Menlo Park, CA.)

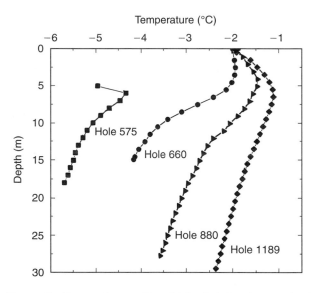

Figure 5 Temperature profiles obtained during the month of May in sub-sea permafrost near Barrow, Alaska, showing the thermal evolution with distance (equivalently time) offshore. Hole designation is the distance (m) offshore and the shoreline erosion rate is about 2.4 m y^{-1}. Sea ice freezes to the seabed within 600 m of shore. (Adapted from Osterkamp TE and Harrison WD (1985) Report UAGR-301. Fairbanks, AK: Geophysical Institute, University of Alaska.)

Nearshore at Prudhoe Bay, $\dot{X}_{\text{top}}$ varies typically from centimeters to tens of centimeters per year while farther offshore it appears to be on the order of millimeters per year. The thickness of the thawed layer at the seabed is typically 10 m to 100 m, although values of less than a meter have been observed. At some sites, the thawed layer is thicker in shallow water and thinner in deeper water.

Sub-sea permafrost also thaws from the bottom by geothermal heat flow once the thermal disturbance of the transgression penetrates there. The approximate thawing rate at the bottom of the ice-bonded permafrost is given by eqn [3].

$$\dot{X}_{\text{bot}} \simeq \frac{J_g}{h} - \frac{J'_f}{h} \qquad [3]$$

J_g is the geothermal heat flow entering the phase boundary from below and J'_f is the heat flow from the phase boundary into the ice-bonded permafrost above. J'_f becomes small within a few millennia except when the permafrost contains unfrozen pore fluids. $\dot{X}_{\text{bot}}$ is typically on the order of centimeters per year. Timescales for thawing at the permafrost table and base are such that several tens of thousands of years may be required to completely thaw a few hundred meters of sub-sea permafrost.

Modeling results and field data indicate that impermeable sediments near the seabed, low T_s, high ice contents, and low J_g favor the survival of ice-bearing sub-sea permafrost during a transgression. Where conditions are favorable, substantial thicknesses of ice-bearing sub-sea permafrost may have survived previous transgressions.

Characteristics

The chemical composition of sediment pore fluids is similar to that of sea water, although there are detectable differences. Salt concentration profiles in thawed coarse-grained sediments at Prudhoe Bay (**Figure 6**) appear to be controlled by processes occurring during the initial phases of submergence. There is evidence for highly saline layers within ice-bonded permafrost near the base of gravels overlying a fine-grained sequence both onshore and offshore. In the Mackenzie Delta region, fluvial sand units deposited during regressions have low salt concentrations (**Figure 7**) except when thawed or when lying under saline sub-sea mud. Fine-grained mud sequences from transgressions have higher salt concentrations. Salts increase the amount of unfrozen pore fluids and decrease the phase equilibrium temperature, ice content, and ice bonding. Thus, the sediment layering observed in the Mackenzie Delta region can lead to unbonded material (clay) between layers of bonded material (fluvial sand).

Thawed sub-sea permafrost is often separated from ice-bonded permafrost by a transition layer of ice-bearing permafrost. The thickness of the ice-bearing layer can be small, leading to a relatively sharp (centimeters scale) phase boundary, or large, leading to a diffuse boundary (meters scale). In general, it appears that coarse-grained soils and low salinities produce a sharper phase boundary and fine-grained soils and higher salinities produce a more diffuse phase boundary.

Sub-sea permafrost consists of a mixture of sediments, ice, and unfrozen pore fluids. Its physical and mechanical properties are determined by the individual properties and relative proportions. Since ice and unfrozen pore fluid are strongly temperature dependent, so also are most of the physical and mechanical properties.

Ice-rich sub-sea permafrost has been found in the Alaskan and Canadian portions of the Beaufort Sea and in the Russian shelf. Thawing of this permafrost can result in differential settlement of the seafloor that poses serious problems for development.

Processes

Submergence

Onshore permafrost becomes sub-sea permafrost upon submergence, and details of this process play a major role in determining its future evolution. The rate at which the sea transgresses over land is determined by rising sea levels, shelf topography, tectonic setting, and the processes of shoreline erosion, thaw settlement, thaw strain of the permafrost, seabed erosion, and sedimentation. Sea levels on the polar continental shelves have increased more than 100 m in the last 20 000 years or so. With shelf widths of 100–600 km, the average shoreline retreat rates would have been about 5 to 30 m y^{-1}, although maximum rates could have been much larger. These average rates are comparable to areas with very rapid shoreline retreat rates observed today on the Siberian and North American shelves. Typical values are 1–6 m y^{-1}.

It is convenient to think of the transition from sub-aerial to sub-sea conditions as occurring in five regions (**Figure 8**) with each region representing different thermal and chemical surface boundary conditions. These boundary conditions are successively applied to the underlying sub-sea permafrost during a transgression or regression. Region 1 is the onshore permafrost that forms the initial condition for sub-sea permafrost. Permafrost surface temperatures range down to about −15°C under current sub-aerial conditions and may have been 8–10°C colder during glacial times. Ground water is generally fresh, although salty lithological units may exist within the permafrost as noted above.

Region 2 is the beach, where waves, high tides, and resulting vertical and lateral infiltration of sea water produce significant salt concentrations in the active layer and near-surface permafrost. The active layer and temperature regime on the beach differ from those on land. Coastal banks and bluffs are a trap for wind-blown snow that often accumulates in insulating drifts over the beach and adjacent ice cover.

Region 3 is the area where ice freezes to the seabed seasonally, generally where the water depth is less than about 1.5–2 m. This setting creates unique thermal boundary conditions because, when the ice freezes into the seabed, the seabed becomes conductively coupled to the atmosphere and thus very cold. During summer, the seabed is covered with shallow, relatively warm sea water. Salt concentrations at the seabed are high during winter because of salt rejection from the growing sea ice and restricted circulation under the ice, which eventually freezes into the seabed. These conditions create highly saline brines that infiltrate the sediments at the seabed.

Region 4 includes the areas where restricted under-ice circulation causes higher-than-normal sea water salinities and lower temperatures over the sediments. The ice does not freeze to the seabed or only freezes to it sporadically. The existence of this region depends on the ice thickness, on water depth, and on flushing processes under the ice. Strong currents or steep bottom slopes may reduce its extent or eliminate it.

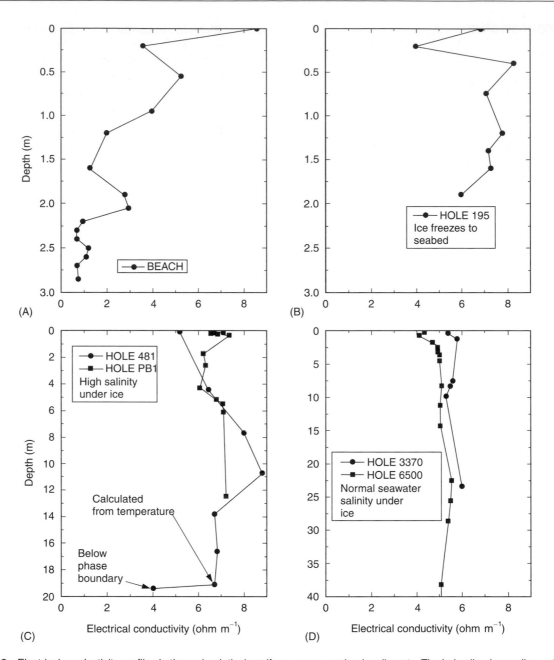

Figure 6 Electrical conductivity profiles in thawed, relatively uniform coarse-grained sediments. The holes lie along a line extending offshore near the West Dock at Prudhoe Bay, Alaska except for PB1, which is in the central portion of Prudhoe Bay. Hole designation is the distance (m) offshore. On the beach (A), concentrations decrease by a factor of 5 at a few meters depth. There are large variations with depth and concentrations may be double that of normal seawater where ice freezes to the seabed (B) and where there is restricted circulation under the ice (C). Farther offshore (D), the profiles tend to be relatively constant with depth with concentrations about the same or slightly greater than the overlying seawater. (Adapted from Iskandar IK *et al.* (1978) *Proceedings, Third International Conference on Permafrost*, Edmonton, Alberta, Canada, pp. 92–98. Ottawa, Ontario: National Research Council.)

The setting for region 5 consists of normal sea water over the seabed throughout the year. This results in relatively constant chemical and thermal boundary conditions.

There is a seasonal active layer at the seabed that freezes and thaws annually in both regions 3 and 4. The active layer begins to freeze simultaneously with the formation of sea ice in shallow water. Brine drainage from the growing sea ice increases the water salinity and decreases the temperature of the water at the seabed because of the requirement for phase equilibrium. This causes partial freezing of the less saline pore fluids in the sediments. Thus, it is not necessary for the ice to contact the seabed for the seabed to freeze. Seasonal changes in the pore fluid salinity show that the partially frozen active layer

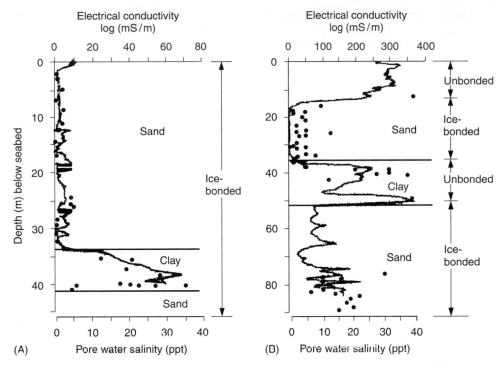

Figure 7 Onshore (A) and offshore (B) (water depth 10 m) electrical conductivity log (lines) and salinity (dots) profiles in the Mackenzie Delta region. The onshore sand–clay–sand sequence can be traced to the offshore site. Sand units appear to have been deposited under sub-aerial conditions during regressions and the clay unit under marine conditions during a transgression. At the offshore site, the upper sand unit is thawed to the 11 m depth. (Adapted with permission from Dallimore SR and Taylor AE (1994). *Proceedings, Sixth International Conference on Permafrost*, Beijing, vol. 1, pp.125–130. Wushan, Guangzhou, China: South China University Press.)

redistributes salts during freezing and thawing, is infiltrated by the concentrated brines, and influences the timing of brine drainage to lower depths in the sediments. These brines, derived from the growth of sea ice, provide a portion or all of the salts required for thawing the underlying sub-sea permafrost in the presence of negative sediment temperatures.

Depth to the ice-bonded permafrost increases slowly with distance offshore in region 3 to a few meters where the active layer no longer freezes to it (**Figure 8**) and the ice-bonded permafrost no longer couples conductively to the atmosphere. This allows the permafrost to thaw continuously throughout the year and depth to the ice-bonded permafrost increases rapidly with distance offshore (**Figure 8**).

The time an offshore site remains in regions 3 and 4 determines the number of years the seabed is subjected to freezing and thawing events. It is also the time required to make the transition from sub-aerial to relatively constant sub-sea boundary conditions. This time appears to be about 30 years near Lonely, Alaska (about 135 km southeast of Barrow) and about 500–1000 years near Prudhoe Bay, Alaska. **Figure 9** shows variations in the mean seabed temperatures with distance offshore near Prudhoe Bay where the shoreline retreat rate is about $1 \, \mathrm{m \, y^{-1}}$.

The above discussion of the physical setting does not incorporate the effects of geology, hydrology, tidal range, erosion and sedimentation processes, thaw settlement, and thaw strain. Regions 3 and 4 are extremely important in the evolution of sub-sea permafrost because the major portion of salt infiltration into the sediments occurs in these regions. The salt plays a strong role in determining T_p and, thus, whether or not thawing will occur.

Heat and Salt Transport

The transport of heat in sub-sea permafrost is thought to be primarily conductive because the observed temperature profiles are nearly linear below the depth of seasonal variations. However, even when heat transport is conductive, there is a coupling with salt transport processes because salt concentration controls T_p. Our lack of understanding of salt transport processes hampers the application of thermal models.

Thawing in the presence of negative seabed temperatures requires that T_p be significantly lower than T_s, so that generally salt must be present for thawing to occur. This salt must exist in the permafrost on submergence and/or be transported from the seabed

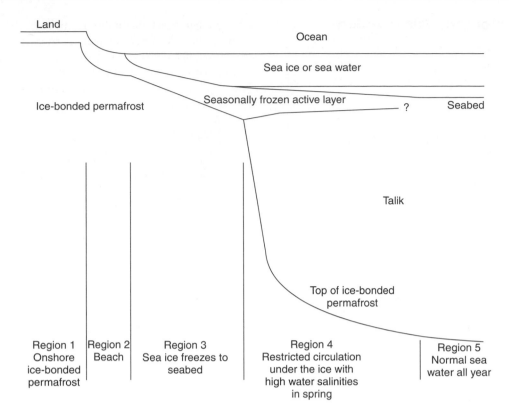

Figure 8 Schematic illustration of the transition of permafrost from sub-aerial to sub-sea conditions. There are five potential regions with differing thermal and chemical seabed boundary conditions. Hole 575 in **Figure 5** is in region 3 and the rest of the holes are in region 4. (Adapted from Osterkamp TE (1975) Report UAGR-234. Fairbanks, AK: Geophysical Institute, University of Alaska.)

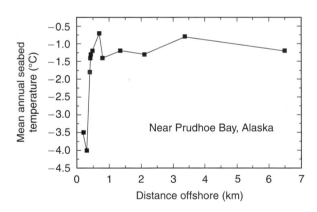

Figure 9 Variation of mean annual seabed temperatures with distance offshore. Along this same transect, about 6 km offshore from Reindeer Island in 17 m of water, the mean seabed temperature was near −1.7°C. Data on mean annual seabed salinities in regions 3 and 4 do not appear to exist. (Adapted from Osterkamp TE and Harrison WD (1985) Report UAGR-301. Fairbanks, AK: Geophysical Institute, University of Alaska.)

to the phase boundary. The efficiency of salt transport through the thawed layer at the seabed appears to be sensitive to soil type. In clays, the salt transport process is thought to be diffusion, a slow process; and in coarse-grained sands and gravels, pore fluid convection, which (involving motion of fluid) can be rapid.

Diffusive transport of salt has been reported in dense overconsolidated clays north of Reindeer Island offshore from Prudhoe Bay. Evidence for convective transport of salt exists in the thawed coarse-grained sediments near Prudhoe Bay and in the layered sands in the Mackenzie Delta region. This includes rapid vertical mixing as indicated by large seasonal variations in salinity in the upper 2 m of sediments in regions 3 and 4 and by salt concentration profiles that are nearly constant with depth and decrease in value with distance offshore in region 5 (**Figures 5–7**). Measured pore fluid pressure profiles (**Figure 10**) indicate downward fluid motion. Laboratory measurements of downward brine drainage velocities in coarse-grained sediments indicate that these velocities may be on the order of $100 \, \mathrm{m \, y^{-1}}$.

The most likely salt transport mechanism in coarse-grained sediments appears to be gravity-driven convection as a result of highly saline and dense brines at the seabed in regions 3 and 4. These brines infiltrate the seabed, even when it is partially frozen, and move rapidly downward. The release of relatively fresh and buoyant water by thawing ice at the phase boundary may also contribute to pore fluid motion.

Occurrence and Distribution

The occurrence, characteristics, and distribution of sub-sea permafrost are strongly influenced by regional and local conditions and processes including the following.

1. Geological (heat flow, shelf topography, sediment or rock types, tectonic setting)
2. Meteorological (sub-aerial ground surface temperatures as determined by air temperatures, snow cover and vegetation)
3. Oceanographic (seabed temperatures and salinities as influenced by currents, ice conditions, water depths, rivers and polynyas; coastal erosion and sedimentation; tidal range)
4. Hydrological (presence of lakes, rivers and salinity of the ground water)
5. Cryological (thickness, temperature, ice content, physical and mechanical properties of the onshore permafrost; presence of sub-sea permafrost that has survived previous transgressions; presence of ice sheets on the shelves)

Lack of information on these conditions and processes over the long timescales required for permafrost to aggrade and degrade, and inadequacies in the theoretical models, make it difficult to formulate reliable predictions regarding sub-sea permafrost. Field studies are required, but field data are sparse and investigations are still producing surprising results indicating that our understanding of sub-sea permafrost is incomplete.

Pechora and Kara Seas

Ice-bonded sub-sea permafrost has been found in boreholes with the top typically up to tens of meters below the seabed. In one case, pure freshwater ice was found 0.3 m below the seabed, extending to at least 25 m. These discoveries have led to difficult design conditions for an undersea pipeline that will cross Baydaratskaya Bay transporting gas from the Yamal Peninsula fields to European markets.

Laptev Sea

Sea water bottom temperatures typically range from $-0.5°C$ to $-1.8°C$, with some values colder than $-2°C$. A 300–850 m thick seismic sequence has been found that does not correlate well with regional tectonic structure and is interpreted to be ice-bonded permafrost. The extent of ice-bonded permafrost appears to be continuous to the 70 m isobath and widespread discontinuous to the 100 m isobath. Depth to the ice-bonded permafrost ranges from 2 to 10 m in water depths from 45 m to the shelf edge. Deep taliks may exist inshore of the 20 m isobath. A shallow sediment core with ice-bonded material at its base was recovered from a water depth of 120 m. Bodies of ice-rich permafrost occur under shallow water at the locations of recently eroded islands and along retreating coastlines.

Bering Sea

Sub-sea permafrost is not present in the northern portion except possibly in near-shore areas or where shoreline retreat is rapid.

Chukchi Sea

Seabed temperatures are generally slightly negative and thermal gradients are negative, indicating ice-bearing permafrost at depth within 1 km of shore near Barrow, Rabbit Creek and Kotzebue.

Alaskan Beaufort Sea

To the east of Point Barrow, bottom waters are typically $-0.5°C$ to $-1.7°C$ away from shore, shoreline erosion rates are rapid (1–10 m y^{-1}) and sediments are thick. Sub-sea permafrost appears to be thicker in the Prudhoe Bay region and thinner west of Harrison Bay to Point Barrow. Ice layers up

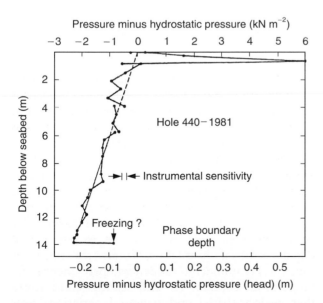

Figure 10 Measured *in situ* pore fluid pressure minus calculated hydrostatic pressure through the thawed layer in coarse-grained sediments at a hole 440 m offshore in May 1981 near Prudhoe Bay, Alaska. High pressure at the 0.54 m depth is probably related to seasonal freezing and the solid line is a least-squares fit to the data below 5.72 m. The negative pressure head gradient (-0.016) indicates a downward component of pore fluid velocity. (Adapted from Swift DW *et al.* (1983) *Proceedings, Fourth International Conference on Permafrost*, Fairbanks, Alaska, vol. 1, pp. 1221–1226. Washington DC: National Academy Press.)

to 0.6 m in thickness have been found off the Saga-vanirktok River Delta.

Surface geophysical studies (seismic and electrical) have indicated the presence of layered ice-bonded sub-sea permafrost. A profile of sub-sea permafrost near Prudhoe Bay (**Figure 11**) shows substantial differences in depth to ice-bonded permafrost between coarse-grained sediments inshore of Reindeer Island and fine-grained sediments farther offshore. Offshore from Lonely, where surface sediments are fine-grained, ice-bearing permafrost exists within 6–8 m of the seabed out to 8 km offshore (water depth 8 m). Ice-bonded material is deeper (~ 15 m). In Elson Lagoon (near Barrow) where the sediments are fine-grained, a thawed layer at the seabed of generally increasing thickness can be traced offshore.

Mackenzie River Delta Region

The layered sediments found in this region are typically fluvial sand and sub-sea mud corresponding to regressive/transgressive cycles (see **Figure 12**). Mean seabed temperatures in the shallow coastal areas are generally positive as a result of warm river water, and negative farther offshore. The thickness of ice-bearing permafrost varies substantially as a result of a complex history of transgressions and regressions, discharge from the Mackenzie River, and possible effects of a late glacial ice cover. Ice-bearing permafrost in the eastern and central Beaufort Shelf exceeds 600 m. It is thin or absent beneath Mackenzie Bay and may be only a few hundred meters thick toward the Alaskan coast. The upper surface of

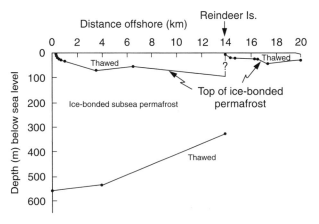

Figure 11 Sub-sea permafrost profile near Prudhoe Bay, Alaska, determined from drilling and well log data. The sediments are coarse-grained with deep thawing inshore of Reindeer Island and fine-grained with shallow thawing farther offshore. Maximum water depths are about 8 m inshore of Reindeer Island and 17 m about 6 km north. (Adapted from Osterkamp TE *et al.* (1985) *Cold Regions Science and Technology* 11: 99–105, 1985, with permission from Elsevier Science.)

ice-bearing permafrost is typically 5–100 m below the seabed and appears to be under control of seabed temperatures and stratigraphy.

The eastern Arctic, Arctic Archipelago and Hudson Bay regions were largely covered during the last glaciation by ice that would have inhibited permafrost growth. These regions are experiencing isostatic uplift with permafrost aggrading in emerging shorelines.

Antarctica

Negative sediment temperatures and positive temperature gradients to a depth of 56 m below the seabed exist in McMurdo Sound where water depth is 122 m and the mean seabed temperature is $-1.9°C$. This sub-sea permafrost did not appear to contain any ice.

Models

Modeling the occurrence, distribution and characteristics of sub-sea permafrost is a difficult task. Statistical, geological, analytical, and numerical models are available. Statistical models attempt to combine geological, oceanographic, and other information into algorithms that make predictions about sub-sea permafrost. These statistical models have not been very successful, although new GIS methods could potentially improve them.

Geological models consider how geological processes influence the formation and development of sub-sea permafrost. It is useful to consider these models since some sub-sea permafrost could potentially be a million years old. A geological model for the Mackenzie Delta region (**Figure 12**) has been developed that provides insight into the nature and complex layering of the sediments that comprise the sub-sea permafrost there.

Analytical models for investigating the thermal regime of sub-sea permafrost include both one- and two-dimensional models. All of the available analytical models have simplifying assumptions that limit their usefulness. These include assumptions of one-dimensional heat flow, stable shorelines or shorelines that undergo sudden and permanent shifts in position, constant air and seabed temperatures that neglect spatial and temporal variations over geological timescales, and constant thermal properties in layered sub-sea permafrost that is likely to contain unfrozen pore fluids. Neglect of topographical differences between the land and seabed, geothermal heat flow, phase change at the top and bottom of the sub-sea permafrost, and salt effects also limits their application. An analytical model exists that addresses the coupling between heat and

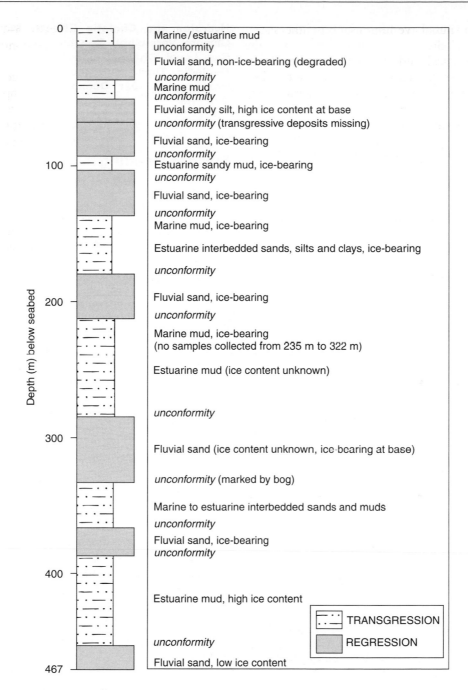

Depth (m) below seabed

0 — Marine / estuarine mud
unconformity
Fluvial sand, non-ice-bearing (degraded)
unconformity
Marine mud
unconformity
Fluvial sandy silt, high ice content at base
unconformity (transgressive deposits missing)

Fluvial sand, ice-bearing
unconformity
100 — Estuarine sandy mud, ice-bearing
unconformity

Fluvial sand, ice-bearing

unconformity
Marine mud, ice-bearing

Estuarine interbedded sands, silts and clays, ice-bearing

unconformity

200 — Fluvial sand, ice-bearing

unconformity

Marine mud, ice-bearing
(no samples collected from 235 m to 322 m)

Estuarine mud (ice content unknown)

unconformity

300 —

Fluvial sand (ice content unknown, ice-bearing at base)

unconformity (marked by bog)

Marine to estuarine interbedded sands and muds
unconformity
Fluvial sand, ice-bearing
unconformity

400 —

Estuarine mud, high ice content

| | TRANSGRESSION |
| | REGRESSION |

unconformity
Fluvial sand, low ice content
467 —

Figure 12 Canadian Beaufort Shelf stratigraphy in 32 m of water near the Mackenzie River Delta. Eight regressive/transgressive fluvial sand/marine mud cycles are shown. It is thought that, except for thawing near the seabed, the ice-bearing sequence has been preserved through time to the present. (Adapted with permission from Blasco S (1995) GSC Open File Report 3058. Ottawa, Canada: Geological Survey of Canada.)

salt transport but only for the case of diffusive transport with simplifying assumptions. Nevertheless, these analytical models appear to be applicable in certain special situations and have shaped much of the current thinking about sub-sea permafrost.

Two-dimensional numerical thermal models have addressed most of the concerns related to the assumptions in analytical models except for salt transport. Models have been developed that address salt transport via the buoyancy of fresh water generated by thawing ice at the phase boundary. Models for the infiltration of dense sea water brines derived from the growth of sea ice into the sediments do not appear to exist.

Successful application of all models is limited because of the lack of information over geological timescales on initial conditions, boundary conditions, material properties, salt transport and the coupling of heat and salt transport processes. There is also a lack of areas with sufficient information and measurements to fully test model predictions.

Nomenclature

h	Volumetric latent heat of the sediments (1 to 2×10^8 J m^{-3})
J_g	Geothermal heat flow entering the bottom phase boundary from below
J_f	Heat flow from the bottom phase boundary into the ice-bonded permafrost above
J_t	Heat flux into the top phase boundary from above
J_f	Heat flux from the top phase boundary into the ice-bonded permafrost below
K	Thermal conductivity of the ice-bonded permafrost (1 to 5 W m^{-1} K^{-1})
t	Time
T_s	Long-term mean temperature at the seabed
T_p	Phase boundary temperature at the top of the ice-bonded permafrost (0 to $-2°$C)
T_e	Phase boundary temperature at the bottom of the ice-bonded permafrost (0 to $-2°$C)
T_g	Long-term mean ground surface temperature during emergence
$X(t)$	Depth to the bottom of ice-bonded permafrost at time, t
$\dot{X}_{top}$	Thawing rate at top of ice-bonded permafrost during submergence
$\dot{X}_{bot}$	Thawing rate at the bottom of ice-bonded permafrost during submergence

See also

Arctic Ocean Circulation. Glacial Crustal Rebound, Sea Levels, and Shorelines. Heat Transport and Climate. Holocene Climate Variability. Methane Hydrates and Climatic Effects. Millennial-Scale Climate Variability. Sea Ice. Sea Ice: Overview. Sea Level Change. Sea Level Variations Over Geologic Time. Sub Ice-Shelf Circulation and Processes. Upper Ocean Heat and Freshwater Budgets.

Further Reading

Dallimore SR and Taylor AE (1994) Permafrost conditions along an onshore–offshore transect of the Canadian Beaufort Shelf. *Proceedings, Sixth International Conference on Permafrost, Beijing, vol. 1, pp 125–130.* South China University Press: Wushan, Guangzhou, China.

Hunter JA, Judge AS, MacAuley HA, *et al.* (1976) *The Occurrence of Permafrost and Frozen Sub-sea Bottom Materials in the Southern Beaufort Sea. Beaufort Sea Project, Technical Report 22.* Ottawa: Geological Survey Canada.

Lachenbruch AH, Sass JH, Marshall BV, and Moses TH Jr (1982) Permafrost, heat flow, and the geothermal regime at Prudhoe Bay, Alaska. *Journal of Geophysical Research* 87(B11): 9301–9316.

Mackay JR (1972) Offshore permafrost and ground ice, Southern Beaufort Sea, Canada. *Canadian Journal of Earth Science* 9(11): 1550–1561.

Osterkamp TE, Baker GC, Harrison WD, and Matava T (1989) Characteristics of the active layer and shallow sub-sea permafrost. *Journal of Geophysical Research* 94(C11): 16227–16236.

Romanovsky NN, Gavrilov AV, Kholodov AL, *et al.* (1998) *Map of predicted offshore permafrost distribution on the Laptev Sea Shelf. Proceedings, Seventh International Conference on Permafrost, Yellowknife, Canada, pp. 967–972.* University of Laval, Quebec: Center for Northern Studies.

Sellmann PV and Hopkins DM (1983) *Sub-sea permafrost distribution on the Alaskan Shelf. Final Proceedings, Fourth International Conference on Permafrost, Fairbanks, Alaska, pp. 75–82.* Washington, DC: National Academy Press.

SATELLITE MEASUREMENTS

SATELLITE OCEANOGRAPHY, HISTORY, AND INTRODUCTORY CONCEPTS

W. S. Wilson, NOAA/NESDIS, Silver Spring, MD, USA
E. J. Lindstrom, NASA Science Mission Directorate, Washington, DC, USA
J. R. Apel[†], Global Ocean Associates, Silver Spring, MD, USA

Published by Elsevier Ltd.

Oceanography from a satellite – the words themselves sound incongruous and, to a generation of scientists accustomed to Nansen bottles and reversing thermometers, the idea may seem absurd.

Gifford C. Ewing (1965)

Introduction: A Story of Two Communities

The history of oceanography from space is a story of the coming together of two communities – satellite remote sensing and traditional oceanography.

For over a century oceanographers have gone to sea in ships, learning how to sample beneath the surface, making detailed observations of the vertical distribution of properties. Gifford Ewing noted that oceanographers had been forced to consider "the class of problems that derive from the vertical distribution of properties at stations widely separated in space and time."

With the introduction of satellite remote sensing in the 1970s, traditional oceanographers were provided with a new tool to collect synoptic observations of conditions at or near the surface of the global ocean. Since that time, there has been dramatic progress; satellites are revolutionizing oceanography. (Appendix 1 provides a brief overview of the principles of satellite remote sensing.)

Yet much remains to be done. Traditional subsurface observations and satellite-derived observations of the sea surface – collected as an integrated set of observations and combined with state-of-the-art models – have the potential to yield estimates of the three-dimensional, time-varying distribution of properties for the global ocean. Neither a satellite nor an *in situ* observing system can do this on its own. Furthermore, if such observations can be collected over

the long term, they can provide oceanographers with an observational capability conceptually similar to that which meteorologists use on a daily basis to forecast atmospheric weather.

Our ability to understand and forecast oceanic variability, how the oceans and atmosphere interact, critically depends on an ability to observe the three-dimensional global oceans on a long-term basis. Indeed, the increasing recognition of the role of the ocean in weather and climate variability compels us to implement an integrated, operational satellite and *in situ* observing system for the ocean now – so that it may complement the system which already exists for the atmosphere.

The Early Era

The origins of satellite oceanography can be traced back to World War II – radar, photogrammetry, and the V-2 rocket. By the early 1960s, a few scientists had recognized the possibility of deriving useful

Figure 1 Thermal infrared image of the US southeast coast showing warmer waters of the Gulf Stream and cooler slope waters closer to shore taken in the early 1960s. While the resolution and accuracy of the TV on *Tiros* were not ideal, they were sufficient to convince oceanographers of the potential usefulness of infrared imagery. The advanced very high resolution radiometer (AVHRR) scanner (see text) has improved images considerably. Courtesy of NASA.

[†] Deceased

oceanic information from the existing aerial sensors. These included (1) the polar-orbiting meteorological satellites, especially in the 10–12-µm thermal infrared band; and (2) color photography taken by astronauts in the Mercury, Gemini, and Apollo manned spaceflight programs. Examples of the kinds of data obtained from the National Aeronautics and Space Administration (NASA) flights collected in the 1960s are shown in **Figures 1** and **2**.

Such early imagery held the promise of deriving interesting and useful oceanic information from space, and led to three important conferences on space oceanography during the same time period.

In 1964, NASA sponsored a conference at the Woods Hole Oceanographic Institution (WHOI) to examine the possibilities of conducting scientific research from space. The report from the conference, entitled *Oceanography from Space*, summarized findings to that time; it clearly helped to stimulate a number of NASA projects in ocean observations and sensor development. Moreover, with the exception of the synthetic aperture radar (SAR), all instruments flown through the 1980s used techniques described in this report. Dr. Ewing has since come to be justifiably regarded as the father of oceanography from space.

A second important step occurred in 1969 when the Williamstown Conference was held at Williams College in Massachusetts. The ensuing Kaula report set forth the possibilities for a space-based geodesy mission to determine the equipotential figure of the Earth using a combination of (1) accurate tracking of satellites and (2) the precision measurement of satellite elevation above the sea surface using radar altimeters. Dr. William Von Arx of WHOI realized the possibilities for determining large-scale oceanic currents with precision altimeters in space. The requirements for measurement precision of 10-cm height error in the elevation of the sea surface with respect to the geoid were articulated. NASA scientists and engineers felt that such accuracy could be achieved in the long run, and the agency initiated the Earth and Ocean Physics Applications Program, the first formal oceans-oriented program to be established within the organization. The required accuracy was not to be realized until 1992 with *TOPEX/Poseidon*, which was reached only over a 25-year

Figure 2 Color photograph of the North Carolina barrier islands taken during the Apollo-Soyuz Mission (AS9-20-3128). Capes Hatteras and Lookout, shoals, sediment- and chlorophyll-bearing flows emanating from the coastal inlets are visible, and to the right, the blue waters of the Gulf Stream. Cloud streets developing offshore the warm current suggest that a recent passage of a cold polar front has occurred, with elevated air–sea evaporative fluxes. Later instruments, such as the coastal zone color scanner (CZCS) on *Nimbus-7* and the SeaWiFS imager have advanced the state of the art considerably. Courtesy of NASA.

period of incremental progress that saw the flights of five US altimetric satellites of steadily increasing capabilities: *Skylab*, *Geos-3*, *Seasat*, *Geosat*, and *TOPEX/Poseidon* (see **Figure 3** for representative satellites).

A third conference, focused on sea surface topography from space, was convened by the National Oceanic and Atmospheric Administration (NOAA), NASA, and the US Navy in Miami in 1972, with 'sea surface topography' being defined as undulations of the ocean surface with scales ranging from

approximately 5000 km down to 1 cm. The conference identified several data requirements in oceanography that could be addressed with space-based radar and radiometers. These included determination of surface currents, Earth and ocean tides, the shape of the marine geoid, wind velocity, wave refraction patterns and spectra, and wave height. The conference established a broad scientific justification for space-based radar and microwave radiometers, and it helped to shape subsequent national programs in space oceanography.

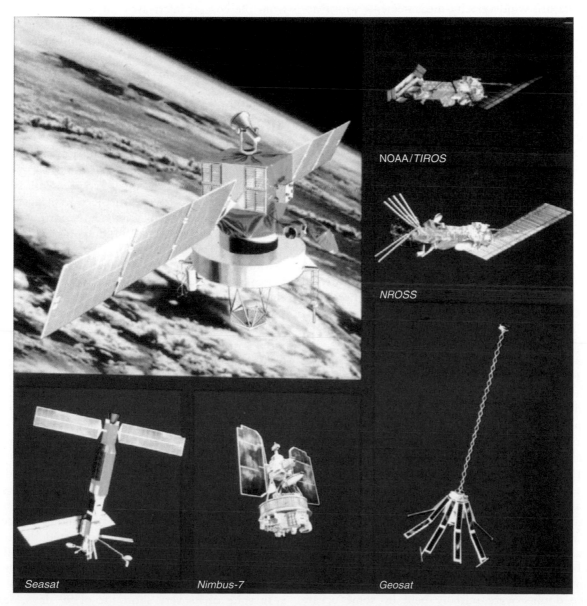

Figure 3   Some representative satellites: (1) *Seasat*, the first dedicated oceanographic satellite, was the first of three major launches in 1978; (2) the *Tiros* series of operational meteorological satellites carried the advanced very high resolution radiometer (AVHRR) surface temperature sensor; *Tiros-N*, the first of this series, was the second major launch in 1978; (3) *Nimbus-7*, carrying the CZCS color scanner, was the third major launch in 1978; (4) *NROSS*, an oceanographic satellite approved as an operational demonstration in 1985, was later cancelled; (5) *Geosat*, an operational altimetric satellite, was launched in 1985; and (6) this early version of *TOPEX* was reconfigured to include the French *Poseidon*; the joint mission *TOPEX/Poseidon* was launched in 1992. Courtesy of NASA.

The First Generation

Two first-generation ocean-viewing satellites, *Skylab* in 1973 and *Geos-3* in 1975, had partially responded to concepts resulting from the first two of these conferences. *Skylab* carried not only several astronauts, but a series of sensors that included the S-193, a radar-altimeter/wind-scatterometer, a long-wavelength microwave radiometer, a visible/infrared scanner, and cameras. S-193, the so-called Rad/Scatt, was advanced by Drs. Richard Moore and Willard Pierson. These scientists held that the scatterometer could return wind velocity measurements whose accuracy, density, and frequency would revolutionize marine meteorology. Later aircraft data gathered by NASA showed that there was merit to their assertions. *Skylab*'s scatterometer was damaged during the opening of the solar cell panels, and as a consequence returned indeterminate results (except for passage over a hurricane), but the altimeter made observations of the geoid anomaly due to the Puerto Rico Trench.

Geos-3 was a small satellite carrying a dual-pulse radar altimeter whose mission was to improve the knowledge of the Earth's marine geoid, and coincidentally to determine the height of ocean waves via the broadening of the short transmitted radar pulse upon reflection from the rough sea surface. Before the end of its 4-year lifetime, *Geos-3* was returning routine wave height measurements to the National Weather Service for inclusion in its Marine Waves Forecast. Altimetry from space had become a clear possibility, with practical uses of the sensor immediately forthcoming. The successes of *Skylab* and *Geos-3* reinforced the case for a second generation of radar-bearing satellites to follow.

The meteorological satellite program also provided measurements of sea surface temperature using far-infrared sensors, such as the visible and infrared scanning radiometer (VISR), which operated at wavelengths near 10 μm, the portion of the terrestrial spectrum wherein thermal radiation at terrestrial temperatures is at its peak, and where coincidentally the atmosphere has a broad passband. The coarse, 5-km resolution of the VISR gave blurred temperature images of the sea, but the promise was clearly there. **Figure 1** is an early 1960s TV image of the southeastern USA taken by the NASA *TIROS* program, showing the Gulf Stream as a dark signal. While doubts were initially held by some oceanographers as to whether such data actually represented the Gulf Stream, nevertheless the repeatability of the phenomenon, the verisimilitude of the positions and temperatures with respect to conventional wisdom, and their own objective judgment finally convinced most workers of the validity of the data. Today, higher-resolution, temperature-calibrated infrared imagery constitutes a valuable data source used frequently by ocean scientists around the world.

During the same period, spacecraft and aircraft programs taking ocean color imagery were delineating the possibilities and difficulties of determining sediment and chlorophyll concentrations remotely. **Figure 2** is a color photograph of the North Carolina barrier islands taken with a hand-held camera, with Cape Hatteras in the center. Shoals and sediment- and chlorophyll-bearing flows emanating from the coastal inlets are visible, and to the right, the blue waters of the Gulf Stream. Cloud streets developing offshore the warm stream suggest a recent passage of a cold polar front and attendant increases in air–sea evaporative fluxes.

The Second Generation

The combination of the early data and advances in scientific understanding that permitted the exploitation of those data resulted in spacecraft sensors explicitly designed to look at the sea surface. Information returned from altimeters and microwave radiometers gave credence and impetus to dedicated microwave spacecraft. Color measurements of the sea made from aircraft had indicated the efficacy of optical sensors for measurement of near-surface chlorophyll concentrations. Infrared radiometers returned useful sea surface temperature measurements. These diverse capabilities came together when, during a 4-month interval in 1978, the USA launched a triad of spacecraft that would profoundly change the way ocean scientists would observe the sea in the future. On 26 June, the first dedicated oceanographic satellite, *Seasat*, was launched; on 13 October, *TIROS-N* was launched immediately after the catastrophic failure of *Seasat* on 10 October; and on 24 October, *Nimbus-7* was lofted. Collectively they carried sensor suites whose capabilities covered virtually all known ways of observing the oceans remotely from space. This second generation of satellites would prove to be extraordinarily successful. They returned data that vindicated their proponents' positions on the measurement capabilities and utility, and they set the direction for almost all subsequent efforts in satellite oceanography.

In spite of its very short life of 99 days, *Seasat* demonstrated the great utility of altimetry by measuring the marine geoid to within a very few meters, by inferring the variability of large-scale ocean surface currents, and by determining wave heights. The wind scatterometer could yield oceanic surface wind velocities equivalent to 20 000 ship observations per day. The scanning multifrequency radiometer also provided wind speed and atmospheric water content data; and the SAR penetrated clouds to show features on the surface of the sea, including surface and internal waves, current boundaries, upwellings, and rainfall patterns. All of these measurements could be extended to basin-wide scales, allowing oceanographers a view of the sea never dreamed of before. *Seasat* stimulated several subsequent generations of ocean-viewing satellites, which outline the chronologies and heritage for the world's ocean-viewing spacecraft. Similarly, the early temperature and color observations have led to successor programs that provide large quantities of quantitative data to oceanographers around the world.

The Third Generation

The second generation of spacecraft would demonstrate that variables of importance to oceanography could be observed from space with scientifically useful accuracy. As such, they would be characterized as successful concept demonstrations. And while both first- and second-generation spacecraft had been exclusively US, international participation in demonstrating the utility of their data would lead to the entry of Canada, the European Space Agency (ESA), France, and Japan into the satellite program during this period. This article focuses on the US effort. Additional background on US third-generation missions covering the period 1980–87 can be found in the series of *Annual Reports for the Oceans Program* (NASA Technical Memoranda 80233, 84467, 85632, 86248, 87565, 88987, and 4025).

Partnership with Oceanography

Up to 1978, the remote sensing community had been the prime driver of oceanography from space and there were overly optimistic expectations. Indeed, the case had not yet been made that these observational techniques were ready to be exploited for ocean science. Consequently, in early 1979, the central task was establishing a partnership with the traditional oceanographic community. This meant involving them in the process of evaluating the performance of *Seasat* and *Nimbus-7*, as well as building an ocean science program at NASA headquarters to complement the ongoing remote sensing effort.

National Oceanographic Satellite System

This partnership with the oceanographic community was lacking in a notable and early false start on the part of NASA, the US Navy, and NOAA – the National Oceanographic Satellite System (NOSS). This was to be an operational system, with a primary and a backup satellite, along with a fully redundant ground data system. NOSS was proposed shortly after the failure of *Seasat*, with a first launch expected in 1986. NASA formed a 'science working group' (SWG) in 1980 under Francis Bretherton to define the potential that NOSS offered the oceanographic community, as well as to recommend sensors to constitute the 25% of its payload allocated for research. However, with oceanographers essentially brought in as junior partners, the job of securing a new start for NOSS fell to the operational community – which it proved unable to do. NOSS was canceled in early 1981. The prevailing and realistic view was that the greater community was not ready to implement such an operational system.

Science Working Groups

During this period, SWGs were formed to look at each promising satellite sensing technique, assess its potential contribution to oceanographic research, and define the requirements for its future flight. The notable early groups were the TOPEX SWG formed in 1980 under Carl Wunsch for altimetry, Satellite Surface Stress SWG in 1981 under James O'Brien for scatterometry, and Satellite Ocean Color SWG in 1981 under John Walsh for color scanners. These SWGs were true partnerships between the remote sensing and oceanographic communities, developing consensus for what would become the third generation of satellites.

Partnership with Field Centers

Up to this time, NASA's Oceans Program had been a collection of relatively autonomous, in-house activities run by NASA field centers. In 1981, an overrun in the space shuttle program forced a significant budget cut at NASA headquarters, including the Oceans Program. This in turn forced a reprioritization and refocusing of NASA programs. This was a blessing in disguise, as it provided an opportunity to initiate a comprehensive, centrally led program – which would ultimately result in significant funding for the oceanographic as well as remote-sensing communities. Outstanding relationships with individuals like Mous Chahine in senior management at the Jet Propulsion Laboratory (JPL) enabled the partnership between NASA headquarters and the two prime ocean-related field centers (JPL and the Goddard Space Flight Center) to flourish.

Partnerships in Implementation

A milestone policy-level meeting occurred on 13 July 1982 when James Beggs, then Administrator of NASA, hosted a meeting of the Ocean Principals Group – an informal group of leaders of the ocean-related agencies. A NASA presentation on opportunities and prospects for oceanography from space was received with much enthusiasm. However, when asked how NASA intended to proceed, Beggs told the group that – while NASA was the sole funding agency for space science and its missions – numerous agencies were involved in and support oceanography. Beggs said that NASA was willing to work with other agencies to implement an ocean satellite program, but that it would not do so on its own. Beggs' statement defined the approach to be pursued in implementing oceanography from space, namely, a joint approach based on partnerships.

Research Strategy for the Decade

As a further step in strengthening its partnership with the oceanographic community, NASA collaborated with the Joint Oceanographic Institutions Incorporated (JOI), a consortium of the oceanographic institutions with a deep-sea-going capability. At the time, JOI was the only organization in a position to represent and speak for the major academic oceanographic institutions. A JOI satellite planning committee (1984) under Jim Baker examined SWG reports, as well as the potential synergy between the variety of oceanic variables which could be measured from space; this led to the idea of understanding the ocean as a system. (From this, it was a small leap to understanding the Earth as a system, the goal of NASA's Earth Observing System (EOS).)

The report of this Committee, *Oceanography from Space*: *A Research Strategy for the Decade, 1985–1995*, linked altimetry, scatterometry, and ocean color with the major global ocean research programs being planned at that time – the World Ocean Circulation Experiment (WOCE), Tropical Ocean Global Atmosphere (TOGA) program, and Joint Global Ocean Flux Study (JGOFS). This strategy, still being followed today, served as a catalyst to engage the greater community, to identify the most important missions, and to develop an approach for their prioritization. Altimetry, scatterometry, and ocean color emerged from this process as national priorities.

Promotion and Advocacy

The *Research Strategy* also provided a basis for promoting and building an advocacy for the NASA program. If requisite funding was to be secured to pay for proposed missions, it was critical that government policymakers, the Congress, the greater oceanographic community, and the public had a good understanding of oceanography from space and its potential benefits. In response to this need, a set of posters, brochures, folders, and slide sets was designed by Payson Stevens of Internetwork Incorporated and distributed to a mailing list which grew to exceed 3000. These award-winning materials – sharing a common recognizable identity – were both scientifically accurate and esthetically pleasing.

At the same time, dedicated issues of magazines and journals were prepared by the community of involved researchers. The first example was the issue of *Oceanus* (1981), which presented results from the second-generation missions and represented a first step toward educating the greater oceanographic community in a scientifically useful and balanced way about realistic prospects for satellite oceanography.

Implementation Studies

Given the SWG reports taken in the context of the *Research Strategy*, the NASA effort focused on the following sensor systems (listed with each are the various flight opportunities which were studied):

- altimetry – the flight of a dedicated altimeter mission, first *TOPEX* as a NASA mission, and then *TOPEX/Poseidon* jointly with the French Centre Nationale d'Etudes Spatiales (CNES);
- scatterometry – the flight of a NASA scatterometer (NSCAT), first on NOSS, then on the *Navy Remote Ocean Observing Satellite* (NROSS), and finally on the *Advanced Earth Observing Satellite* (ADEOS) of the Japanese National Space Development Agency (NASDA);
- visible radiometry – the flight of a NASA color scanner on a succession of missions (NOSS, *NOAA-H/-I*, *SPOT-3* (*Système Pour l'Observation de la Terre*), and *Landsat-6*) and finally the purchase of ocean color data from the Sea-viewing Wide Field-of-view Sensor (SeaWiFS) to be flown by the Orbital Sciences Corporation (OSC);
- microwave radiometry – a system to utilize data from the series of special sensor microwave imager (SSMI) radiometers to fly on the Defense Meteorological Satellite Program satellites;
- SAR – a NASA ground station, the Alaska SAR Facility, to enable direct reception of SAR data from the *ERS-1/-2*, *JERS-1*, and *Radarsat* satellites of the ESA, NASDA, and the Canadian Space Agency, respectively.

New Starts

Using the results of the studies listed above, the Oceans Program entered the new start process at NASA headquarters numerous times attempting to secure funds for implementation of elements of the third generation. *TOPEX* was first proposed as a NASA mission in 1980. However, considering limited prospects for success, partnerships were sought and the most promising one was with the French. CNES initially proposed a mission using a *SPOT* bus with a US launch. However, NASA rejected this because *SPOT*, constrained to be Sun-synchronous, would alias solar tidal components. NASA proposed instead a mission using a US bus capable of flying in a non-Sun-synchronous orbit with CNES providing an *Ariane* launch. The NASA proposal was accepted for study in fiscal year (FY) 1983, and a new start was finally secured for the combined *TOPEX/Poseidon* in FY 1987.

In 1982 when the US Navy first proposed *NROSS*, NASA offered to be a partner and provide a scatterometer. The US Navy and NASA obtained new starts for both *NROSS* and NSCAT in FY 1985. However, *NROSS* suffered from a lack of strong support within the navy, experienced a number of delays, and was finally terminated in 1987. Even with this termination, NASA was able to keep NSCAT alive until establishing the partnership with NASDA for its flight on their *ADEOS* mission.

Securing a means to obtain ocean color observations as a follow-on to the coastal zone color scanner (CZCS) was a long and arduous process, finally coming to fruition in 1991 when a contract was signed with the OSC to purchase data from the flight of their SeaWiFS sensor. By that time, a new start had already been secured for NASA's EOS, and ample funds were available in that program for the SeaWiFS data purchase.

Finally, securing support for the Alaska SAR Facility (now the Alaska Satellite Facility to reflect its broader mission) was straightforward; being small in comparison with the cost of flying space hardware, its funding had simply been included in the new start that NSCAT obtained in FY 1985. Also, funding for utilization of SSMI data was small enough to be covered by the Oceans Program itself.

Implementing the Third Generation

With the exception of the US Navy's *Geosat*, these third-generation missions would take a very long time to come into being. As seen in **Table 1**, *TOPEX/Poseidon* was launched in 1992 – 14 years after *Seasat*; NSCAT was launched on *ADEOS* in 1996 – 18 years

after *Seasat*; and SeaWiFS was launched in 1997 – 19 years after *Nimbus-7*. (In addition to the missions mentioned in **Table 1**, the Japanese *ADEOS-1* included the US NSCAT in its sensor complement, and the US *Aqua* included the Japanese advanced microwave scanning radiometer (AMSR); the United States provided a launch for the Canadian *RADARSAT-1*.) In fact, these missions came so late that they had limited overlap with the field phases of the major ocean research programs (WOCE, TOGA, and JGOFS) they were to complement. Why did it take so long?

Understanding and Consensus

First, it took time to develop a physically unambiguous understanding of how well the satellite sensors actually performed, and this involved learning to cope with the data – satellite data rates being orders of magnitude larger than those encountered in traditional oceanography. For example, it was not until 3 years after the launch of *Nimbus-7* that CZCS data could be processed as fast as collected by the satellite. And even with only a 3-month data set from *Seasat*, it took 4 years to produce the first global maps of variables such as those shown in **Figure 4**.

In evaluating the performance of both *Seasat* and *Nimbus-7*, it was necessary to have access to the data. *Seasat* had a free and open data policy; and after a very slow start, the experiment team concept (where team members had a lengthy period of exclusive access to the data) for the *Nimbus-7* CZCS was replaced with that same policy. Given access to the data, delays were due to a combination of sorting out the algorithms for converting the satellite observations into variables of interest, as well as being constrained by having limited access to raw computing power.

In addition, the rationale for the third-generation missions represented a major paradigm shift. While earlier missions had been justified largely as demonstrations of remote sensing concepts, the third-generation missions would be justified on the basis of their potential contribution to oceanography. Hence, the long time it took to understand sensor performance translated into a delay in being able to convince traditional oceanographers that satellites were an important observational tool ready to be exploited for ocean science. As this case was made, it was possible to build consensus across the remote sensing and oceanographic communities.

Space Policy

Having such consensus reflected at the highest levels of government was another matter. The *White House*

Table 1   Some major ocean-related missions

Year	USA	Russia	Japan	Europe	Canada	Other
1968		Kosmos 243				
	Nimbus-3					
1970	Nimbus-4	Kosmos 384				
1972	Nimbus-5					
1974	Skylab					
	Nimbus-6, Geos-3					
1976						
1978	Nimbus-7, Seasat					
		Kosmos 1076				
1980		Kosmos 1151				
1982						
		Kosmos 1500				
1984		Kosmos 1602				
	Geosat					
1986		Kosmos 1776				
		Kosmos 1870	MOS 1A			
1988		OKEAN 1				
1990		OKEAN 2	MOS 1B			
		Almaz-1, OKEAN 3		ERS-1		
1992	Topex/Poseidon[a]		JERS-1			
1994		OKEAN 7				
		OKEAN 8		ERS-2	RADARSAT-1	
1996			ADEOS-1			
	SeaWiFS		TRMM[b]			
1998	GFO					
	Terra, QuikSCAT	OKEAN-O #1				OCEANSAT-1[i]
2000				CHAMP[c]		
		Meteor-3 #1		Jason-1[e]		
2002	Aqua; GRACE[d]			ENVISAT		HY-1A[j]
	WINDSAT		ADEOS-2			
2004	ICESat	SICH-1M[f]				
				CRYOSat		
2006			ALOS	GOCE, MetOp-1		HY-1B[j]
		Meteor-3M #2		SMOS	RADARSAT-2	OCEANSAT-2[i]
2008				OSTM/Jason-2[g]		
	NPP; Aquarius[h]			CryoSat-2		HY-1C, HY-2A[j]
2010			GCOM-W	Sentinel-3, MetOp-2		OCEANSAT-3[i]

[a] US/France TOPEX/Poseidon.
[b] Japan/US TRMM.
[c] German CHAMP.
[d] US/German GRACE.
[e] France/US Jason-1.
[f] Russia/Ukraine Sich-1M.
[g] France/US Jason-2/OSTM.
[h] US/Argentina Aquarius.
[i] India OCEANSAT series.
[j] China HY series.
Updated version of similar data in Wilson WS, Fellous JF, Kawamura H, and Mitnik L (2006) A history of oceanography from space. In: Gower JFR (ed.) Manual of Remote Sensing, Vol. 6: Remote Sensing of the Environment, pp. 1–31. Bethesda, MD: American Society for Photogrammetry and Remote Sensing.

Fact Sheet on US Civilian Space Policy of 11 October 1978 states, "… emphasizing space applications … will bring important benefits to our understanding of earth resources, climate, weather, pollution … and provide for the private sector to take an increasing responsibility in remote sensing and other applications." *Landsat* was commercialized in 1979 as part of this space policy. As Robert Stewart explains, "Clearly the mood at the presidential level was that earth remote sensing, including the oceans, was a practical space application more at home outside the scientific community. It took almost a decade to get an

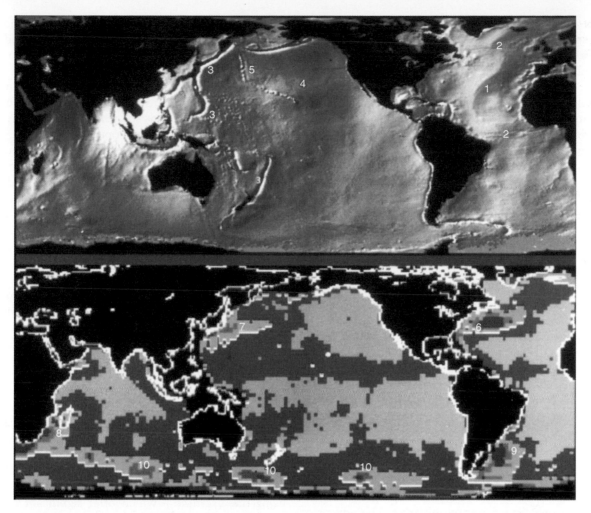

Figure 4 Global sea surface topography *c.* 1983. This figure shows results computed from the 70 days of *Seasat* altimeter data in 1978. Clearly visible in the mean sea surface topography, the marine geoid (upper panel), are the Mid-Atlantic Ridge (1) and associated fracture zones (2), trenches in the western Pacific (3), the Hawaiian Island chain (4), and the Emperor seamount chain (5). Superimposed on the mean surface is the time-varying sea surface topography, the mesoscale variability (lower panel), associated with the variability of the ocean currents. The largest deviations (10–25 cm), yellow and orange, are associated with the western boundary currents: Gulf Stream (6), Kuroshio (7), Agulhas (8), and Brazil/Falkland Confluence (9); large variations also occur in the West Wind Drift (10). Courtesy of NASA.

understanding at the policy level that scientific needs were also important, and that we did not have the scientific understanding necessary to launch an operational system for climate." The failures of NOSS, and later *NROSS*, were examples of an effort to link remote sensing directly with operational applications without the scientific underpinning.

The view in Europe was not dissimilar; governments felt that cost recovery was a viable financial scheme for ocean satellite missions, that is, the data have commercial value and the user would be willing to pay to help defray the cost of the missions.

Joint Satellite Missions

It is relatively straightforward to plan and implement missions within a single agency, as with NASA's

space science program. However, implementing a satellite mission across different organizations, countries, and cultures is both challenging and time-consuming. An enormous amount of time and energy was invested in studies of various flight options, many of which fizzled out, but some were implemented. With the exception of the former Soviet Union, NASA's third-generation missions would be joint with each nation having a space program at that time, as well as with a private company.

The *Geosat* Exception

Geosat was the notable US exception, having been implemented so quickly after the second generation. It was approved in 1981 and launched in 1985 in order to address priority operational needs on the part of the US Navy. During the second half of its mission, data would become available within 1–2 days. As will be discussed below, *Geosat* shared a number of attributes with the meteorological satellites: it had a specific focus; it met priority operational needs for its user; experience was available for understanding and using the observations; and its implementation was done in the context of a single organization.

Challenges Ahead

Scientific Justification

As noted earlier, during the decade of the 1980s, there was a dearth of ocean-related missions in the United States, it being difficult to justify a mission based on its contribution to ocean science. Then later in that decade, NASA conceived of the EOS and was able to make the case that Earth science was sufficient justification for a mission. Also noted earlier, ESA initially had no appropriate framework for Earth science missions, and a project like *ERS-1* was pursued under the assumption that it would help develop commercial and/or operational applications of remote sensing of direct societal benefit. Its successor, *ERS-2*, was justified on the basis of needing continuity of SAR coverage for the land surface, rather than the need to monitor ocean currents. And the *ENVISAT* was initially decided by ESA member states as part of the Columbus program of the International Space Station initiative. The advent of an Earth Explorer program in 1999 represented a change in this situation. As a consequence, new Earth science missions – *GOCE*, *CryoSat*, and *SMOS* – all represent significant steps forward.

This ESA program, together with similar efforts at NASA, are leading to three sets of ground-breaking scientific missions which have the potential to significantly impact oceanography. *GOCE* and *GRACE* will contribute to an improved knowledge of the Earth's gravity field, as well as the mass of water on the surface of the Earth. *CryoSat-2* and *ICESat* will contribute to knowledge of the volume of water locked up in polar and terrestrial ice sheets. Finally, *SMOS* and *Aquarius* will contribute to knowledge of the surface salinity field of the global oceans. Together, these will be key ingredients in addressing the global water cycle.

Data Policy

The variety of missions described above show a mix of data policies, from full and open access without any period of exclusive use (e.g., *TOPEX/Poseidon*, *Jason*, and *QuikSCAT*) to commercial distribution (e.g., real-time SeaWiFS for nonresearch purposes, *RADARSAT*), along with a variety of intermediate cases (e.g., *ERS*, *ALOS*, and *ENVISAT*). From a scientific perspective, full and open access is the preferred route, in order to obtain the best understanding of how systems perform, to achieve the full potential of the missions for research, and to lay the most solid foundation for an operational system. Full and open access is also a means to facilitate the development of a healthy and competitive private sector to provide value-added services. Further, if the international community is to have an effective observing system for climate, a full and open data policy will be needed, at least for that purpose.

In Situ Observations

Satellites have made an enormous contribution enabling the collection of *in situ* observations from *in situ* platforms distributed over global oceans. The Argos (plural spelling; not to be confused with *Argo* profiling floats) data collection and positioning system has flown on the NOAA series of polar-orbiting operational environmental satellites continuously since 1978. It provides one-way communication from data collection platforms, as well as positioning of those platforms. While an improved Argos capability (including two-way communications) is coming with the launch of *MetOp-1* in 2006, oceanographers are looking at alternatives – Iridium being one example – which offer significant higher data rates, as well as two-way communications.

In addition, it is important to note that the Intergovernmental Oceanographic Commission and the World Meteorological Organization have established the Joint Technical Commission for Oceanography and Marine Meteorology (JCOMM) to bring a focus to the collection, formatting, exchange, and archival of data collected at sea, whether they be oceanic or atmospheric. JCOMM has established a center, JCOMMOPS, to serve as the specific institutional focus to harmonize the national contributions of *Argo* floats, surface-drifting buoys, coastal tide gauges, and fixed and moored buoys. JCOMMOPS will play an important role helping contribute to 'integrated observations' described below.

Integrated Observations

To meet the demands of both the research and the broader user community, it will be necessary to focus, not just on satellites, but 'integrated' observing systems. Such systems involve combinations of satellite and *in situ* systems feeding observations into data-processing systems capable of delivering a comprehensive view of one or more geophysical variables (sea level, surface temperature, winds, etc.).

Three examples help illustrate the nature of integrated observing systems. First, consider global sea level rise. The combination of the *Jason-1* altimeter, its precision orbit determination system, and the suite of precision tide gauges around the globe allow scientists to monitor changes in volume of the oceans. The growing global array of *Argo* profiling floats allows scientists to assess the extent to which those changes in sea level are caused by changes in the temperature and salinity structure of the upper ocean. *ICESat* and *CryoSat* will provide estimates of changes in the volume of ice sheets, helping assess the extent to which their melting contributes to global sea level rise. And *GRACE* and *GOCE* will provide estimates of the changes in the mass of water on the Earth's surface. Together, systems such as these will enable an improved understanding of global sea level rise and, ultimately, a reduction in the wide range of uncertainty in future projections.

Second, global estimates of vector winds at the sea surface are produced from the scatterometer on *QuikSCAT*, a global array of *in situ* surface buoys, and the Seawinds data processing system. Delivery of this product in real time has significant potential to improve marine weather prediction. The third example concerns the *Jason-1* altimeter together with *Argo*. When combined in a sophisticated data assimilation system – using a state-of-the-art ocean model – these data enable the estimation of the physical state of the ocean as it changes through time. This information – the rudimentary 'weather map' depicting the circulation of the oceans – is a critical component of climate models and provides the fundamental context for addressing a broad range of issues in chemical and biological oceanography.

Transition from Research to Operations

The maturing of the discipline of oceanography includes the development of a suite of global oceanographic services being conducted in a manner similar to what exists for weather services. The delivery of these services and their associated informational products will emerge as the result of the successes in ocean science ('research push'), as well as an increasing demand for ocean analyses and forecasts from a variety of sectors ('user pull').

From the research perspective, it is necessary to 'transition' successfully demonstrated ('experimental') observing techniques into regular, long-term, systematic ('operational') observing systems to meet a broad range of user requirements, while maintaining the capability to collect long-term, 'research-quality'

observations. From the operational perspective, it is necessary to implement proven, scientifically sound, cost-effective observing systems – where the uninterrupted supply of real-time data is critical. This is a big challenge to be met by the space systems because of the demand for higher reliability and redundancy, at the same time calling for stringent calibration and accuracy requirements. Meeting these sometimes competing, but quite complementary demands will be the challenge and legacy of the next generation of ocean remote-sensing satellites.

Meteorological Institutional Experience

With the launch in 1960 of the world's first meteorological satellite, the polar-orbiting *Tiros-1* carrying two TV cameras, the value of the resulting imagery to the operational weather services was recognized immediately. The very next year a National Operational Meteorological System was implemented, with NASA to build and launch the satellites and the Weather Bureau to be the operator. The feasibility of using satellite imagery to locate and track tropical storms was soon demonstrated, and by 1969 this capability had become a regular part of operational weather forecasting. In 1985, Richard Hallgren, former Director of the National Weather Service, stated, "the use of satellite information simply permeates every aspect of the [forecast and warning] process and all this in a mere 25 years." In response to these operational needs, there has been a continuing series of more than 50 operational, polar-orbiting satellites in the United States alone!

The first meteorological satellites had a specific focus on synoptic meteorology and weather forecasting. Initial image interpretation was straightforward (i.e., physically unambiguous), and there was a demonstrated value of resulting observations in meeting societal needs. Indeed, since 1960 satellites have ensured that no hurricane has gone undetected. In addition, the coupling between meteorology and remote sensing started very near the beginning. An 'institutional mechanism' for transition from research to operations was established almost immediately. Finally, recognition of this endeavor extended to the highest levels of government, resulting in the financial commitment needed to ensure success.

Oceanographic Institutional Issue

Unlike meteorology where there is a National Weather Service in each country to provide an institutional focus, ocean-observing systems have multiple performer and user institutions whose interests must be reconciled. For oceanography, this is a significant challenge working across 'institutions', where the *in situ* research is in one or more agencies, the space research and development is in another, and operational activities in yet another – with possibly separate civil and military systems. In the United States, the dozen agencies with ocean-related responsibilities are using the National Oceanographic Partnership Program and its Ocean.US Office to provide a focus for reconciling such interests. In the United Kingdom, there is the Interagency Committee on Marine Science and Technology.

In France, there is the Comité des Directeurs d'Organismes sur l'Océanographie, which gathers the heads of seven institutions interested in the development of operational oceanography, including CNES, meteorological service, ocean research institution, French Research Institute for Exploitation of the Sea (IFREMER), and the navy. This group of agencies has worked effectively over the past 20 years to establish a satellite data processing and distribution system (AVISO), the institutional support for a continuing altimetric satellite series (*TOPEX/Poseidon, Jason*), the framework for the French contribution to the *Argo* profiling float program (CORIOLIS), and to create a public corporation devoted to ocean modeling and forecasting, using satellite and *in situ* data assimilation in an

operational basis since 2001 (Mercator). This partnership could serve as a model in the effort to develop operational oceanography in other countries. Drawing from this experience working together within France, IFREMER is leading the European integrated project, MERSEA, aimed at establishing a basis for a European center for ocean monitoring and forecasting.

Ocean Climate

If we are to adequately address the issue of global climate change, it is essential that we are able to justify the satellite systems required to collect the global observations 'over the long term'. Whether it be global sea level rise or changes in Arctic sea ice cover or sea surface temperature, we must be able to sustain support for the systems needed to produce climate-quality data records, as well as ensure the continuing involvement of the scientific community.

Koblinsky and Smith have outlined the international consensus for ocean climate research needs and identified the associated observational requirements. In addition to their value for research, we are compelled by competing interests to demonstrate the value of such observations in meeting a broad range of societal needs. Climate observations pose

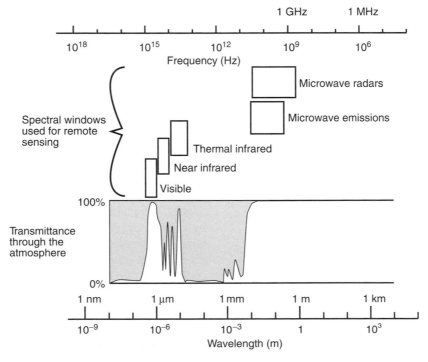

Figure 5 The electromagnetic spectrum showing atmospheric transmitance as a function of frequency and wavelength, along with the spectral windows used for remote sensing. Microwave bands are typically defined by frequency and the visible/infrared by wavelength. Adapted from Robinson IS and Guymer T (1996) Observing oceans from space. In: Summerhayes CP and Thorpe SA (eds.) *Oceanography: An Illustrated Guide*, pp. 69–87. Chichester, UK: Wiley.

challenges, since they require operational discipline to be collected in a long-term systematic manner, yet also require the continuing involvement of the research community to ensure their scientific integrity, and have impacts that may not be known for decades (unlike observations that support weather forecasting whose impact can be assessed within a matter of hours or days). Together, the institutional and observational challenges for ocean climate have been difficult to surmount.

International Integration

The paper by the Ocean Theme Team prepared under the auspices of the Integrated Global Observing Strategy (IGOS) Partnership represents how the space-faring nations are planning for the collection of global ocean observations. IGOS partners include the major global research program sponsors, global observing systems, space agencies, and international organizations.

On 31 July 2003, the First Earth Observations Summit – a high-level meeting involving ministers from over 20 countries – took place in Washington, DC, following a recommendation adopted at the G-8 meeting held in Evian the previous month; this summit proposed to 'plan and implement' a Global Earth Observation System of Systems (GEOSS). Four additional summits have been held, with participation having grown to include 60 nations and 40 international organizations; the GEOSS process provides the political visibility – not only to implement the plans developed within the IGOS Partnership – but to do so in the context of an overall Earth observation framework. This represents a remarkable opportunity to develop an improved understanding of the oceans and their influence on the Earth system, and to contribute to the delivery of improved oceanographic products and services to serve society.

Appendix 1: A Brief Overview of Satellite Remote Sensing

Unlike the severe attenuation in the sea, the atmosphere has 'windows' in which certain electromagnetic (EM) signals are able to propagate. These windows, depicted in **Figure 5**, are defined in terms

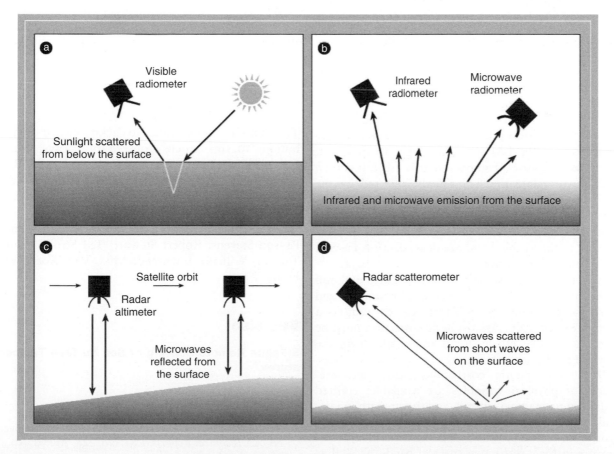

Figure 6 Four techniques for making oceanic observations from satellites: (a) visible radiometry, (b) infrared and microwave radiometry, (c) altimetry, and (d) scatterometry. Adapted from Robinson IS and Guymer T (1996) Observing oceans from space. In: Summerhayes CP and Thorpe SA (eds.) *Oceanography: An Illustrated Guide*, pp. 69–87. Chichester, UK: Wiley.

Passive sensors (radiometers)			Active sensors (microwave radars)		
Sensor type					
Visible	Infrared	Microwave	Altimetry	Scatterometry	SAR
Measured physical variable					
Solar radiation backscattered from beneath the sea surface	Infrared emission from the sea surface	Microwave emission from the sea surface	Travel time, shape, and strength of reflected pulse	Strength of return pulse when illuminated from different directions	Strength and phase of return pulse
Applications					
Ocean color; chlorophyll; primary production; water clarity; shallow-water bathymetry	Surface temperature; ice cover	Ice cover, age and motion; sea surface temperature; wind speed	Surface topography for geostrophic currents and tides; bathymetry; oceanic geoid; wind and wave conditions	Surface vector winds; ice cover	Surface roughness at fine spatial scales; surface and internal wave patterns; bathymetric patterns; ice cover and motion

Figure 7   Measured physical variables and applications for both passive and active sensors, expressed as a function of sensor type.

of atmospheric transmittance – the percentage of an EM signal which is able to propagate through the atmosphere – expressed as a function of wavelength or frequency.

Given a sensor on board a satellite observing the ocean, it is necessary to understand and remove the effects of the atmosphere (such as scattering and attenuation) as the EM signal propagates through it. For passive sensors (**Figures 6(a)** and **6(b)**), it is then possible to relate the EM signals collected by the sensor to the associated signals at the bottom of the atmosphere, that is, the natural radiation emitted or reflected from the sea surface. Note that passive sensors in the visible band are dependent on the Sun for natural illumination.

Active sensors, microwave radar (**Figures 6(c)** and **6(d)**), provide their own source of illumination and have the capability to penetrate clouds and, to a certain extent, rain. Atmospheric correction must be done to remove effects for a round trip from the satellite to the sea surface.

With atmospheric corrections made, measurements of physical variables are available: emitted radiation for passive sensors, and the strength, phase, and/or travel time for active sensors. **Figure 7** shows typical measured physical variables for both types of sensors in their respective spectral bands, as well as applications or derived variables of interest – ocean color, surface temperature, ice cover, sea level, and

surface winds. The companion articles on this topic address various aspects of **Figure 7** in more detail, so only this general overview is given here.

Acknowledgments

The authors would like to acknowledge contributions to this article from Mary Cleave, Murel Cole, William Emery, Michael Freilich, Lee Fu, Rich Gasparovic, Trevor Guymer, Tony Hollingsworth[†], Hiroshi Kawamura, Michele Lefebvre, Leonid Mitnik, Jean-François Minster, Richard Moore, William Patzert, Willard Pierson[†], Jim Purdom, Keith Raney, Payson Stevens, Robert Stewart, Ted Strub, Tasuku Tanaka, William Townsend, Mike Van Woert, and Frank Wentz.

See also

Satellite Remote Sensing of Sea Surface Temperatures.

Further Reading

Apel JR (ed.) (1972) Sea surface topography from space. *NOAA Technical Reports: ERL No. 228, AOML No. 7.* Boulder, CO: NOAA.

Cherny IV and Raizer VY (1998) *Passive Microwave Remote Sensing of Oceans.* Chichester, UK: Praxis.

Committee on Earth Sciences (1995) *Earth Observations from Space: History, Promise, and Reality*. Washington, DC: Space Studies Board, National Research Council.

Ewing GC (1965) Oceanography from space. *Proceedings of a Conference held at Woods Hole, 24–28 August 1964. Woods Hole Oceanographic Institution Ref. No. 65-10*. Woods Hole, MA: WHOI.

Fu L-L, Liu WT, and Abbott MR (1990) Satellite remote sensing of the ocean. In: Le Méhauté B (ed.) *The Sea, Vol. 9: Ocean Engineering Science*, pp. 1193–1236. Cambridge, MA: Harvard University Press.

Guymer TH, Challenor PG, and Srokosz MA (2001) Oceanography from space: Past success, future challenge. In: Deacon M (ed.) *Understanding the Oceans: A Century of Ocean Exploration*, pp. 193–211. London: UCL Press.

JOI Satellite Planning Committee (1984) *Oceanography from Space: A Research Strategy for the Decade, 1985–1995*, parts 1 and 2. Washington, DC: Joint Oceanographic Institutions.

Kaula WM (ed.) (1969) The terrestrial environment: solid-earth and ocean physics. *Proceedings of a Conference held at William College, 11–21 August 1969, NASA CR-1579*. Washington, DC: NASA.

Kawamura H (2000) Era of ocean observations using satellites. *Sokko-Jiho* 67: S1–S9 (in Japanese).

Koblinsky CJ and Smith NR (eds.) (2001) *Observing the Oceans in the 21st Century*. Melbourne: Global Ocean Data Assimilation Experiment and the Bureau of Meteorology.

Masson RA (1991) *Satellite Remote Sensing of Polar Regions*. London: Belhaven Press.

Minster JF and Lefebvre M (1997) *TOPEX/Poseidon* satellite altimetry and the circulation of the oceans. In: Minster JF (ed.) *La Machine Océan*, pp. 111–135 (in French). Paris: Flammarion.

Ocean Theme Team (2001) *An Ocean Theme for the IGOS Partnership*. Washington, DC: NASA. http://www.igospartners.org/docs/theme_reports/IGOS-Oceans-Final-0101.pdf (accessed Mar. 2008).

Purdom JF and Menzel WP (1996) Evolution of satellite observations in the United States and their use in meteorology. In: Fleming JR (ed.) *Historical Essays on Meteorology: 1919–1995*, pp. 99–156. Boston, MA: American Meteorological Society.

Robinson IS and Guymer T (1996) Observing oceans from space. In: Summerhayes CP and Thorpe SA (eds.) *Oceanography: An Illustrated Guide*, pp. 69–87. Chichester, UK: Wiley.

Victorov SV (1996) *Regional Satellite Oceanography*. London: Taylor and Francis.

Wilson WS (ed.) (1981) *Special Issue: Oceanography from Space. Oceanus* 24: 1–76.

Wilson WS, Fellous JF, Kawamura H, and Mitnik L (2006) A history of oceanography from space. In: Gower JFR (ed.) *Manual of Remote Sensing, Vol. 6: Remote Sensing of the Environment*, pp. 1–31. Bethesda, MD: American Society for Photogrammetry and Remote Sensing.

Wilson WS and Withee GW (2003) A question-based approach to the implementation of sustained, systematic observations for the global ocean and climate, using sea level as an example. *MTS Journal* 37: 124–133.

Relevant Websites

http://www.aviso.oceanobs.com
– AVISO.

http://www.coriolis.eu.org
– CORIOLIS.

http://www.eohandbook.com
– Earth Observation Handbook, CEOS.

http://www.igospartners.org
– IGOS.

http://wo.jcommops.org
– JCOMMOPS.

http://www.mercator-ocean.fr
– Mercator Ocean.

SATELLITE PASSIVE-MICROWAVE MEASUREMENTS OF SEA ICE

C. L. Parkinson, NASA Goddard Space Flight Center, Greenbelt, MD, USA

Published by Elsevier Ltd.

Introduction

Satellite passive-microwave measurements of sea ice have provided global or near-global sea ice data for most of the period since the launch of the *Nimbus 5* satellite in December 1972, and have done so with horizontal resolutions on the order of 25–50 km and a frequency of every few days. These data have been used to calculate sea ice concentrations (percent areal coverages), sea ice extents, the length of the sea ice season, sea ice temperatures, and sea ice velocities, and to determine the timing of the seasonal onset of melt as well as aspects of the ice-type composition of the sea ice cover. In each case, the calculations are based on the microwave emission characteristics of sea ice and the important contrasts between the microwave emissions of sea ice and those of the surrounding liquid-water medium.

The passive-microwave record is most complete since the launch of the scanning multichannel microwave radiometer (SMMR) on the *Nimbus 7* satellite in October 1978; and the SMMR data and follow-on data from the special sensor microwave imagers (SSMIs) on satellites of the United States Defense Meteorological Satellite Program (DMSP) have been used to determine trends in the ice covers of both polar regions since the late 1970s. The data have revealed statistically significant decreases in Arctic sea ice coverage and much smaller magnitude increases in Antarctic sea ice coverage.

Background on Satellite Passive-Microwave Sensing of Sea Ice

Rationale

Sea ice is a vital component of the climate of the polar regions, insulating the oceans from the atmosphere, reflecting most of the solar radiation incident on it, transporting cold, relatively fresh water toward equator, and at times assisting overturning in the ocean and even bottom water formation through its rejection of salt to the underlying water. Furthermore, sea ice spreads over vast distances, globally

covering an area approximately the size of North America, and it is highly dynamic, experiencing a prominent annual cycle in both polar regions and many short-term fluctuations as it is moved by winds and waves, melted by solar radiation, and augmented by additional freezing. It is a major player in and indicator of the polar climate state and has multiple impacts on all levels of the polar marine ecosystems. Consequently it is highly desirable to monitor the sea ice cover on a routine basis. In view of the vast areal coverage of the ice and the harsh polar conditions, the only feasible means of obtaining routine monitoring is through satellite observations. Visible, infrared, active-microwave, and passive-microwave satellite instruments are all proving useful for examining the sea ice cover, with the passive-microwave instruments providing the longest record of near-complete sea ice monitoring on a daily or near-daily basis.

Theory

The tremendous value of satellite passive-microwave measurements for sea ice studies results from the combination of the following four factors:

1. Microwave emissions of sea ice differ noticeably from those of seawater, making sea ice generally readily distinguishable from liquid water on the basis of the amount of microwave radiation received by the satellite instrument. For example, **Figure 1** presents color-coded images of the data from one channel on a satellite passive-microwave instrument, presented in units (termed 'brightness temperatures') indicative of the intensity of emitted microwave radiation at that channel's frequency, 19.4 GHz. The ice edge, highlighted by the broken white curve, is readily identifiable from the brightness temperatures, with open-ocean values of 172–198 K outside the ice edge and sea ice values considerably higher, predominantly greater than 230 K, within the ice edge.

2. The microwave radiation received by Earth-orbiting satellites derives almost exclusively from the Earth system. Hence, microwave sensing does not require sunlight, and the data can be collected irrespective of the level of daylight or darkness. This is a major advantage in the polar latitudes, where darkness lasts for months at a time, centered on the winter solstice.

(a)   (b)

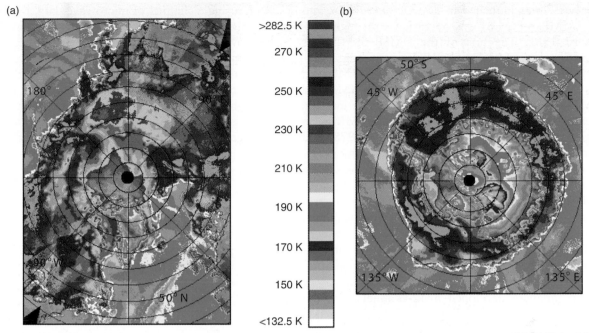

Figure 1 Late-winter brightness temperature images of 19.4-GHz vertically polarized (19.4 V) data from the DMSP SSMI for (a) the north polar region on 15 Mar. 1998 and (b) the south polar region on 15 Sep. 1998, showing near-maximum sea ice coverage in each hemisphere. The broken white curve has been added to indicate the location of the sea ice edge. Black signifies areas of no data; the absence of data poleward of 87.6° latitude results from the satellite's near-polar orbit and is consistent throughout the SSMI data set.

3. Many of the microwave data are largely unaffected by atmospheric conditions, including the presence or absence of clouds. Storm systems can produce atmospheric interference, but, at selected wavelengths, the microwave signal from the ice–ocean surface can pass through most nonprecipitating clouds essentially unhindered. Hence, microwave sensing of the surface does not require cloud-free conditions.

4. Satellite passive-microwave instruments can obtain a global picture of the sea ice cover at least every few days with a resolution of 50 km or better, providing reasonable spatial resolution and extremely good temporal resolution for most large-scale or climate-related studies.

Satellite Passive-Microwave Instruments

The first major satellite passive-microwave imager was the electrically scanning microwave radiometer (ESMR) launched on the *Nimbus 5* satellite of the United States National Aeronautics and Space Administration (NASA) in December 1972, preceded by a nonscanning passive-microwave radiometer launched on the Russian Cosmos satellite in September 1968. The ESMR was a single-channel instrument recording radiation at a wavelength of 1.55 cm and corresponding frequency of 19.35 GHz. It collected good-quality data for much of the 4-year period from January 1973 through December 1976, although with some major data gaps, including one that lasted for 3 months, from June through August 1975. Being a single-channel instrument, it did not allow some of the more advanced studies that have been done with subsequent instruments, but its flight was a highly successful proof-of-concept mission, establishing the value of satellite passive-microwave technology for observing the global sea ice cover and other variables. The ESMR data were used extensively in the determination and analysis of sea ice conditions in both the Arctic and the Antarctic over the 4 years 1973–76. Emphasis centered on the determination of ice concentrations (percent areal coverages of ice) and, based on the ice concentration results, the calculation of ice extents (integrated areas of all grid elements with ice concentration ≥15%). This 4-year data set established key aspects of the annual cycles of the polar sea ice covers, including the nonuniformity of the growth and decay seasons and the marked interannual differences even within a 4-year data set.

The *Nimbus 5* ESMR was followed by a less successful ESMR on the *Nimbus 6* satellite and then by the more advanced 10-channel SMMR on board NASA's *Nimbus 7* satellite and a sequence of seven-channel SSMIs on board satellites of the DMSP.

Nimbus 7 was launched in late October 1978, and the SMMR on board was operational through mid-August 1987. The first of the DMSP SSMIs was operational as of early July 1987, providing a welcome data overlap with the *Nimbus 7* SMMR and thereby allowing intercalibration of the SMMR and SSMI data sets. SSMIs continue to operate into the twenty-first century. There was also an SMMR on board the short-lived *Seasat* satellite in 1978; and there was a two-channel microwave scanning radiometer (MSR) on board the Japanese Marine Observation Satellites starting in February 1987. Each of these successor satellite passive-microwave instruments, after the ESMR, has been multichannel, allowing both an improved accuracy in the ice concentration derivations and the calculation of additional sea ice variables, including ice temperature and the concentrations of separate ice types.

The Japanese developed a 12-channel advanced microwave scanning radiometer (AMSR) for the Earth Observing System's (EOS) *Aqua* satellite (formerly named the *PM-1* satellite), launched by NASA in May 2002, and for the Advanced Earth Observing Satellite II (*ADEOS-II*), launched by the Japanese National Space Development Agency in December 2002. *ADEOS-II* prematurely ceased operations in October 2003, but the *Aqua* AMSR, labeled AMSR-E in recognition of its place in the EOS and to distinguish it from the *ADEOS-II* AMSR, has collected a multiyear data record. The AMSR-E has a major advantage over the SMMR and SSMI instruments in allowing sea ice measurements at a higher spatial resolution (12–25 km vs. 25–50 km for the major derived sea ice products). It furthermore has an additional advantage over the SMMR in having channels at 89 GHz in addition to its lower-frequency channels, at 6.9, 10.7, 18.7, 23.8, and 36.5 GHz.

Sea Ice Determinations from Satellite Passive-Microwave Data

Sea Ice Concentrations

Ice concentration is among the most fundamental and important parameters for describing the sea ice cover. Defined as the percent areal coverage of ice, it is directly critical to how effectively the ice cover restricts exchanges between the ocean and the atmosphere and to how much incoming solar radiation the ice cover reflects. Ice concentration is calculated at each ocean grid element, for whichever grid is being used to map or otherwise display the derived satellite products. A map of ice concentrations presents the areal distribution of the ice cover, to the resolution of the grid.

With a single channel of microwave data, taken at a radiative frequency and polarization combination that provides a clear distinction between ice and water, approximate sea ice concentrations can be calculated by assuming a uniform radiative brightness temperature TB_w for water and a uniform radiative brightness temperature TB_I for ice, with both brightness temperatures being appropriate for the values received at the altitude of the satellite, that is, incorporating an atmospheric contribution. Assuming no other surface types within the field of view, the observed brightness temperature TB is given by

$$TB = C_w TB_w + C_I TB_I \qquad [1]$$

C_w is the percent areal coverage of water and C_I is the ice concentration. With only the two surface types, $C_w + C_I = 1$, and eqn [1] can be expressed as

$$TB = (1 - C_I)TB_w + C_I TB_I \qquad [2]$$

This is readily solved for the ice concentration:

$$C_I = \frac{TB - TB_w}{TB_I - TB_w} \qquad [3]$$

Equation [3] is the standard equation used for the calculation of ice concentrations from a single channel of microwave data, such as the data from the ESMR instrument. A major limitation of the formulation is that the polar ice cover is not uniform in its microwave emission, so that the assumption of a uniform TB_I for all sea ice is only a rough approximation, far less justified than the assumption of a uniform TB_w for seawater, although that also is an approximation.

Multichannel instruments allow more sophisticated, and generally more accurate, calculation of the ice concentrations. They additionally allow many options as to how these calculations can be done. To illustrate the options, two algorithms will be described, both of which assume two distinct ice types, thereby advancing over the assumption of a single ice type made in eqns [1]–[3] and being particularly appropriate for the many instances in which two ice types dominate the sea ice cover. For this approximation, the assumption is that the field of view contains only water and two ice types, type 1 ice and type 2 ice (see the section 'Sea ice types' for more information on ice types), and that the three surface types have identifiable brightness temperatures, TB_w, TB_{I1}, and TB_{I2}, respectively. Labeling the concentrations of the two ice types as C_{I1} and C_{I2}, respectively, the percent coverage of water is $1 - C_{I1} - C_{I2}$, and the integrated observed brightness temperature is

$$TB = (1 - C_{I1} - C_{I2})TB_w + C_{I1}TB_{I1} + C_{I2}TB_{I2} \quad [4]$$

With two channels of information, as long as appropriate values for TB_w, TB_{I1}, and TB_{I2} are known for each of the two channels, eqn [4] can be used individually for each channel, yielding two linear equations in the two unknowns C_{I1} and C_{I2}. These equations are immediately solvable for C_{I1} and C_{I2}, and the total ice concentration C_I is then given by

$$C_I = C_{I1} + C_{I2} \quad [5]$$

Although the scheme described in the preceding paragraph is a marked advance over the use of a single-channel calculation (eqn [3]), most algorithms for sea ice concentrations from multichannel data make use of additional channels and concepts to further improve the ice concentration accuracies. A frequently used algorithm (termed the NASA Team algorithm) for the SMMR data employs three of the 10 SMMR channels, those recording horizontally polarized radiation at a frequency of 18 GHz and vertically polarized radiation at frequencies of 18 and 37 GHz. The algorithm is based on both the polarization ratio (PR) between the 18-GHz vertically polarized data (abbreviated 18 V) and the 18-GHz horizontally polarized data (18 H) and the spectral gradient ratio (GR) between the 37-GHz vertically polarized data (37 V) and the 18-V data. PR and GR are defined as:

$$PR = \frac{TB(18\ V) - TB(18\ H)}{TB(18\ V) + TB(18\ H)} \quad [6]$$

$$GR = \frac{TB(37\ V) - TB(18\ V)}{TB(37\ V) + TB(18\ V)} \quad [7]$$

Substituting into eqns [6] and [7] expanded forms of $TB(18\ V)$, $TB(18\ H)$, and $TB(37\ V)$ obtained from eqn [4], the result yields equations for PR and GR in the two unknowns C_{I1} and C_{I2}. Solving for C_{I1} and C_{I2} yields two algebraically messy but computationally straightforward equations for C_{I1} and C_{I2} based on PR, GR, and numerical coefficients determined exclusively from the brightness temperature values assigned to water, type 1 ice, and type 2 ice for each of the three channels (these assigned values are termed 'tie points' and are determined empirically). These are the equations that are then used for the calculation of the concentrations of type 1 and type 2 ice once the observations are made and are used to calculate PR and GR from eqns [6] and [7]. The total ice concentration C_I is then obtained from eqn [5]. The use of PR and GR in this formulation reduces the impact of ice temperature variations on the ice

concentration calculations. This algorithm is complemented by a weather filter that sets to 0 all ice concentrations at locations and times with a GR value exceeding 0.07. The weather filter eliminates many of the erroneous calculations of sea ice presence arising from the influence of storm systems on the microwave data.

For the SSMI data, the same basic NASA Team algorithm is used, although 18 V and 18 H in eqns [6] and [7] are replaced by 19.4 V and 19.4 H, reflecting the placement on the SSMI of channels at a frequency of 19.4 GHz rather than the 18-GHz channels on the SMMR. Also, because the data from the 19.4-GHz channels tend to be more contaminated by water vapor absorption/emission and other weather effects than the 18-GHz data, the weather filter for the SSMI calculations incorporates a threshold level for the GR calculated from the 22.2-GHz vertically polarized data and 19.4-V data as well as a threshold for the GR calculated from the 37 V and 19.4-V data. To illustrate the results of this ice concentration algorithm, **Figure 2** presents the derived sea ice concentrations for 15 March 1998 in the Northern Hemisphere and for 15 September 1998 in the Southern Hemisphere, the same dates as used in **Figure 1**.

As mentioned, there are several alternative ice concentration algorithms in use. Contrasts from the NASA Team algorithm just described include: use of different microwave channels; use of regional tie points rather than hemispherically applicable tie points; use of cluster analysis on brightness temperature data, without PR and GR formulations; use of iterative techniques whereby an initial ice concentration calculation leads to refined atmospheric temperatures and opacities, which in turn lead to refined ice concentrations; use of iterative techniques involving surface temperature, atmospheric water vapor, cloud liquid water content, and wind speed; incorporation of higher-frequency data to reduce the effects of snow cover on the computations; and use of a Kalman filtering technique in conjunction with an ice model. The various techniques tend to obtain very close overall distributions of where the sea ice is located, although sometimes with noticeable differences (up to 20%, and on occasion even higher) in the individual, local ice concentrations. The differences can often be markedly reduced by adjustment of tunable parameters, such as the algorithm tie points, in one or both of the algorithms being compared. However, the lack of adequate ground data often makes it difficult to know which tuning is most appropriate or which algorithm is yielding the better results. To help resolve the uncertainties, *in situ* and aircraft measurements are being made in both the Arctic and the

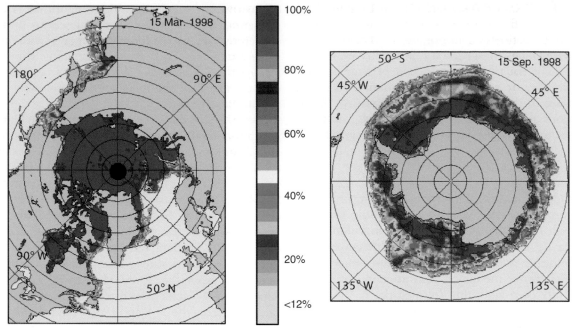

Figure 2 North and south polar sea ice concentration images for 15 Mar. 1998 and 15 Sep. 1998, respectively. The ice concentrations are derived from the data of the DMSP SSMI, including the 19.4-V data depicted in **Figure 1**.

Antarctic to help validate and improve the AMSR-E sea ice products.

Whichever ice concentration algorithm is used, the result provides estimates of ice coverage, not ice thickness. In cases where ice thickness data are also available, the combination of ice concentration and ice thickness allows the calculation of ice volume. Ice thickness data, however, are quite limited both spatially and over time. Furthermore, they are generally not derived from the passive-microwave observations and so are not highlighted in this article. The limited ice thickness data traditionally have come from *in situ* and submarine measurements, although some recent data are also available from satellite radar altimetry and, since the January 2003 launch of the Ice, Cloud and land Elevation Satellite (*ICE-Sat*), from satellite laser altimetry. The laser technique appears promising, although the laser on board ICESat operates only for short periods, and hence the full value of the technique will not be realized until a new laser is launched with a longer operational lifetime. Average sea ice thickness in the Antarctic is estimated to be in the range 0.4–1.5 m, and average sea ice thickness in the Arctic is estimated to be in the range 1.5–3.5 m.

Sea Ice Extents and Trends

Sea ice extent is defined as the total ocean area of all grid cells with sea ice concentration of at least 15% (or, occasionally, an alternative prescribed minimum percentage). Sea ice extents are now routinely calculated for the north polar region as a whole, for the south polar region as a whole, and for each of several subregions within the two polar domains, using ice concentration maps determined from satellite passive-microwave data.

A major early result from the use of satellite passive-microwave data was the detailed determination of the seasonal cycle of ice extents in each hemisphere. Incorporating the interannual variability observed from the 1970s through the early twenty-first century, the Southern Ocean ice extents vary from about $2\text{–}4 \times 10^6\,\text{km}^2$ in February to about $17\text{–}20 \times 10^6\,\text{km}^2$ in September, and the north polar ice extents vary from about $5\text{–}8 \times 10^6\,\text{km}^2$ in September to about $14\text{–}16 \times 10^6\,\text{km}^2$ in March. The exact timing of minimum and maximum ice coverage and the smoothness of the growth from minimum to maximum and the decay from maximum to minimum vary noticeably among the different years.

As the data sets lengthened, a major goal became the determination of trends in the ice extents and the placement of these trends in the context of other climate variables and climate change. Because of the lack of a period of data overlap between the ESMR and the SMMR, matching of the ice extents derived from the ESMR data to those derived from the SMMR and SSMI data has been difficult and uncertain. Consequently, most published results regarding trends found from the SMMR and SSMI data sets do not include the ESMR data.

The SMMR/SSMI record from late 1978 until the early twenty-first century indicates an overall decrease in Arctic sea ice extents of about 3% per decade and an overall increase in Antarctic sea ice extents of about 1% per decade. The Arctic ice decreases have received particular attention because they coincide with a marked warming in the Arctic and likely are tied closely to that warming. Although the satellite data reveal significant interannual variability in the ice cover, even as early as the late 1980s it had become clear from the satellite record that the Arctic as a whole had lost ice since the late 1970s. The picture was mixed regionally, with some regions having lost ice and others having gained ice; and with such a short data record, there was a strong possibility that the decreases through the 1980s were part

of an oscillatory pattern and would soon reverse. However, although the picture remained complicated by interannual variability and regional differences, the Arctic decreases overall continued (with fluctuations) through the 1990s and into the twenty-first century, and by the middle of the first decade of the twenty-first century no large-scale regions of the Arctic showed overall increases since late 1978. Moreover, by the early twenty-first century the Arctic sea ice decreases were apparent in all seasons of the year, and the decreases in ice extent found from satellite data were complemented by decreases in ice thickness found from submarine and *in situ* data.

The satellite-derived Arctic sea ice decreases are illustrated in **Figure 3** with 26-year March and September time series for the Northern Hemisphere sea

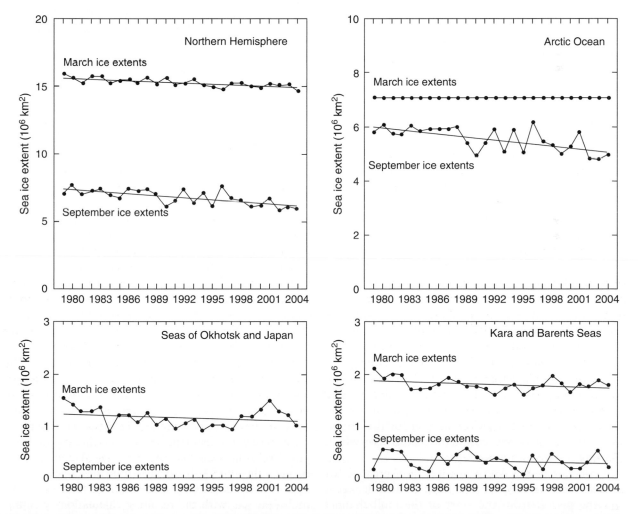

Figure 3 Time series of monthly average 1979–2004 March and September sea ice extents for the Northern Hemisphere and the following three regions within the Northern Hemisphere: the Arctic Ocean, the Seas of Okhotsk and Japan, and the Kara and Barents Seas. All ice extents are derived from the *Nimbus 7* SMMR and DMSP SSMI satellite passive-microwave data. The trend lines are linear least squares fits through the data points, and the slopes of the trend lines for the Northern Hemisphere total are $-29\,500 \pm 5400\,\mathrm{km^2\,yr^{-1}}$ ($-1.9 \pm 0.35\%$ per decade) for the March values and $-51\,300 \pm 10\,800\,\mathrm{km^2\,yr^{-1}}$ ($6.9 \pm 1.5\%$ per decade) for the September values. For the Seas of Okhotsk and Japan, all the September ice extents are $0\,\mathrm{km^2}$, as the ice cover fully disappeared from these seas by the end of summer in each of the 26 years.

ice cover as a whole and for three regions within the Northern Hemisphere ice cover, those being the Arctic Ocean, the Seas of Okhotsk and Japan, and the Kara and Barents Seas. March and September are typically the months of maximum and minimum Northern Hemisphere sea ice coverage. Among the regional and seasonal differences visible in the plots, the Arctic Ocean shows little or no variation in March ice extents because of consistently being fully covered with ice in March, to at least 15% ice coverage in each grid square, but shows noticeable fluctuations and overall decreases in September ice coverage. In contrast, the region of the Seas of Okhotsk and Japan has no variability in September, because of having no ice in any of the Septembers, but for March shows marked fluctuations and slight (not statistically significant) overall decreases, interrupted by prominent increases from 1994 to 2001. The Kara and Barents Seas exhibit significant variability in both the March and September values, although with slight (not statistically significant) overall decreases for both months (**Figure 3**). The March and September ice extent decreases for the Northern Hemisphere as a whole are both statistically significant at the 99% level, as are the September decreases for the Arctic Ocean region.

The lack of uniformity in the Arctic sea ice losses (e.g., **Figure 3**) complicates making projections into the future. Nonetheless, in the early twenty-first century, several scientists have offered projections in light of the expectation of continued warming of the climate system. These projections – some based on extrapolation from the data, others on computer modeling – suggest continued Arctic sea ice decreases, with the effects of warming dominating over oscillatory behavior and other fluctuations. Some studies project a totally ice-free late summer Arctic Ocean by the end of the century, the middle of the century, or even, in one projection, by as early as the year 2013. An ice-free Arctic would greatly ease shipping but would also have multiple effects on the Arctic climate (by seasonally eliminating the highly reflective and insulating ice cover) and on Arctic ecosystems (by seasonally removing both the habitat for the organisms that live within the ice and the platform on which a variety of polar animals depend).

In contrast to the Arctic sea ice, the Antarctic sea ice cover as a whole does not show a warming signal over the period from the start of the multichannel satellite passive-microwave record, in late 1978, through the end of 2004. Over this period, some regions in the Antarctic experienced overall ice cover increases and other regions experienced overall ice cover decreases, with the hemispheric 1% per decade ice extent increases incorporating contrasting regional conditions. In **Figure 4**, the 26-year February (generally the month of Antarctic sea ice minimum) and September (generally the month of Antarctic sea ice maximum) ice extent time series are plotted for the Southern Hemisphere total, the Weddell Sea, the Bellingshausen and Amundsen Seas, and the Ross Sea. The region of the Bellingshausen and Amundsen Seas shows statistically significant (99% confidence level) February sea ice decreases but shows very slight (not statistically significant) September sea ice increases. The Weddell Sea instead shows statistically significant (95% confidence level) increases in February ice coverage and slight (not statistically significant) decreases in September ice coverage, while the Ross Sea shows ice increases in both months, with the February increases being statistically significant at the 99% confidence level (**Figure 4**). In line with the mixed pattern of ice extent increases and decreases, the Antarctic has also experienced a mixed pattern of temperature increases and decreases, with the Antarctic Peninsula (separating the Bellingshausen and Weddell seas) being the one region of the Antarctic with a prominent warming signal.

Sea Ice Types

The sea ice covers in both polar regions are mixtures of various types of ice, ranging from individual ice crystals to coherent ice floes several kilometers across. Common ice types include frazil ice (fine spicules of ice suspended in water); grease ice (a soupy layer of ice that looks like grease from a distance); slush ice (a viscous mass formed from a mixture of water and snow); nilas (a thin elastic sheet of ice 0.01–0.1-m thick); pancake ice (small, roughly circular pieces of ice, 0.3–3 m across and up to 0.1-m thick); first-year ice (ice at least 0.3-m thick that has not yet undergone a summer melt period); and multiyear ice (ice that has survived a summer melt).

Because different ice types have different microwave emission characteristics, once these differences are understood, appropriate satellite passive-microwave data can be used to distinguish ice types. The ice types most frequently distinguished with such data are first-year ice and multiyear ice in the Arctic Ocean. In fact, the NASA Team algorithm described earlier was initially developed specifically for first-year and multiyear ice, with the resulting calculations yielding the concentrations, C_{I1} and C_{I2}, of those two ice types. First-year and multiyear ice are distinguishable in their microwave signals because the summer melt process drains some of the salt content downward through the ice, reducing the salinity of the upper layers of the ice and thereby changing the microwave emissions; these

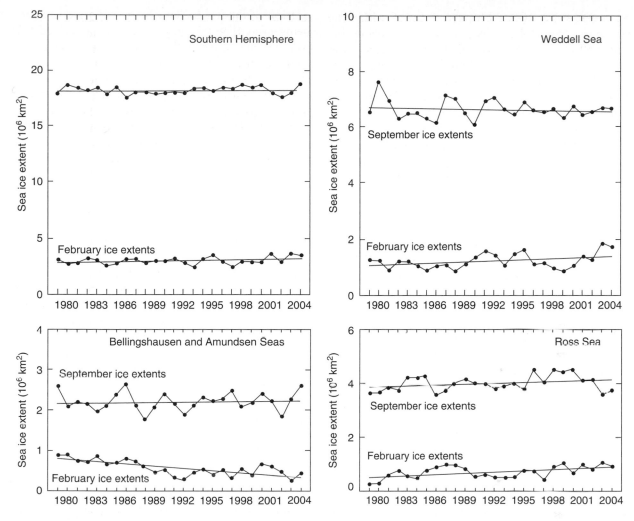

Figure 4 Time series of monthly average 1979–2004 February and September sea ice extents for the Southern Hemisphere and the following three regions within the Southern Hemisphere: the Weddell Sea, the Bellingshausen and Amundsen Seas, and the Ross Sea. All ice extents are derived from the *Nimbus 7* SMMR and DMSP SSMI satellite passive-microwave data. The trend lines are linear least squares fits through the data points, and the slopes of the trend lines for the Southern Hemisphere total are $14\,100 \pm 8000\,\text{km}^2\,\text{yr}^{-1}$ ($5.0 \pm 2.8\%$ per decade) for the February values and $7000 \pm 8200\,\text{km}^2\,\text{yr}^{-1}$ ($0.4 \pm 0.5\%$ per decade) for the September values.

changes are dependent on the frequency and polarization of the radiation. To illustrate the differences, **Figure 5** presents a plot of the tie points employed in the NASA Team algorithm for the Arctic ice. Tie points were determined empirically and are included for each of the three SSMI channels used in the calculation of ice concentrations prior to the application of the weather filter. The plot shows that while the transition from first-year to multiyear ice lowers the brightness temperatures for each of the three channels, the reduction is greatest for the 37-V data and least for the 19.4-V data. The plot further reveals that the polarization PR (eqn [6], revised for 19.4 GHz rather than 18-GHz data) is larger for multiyear ice than for first-year ice, and larger for

water than for either ice type. Furthermore, the GR (eqn [7], revised for 19.4-GHz data) is positive for water, slightly negative for first-year ice, and considerably more negative for multiyear ice (**Figure 5**). The differences allow the sorting out, either through the calculation of C_{I1} and C_{I2} as described earlier or through alternative algorithms, of the first-year ice and multiyear ice percentages in the satellite field of view.

Other Sea Ice Variables: Season Length, Temperature, Melt, Velocity

Although ice concentrations, ice extents, and, to a lesser degree, ice types have been the sea ice variables

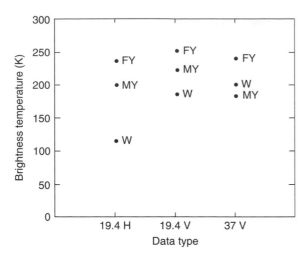

Figure 5 Typical brightness temperatures of first-year ice (FY), multiyear ice (MY), and liquid water (W) at three channels of SSMI data from the DMSP F13 satellite, specifically those for 19.4-GHz horizontally polarized data (19.4 H), 19.4-GHz vertically polarized data (19.4 V), and 37-GHz vertically polarized data (37 V). These are the values used as tie points for the Arctic calculations in the NASA Team algorithm described in the text. Data from Cavalieri DJ, Parkinson CL, Gloersen P, Comiso JC, and Zwally HJ (1999) Deriving long-term time series of sea ice cover from satellite passive-microwave multisensor data sets. *Journal of Geophysical Research* 104(C7): 15803–15814.

most widely calculated and used from satellite passive-microwave data, several additional variables have also been obtained from these data, including the length of the sea ice season, sea ice temperature, sea ice melt, and sea ice velocity. The length of the sea ice season for any particular year is calculated directly from that year's daily maps of sea ice concentrations, by counting, at each grid element, the number of days with ice coverage of at least some predetermined (generally 15% or 30%) ice concentration. Trends in the length of the sea ice season from the late 1970s to the early twenty-first century show coherent spatial patterns in both hemispheres, with a predominance of negative values (shortening of the sea ice season) in the Northern Hemisphere and in the vicinity of the Antarctic Peninsula and a much lesser predominance of positive values in the rest of the Southern Hemisphere's sea ice region, consistent with the respective hemispheric trends in sea ice extents.

The passive-microwave-based ice temperature calculations generally depend on the calculated sea ice concentrations, empirically determined ice emissivities, a weighting of the water and ice temperatures within the field of view, and varying levels of sophistication in incorporating effects of the polar atmosphere and the presence of multiple ice types. The derived temperature is not the surface temperature but the temperature of the radiating

portion of the ice, for whichever radiative frequency being used. The ice temperature fields derived from passive-microwave data complement those derived from satellite infrared data, which have the advantages of generally having finer spatial resolution and of more nearly approaching surface temperatures but have the disadvantage of greater contamination by clouds. The passive-microwave and infrared data are occasionally used together, iteratively, for an enhanced ice temperature product.

The seasonal melting of portions of the sea ice and its overlying snow cover generally produces marked changes in microwave emissions, first increasing the emissions as liquid water emerges in the snow, then decreasing the emissions once the snow has melted and meltwater ponds cover the ice. Because of the emission changes, these events on the ice surface frequently become detectable through time series of the satellite passive-microwave data. The onset of melt in particular can often be identified, and hence yearly maps can be created of the dates of melt onset. Melt ponds, however, present greater difficulties, as they can have similar microwave emissions to those of the water in open spaces between ice floes, so that a field of heavily melt-ponded ice can easily be confused in the microwave data with a field of low-concentration ice. The ambiguities can be reduced through analysis of the passive-microwave time series and comparisons with active-microwave, visible, and infrared data. Still, because of these complications under melt conditions, the passive-microwave-derived ice concentrations tend to have larger uncertainties for summertime ice than for wintertime ice.

The calculation of sea ice velocities from satellite data has in general relied upon data with fine enough resolution to distinguish individual medium-sized ice floes, such as visible data and active-microwave data rather than the much coarser resolution passive-microwave data. However, in the 1990s, several groups devised methods of determining ice velocity fields from passive-microwave data, some using techniques based on cross-correlation of brightness temperature fields and others using wavelet analysis. These techniques have yielded ice velocity maps on individual dates for the entire north and south polar sea ice covers. Comparisons with buoy and other data have been quite encouraging regarding the potential of using the passive-microwave satellite data for long-term records and monitoring of ice motions.

Looking toward the Future

Monitoring of the polar sea ice covers through satellite passive-microwave technology is ongoing with the operational SSMI instruments on the DMSP satellites and the Japanese AMSR-E instrument on NASA's Aqua satellite. Both Japan and the United States anticipate launching additional passive-microwave instruments to maintain an uninterrupted satellite passive-microwave data record. The resulting lengthening sea ice records should continue to provide an improved basis with which scientists can examine trends in the sea ice cover and interactions between the sea ice and other elements of the climate system. For instance, the lengthened records will be essential to answering many of the questions raised concerning whether the negative overall trends found in Arctic sea ice extents for the first quarter century of the SMMR/SSMI record will continue and how these trends relate to temperature trends, in particular to climate warming, and to oscillations within the climate system, in particular the North Atlantic Oscillation, the Arctic Oscillation, and the Southern Oscillation. In addition to covering a longer period, other expected improvements in the satellite passive-microwave record of sea ice include further algorithm developments, following additional analyses of the microwave properties of sea ice and liquid water. Such analyses are likely to lead both to improved algorithms for the variables already examined and to the development of techniques for calculating additional sea ice variables from the satellite data.

Glossary

Brightness temperature Unit used to express the intensity of emitted microwave radiation received by the satellite, presented in temperature units (K) following the Rayleigh–Jeans approximation to Planck's law, whereby the radiation emitted from a perfect emitter at microwave wavelengths is proportional to the emitter's physical temperature.

Sea ice concentration Percent areal coverage of sea ice.

Sea ice extent Integrated area of all grid elements with sea ice concentration of at least 15%.

See also

Abrupt Climate Change. Antarctic Circumpolar Current. Arctic Ocean Circulation. Ice–Ocean Interaction. Millennial-Scale Climate Variability. North Atlantic Oscillation (NAO). Satellite Oceanography, History and Introductory Concepts. Sea Ice. Sea Ice: Overview.

Further Reading

Barry RG, Maslanik J, Steffen K, et al. (1993) Advances in sea-ice research based on remotely sensed passive microwave data. *Oceanography* 6(1): 4–12.

Carsey FD (ed.) (1992) *Microwave Remote Sensing of Sea Ice*. Washington, DC: American Geophysical Union.

Cavalieri DJ, Parkinson CL, Gloersen P, Comiso JC, and Zwally HJ (1999) Deriving long-term time series of sea ice cover from satellite passive-microwave multisensor data sets. *Journal of Geophysical Research* 104(C7): 15803–15814.

Comiso JC, Yang J, Honjo S, and Krishfield RA (2003) Detection of change in the Arctic using satellite and *in situ* data. *Journal of Geophysical Research* 108(C12): 3384 (doi:10.1029/2002JC001347).

Gloersen P, Campbell WJ, Cavalieri DJ, et al. (1992) *Arctic and Antarctic Sea Ice, 1978–1987: Satellite Passive-Microwave Observations and Analysis*. Washington, DC: National Aeronautics and Space Administration.

Jeffries MO (ed.) (1998) *Antarctic Sea Ice: Physical Processes, Interactions and Variability*. Washington, DC: American Geophysical Union.

Johannessen OM, Bengtsson L, Miles MW, et al. (2004) Arctic climate change: Observed and modelled temperature and sea-ice variability. *Tellus* 56A(4): 328–341.

Kramer HJ (2002) *Observation of the Earth and Its Environment*, 4th edn. Berlin: Springer.

Lubin D and Massom R (2006) *Polar Remote Sensing, Vol. 1: Atmosphere and Oceans*. Berlin: Springer-Praxis.

Parkinson CL (1997) *Earth from Above: Using Color-Coded Satellite Images to Examine the Global Environment*. Sausalito, CA: University Science Books.

Parkinson CL (2004) Southern Ocean sea ice and its wider linkages: Insights revealed from models and observations. *Antarctic Science* 16(4): 387–400.

Smith WO, Jr. and Grebmeier JM (eds.) (1995) *Arctic Oceanography: Marginal Ice Zones and Continental Shelves*. Washington, DC: American Geophysical Union.

Thomas DN and Dieckmann GS (eds.) (2003) *Sea Ice: An Introduction to Its Physics, Chemistry, Biology and Geology*. Oxford, UK: Blackwell Science.

Ulaby FT, Moore RK, and Fung AK (eds.) (1986) *Monitoring sea ice*. In: *Microwave Remote Sensing: Active and Passive, Vol. III: From Theory to Applications*, pp. 1478–1521. Dedham, MA: Artech House.

Walsh JE, Anisimov O, Hagen JOM, et al. (2005) Cryosphere and hydrology. In: Symon C, Arris L, and Heal B (eds.) *Arctic Climate Impact Assessment*, pp. 183–242. Cambridge, UK: Cambridge University Press.

Relevant Websites

http://www.awi-bremerhaven.de
 – Alfred Wegener Institute for Polar and Marine Research.

http://www.antarctica.ac.uk
  – Antarctica, British Antarctic Survey.
http://www.arctic.noaa.gov
  – Arctic Change, NOAA Arctic Theme Page.
http://www.aad.gov.au
  – Australian Antarctic Division.
http://www.dcrs.dtu.dk
  – Danish Center for Remote Sensing.
http://www.jaxa.jp
  – Japan Aerospace Exploration Agency.

http://www.spri.cam.ac.uk
  – Scott Polar Research Institute.
http://www.nasa.gov
  – US National Aeronautics and Space Administration.
http://www.natice.noaa.gov
  – US National Ice Center.
http://nsidc.org
  – US National Snow and Ice Data Center.

SATELLITE REMOTE SENSING OF SEA SURFACE TEMPERATURES

P. J. Minnett, University of Miami, Miami, FL, USA

Introduction

The ocean surface is the interface between the two dominant, fluid components of the Earth's climate system: the oceans and atmosphere. The heat moved around the planet by the oceans and atmosphere helps make much of the Earth's surface habitable, and the interactions between the two, that take place through the interface, are important in shaping the climate system. The exchange between the ocean and atmosphere of heat, moisture, and gases (such as CO_2) are determined, at least in part, by the sea surface temperature (SST). Unlike many other critical variables of the climate system, such as cloud cover, temperature is a well-defined physical variable that can be measured with relative ease. It can also be measured to useful accuracy by instruments on observation satellites.

The major advantage of satellite remote sensing of SST is the high-resolution global coverage provided by a single sensor, or suite of sensors on similar satellites, that produces a consistent data set. By the use of onboard calibration, the accuracy of the time-series of measurements can be maintained over years, even decades, to provide data sets of relevance to research into the global climate system. The rapid processing of satellite data permits the use of the global-scale SST fields in applications where the immediacy of the data is of prime importance, such as weather forecasting – particularly the prediction of the intensification of tropical storms and hurricanes.

Measurement Principle

The determination of the SST from space is based on measuring the thermal emission of electromagnetic radiation from the sea surface. The instruments, called radiometers, determine the radiant energy flux, B_λ, within distinct intervals of the electromagnetic spectrum. From these the brightness temperature (the temperature of a perfectly emitting 'black-body' source that would emit the same radiant flux) can be calculated by the Planck equation:

$$B_\lambda(T) = 2hc^2\lambda^{-5}\left(e^{hc/(\lambda kT)} - 1\right)^{-1} \quad [1]$$

where h is Planck's constant, c is the speed of light in a vacuum, k is Boltzmann's constant, λ is the wavelength and T is the temperature. The spectral intervals (wavelengths) are chosen where three conditions are met: (1) the sea emits a measurable amount of radiant energy, (2) the atmosphere is sufficiently transparent to allow the energy to propagate to the spacecraft, and (3) current technology exists to build radiometers that can measure the energy to the required level of accuracy within the bounds of size, weight, and power consumption imposed by the spacecraft. In reality these constrain the instruments to two relatively narrow regions of the infrared part of the spectrum and to low-frequency microwaves. The infrared regions, the so-called atmospheric windows, are situated between wavelengths of $3.5-4.1\,\mu m$ and $8-12\,\mu m$ (**Figure 1**); the microwave measurements are made at frequencies of 6–12 GHz.

As the electromagnetic radiation propagates through the atmosphere, some of it is absorbed and scattered out of the field of view of the radiometer, thereby attenuating the original signal. If the attenuation is sufficiently strong none of the radiation from the sea reaches the height of the satellite, and such is the case when clouds are present in the field of view of infrared radiometers. Even in clear-sky conditions a significant fraction of the sea surface emission is absorbed in the infrared windows. This energy is re-emitted, but at a temperature characteristic of that height in the atmosphere. Consequently the brightness temperatures measured through the clear atmosphere by a spacecraft radiometer are cooler than would be measured by a similar device just above the surface. This atmospheric effect, frequently referred to as the temperature deficit, must be corrected accurately if the derived sea surface temperatures are to be used quantitatively.

Infrared Atmospheric Correction Algorithms

The peak of the Planck function for temperatures typical of the sea surface is close to the longer wavelength infrared window, which is therefore well suited to SST measurement (**Figure 1**). However, the

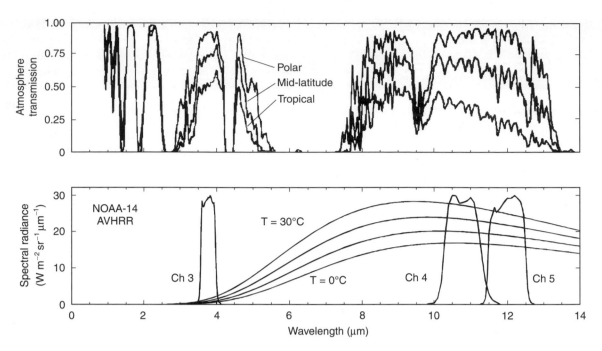

Figure 1 Spectra of atmospheric transmission in the infrared (wavelengths 1–14 μm) calculated for three typical atmospheres from diverse parts of the ocean; polar, mid-latitude and tropical with integrated water vapor content of 7 kg m^{-2} (polar), 29 kg m^{-2} (mid-latitude) and 54 kg m^{-2} (tropical). Regions where the transmission is high are well suited to satellite remote sensing of SST. The lower panel shows the electromagnetic radiative flux for four sea surface temperatures (0, 10, 20, and 30°C) with the relative spectral response functions for channels 3, 4, and 5 of the AVHRR on the NOAA-14 satellite. The so-called 'split-window' channels, 4 and 5, are situated where the sea surface emission is high, and where the atmosphere is comparatively clear but exhibits a strong dependence on atmospheric water vapor content.

main atmospheric constituent in this spectral interval that contributes to the temperature deficit is water vapor, which is very variable both in space and time. Other molecular species that contribute to the temperature deficit are quite well mixed throughout the atmosphere, and therefore inflict a relatively constant temperature deficit that is simple to correct.

The variability of water vapor requires an atmospheric correction algorithm based on the information contained in the measurements themselves. This is achieved by making measurements at distinct spectral intervals in the windows when the water vapor attenuation is different. These spectral intervals are defined by the characteristics of the radiometer and are usually referred to as bands or channels (**Figure 1**). By invoking the hypothesis that the difference in the brightness temperatures measured in two channels, i and j, is related to the temperature deficit in one of them, the atmospheric correction algorithm can be formulated thus:

$$SST_{ij} - T_i = f(T_i - T_j) \qquad [2]$$

where SST_{ij} is the derived SST and T_i, T_j are the brightness temperatures in channels i, j.

Further, by assuming that the atmospheric attenuation is small in these channels, so that the radiative transfer can be linearized, and that the channels are spectrally close so that Planck's function can be linearized, the algorithm can be expressed in the very simple form:

$$SST_{ij} = a_o + a_i T_i + a_j T_j \qquad [3]$$

where are a_o, a_i, and a_j are coefficients. These are determined by regression analysis of either coincident satellite and *in situ* measurements, mainly from buoys, or of simulated satellite measurements derived by radiative transfer modeling of the propagation of the infrared radiation from the sea surface through a representative set of atmospheric profiles.

The simple algorithm has been applied for many years in the operational derivation of the sea surface from measurements of the Advanced Very High Resolution Radiometer (AVHRR, see below), the product of which is called the multi-channel SST (MCSST), where i refers to channel 4 and j to channel 5.

More complex forms of the algorithms have been developed to compensate for some of the short-comings of the linearization. One such widely applied algorithm takes the form:

$$SST_{ij} = b_o + b_1 T_i + b_2(T_i - T_j)SST_r \\ + b_3(T_i - T_j)(\sec\theta - 1) \qquad [4]$$

where SST_r is a reference SST (or first-guess temperature), and θ is the zenith angle to the satellite radiometer measured at the sea surface. When applied to AVHRR data, with i and j referring to channels 4 and 5 derived SST is called the nonlinear SST (NLSST). A refinement is called the Pathfinder SST (PFSST) in the research program designed to post-process AVHRR data over a decade or so to provide a consistent data set for climate research. In the PFSST, the coefficients are derived on a monthly basis for two different atmospheric regimes, distinguished by the value of the T_4–T_5 differences being above or below 0.7 K, by comparison with measurements from buoys.

The atmospheric correction algorithms work effectively only in the clear atmosphere. The presence of clouds in the field of view of the infrared radiometer contaminates the measurement so that such pixels must be identified and removed from the SST retrieval process. It is not necessary for the entire pixel to be obscured, even a portion as small as 3–5%, dependent on cloud type and height, can produce unacceptable errors in the SST measurement. Thin, semi-transparent cloud, such as cirrus, can have similar effects to subpixel geometric obscuration by optically thick clouds. Consequently, great attention must be paid in the SST derivation to the identification of measurements contaminated by even a small amount of clouds. This is the principle disadvantage to SST measurement by spaceborne infrared radiometry. Since there are large areas of cloud cover over the open ocean, it may be necessary to composite the cloud-free parts of many images to obtain a complete picture of the SST over an ocean basin.

Similarly, aerosols in the atmosphere can introduce significant errors in SST measurement. Volcanic aerosols injected into the cold stratosphere by violent eruptions produce unmistakable signals that can bias the SST too cold by several degrees. A more insidious problem is caused by less readily identified aerosols at lower, warmer levels of the atmosphere that can introduce systematic errors of a much smaller amplitude.

Microwave Measurements

Microwave radiometers use a similar measurement principle to infrared radiometers, having several spectral channels to provide the information to correct for extraneous effects, and black-body calibration targets to ensure the accuracy of the measurements. The suite of channels is selected to include sensitivity to the parameters interfering with the SST measurements, such as cloud droplets and surface wind speed, which occurs with microwaves at higher frequencies. A simple combination of the brightness temperature, such as eqn [2], can retrieve the SST.

The relative merits of infrared and microwave radiometers for measuring SST are summarized in **Table 1**.

Characteristics of Satellite-derived SST

Because of the very limited penetration depth of infrared and microwave electromagnetic radiation in sea water the temperature measurements are limited to the sea surface. Indeed, the penetration depth is typically less than 1 mm in the infrared, so that temperature derived from infrared measurements is characteristic of the so-called skin of the ocean. The skin temperature is generally several tenths of a degree cooler than the temperature measured just below, as a result of heat loss from the ocean to atmosphere. On days of high insolation and low wind speed, the absorption of sunlight causes an increase in near surface temperature so that the water just below the skin layer is up to a few degrees warmer than that measured a few meters deeper, beyond the influence of the diurnal heating. For those people interested in a temperature characteristic of a depth of a few meters or more, the decoupling of the skin and deeper, bulk temperatures is perceived as a disadvantage of using satellite SST. However, algorithms generated by comparisons between satellite

Table 1 Relative merits of infrared and microwave radiometers for sea surface temperature measurement

Infrared	Microwave
Good spatial resolution (~1 km)	Poor spatial resolution (~50 km)
Surface obscured by clouds	Clouds largely transparent, but measurement perturbed by heavy rain
No side-lobe contamination	Side-lobe contamination prevents measurements close to coasts or ice
Aperture is reasonably small; instrument can be compact for spacecraft use	Antenna is large to achieve spatial resolution from polar orbit heights (~800 km above the sea surface)
4 km resolution possible from geosynchronous orbit; can provide rapid sampling data	Distance to geosynchronous orbit too large to permit useful spatial resolution with current antenna sizes

Table 2 Spectral characteristics of current and planned satellite-borne infrared radiometers

AVHRR		ATSR		MODIS		OCTS		GLI	
λ (μm)	NEΔT (K)	λ (μm)	NEΔT (K)	λ (μm)	NEΔT (K)	λ (μm)	NEΔT (K)	λ (μm)	NEΔT (K)
3.75	0.12	3.7	0.019	3.75	0.05	3.7	0.15	3.715	<0.15
				3.96	0.05				
				4.05	0.05				
				8.55	0.05	8.52	0.15	8.3	<0.1
10.5	0.12	10.8	0.028	11.03	0.04	10.8	0.15	10.8	<0.1
11.5	0.12	12.0	0.025	12.02	0.04	11.9	0.15	12	<0.1

and *in situ* measurements from buoys include a mean skin effect masquerading as part of the atmospheric effect, and so the application of these results in an estimate of bulk temperatures.

The greatest advantage offered by satellite remote sensing is, of course, coverage. A single, broad-swath, imaging radiometer on a polar-orbiting satellite can provide global coverage twice per day. An imaging radiometer on a geosynchronous satellite can sample much more frequently, once per half-hour for the Earth's disk, or smaller segments every few minutes, but the spatial extent of the data is limited to that part of the globe visible from the satellite.

The satellite measurements of SST are also reasonably accurate. Current estimates for routine measurements show absolute accuracies of ± 0.3 to ± 0.5 K when compared to similar measurements from ships, aircraft, and buoys.

Spacecraft Instruments

All successful instruments have several attributes in common: a mechanism for scanning the Earth's surface to generate imagery, good detectors, and a mechanism for real-time, in-flight calibration. Calibration involves the use of one or more black-body calibration targets, the temperatures of which are accurately measured and telemetered along with the imagery. If only one black-body is available a measurement of cold dark space provides the equivalent of a very cold calibration target. Two calibration points are needed to provide in-flight calibration; nonlinear behavior of the detectors is accounted for by means of pre-launch instrument characterization measurements.

The detectors themselves inject noise into the data stream, at a level that is strongly dependent on their temperature. Therefore, infrared radiometers require cooled detectors, typically operating from 78 K ($-195°C$) to 105 K ($-168°C$) to reduce the noise

equivalent temperature difference (NEΔT) to the levels shown in **Table 2**.

The Advanced Very High Resolution Radiometer (AVHRR)

The satellite instrument that has contributed the most to the study of the temperature of the ocean surface is the AVHRR that first flew on TIROS-N launched in late 1978. AVHRRs have flown on successive satellites of the NOAA series from NOAA-6 to NOAA-14, with generally two operational at any given time. The NOAA satellites are in a near-polar, sun-synchronous orbit at a height of about 780 km above the Earth's surface and with an orbital period of about 100 min. The overpass times of the two NOAA satellites are about 2.30 a.m. and p.m. and about 7.30 a.m. and p.m. local time. The AVHRR has five channels: 1 and 2 at ~ 0.65 and $\sim 0.85 \mu m$ are responsive to reflected sunlight and are used to detect clouds and identify coastlines in the images from the daytime part of each orbit. Channels 4 and 5 (**Table 2** and **Figure 1**) are in the atmospheric window close to the peak of the thermal emission from the sea surface and are used primarily for the measurement of sea surface temperature. Channel 3, positioned at the shorter wavelength atmospheric window, is responsive to both surface emission and reflected sunlight. During the nighttime part of each orbit, measurements of channel 3 brightness temperatures can be used with those from channels 4 and 5 in variants of the atmospheric correction algorithm to determine SST. The presence of reflected sunlight during the daytime part of the orbit prevents much of these data from being used for SST measurement. Because of the tilting of the sea surface by waves, the area contaminated by reflected sunlight (sun glitter) can be quite extensive, and is dependent on the local surface wind speed. It is limited to the point of specular reflection only in very calm seas.

The images in each channel are constructed by scanning the field of view of the AVHRR across the

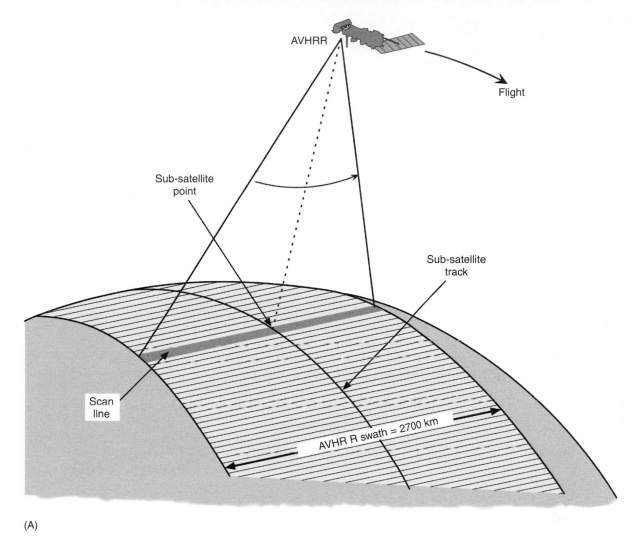

(A)

Figure 2 Scan geometries of AVHRR (A) and ATSR (B). The continuous wide swath of the AVHRR is constructed by linear scan lines aligned across the direction of motion of the subsatellite point. The swaths of the ATSR are generated by an inclined conical scan, which covers the same swath through two different atmospheric path lengths. The swath is limited to 512 km by geometrical constraints. Both radiometers are on sun-synchronous, polar-orbiting satellites.

Earth's surface by a mirror inclined at 45° to the direction of flight (**Figure 2A**). The rate of rotation, 6.67 Hz, is such that successive scan lines are contiguous at the surface directly below the satellite. The width of the swath (~2700 km) means that the swaths from successive orbits overlap so that the whole Earth's surface is covered without gaps each day.

The Along-Track Scanning Radiometer (ATSR)

An alternative approach to correcting the effects of the intervening atmosphere is to make a brightness temperature measurement of the same area of sea surface through two different atmospheric path lengths. The pairs of such measurements must be made in quick succession, so that the SST and atmospheric conditions do not change in the time interval. This approach is that used by the ATSR, two of which have flown on the European satellites ERS-1 and ERS-2.

The ATSR has infrared channels in the atmospheric windows comparable to those of AVHRR, but the rotating scan mirror sweeps out a cone inclined from the vertical by its half-angle (**Figure 2B**). The field of view of the ATSR sweeps out a curved path on the sea surface, beginning at the point directly below the satellite, moving out sideways and forwards. Half a mirror revolution later, the field of view is about 900 km ahead of the sub-satellite track in the center of the 'forward view'. The path of the field of view returns to the sub-satellite point, which,

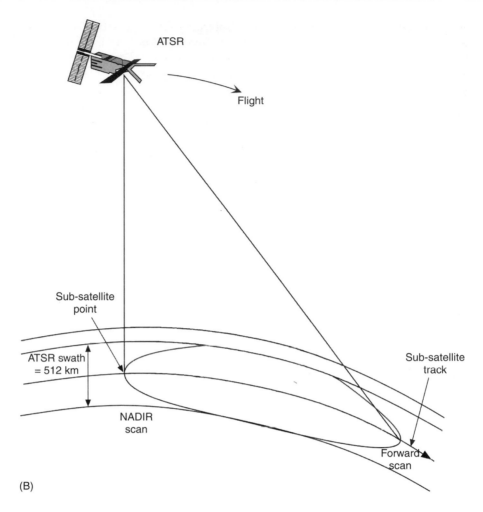

ATSR

Flight

Sub-satellite point

ATSR swath = 512 km

Sub-satellite track

NADIR scan

Forward scan

(B)

Figure 2 continued

during the period of the mirror rotation, has moved 1 km ahead of the starting point. Thus the pixels forming the successive swaths through the nadir point are contiguous. The orbital motion of the satellite means that the nadir point overlays the center of the forward view after about 2 min. The atmospheric path length of the measurement at nadir is simply the thickness of the atmosphere, whereas the slant path to the center of the forward view is almost double that, resulting in colder brightness temperatures. The differences in the brightness temperatures between the forward and nadir swaths are a direct measurement of the effect of the atmosphere and permit a more accurate determination of the sea surface temperature. The atmospheric correction algorithm takes the form:

$$SST = c_o + \sum_i c_{n,i} T_{n,i} + \sum_i c_{f,i} T_{f,i} \qquad [5]$$

where the subscripts n and f refer to measurements from the nadir and forward views, i indicates two or three atmospheric window channels and the set of c are coefficients. The coefficients, derived by radiative transfer simulations, have an explicit latitudinal dependence.

Accurate calibration of the brightness temperatures is achieved by using two onboard black-body cavities, situated between the apertures for the nadir and forward views such that they are scanned each rotation of the mirror. One calibration target is at the spacecraft ambient temperature while the other is heated, so that the measured brightness temperatures of the sea surface are straddled by the calibration temperatures.

The limitation of the simple scanning geometry of the ATSR is a relatively narrow swath width of 512 km. The ERS satellites have at various times in their missions been placed in orbits with repeat patterns of 3, 35, and 168 days, and given the narrow ATSR swath, complete coverage of the globe has been possible only for the 35 and 186 day cycles. This disadvantage is offset by the intended improvement in absolute accuracy of the atmospheric correction, and of its better insensitivity to aerosol effects.

The Moderate Resolution Imaging Spectroradiometer (MODIS)

The MODIS is a 36-band imaging radiometer on the NASA Earth Observing System (EOS) satellites *Terra*, launched in December 1999, and *Aqua*, planned for launch by late 2001. MODIS is much more complex than other radiometers used for SST measurement, but uses the same atmospheric windows. In addition to the usual two bands in the 10–12 μm interval, MODIS has three narrow bands in the 3.7–4.1 μm windows, which, although limited by sun-glitter effects during the day, hold the potential for much more accurate measurement of SST during the night. Several of the other 31 bands of MODIS contribute to the SST measurement by better identification of residual cloud and aerosol contamination.

The swath width of MODIS, at 2330 km, is somewhat narrower than that of AVHRR, with the result that a single day's coverage is not entire, but the gaps from one day are filled in on the next. The spatial resolution of the infrared window bands is 1 km at nadir.

The GOES Imager

SST measurements from geosynchronous orbit are made using the infrared window channels of the GOES Imager. This is a five-channel instrument that remains above a given point on the Equator. The image of the Earth's disk is constructed by scanning the field of view along horizontal lines by an oscillating mirror. The latitudinal increments of the scan line are done by tilting the axis of the scan mirror. The spatial resolution of the infrared channels is 2.3 km (east–west) by 4 km (north–south) at the subsatellite point. There are two imagers in orbit at the same time on the two GOES satellites, covering the western Atlantic Ocean (GOES-East) and the eastern Pacific Ocean (GOES-West). The other parts of the global oceans visible from geosynchronous orbit are covered by three other satellites operated by Japan, India, and the European Meteorological Satellite organization (Eumetsat). Each carries an infrared imager, but with lesser capabilities than the GOES Imager.

TRMM Microwave Imager (TMI)

The TMI is a nine-channel microwave radiometer on the Tropical Rainfall Measuring Mission satellite, launched in 1997. The nine channels are centered at five frequencies: 10.65, 19.35, 21.3, 37.0, and 85.5 GHz, with four of them being measured at two polarizations. The 10.65 GHz channels confer a sensitivity to SST, at least in the higher SST range found in the tropics, that has been absent in microwave radiometers since the SMMR (Scanning Multifrequency Microwave Radiometer) that flew on the short-lived Seasat in 1978 and on Nimbus-7 from 1978 to 1987. Although SSTs were derived from SMMR measurements, these lacked the spatial resolution and absolute accuracy to compete with those of the AVHHR. The TMI complements AVHRR data by providing SSTs in the tropics where persistent clouds can be a problem for infrared retrievals. Instead of a rotating mirror, TMI, like other microwave imagers, uses an oscillating parabolic antenna to direct the radiation through a feed-horn into the radiometer.

The swath width of TMI is 759 km and the orbit of TRMM restricts SST measurements to within 38.5° of the equator. The beam width of the 10.65 GHz channels produces a footprint of 37×63 km, but the data are over-sampled to produce 104 pixels across the swath.

Applications

With absolute accuracies of satellite-derived SST fields of ~ 0.5 K or better, and even smaller relative uncertainties, many oceanographic features are resolved. These can be studied in a way that was hitherto impossible. They range from basin-scale perturbations to frontal instabilities on the scales of tens of kilometers. SST images have revealed the great complexity of ocean surface currents; this complexity was suspected from shipboard and aircraft measurements, and by acoustically tacking neutrally buoyant floats. However, before the advent of infrared imagery the synoptic view of oceanic variability was elusive, if not impossible.

El Niño

The El Niño Southern Oscillation (ENSO) phenomenon has become a well-known feature of the coupled ocean–atmosphere system in terms of perturbations that have a direct influence on people's lives, mainly by altering the normal rainfall patterns causing draughts or deluges – both of which imperil lives, livestock, and property.

The normal SST distribution in the topical Pacific Ocean is a region of very warm surface waters in the west, with a zonal gradient to cooler water in the east; superimposed on this is a tongue of cool surface water extending westward along the Equator. This situation is associated with heavy rainfall over the western tropical Pacific, which is in turn associated with lower level atmospheric convergence and deep atmospheric convection. The atmospheric convergence and convection are part of the large-scale global circulation. The warm area of surface water, enclosed by the 28°C isotherm, is commonly referred to as the 'Warm Pool' and in the normal situation is confined to the western part of the tropical Pacific. During an El Niño event the warm surface water, and associated convection and rainfall, migrate eastward perturbing the global atmospheric circulation. El Niño events occur up to a few times per decade and are of very variable intensity. Detailed knowledge of the shape, area, position, and movement of the Warm Pool can be provided from satellite-derived SST to help study the phenomenon and forecast its consequences.

Figure 3 shows part of the global SST fields derived from the Pathfinder SST algorithm applied to AVHRR measurements. The tropical Pacific SST field in the normal situation (December 1993) is shown in the upper panel, while the lower panel shows the anomalous field during the El Niño event of 1997–98. This was one of the strongest El Niños on record, but also the best documented and forecast. Seasonal

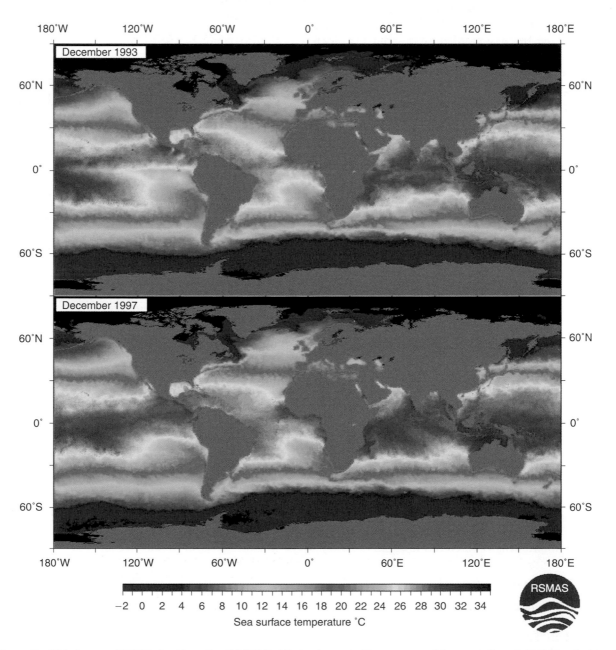

Figure 3 Global maps of SST derived from the AVHRR Pathfinder data sets. These are monthly composites of cloud-free pixels and show the normal situation in the tropical Pacific Ocean (above) and the perturbed state during an El Niño event (below).

predictions of disturbed patterns of winds and rainfall had an unprecedented level of accuracy and provided improved useful forecasts for agriculture in many affected areas. Milder than usual hurricane and tropical cyclone seasons were successfully forecast, as were much wetter winters and severe coastal erosion on the Pacific coasts of the Americas.

Hurricane Intensification

The Atlantic hurricane season in 1999 was one of the most damaging on record in terms of land-falling storms in the eastern USA, Caribbean, and Central America. Much of the damage was not a result of high winds, but of torrential rainfall. Accurate forecasting of the path and intensity of these land-falling storms is very important, and a vital component of this forecasting is detailed knowledge of SST patterns in the path of the hurricanes. The SST is indicative of the heat stored in the upper ocean that is available to intensify the storms, and SSTs of $>26°C$ appear to be necessary to trigger the intensification of the hurricanes. Satellite-derived SST maps are used in the prediction of the development of storm propagation across the Atlantic Ocean from the area off Cape Verde where atmospheric disturbances spawn the nascent storms. Closer to the USA and Caribbean, the SST field is important in determining the sudden intensification that can occur just before landfall. After the hurricane has passed, they sometimes leave a wake of cooler water in the surface that is readily identifiable in the satellite-derived SST fields.

Frontal Positions

One of the earliest features identified in infrared images of SST were the positions of ocean fronts, which delineate the boundaries between dissimilar surface water masses. Obvious examples are western boundary currents, such as the Gulf Stream in the Atlantic Ocean (**Figure 4**) and the Kuroshio in the Pacific Ocean, both of which transport warm surface water poleward and away from the western coastlines. In the Atlantic, the path of the warm surface water of the Gulf Stream can be followed in SST images across the ocean, into the Norwegian Sea, and into the Arctic Ocean. The surface water loses heat to the atmosphere, and to adjacent cooler waters on this path from the Gulf of Mexico to the Arctic, producing a marked zonal difference in the climates of the opposite coasts of the Atlantic and Greenland-Norwegian Seas. Instabilities in the fronts at the sides of the currents have been revealed in great detail in the SST images. Some of the large-scale instabilities can lead to loops on scales of a few tens to hundreds of kilometers that can become

'pinched off' from the flow and evolve as independent features, migrating away from the currents. When these occur on the equator side of the current these are called 'Warm Core Rings' and can exist for many months; in the case of the Gulf Stream these can propagate into the Sargasso Sea.

Figure 5 shows a series of instabilities along the boundaries of the Equatorial current system in the Pacific Ocean. The extent and structure of these features were first described by analysis of satellite SST images.

Coral Bleaching

Elevated SSTs in the tropics have adverse influences on living coral reefs. When the temperatures exceed the local average summertime maximum for several days, the symbiotic relationship between the coral polyps and their algae breaks down and the reef-building animals die. The result is extensive areas where the coral reef is reduced to the skeletal structure without the living and growing tissue, giving the reef a white appearance. Time-series of AVHRR-derived SST have been shown to be valuable predictors of reef areas around the globe that are threatened by warmer than usual water temperatures. Although it is not possible to alter the outcome, SST maps have been useful in determining the scale of the problem and identifying threatened, or vulnerable reefs.

The 'Global Thermometer'

Some of the most pressing problems facing environmental scientists are associated with the issue of global climate change: whether such changes are natural or anthropogenic, whether they can be forecast accurately on regional scales over decades, and whether undesirable consequences can be avoided. The past decade has seen many air temperature records being surpassed and indeed the planet appears to be warming on a global scale. However, the air temperature record is rather patchy in its distribution, with most weather stations clustered on Northern Hemisphere continents.

Global SST maps derived from satellites provide an alternative approach to measuring the Earth's temperature in a more consistent fashion. However, because of the very large thermal inertia of the ocean (it takes as much heat to raise the temperature of only the top meter of the ocean through one degree as it does for the whole atmosphere), the SST changes indicative of global warming are small. Climate change forecast models indicate a rate of temperature increase of only a few tenths of a degree per decade, and this is far from certain because of our incomplete understanding of how the climate

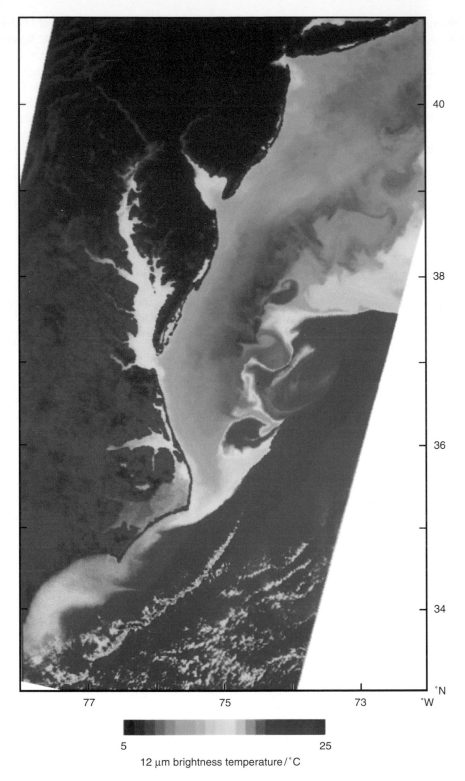

5   25
12 µm brightness temperature / °C

Figure 4 Brightness temperature image derived from the measurements of the ATSR on a nearly cloud-free day over the eastern coast of the USA. The warm core of the Gulf Stream is very apparent; it departs from the coast at Cape Hatteras. The cool, shelf water from the north entrains the warmer outflows from the Chesapeake and Delaware Bays. The north wall of the Gulf Stream reveals very complex structure associated with frontal instabilities that lead to exchanges between the Gulf Stream and inshore waters. The small-scale multicolored patterns over the warm Gulf Stream waters to the south indicate the presence of cloud. This image was taken at 15.18 UTC on 21 May 1992, and is derived from nadir view data from the 12 µm channel. (Generated from data © NERC/ESA/RAL/BNSC, 1992)

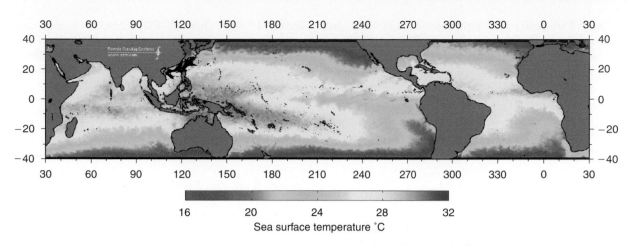

Figure 5 Tropical SSTs produced by microwave radiometer measurements from the TRMM (Tropical Rainfall Measuring Mission) Microwave Imager (TMI). This is a composite generated from data taken during the week ending December 22, 1999. The latitudinal extent of the data is limited by the orbital geometry of the TRMM satellite. The measurement is much less influenced by clouds than those in the infrared, but the black pixels in parts of the oceans where there are no islands indicate areas of heavy rainfall. The image reveals the cold tongue of surface water along the Equator in the Pacific Ocean and cold water off the Pacific coast of South America, indicating a non-El Niño situation. Note that the color scale is different from that used in **Figure 3**. The image was produced by Remote Sensing Systems, sponsored in part by NASA's Earth Science Information Partnerships (ESIP) (a federation of information sites for Earth science); and by the NOAA/NASA Pathfinder Program for early EOS products; principal investigator: Frank Wentz.

system functions, especially in terms of various feedback factors such as those involving changes in cloud and aerosol properties. Such a rate of temperature increase will require SST records of several decades length before the signal, if present, can be unequivocally identified above the uncertainties in the accuracy of the satellite-derived SSTs. Furthermore, the inherent natural variability of the global SST fields tends to mask any small, slow changes. Difficult though this task may be, global satellite-derived SSTs are an important component in climate change research.

Air-sea Exchanges

The SST fields play further indirect roles in the climate system in terms of modulating the exchanges of heat and greenhouse gases between the ocean and atmosphere. Although SST is only one of several variables that control these exchanges, the SST distributions, and their evolution on seasonal timescales can help provide insight into the global patterns of the air–sea exchanges. An example of this is the study of tropical cloud formation over the ocean, a consequence of air–sea heat and moisture exchange, in terms of SST distributions.

Future Developments

Over the next several years continuing improvement of the atmospheric correction algorithms can be anticipated to achieve better accuracies in the derived

SST fields, particularly in the presence of atmospheric aerosols. This will involve the incorporation of information from additional spectral channels, such as those on MODIS or other EOS era satellite instruments. Improvements in SST coverage, at least in the tropics, can be expected in areas of heavy, persistent cloud cover by melding SST retrievals from high-resolution infrared sensors with those from microwave radiometers, such as the TMI.

Continuing improvements in methods of validating the SST retrieval algorithms will improve our understanding of the error characteristics of the SST fields, guiding both the appropriate applications of the data and also improvements to the algorithms.

On the hardware front, a new generation of infrared radiometers designed for SST measurements will be introduced on the new operational satellite series, the National Polar-Orbiting Environmental Satellite System (NPOESS) that will replace both the civilian (NOAA-n) and military (DMSP, Defense Meteorological Satellite Program) meteorological satellites. The new radiometer, called VIIRS (the Visible and Infrared Imaging Radiometer Suite), will replace the AVHRR and MODIS. The prototype VIIRS will fly on the NPP (NPOESS Preparatory Program) satellite scheduled for launch in late 2005. At present, the design details of the VIIRS are not finalized, but the physics of the measurement constrains the instrument to use the same atmospheric window channels as previous and current instruments, and have comparable, or better, measurement accuracies.

The ATSR series will continue with at least one more model, called the Advanced ATSR (AATSR) to fly on Envisat to be launched in 2001. The SST capability of this will be comparable to that of its predecessors.

Thus, the time-series of global SSTs that now extends for two decades will continue into the future to provide invaluable information for climate and oceanographic research.

See also

Carbon Dioxide (CO₂) Cycle. El Niño Southern Oscillation (ENSO). Evaporation and Humidity. Heat and Momentum Fluxes at the Sea Surface. Ocean Circulation. Satellite Oceanography, History and Introductory Concepts.

Further Reading

Barton IJ (1995) Satellite-derived sea surface temperatures: Current status. *Journal of Geophysical Research* 100: 8777–8790.

Gurney RJ, Foster JL, and Parkinson CL (eds.) (1993) *Atlas of Satellite Observations Related to Global Change*. Cambridge: Cambridge University Press.

Ikeda M and Dobson FW (1995) *Oceanographic Applications of Remote Sensing*. London: CRC Press.

Kearns EJ, Hanafin JA, Evans RH, Minnett PJ, and Brown OB (2000) An independent assessment of Pathfinder AVHRR sea surface temperature accuracy using the Marine-Atmosphere Emitted Radiance Interferometer (M-AERI). Bulletin of the American Meteorological Society. 81: 1525–1536.

Kidder SQ and Vonder Haar TH (1995) *Satellite Meteorology: An Introduction*. London: Academic Press.

Legeckis R and Zhu T (1997) Sea surface temperature from the GEOS-8 geostationary satellite. *Bulletin of the American Meteorological Society* 78: 1971–1983.

May DA, Parmeter MM, Olszewski DS, and Mckenzie BD (1998) Operational processing of satellite sea surface temperature retrievals at the Naval Oceanographic Office. *Bulletin of the American Meteorological Society*, 79: 397–407.

Robinson IS (1985) *Satellite Oceanography: An Introduction for Oceanographers and Remote-sensing Scientists*. Chichester: Ellis Horwood.

Stewart RH (1985) *Methods of Satellite Oceanography*. Berkeley, CA: University of California Press.

Victorov S (1996) *Regional Satellite Oceanography*. London: Taylor and Francis.

PALEOCLIMATE MARINE RECORD

MILLENNIAL-SCALE CLIMATE VARIABILITY

J. T. Andrews, University of Colorado, Boulder, CO, USA

Introduction

Analysis of Quaternary marine sediment cores has changed in emphasis several times over the last four decades. In particular, this has involved a change in focus from variations in proxy records on orbital or Milankovitch timescales (with recurring periodicities of *c.* 20 000, 40 000, and 100 000 years), to an interest in the sub-Milankovitch variability (**Figure 1**). In turn, this has frequently meant a change in the length of the record from several million years, to several tens of thousands of years (often the last glacial/deglacial cycle which extended from 120 000 years ago to the present). It has also meant an increased interest in sites with high rates of sediment accumulation ($\gg 10 \, \text{cm ky}^{-1}$).

Although not precisely defined, the term 'millennial-scale climate variability' is usually considered to cover events with periods of between 1000 and 10 000 years. Evidence for abrupt, millennial-scale changes in ocean sediments (**Figures 1(a)** and **1(b)**) has resulted in a paradigm shift. The role of the oceans in abrupt climate forcing is now considered to be paramount, whereas under the Milankovitch scenario the role of the oceans was frequently considered subordinate to changes on land associated with the growth and decay of the large Quaternary ice sheets (**Figure 2**), which were principally driven by changes in high-latitude, Northern Hemisphere insolation (e.g., **Figure 1(d)**).

History

The ability to undertake millennial-scale climate reconstructions from marine sediments was conditioned by several requirements, which could not be met until the 1980s and early 1990s. An underlying rationale for this interest was that of the results from

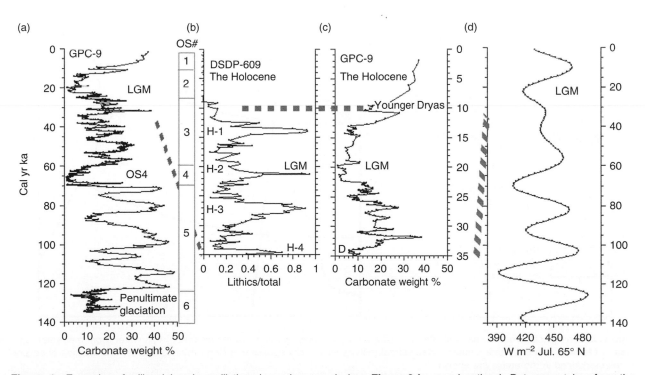

Figure 1 Examples of millennial-scale oscillations in marine records (see **Figure 2** for core locations). Data were taken from the National Oceanographic and Atmospheric Administration Paleoclimate Database (http://www.ncdc.noaa.gov/paleo). Age is given in thousands of years ago (ka) and the marine oxygen isotope stages (OS) are shown. The temporal location of Heinrich events (H-1–H-3), the Last Glacial Maximum (LGM), and the dramatic Younger Dryas cold event are shown. On the right-hand side is the summer insolation at 65° N for July. Notice the absence of millennial-scale variability in the Milankovitch forcing of global climate although the main peaks and troughs are picked out in the carbonate record (left-hand column).

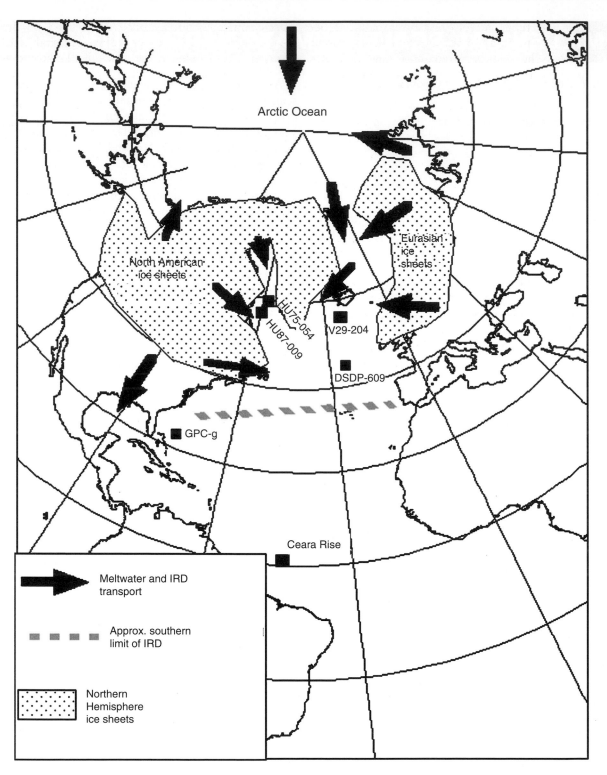

Figure 2 The glacial world of the Last Glacial Maximum (LGM) showing routes for the export of fresh water (from meltwater, e.g., Gulf of Mexico), and the major iceberg-rafted detrital (IRD) sources for deposition in North Atlantic basins. The location of several cores mentioned in the article are shown.

the Greenland and Antarctic ice core records which showed remarkably rapid oscillations in a variety of proxies for the last 40 000–80 000 years. The advent of accelerator mass spectrometry (AMS) ^{14}C dating of small (2–10 mg) samples of foraminifera allowed many of the world's deep-sea and shelf sediments to be directly dated up to a limit of between 30 000 and 40 000 years ago. Because of the small sample size required and the relatively fast turnaround, it became possible to obtain many dates on a core, and in some cases the density approached one date per thousand years. This technology also meant that sediment cores from environments with high rates of sediment accumulation could now be successfully dated; thus a variety of sediment environments from the 'drifts' around the North Atlantic, to glaciated shelves and fiords, could now be studied. In these environments, sediment accumulation rates (SARs) were often greater than $20 \, cm \, ky^{-1}$ and could reach rates as high as $2 \, m \, ky^{-1}$. In areas with these very high rates of sediment accumulation there was a clear need for improved coring technology such that cores of tens of meters in length could be obtained. Giant piston cores were thus developed with recoveries in the range of 20–60 m. An example is the Calypso system deployed from the French research vessel *Marion Dufresne* and employed as part of the IMAGES (International Marine Past Global Change) program. This allowed for very high-resolution studies (SAR of $\geq 1 \, m \, ky^{-1}$, decadal resolution) if the cores recovered sediments with basal dates of 10 000 BP. If the cores were extracted from areas with more modest rates of sediment accumulation (>10 and $<100 \, cm \, ky^{-1}$), then millennial-scale studies would have been possible. However, one problem with longer temporal records was that they recovered sediments with ages greater than the radiocarbon limit of *c.* 45 ka. Records that extended back into marine or oxygen isotope stages 4 and 5 (**Figure 1**, OS column) could not be numerically dated *per se*, but their chronology had to be derived by correlation with other records. In some important but rare cases, the marine sediment is annually laminated and a chronology can be developed by counting the varves.

The rationale for conducting high-resolution, millennial-scale studies of marine sediments has been largely driven by the need to ascertain if the abrupt climate changes recorded in the polar ice sheet records, particularly the millennial-scale Dansgaard–Oeschger (D–O) events, were evident in the ocean system (**Figures 1(b)** and **1(c)**).

Examples of Millennial-scale Oceanographic Proxy Records

In the last decade, the number of papers on millennial-scale ocean variability has increased substantially. In all cases, some property of the sediment is measured and climatic variability deduced. The most-documented proxies for millennial changes in ocean climate and hydrography are: (1) changes in the noncarbonated sand size (<2000 and $>63 \, \mu m$) fraction, the so-called iceberg-rafted detrital (IRD) fraction; (2) changes in the $\delta^{18}O$ of planktonic and benthic foraminifera which reflect both changes in the global ice volume, temperature, and meltwater volume; (3) changes in the $\delta^{13}C$ of marine carbonates which is a measure of productivity and water mass history and is used to trace variations in the production and circulation patterns of bottom water; (4) changes in the composition of faunas or floras which reflect the response of the biota to oceanographic changes; and (5) changes in the geochemical properties of the inorganic shell of organisms, or bulk sediment, which can be calibrated against climatic variables, such as sea surface temperature (SST). Usually, more than one of these parameters are measured, or a ratio, for example, lithics/(lithics + foraminifera), is used to develop a scenario of oceanic climate variability (**Figure 1(b)**).

Iceberg-rafted Detrital (Heinrich) Events

Heinrich's seminal paper on the occurrence, in cores off Portugal, of discrete IRD peaks during the last 60 000 years or so resulted in a wealth of data and hypotheses about 'Heinrich events'. It is now believed that 'armadas of icebergs' were released on a quasi-periodic basis into the North Atlantic, with the major source area being the Hudson Strait. Hudson Strait is a large, deep trough, which drained a substantial fraction ($2–4 \times 10^6 \, km^2$) of the interior of the Laurentide Ice Sheet (**Figure 2**). In the Labrador Sea, and the areas south of Greenland and toward Europe, evidence for these armadas is dramatically visible in many parameters, but especially in the changes of the detrital carbonate content of the cores, derived from the erosion of the Paleozoic limestone that outcrops on the floors of Hudson Strait and Hudson Bay (**Figure 3**).

These dramatic sedimentological events have been termed H-0, H-1, etc., and have the following radiocarbon ages (years ago): H-0 = 10 000–11 000; H-1 = 14 500±; H-2 = 20 500±; H-3 = 27 000±; H-4 = 34 000± (**Figures 1** and **3**). Older H-events, H-5 and H-6, lie beyond the limits of ^{14}C dating but have inferred ages of 48 000 and 60 000 years ago. Because of the ±error in the ^{14}C dates, the duration of each of these IRD intervals is not well defined. Available dates indicate that they persisted for a few hundred to about a thousand years (**Figure 1(b)**) and have a quasi-periodic return interval averaging *c.* 6 ky.

Studies in the North Atlantic indicate that during H-events there are coeval changes in other parameters,

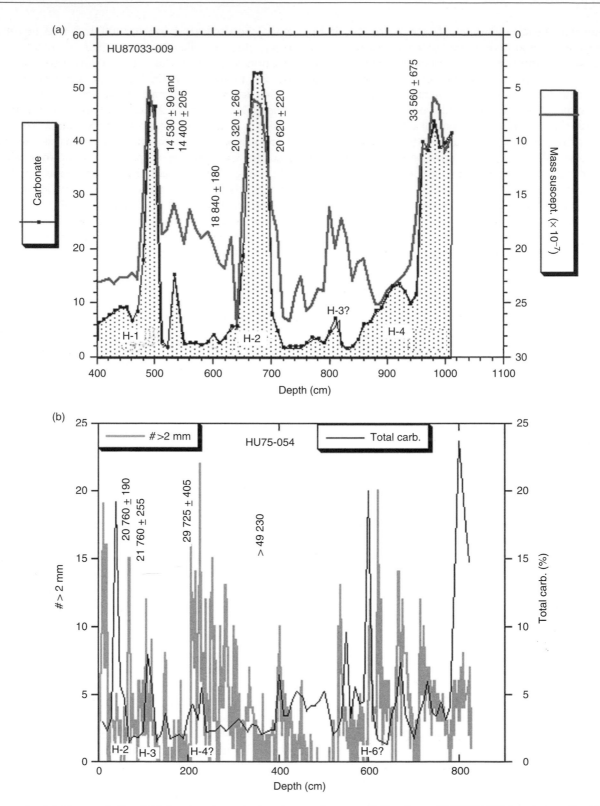

Figure 3 (a) Changes in the detrital carbonate content and magnetic susceptibility of core HU87033-009 from just north of the Hudson Strait outlet (see **Figure 2**). Note that the scale for mass magnetic susceptibility ($\times 10^{-7}\,\mathrm{m}^3\,\mathrm{kg}^{-1}$) is reversed because in this area the magnetite concentrations are diluted by the input of diamagnetic detrital carbonate. (b) Detrital carbonate and clasts $>2\,\mathrm{mm}$ in HU75-054 from south of Davis Strait, northwestern North Atlantic (**Figure 2**). Note that the agreement between detrital carbonate events (primarily a measure of North Atlantic Heinrich events) and coarse ice-rafted detritus is far from perfect.

with planktonic foraminiferal assemblages decreasing in numbers per gram but also becoming nearly entirely polar in composition. Benthic foraminifera show strongly decreased productivity. At the same time, the stable oxygen isotopic composition suggests an increase in surface meltwater.

Controversy exists on the regional extent of H-events and the underlying mechanism(s) for discrete IRD events. Dating is clearly a critical issue as these millennial-scale events are of short duration and often date from times (<20 000 years ago) when the errors of the radiocarbon dates are measured in one to several hundred years. In addition, efforts to correlate millennial-scale oceanic H-events with abrupt events on land (or in ice cores) face the problem of correcting the marine dates for both changes in the ocean reservoir correction and ^{14}C production rates. Thus ice core/ocean record correlations are often based on fitting the 'wiggles' of the proxies from the two systems.

The extent of IRD events coeval with the main North Atlantic belt of iceberg-rafted materials (**Figure 2**) is a matter of debate. In the Nordic Seas, in the Labrador Sea, and in Baffin Bay, the IRD signal in cores is pervasive during the last glacial cycle and cannot be used *per se* to identify H-events (**Figure 3(b)**). In contrast, in the Labrador Sea, H-events are easily distinguished by the dramatic increase in detrital carbonate during these abrupt events (**Figures 3(a)** and **3(b)**). These data are not surprising; in areas close to ice margins, the rafting of sediments in icebergs would be a persistent transport mechanism whereas the collapse or surge of a major outlet might be distinguished by an abrupt change in sediment provenance. It is also uncertain as to what extent small changes in IRD have any significance given the strong, stochastic nature of iceberg/sediment relationships.

The origins of these millennial-scale changes in ice sheet dynamics is considered to be attributable to either inherent glaciological mechanisms associated with changes at the bed of these former ice sheets, or alternatively researchers have argued that they represent climate forcing. The main issue of concern for glaciologists is how atmospheric processes could translate to the bed of large ice sheets at a rate compatible with millennial-scale variations. No plausible mechanism has been discovered. On the other hand, if coeval H-events are seen outside the North Atlantic, which has been suggested, then mechanisms are needed to transfer the process from a regional scale to a global scale.

Two mechanisms have been invoked. They are not mutually exclusive. In the first scenario, it is posited that a collapse of the North American Ice Sheet during an H-event causes a rapid rise in sea level of 1–5 m, which then triggers instabilities in other ice sheet margins, which have advanced toward the shelf break (in areas such as Norway, Greenland, and Iceland). As yet it is unclear whether these events around the North Atlantic affected the grounded margins of the Antarctic Ice Sheet and the West Antarctic Ice Sheet in particular. In the second scenario, the collapse of the Northern Hemisphere ice sheets, or the Laurentide (North American) Ice Sheet in particular, would result in the transport of large volumes of fresh water, in the form of icebergs and basal meltwater, to the North Atlantic. Isotopic changes in the $\delta^{18}O$ of planktonic foraminifera certainly occur during H-events (although foraminifera often disappear from the sediment during H-events, hence detailed records are sparse) and indicate that $\delta^{18}O$ values get lower, indicating the presence of a surface, low-salinity layer. If these waters are advected toward sites of vertical convection in the North Atlantic then both theory and observations indicate that this process will turn off or curtail the global thermohaline circulation. Thus the next question is whether sites beyond the normal limits of iceberg rafting and direct glacial impact show any evidence of millennial-scale climate oscillations in either the surface or deep waters.

Other Millennial-scale Proxies

A variety of proxy data from deep-sea sediment drifts in the western North Atlantic, south of 35° N (**Figure 2**), indicate substantial, millennial-scale changes in the deep ocean (**Figures 1(a)** and **1(c)**). Changes in the $CaCO_3$ content of the sediment reflect the integration of carbonate production in surface waters, carbonate dissolution, resuspension and transport of continental margin sediments, and dilution with glacially and fluvially derived terrigenous sediments from the Canadian Maritimes and Eastern Canadian Arctic. Core GPC-9 was taken from a water depth of 4758 m on the Bahamas Outer Ridge. The $CaCO_3$ record spectacularly captures oscillations in this proxy, which range over 48%, from lows during marine oxygen isotope stages (OS) 2 and 4 of ~2% to peaks during OS 5 of *c.* 50% (**Figure 1(a)**). It was also shown that these oscillations were also evident in the stable isotopic composition of foraminifera, although the isotopic changes tended to lead the carbonate fluctuations by 1000 years or so during carbonate event D. Changes in the $\delta^{13}C$ may be linked with changes in the thermohaline circulation, such that a significant reduction in the formation of North Atlantic Deep

Water (NADW) is indicated by low (δ^{13}C) in two benthic foraminifera genera.

More recent work has concentrated on the events during OS 3 (**Figure 1**), as this was a period of extreme and abrupt oscillations in the Greenland ice core records. This interval includes Heinrich events 3, 4, and 5 (i.e., between ~31 600 and 47 800) calendar years ago (**Figure 1(b)**). Variations in δ^{13}C of planktonic foraminifera along a transect from the south of Iceland (c. 60° N) to the Ceara Rise (c. 5° N) (**Figure 2**) have been examined. Cores from different water depths along the transect were used to reconstruct changes in water mass on millennial timescales. A critical question is the relationship between changes in the deep-sea circulation and ventilation and H-events. Is there a cause-and-effect relationship such that H-events result in a response in ocean circulation? Because these sites are outside the IRD belt of the North Atlantic, the correlation between actual IRD or carbonate-rich H-events (e.g., **Figure 3**) and ocean geochemical responses relies on the quality of the chronology. Within OS 3, the errors on AMS ^{14}C dates are frequently between ± 300 and ± 500 years; hence the issue of a direct correlation to events lasting a mere 1000 years is of concern. However, the δ^{13}C records from the Ceara Rise indicate that cold, relatively fresh Antarctic Bottom Water (AABW, lower δ^{13}C), which underlies the warmer and saltier NADW (higher δ^{13}C), thickened by a factor of 2. The thickening of the AABW at the site began 'several thousands of years' prior to each H-event and extended for 'several thousand years' after each event. These intervals of expanded AABW were times of reduced NADW production. These intervals of reduced NADW production and associated reduction in the thermohaline circulation cannot be directly caused by ice sheet collapse and the presence of a freshwater cap over the northern North Atlantic.

Further, in a core from the Bermuda Rise (near GPC-9) (**Figure 2**), reconstructed SST fluctuations of 2–5 °C have been shown. These SST estimates could be mapped directly onto the δ^{18}O oscillations from the ice cores at the Greenland Summit.

Millennial-scale Events in the Last 12 000 Years (The Holocene)

A critical question for society is whether such rapid millennial-scale oscillations continued during the 'Postglacial' or Holocene period of the last 10 000 radiocarbon years (about 12 000 calendar years), and if so were they too associated with changes in the thermohaline circulation and episodic ice-rafting events? In general, data from the Greenland ice cores indicate that climatic variability was substantially reduced during the Holocene. Temperature reconstructions from borehole and isotopic measurements indicate that temperatures at the summit of the ice sheet warmed dramatically by 16 °C at the onset of the Holocene. Over the last 10 000 years there have been temperature variations of c. 2–3 °C, and in the last 5000 years these are superimposed on a gradual, long-term temperature decrease. Based on the chronology of Holocene glacial readvances, a 2500± year cycle has been advocated.

It is only in the last few years or so that researchers in the marine community have focused on producing high-resolution records from this most recent interval of Earth's history. Cores have often been selected that have sufficiently high rates of sediment accumulation that sampling can resolve multicentury-, even multidecadal-, scale events. It has been proposed that variations in the numbers of hematite-stained quartz grains at sites in the North Atlantic reflective a pervasive 1470-year 'beat' during the Holocene that are linked with variations in solar activity. Cycles with a similar periodicity have been reported from a variety of sedimentary archives including silt size (as a measure of current speed), sediment color, and the amount and composition of the sand fraction. Although the ~1470-year cycles have been attributed to iceberg-rafting events, their magnitude in the records is not remotely at the scale of the Heinrich events. Indeed the 'pervasive' nature of this ice-rafting signal has been questioned on the basis of quantitative studies of quartz and plagioclase weight percentage data at sites from North Iceland and the Vorring Plateau.

Discussion: Importance and Mechanisms

In the 1970s and 1980s, a common view of the global climate system on scales from 1 to 10^6 years was that there were systematic changes associated with the Milankovitch orbital variations, which effected insolation. Evaluation of changes in the global ice volume indicated dominant periodicities of 41 000 and c. 20 000 years, and in the last 0.7 million years a 100 000-year cycle became evident. At higher frequencies, the spectra of climatic variability was essentially blank between 20 000-year and the 22-year sunspot cycle. This absence of recurring periodicities suggested that global climate change within this range had no obvious or repetitive forcing function. The advent of the successful ice-coring programs, especially the Greenland Ice Sheet boreholes, and the subsequent development of well-dated, multiproxy records of the atmosphere, led to a search for recurring frequencies between 1/20 000 and 1/22 cycles per year. This analysis suggested that there is a 6000–7000-year periodicity, which is approximately the same as the interval between the successive Heinrich events. Because of the largely unknown errors connected with the value of the ocean reservoir correction, and the conversion from radiocarbon years to sidereal years, the spacing between H-0 and H-4 was c. 5000, 8000, 7000, and 8000 years with uncertainties of several hundred years. However, each H-cycle was composed of several higher-frequency events, the D–O cycles, which had a recurrence interval of around 2000 years. In detailed records from cores in the North Atlantic, a series of D–O events are bundled with bounding H-events. These 'packages' show an overall saw-tooth decrease in warm surface water indicators over the course of a cycle, with a final abrupt and extreme minimum, which marks the onset of an H-event. This was rapidly followed by a dramatic rise in the warm-water proxies. Broecker referred to this pattern as a 'Bond cycle'. The prevailing wisdom calls for these oscillations to be associated with changes in the thermohaline circulation, but there is the 'chicken or egg' syndrome. Changes in the thermohaline circulation are usually associated with changes in the saltiness of the surface waters. Thus the dramatic collapse of a large ice sheet, and the subsequent export of fresh water in the form of meltwater plumes and icebergs, is a legitimate mechanism for curtailing convective overturning at sites in the northern North Atlantic.

An important question, presently unanswered, then becomes that how these changes are transmitted rapidly and at the millennial scale, synchronously through the atmosphere (to account for the observed rapid changes in ice sheet isotopes and precipitation chemistry), and within the oceans. There are several lines of evidence to suggest that one way in which the ocean circulation compensates for changes in the 'deep' thermohaline circulation is by the increased production of what has been termed 'glacial intermediate water'.

There is a dramatic decrease in the variability of most climate proxies in ocean sediments over the last 11 000 cal years. H- and D–O events characterize marine oxygen isotope stages 2, 3, and 4 (**Figure 1**) when the Earth was marked by extensive glaciation and sea levels were lowered between 40 and 110 m. High-resolution sampling of marine cores from deep ocean basins and continental margins which span one to several thousand of years indicate that changes in oceanography have taken place at millennial timescales over the present interglacial period.

A key question is whether all proxies will record the same oscillations? The notion that there are thresholds in the climate system suggests that not all events may be archived in marine sedimentary records. An example is the Holocene ice rafting of sediments. In some parts of the world oceans, measurable quantities of sediment could be rafting to ice-distal locations on and in sea ice. However, the sediment burden in sea ice is relatively light and, furthermore, the thickness of a typical multiyear pan of sea ice is measured in meters to a few tens of meters versus hundreds of meters for true icebergs. Hence, melting and erosion of sea ice results in a limited transport of sea-ice-rafted sediment when compared to iceberg rafting. Most of the North Atlantic's margins and offshore basins have seen a massive reduction in IRD following the retreat and disappearance of late Quaternary ice sheets (**Figures 2** and **4**). In today's world (**Figure 4**), the distribution of IRD-rich sediments is primarily restricted to the Greenland shelves and the eastern Canadian Arctic (Baffin Island and Labrador) margins, therefore even small traces of sand-size minerals distal to these areas may indicate intervals of iceberg rafting. However, the threshold in question is the presence of tidewater calving glaciers in the Greenland fiords. Observations from Greenland indicate that the ice sheet was well behind its present margin by 6000 years ago and probably by 7000–8000 years ago. Even on the East Greenland margin, which today is 'well traveled' by icebergs, sediments deposited between 6000 and 8000 years ago are largely devoid of IRD, whereas between 5000 and 6000 years ago the iceberg rafting of coarse, clastic sediments becomes a pervasive depositional process.

Modern observations, however, do indicate that the production of Intermediate Atlantic Water is sensitive to modern-day atmospheric and oceanographic processes. The Great Salinity Anomaly of the late 1960s and early 1970s (depending on location) was the result of an excess freshwater output from the Arctic Ocean (as sea ice) (**Figure 4**). This pool of relatively fresh, surface water, caused dramatic cooling of the water column off North Iceland (by 5 °C), and as it moved into the Labrador Sea it caused a cessation in convective overturning. This resulted in a temperature drop of ∼2 °C on the west Greenland and Canadian margins.

Because another salinity anomaly occurred in the early 1980s, this time sourced from the Hudson Bay/Labrador Sea region, there certainly appear to be mechanisms within the present climate system that are capable of generating rather severe and abrupt oceanographic changes. The question is whether the processes responsible for multidecadal climatic

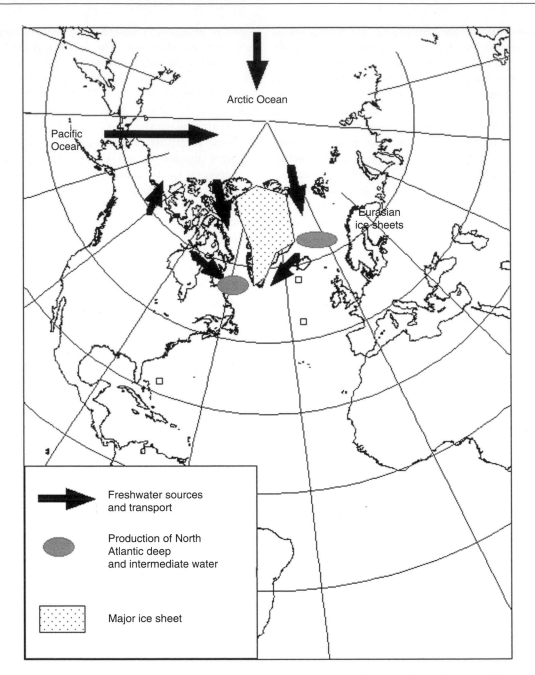

Figure 4 Major sources of salinity events and location of convection areas in the North Atlantic during the present 'interglacial' world. Sources include the influx of freswater from the Pacific Ocean via the Bering Strait, river runoff into the Arctic Ocean, and the export of sea ice from the Arctic Ocean via Fram Strait and the Canadian High Arctic channels. Tidewater calving margins around the Greenland Ice Sheet lead to the calving of about 350 km^3 of ice per year.

variability can be scaled up, so that the processes persist and produce millennial oscillations in ocean records.

Conclusions

Millennial-scale changes have become an accepted reality in the climate system. Initial research concentrated on the massive changes associated with the discharge of sediments and water into the North Atlantic Ocean during the last glacial cycle (marine oxygen isotope stages 2–5) (**Figures 1–3**). However, high-resolution studies of our deglacial world (**Figure 4**) appear to indicate that similarly spaced but subdued events persist but with very different boundary conditions. A number of publications have also demonstrated that millennial-scale changes in various proxy records are a feature of ocean sediments over at least the last 500 000 years.

The work from tropical and subtropical sites (**Figure 2**) indicates that Heinrich events have manifestations in ocean reconstructions, which belie a simple association with ice sheet instability and collapse. It is far from clear how oceanographic and atmospheric changes are transmitted to the bed of large ice streams, and there is indeed disagreement as to whether the collapse of Northern Hemisphere ice sheets (**Figure 2**) was regionally coeval or whether the collapses are linked temporally by a mechanism, such as rapid changes in relative sea level. It has, however, been observed that the routing of fresh water (**Figure 4**) can have dramatic effects, even in the present world, and the key may well lie in a better understanding of the role of the ocean thermohaline circulation system in the global climate system.

See also

Cenozoic Climate – Oxygen Isotope Evidence. Ocean Circulation: Meridional Overturning Circulation.

Further Reading

Alley RB, Clark PU, Keigwin LD, and Webb RS (1999) Making sense of millennial-scale climate change. In: Clark PU, Webb RS, and Keigwin LD (eds.) *Mechanisms of Global Climate Change at Millennial Time Scales*, pp. 301–312. Washington, DC: American Geophysical Union.

Anderson DM (2001) Attenuation of millennial-scale events by bioturbation in marine sediments. *Paleoceanography* 16: 352–357.

Andrews JT (1998) Abrupt changes (Heinrich events) in late Quaternary North Atlantic marine environments: A history and review of data and concepts. *Journal of Quaternary Science* 13: 3–16.

Andrews JT, Hardardottir J, Stoner JS, Mann ME, Kristjansdottir GB, and Koc N (2003) Decadal to millennial-scale periodicities in North Iceland shelf sediments over the last 12,000 cal yrs: Long-term North Atlantic oceanographic variability and solar forcing. *Earth and Planetary Science Letters* 210: 453–465.

Bond G, Kromer B, Beer J, *et al.* (2001) Persistent solar influence on North Atlantic climate during the Holocene. *Science* 294: 2130–2136.

Bond GC and Lotti R (1995) Iceberg discharges into the North Atlantic on millennial time scales during the last glaciation. *Science* 267: 1005–1009.

Broecker WS (1997) Thermohaline circulation, the Achilles heel of our climate system will man-made CO_2 upset the current balance? *Science* 278: 1582–1588.

Broecker WS, Bond G, McManus J, Klas M, and Clark E (1992) Origin of the northern Atlantic's Heinrich events. *Climatic Dynamics* 6: 265–273.

Clarke GKC, Marshall SJ, Hillaire-Marcel C, Bilodeau G, and Veiga-Pires CA (1999) Glaciological perspective on Heinrich events. In: Clark PU, Webb RS, and Keigwin LD (eds.) *Mechanisms of Global Climate Change at Millennial Time Scales*, pp. 243–262. Washington, DC: American Geophysical Union.

Clemens SC (2005) Millennial-band climate spectrum resolved and linked to centennial-scale solar cycles. *Quaternary Science Reviews* 24: 521–531.

Curry WB, Marchitto TM, McManus JF, Oppo DW, and Laarkamp KL (1999) Millennial-scale changes in the ventilation of the thermocline, intermediate, and deep waters of the glacial North Atlantic. In: Clark PU, Webb RS, and Keigwin LD (eds.) *Mechanisms of Global Climate Change at Millennial Time Scales*, pp. 59–76. Washington, DC: American Geophysical Union.

Heinrich H (1988) Origin and consequences of cyclic ice rafting in the Northeast Atlantic Ocean during the past 130 000 years. *Quaternary Research* 29: 143–152.

Hughen KA, Overpeck JT, and Lehman SJ (1998) Deglacial changes in ocean circulation from an extended radiocarbon calibration. *Nature* 391: 65–68.

Keigwin LD and Jones GA (1994) Western North Atlantic evidence for millennial-scale changes in ocean circulation and climate. *Journal of Geophysical Research* 99: 12397–12410.

Lowell TV, Heusser CJ, and Andersen BG (1995) Inter-hemispheric correlation of late Pleistocene glacial events. *Science* 269: 1541–1549.

McManus JR, Oppo DW, and Cullen JL (1999) A 0.5 million-year record of millennial-scale climate variability in the North Atlantic. *Science* 283: 971–975.

Moros M, Andrews JT, Eberl DD, and Jansen E (2006) The Holocene history of drift ice in the northern North Atlantic: Evidence for different spatial and temporal modes. *Palaeoceanography* 21 (doi:10.1029/2005PA001214).

Sachs JP and Lehman SJ (1999) Subtropical North Atlantic temperatures 60 000 to 30 000 years ago. *Science* 286: 756–759.

Thomas E, Booth L, Maslin M, and Shackelton NJ (1995) Northeastern Atlantic benthic foraminifera during the last 45 000 years: Changes in productivity seen from the bottom up. *Paleoceanography* 10: 545–562.

Relevant Websites

http://www.ncdc.noaa.gov/paleo
  – NOAA Paleoclimatology Program, NOAA.

OXYGEN ISOTOPES IN THE OCEAN

K. K. Turekian, Yale University, New Haven, CT, USA

$$\left[\frac{{}^{18}O/{}^{16}O_{sample}}{{}^{18}O/{}^{16}O_{standard}} - 1\right] \times 1000$$

The oxygen isotope signature of sea water varies as a function of the processing of water in the oceanic cycle. The two chemical parameters, salinity and oxygen isotope ratio, are distinctive for various water types. The oxygen-18 to oxygen-16 ratio is represented in comparison to a standard. The notation is δO^{18} which is defined as follows:

Figure 1 shows the results for the world oceans from Craig and Gordon (1965). During the GEOSECS program the oxygen isotope ratios of seawater samples were also determined. The features resemble those in **Figure 1**. The GEOSECS data are available in the shore-based measurements volume of the GEOSECS Atlas (1987) published by the US National Science Foundation.

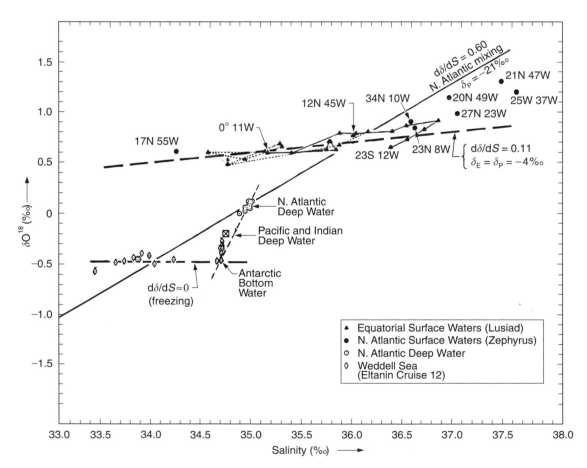

Figure 1 Oxygen-18–salinity relationships in Atlantic surface and deep waters. δ_E and δ_P refer to the isotopic composition of evaporating vapor and precipitation, respectively. (From Craig and Gordon, 1965.)

The tracking of fresh water from streams draining into the ocean at different latitudes has been used to study several coastal oceanic regimes. The work of Fairbanks (1982) is one of the earliest of these efforts.

See also

Cenozoic Climate – Oxygen Isotope Evidence.

Further Reading

Craig H and Gordon LI (1965) Deuterium and oxygen-18 variations in the ocean and marine atmosphere. *Stable Isotopes in Oceanographic Studies and Paleotemperatures, Consiglio Nazionale Delle Ricerche, Laboratorio di Geologia Nucleare-Pisa, 122 pp. (Also in Symposium on Marine Chemistry, Publ. 3., Kingston, Graduate School of Oceanography, University of Rhode Island,* 277–374).

Fairbanks RG (1982) The origin of continental shelf and slope water in the New York Bight and Gulf of Maine: evidence from $H_2^{18}O/H_2^{16}O$ ratio measurements. *Journal of Geophysical Research* 87: 5796–5808.

GEOSECS Atlantic, Pacific, and Indian Ocean Expeditions, Volume 7, Shorebased Data and Graphics (1987) National Science Foundation. 200pp.

CENOZOIC CLIMATE – OXYGEN ISOTOPE EVIDENCE

J. D. Wright, Rutgers University, Piscataway, NJ, USA

Discoveries of fossil remains of 50 million year old alligators on Ellesmere Island and 30–40 million year-old forests on Antarctica contrast sharply with our present vision of polar climates. These are not isolated discoveries or quirks of nature. An ever-growing body of faunal, floral, and geochemical evidence shows that the first half of the Cenozoic Era was much warmer than the present time. What maintained such a warm climate and could it be an analog for future global warming? To address these and other questions, one needs more than a qualitative estimate of planetary temperatures. Quantitative temperature estimates (both magnitudes and rates of change) are required to depict how the Earth's climate has changed through time. One of the most powerful tools used to reconstruct past climates during the Cenozoic (the last 65 million years of Earth's history) is the analysis of oxygen isotopes in the fossil shells of marine organisms. The calcium carbonate shells of the protist foraminifera are the most often analyzed organisms because the different species are distributed throughout surface (planktonic) and deep (benthic) marine environments.

Oxygen Isotope Systematics

The stable isotopes of oxygen used in paleooceanographic reconstructions are ^{16}O and ^{18}O. There are about 500 ^{16}O atoms for every ^{18}O atom in the ocean/atmosphere environment. During the 1940s, Harold Urey at the University of Chicago predicted that the $^{18}O/^{16}O$ ratio in calcite ($CaCO_3$) should vary as a function of the temperature at which the mineral precipitated. His prediction spurred on experiments by himself and others at the University of Chicago who measured $^{18}O/^{16}O$ ratios in $CaCO_3$ precipitated in a wide range of temperatures, leading to the use of stable oxygen isotope measurements as a paleothermometer.

To determine oxygen isotopic ratios, unknown $^{18}O/^{16}O$ ratios are compared to the known $^{18}O/^{16}O$ ratio of a standard. The resulting values are expressed in delta notation, $\delta^{18}O$, where:

$$\delta^{18}O = \frac{^{18}O/^{16}O_{sample} - ^{18}O/^{16}O_{standard}}{^{18}O/^{16}O_{standard}} \times 1000 \quad [1]$$

Carbonate samples are reacted in phosphoric acid to produce CO_2. To analyze water samples, CO_2 gas is equilibrated with water samples at a constant temperature. Given time, the CO_2 will isotopically equilibriate with the water. For both the carbonate and water samples, the isotopic composition of CO_2 gas is compared with CO_2 gas of known isotopic composition. There are two standards for reporting $\delta^{18}O$ values. For carbonate samples, the reference standard is PDB, which was a crushed belemnite shell (*Belemnitella americana*) from the Peedee formation of Cretaceous age in South Carolina. The original PDB material has been exhausted, but other standards have been calibrated to PDB and are used as an intermediate reference standard through which a PDB value can be calculated. For measuring the isotopic composition of water samples, Standard Mean Ocean Water (SMOW) is used as the reference. The SMOW reference was developed so that its $\delta^{18}O_{water}$ value is 0.0‰ (parts per thousand) and approximates the average oxygen isotopic composition of the whole ocean. Deep ocean $\delta^{18}O_{water}$ values are close to the SMOW value, ranging from −0.2 to 0.2‰. In contrast, surface ocean $\delta^{18}O_{water}$ values exhibit a much greater variability, varying between −0.5 and +1.5‰.

Oxygen Isotope Paleothermometry

Early studies into the natural variations in oxygen isotopes led to the development of a paleotemperature equation. The temperature during the precipitation of calcite can be estimated by measuring the $\delta^{18}O$ value in calcite-secreting organisms (foraminifera, corals, and mollusks) and the value of the water in which the organisms live. The various paleotemperature equations all follow the original proposed by Sam Epstein and his colleagues (University of Chicago):

$$T = 16.5 - 4.3 \times \left(\delta^{18}O_{calcite} - \delta^{18}O_{water}\right) \\ + 0.14 \times \left(\delta^{18}O_{calcite} - \delta^{18}O_{water}\right)^2 \quad [2]$$

where T and $\delta^{18}O_{water}$ are the temperature (°C) and oxygen isotope value of the water in which the organism lived[1] and $\delta^{18}O_{calcite}$ is the oxygen isotope value of calcite measured in the mass spectrometer.

Eqn [2] shows that the changes in $\delta^{18}O_{calcite}$ are a function of the water temperature and $\delta^{18}O_{water}$ value. A one-to-one relationship between $\delta^{18}O_{calcite}$ and $\delta^{18}O_{water}$ values dictates that a change in the $\delta^{18}O_{water}$ term will cause a similar change in the measured $\delta^{18}O_{calcite}$ value. However, an inverse relationship between $\delta^{18}O_{calcite}$ and T changes dictates that for every 1°C increase in temperature, there is a 0.23‰ decrease in the measured $\delta^{18}O_{calcite}$ value. These relationships enable us to interpret $\delta^{18}O_{calcite}$ changes generated from foraminifera, corals, and mollusks. For many years, the convention was to plot $\delta^{18}O_{calcite}$ values with the axis reversed (higher values to the left or bottom) so that $\delta^{18}O$ records reflect climate changes (e.g., colder to the left or bottom). More recently, there has been a trend among some scientists to plot $\delta^{18}O_{calcite}$ values without reversing the axis.

The paleotemperature equation contains two unknowns (temperature, $\delta^{18}O_{water}$). Although temperature is the main target in reconstructions, one cannot ignore the $\delta^{18}O_{water}$ term. In the modern ocean, the equator-to-pole gradient measured in planktonic foraminifera $\delta^{18}O_{calcite}$ values is ~5.0‰ and largely reflects the temperature gradient (~28°C). However, if temperature were the sole influence on modern $\delta^{18}O_{calcite}$ values, the equator-to-pole gradient would be ~6.5‰ (28°C × 0.23‰/°C). The attenuated $\delta^{18}O_{calcite}$ gradient measured in planktonic foraminifera reflects the surface ocean $\delta^{18}O_{water}$ variability. Therefore, a key to using $\delta^{18}O_{calcite}$ records as indicators of past climates is to understand the hydrographic parameters that produce the modern $\delta^{18}O_{calcite}$ gradient. For instance, ignoring the $\delta^{18}O_{water}$ term results in a 5–6°C underestimation compared to the observed temperature gradient. This occurs largely because tropical temperature estimates will be too cold (~4°C) whereas polar estimates will be warm (~1–2°C).

$\delta^{18}O$ Variation in the Natural Environment

$\delta^{18}O_{water}$ values in the ocean/atmosphere system vary both spatially and temporally because fractionation between the $H_2^{18}O$ and $H_2^{16}O$ molecules is temperature-dependent in the hydrologic cycle and follows the Rayleigh Distillation model (**Figure 1**). In general, water vapor evaporates at low latitudes and

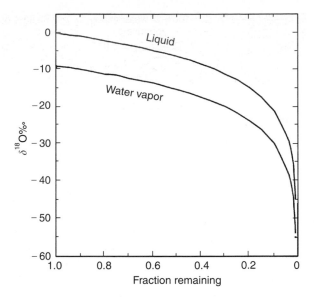

Figure 1 Rayleigh distillation model showing the effects of evaporation and precipitation on the $\delta^{18}O$ values in the vapor and liquid phases. The initial conditions are a temperature of 25°C and $\delta^{18}O_{water}$ value of 0‰. This model also assumes that it is a closed system, meaning that water vapor is not added once the cloud moves away from the source regions. As clouds lose moisture, fractionation during the condensation further lowers the $\delta^{18}O_{water}$ value in the water vapor.

precipitates at higher latitudes. Fractionation during evaporation concentrates the lighter $H_2^{16}O$ molecule in the water vapor, leaving the water enriched in $H_2^{18}O$ and $H_2^{16}O$. On average, the $\delta^{18}O_{water}$ value of water vapor is 9‰ lower than its source water (**Figure 1**). Fractionation during condensation concentrates the $H_2^{18}O$ molecules in the precipitation (rain/snow) by ~9‰. Therefore, if all of the water evaporated in the tropics rained back into the tropical oceans, there would be no net change in the $\delta^{18}O_{water}$ term. However, some water vapor is transported to higher latitudes. If the clouds remain a closed system (i.e., mid-to-high latitude evaporation does not influence the $\delta^{18}O_{water}$ value in the clouds[2]), then precipitation will further deplete the clouds (water vapor) in $H_2^{18}O$ relative to $H_2^{16}O$. Consequently, the $\delta^{18}O$ value of water vapor decreases from the original value as water vapor condenses into precipitation (**Figure 1**) and the cloud that formed from the evaporation in the tropics will eventually lose moisture, fractionating the $\delta^{18}O_{water}$ value of the remaining water vapor (**Figures 1 and 2**).

---

[1] $\delta^{18}O$ calcareous deposits are commonly reported relative to a carbonate standard, PDB (Peedee belemnite), and not SMOW (Standard Mean Ocean Water). PDB is 22‰ relative to SMOW.

[2] Many island or coastal regions have significantly higher $\delta^{18}O_{water}$ values relative to continental locations at similar latitudes. This occurs because local evaporation increases the $\delta^{18}O_{water}$ values, thus resetting the initial conditions for Rayleigh distillation to occur.

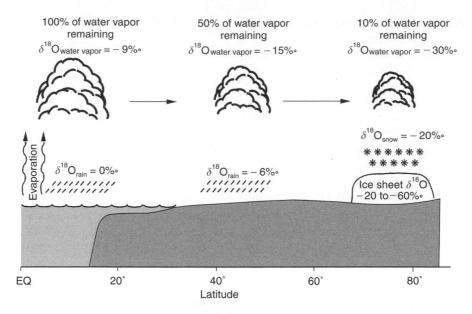

Figure 2 Illustration of the Rayleigh distillation process on $\delta^{18}O$ values as clouds move over land and into the polar regions. Decreasing air temperatures cause moisture to rain/snow out of the cloud. Fractionation of the oxygen isotopes during condensation further decreases values. By the point that a cloud reaches the high latitudes, less than 10% of the original water vapor remains. Snowfall on Antarctica has values between -20 and $-60\permil$. The average $\delta^{18}O$ value for ice on Antarctica is $\sim -40\permil$.

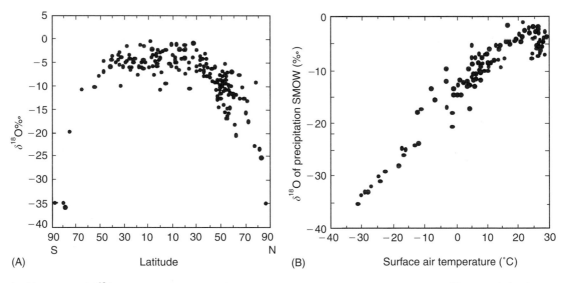

Figure 3 Mean annual $\delta^{18}O$ water of precipitation (rain/snow) versus mean annual temperatures. The correlation between $\delta^{18}O$ values and latitude (A) is a function of temperature (B). The rainout/fraction of water remaining, and hence the fraction of $\delta^{18}O$ values, is determined by the cloud temperatures. Latitude is the dominant effect shown here. The scatter among sites at similar latitude results from elevation differences as well as differences in the distance from the ocean.

By the time 50% of the initial moisture precipitates, the $\delta^{18}O$ value of the water vapor will be $\sim -15\%$, while precipitation will be $\sim -6\permil$. Once the cloud reaches the poles, over 90% of the initial water vapor will have been lost, producing $\delta^{18}O$ values of snow less than $-20\permil$. Snow at the South Pole approaches $-60\permil$. There is a strong relationship between $\delta^{18}O$ values in precipitation and air mass temperatures because air temperature dictates how much water vapor it can hold, and the $\delta^{18}O$ values of the precipitation is a function of the amount of water remaining in the clouds (**Figure 3**).

Spatial Variations in $\delta^{18}O_{water}$ of Modern Sea Water

The evaporation/precipitation process that determines the $\delta^{18}O_{water}$ values of precipitation (e.g., **Figure 1**) also controls the $\delta^{18}O_{water}$ values in regions in the ocean. At any one time, the volume of water being transported through the hydrologic cycle (e.g., atmosphere, lakes, rivers, and groundwater) is small compared to the volume of water in the oceans (1:130). Therefore, the hydrologic cycle can influence the whole ocean $\delta^{18}O_{water}$ value only by creating a new or enlarging an existing reservoir (e.g., glacier/ice sheets). In contrast, evaporation/precipitation processes will change the $\delta^{18}O_{water}$ and salinity values in the surface waters because only the thin surface layer of the ocean communicates with the atmosphere. As noted above, the process of evaporation enriches surface water in $H_2^{18}O$ molecules and salt because the water vapor is enriched in $H_2^{16}O$ molecules. For this reason, high salinity sea water has a high $\delta^{18}O_{water}$ value and vice versa. More specifically, tropical and subtropical surface water $\delta^{18}O_{water}$ values are $\sim 1‰$ higher than mean ocean water values (**Figure 4**). Interestingly, subtropical $\delta^{18}O_{water}$ values are higher than tropical values even though evaporation is higher in the tropics. Atmospheric circulation patterns produce intense rainfall in the tropics to offset some of the evaporation, whereas very little rain falls in the subtropical regions. Because evaporation minus precipitation (E – P) is greater in the subtropics, these regions have higher salinity and $\delta^{18}O_{water}$ values. In contrast, subpolar and polar regions have greater precipitation than evaporation; hence, high-latitude surface waters have low salinity and $\delta^{18}O_{water}$ values that approach $-0.5‰$ (**Figure 4**).

Temporal Variations

Variations in the amount of water stored on land through time, usually in the form of ice, can have a significant effect on the mean ocean $\delta^{18}O_{water}$ value, and hence, the marine $\delta^{18}O_{calcite}$ record. At present, high-latitude precipitation returns to the oceans through summer ice/snow melting. During glacial periods, snow and ice accumulate into large ice sheets. Because the difference in ice sheet and mean ocean values is large ($\delta^{18}O_{ice} = -35$ to $-40‰$ vs. $\delta^{18}O_{water}$ mean ocean $= \sim 0‰$), ice sheet fluctuations are reflected in mean oceanic $\delta^{18}O_{water}$ values. This relationship can be illustrated by examining how the mean ocean $\delta^{18}O_{water}$ value increased during the last glacial maximum (LGM) relative to the present (**Figure 5**). During the LGM, water stored in continental ice lowered global sea level by 120 m, removing $\sim 3\%$ of the ocean's volume. Thus, the mean ocean $\delta^{18}O_{water}$ value increased by 1.2‰ during the LGM relative to the present (**Figure 5**).

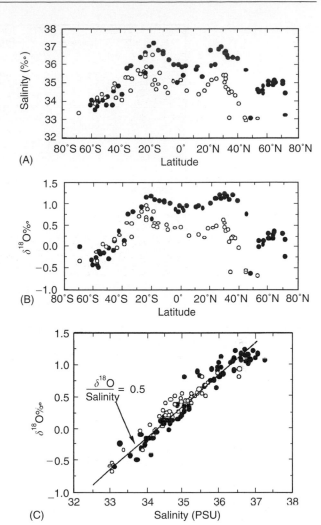

Figure 4 The salinity (A) and $\delta^{18}O_{water}$ values (B) measured in the open Atlantic (●) and Pacific (○) Oceans. Note the higher values in the tropical and subtropical region relative to the subpolar and polar regions. Evaporation and precipitation/runoff processes produce similar patterns in salinity and $\delta^{18}O_{water}$ values which is illustrated by the linearity in the $\delta^{18}O$ versus linity plot (C). The ocean-to-ocean difference between the Atlantic and Pacific results from a net transfer of fresh water from the Atlantic to the Pacific.

Pleistocene Oxygen Isotope Variations

The first systematic downcore examination of the marine stable isotope record was made by Cesaré Emiliani during the 1950s on $\delta^{18}O_{calcite}$ records generated from planktonic foraminifera in Caribbean deep-sea cores. Emiliani recognized the cyclic pattern of low and high $\delta^{18}O_{calcite}$ values and concluded that these represented glacial–interglacial intervals. Emiliani identified the seven most recent climate cycles and estimated that they spanned the last 280 000 years. (Current age estimates indicate that the duration of the cycles is approximately

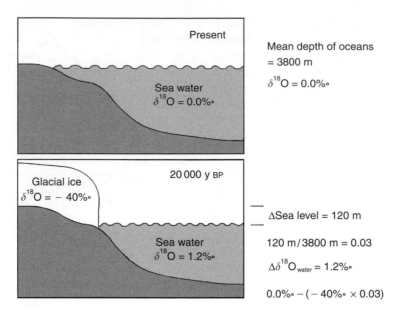

Figure 5 The effect of building or removing large ice sheets on the $\delta^{18}O$ composition of the ocean can be significant. The removal of 3% of the ocean's water during the last glacial maximum lowered sea level by 120 m. The $\delta^{18}O$ difference between the ocean and the ice is 40‰, causing a whole ocean $\delta^{18}O$ change of 1.2‰. The reverse process occurs during the melting of large ice sheets. If the Antarctic and Greenland ice were to melt, then sea level would rise $\sim$70 m. The volume of water stored in these ice sheets is equivalent to $\sim$2% of the water in the ocean. Therefore, the mean $\delta^{18}O$ value of the ocean would decrease by 0.7–0.8‰ (relative to PDB).

525 000 years.) To apply the paleotemperature equation to these records, Emiliani estimated that ice sheet-induced ocean $\delta^{18}O_{water}$ variability was relatively small, 0.3‰. (As shown above, the maximum glacial–interglacial ice sheet signal was closer to 1.2‰.) Therefore, most of the $\delta^{18}O_{calcite}$ variability between glacial and interglacial intervals represented temperature changes of 5–10°C. Emiliani divided the $\delta^{18}O_{calcite}$ record into warm stages (designated with odd numbers counting down from the Holocene) and cold stages (even numbers). Hence, 'Isotope Stage 1' refers to the present interglacial interval and 'Isotope Stage 2' refers to the LGM (**Figure 6**). During the 1960s and 1970s, many argued that most of the glacial to interglacial difference in $\delta^{18}O_{calcite}$ values resulted from ice volume changes. Nicholas Shackleton of Cambridge University made the key observation that benthic foraminiferal $\delta^{18}O$ values show a glacial to interglacial difference of $\sim$1.8‰. If the ice volume contribution was only 0.3‰ as argued by Emiliani, then the deep ocean temperatures would have been 6–7°C colder than the present temperatures of 0–3°C. Sea water freezes at -1.8°C, precluding Emiliani's 'low' ice volume estimate. By the early 1970s, numerous $\delta^{18}O$ records had been generated and showed a cyclic variation through the Pleistocene and into the late Pliocene. One hundred oxygen isotope stages, representing

50 glacial–interglacial cycles, have been identified for the interval since 2.6 million years ago (Ma) (**Figure 6**).

Cenozoic $\delta^{18}O$ Records

The first Cenozoic $\delta^{18}O$ syntheses based on foraminiferal $\delta^{18}O$ records were produced during the mid-1970s. Nicholas Shackleton and James Kennett produced a composite benthic $\delta^{18}O$ record for the Cenozoic from cores to the south of Australia. A second group led by Samuel Savin generated low-latitude planktonic and benthic foraminiferal $\delta^{18}O$ syntheses. Both records are important to understanding Cenozoic climate changes. Benthic foraminiferal records best reflect global temperature and ice volume changes. Additional advantages of the benthic foraminiferal composite include: (1) deep-ocean temperatures are more uniform with respect to horizontal and vertical gradients; (2) deep-ocean $\delta^{18}O_{water}$ values are less variable compared to the large surface water changes; (3) the deep ocean approximates high-latitude surface water conditions where deep waters originated during the Cenozoic (i.e., Antarctica, northern North Atlantic); and (4) many benthic foraminifera taxa are long-lived so that one species can be used to construct records spanning several millions of years in contrast to planktonic taxa which have shorter

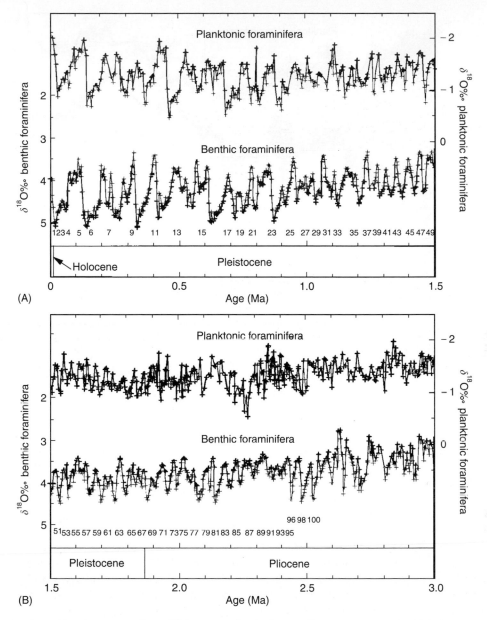

Figure 6 Planktonic and benthic foraminiferal $\delta^{18}O$ records for the last 3 million years. Note the high frequency signals in the records. For the interval between 3 and 1 Ma, a 40 000 year cycle dominates the records. After 1 Ma, the beat changes to a 100 000 year cycle and the amplitudes increase (relative to PDB).

durations and require records to be spliced together from several species.

Low-latitude planktonic foraminiferal $\delta^{18}O$ records are good proxies for tropical sea surface temperatures. Tropical temperatures are an important component of the climate system because they influence evaporation, and hence, total moisture in the atmosphere. Planktonic and benthic foraminiferal $\delta^{18}O$ comparison allows one to assess equator-to-pole as well as vertical temperature gradients during the Cenozoic, and thus, to determine planetary temperature changes. Finally, much of the climatic change in the last 65 million years has been ascribed

to poleward heat transport or greenhouse gas fluctuations. General circulation models indicate that each mechanism should produce different temperature patterns that can be approximated with the planktonic and benthic $\delta^{18}O$ records.

The first benthic $\delta^{18}O$ syntheses generated, as well as more recent compilations, show the same long-term patterns. After the Cretaceous–Tertiary (K/T) boundary events, deep-water $\delta^{18}O$ values remained relatively constant for the first 7 million years of the Paleocene (**Figure 7A**). At 58 Ma, benthic foraminiferal $\delta^{18}O$ values began a decrease over the next 6 My that culminated during the early Eocene

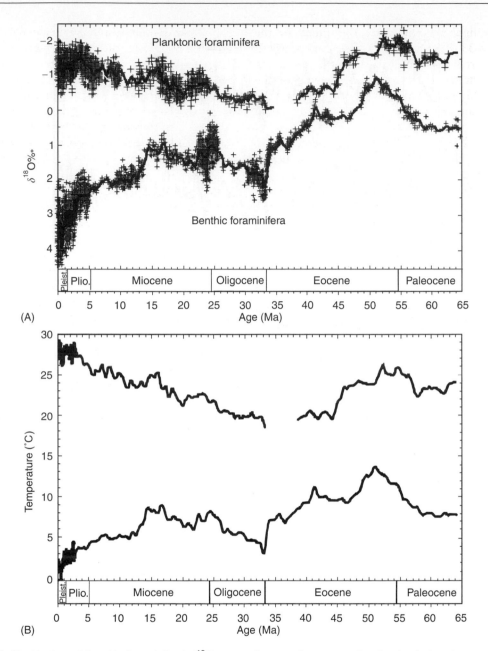

Figure 7 (A) Planktonic and benthic foraminiferal $\delta^{18}O$ composite records representing the tropical surface and deep ocean conditions (relative to PDB). The thick line through both records was generated using a 1 million year Gaussian filter. (B) Temperature estimates based on planktonic and benthic records and ice volume estimates discussed in the text.

with the lowest recorded value ($-0.5‰$) of thc Cenozoic. Following this minimum at 52 Ma, $\delta^{18}O$ values increased by 5.5‰, recording maximum values ($\sim 5‰$) during the glacial intervals of Pleistocene (**Figure 6**). The first part of this long-term change was a gradual increase of 2‰ through the end of the Eocene (52–34 Ma). The remainder of the increase was accomplished through large steps at the Eocene/Oligocene boundary (~ 33.5 Ma), during the middle Miocene (ca. 15–13 Ma) and late Pliocene (ca. 3.2–2.6 Ma). After 2.6 Ma, the amplitude of the high-

frequency signal increased to>1‰, reaching 1.8‰ over the past 800 thousand years.

Planktonic and benthic foraminiferal $\delta^{18}O$ values co-varied in general during the early Cenozoic (6.5–34 Ma). Values averaged about $-1‰$ between 65 and 58 Ma, before decreasing to $-2.5‰$ during the early Eocene, recording the lowest values of the Cenozoic (**Figure 7A**). From 52 to 33 Ma, planktonic foraminiferal values increased by 2‰. In spite of a break in the latest Eocene record, it appears that the tropical ocean differed from the deep ocean across the Eocene/Oligocene boundary. For much of the Oligocene ($\sim$33–25 Ma), planktonic foraminiferal $\delta^{18}O$ values remained unusually high, averaging $-0.5‰$. Beginning around the Oligocene/Miocene boundary ($\sim$25 Ma), planktonic foraminiferal $\delta^{18}O$ values began a long-term decrease, culminating in the Pleistocene with average values of $-1.5‰$. In contrast, the benthic $\delta^{18}O$ record permanently changed during the middle Miocene $\delta^{18}O$ shift and late Pliocene increase.

Apportioning the $\delta^{18}O$ changes recorded by the benthic and planktonic foraminifera between temperature and ice volume changes requires knowledge of, or reasonable estimates for, one of these parameters. One promising tool that may help discriminate between each effect is the Mg/Ca ratio measured in benthic foraminifera. Initial studies using Mg/Ca ratios confirmed the long-term temperature changes during the Cenozoic calculated using the $\delta^{18}O$ record and other climate proxies. If verified, this record implies that small ice sheets grew during the middle and late Eocene and fluctuated in size throughout the Oligocene to Miocene. At present, the Mg/Ca record lacks the resolution for key intervals and still requires verification of interspecies offsets before it can be applied unequivocally to isolate the ice volume-induced $\delta^{18}O_{water}$ component in the foraminiferal $\delta^{18}O$ records. For the discussion that follows, glaciological evidence is used to estimate the ice volume/$\delta^{18}O_{water}$ variations.

The Greenhouse World

The oldest unequivocal evidence for ice sheets on Antarctica, ice-rafted detritus (IRD) deposited by icebergs in the ocean, places the first large ice sheet in the earliest Oligocene. Thus, it is reasonable to assume that ice sheets were small to absent and that surface and deep-water temperature changes controlled much if not all the $\delta^{18}O$ change prior to 34 Ma. The modern Antarctic and Greenland ice sheets lock up $\sim$2% of the total water in the world's ocean. If melted, these ice sheets would raise global sea level by $\sim$70–75 m and mean ocean $\delta^{18}O_{water}$ value would decrease to $-0.9‰$ PDB (see above). One can then apply eqn [2] to the benthic and planktonic foraminiferal $\delta^{18}O$ records to estimate deep- and surface-ocean temperatures for the first half of the Cenozoic (c. 65–34 Ma).

During the early to middle Paleocene, deep-water temperatures remained close to 10°C (**Figure 7B**). The 1‰ decrease between 58 and 52 Ma translates into a deep-water warming of 4°C, reaching a high of 14°C. This is in sharp contrast to the modern

deep-water temperatures, which range between 0 and 3°C. Following the peak warmth at 52 Ma, the 2‰ increase in benthic foraminiferal values indicates that the deep waters cooled by 7°C and were 7°C by the end of the Eocene. If small ice sheets existed during the Paleocene and Eocene, then temperature estimates would be on the order of 1°C warmer than those calculated for the ice-free assumption. (Some data indicate that smaller ice sheets may have existed on the inland parts of Antarctica during the late Eocene. However, these were not large enough to deposit IRD in the ocean. Therefore, their effect on the $\delta^{18}O$ values of the ocean was probably less than 0.3‰.)

Tropical surface water temperatures warmed from 22 to 24°C, based on eqn [2], at the beginning of the Cenozoic to 28°C during the early Eocene (52 Ma; **Figure 7B**). The higher estimate is similar to temperatures in the equatorial regions of the modern oceans. Planktonic foraminiferal $\delta^{18}O$ values recorded a long-term increase of by 2‰ (-2.5 to $-0.5‰$) through the remainder of the Eocene. Just prior to the Eocene/Oligocene boundary, tropical surface water temperatures were $\sim$21°C, ending the long-term tropical cooling of 7°C from 52 to 34 Ma.

The Ice House World of the Last 33 Million Years

As mentioned above, southern ocean cores contain IRD at and above the Eocene/Oligocene boundary. Widely distributed IRD and glacial tills on parts of the Antarctic continental margin representing the Oligocene to Recent mark the onset of large ice sheets. Whether these sediments represent persistent or periodic ice cover is uncertain. At least some ice was present on Antarctica during the Oligocene to early Miocene. The Antarctic ice sheet has been a fixture since the middle Miocene ($\sim$15 Ma). Our record of Northern Hemisphere ice sheets suggests that they were small or nonexistent prior to the late Pliocene. For the purpose of estimating surface and deep temperatures, an ice volume estimate slightly lower than the modern will be applied for the interval that spans from the Oligocene into the middle Miocene (33–15 Ma). For the interval between 15 and 3 Ma, ice volumes were probably similar to those of today. From 3 Ma, ice volumes ranged between the modern and LGM. Using these broad estimates for ice volumes, mean ocean $\delta^{18}O_{water}$ values for those three intervals were -0.5, -0.22, and 0.4‰ PDB, respectively. The 0.4‰ estimate reflects the average between the maximum and minimum conditions during the Plio-Pleistocene. As noted above, the largest portion of the high frequency signal is controlled by ice volume changes.

The benthic foraminiferal $\delta^{18}O$ increase at the Eocene/Oligocene boundary occurred rapidly ($\sim 10\,000$ years; **Figure 8A**). At the peak of the Eocene/Oligocene boundary event, benthic foraminifera recorded $\delta^{18}O$ values similar to modern values. Using the ice volume assumption from above, deepwater temperatures approached modern deep-ocean temperatures ($3°C$). This marks an important transition from the relatively warm oceans of the Paleocene and Eocene to the cold deep waters of the Oligocene to present. This switch to a cold ocean where bottom waters formed at near-freezing temperatures heralded the development of the psychrosphere. Following the Eocene/Oligocene boundary, deep-water temperatures began a long-term warming over the next 18 million years (33–15 Ma). The coldest deep-water temperatures of $3°C$ were recorded at 33 Ma, while temperatures reached $9°C$ at ~ 25 and ~ 15 Ma (**Figure 7B**).

There is a gap in the planktonic foraminiferal $\delta^{18}O$ record for the latest Eocene that hampers our assessment of tropical response during Eocene/Oligocene climate event. However, it is clear from the data that do exist that the planktonic response across the Eocene/Oligocene boundary differed from the benthic response. The planktonic foraminiferal $\delta^{18}O$ values for the early Oligocene are similar to late Eocene values, whereas the benthic values recorded a 1.5‰ increase. Planktonic foraminiferal records from other regions that span the Eocene/Oligocene boundary indicate that the surface water $\delta^{18}O$ increase was on the order of 0.5‰. This change is approximately equal to the effect of the modern Antarctic ice sheet. Combined with the physical evidence, it seems probable that the planktonic foraminiferal $\delta^{18}O$ increase at the Eocene/Oligocene boundary recorded the ice volume influence with little temperature effect. Therefore, tropical surface temperatures remained around $22°C$ while the deep ocean cooled during

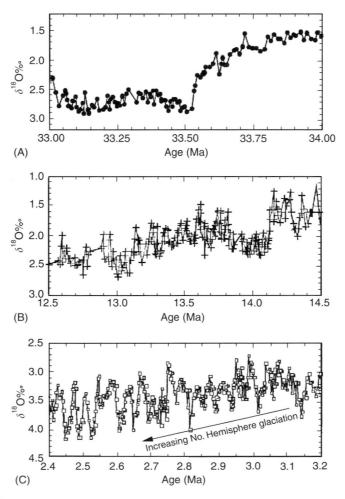

(A)

(B)

(C)

Figure 8 High-resolution $\delta^{18}O$ records representing the Eocene/Oligocene boundary (A), middle Miocene (B), and late Pliocene (C) $\delta^{18}O$ shifts (relative to PDB).

this $\delta^{18}O$ shift. Following the boundary event, planktonic foraminifera $\delta^{18}O$ record during the Oligocene and early Miocene mirrored the benthic record in many respects. For much of the Oligocene and early Miocene, the absolute values are close to $-0.5‰$, which translates into a temperature estimate of $21°C$ (**Figure 7B**). By 15 Ma, tropical surface waters had warmed to $26°C$.

The middle Miocene $\delta^{18}O$ shift represents an increase of $1.5‰$ in the benthic record between 15 and 13 Ma. This transition is composed of two sharp increases around 14 and 13 Ma (**Figure 8B**). These $\delta^{18}O$ steps occurred in less than 200 000 years with each recording an increase of $\sim 1‰$ followed by a small decrease. During these two shifts, deep waters cooled from 9 to $5°C$. The planktonic foraminiferal $\delta^{18}O$ record from 15 to 13 Ma shows two increases as recorded in the benthic foraminiferal record (**Figure 7**). However, it does not show the large permanent shift recorded by the benthic foraminifera, indicating a small cooling from 26 to $24°C$. From 13 to 3 Ma, the deep ocean cooled slightly from 5 to $3°C$ while the surface waters warmed from 24 to $26°C$ (**Figure 7B**).

The last of the large $\delta^{18}O$ steps in the Cenozoic was recorded during the late Pliocene from 3.2 to 2.6 Ma. This 'step' is better characterized as a series of $\delta^{18}O$ cycles with increasing amplitudes and values over this interval (**Figures 6 and 8C**). The cycles have been subsequently determined to be 40 000 year cycles related to variations in the solar radiation received in the high latitudes. This interval ushered in the large-scale Northern Hemisphere ice sheets that have since dominated Earth's climate. At 2.6 Ma, the first IRD was deposited in the open North Atlantic and was coeval with the $\delta^{18}O$ maximum. Prior to 2.6 Ma, IRD was confined to the marginal basins to the north, Greenland's and Iceland's continental margins. Subsequent $\delta^{18}O$ maxima were associated with IRD. Between 2.6 and 1 Ma, large Northern Hemisphere ice sheets waxed and waned on the 40 000 year beat. Beginning around 1 Ma, the ice sheets increased in size and switched to a 100 000 year beat (**Figure 6**). During this interval, deep-water temperatures remained similar to those in the modern ocean (0 to $3°C$).

The planktonic foraminiferal $\delta^{18}O$ response during the late Pliocene event shows the cyclic behavior, but not the overall increase recorded by the benthic foraminifera. As with the middle Miocene $\delta^{18}O$ shift, the late Pliocene increase represents the cyclic build-up of ice sheets accompanied by deep water cooling. The tropical surface water temperatures, however, varied between 26 and $28°C$.

Mechanisms for Climate Change

Most climate change hypotheses for the Cenozoic focus on either oceanic heat transport and/or greenhouse gas concentrations. Each mechanism produces different responses in the equatorial-to-pole and surface-to-deep temperature gradients. An increase in the meridional heat transport generally cools the tropics and warms the poles. If poleward transport of heat decreases, then the tropics will warm and the poles will cool. Variations in greenhouse gas concentrations should produce similar changes in both the tropical and polar regions.

Tropical surface water and deep-ocean records covaried for the first part of the Cenozoic. The warming and subsequent cooling between 65 and 34 Ma are most often ascribed to changing greenhouse gas concentrations. The interval of warming that began around 58 Ma and peaked at 52 Ma coincided with the release of large amounts of CO_2 into the atmosphere as a consequence of tectonic processes. The eruption of the Thulean basalts in the northeastern Atlantic Ocean began during the Paleocene and peaked around 54 Ma. It is also recognized that there was a large-scale reorganization of the mid-ocean ridge hydrothermal system which began during the late Paleocene and extended into the Eocene. Both tectonic processes accelerate mantle degassing which raises atmospheric levels of CO_2. Recently, evidence for another potentially large CO_2 reservoir was found along the eastern continental margin of North America. Methane hydrates frozen within the sediments appear to have released catastrophically at least once and possibly multiple times during the latest Paleocene and early Eocene (~ 58–52 Ma). One or all of these sources could have contributed to the build-up of greenhouse gases in the atmosphere between 58 and 52 Ma. Following the thermal maximum, the long-term cooling in both the surface and deep waters implies that greenhouse gas concentrations slowly decreased. Proxies for estimating pCO_2 concentrations ($\delta^{13}C$ fractionation within organic carbon and boron isotopes) are still being developed and refined. However, preliminary indications are that atmospheric pCO_2 levels were high (>1000 p.p.m.) during the early Eocene, dropped to ~ 400–500 p.p.m. during the middle to late Eocene, and reached late Pleistocene concentrations (200–300 p.p.m.) by the early Oligocene.

The deep-water temperature cooling across the Eocene/Oligocene boundary (**Figure 7B**) was not accompanied by tropical cooling, and resulted from the first step in the thermal isolation of Antarctica. In modern ocean, the Antarctic Circumpolar Current is a vigorous surface-to-bottom current that provides an effective barrier to southward-flowing warm surface waters. The development of this current during the Cenozoic hinged on the deepening for the Tasman Rise and opening of the Drake Passage. Recent drilling indicates that marine connections developed across the Tasman Rise at or near the Eocene/Oligocene boundary (33.5 Ma). Tectonic constraints on the separation of the Drake Passage are less precise. Estimates range from 35 to 22 Ma for the opening of this gateway. The uncertainty lies in the tectonic complexity of the region and what constitutes an effective opening for water to flow through. The climatic consequence of creating a circumpolar flow was to thermally isolate Antarctica and promote the growth of the Antarctic ice sheet. As noted above, the first large ice sheet grew at the beginning at the Eocene/Oligocene boundary.

The most notable divergence in the $\delta^{18}O$ records occurred during the middle Miocene (~ 15 Ma). For the first time during the Cenozoic, the tropical surface and deep waters recorded a clear divergence in $\delta^{18}O$ values, a trend that increased in magnitude and reached a maximum in the modern ocean. Any poleward transport of heat appears to have been effectively severed from Antarctica by 15 Ma, promoting further cooling. On the other hand, the tropics have been warming over the past 15 My. A combination of different factors fueled this warming. First, less heat was being transported out of the low- and mid-latitude regions to the high southern latitudes. Second, the opening of the Southern Hemisphere gateways that promoted the formation of the circumpolar circulation led to the destruction of the circumequatorial circulation. The effects of the closure of the Tethys Ocean (predecessor to the Mediterranean), shoaling of the Panamanian Isthmus (4.5–2.6 Ma), and constriction in the Indonesian Passage (~ 3 Ma to present) allowed the east-to-west flowing surface waters in the tropics to 'pile' up and absorb more solar radiation. A consequence of the equatorial warming and high-latitude cooling was an increase in the equator-to-pole temperature gradient. As the gradient increased, winds increased, promoting the organization of the surface ocean circulation patterns that persist today.

Some Caveats

A concern in generating marine isotope records is that the isotopic analyses should be made on the same species. This is important because $\delta^{18}O_{calcite}$ values can vary among the different species of organisms. Coexisting taxa of benthic foraminifera record $\delta^{18}O$ values that can differ by as much as 1‰.

In planktonic foraminifera, variations between species can be as great as 1.5‰. For both the planktonic and benthic foraminifera, interspecific differences are as large as the glacial–interglacial signal. These interspecific $\delta^{18}O_{calcite}$ variations are often ascribed to a vital effect or kinetic fractionation of the oxygen isotopes within the organism. However, some of the differences in the planktonic taxa results from different seasonal or depth habitats and therefore provides important information about properties in the upper part of the water column. It is noteworthy that the first $\delta^{18}O$ syntheses were based on mixed species analyses and yet basic features captured in these curves still persist today. This attests to the robustness of these records and method for reconstructing climate changes in the ocean.

The high-frequency signal that dominates the late Pliocene to Pleistocene records is also present in the Miocene and Oligocene intervals. The cloud of points about the mean shown in **Figure 7** reflects records that were sampled at a resolution sufficient to document the high frequency signal. For the interval between 35 and 1 Ma, the benthic foraminiferal $\delta^{18}O$ record has a 40 000 year frequency superimposed on the long-term means that are represented by the smoothed line. The origin of the 40 000 year cycles lies in variations in the tilt of the earth's axis that influences the amount of solar radiation received in the high latitudes. This insolation signal is transmitted to the deep ocean because the high latitudes were the source regions for deep waters during much, if not all, of the Cenozoic. The record prior to 35 Ma is unclear with regard to the presence or absence of 40 000 year cycles.

See also

Holocene Climate Variability. Oxygen Isotopes in the Ocean. Sea Ice: Overview. Sub Ice-Shelf Circulation and Processes.

Further Reading

Craig H (1957) Isotopic standards for carbon and oxygen and correction factors for mass spectrometric analysis of carbon dioxide. *Geochemica et Cosmochemica Acta* 12: 133–149.

Craig H (1965) The measurement of oxygen isotope paleotemperatures. In: Tongiorgi E (ed.) *Stable Isotopes in Oceanographic Studies and Paleotemperatures*, pp. 161–182. Spoleto: Consiglio Nazionale delle Ricerche, Laboratorio di Geologica Nucleare, Pisa.

Craig H and Gordon LI (1965) Deuterium and oxygen-18 variations in the oceans and marine atmosphere.

In: Tongiorgi E (ed.) *Stable Isotopes in Oceanographic Studies and Paleotemperatures*, pp. 1–122. Spoleto: Consiglio Nazionale delle Ricerche, Laboratorio di Geologica Nucleare, Pisa.

Emiliani C (1955) Pleistocene temperatures. *Journal of Geology* 63: 539–578.

Epstein S, Buchsbaum R, Lowenstam H, and Urey HC (1953) Revised carbonate-water temperature scale. *Bulletin of the Geological Society of America* 64: 1315–1326.

Fairbanks RG, Charles CD, and Wright JD (1992) Origin of Melt Water Pulses. In: Taylor RE, Long A, and Kra RS (eds.) *Radiocarbon After Four Decades*, pp. 473–500. New York: Springer-Verlag.

Imbrie J, Hays JD, Martinson DG, *et al.* (1984) The orbital theory of Pleistocene climate: support from a revised chronology of the marine $\delta^{18}O$ record. In: Berger AL, Imbrie J, Hays JD, Kukla G, and Saltzman B (eds.) *Milankovitch and Climate, part I*, pp. 269–305. Dordrecht: Reidel.

Lear CH, Elderfield H, and Wilson PA (1999) Cenozoic deep-sea temperatures and global ice volumes from Mg/Ca in benthic foraminiferal calcite. *Science* 287: 269–272.

Miller KG, Fairbanks RG, and Mountain GS (1987) Tertiary oxygen isotope synthesis, sea-level history, and continental margin erosion. *Paleoceanography* 2: 1–19.

Miller KG, Wright JD, and Fairbanks RG (1991) Unlocking the Ice House: Oligocene–Miocene oxygen isotopes, eustasy, and margin erosion. *Journal of Geophysical Research* 96: 6829–6848.

Pagani M, Arthur MA, and Freeman KH (1999) Miocene evolution of atmospheric carbon dioxide. *Paleoceanography* 14: 273–292.

Palmer MR, Pearson PN, and Cobb SJ (1998) Reconstructing past ocean pH-depth profiles. *Science* 282: 1468–1471.

Pearson PN and Palmer MR (1999) Middle Eocene seawater pH and atmospheric carbon dioxide concentrations. *Science* 284: 1824–1826.

Rozanski K, Araguas-Araguas L and Gonfiantini R (1993) Isotopic patterns in modern global precipitation. In: Swart PK, McKenzie J and Savin S (eds) *Climate Change in Continental Isotopic Records*. Geophysical Monograph 78, pp. 1–35. Washington, DC: American Geophysical Union.

Rye DM and Sommer MA (1980) Reconstructing paleotemperature and paleosalinity regimes with oxygen isotopes. In: Rhoads DC and Lutz RA (eds.) *Skeletal Growth of Aquatic Organisms*, pp. 162–202. New York: Plenum.

Savin SM, Douglas RG, and Stehli FG (1975) Tertiary marine paleotemperatures. *Geological Society of America Bulletin* 86: 1499–1510.

Shackleton NJ (1967) Oxygen isotope analyses and Pleistocene temperatures re-assessed. *Nature* 215: 115–117.

Shackleton NJ, Berger A, and Peltier WR (1990) An alternative astronomical calibration of the Lower Pleistocene time scale based on ODP Site 677. *Transactions of the Royal Society of Edinburgh, Earth Science* 81: 251–261.

Shackleton NJ and Kennett JP (1975) Paleotemperature history of the Cenozoic and initiation of Antarctic glaciation. Oxygen and carbon isotopic analysis in DSDP Sites 277, 279, and 281. *Initial Report Deep Sea Drilling Project* 29: 743–755.

Shackleton NJ and Opdyke ND (1973) Oxygen isotope and paleomagnetic stratigraphy of equatorial Pacific core V28-238. Oxygen isotope temperatures and ice volumes on a 10^5 year and 10^6 year scale. *Quaternary Research* 3: 39–55.

Tiedemann RM, Sarnthein M, and Shackleton NJ (1994) Astronomic calibration for the Pliocene Atlantic $\delta^{18}O$ and dust flux records of Ocean Drilling Program Site 659. *Paleoceanography* 9: 619–638.

Urey HC (1947) The thermodynamic properties of isotopic substances. *Journal of the Chemical Society* pp. 562–581.

DETERMINATION OF PAST SEA SURFACE TEMPERATURES

M. Kucera, Eberhard Karls Universität Tübingen, Tübingen, Germany

Introduction

Temperature is one of the most striking aspects of the climate system. Its distribution on the surface of the Earth determines regional climate and habitability and its changes through time mirror the climatic evolution of our planet. The surface ocean represents the main reservoir of latent heat with ocean currents being the major means of heat redistribution at the Earth's surface. In addition to the position of currents, sea surface temperature also reflects vertical mixing in the ocean, where upwelling brings deep cold waters to the surface.

Clearly, knowledge of past variations in sea surface temperature (SST) and its spatial distribution is essential for the assessment of current warming trends as well as for the general understanding of the dynamic processes in the ocean and their links with global climate. Importantly, quantitative data on spatial and temporal distribution of past SST are instrumental for validation of numerical climate models. Unfortunately, direct instrumental measurements of SST are only available for the last two centuries. This time window is too short to assess the nature of natural variability and understand how the ocean and atmosphere systems behave under different climatic regimes. To know when, why, by how much, and how fast SST changes, scientists need long, continuous records spanning centuries to millions of years.

Fortunately, SST affects a range of metabolic and thermodynamic processes that leave a distinct signature in biological materials. Some of these materials are preserved in the geological record and the signals locked in them can be decoded to estimate SST variation in the past. Unsurprisingly, reconstructing the temperature history of the surface ocean became one of the major tasks in geosciences and the last 50 years have seen the establishment of quantitative paleothermometry as the backbone of paleoceanography and paleoclimatology. Among the main achievements of SST paleothermometry are the reconstruction of the ocean surface during the Last Glacial Maximum by the CLIMAP (Climate: Long Range Investigation Mapping and Prediction) group (**Figure 1**), the recognition of the pattern, rate, and magnitude of ocean cooling and warming during glacial cycles, and the determination of tropical and polar temperatures during supergreenhouse climates of the Mesozoic and early Cenozoic.

Characteristics of Sea Surface Temperatures

Before explaining the various paleothermometry methods, it is important to understand what is actually meant by SST. The surface ocean derives its temperature from solar insolation, whereas the temperature of the deep waters is relatively constant, preserving the SST signature at the place of their origin. At present, deep waters in the oceans are formed chiefly in polar and subpolar regions and their temperatures range between 0 and 4 °C. The resulting thermal gradient (**Figure 2**) implies that only a relatively shallow upper layer of the ocean retains the temperature of the ocean's surface. The 'mixed layer' is homogenized by wave action and turbulence. Its depth is typically 50–100 m (**Figure 3**), although in summer during calm weather and due to intense heating, it can be only a few meters thin. Importantly, the mixed layer can be taken to roughly correspond to the photic zone, which means that phytoplankton and photosymbiotic organisms like corals and some planktonic foraminifera are likely to record the SST, whereas some zooplankton may live below the mixed layer, recording the temperature of the thermocline rather than the surface ocean (**Figure 2**).

Due to the high thermal capacity of water, the range of surface ocean temperature is smaller than that of the atmosphere. At present, the coldest surface waters in polar oceans reach the freezing point of seawater at −1.7 °C while the highest surface temperatures of 32 °C are recorded in the Western Pacific Warm Pool. Seasonal variations in SST are also relatively muted, which makes the SST a good recorder of the mean state of the climate system. However, the magnitude of the seasonal signal (**Figure 3**) exceeds the precision of most methods for past SST determination and the difficulty in seasonal attribution of SST reconstructions is one of the main caveats in paleothermometry.

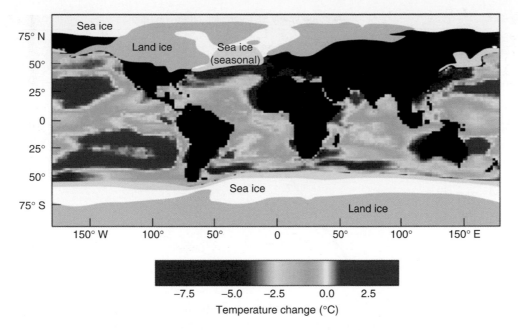

Figure 1 The CLIMAP reconstruction of SST during the Last Glacial Maximum (*c.* 21 000 years ago), expressed as the difference between present-day and glacial values. The map represents one of the milestones of paleoceanography. It was derived from abundances of microfossil species in sediment cores using the transfer function method. Reproduced from Mix A, Bard E, and Schneider R (2001) Environmental processes of the ice age: Land, oceans, glaciers (EPILOG). *Quaternary Science Reviews* 20: 627–657, with permission from Elsevier. Data from CLIMAP Project Members (1976) The surface of the ice-age Earth. *Science* 191: 1131–1137.

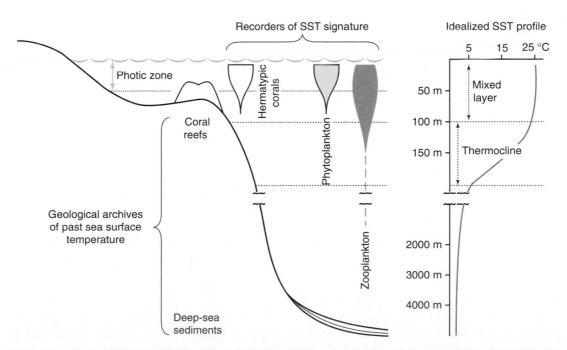

Figure 2 An idealized scheme of the habitats of SST signal carriers and the occurrence of geological archives of SST. Hermatypic corals and phytoplankton grow only in the photic zone, which normally overlaps with the mixed layer. Some zooplankton species live below the photic zone and record thermocline temperatures rather than SST.

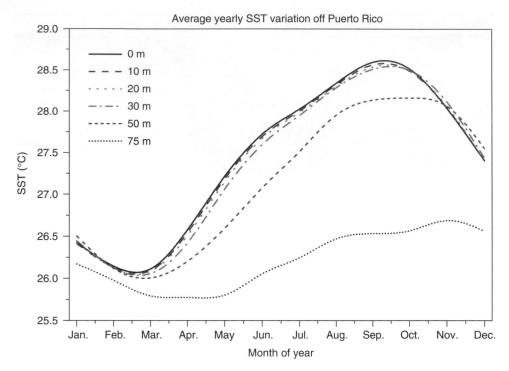

Figure 3 A typical tropical mean annual SST record for different depths. The temperature of the Caribbean Sea off Puerto Rico varies by >2 °C throughout the year; the thickness of the mixed layer varies between >50 m in winter and *c.* 30 m in summer. Data from the Comprehensive Ocean–Atmosphere Data Set (COADS).

Methods for Determination of Past Sea Surface Temperatures

Nature of the Sea Surface Temperatures Signal in Geological Materials

Many properties of marine organisms are affected by their chemical and physical environment. For SST reconstructions, the most significant are kinetic processes that alter the rate of incorporation of trace elements and isotopes into biominerals, metabolic regulation of the composition of cellular membranes, and temperature-driven changes in species abundance (**Table 1**). These various signals locked in fossil materials are not directly related to individual environmental parameters. Therefore, a whole branch of paleoceanography has been established to develop and test proxies – recipes and algorithms describing ways of how to relate measurements and observations made on fossils and other geological material to past environmental variables. Paleothermometry is the branch of paleoceanography devoted to the development of proxies for the reconstruction of past ocean temperatures.

At present, only two archives have the capacity to provide long and continuous records of past SST: marine sediments and coral colonies (**Figure 2**). Deep-sea sediments are particularly suitable as SST archives because they are often undisturbed and accumulate continuously for millions of years. Marine sediments are sampled by coring or drilling and SST signals are extracted from mineral or organic remains of planktonic microorganisms that lived in the surface ocean above the site of deposition. Strictly speaking, the SST signal is not *in situ*, because the plankton remains were transported to the seafloor by sinking through several kilometers of overlaying water. The degree of potential lateral displacement of the signal depends on the size of the signal carrier: small organic-walled microfossils and organic compounds are more easily transported than shells of planktonic foraminifera (**Figure 4**) or microfossils carried down in fecal pellets. SST records in marine sediments have typically centennial and rarely decadal resolution.

The skeletons of hermatypic corals grow by accretion and like tree rings they record with high resolution the conditions of the surface ocean. Hermatypic corals are restricted to the tropical oceans with normal salinity where water temperature exceeds 25 °C. The vast majority of reef-building corals (such as the genus *Porites*) harbor photosymbiotic algae and are thus limited to the photic zone (**Figure 2**). Individual coral records recovered from cores drilled into living or fossil submerged coral heads are typically only a few

Table 1 Parameters of the main methods used to determine past sea surface temperatures

Method	Substrate	SST range[a]	Typical standard error (°C)	Temporal applicability	Regional applicability
Oxygen isotopes in planktonic foraminifera	Foraminiferal calcite	−2 °C to no upper limit	0.5[b]	0–120 My[c]	All oceans
Mg/Ca in planktonic foraminifera	Foraminiferal calcite	~5 °C to no upper limit	1	0–120 My[d]	All oceans, best performance in Tropics
Oxygen isotopes and Sr/Ca in corals	Sclearctinian aragonite	Mostly tropical waters, no upper limit	0.5[b]	0–130 ky	Mainly Tropics
Alkenone unsaturation (U_{37}^{k})	Haptophyte alkenones in sediments	Best performance between 5 and 27 °C	1.5	0–5 My[e]	All oceans except those of polar regions
Tetraether index (TEX$_{86}$)	Crenarchaeotal membrane lipids in sediments	0–28 °C >28 °C[h]	2	0–120 My[f]	All oceans, possibly also lakes
Transfer functions	Planktonic foraminifera	5–30 °C	1–1.5	0–500 ky[g]	All oceans except those of polar regions
	Radiolaria	0–30 °C	1–1.5	0–500 ky[g]	All oceans except those of polar regions
	Diatoms	−2 to 18 °C	1–1.5	0–500 ky[g]	Polar to transitional waters
	Dinoflagellate cysts	−2 to 22 °C	1.5–2	0–500 ky[g]	All oceans except those of deep oligotrophic basins

[a] Mean annual SST.
[b] Provided the isotopic and trace element composition of seawater is known.
[c] In sediments older than c. 25 My in pristine preservation only.
[d] In sediments older than c. 5 My, unknown species offsets and Mg content of the ocean may cause significant bias.
[e] Ecology of pre-Quaternary alkenone-producing haptophytes is unknown.
[f] Possibly older, if well-preserved lipids are found.
[g] Estimate based on the age of the modern planktonic foraminifer fauna; older applications are problematic.
[h] Several alternative calibrations are used for SST > 28 °C

centuries long. However, depending on the growth rate, they can provide yearly or even monthly resolution and data from several heads can be stacked to develop longer, regional records.

Paleothermometers Based on Chemical Signals in Biominerals

Oxygen isotopes in planktonic foraminifera Measurement of oxygen isotopic composition of marine biogenic carbonates provided geoscientists and paleoceanographers with the first quantitative paleothermometer. The method is based on the discovery by the American physicist and Nobel Prize laureate Harold Urey that during precipitation of calcite from seawater, a thermodynamic fractionation occurs between the two main stable isotopes of oxygen, ^{16}O and ^{18}O, and that this fractionation is a logarithmic function of temperature. The technique was originally calibrated on mollusk shells, but since

the pioneering work of Cesare Emiliani in the 1950s, planktonic foraminifera (**Figure 4**) became the prime target for oxygen isotope paleothermometry in oceanic sediments. The oxygen isotopic composition of foraminiferal shells is determined by mass spectrometry of CO_2 evolved by acid digestion of foraminiferal calcite. Although modern analytical methods allow measurements of oxygen isotopes in single shells, a standard analysis requires about 10–20 specimens, in order to obtain a representative average of the conditions integrated in a fossil sample. The oxygen isotopic composition of foraminiferal calcite is expressed as a deviation, in per mille, from a standard (normally V-PDB):

$$\delta^{18}O = \left[\frac{\left({}^{18}O/{}^{16}O \right)_{sample} - \left({}^{18}O/{}^{16}O \right)_{standard}}{\left({}^{18}O/{}^{16}O \right)_{standard}} \right] \times 10^3$$

[1]

Figure 4 Shells of planktonic foraminifera from a Caribbean surface sediment sample. Foraminifera are the main substrate for geochemical and transfer function paleothermometry. The shells have been extracted by washing and sieving of the sediments through a 0.15-mm mesh. The image is 4 mm across.

The determination of past SST from oxygen isotopes requires *a priori* knowledge of the isotopic composition of ambient seawater. When this is known, empirical calibrations from culture experiments or surface sediments can be used to derive paleotemperature equations for different species and SST ranges. These equations usually take on a second-order polynomial form; the Shackleton calibration is perhaps the most commonly used:

$$T_{\text{calcification}}[^{\circ}C] = 16.9 - 4.38$$
$$\times \left(\delta^{18}O_{\text{calcite}} - \delta^{18}O_{\text{seawater}}\right)$$
$$+ 0.1 \times \left(\delta^{18}O_{\text{calcite}} - \delta^{18}O_{\text{seawater}}\right)^2$$
$$[2]$$

Given the high analytical precision of oxygen isotope measurements ($<0.1‰$), the method allows SST reconstruction with the precision of 0.5–1 °C. However, the biomineralization of calcite by foraminifera is affected by a range of secondary factors and the $\delta^{18}O$ composition of seawater in the past must be estimated from a number of assumptions.

The two main factors influencing $\delta^{18}O_{\text{seawater}}$ are the global ice volume and the local precipitation/evaporation (P/E) balance. Due to fractionation during water phase transitions, continental ice is depleted in the heavy isotope, leaving the seawater isotopically heavy. Similarly, during evaporation, the light isotope preferentially enters the gaseous phase, making water vapor and precipitation isotopically light. The P/E effect is also known as the salinity effect, because seawater salinity is affected by the P/E balance in a similar way as its oxygen isotopic composition. The combined effect of ice volume and P/E balance on $\delta^{18}O$ in foraminifera can be substantially larger than that of SST, making paleotemperature reconstructions from foraminiferal $\delta^{18}O$ rather difficult. In fact, the $\delta^{18}O$ method is now used predominantly to reconstruct the isotopic composition of seawater and, if available, other paleothermometers are preferred.

Planktonic foraminifera are heterotrophic zooplankton and their habitat is not limited to the surface mixed layer (**Figure 2**). Each species calcifies at specific depths, often below the mixed layer, and maximum production occurs at specific times of the year. In addition, the presence of symbiotic algae in some species offsets the $\delta^{18}O$ signal, as do changes in growth rate, addition of gametogenic calcite at depth, carbonate ion concentration in seawater, and various other 'vital effects'. Dissolution appears to increase $\delta^{18}O$ in foraminifera by about 0.2‰ per kilometer water depth, but this effect is relatively easy to contain. On geological scales, foraminiferal calcite is gradually recrystallized during burial in the sediment and the secondary calcite assumes the isotopic composition of inorganic precipitate from pore water fluid. Foraminiferal $\delta^{18}O$ paleothermometry in Cretaceous and Paleogene sediments appears only feasible on pristinely preserved shells recovered from clay-rich sediments. Although the method is no longer seen as the prime SST paleothermometer, the development of independent methods based on the same signal carrier (Mg/Ca, transfer functions) provides an exciting opportunity to subtract the SST contribution to foraminiferal $\delta^{18}O$ and reconstruct more precisely than before the $\delta^{18}O$ of seawater.

Mg/Ca in planktonic foraminifera   The amount of trace elements incorporated into inorganically precipitated calcite depends chiefly on their concentration in the solution from which the mineral grows. However, the rate of cation substitution in calcite also depends on temperature. This latter effect is particularly pronounced for magnesium and forms the basis of the Mg/Ca paleothermometer. The substitution of magnesium for calcium in calcite is endothermic and both thermodynamic calculations and experimental data confirm that Mg/Ca of inorganically precipitated calcite increases by 3% per °C. A similar phenomenon is observed in biologically precipitated calcite produced by a range of organisms, although the actual relationship between Mg/Ca and SST is often offset from the thermodynamic prediction.

In oceanic sediments, shells of planktonic foraminifera offer the best source of calcite for Mg/Ca paleothermometry. Planktonic foraminiferal shells are composed of low-magnesium calcite and contain several times less magnesium ($<1\%$ Mg) compared to inorganic calcite precipitated at the same conditions. The response of foraminiferal Mg/Ca to increasing SST is 9% per °C, that is, 3 times larger than the thermodynamic prediction. The reason for the latter effect is not known, although it is likely to be related to the lower Mg/Ca content of foraminiferal calcite. No matter what its cause, the larger response of foraminiferal Mg/Ca to SST change is extremely important for paleothermometry as it increases the sensitivity of the method and reduces the uncertainty of Mg/Ca-derived SST reconstructions.

Following the thermodynamic prediction, the relationship between Mg/Ca in foraminiferal calcite and calcification temperature is expressed as an exponential function:

$$\text{Mg/Ca}\left[\text{mmol} \times \text{mol}^{-1}\right] = b \times e^{m \times T_{\text{calcification}}[°C]} \quad [3]$$

After a rigorous cleaning to remove Mg from adhering sediment, organic matter, and secondary precipitates, the magnesium content of foraminiferal shells is routinely determined by inductively coupled plasma mass spectrometry (ICP-MS) or atomic emission spectrometry (ICP-AES). Both techniques provide a precision of Mg/Ca ratio measurements of the order 0.5%. Data from laboratory cultures as well as calibrations from sediments and water-column samples converge on the value of m being between 10.7% and 8.8%. The pre-exponential term b appears more variable, ranging between 0.30 and 0.53. The parameters of the equation vary depending on which species is analyzed and how the calcification temperature is determined. An extensive calibration based on foraminifera from sediment traps (**Figure 5**) yielded the following conversion

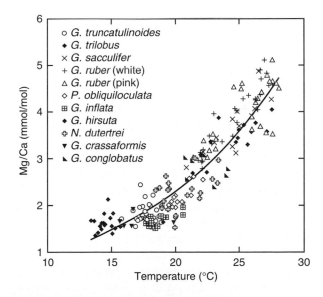

Figure 5   The relationship between calcification temperature (inferred from oxygen isotopes) and Mg/Ca in a range of planktonic foraminifera species from a Bermuda sediment trap. For most species, the global calibration expressed in eqn [4] can be used, yielding a standard error of 1.5 °C. Modified from Barker S, Cacho I, Benway HM, and Tachikawa K (2005) Planktonic foraminiferal Mg/Ca as a proxy for past oceanic temperatures: A methodological overview and data compilation for the Last Glacial Maximum. *Quaternary Science Reviews* 24: 821–834. Data from Anand P, Elderfield H, and Conte MH (2003) Calibration of Mg/Ca thermometry in planktonic foraminifera from a sediment trap time series. *Paleoceanography* 18: 1050, with permission from Elsevier.

equation, which appears to be applicable globally and across a range of species:

$$T_{\text{calcification}} [^{\circ}\text{C}] = \left(\frac{1}{0.09}\right)$$
$$\times \ln\left(\frac{\text{Mg/Ca}\left[\text{mmol} \times \text{mol}^{-1}\right]}{0.38}\right) \quad [4]$$

The standard error of foraminiferal Mg/Ca calibrations is typically around 1 °C. Because of the exponential nature of the conversion equations, the sensitivity of the Mg/Ca paleothermometer decreases with temperature. The method is thus less suitable for SST reconstructions in subpolar and polar waters, whereas its sensitivity is higher in the Tropics where it ought to yield robust reconstructions even when past tropical SST was higher than today.

As discussed in the previous section, planktonic foraminifera migrate through the water column during life and different parts of the shell calcify at different depths, often within or even below the thermocline. In order to circumvent this problem, Mg/Ca paleothermometry generally focuses on surface-dwelling species, which spend most of their life within the mixed layer: *Globigerinoides ruber*, *Globigerinoides sacculifer*, and *Globigerina bulloides*. The sedimentary Mg/Ca signal is typically derived by a pooled analysis of 10–50 shells and is therefore biased toward the season of highest production of the analyzed species.

The effect of salinity on foraminiferal Mg/Ca (for the range of normal marine conditions) is believed to correspond to 0.6–0.8 °C per psu; the effect of surface ocean pH is *c.* –0.6 °C per 0.1 pH units. Both these effects are relatively small and affect Mg/Ca SST reconstructions only in extreme environments or on geological timescales. On these timescales, the possibility of changes in oceanic Mg/Ca must be considered as well. By far the largest secondary effect on Mg in foraminiferal shells is that of carbonate dissolution. Mg is preferentially removed from calcite during dissolution, and in foraminifera sinking through the water column and on the seafloor, this effect translates to about 0.4–0.6 °C loss per km or 2.8 °C loss per km of lysocline shift. With some effort, the influence of carbonate dissolution on Mg/Ca SST reconstructions can be quantified and it is generally assumed that in normal marine sediments, the maximum bias on the three commonly used species is of the order of 0.5 °C.

Foraminiferal Mg/Ca paleothermometry is analytically relatively straightforward and offers the possibility to determine past SST from the same phase as oxygen isotopes. This provides a powerful means for subtracting the temperature signal obtained from Mg/Ca measurements from the $\delta^{18}O$ signal in foraminifera, allowing reconstructions of past seawater isotopic composition and assessment of phase relationships between SST and ice sheet dynamics in the past. The greatest advantage of the Mg/Ca method is its high sensitivity at warm SST with no upper limit to reconstructed SST values. Therefore, Mg/Ca in planktonic foraminifera is currently considered the best paleothermometer for the Tropics.

Sr/Ca and oxygen isotopes in hermatypic corals

Skeletons of hermatypic corals also offer a possibility to obtain long continuous records of past SST variability. Unlike planktonic foraminifera, scleractinian corals produce aragonite, which is the rhombohedric variety of calcium carbonate. Elemental and isotopic substitutions in aragonite are also controlled by temperature and both the $\delta^{18}O$ and Sr/Ca of coral skeletons record SST. The $\delta^{18}O$ signal in scleractinian aragonite follows the slope of the relationship derived from inorganic precipitation experiments, but the curve is offset from equilibrium. The offset is presumably due to biological 'vital effects' such as growth rate or symbiont activity, and it appears to vary even within individual coral colonies. In order to convert coral $\delta^{18}O$ to past SST, the $\delta^{18}O$ of seawater must be known. The combined influence of the vital effect and uncertainties in the determination of seawater $\delta^{18}O$ make the coral $\delta^{18}O$ signal very difficult to deconvolve and the signal is thus often interpreted as a generalized 'climate' record, rather than SST.

The Sr content of inorganically precipitated aragonite is an exponential function of temperature, but the exponential constant is small (–0.45% per °C), and within the range of marine tropical SST, the relationship is often described as a linear function. The substitution reaction is exothermic, favoring Sr substitution at lower SST. Coralline Sr/Ca is determined by atomic absorption spectrophotometry (AAS) and the standard error of empirical calibrations is around 0.5 °C. However, recent studies indicate that the Sr content of coralline aragonite also responds to a number of secondary effects including symbiont activity, and the slope of empirical calibrations shows a large variability. In addition, the Sr/Ca paleothermometer may be affected on glacial/interglacial and longer timescales by secular changes of the Sr content of seawater.

In comparison with calcite, aragonite is much more susceptible to dissolution and alteration, especially when exposed to meteoric waters. Both the

δ^{18}O and Sr/Ca signals are heavily altered during this process and the applicability of coral paleothermometry on uplifted reefs is severely reduced. The best coral climatic records are therefore derived from submerged corals. Coral paleothermometry is obviously limited to the Tropics, where hermatypic corals occur, and it provides an important means of studying tropical climate variability. Due to the high resolution of the records, climate data extracted from corals have provided fascinating insights into decadal dynamics of tropical oceans. It is, however, not always possible to separate the SST component in the δ^{18}O and Sr/Ca signals in coral skeletons and the method is therefore more useful when seen as recording the combined effect of SST, P/E balance, and seawater chemistry.

Paleothermometers Based on Organic Biomarkers in Sediments

Alkenone unsaturation $(U_{37}^{k'})$ The alkenone unsaturation index $U_{37}^{k'}$ (U^k standing for 'unsaturated ketones') is the main organic biomarker proxy for SST. It is based on the measurement of the relative abundance in marine sediments of unbranched long-chained methyl ketones with 37 carbon atoms (C_{37}) and a variable number of double bonds. C_{37} alkenones, together with C_{38} and C_{39} alkenones, their fatty acid methyl esters (alkenoates), and alkenes, are specific for certain species of haptophyte algae, including one of the main primary producers in the oceans, the bloom-forming *Emiliania huxleyi* (**Figure 6**). These molecules are chemically inert and survive transport throughout the water column as well as burial in marine sediments, where they are known to persist for millions of years, in concentrations ranging typically between 0.1 and 10 ppm. The abundance of the individual compounds is determined from total lipid extracts or purified lipid fractions by high-temperature elution gas chromatography. Each analysis requires between 1 and 10 g of dry sediment. Current technology allows determination of the unsaturation index in all marine sediments, except where virtually no organic matter is preserved.

The function of the long-chained alkenones is not fully understood. The high abundance of these compounds in the haptophyte algae (up to 10% of cellular carbon) indicates that they play a significant role in the cellular structure. It has been suggested that alkenones may serve as storage molecules or that they may be incorporated into the cellular membrane where they act as fluidity regulators. The latter hypothesis links alkenone unsaturation directly with growth temperature and provides a mechanistic explanation for the striking relationship between the degree of unsaturation and temperature, observed in C_{37} alkenone extracts from surface sediment samples and from laboratory cultures. The degree of unsaturation has been originally expressed as a ratio between the quantities of the di- ($C_{37:2}$), tri- ($C_{37:3}$), and tetra- ($C_{37:4}$) unsaturated C_{37} alkenones:

$$U_{37}^{k} = \frac{[C_{37:2} - C_{37:4}]}{[C_{37:2} + C_{37:3} + C_{37:4}]} \qquad [5]$$

It was later shown that there is no benefit in including the $C_{37:4}$ alkenone and the currently used version of the index is marked by a prime:

$$U_{37}^{k'} = \frac{[C_{37:2}]}{[C_{37:2} + C_{37:3}]} \qquad [6]$$

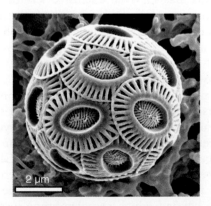

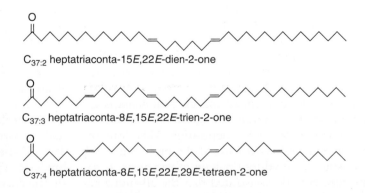

$C_{37:2}$ heptatriaconta-15*E*,22*E*-dien-2-one

$C_{37:3}$ heptatriaconta-8*E*,15*E*,22*E*-trien-2-one

$C_{37:4}$ heptatriaconta-8*E*,15*E*,22*E*,29*E*-tetraen-2-one

Figure 6 Scanning electron micrograph (SEM; in false color) of a single cell of the alkenone-producing haptophyte alga *Emiliania huxleyi*. The cell is covered by a mineral skeleton (coccosphere), consisting of isolated interlocking plates (coccoliths). The structures of the three unsaturated alkenones produced by this species and used to calculate the U_{37}^{k} paleotemperature index are shown to the right. SEM image courtesy of Dr. Marcus Geisen and Dr. Claudia Sprengel, Alfred Wegener Institute for Polar and Marine Research, Bremerhaven, Germany.

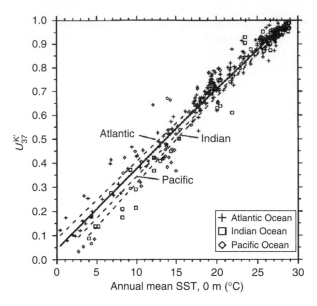

Figure 7 Empirical calibration of mean annual SST at 0-m depth and the alkenone unsaturation index $U^{k'}_{37}$ determined in 149 globally distributed surface sediment samples. Thick line shows the global regression as expressed in eqn [7]; standard error of the regression is 1.5 °C. Reproduced from Müller PJ, Kirst G, Ruhland G, von Storch I, and Rosell-Melé A (1998) Calibration of the alkenone paleotemperature index $U^{k'}_{37}$ based on core-tops from the eastern South Atlantic and the global ocean (60° N–60° S). *Geochimica et Cosmochimica Acta* 62: 1757–1772, with permission from Elsevier.

This modified index shows a strong and consistent linear relationship with SST (**Figure 7**). Empirical calibrations in surface sediment samples yielded the following conversion equation between the unsaturation index and annual mean SST at 0-m depth:

$$\text{SST}_{0\,m}[^\circ C] = \left(U^{k'}_{37} - 0.044 \right)/0.033 \qquad [7]$$

The form of this equation has been confirmed by laboratory culture studies as well as measurements of particulate organic matter in the water column. The standard error of the empirical calibration is about 1.5 °C and the method appears to work best in the SST range between 5 and 27 °C.

The alkenone content of marine sediments is the result of many years of integration of alkenone production over the site of deposition. Therefore, the signal should be biased toward the preferred depth and season of growth of the haptophyte algae. It has been repeatedly demonstrated that the alkenone unsaturation index measured in surface sediments shows a stronger correlation with annual mean SST at 0-m depth than for any other depth and season. Consequently, past SST reconstructions based on alkenone unsaturation in fossil sediments are interpreted as reflecting annual mean values at the ocean surface.

Like all other algae, haptophytes are dependent on photosynthesis and their growth occurs within the photic zone (**Figure 2**). Therefore, the alkenone unsaturation signal records the conditions within or near the mixed layer and this conjecture ought to remain valid for as long as the haptophytes were the sole producers of the alkenones. The main production of haptophyte algae occurs in spring or in autumn. The temperature of these seasons is comparable to the annual average, explaining the best fit of sedimentary unsaturation values with mean annual SST. However, in the absence of information on the ecology of the alkenone producers in the past, interpretations of alkenone unsaturation values in fossil sediments as reflecting past SST for a specific season remain speculative. The modern alkenone-producing species evolved during the late Quaternary (within the last 1 My) and ecological shifts could be significant for the interpretation of older alkenone signals, particularly at high latitudes where seasonal differences in SST are large.

Apart from changes to vertical and seasonal production of the alkenones, the unsaturation signal appears robust to the usual sources of bias. Alkenones are recalcitrant to diagenesis and reworking of older alkenones has been found significant only in rare circumstances, such as in the polar waters where the primary production by the haptophytes was low, particularly during glacial times. Similarly, there is little evidence that the signal is affected by other environmental parameters such as productivity or salinity (within the range of normal oceanic values). To conclude, the fortuitous discovery by the Bristol organic geochemistry group in 1986 of the strong relationship between alkenone unsaturation and SST has survived extensive scrutiny and the unsaturation index remains one of the most robust means for determination of past SST in marine sediments.

Crenarchaeotal membrane lipids (TEX$_{86}$) The relatively novel biomarker paleothermometer TEX$_{86}$ is based on the abundance of different types of crenarchaeotal membrane lipids called glycerol dialkyl glycerol tetraethers (GDGTs) in sediments. The presence of cyclopentane rings in the GDGT lipids of hyperthermophilic Crenarchaeota increases the thermal transition point of the membrane (the temperature beyond which the membrane loses its biological functionality) and the number of such rings in crenarchaeotal GDGT is known from laboratory cultures to increase with temperature. The kingdom Crenarchaeota of the domain Archaea was long believed to include exclusively hyper-thermophilic organisms, but recent studies showed that one group of Crenarchaeota is abundant in the upper water column of the oceans, where it may make up a significant portion of the picoplankton. Oceanic crenarchaeota also synthesize GDGT lipids and the proportions of four types of GDGT molecules with 86 carbon atoms and different numbers of cyclopentane rings correlates with SST (**Figure 8**).

Empirical calibration of GDGT abundance in surface sediments led to the definition of the TEX_{86}

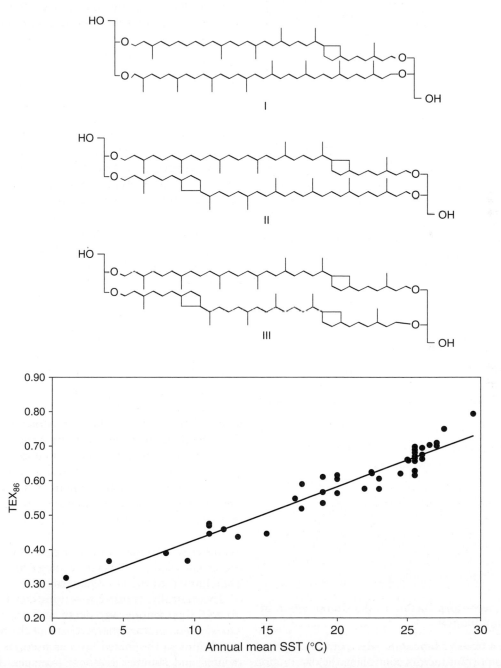

Figure 8 Empirical calibration of mean annual SST at 0 m depth and the crenarchaeotal membrane lipid index TEX_{86} determined in 40 globally distributed surface sediment samples. The thick lines show the linear regression expressed in eqn [9]; the standard error is 2 °C. The structure of the GDGTs is shown; roman numerals refer to the number of cyclopentane rings. The structure of the $GDGT_{IV}$ is not yet known. Modified from Schouten S, Hopmans EC, Schefuß E, and Sinninghe Damsté JS (2002) Distributional variations in marine crenarchaeotal membrane lipids: A new tool for reconstructing ancient sea water temperatures? *Earth and Planetary Science Letters* 204: 265–274, with permission from Elsevier.

'tetraether index of lipids with 86 carbon atoms':

$$TEX_{86}$$
$$= \frac{([GDGT_{II}] + [GDGT_{III}] + [GDGT_{IV'}])}{([GDGT_{I}] + [GDGT_{II}] + [GDGT_{III}] + [GDGT_{IV'}])}$$
[8]

This index is highly correlated with annual mean SST at 0-m depth and for SST range 0–28 °C the following conversion can be used:

$$SST_{0\,m}[°C] = (TEX_{86} - 0.28)/0.015 \quad [9]$$

The abundance of the different GDGT types (**Figure 8**) is determined from total lipid extracts by high-performance liquid chromatography/atmospheric pressure positive ion chemical ionization mass spectrometry (HPLC/APCI-MS). The combined uncertainty in TEX_{86} SST reconstructions due to the standard error of the calibration and the analytical error is less than 2 °C.

The TEX_{86} technique was first described in 2002 and there are many aspects of the method that require further investigation. The effects of secondary parameters like salinity and productivity are not known and it is not clear why the index correlates best with annual mean SST at 0-m depth, when the GDGT producing Crenarchaeota are known to inhabit the top 500 m of the water column. The abundance of marine Crenarchaeota is negatively correlated with that of algal phytoplankton and the GDGT lipids deposited on the seafloor should represent different seasons than those when algal blooms occur. On the other hand, the application of the TEX_{86} technique is not restricted to the oceanic environments. First results indicate that a similar relationship between SST and GDGT abundance also holds for lake sediments. Like alkenones, the TEX_{86} proxy can be used in carbonate-free sediments, and it additionally offers the possibility of reconstructing SST in very old sediments. The GDGT-producing Crenarchaeota are most likely ancient and measurable amounts of the GDGT lipids have been found in sediments older than 100 My.

Paleothermometers Based on the Composition of Fossil Assemblages

Principles of transfer functions Temperature is one of the most important factors controlling the distribution and abundance of species of marine phyto- and zooplankton (**Figure 9**). Assemblage composition in those groups of plankton that leave a fossil record can therefore be used as a means to reconstruct past SST variation. In order to obtain quantitative, calibrated reconstructions of past SST, microfossil assemblages are extracted from surface sediments, abundance of individual species is determined, and the data are linked to a long-term average SST above the site of their deposition. The relationship between the assemblage composition and SST is described in the form of the so-called transfer functions. Transfer functions are empirically calibrated mathematical formulas or algorithms that serve to optimally extract the general relationship between microfossil assemblage composition in sediment samples and environmental conditions in the surface ocean. Assuming that SST changes in the past were manifested mainly through zonal redistribution of microfossil assemblages, a transfer function derived from modern sediments can then be used to convert assemblage composition data from fossil samples to SST reconstructions (**Figure 10**). Unfortunately, this assumption is not always met and many fossil samples yield microfossil assemblages without analogs in the modern ocean. Such assemblages may reflect secondary alteration or reworking as well as no-analog oceanographic conditions in the past. No-analog assemblages are typically easy to recognize, but their use for SST reconstructions is always problematic.

The potential of the transfer function technique in providing calibrated paleoenvironmental reconstructions has been realized in 1976 by the CLIMAP project in the seminal paper on SSTs of the Last Glacial Maximum. Ever since, transfer functions have become a standard method in paleothermometry, applicable in virtually all marine sediments and extendable to variables other than SST. The calibration of a transfer function requires a large database of census counts from surface sediments, including typically 100–1000 samples with counts of 20–50 species. The data set may be global or regional. Like any other empirical calibration, transfer functions rely on a number of assumptions (**Figure 10**). Unlike geochemical paleothermometers, transfer functions are robust to mild diagenesis, require no cleaning or chemical extraction, and can be relatively cheap and fast. However, they rely on the exact knowledge of species ecology and their use is thus limited to the late Quaternary (~ 0.1–0.5 My).

Theoretically, transfer functions can be calibrated to SST representing any depth and season. In most cases, a temperature characteristic of the mean annual conditions in the mixed layer is used, as well as the winter and summer seasonal averages. However, a simultaneous reconstruction of summer and winter SST by the transfer function method does not equate to the reconstruction of a seasonal contrast in the past. The simultaneous reconstruction of several SST representations is only possible because all SST

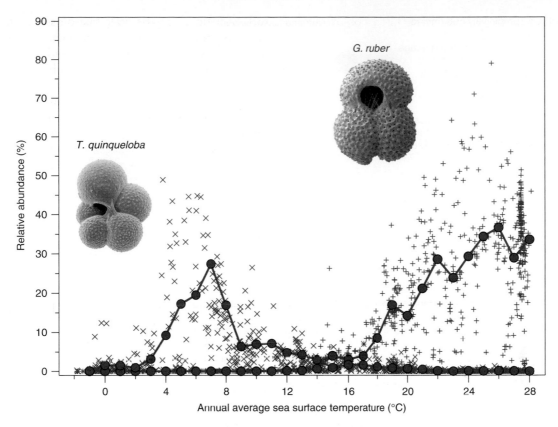

Figure 9 Relative abundance of two species of planktonic foraminifera (*Turborotalita quinqueloba* and *Globigerinoides ruber*) vs. mean annual SST at 10-m depth in 863 surface sediment samples from the North Atlantic (10° S to 90° N). Thick lines show average abundance per 1 °C increment. Note the strong and distinctly different SST response of the two species. Scanning electron micrographs of the two species are not to scale. Data from the MARGO database (http://www.pangaea.de).

representations in the present-day ocean are extremely highly correlated to each other. Transfer functions can only produce reliable SST reconstructions within the limits of the calibration range. The method is therefore most sensitive in the middle of the SST range and weaker at the SST extremes. Prediction errors are estimated by validation where a portion of the calibration database is held back, the transfer function is then applied on this validation data set, and the SST 'reconstructions' are compared with the actual values. If the calibration data set is small, the leaving-one-out method may be used where one sample is held back at a time and a new transfer function is produced. The prediction error is then calculated as the average of the prediction errors for the individual samples.

The form of the transfer function may vary considerably. A simple linear relationship is expressed by the weighted average method, where past SST is calculated as the weighted average of the abundances of *n* species:

$$\text{SST} = \sum_{i=1}^{n} (p_i \times \text{SST}_i) \bigg/ \sum_{i=1}^{n} p_i \qquad [10]$$

where p_i is the proportion of species i and SST_i is the 'optimal temperature' for species i determined empirically from the calibration data set. The performance of the weighted average method has been improved by including partial least squares regression. The classical method by Imbrie and Kipp is mathematically related. It involves transformation of the census counts into a number of artificial 'assemblages' by means of factor analysis, followed by a multiple regression of the assemblage scores (and their cross products and squares) onto SST.

The modern analog technique (MAT) is a variant of the *k*-nearest neighbor regression. This method searches the calibration data set for samples with assemblages that most resemble the fossil assemblage. To identify the best analogs in the calibration data set, the square chord distance measure is used:

$$d_{ij} = \sum_{k=1}^{p} \left[(x_{ik})^{1/2} - (x_{jk})^{1/2} \right]^2 \qquad [11]$$

where d_{ij} is the dissimilarity coefficient between sample i and sample j, x_k is the percentage of species k in a sample, and p is the total number of species in

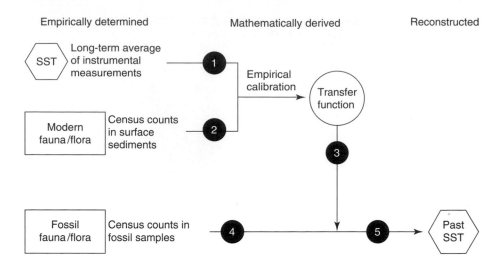

Assumptions of paleotemperature transfer functions:

1 Species abundances in the calibration data set are systematically related to sea surface temperature (SST).

2 SST is a significant determinant in the ecological system represented by the calibration data set or is at least strongly correlated to such a determinant.

3 The mathematical method used to derive the transfer function provides an adequate model of the ecological system and yields a calibration with sufficient predictive power.

4 Species in fossil samples all have modern counterparts and their ecological responses to SST have not changed significantly.

5 The joint distribution of SST with other environmental parameters acting on the ecological system in the calibration data set is the same as in the fossil sample.

Figure 10   The principles of the transfer function paleothermometer, showing the five main assumptions on which the method is based.

the calibration data set. Past SST is then reconstructed from the physical properties recorded in the best modern analog samples:

$$SST = \sum_{i=1}^{m}(s_i \times SST_i) \Big/ \sum_{i=1}^{m} s_i \qquad [12]$$

where SST_i is the SST value in ith of the m best analog samples and s_i is a similarity coefficient between the fossil sample and the ith best analog sample:

$$s_{ij} = \sum_{k=1}^{p} \sqrt{x_{ik} \times x_{jk}} \qquad [13]$$

The MAT approach has been further modified in the SIMMAX technique which weighs the best analogs additionally by their geographical distance from the fossil sample. The revised analog method (RAM) expands the calibration data set by interpolation in the space of the reconstructed environmental parameters. It also includes an algorithm for flexible selection of the number of best analogs.

Increasingly more complex mathematical methods are being used to extract the relationship between species counts and SST, including artificial intelligence algorithms and Bayesian techniques. The main issue in further development of transfer functions is how to avoid overfitting and extract the general signal. A useful approach appears to be that of comparison of the differences among SST reconstructions derived by different mathematical methods.

Microfossil groups used for transfer functions   Four microfossil groups are commonly used for transfer function paleothermometry. By far the most commonly used are planktonic foraminifera (**Figure 4**). The foraminifera are isolated from marine sediments by washing and sieving and the determination of foraminiferal assemblage census typically involves counting under a binocular microscope of 300–500 specimens in random splits of the $>0.150\,mm$ fraction. Foraminiferal transfer functions are applicable in all marine sediments deposited above the calcite lysocline, apart from polar waters where the dominance of a single species (*Neogloboquadrina pachyderma*) causes lack of resolution (**Figure 11**).

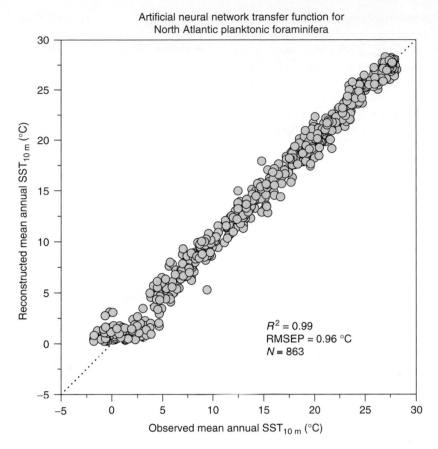

Figure 11 Comparison of actual SST values with transfer function estimates for a calibration of census counts of 26 foraminiferal species in 863 North Atlantic (10° S to 90° N) surface sediment samples. The transfer function was derived by training of back-propagation artificial neural networks. The root mean square error of prediction of the calibration is 1 °C; the calibration breaks down in polar samples (SST<3 °C) due to the dominance of a single species. Modified from Kucera M, Weinelt M, Kiefer T, *et al.* (2005) Reconstruction of sea-surface temperatures from assemblages of planktonic foraminifera: Multi-technique approach based on geographically constrained calibration datasets and its application to glacial Atlantic and Pacific Oceans. *Quaternary Science Reviews* 24: 951–998. Data from the MARGO database (http://www.pangaea.de).

Typically, annual, summer, and winter SST are reconstructed simultaneously. Both regional and global calibrations exist, all showing prediction error estimates in the range of 1–1.5 °C. The approach has been successfully used throughout the last 0.5 My; applications to sediments as old as the mid-Pliocene exist, but must be viewed with utmost caution due to evolutionary modifications of species ecology.

In sediments affected by carbonate dissolution, in the tropical Pacific and in polar oceans, the siliceous skeletons of diatoms (Chrysophyceae) and radiolarians are the prime target for transfer functions. The siliceous skeletons of diatoms and radiolarians are concentrated using chemical maceration of the sediment. The residue is mounted on glass slides and typically 300–500 specimens are counted under light microscope. Radiolaria are eurybathyal heterotrophic protists and their production maximum occurs in tropical and subpolar/transitional waters. The maximum production of the diatom algae is in polar waters and high-productivity upwelling regions. Diatoms are particularly suitable for SST reconstruction in polar waters of the Southern Hemisphere. As light-dependent phytoplankton, their production during polar winter ceases and diatom transfer functions are therefore commonly used to reconstruct summer SST only. Unlike foraminifera, diatoms are diverse even in the coldest waters and can be used to reconstruct SST up to the freezing point of seawater. In fact, diatom transfer functions can be even used to reconstruct the extent of permanent sea ice. The prediction error of diatom transfer functions are estimated to be around 1 °C; slightly higher values are reported for radiolarians. Both groups have been used for paleothermometry only in the late Quaternary (last 0.5 My).

Cysts of dinoflagellate algae are made of resistant biopolymers and can be extracted from sediments by chemical maceration involving hydrofluoric acid. The residue is mounted on glass slides and preferably 300, but sometimes less, specimens are counted. Dinoflagellates are normally phototrophic, although some species are mixotrophic or heterotrophic. Cyst-producing forms live typically in coastal waters and the use of dinocyst transfer functions in open-ocean settings is therefore somewhat limited. Large calibration data sets exist only for the Northern Hemisphere and dinocyst transfer functions are particularly useful in the high latitudes. Like diatoms, dinoflagellates can be used to reconstruct SST up to the freezing point of seawater and even to detect the presence of permanent sea ice. The prediction error of dinocyst transfer functions are estimated at 1.5–2 °C and the method has been used only in the late Quaternary (last 0.5 My). Apart from the four microfossil groups mentioned above, coccolithophore algae have been used for transfer functions in the past, but due to taxonomic uncertainties, this method has been largely abandoned.

The transfer function approach has seen a recent revival, fueled mainly by the development and application of advanced computational techniques. Despite its caveats and limitation, the method is less sensitive to shifts in production season of the microfossils, is taphonomically more robust, and offers a methodologically independent validation of geochemical paleothermometers.

Conclusion

The methods used to reconstruct past SST depend on analytical instrumental precision and computational power and their development has accelerated in the past few decades hand in hand with the technological progress. After the first attempts to quantify the magnitude of past SST change, the science is now focusing on the reduction of uncertainties in SST reconstructions, on a better understanding of the origin of the SST signal and on the search for new types of SST signatures. Existing methods are being tested on new substrates, such as oxygen isotopes in diatom opal and trace elements in monospecific isolates of coccolithophore calcite. Modern instrumentation is opening up new chemical and isotopic systems and the analysis of calcium isotopes in planktonic foraminifera appears a promising avenue of research. Finally, physical properties of microfossils other than assemblage composition are being studied and it appears that size of planktonic foraminifera and morphology of coccoliths (such as *Gephyrocapsa oceanica*) could be used to reconstruct past SST.

See also

Land–Sea Global Transfers. Past Climate from Corals. Plankton and Climate. Satellite Remote Sensing of Sea Surface Temperatures.

Further Reading

Anand P, Elderfield H, and Conte MH (2003) Calibration of Mg/Ca thermometry in planktonic foraminifera from a sediment trap time series. *Paleoceanography* 18: 1050.

Bard E (2001) Comparison of alkenone estimates with other paleotemperature proxies. *Geochemistry Geophysics Geosystems* 2 (doi:10.1029/2000GC000050).

Barker S, Cacho I, Benway HM, and Tachikawa K (2005) Planktonic foraminiferal Mg/Ca as a proxy for past oceanic temperatures: A methodological overview and data compilation for the Last Glacial Maximum. *Quaternary Science Reviews* 24: 821–834.

Bemis BE, Spero HJ, Bijma J, and Lea DW (1998) Reevaluation of the oxygen isotopic composition of planktonic foraminifera: Experimental results and revised paleotemperature equations. *Paleoceanography* 13: 150–160.

Birks HJB (1995) Quantitative palaeoenvironmental reconstructions. In: Maddy D and Brew JS (eds.) *Technical Guide 5: Statistical Modelling of Quaternary Science Data*, pp. 161–254. Cambridge, MA: Quaternary Research Association.

CLIMAP, Project Members (1976) The surface of the ice-age Earth. *Science* 191: 1131–1137.

de Vernal A, Eynaud F, Henry M, *et al.* (2005) Reconstruction of sea-surface conditions at middle to high latitudes of the Northern Hemisphere during the Last Glacial Maximum (LGM) based on dinoflagellate cyst assemblages. *Quaternary Science Reviews* 24: 897–924.

Gagan MK, Ayliffe LK, Beck JW, *et al.* (2000) New views of tropical paleoclimates from corals. *Quaternary Science Reviews* 19: 45–64.

Gersonde R, Crosta X, Abelmann A, and Armand L (2005) Sea surface temperature and sea ice distribution of the Southern Ocean at the EPILOG Last Glacial Maximum – a circum-Antarctic view based on siliceous microfossil records. *Quaternary Science Reviews* 24: 869–896.

Herbert TD (2006) Alkenone paleotemperature determinations. In: Elderfield H (ed.) *Treatise on Geochemistry, Vol. 6: The Oceans and Marine Geochemistry*, pp. 391–432. Oxford, UK: Elsevier.

Kucera M, Weinelt M, Kiefer T, *et al.* (2005) Reconstruction of sea-surface temperatures from assemblages of planktonic foraminifera: Multi-technique approach based on geographically constrained calibration datasets and its application to glacial Atlantic and Pacific Oceans. *Quaternary Science Reviews* 24: 951–998.

Lea DW (2006) Elemental and isotopic proxies of past ocean temperatures. In: Elderfield H (ed.) *Treatise on*

Geochemistry, Vol. 6: The Oceans and Marine Geochemistry, pp. 365–390. Oxford, UK: Elsevier.

Mix A, Bard E, and Schneider R (2001) Environmental processes of the ice age: Land, oceans, glaciers (EPILOG). *Quaternary Science Reviews* 20: 627–657.

Müller PJ, Kirst G, Ruhland G, von Storch I, and Rosell-Melé A (1998) Calibration of the alkenone paleotemperature index $U_{37}^{k'}$ based on core-tops from the eastern South Atlantic and the global ocean (60°N–60°S). *Geochimica et Cosmochimica Acta* 62: 1757–1772.

Schouten S, Hopmans EC, Schefuß E, and Sinninghe Damsté JS (2002) Distributional variations in marine crenarchaeotal membrane lipids: A new tool for reconstructing ancient sea water temperatures? *Earth and Planetary Science Letters* 204: 265–274.

Waelbroeck C, Mulitza S, Spero H, Dokken T, Kiefer T, and Cortijo E (2005) A global compilation of late Holocene planktonic foraminiferal $\delta^{18}O$: Relationship between surface water temperature and $\delta^{18}O$. *Quaternary Science Reviews* 24: 853–868.

Relevant Website

http://margo.pangaea.de
  – MARGO at PANGAEA; the MARGO project houses a database of geochemical and microfossil proxy results for the Last Glacial Maximum, numerous calibration data sets, software tools, and publications.

PLIO-PLEISTOCENE GLACIAL CYCLES AND MILANKOVITCH VARIABILITY

K. H. Nisancioglu, Bjerknes Centre for Climate Research, University of Bergen, Bergen, Norway

Introduction

A tremendous amount of data on past climate has been collected from deep-sea sediment cores, ice cores, and terrestrial archives such as lake sediments. However, several of the most fundamental questions posed by this data remain unanswered. In particular, the Plio-Pleistocene glacial cycles which dominated climate during the past ~2.8 My have puzzled scientists. More often than not during this period large parts of North America and northern Europe were covered by massive ice sheets up to 3 km thick, which at regular intervals rapidly retreated, giving a sea level rise of as much as 120 m.

The prevalent theory is that these major fluctuations in global climate, associated with the glacial cycles, were caused by variations in insolation at critical latitudes and seasons. In particular, ice sheet growth and retreat is thought to be sensitive to high northern-latitude summer insolation as proposed by Milankovitch in his original astronomical theory.

Brief History of the Astronomical Theory

Long before the first astronomical theory of the ice ages, the people of northern Europe had been puzzled by the large erratic boulders scattered a long way from the Alpine mountains where they originated. Based on these observations the Swiss geologist and zoologist Louis Agassiz presented his ice age theory at a meeting of the Swiss Society of Natural Sciences in Neuchatel in 1837, where he claimed that the large boulders had been transported by Alpine glaciers covering most of Switzerland in a past ice age.

A few years later, the French mathematician Joseph Alphonse Adhemar was the first to suggest that the observed ice ages were controlled by variations in the Earth's orbit around the Sun. At this point it was known that there had been multiple glaciations, and Adhemar proposed that there had been alternating ice ages between the North and the South Pole following the precession of the equinoxes.

Indeed, the winter is warmer when the Earth is at the point on its orbit closest to the Sun, and colder when the Earth is furthest from the Sun. Adhemar correctly deduced that the precession of the equinoxes had a period of approximately 21 000 years, giving alternating cold and warm winters in the two hemispheres every 10 500 years.

In 1864, James Croll expanded on the work by Adhemar and described the influence of changing eccentricity on the precession of the equinoxes. He assumed that winter insolation controlled glacial advances and retreats, and determined that the precession of the equinoxes played an important role in regulating the amount of insolation received during winter. Based on this, he estimated that the last ice age lasted from about 240 000 to 80 000 years ago. Croll was aware of the fact that the amplitude of the variations in insolation was relatively small, and introduced the concept of positive feedbacks due to changing surface snow and ice cover as well as changes in atmosphere and ocean circulation.

In parallel to the work of Croll, geologists in Europe and America found evidence of multiple glacial phases separated by interglacial periods with milder climate similar to that of the present day, or even warmer. These periodic glaciations were consistent with Croll's astronomical theory. However, most geologists abandoned his theory after mounting evidence from varved lake sediments in Scandinavia and North America showed that the last glacial period ended as late as 15 000 years ago, and not 80 000 years ago as suggested by Croll.

Milankovitch's Astronomical Theory of the Ice Ages

Following Croll's astronomical theory there was a period where scientists such as Chamberlin and Arrhenius tried to explain the ice ages by natural variations in the atmospheric content of carbon dioxide. The focus of the scientific community on an astronomical cause of the ice ages was not renewed until the publication of Milankovitch's theory in a textbook on climate by the well-known geologists Wladimir Köppen and Alfred Wegener in 1924. This was the first comprehensive astronomical theory of the Pleistocene glacial cycles, including detailed calculations of the orbitally induced changes in insolation.

Milutin Milankovitch was of Serbian origin, born in 1879. He obtained his PhD in Vienna in 1904 and was later appointed Professor of Applied Mathematics at the University of Belgrade. He was captured during World War I, but allowed to work at the Hungarian Academy of Sciences, where he completed his calculation of the variations of the orbital parameters of the Earth and their impact on insolation and climate. Milankovitch's basic idea was that at times of reduced summer insolation, snow and ice could persist at high latitudes through the summer melt season. At the same time, the cool summer seasons were accompanied by mild winter seasons leading to enhanced winter accumulation of snow. When combined, reduced summer melt and a slight increase in winter accumulation, enhanced by a positive snow albedo feedback, could eventually lead to full glacial conditions.

During World War II, Milankovitch worked on a complete revision of his astronomical theory which was published as the *Kanon der Erdbestrahlung* in 1941. However, the scientific establishment was critical of Milankovitch, and his theory was largely rejected until the early 1970s. By this time, great advances in sediment coring, deep-sea drilling, and dating techniques had made it possible to recover climate records covering the last 500 000 years. By studying the variations in oxygen isotopes of foraminifera in deep-sea sediment cores as well as by reconstructing past sea level from terraces of fossil coral reefs, new support was emerging for an astronomical phasing of the glacial cycles. The oxygen isotope data from the long deep-sea cores presented in a paper in 1976 by Hays *et al.* were considered as proof of the Milankovitch theory, as they showed cycles with lengths of roughly 20 000 and 40 000 years as well as 100 000 years in agreement with Milankovitch's original calculations.

Orbital Parameters and Insolation

The Earth's orbit around the Sun is an ellipse where the degree to which the orbit departs from a circle is measured by its eccentricity (e). The point on the orbit closest to the Sun is called the perihelion, and the point most distant from the Sun the aphelion (**Figure 1**). If the distance from the Earth to the Sun is r_p at perihelion, and r_a at aphelion, then the eccentricity is defined as $e = (r_a - r_p)/(r_a + r_p)$.

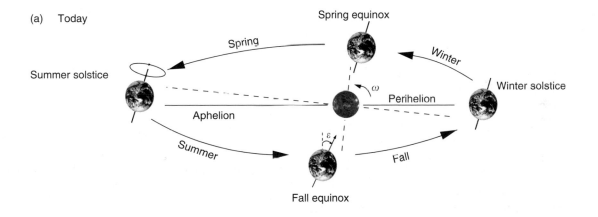

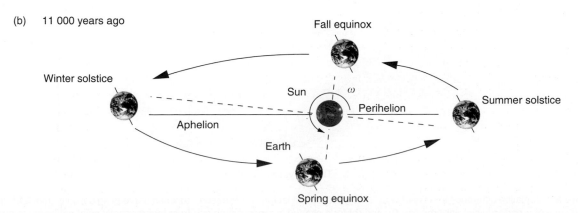

Figure 1 Sketch of the Earth's orbit around the Sun today and at the end of the last glacial cycle (11 000 years ago), showing the positions of the solstices and equinoxes relative to perihelion. The longitude of perihelion (ω) is measured as the angle between the line to the Earth from the Sun at spring equinox and the line to the Earth at perihelion.

Variations in the eccentricity of the Earth's orbit follow cycles of 100 000 and 400 000 years giving a change in annual mean insolation on the order of 0.2% or less. This change in insolation is believed to be too small to produce any notable effect on climate.

A more significant change in insolation is caused by variations in the seasonal and latitudinal distribution of insolation due to obliquity. Obliquity (ε) is the angle between Earth's axis of rotation and the normal to the Earth's plane around the Sun (Figure 1). This angle is 23.5° today, but varies between values of 22.1° and 24.5° with a period of 41 000 years. A decrease in obliquity decreases the seasonal insolation contrast, with the largest impact at high latitudes. At the same time, annual mean insolation at high latitudes is decreased compared to low latitudes. An example of the effect of obliquity variations on seasonal insolation is shown in Figure 2(a). During times when obliquity is small, high-latitude summertime insolation decreases, whereas mid-latitude wintertime insolation increases. The magnitude of the change in high-latitude summer insolation due to obliquity variations can be as large as 10%.

The third and last variable affecting insolation is the longitude of perihelion (ω). This parameter is defined as the angle between the line to the Earth from the Sun at spring equinox and the line to the Earth at perihelion (Figure 1). It determines the direction of the Earth's rotational axis relative to the orientation of the Earth's orbit around the Sun, thereby giving the position of the seasons on the orbit relative to perihelion. Changes in the longitude of perihelion result in the Earth being closest to the Sun at different times of the year. Today, the Earth is closest to the Sun in early January, or very near winter solstice in the Northern Hemisphere. All other things being equal, this will result in relatively warm winter and cool summer seasons in the Northern Hemisphere, whereas the opposite is the case in the Southern Hemisphere. At the time of the last deglaciation, 11 000 years ago the Earth was closest to the Sun at summer solstice, resulting in extra warm summers and cool winters in the Northern Hemisphere. An example of the effect of changes in precession on seasonal insolation is shown in Figure 2(b).

If the Earth's orbit were a circle, the distance to the Sun would remain constant at all times of the year and it would not make any difference where on the orbit the seasons were positioned. Therefore, the impact of variations in the longitude of perihelion depends on the eccentricity of the Earth's orbit and is described by

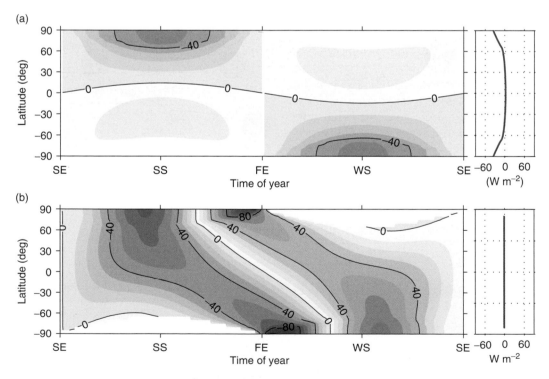

Figure 2 Insolation difference in units of W m^{-2} as a function of latitude and season: (a) when decreasing obliquity from 24.5° to 22° in the case of a perfectly circular orbit ($e=0$); and (b) for a change in precession going from summer solstice at perihelion to summer solstice at aphelion while keeping obliquity at today's value ($\varepsilon=23.5$°) and using a mean value for eccentricity ($e=0.03$). The annual mean insolation difference is shown to the right of each figure and the seasons are defined as follows: FE, fall equinox; SE, spring equinox; SS, summer solstice; WS, winter solstice.

the precession parameter ($e \sin \omega$). The combined effect of eccentricity and longitude of perihelion can give changes in high-latitude summer insolation on the order of 15% and varies with periods of 19 000 and 23 000 years, but is modulated by the longer-period variations in eccentricity. **Figure 3** shows the variations in obliquity (ε), eccentricity (e), and the precession parameter ($e \sin \omega$).

Plio-Pleistocene Glacial Cycles

Some of the longest continuous records of past climate come from deep-sea sediment cores. Ocean sediments are laid down over time, and by drilling into the seafloor, layered sediment cores can be extracted containing valuable information about the conditions at the time when the layers were formed. By studying the relative abundance of oxygen isotopes in shells of tiny marine organisms (foraminifera) found in the sediments, it is possible to estimate the amount of water tied up in the continental ice sheets and glaciers. This is because water molecules containing the lighter isotope of oxygen (^{16}O) are more readily evaporated and transported from the oceans to be deposited as ice on land. Thus, leaving the ocean water enriched with the heavy oxygen isotope (^{18}O) during glacial periods. However, the fractionation of the oxygen isotopes when forming the shells of the foraminifera also depends on the surrounding water temperature: low water temperature gives higher $\delta^{18}O$ values (the ratio of ^{18}O and ^{16}O relative to a standard). Therefore records of $\delta^{18}O$ are a combination of ice volume and temperature. By analyzing benthic foraminifera living on the seafloor where the ocean is very cold, and could not have been much colder during glacial times, the contribution of temperature variations to the $\delta^{18}O$ value is reduced.

The benthic $\delta^{18}O$ ice volume record of Hays et al. from 1976 was one of the very first continuous records of the late Pleistocene extending back to the Brunhes–Matuyama magnetic reversal event (780 000 years ago), making it possible to construct a timescale by assuming linear accumulation rates. Analysis of the data showed cycles in ice volume with periods of about 20 000 years and 40 000 years, with a particularly strong cycle with a period of roughly 100 000 years. Later studies extended the record past the Brunhes–Matuyama reversal, showing that the late Pliocene (3.6–1.8 Ma) and early Pleistocene records (1.8–0.8 Ma) were dominated by smaller-amplitude cycles with a period of 41 000 years, rather than the large 100 000 years cycles of the late Pleistocene (0.8–0 Ma).

Many records generated since this time have confirmed these early observations, namely:

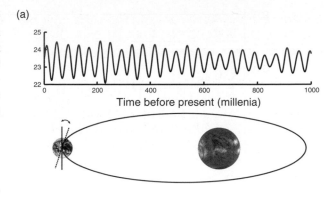

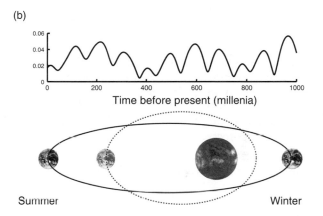

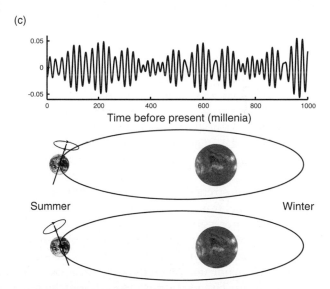

Figure 3 The three most important cycles regulating insolation on Earth are obliquity, eccentricity, and precession: (a) obliquity, or tilt of the Earth's axis varies with a period of 41 000 years; (b) eccentricity of the Earth's orbit varies with periods of 100 000 years and 400 000 years; and (c) precession of the equinoxes has a dominant period of 21 000 years and is modulated by eccentricity.

1. from about 3 to 0.8 Ma, the main period of ice volume change was 41 000 years, which is the dominant period of orbital obliquity;
2. after about 0.8 Ma, ice sheets varied with a period of roughly 100 000 years and the amplitude of

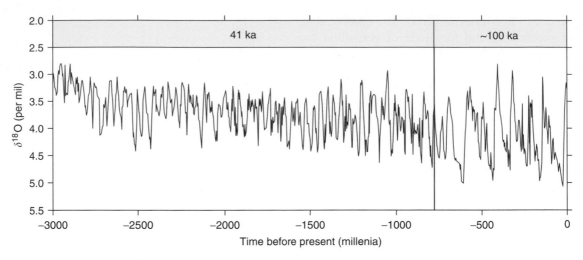

Figure 4 Benthic foraminiferal $\delta^{18}O$ ice volume record from the North Atlantic plotted to a paleomagnetic timescale covering the last 3 My. The transition from a dominant 41 000 to a 100 000-year periodicity in ice volume occurs close to the Brunhes–Matuyama magnetic reversal event ($\sim$780 000 years ago).

oscillations in $\delta^{18}O$ increases, implying growth of larger ice sheets.

The long benthic $\delta^{18}O$ record from a deep-sea sediment core extracted from the North Atlantic shown in **Figure 4** illustrates both of these points. This isotope record is plotted with a paleomagnetic timescale determined by the depth of magnetic field reversals recorded by ferromagnetic grains in the sediment core. Using this simple timescale, which is not biased by orbital tuning, one can clearly observe the 41 000-year periodicity of the late Pliocene and early Pleistocene (3.0–0.8 My), as well as the dominance of the stronger $\sim$100 000-year periodicity of the late Pleistocene (last 800 000 years).

Note that the main periods of orbital precession (19 000 and 23 000 years) are of less importance in the benthic ice volume record, whereas it is known that they increase in strength after about 800 000 years (the mid-Pleistocene transition). The lack of an imprint from orbital precession in the early part of the record and the reason for the dominance of roughly 100 000 years periodicity in the recent part of the record are some of the major unanswered questions in the field.

Only eccentricity varies with periods matching the roughly 100 000 years periods observed in the late Pleistocene. Although eccentricity is the only orbital parameter which changes the annual mean global insolation received on Earth, it has a very small impact. This was known to Croll and Milankovitch, who saw little direct importance in variations in eccentricity and assumed that changes in precession and obliquity would dominate climate by varying the amount of seasonal, rather than annual mean insolation received at high latitudes. Milankovitch postulated that the total amount of energy received from

the Sun during the summer at high northern latitudes is most important for controlling the growth and melt of ice. To calculate this insolation energy, he divided the year into two time periods of equal duration, where each day of the summer season received more insolation than any day of the winter season. The seasons following these requirements were defined as the caloric summer and caloric winter half-years. These caloric half-years are of equal duration through time and the amount of insolation energy received in each can be compared from year to year.

For Milankovitch's caloric summer half-year insolation (**Figure 5**), obliquity (ε) dominates at high latitudes ($>65°$ N), whereas climatic precession ($e \sin \omega$) dominates at low latitudes ($<55°$ N). In the mid-latitudes ($\sim 55 - 65°$ N), the contribution by obliquity and climatic precession are of similar magnitude. In the Southern Hemisphere, variations in caloric half-year insolation due to obliquity are in phase with the Northern Hemisphere and could potentially amplify the global signal, whereas variations due to climatic precession are out of phase. By taking into account the positive snow albedo feedback, Milankovitch used his caloric insolation curves to reconstruct the maximum extent of the glacial ice sheets back in time (**Figure 6**).

Milankovitch's predicted cold periods occurred roughly every 40 000–80 000 years, which fit reasonably well with the glacial advances known to geologists at that time. However, as the marine sediment core data improved, it became clear that the last several glacial periods were longer and had a preferred period of roughly 100 000 years (**Figure 4**), which was not consistent with Milankovitch's original predictions. Based on these observations, and without knowledge of the 41 000 years cycles of the

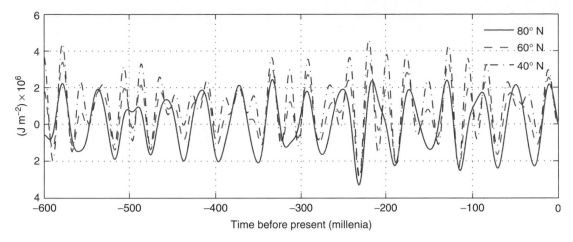

Figure 5 Caloric summer half-year insolation following Milankovitch's definition plotted for the latitudes of 40° N, 60° N, and 80° N. Caloric half-years are periods of equal duration where each day of the summer half-year receives more insolation than any day of the winter half-year.

early Pliocene and late Pleistocene, scientists reasoned that climatic precession and its modulation by eccentricity must play the leading role in past climates. Following Andre Berger and others, who recalculated and improved the records of orbital insolation, most researchers replaced the caloric half-year insolation as a driver of glacial climate by mid-month, or monthly mean insolation, for example, June or July at 65° N (**Figure 7**).

As can be seen from **Figure 7**, monthly mean insolation is dominated by precession. As insolation time series at a given time of the year (e.g., June or July) are in phase across all latitudes of the same hemisphere, the proxy records could be compared equally well with insolation from other latitudes than the typical choice of 65° N shown here. This means that any direct response of climate at high latitudes to monthly or daily insolation requires a strong presence of precession in the geologic record. Although both the frequencies of precession and obliquity are clearly found in the proxy records, a simple linear relationship between summer insolation and glacial cycles is not possible. This is particularly true for the main terminations spaced at roughly 100 000 years, which must involve strongly nonlinear mechanisms.

The strong positive feedback on global climate caused by greenhouse gases, such as CO_2, was pointed out as early as 1896 by the Swedish physical chemist Svante Arrhenius. Shortly thereafter, the American geologist Thomas Chamberlin suggested a possible link between changing levels of CO_2 and glacial cycles. From the long ice cores extracted from Antarctica, covering the last 740 000 years, it is now known that atmospheric levels of CO_2 closely follow the glacial temperature record (**Figure 8**). Although

greenhouse gases, such as CO_2, cannot explain the timing and rapidity of glacial terminations, the changing levels of atmospheric greenhouse gases clearly contributed by amplifying the temperature changes observed during the glacial cycles.

Modeling the Glacial Cycles

Following the discovery of the orbital periods in the proxy records, a considerable effort has gone into modeling and understanding the physical mechanisms involved in the climate system's response to variations in insolation and changes in the orbital parameters. In this work, which requires modeling climate on orbital timescales ($> 10 000$ years), the typical general circulation models (GCMs) used for studying modern climate and the impact of future changes in greenhouse gases require too much computing power. These GCMs can be used for simulations covering a few thousand years at most, but provide valuable equilibrium simulations of the past climates, such as the Last Glacial Maximum (LGM).

Instead of the GCMs, it has been common to use Energy Balance Models (EBMs) to study changes in climate on orbital timescales. These types of models can be grouped into four categories: (1) annual mean atmospheric models; (2) seasonal atmospheric models with a mixed layer ocean; (3) Northern Hemisphere ice sheet models; and (4) coupled climate–ice sheet models, which in some cases include a representation of the deep ocean.

Studies with the first type of simple climate models were pioneered by the early work of Budyko in the 1960s, who investigated the sensitivity of climate to changes in global annual mean insolation. However, changes in the Earth's orbital parameters result in a

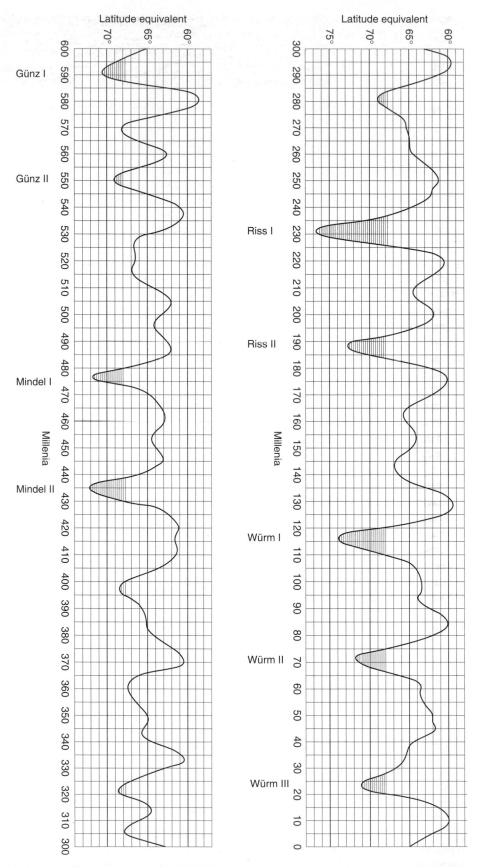

Figure 6 Milankovitch's reconstructed maximum glacial ice extent for the past 600 000 years. From Milankovitch M (1998) *Canon of Insolation and the Ice-Age Problem* (orig. publ. 1941). Belgrade: Zavod za Udzbenike I Nastavna Sredstva.

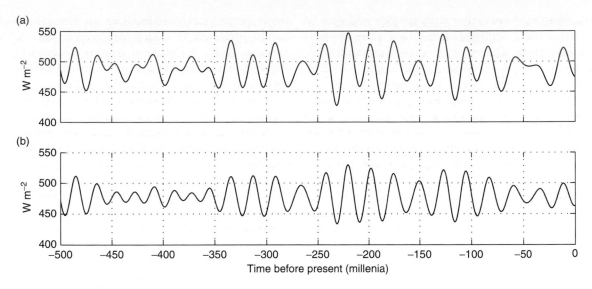

(a)

(b)

Figure 7 Summer solstice insolation at (a) 65° N and (b) 25° N for the past 500 000 years.

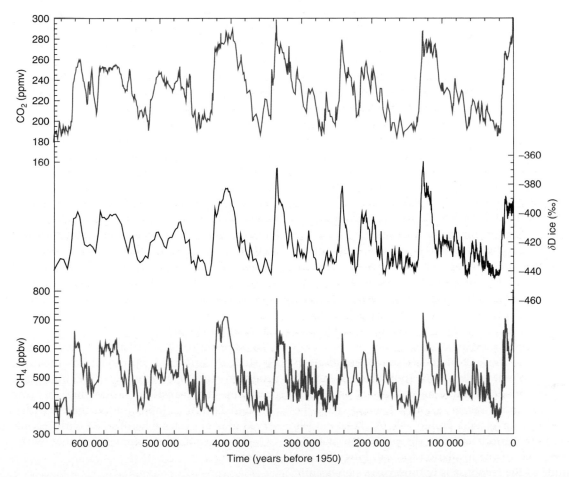

Figure 8 Variations in deuterium (δD; black), a proxy for local temperature, and atmospheric concentrations of the greenhouse gases carbon dioxide (CO_2; red), and methane (CH_4; green), from measurements of air trapped within Antarctic ice cores. Data from Spahni R, Chappellaz J, Stocker TF, *et al.* (2005) Atmospheric methane and nitrous oxide of the late Pleistocene from Antarctica ice cores. *Science* 310: 1317–1321 and Siegenthaler U, Stocker TF, Monnin E, *et al.* (2005) Stable carbon cycle-climate relationship during the late Pleistocene. *Science* 310: 1313–1317.

redistribution of insolation with latitude and time of the year, with a negligible impact on global annual mean insolation. Therefore, annual mean models are not adequate when investigating the impact of orbital insolation on climate, as they cannot capture the parts of the insolation variations which are seasonal and translate them into long-term climate change.

The second type of models includes a representation of the seasonal cycle, and has been used to investigate the orbital theory of Milankovitch. In this case, the seasonal variations in orbital insolation are resolved. However, as for the first type of models, past changes in ice cover are assumed to follow the simulated variations in the extent of perennial snow. This approach assumes that ice cover and the powerful ice albedo feedback are governed only by temperature, as the extent of snow in these models is fixed to the latitude with a temperature of $0\,^\circ C$. In reality, the growth and decay of land-based ice sheets are governed by the balance of accumulation and ablation. Therefore, when investigating changes in ice cover, it is necessary to include an appropriate representation of the dynamics and mass balance of ice sheets in the model.

The third type of models improves upon this by focusing on modeling past changes in mass balance and size of typical Northern Hemisphere ice sheets, such as the Laurentide. This type of studies was initiated by Weertman in the 1960s who used simple ice sheet models, forced by a prescribed distribution of accumulation minus ablation, to predict ice thickness versus latitude. These models do not calculate the atmospheric energy balance in order to estimate snowfall and surface melt; instead, changes to the prescribed distribution of net accumulation follow variations in mean summer insolation.

The fourth type of models include zonal mean seasonal climate models coupled to the simple Weertman-type ice sheet model, as well as earth models of intermediate complexity (EMICs) coupled to a dynamic ice sheet. These models give a more realistic representation of the climate as compared with the simpler models.

Partly due to the lack of good data on variations in global ice volume older than about half a million years, most model studies have focused on understanding the more recent records dominated by the $\sim 100\,000$ years glacial cycles. All of these models respond with periods close to the precession and obliquity periods of the insolation forcing. However, the amplitude of the response is in most cases significantly smaller than what is observed in the proxy records. At the same time, the dominant $\sim 100\,000$ year cycles of the ice volume record, characterized by rapid deglaciations, are only found when including a time lag in

the response of the model. Such an internal time lag can be produced by taking into account bedrock depression under the load of the ice, or by adding a parametrization of ice calving into proglacial lakes, or marine incursions at the margin of the ice sheet. Alternatively, the $\sim 100\,000$-year cycles have been explained as free, self-sustained oscillations, which might be phase-locked to oscillations in orbital insolation.

One of the very few model studies that have investigated variations in ice volume before the late Pleistocene transition ($\sim 800\,000$ years ago) used a two-dimensional climate model developed at Louvain-la-Neuve in Belgium. It falls within the definition of an EMIC and includes a simple atmosphere coupled to a mixed layer ocean, sea ice and ice sheets. By forcing this model with insolation and steadily decreasing atmospheric CO_2 concentrations, the model reproduces some of the characteristics of the ice volume record. The 41 000-year periodicity is present in the simulated ice volume for most of the past 3 My and the strength of the 100 000-year signal increases after about 1 My. However, a longer 400 000-year year period is also present and often dominates the simulated Northern Hemisphere ice volume record.

This nicely illustrates the remaining questions in the field. It is expected that models responding to the 100 000-year period will also respond to the longer 400 000-year period of eccentricity. However, this later period is not present in the ice volume record. At the same time, the late Pleistocene transition from a dominance of 41 to $\sim 100\,000$-year period oscillations in ice volume is not well understood. Explanations for the transition which have been tested in models are: a steady decrease in CO_2 forcing and its associated slow global cooling; or a shift from a soft to a hard sediment bed underlying the North American ice sheet through glacial erosion and exposure of unweathered bedrock. Neither of these changes are in themselves abrupt, but could cause a transition in the response of the ice sheets to insolation as the ice sheets grew to a sufficiently large size.

In addition to the challenge of modeling the mid-Pleistocene transition, no model has successfully reproduced the relatively clean 41 000-year cycles preceding the transition. Following the transition, the models only exhibit a good match with the observed glacial cycles when forced with reconstructed CO_2 from Antarctic ice cores together with orbital insolation.

Summary

The Plio-Pleistocene glacial cycles represent some of the largest and most significant changes in past climate, with a clear imprint in terrestrial and marine proxy records. Many of the physical mechanisms driving these large cycles in ice volume are not well understood. However, the pursuit to explain these climate changes has greatly advanced our understanding of the climate system and its future response to man-made forcing. New and better resolved proxy records will improve our spatial and temporal picture of the glacial cycles. Together with the advent of comprehensive climate models able to simulate longer periods of the glacial record, scientists will be able to better resolve the interaction of the atmosphere, ocean, biosphere, and ice sheets and the mechanisms linking them to the astronomical forcing.

Nomenclature

e	eccentricity
r_a	aphelion
r_p	perihclion
$\delta^{18}O$	oxygen isotope ratio (ratio of ^{18}O and ^{16}O relative to a standard)
ε	obliquity
ω	longitude of perihelion

See also

Monsoons, History of. Oxygen Isotopes in the Ocean. Satellite Remote Sensing of Sea Surface Temperatures. Stable Carbon Isotope Variations in the Ocean.

Further Reading

Bard E (2004) Greenhouse effect and ice ages: Historical perspective. *Comptes Rendus Geoscience* 336: 603–638.

Berger A, Li XS, and Loutre MF (1999) Modelling Northern Hemisphere ice volume over the last 3 Ma. *Quaternary Science Reviews* 18: 1–11.

Budyko MI (1969) The effect of solar radiation variations on the climate of the Earth. *Tellus* 5: 611–619.

Crowley TJ and North GR (1991) *Paleoclimatology*. New York: Oxford University Press.

Hays JD, Imbrie J, and Shackleton NJ (1976) Variations in the Earth's orbit: Pacemakers of the ice ages. *Science* 194: 1121–1132.

Imbrie J and Imbrie KP (1979) *Ice Ages, Solving the Mystery*. Cambridge, MA: Harvard University Press.

Köppen W and Wegener A (1924) *Die Klimate Der Geologischen Vorzeit*. Berlin: Gebrüder Borntraeger.

Milankovitch M (1998) *Canon of Insolation and the Ice-Age Problem* (orig. publ. 1941). Belgrade: Zavod za Udzbenike 1 Nastavna Sredstva.

Paillard D (2001) Glacial cycles: Toward a new paradigm. *Reviews of Geophysics* 39: 325–346.

Saltzman B (2002) *Dynamical Paleoclimatology*. San Diego, CA: Academic Press.

Siegenthaler U, Stocker TF, Monnin E, *et al.* (2005) Stable carbon cycle–climate relationship during the late Pleistocene. *Science* 310: 1313–1317.

Spahni R, Chappellaz J, Stocker TF, *et al.* (2005) Atmospheric methane and nitrous oxide of the late Pleistocene from Antarctica ice cores. *Science* 310: 1317–1321.

Weertman J (1976) Milankovitch solar radiation variations and ice age ice sheet sizes. *Nature* 261: 17 20.

PALEOCEANOGRAPHY

E. Thomas, Yale University, New Haven, CT, USA

Introduction: The Relevance of Paleoceanography

Paleoceanography encompasses (as its name implies) the study of 'old oceans', that is, the oceans as they were in the past. In this context, 'the past' ranges from a few decades through centennia and millennia ago to the very deep past, millions to billions of years ago. In reconstructing oceans of the past, paleoceanography needs to be highly interdisciplinary, encompassing aspects of all topics in this encyclopedia, from plate tectonics (positions of continents and oceanic gateways, determining surface and deep currents, thus influencing heat transport) through biology and ecology (knowledge of present-day organisms needed in order to understand ecosystems of the past), to geochemistry (using various properties of sediment, including fossil remains, in order to reconstruct properties of the ocean waters in which they were formed).

Because paleoceanography uses properties of components of oceanic sediments (physical, chemical, and biological) in order to reconstruct various aspects of the environments in these 'old oceans', it is limited in its scope and time resolution by the sedimentary record. Ephemeral ocean properties cannot be measured directly, but must be derived from proxies. For instance, several types of proxy data make it possible to reconstruct such ephemeral properties as temperature (*see* Determination of Past Sea Surface Temperatures) and nutrient content of deep and surface waters at various locations in the world's oceans, and thus obtain insights into past thermohaline circulation patterns, as well as patterns of oceanic primary productivity. The information on ocean circulation can be combined with information on planktic and benthic microfossils, allowing a view of interactions between fluctuations in oceanic environments and oceanic biota, on short but also on evolutionary timescales.

Paleoceanography offers information that is available from no other field of study: a view of a world alternative to, and different from, our present world, including colder worlds ('ice ages'; *see* Plio-Pleistocene Glacial Cycles and Milankovitch Variability) as well as warmer worlds (*see* Paleoceanography: the Greenhouse World). Paleoceanographic data thus serve climate modelers in providing data on boundary conditions of the ocean–atmosphere system very different from those in the present world. In addition, paleoceanography provides information not just on climate, but also on climate changes of the past, on their rates and directions, and possible linkages (or lack thereof) to such factors as atmospheric $p\mathrm{CO_2}$ levels (*see* Plio-Pleistocene Glacial Cycles and Milankovitch Variability) and the location of oceanic gateways and current patterns. Paleoceanography therefore enables us to gauge the limits of uniformitarianism: in which aspects is the present world indeed a guide to the past, in which aspects is the ocean–atmosphere system of the present world with its present biota just a snapshot, proving information only on one possible, but certainly not the only, stable mode of the Earth system? How stable are such features as polar ice caps, on timescales varying from decades to millions of years? How different were oceanic biota in a world where deep-ocean temperatures were 10–12 °C rather than the present (almost ubiquitous) temperatures close to freezing? Such information is relevant to understanding the climate variability of the Earth on different timescales, and modeling possible future climate change ('global warming'), as recognized by the incorporation of a paleoclimate chapter in the *Fourth Assessment Report of the Intergovernmental Panel on Climate Change* (2007).

Paleoceanography thus enables us to use the past in order to gain information on possible future climatic and biotic developments: the past is the key to the future, just as much and maybe more than the present is the key to the past.

Paleoceanography: Definition and History

Paleoceanography is a relatively recent, highly interdisciplinary, and strongly international field of science. International Conferences on Paleoceanography (ICP) have been held every 3 years since 1983, and the locations of these meetings reflect the international character of the paleoceanographic research community (**Table 1**). The flagship journal of paleoceanographic research, *Paleoceanography*, was first published by the American Geophysical Union in March 1986. In the editorial in its first volume, its target was defined as follows by editor J.P. Kennett:

Paleoceanography publishes papers dealing with the marine sedimentary record from the present ocean basins and

Table 1 International paleoceanographic conferences (ICP meetings)

Meeting	Year	Location
ICP1	1983	Zurich, Switzerland
ICP2	1986	Woods Hole, USA
ICP3	1989	Cambridge, UK
ICP4	1992	Kiel, Germany
ICP5	1995	Halifax, Canada
ICP6	1998	Lisbon, Portugal
ICP7	2001	Sendai, Japan
ICP8	2004	Biarritz, France
ICP9	2007	Shanghai, China

margins and from exposures of ancient marine sediments on the continents. An understanding of past oceans requires the employment of a wide range of approaches including sedimentology; stable isotope geology and other areas of geochemistry; paleontology; seismic stratigraphy; physical, chemical, and biological oceanography; and many others. The scope of this journal is regional and global, rather than local, and includes studies of any geologic age (Precambrian to Quaternary, including modern analogs). Within this framework, papers on the following topics are to be included: chronology, stratigraphy (where relevant to correlation of paleoceanographic events), paleoreconstructions, paleoceanographic modeling, paleocirculation (deep, intermediate, and shallow), paleoclimatology (e.g., paleowinds and cryosphere history), global sediment and geochemical cycles, anoxia, sea level changes and effects, relations between biotic evolution and paleoceanography, biotic crises, paleobiology (e.g., ecology of 'microfossils' used in paleoceanography), techniques and approaches in paleoceanographic inferences, and modern paleoceanographic analogs.

Perusal of the volumes of the journal published since demonstrates that it indeed covers the full range of topics indicated above.

What is the shared property of all these papers? They deal with various aspects of data generation or modeling using information generated from the sedimentary record deposited in the oceans, including such materials as carbonates, clays, and authigenic minerals, as well as carbonate, phosphate, and opaline silica secreted by marine organisms, including for instance the calcium carbonate secreted by corals (*see* Past Climate from Corals). Sediments recovered from now-vanished oceans as well as sediments recovered from the present oceans are included, but paleoceanography as a distinct field of study is tightly linked to recovery of sediment cores from the ocean floor, deposited onto oceanic basement. Such sediments represent times from which oceanic crust is still in existence in the oceans (i.e., has not been subducted), which is about the last 200 My of Earth history.

Paleoceanography thus is a young field of scientific endeavor because the recovery of oceanic sediments in

cores started in earnest only in the 1950s, long after the *Challenger* Expedition (1872–76), usually seen as the initiation of modern oceanography. Little research was conducted on ocean sediments until the late 1940s and 1950s, although some short cores were recovered by the German South Polar Expedition (1901–03), the German *Meteor* Expedition (1925–27), the English *Discovery* Expedition (1925–27), and Dutch *Snellius* Expedition (1929–30). This lack of information on oceanic sediments is documented by the statement in the book *The Oceans* by H. Sverdrup, N. Johnson, and R. Fleming published in 1942:

> From the oceanographic point of view, the chief interest in the topography of the seafloor is that it forms the lower and lateral boundaries of the water.

This situation quickly changed in the years after the World War II, when funding for geology and geophysics increased, and new technology became available to recover material from the seafloor. Sediment cores collected by the *Challenger* in 1873 reached a maximum length of 2 ft (**Figure 1**), and little progress in core collection was made in more than 50 years, so that cores collected by the *Meteor* (1925–27) reached only 3 ft, as described by H. Petterson in the *Proceedings of the Royal Society of London* in 1947. In 1936, the American geophysicist C.S. Piggott obtained a 10-ft core using an exploding charge to drive a coring tube into the sediment, and in the early 1940s H. Petterson and B. Kullenberg developed the prototype of the piston corer. Piston corers were first used extensively on the Swedish Deep-Sea Expedition (1946–47) using the *Albatross*, the first expedition to focus "on the bottom deposits, their chemistry, stratigraphy, etc.," and thus arguably the first paleoceanographic expedition.

A major expansion occurred in US oceanographic institutions in the postwar years, and piston corers were used extensively by Woods Hole Oceanographic Institution (Woods Hole, Massachusetts) with its research vessel *Atlantis*, the first American ship built for sea research (1931–66), Scripps Institution of Oceanography (La Jolla, California) with the *E.W. Scripps* (1937–55), and the Lamont Geological Observatory of Columbia University (Palisades, New York), now the Lamont-Doherty Earth Observatory, with the *Vema* (1953–81). In 1948, only about 100 deep-sea cores existed, and by 1956 the *Vema* alone had collected 1195 cores.

Before the Swedish Deep-Sea Expedition, only limited estimates of the sedimentation rates in the deep ocean were available, described by H. Petterson in 1947 as "the almost unknown chronology of the oceans" (**Figure 1**). The development of radiocarbon dating of carbonates in combination with geochemical

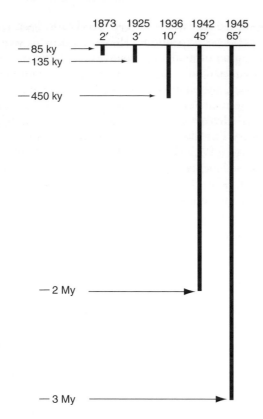

Figure 1 Lengths of sediments cores (in ft) obtained by the *Challenger* (1873), the *Meteor* (1925), Piggott's explosive-driven coring device (1936), and piston cores obtained by corers constructed by H. Petterson and B. Kullenberg, Institute of Oceanography, Göteborg, Sweden (1942–45). Adapted from Petterson H (1947) A Swedish deep-sea expedition. *Proceedings of the Royal Society of London, Series B: Biological Sciences* 134(876): 399–407, figure 5.

determination of the titanium content of sediments as a tracer for clay content led to estimates of sedimentation rates by G. Arrhenius, G. Kjellberg, and W.L. Libby in 1951, and of differences in sedimentation rates in glacial and interglacial times by W.S. Broecker, K.K. Turekian, and B.C. Heezen in 1958.

Cores collected by the *Albatross* during the Swedish Deep-Sea Expedition as well as the many later expeditions of the *Atlantis* and *Vema* were used extensively in seminal paleoceanographic studies of the Pleistocene ice ages, including those of the variation of calcium carbonate accumulation by G. Arrhenius and E. Olausson, and of the variations in populations of pelagic microorganisms (foraminifera) by F. Phleger, F. Parker, and J.F. Peirson published in 1953, as well as D.B. Erickson and G. Wollin published in 1956. The oxygen isotopic method to reconstruct past oceanic temperatures using carbonate sediments, as outlined by H. Urey in a paper in *Science* in 1948, was first applied to oceanic microfossils by C. Emiliani, in order to reconstruct ocean surface temperatures during the Pleistocene ice ages.

It turned out that the oxygen isotope record combined signals of temperature change as well as ice volume (*see* Cenozoic Climate – Oxygen Isotope Evidence and Determination of Past Sea Surface Temperatures), but the method has become widely established since those early days. Oxygen isotope analysis was the major tool in the first attempt to look at paleoclimate globally, in the CLIMAP (Climate/Long Range Investigation Mappings and Predictions) project in the early 1970s, which led to the documentation that major, long-term changes in past climate are associated with variations in the geometry of the Earth's orbit, as described by J. Hays, J. Imbrie, and N.J. Shackleton in 1976 in *Science* (*see* Plio-Pleistocene Glacial Cycles and Milankovitch Variability). Oxygen isotope analysis is still one of the 'workhorses' of paleoceanographic research, as documented in the review of global climate of the last 65 My by J.C. Zachos and others in *Science* (2001).

Most material older than the last few hundred thousands of years became available in much longer cores which could be recovered only after the start of drilling by the Deep Sea Drilling Project (DSDP) in 1968, an offshoot of a 1957 suggestion by W. Munk (Scripps Institution of Oceanography) and H. Hess (Princeton University) to drill deeply into the Earth and penetrate the crust–mantle boundary. In 1975, various countries joined the United States of America in the International Phase of Ocean Drilling, and DSDP became a multinational enterprise. Between 1968 and 1983, the drill ship *Glomar Challenger* recovered more than 97 km of core at 624 drill sites. Cores recovered by DSDP provided the material for the first paper using oxygen isotope data to outline Cenozoic climate history and the initiation of ice sheets on Antarctica, published in the *Initial Reports of the Deep Sea Drilling Project* in 1975 by N.J. Shackleton and J.P. Kennett.

DSDP's successor was the Ocean Drilling Program (1983–2003), which recovered more than 222 km sediment at 652 sites with its vessel *Joides Resolution*. These two programs were succeeded by the present Integrated Ocean Drilling Program (IODP), which plans to greatly expand the reach of the previous programs by using multiple drilling platforms, including riser drilling by the Japanese-built vessel *Chikyu*, riserless drilling by a refitted *Joides Resolution*, and mission-specific drilling using various vessels.

Paleoceanographic Techniques

Over the last 30 years there has been an explosive development of techniques for obtaining information from oceanic sediments (**Figure 2**). One of the limitations of paleoceanographic research on samples from ocean cores is the limited size of each sample. However, a positive result of this limited availability is that researchers from different disciplines are forced to work closely together, leading to the generation of independent proxy records on the same sample set, thus integrating various aspects of chemical and biotic change over time. Proxies are commonly measured on carbonate shells of pelagic and benthic microorganisms (*see* Benthic Foraminifera and Coccolithophores), thus providing records from benthic and several planktonic environments (surface, deep thermocline).

Methods used since the first paleoceanographic core studies include micropaleontology, with the most commonly studied fossil groups including pelagic calcareous and siliceous-walled heterotroph protists, organic-walled cysts of heterotroph and autotroph dinoflagellates, siliceous-walled (Diatoms) and calcareous-walled autotroph protists (*see* Coccolithophores), benthic protists (*see* Benthic Foraminifera), and microscopic metazoa, Ostracods (Crustacea). Micropaleontology is useful in biostratigraphic correlation as well as in its own right, providing information on evolutionary processes and their linkage (or lack thereof) to climate change, as well as on changes in oceanic productivity.

Classic stable isotopes used widely and commonly in paleoceanographic studies include those of oxygen (*see* Cenozoic Climate – Oxygen Isotope Evidence) and carbon (*see* Cenozoic Oceans – Carbon Cycle Models). Carbon isotope records are of prime interest in investigations of deep oceanic circulation (*see* Ocean Circulation: Meridional Overturning Circulation) and of oceanic productivity. In addition to these classic paleoceanographic proxies, many new methods of investigation have been and are being developed using different geochemical (isotope, trace element, organic geochemical) proxies (*see* Determination of Past Sea Surface Temperatures). Many more proxies are in development, including proxies on different organic compounds, to investigate aspects of global biogeochemical cycles, biotic evolution and productivity, and thermohaline circulation patterns (*see* Ocean Circulation: Meridional Overturning Circulation) (**Figure 2**).

Techniques to correlate the age of features in sediment records recovered at different locations and to assign numerical ages to sediment samples are of

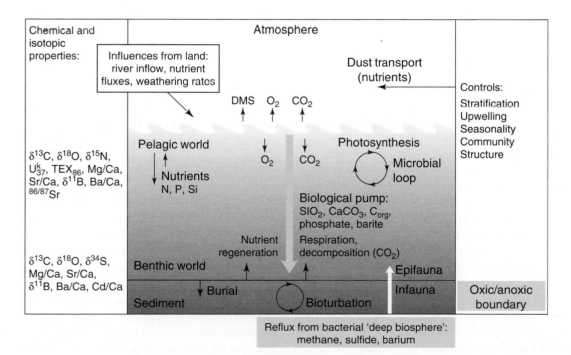

Figure 2 Linkages between the marine biosphere and global biogeochemical cycles, as well as to various proxies used in paleoceanographic studies. The proxies include isotope measurements (indicated by lower case deltas, followed by the elements and the heavier isotope), elemental ratios, and organic geochemical temperature proxies (e.g., U^k_{37}', an alkenone ratio in which the 37 refers to the carbon number of the alkenones; TEX_{86}, a proxy based on the number of cyclopentane rings in sedimentary membrane lipids derived from marine crenarchaeota). DMS is dimethylsulfide, a sulfur compound produced by oceanic phytoplankton. The various proxies can be used to trace changes in ocean chemistry (including alkalinity), temperature, and productivity, as well as changes in the reservoirs of the carbon cycle. Adapted from Pisias NG and Delaney ML (eds.) (1999) *COMPLEX: Conference of Multiple Platform Exploration of the Ocean*. Washington, DC: Joint Oceanographic Institutions.

the utmost importance to be able to estimate rates of deposition of sediments and their components. Correlation between sediment sections is commonly achieved by biostratigraphic techniques, which cannot directly provide numerical age estimates, and are commonly limited to a resolution of hundreds of thousands of years. Techniques used in numerical dating include the use of various radionuclides, the correlation of sediment records to the geomagnetic polarity timescale, and the more recently developed techniques of linking high-resolution records of variability in sediment character (e.g., color, magnetic susceptibility, density, and sediment composition) to variability in climate caused by changes in the Earth's orbit and thus energy supplied by the sun to the Earth's surface at specific latitudes (see Paleoceanography: Orbitally Tuned Timescales). Such an orbitally tuned timescale has been fully developed for the last 23 My of Earth history, with work in progress for the period of 65–23 Ma. Remote sensing techniques are being used increasingly in order to characterize sediment *in situ* in drill holes, even if these sediments have not been recovered, and to establish an orbital chronology even under conditions of poor sediment recovery.

Contributions of Paleoceanographic Studies

Paleoceanographic studies have contributed to a very large extent to our present understanding that the Earth's past environments were vastly different from today's, and that changes have occurred on many different timescales. Paleoceanographic studies have been instrumental in establishing the fact that climate change occurred rapidly and stepwise rather than gradually, whether in the establishment of the Antarctic ice cap on timescales of tens to hundreds of thousands of years rather than millions of years, or in the ending of a Pleistocene ice age on a timescale of decades rather than tens of thousands of years (see Abrupt Climate Change and Millennial-Scale Climate Variability).

Unexpected paleoceanographic discoveries over the last few years include the presence of large amounts of methane hydrates (clathrates), in which methane is trapped in ice in sediments along the continental margins, in quantities potentially larger than the total global amount of other fossil fuels. The methane in gas hydrates may become a source of energy, with exploratory drilling advanced furthest in the waters off Japan and India. Such exploration and use of gas hydrates might, however, lead to rapid global warming if drilling and use of methane hydrates inadvertently lead to uncontrolled destabilization. Destabilization of methane hydrates may have occurred in the past as a result of changes in sea level (see Sea Level Change and Sea Level Variations Over Geologic Time) and/or changes in thermohaline circulation and subsequent changes in deep-ocean temperature (see Methane Hydrates and Climatic Effects). Dissociation of gas hydrates and subsequent oxidation of methane in the atmosphere or oceans have been speculated to have played a role in the ending of ice ages, and in a major upheaval in the global carbon cycle and global warming (see Methane Hydrates and Climatic Effects). The influence on global climate and the global carbon cycle of methane hydrate reservoirs (with their inherent capacity to dissociate on timescales of a few thousand years at most) is as yet not well understood or documented (see Carbon Cycle and Ocean Carbon System, Modeling of).

Most methane hydrates are formed by bacterial action upon organic matter, and another unexpected discovery was that of the huge and previously unknown microbial biomass in seawater, also found buried deep in the sediments. Fundamental issues such as the conditions that support and limit this biomass are still not understood and neither are their linkage to the remainder of the oceanic biosphere and the role of chemosynthesis and chemosymbiosis in the deep oceanic food supply and the global carbon cycle.

In contrast to these unexpected discoveries, paleoceanographers of a few decades ago could have predicted at least in part our increased knowledge of aspects of climate change such as the patterns of change in sea level at various timescales (see Sea Level Variations Over Geologic Time). The suddenness and common occurrence of rapid climate change events in Earth history, however, was unpredicted. On timescales of millions of years, the Earth's climate was warm globally during most of the Cretaceous and the early part of the Cenozoic (65–35 Ma), and the Earth had no large polar ice caps reaching sea level (see Paleoceanography: the Greenhouse World). Drilling in the Arctic Ocean, for instance, established that average summer surface water temperatures might have reached up to 18 °C. The use of climate models has assisted in understanding such a warm world (see Paleoceanography, Climate Models in), but the models still cannot fully reproduce the necessary efficient heat transport to high latitudes at extremely low latitudinal temperature gradients.

Paleoceanographic research has provided considerable information on the major biogeochemical cycles over time (*see* Cenozoic Oceans – Carbon Cycle Models). Geochemical models of the carbon cycle rely on carbon isotope data on bulk carbonates and on planktonic and benthic foraminifera in order to evaluate transfer of carbon from one reservoir (e.g., organic matter, including fossil fuel; methane hydrates) to another (e.g., limestone, the atmosphere, and dissolved carbon in the oceans). Information on pelagic carbonates and their microfossil content as well as stable isotope composition has assisted in delineating the rapidity and extent of the extinction at the end of the Cretaceous in the marine realm. Sedimentological data at many locations have documented the large-scale failure of the western margin of the Atlantic Ocean, with large slumps covering up to half of the basin floor in the North Atlantic as the result of the asteroid impact on the Yucatan Peninsula.

One of the major successes of paleoceanographic research has been the establishment of the nature and timing of polar glaciation during the Cenozoic cooling (*see* Cenozoic Climate – Oxygen Isotope Evidence and Paleoceanography: the Greenhouse World). After a prolonged period of polar cooling in the middle to late Eocene, the East Antarctic Ice Sheet became established during a period of rapid ice volume growth (< 100 ky) in the earliest Oligocene, *c.* 33.5 Ma. This establishment was followed by times of expansion and contraction of the ice sheet, and the West Antarctic Ice Sheet may have started to grow at *c.* 14 Ma. Until recently it was argued that the Northern Hemispheric Ice Sheets formed much later: these ice sheets increased in size around 3 Ma (in the Pliocene), and ever since have contracted and expanded on orbital timescales (*see* Plio-Pleistocene Glacial Cycles and Milankovitch Variability). There is now considerable evidence that the polar ice sheets in the Northern Hemisphere also became established, at least in part, in the earliest Oligocene, that is, at a similar time as the Southern Hemisphere ice sheets.

The cause(s) of the long-term Cenozoic cooling are not fully known. Possible long-term drivers of climate include the opening and closing of oceanic gateways, which direct oceanic heat transport (*see* Heat Transport and Climate). Such changes in gateway configuration include the opening of the Tasman Gateway and Drake Passage which made the Antarctic Circumpolar Current possible (*see* Antarctic Circumpolar Current), and closing of the Isthmus of Panama which ended the flow of equatorial currents from the Atlantic into the Pacific (*see* and Pacific Ocean Equatorial Currents). Evidence has accumulated, however, that changes in atmospheric CO_2 levels may have been more important

than changes in gateway configuration. For instance, long-term episodes of global warmth (in timescales of millions of years) may have been sustained by CO_2 emissions from large igneous provinces , and short-term global warming (on timescales of ten to one or two hundred thousands of years) could have been triggered by the release of greenhouse gases from methane hydrate dissociation or burning/oxidation or organic material (including peat). Decreasing atmospheric CO_2 levels due to decreasing volcanic activity and/or increased weathering intensity have been implicated in the long-term Cenozoic cooling, and high-resolution paleoceanographic records show that climate change driven by changes in atmospheric CO_2 levels may have been modulated by changes in insolation caused by changes in orbital configuration. Recognition of variability in climatic signals in sediments at orbital frequencies has led to major progress in the establishment of orbitally tuned timescales throughout the Cenozoic (*see* Paleoceanography: Orbitally Tuned Timescales).

Paleoceanographic research has led to greatly increased understanding of the Plio-Pleistocene Ice Ages as being driven by changes in the Earth's orbital parameters (*see* Paleoceanography: Orbitally Tuned Timescales and Plio-Pleistocene Glacial Cycles and Milankovitch Variability), and the correlation of data from oceanic sediments to records from ice cores on land. Orbital forcing is the 'pacemaker of the ice ages', but it is not yet fully understood how feedback processes magnify the effects of small changes in insolation into the major climate swings of the Plio-Pleistocene. Ice-core data and carbon isotope data for marine sediments show that changes in atmospheric CO_2 levels play a role. We also do not yet understand why the amplitude of these orbitally driven climate swings increased at *c.* 0.9 Ma, and why the dominant periodicity of glaciation switched from 40 000 (obliquity) to 100 000 (eccentricity) years at that time, the 'mid-Pleistocene revolution'. Effects of glaciation at low latitudes, including changes in upwelling, productivity, and monsoonal activity, are only beginning to be documented (*see* Monsoons, History of).

Great interest has been generated by the information on climate change at shorter timescales than 20 ky, the duration of the shortest orbital cycle, precession. Such climate variability includes the millennial-scale changes that occurred during the glacial periods (*see* Millennial-Scale Climate Variability) and the climate variations of lesser amplitudes that have occurred since the last deglaciation (*see* Holocene Climate Variability). These research efforts are beginning to provide information on timescales that are close to the human timescale, on such topics as abrupt climate change (*see* Abrupt Climate Change), and the fluctuations in intensity and occurrence of the El Niño–Southern Oscillation (*see* El Niño Southern Oscillation (ENSO)) and the North Atlantic Oscillation (NAO) (*see* North Atlantic Oscillation (NAO)), during overall colder and warmer periods of the Earth history.

The Future of Paleoceanography

Past progress in paleoceanography has been linked to advances in technology since the invention of the piston corer in the 1940s. In the early to mid-1980s, paleoceanographic studies appeared to reach a plateau, with several review volumes published on, for example, Plio-Pleistocene ice ages (CLIMAP), the oceanic lithosphere, and the global carbon cycle, as well as a textbook *Marine Geology* by J.P. Kennett. Paleoceanographic research, however, has benefited from research programs in the present oceans (such as the Joint Global Ocean Flux Program, JGOFS), and new technology has spurred major research activity. The extensive use and improvement of the hydraulic piston corer by DSDP/ODP/IODP led to recovery of minimally disturbed soft sediment going back in age through the Cenozoic, making high-resolution studies possible, including paleoceanographic and stratigraphic studies of sediment composition using the X-ray fluorescence (XRF) scanner. Progress in computers led to strongly increased use of paleoceanographic data in climate modeling, and to increased possibilities of remote sensing in drill holes. Developments in mass spectrometry led to the possibility to measure isotopes and trace elements in very small samples, such as those recovered in deep-sea cores, and new proxies continue to be developed, including proxies (e.g., on levels of oxygenation and temperature) using organic geochemical methods.

The IODP started in 2003 and is scheduled to start drilling in 2008 with three different platforms. Drilling activity has become integrated with that of the French research vessel *Marion Dufresne*, which has recovered many long piston cores (several tens of meters) for studies covering the last few hundred thousand years of the Earth history, in the International Marine Global Change Study (IMAGES) program. Examples of drilling by alternative platforms include the drilling in the Arctic Ocean, one of the frontiers in ocean science, and drilling in shallower regions than accessible to the drilling vessel *Joides Resolution*, such as coral drilling in Tahiti for studies of Holocene climate and rates of sea level rise. If these ambitious programs are carried out, we can expect to learn much about the working of the Earth system of lithosphere–ocean–atmosphere–biosphere, specifically about the sensitivity of the climate system, about the controls on the long-term evolution of this sensitivity, and about the complex interaction of the biospheric, lithospheric, oceanic, and atmospheric components of the Earth system at various timescales.

See also

Abrupt Climate Change. Benthic Foraminifera. Carbon Cycle. Cenozoic Climate – Oxygen Isotope Evidence. Cenozoic Oceans – Carbon Cycle Models. Coccolithophores. Determination of Past Sea Surface Temperatures. El Niño Southern Oscillation (ENSO). Heat Transport and Climate. Holocene Climate Variability. Methane Hydrates and Climatic Effects. Millennial-Scale Climate Variability. Monsoons, History of. North Atlantic Oscillation (NAO). Ocean Carbon System, Modeling of. Ocean Circulation: Meridional Overturning Circulation. Pacific Ocean Equatorial Currents. Paleoceanography, Climate Models in. Paleoceanography: Orbitally Tuned Timescales. Paleoceanography: the Greenhouse World. Past Climate from Corals. Plio-Pleistocene Glacial Cycles and Milankovitch Variability. Sea Level Change. Sea Level Variations Over Geologic Time.

Further Reading

Elderfield H (2004) The Oceans and Marine Geochemistry. In: Holland HD and Turekian KK (eds.) *Treatise on Geochemistry*, vol. 6. Amsterdam: Elsevier.

Gradstein F, Ogg J, and Smith AG (2004) *A Geologic Time Scale*. Cambridge, UK: Cambridge University Press.

Hillaire-Marcel C and de Vernal A (2007) *Proxies in Late Cenozoic Paleoceanography*. Amsterdam: Elsevier.

National Research Council, Ocean Sciences Board (2000) *50 years of Ocean Discovery*. Washington, DC: National Academies Press.

Oceanography (2006) A Special Issue on The Impact of the Ocean Drilling Program. *Oceanography* 19(4).

Olausson E (1996) *The Swedish Deep-Sea Expedition with the 'Albatross' 1947–1948: A Summary of Sediment Core Studies*. Göteborg, Sweden: Novum Grafiska AB.

Petterson H (1947) A Swedish deep-sea expedition. *Proceedings of the Royal Society of London, Series B: Biological Sciences* 134(876): 399–407.

Pisias NG and Delaney ML (eds.) (1999) *COMPLEX: Conference of Multiple Platform Exploration of the Ocean*. Washington, DC: Joint Oceanographic Institutions.

Sarmiento JL and Gruber N (2006) *Ocean Biogeochemical Dynamics*. Princeton, NJ: Princeton University Press.

Relevant Websites

http://www.andrill.org
 – ANDRILL: Antarctic drilling.

http://www.ecord.org
 – European Consortium for Ocean Research Drilling (ECORD).

http://www.iodp.org
 – Integrated Ocean Drilling Program (IODP).

http://www.ipcc-wg1.ucar.edu
 – Intergovernmental Panel on Climate Change, Working Group 1: The Physical Basis of Climate Change.

http://www.ngdc.noaa.gov
 – National Geophysical Data Center (Marine Geology and Geophysics).

http://www.images-pages.org
 – The International Marine Past Global Change Study (IMAGES).

PALEOCEANOGRAPHY, CLIMATE MODELS IN

W. W. Hay, Christian-Albrechts University, Kiel, Germany

Introduction

Climate models have been applied to the investigation of ancient climates for about 25 years. The early investigations used simple energy balance models, but since the 1970s increasingly more elaborate atmospheric general circulation models (AGCMs) have been applied to paleoclimate problems. It is only within the last decade that climate model results have been used to drive ocean circulation models to simulate the behavior of ancient oceans. Similarly, increasingly more complex ocean models have been used since the 1970s to explore the effects of changing boundary conditions on ocean circulation and climate. The early paleo-ocean models were driven by modern winds and hydrologic cycle data; the more recent simulations use paleoclimate-model-generated data. In some of the early models sea surface temperatures were specified, but this locks the model to a particular solution. Runoff from land has only been included in the most recent models. Experiments with coupled atmosphere–ocean models are just beginning.

Paleo-ocean modeling has been directed largely towards trying to understand two major problems related to possible changes in ocean heat transport with global climatic implications: (1) the origin of the Late Cenozoic glacial ages, with special attention to the initiation and cessation or slowdown of the 'Global Conveyor System,' and (2) the warm climates of the Late Mesozoic and Eocene, with low meridional temperature gradients. One group of modeling exercises has concentrated on the effects of interocean connections, particularly on the effects of opening and closing of gateways between the major ocean basins on the global thermohaline circulation system. Another major set of investigations has focused on ocean surface and thermohaline circulation on a warm Earth, with higher concentrations of atmospheric CO_2 and the very different paleogeography of the Late Mesozoic.

Ocean Gateways and Climate

The effects of interocean gateways on ocean circulation and the global climate system have been a matter of speculation since the late 1960s and of experimentation via box models and numerical simulation since the 1980s. **Figure 1** shows the major gateways that have opened or closed during the Cenozoic. In most instances the timing of opening or closing of gateways has been deduced from the regional plate tectonic context, but has an uncertainty of a few million years. In some cases, such as the opening of passages between Australia, South America and Antarctica the plate tectonic record is ambiguous and the timing is inferred from climate change in the surrounding regions. For a few passages, such as Panama and Gibraltar, the timing is known to within a few tens of thousands of years because the resulting oceanographic changes are reflected in ocean floor sediments on either side of the gateway. Until now it has been possible to explore the importance of only a few of the most critical gateways for the global climate system. Most of these investigations have concentrated on exploring the role of salinity in forcing the thermohaline circulation to develop an understanding of the behavior of the 'Great Conveyor.' The term 'Great Conveyor' is used to describe the global thermohaline circulation that starts with sinking of cold saline water in the Norwegian–Greenland Sea and continues southward as the North Atlantic Deep Water (NADW). On reaching the Southern Ocean, it is supplemented by water sinking in the region of the Weddell Sea. The deep water then flows into the Indian and Pacific Oceans where it returns to the surface. The return surface flow is then from the Pacific through the Indonesian Passages and across the Indian Ocean, then around South Africa into the South Atlantic and finally northward to the Norwegian–Greenland Sea. Paleoceanographic data show that the production of NADW has slowed or even stopped from time to time. This has major implications for the paleoclimate of the continents surrounding the North Atlantic because it is the sinking of water in the Norwegian–Greenland Sea that draws warm surface waters northward along the western margin of Europe and Scandinavia.

It has been argued on the basis of simple model calculations that the Atlantic acts as a salt oscillator. The southward flow of NADW is thought to export salt from the North Atlantic. When so much salt has

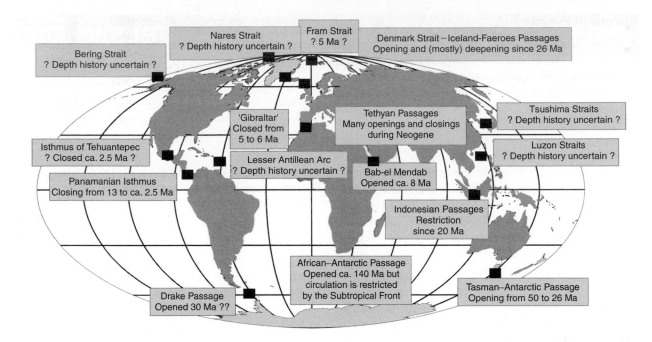

Figure 1 Interocean gateways which have opened or closed during the Cenozoic (since 65 Ma).

been exported that the surface waters are no longer so saline, deep water formation in the Norwegian–Greenland Sea will stop, shutting down the Global Conveyor. The Global Conveyor is revived by the loss of water from the Atlantic to the Pacific through the atmosphere across Central America. This increases the salinity of the North Atlantic to levels where the deep water formation is initiated again.

It has also been suggested that the saline Mediterranean outflow is an important forcing factor controlling the generation of NADW. Some of the Mediterranean outflow water becomes entrained in the surface waters crossing the Iceland–Scotland Ridge, raising the salinity to a level that can promote sinking as the water cools.

The salinity control hypothesis has been questioned by authors who argue that the global thermohaline circulation system is driven by Southern Hemisphere winds, and that the salinity difference between the Atlantic and Pacific plays only a minor role. This conclusion is based on an idealized ocean general circulation model (OGCM) using the simple geometry of two polar continents connected by a long thin meridional land mass. The model has no salinity forcing or sea ice formation, and no separate Atlantic Basin. The ocean model is coupled to a simple energy balance atmosphere with imposed zonal winds. Removing a piece of the land mass at the latitude of the Drake Passage induces an interhemispheric conveyor, similar to the modern Great Conveyor, which warms the entire Northern Hemisphere by several degrees. In this case, the

conveyor is driven by the Southern Hemisphere winds and salinity plays no role at all.

Both of these models are great oversimplifications of the real world. They raise the question of whether the same results will be obtained with simulations having more realistic geographic boundary conditions, wind forcing, and radiative balance.

Drake Passage

Paleoceanographers have assumed that the opening of the Tasman–Antarctic Passage and subsequent opening of the Drake Passage led to isolation of the Antarctic continent. It has been argued that this resulted in a sharp meridional thermal gradient across the Southern Ocean and promoted the glaciation of the Antarctic continent.

The importance of the Drake Passage for the modern thermohaline circulation has been recognized by physical oceanographers since 1971 when scientists at the Geophysical Fluid Dynamics Laboratory in Princeton, New Jersey used an early OGCM to experiment with a closed Drake Passage. They found that closure of the passage would increase the outflow of Antarctic Bottom Water (AABW) to the world ocean. The same result has been obtained for the effect of closure of the Drake Passage using other ocean models. It has been concluded that the opening of the Drake Passage was responsible for creating the circulation system we observe today.

Isthmus of Panama

Another important event affecting Earth's climate history was the closure of the connection between the Atlantic and Pacific across Central America. It has long been suspected that this paleographic change set the stage for the Northern Hemisphere glaciation.

Workers at the Max Planck Institute for Meteorology in Hamburg, explored the effects of an open Central American Passage using a model with a $3.5° \times 3.5°$ grid with 11 vertical levels, realistic bottom topography, and a full seasonal cycle. Although it does not simulate mesoscale eddies, the Hamburg OGCM does well at simulating the present surface and deep circulation of the ocean. A Central American Passage was opened between Yucatan and South America, and a sill depth of 2711 m was specified. The simulation was forced by the modern observed wind stress field and air temperatures. The large-scale low-latitude interchanges of waters between the Atlantic and Pacific had dramatic effects. The regional slope on the ocean surface from the western Pacific to the Norwegian–Greenland Sea was reduced. Flow through the Central American Passage was 1 Sv ($1 \text{ Sv} = 10^6 \text{ m}^3 \text{ s}^{-1}$) of wind-driven surface water from Atlantic to Pacific and 10 Sv of subsurface water driven by the hydrostatic head from Pacific to Atlantic. Flow through the Bering Strait, presently from the Pacific to the Arctic, was reversed, with a low salinity outflow from the Arctic into the North Pacific. There was little effect on flow through the Drake Passage. The salinity difference between the Atlantic and Pacific was greatly reduced. The production of NADW ceased, and flow of the Gulf Stream was reduced whereas flow of the Brazil Current was enhanced. The outflow of AABW into the world ocean increased about 25% with the closed Central American Passage. Southward oceanic heat transport also increased by about 25%, as more warm water from the South Atlantic was drawn southward to replace the surface water sinking in the Weddell Sea to form AABW. These changes would have major implications for global climates, making Europe colder and providing a larger snow source for the Antarctic. The results of this experiment did much to stimulate further studies.

The extreme sensitivity of the Atlantic thermohaline circulation to an open or closed Panama Strait has not been borne out by more realistic coupled atmosphere–ocean models (AOGCM). A subsequent experiment found a critical threshold for water transport through the passage affecting formation of NADW. Flows up to 5 Sv from the Pacific to the Atlantic through the passage still allow NADW to form, but at a rate only about 50% that of the present day. Greater flows shut down the NADW production.

In contrast, a recent AOGCM simulation using the recently developed Fanning and Weaver atmospheric energy-moisture balance model indicated that NADW formation could take place even with a fully open passage through Panama.

Combined Opening and Closure of the Drake and Panama Passages

One set of experiments has combined the effect of closing the Drake Passage and opening Panama. Two scenarios were examined, a closed Drake Passage and closed Central American Passage, and a closed Drake Passage with an open Central American Passage. Atmospheric forcing of the ocean model was with present day winds. The results are summarized in **Figure 2**.

During the past 60 million years the Drake and Central American Passages have not been closed at the same time. Nevertheless, it is an interesting experiment to examine the effect of a pole-to-pole meridional seaway connected to the world ocean through a high latitude passage on the east (between Africa and Antarctica). The major effect of this configuration was to increase the outflow of AABW to the world ocean by a factor of four and suppress NADW production. The net effect is a doubling of the global rate of thermohaline overturning. Enhanced upwelling of AABW in the northern Atlantic reduced the salinity of the surface waters there, preventing formation of deep water.

The scenario with a closed Drake Passage and open Central American Passage corresponds to the paleogeography of the Atlantic from mid-Cretaceous through most of the Oligocene. The result of the experiment was similar to that closing both passages, except that the increase in AABW was less, about 3.2 times that at present. Production of NADW was suppressed, but the overall thermohaline circulation increased. In the South Atlantic, the deep waters were cooler and the thermocline weaker, promoting overturning and cooling of the surface waters. The flow of the Brazil Current was increased and the flow of the North Equatorial Current along the northern margin of South America was reduced. There was reduced equatorward transport in the eastern South Pacific and reduced flow of the Pacific western boundary current. The most controversial result of this experiment was that analysis of the surface heat flux suggested that the opening of the Drake Passage did not result in temperature changes large enough to have triggered Antarctic glaciation.

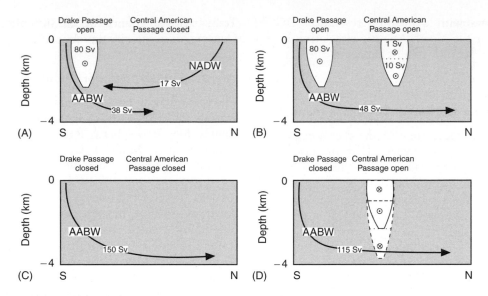

Figure 2 Pacific to Atlantic flow with open and closed Panama and Drake Passages. (After Mikalojewicz et al. 1993.) ⊙ = flow from Pacific to Atlantic. ⊗ = flow from Atlantic to Pacific. Volume fluxes in Sverdrups (Sv).

The effects of different widths of the Central American Passage and different sill depths in the Drake and Central American Passages were also investigated. It was found that widening the low latitude passage by removing all of Central America has little effect. A shallow sill (700 m) in the Drake Passage produced effects intermediate between experiments with deep sill (2700 m) used for the control run and the closed Drake Passage run. An experiment was also conducted with a deep (2700 m) Drake Passage and a very deep (4100 m) Central American Passage. The very deep Central American Passage allowed AABW flowing northward from the strong source in the Weddell Sea to enter the Pacific Basins, resulting in a reversal of the deep circulation in the Pacific. The flow of the western boundary current in the South Pacific reversed from northward (present) to southward.

Another experiment with the idealized OGCM explored the effect of making a gap in the northern hemisphere in the latitude band of the easterly trade winds. This corresponds roughly to a closed Drake Passage and an open Panama Passage. It was found that this produced an overturning circulation that cooled the tropics and warmed the high latitudes.

Gibraltar

Although it has been proposed that the saline outflow from the Mediterranean may play a critical role in the formation of NADW, ocean models do not support this idea. The influence of the Mediterranean outflow on the thermohaline circulation in the Atlantic has been investigated using modern boundary conditions in OGCMs. The opening and closing of the Straits of Gibraltar have been modeled by turning a 'salt source' at Gibraltar on and off.

One simulation indicates that open Straits of Gibraltar have only a minor effect on NADW production. Compared with closed Straits, the open Straits scenario increased the export of NADW from the North Atlantic by about 1 Sv.

Indonesian–Malaysian Passages

The connection between the Pacific and Indian Oceans through the Indonesian region is a critical part of the Great Conveyor. The passages through this region have become increasingly restricted as the Australian Block moves northward and collides with south-east Asia. Unfortunately, few model experiments have explored the effect on global ocean circulation.

OGCM sensitivity experiments suggest that closing these passages up to the level of the thermocline would cool the Indian Ocean about 1.5°C and warm the Pacific Ocean about 0.5°C. The model results also indicate that the influence of these passages on the production of NADW is very small.

Denmark Strait

The Denmark Strait, between Greenland and Iceland, is thought to play a critical role in the formation of NADW. The main body of cold saline intermediate water from the Norwegian–Greenland Sea flows into the North Atlantic through the Denmark Strait. In the North Atlantic it mixes with other waters to form NADW.

Experiments with a medium resolution OGCM indicated that the North Atlantic circulation is very sensitive to relatively small changes in the depth of the Denmark Strait.

Implications for the Global Climate Systems

Changes in the magnitude of the thermohaline circulation of the scale suggested by the circulation experiments exploring changes in interocean gateways have profound implications for the global climate system. On a planetary scale, the ocean and atmosphere carry roughly equal amounts of energy poleward, but their relative importance varies with latitude. The ocean dominates by a factor of two at low latitudes, and the atmosphere dominates by a similar amount at high latitudes. The present poleward ocean heat transport is estimated to reach a maximum of about 3.5×10^{15} W at 25°N and 2.7×10^{15} W at 25°S. The subtropical convergences act as barriers to poleward heat transport by the ocean. Only where deep water forms at high latitudes are warm subtropical waters drawn across the frontal systems to higher latitudes to replace the sinking waters.

If the entire thermohaline circulation were involved in the heat exchange between the polar and tropical regions, there would be an equatorward flow of 55 Sv warmed by about 15°C as it returns to the surface through diffuse upwelling. This is equal to an energy transport of 3.5×10^{15} W, about half the total surface ocean transport at low latitudes. Clearly, the transport of heat by the thermohaline system is an important component of the ocean transport system. The great increases in production of polar bottom waters indicated by the closed Drake Passage models suggest that in earlier times ocean heat transport may have dominated the earth's energy redistribution system.

Ocean Circulation on a Warm Earth

During the Late Cretaceous and Eocene, the Earth was characterized by warm polar regions with mean annual temperatures well above freezing and perhaps as warm as 10°C. A major controversy has arisen as to whether increased ocean heat transport played a role in creating these conditions. It has been suggested that there may have been a reversal of the thermohaline circulation, with sinking of warm saline waters in low latitudes and upwelling of warm waters in the polar regions. A secondary question has concerned the effect of the very different paleogeography, with the low latitude circumglobal Tethys seaway connecting the Atlantic and Pacific Oceans, on ocean circulation.

The Ocean Heat Transport Problem

The role of ocean heat transport in producing global warmth has been assessed by using the US National Center for Atmospheric Research (NCAR) Community Climate Model 1 (CCM1) AGCM with present-day geography, but with a swamp ocean. Swamp oceans were used in many early climate models; they have no heat storage capacity and do not transport heat, but serve as a source of moisture. Simulating a world without ocean currents, it was found that polar temperatures were 5–15°C warmer than observed today. Tropical temperatures were about 2°C less. An experiment with the Princeton model used idealized geography, a single ocean and a wedge-shaped continent extending from pole to pole for two AGCM experiments, one with and one without ocean heat transport. The experiment with ocean heat transport produced surface air temperatures poleward of 50° that were 20°C warmer than in the experiment without ocean heat transport. Subsequently, the meridional ocean heat transport required to maintain polar ocean surface temperatures at 10°C was determined. If the ocean were solely responsible, such warm polar temperatures would require implausible ocean heat transports of 2.3×10^{15} W across the Arctic Circle, implying water volume fluxes in excess of 100 Sv, or over twice the flow of the Gulf Stream today.

In a discussion of the role of ocean heat transport in paleoclimatology it has been argued that theoretical arguments as well as simple models suggest that the total poleward transport of energy is not dependent on the structure of either the atmosphere or ocean. Changes in one transport mechanism tend to be compensated by a change in the opposite sense in the other, so that the total transport remains constant. Apparently only extreme changes in the geography of the polar regions can cause uncompensated changes in poleward heat transport. By using the NCAR CCM1 AGCM linked with a slab mixed-layer ocean model, three experiments reflecting different amounts of ocean heat transport have been conducted. Doubling the present flux and reducing the flux to zero resulted in global average temperatures of 14.4 and 15.9°C, respectively compared with the present 15°C. The major effect of changing the ocean heat flux through this range was to lower tropical temperatures 5°C while making only a small increase in polar temperatures.

By using NASA's Goddard Institute of Space Science (GISS) AGCM with slab ocean to calculate the ocean heat transport required to achieve specific sea surface temperatures it was concluded that a 46% increase in ocean heat transport was required to produce a planetary warming of 6°C, which was assumed to be typical for the Jurassic, and that a 68% increase could produce a warming of 6.5°C, assumed to be typical for the Cretaceous. The reduced sea ice formation and ice–albedo feedback in response to the increased poleward heat transport were considered critical factors.

Surface and Thermohaline Circulation

In 1972 the first attempt was made to simulate Late Cretaceous ocean circulation using an analog model. An experiment was carried out using a large water-filled rotating dish set up with continental blocks in a Cretaceous configuration, with wind stress applied to the surface by fans, assuming a poleward displacement of the winds by 10–15°. This analog model produced large anticyclonic gyres in the Pacific and east to west flow through the Tethys, a pattern widely reproduced in paleo-oceanographic studies (**Figure 3**).

In the late 1980s an OGCM was used to perform experiments on ocean circulation with mid-Cretaceous paleogeography. Two of the models contrasted the effects of the different temperatures and winds generated by AGCMs assuming present day and four times present day atmospheric CO_2 concentrations. The present-day CO_2 experiment showed eastward flow across the Gulf of Mexico, westward flow through the Caribbean, a broad west to east current through the western Atlantic, and a clockwise anticyclonic gyre between Eurasia and Africa. Strong western boundary currents developed along the margins of Asia, Africa, and India. The current off north-east Asia continued into the Arctic to a high-latitude site of deep water formation. The thermohaline circulation indicated a strong polar source north of east Asia, and a weak source in the eastern Tethys. For the four-times present CO_2 experiment surface currents in the tropics and subtropics were similar to those in the first experiment, but the flow of the western boundary current off north-east Asia no longer continued into the Arctic. The thermohaline circulation was completely different, lacking a polar source of deep water, but having a strong source in the eastern Tethys which produced a deep flow south and eastward into the Pacific basin.

A simulation of Late Cretaceous ocean circulation using higher spatial resolution and driven by a later version of the GENESIS Earth System Model suggested that low latitude deep water formation may

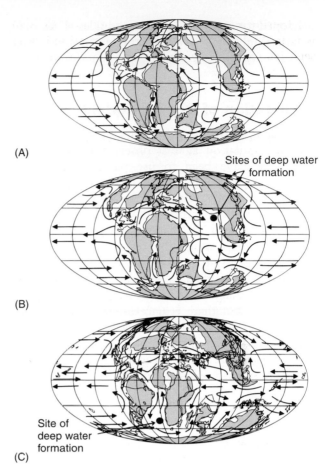

(A)

Sites of deep water formation

(B)

Site of deep water formation

(C)

Figure 3 Three models of Late Cretaceous Ocean circulation. (A) After the analog model of Luyendyk *et al.* (1972) (B) After the OGCM of Barron *et al.* (1993). (C) After the OGCM of Brady *et al.* (1998). Note the different directions of flow through the Tethyan circumglobal seaway.

be more complex than had been thought. The model suggested four potential sites of warm saline dense water formation along the borders of the Tethys and off western South America. However, at three of these sites the water sank only to upper intermediate levels. However, at a site north of the equator on the west African margin the water sank first to an intermediate level, then spread south to the subtropical front where it sank further into the deep ocean interior and flowed out through the Indian Ocean into the Pacific.

Models of Seaways

There have been some model simulations of circulation in shallow continental-scale ocean passages, such as the Cretaceous Western Interior Seaway of North America and the Miocene Tethys–Paratethys connection. These simulations are not based on general ocean models, but use circulation models developed for study of estuaries and tides. They may include tidal currents and simulate flows resulting from salinity differences induced by runoff from land.

Verification of Ocean Models

It is not easy to test the results of a paleoceanographic simulation against observations. The simplest tests have been to use palaeontologic and sedimentologic data to confirm or reject the direction of major current systems. Much geologic evidence suggests that flow through the Tethys was east to west, but some model simulations indicate that it was west to east, others that it was east to west.

Several recent attempts have been made to compare oxygen isotopic data with the temperatures and salinities predicted by the paleo-oceanographic simulations, but there is uncertainty about how to correct for the isotopic differences due to salinity because of the complexity of the relation to the hydrologic cycle. The solution will ultimately lie in incorporating isotopic fractionation into the ocean models.

Conclusions

The use of models to investigate the relations between ancient climates and ocean circulation is still in its infancy. There have been few comparisons of different models using the same boundary conditions. The results produced by different OGCMs for similar boundary conditions are often significantly different. In spite of the different model results it has become clear that the opening and closing of passages between the major ocean basins have had a major effect on the Earth's climate. Attempts to simulate ocean circulation on a warm Earth indicate that sites of deep water formation may shift to low latitudes when atmospheric CO_2 concentrations are high.

See also

Abrupt Climate Change. Cenozoic Climate – Oxygen Isotope Evidence. Heat Transport and Climate.

Further Reading

Barron EJ, Peterson W, Thompson S, and Pollard D (1993) Past climate and the role of ocean heat transport: model simulations for the Cretaceous. *Paleoceanography* 8: 785–798.

Bice KL, Sloan LC, and Barron EJ (2000) Comparison of early Eocene isotopic paleotemperatures and three-dimensional OGCM temperature field: the potential for use of model-derived surface water $\delta^{18}O$. In: Huber BT, MacLeod KG, and Wing SL (eds.) *Warm Climates in Earth History*, pp. 79–131. Cambridge: Cambridge University Press.

Brady EC, DeConto RM, and Thompson SL (1998) Deep water formation and poleward ocean heat transport in the warm climate extreme of the Cretaceous (80 Ma). *Geophysical Research Letters* 25: 4205–4208.

Broecker WS, Bond G, and Klas M (1990) A salt oscillator in the glacial Atlantic? 1. The concept. *Paleoceanography* 5: 469–477.

Bush ABG (1997) Numerical simulation of the Cretaceous Tethys circum global current. *Science* 275: 807–810.

Covey C and Barron E (1988) The role of ocean heat transport in climatic change. *Earth-Science Reviews* 24: 429–445.

Ericksen MC and Slingerland R (1990) Numerical simulation of tidal and wind-driven circulation in the Cretaceous interior seaway of North America. *Geological Society of America Bulletin* 102: 1499–1516.

Fanning AF and Weaver AJ (1996) An atmospheric energy-moisture balance model: climatology, interpentadal climate change and coupling to an ocean general circulation model. *Journal of Geophysical Research* D101: 15111–15128.

Hay WW (1996) Tectonics and climate. *Geologische Rundschau* 85: 409–437.

Kump LR and Slingerland RL (1999) Circulation and stratification of the early Turonian western interior seaway: sensitivity to a variety of forcings. In: Barrera E and Johnson CC (eds.) *Evolution of the Cretaceous Ocean-Climate System*, pp. 181–190. Boulder, co: Geological Society of America.

Luyendyk B, Forsyth D, and Phillips J (1972) An experimental approach to the paleocirculation of ocean surface waters. *Geological Society of America Bulletin* 83: 2649–2664.

Maier-Reimer E, Mikolajewicz U, and Crowley TJ (1990) Ocean general circulation model sensitivity experiment with an open Central American Isthmus. *Paleoceanography* 5: 349–366.

Martel AT, Allen PA, and Slingerland R (1994) Use of tidal-circulation modeling in paleogeographical studies: an example from the Tertiary of the alpine perimeter. *Geology* 22: 925–928.

Mikolajewicz U and Crowley TJ (1997) Response of a coupled ocean/energy balance model to restricted flow through the central American isthmus. *Paleoceanography* 12: 429–441.

Mikolajewicz U, Maier-Reimer E, Crowley TJ, and Kim KY (1993) Effect of Drake and Panamanian gateways on the circulation of an ocean model. *Paleoceanography* 8: 409–426.

Murdock TQ, Weaver AJ, and Fanning AF (1997) Paleoclimatic response of the closing of the Isthmus of Panama in a coupled ocean-atmosphere model. *Geophysical Research Letters* 24: 253–256.

Poulsen CJ, Barron EJ, Peterson WH, and Wilson PA (1999) A reinterpetation of mid-Cretaceous shallow marine temperatures through model-data comparison. *Paleoceanography* 14: 679–697.

Rahmstorf S (1998) Influence of Mediterranean outflow on climate. *EOS Transactions of the American Geophysical Union* 79: 281–282.

Rind D and Chandler M (1991) Increased ocean heat transports and warmer climate. *Journal of Geophysical Research* 96: 7437–7461.

Roberts MJ and Wood RA (1997) Topographic sensitivity studies with a Bryan-Cox-type ocean model. *Journal of Physical Oceanography* 27: 823–836.

Schmidt GA (1999) Forward modeling of carbonate proxy data from planktonic foraminifera using oxygen isotope tracers in a global ocean model. *Paleoceanography* 14: 482–497.

Toggweiler JR and Samuels B (1993) Is the magnitude of the deep outflow from the Atlantic Ocean actually governed by Southern Hemisphere winds? In: Heimann M (ed.) *The Global Carbon Cycle*, vol. 15, pp. 303–331. Berlin: Springer–Verlag.

Toggweiler R and Samuels B (1998) *Energizing the ocean's large-scale circulation for climate change.* Lisbon: 6th International Conference on Paleoceanography.

Washington WM and Meehl GA (1983) General circulation model experiments on the climatic effects due to a doubling and quadrupling of carbon dioxide concentrations. *Journal of Geophysical Research* 88: 6600–6610.

PALEOCEANOGRAPHY: ORBITALLY TUNED TIMESCALES

T. D. Herbert, Brown University, Providence, RI, USA

Introduction

Geologists rely on a variety of 'clocks' built into sediments to place paleoenvironmental events into a time frame. These include radiometric decay systems, annual banding in trees, corals, and some marine and lake sediments, and, increasingly, the correlation of isotopic, geochemical, and paleontological variations to pacing supplied by changes in the Earth's orbit. Variations in three parameters of the orbital system – eccentricity, obliquity, and precession – cause solar insolation to vary over the Earth as a function of latitude, season, and time, and hence cause global changes in climate. Because the timing of orbital changes can be calculated very precisely over the past 30 My, and because their general character can be deduced for much longer intervals of geological time, orbital variations provide a template by which paleoceanographers can fix paleoclimatic variations to geological time. Paleoceanographers now commonly assign either numerical ages or elapsed time to sediment records by optimizing the fit of sedimentary variations to a model of orbital forcing, a process referred to as 'orbital tuning'. Although orbital tuning was first developed to create better timescales for studying the late Pleistocene (approximately the last 400 ky) Ice Ages, it now finds applications to dating sediments and estimating sedimentary fluxes at least into the Mesozoic Era (65–210 Ma).

Orbital tuning came about because of the difficulty in assigning ages to long, continuous records of climate change that became available with ocean sediment coring. The simplest assumption for scaling time to stratigraphic in sediments – that sediment accumulates at a constant rate over long spans of time – is not likely to be true. Sediment compacts with burial, so that a layer 1-cm thick at 50-m burial depth represents significantly more time than the same centimeter of highly porous material at the top of the sediment column. Furthermore, we suspect that climate itself influences the rate of marine sediment accumulation over time, either by varying the production and preservation of biogenic components, or the supply of detrital materials from land. Random factors such as small-scale erosion or excess deposition surely occur as well. Some of these variations in deposition rate can be documented with radiometric systems such as ^{14}C and uranium disequilibrium series. Unfortunately, the radiocarbon clock cannot extend much past 4×10^4 years, and uranium series methods, which have more restrictive conditions to work well, extend to perhaps 3×10^5 years. Strata that contain minerals with radiometric systems that date the age of the sediment (e.g., $^{40}Ar/^{39}Ar$ and other systems in appropriate minerals) are few and far between. Other techniques such as paleomagnetic stratigraphy provide valuable age constraints, but at resolution of 10^5–10^6 years (the frequency of magnetic reversals).

The end product of orbital tuning is a mapping function of stratigraphic position (depth scale in the sediment column) to time, using criteria discussed at more length below (**Figure 1**). Paleobiological and paleoceanographic patterns can be placed into a time frame perhaps accurate to a few thousand years (ky). Time-series studies benefit enormously, since removing the distorting effects of variations in sedimentation rate results in much 'cleaner' frequency spectra. Equally important, the large improvement in time resolution offered by orbital tuning in comparison to other stratigraphic techniques allows paleoceanographers to measure the dynamics of sedimentary records with far more precision than the constant sedimentation rate assumption. Did events recorded in different sedimentary locations occur simultaneously or with an age progression? How long did fundamental biological and geochemical turnovers in Earth history take to occur? What are the precise durations of magnetic polarity intervals? How do marine sediment fluxes vary with climatic state? Orbital tuning methods have addressed these and other questions over the last two decades.

Orbital Parameters and Potential Age Resolution by Orbital Tuning

Gravitational interactions with the other planets and the Earth's moon perturb the orbit of our planet in three modes that affect the solar radiation received by the Earth as a function of latitude and season. Variations in 'eccentricity' describe the quasiperiodic evolution of the Earth's orbit around the Sun from more circular to more elliptical. Orbital

ODP 926B

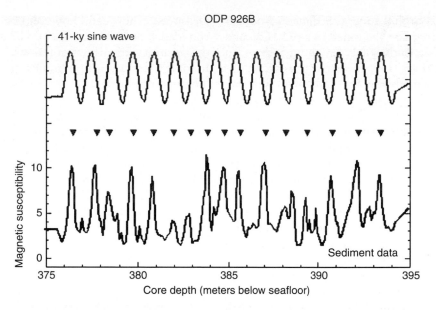

Figure 1 Example of how a sedimentary signal (magnetic susceptibility, a carbonate proxy) recorded as a function of depth below seafloor can be 'tuned' to an orbital chronometer. In this case, Shackleton and colleagues tuned the record of early Miocene sedimentation (lower curve) to a 41-ky obliquity cycle (upper curve). The triangles indicate position of consecutive obliquity cycles identified at ODP Site 926B. Note that the spacing of cycles is not constant, an indication of variation in deposition rate. Note also the presence of smaller higher-frequency features caused by precession (mean period of about 21 ky). Absolute ages can be determined by fitting the amplitude envelopes of the obliquity and precessional signals measured in the sediments to similar features in long-term numerical calculations of the orbital terms. Data from Shackleton NJ, Crowhurst SJ, Weedon GP, and Laskar J (1999) Astronomical calibration of Oligocene–Miocene time. *Philosophical Transactions of the Royal Society London, Series A* 357: 1907–1929, courtesy of N. J. Shackleton.

eccentricity evolves in a complex manner, but has significant terms at about 95, 125, 405, 2000, and 2800 ky. The two shorter periods are close enough together to produce a beat pattern with a mean period of 109 ky, and the two longer terms produce a beat pattern at 2425 ky. Eccentricity is the only orbital cycle to alter the mean annual insolation, but its direct effect is very small. Axial 'tilt' ('obliquity') varies around its mean value of 23.5° by about 1.5° in a nearly sinusoidal pattern with a modern period of 41 ky. The obliquity cycle causes insolation to be distributed poleward during intervals of higher than average tilt, and equatorward during intervals of lower obliquity. Climatic 'precession' affects the summer–winter insolation contrast, and unlike the other two orbital parameters, acts in opposite sign between the hemispheres. The precessional index is modulated by eccentricity, so that its amplitude bears the imprint of the *c.* 109-, 405-, and 2425-ky eccentricity cycles mentioned above. Climatic precession has a mean modern period of 21.2 ky, but, because of the eccentricity amplitude modulation, it can range in repeat time from 14 to 28 ky. Its Fourier series representation has concentrations of variance at 1/23 and 1/19 ky^{-1}.

Long-term secular changes in the Earth–Moon and solar system influence the Earth's orbital parameters on timescales >1Ma. The observed slowing of Earth rotation rate due to tidal friction requires that the periods of obliquity and precessional cycles have gradually decreased as we move back in time. The predicted shortenings (relative to the Present) are less than 1% through the Cenozoic, but amount to several percent in the Mesozoic and more into the Paleozoic. One should also note that because the rate of tidal dissipation has probably not been constant over time, we do not have a good model for how obliquity and precessional periods have evolved over the entire course of the Earth–Moon system. A second significant uncertainty in using orbital variations as a time template comes from the weakly chaotic motion of the solar system. Any numerical solution of the orbital system will show an exponential dependence on initial conditions. There is therefore a limit in time beyond which one may not calculate the phases of the orbital cycles with confidence. This does not mean that one has no idea of the behavior of the orbital system, but rather that one has to rely on average statistical properties, or unusually stable elements of the orbit (the 405-ky eccentricity cycle is believed to be one such case) to perform orbital tuning in sediments older than about 30 Ma.

The fundamental limit on the accuracy of orbital calculations thus imposes a twofold division of Earth time amenable to orbital tuning. Sediments younger than 30 My can in theory be tuned to a precision and accuracy which depends solely on the match to an orbital template. In theory, the error of such an approach depends largely on the unknown time lag between orbital change and the climate change recorded by sedimentary proxies. Such response times could vary from nearly instantaneous (adjustments of ocean surface temperature and hydrology) to ky scale (growth and decay of large ice sheets, changes in the deep-ocean circulation, variations in greenhouse gas content).

Orbital tuning in older sediments functions to measure 'elapsed time' in a sediment record, since one can no longer unequivocally associate a sedimentary feature with a precisely known time value of the Earth's orbit. The latter approach resembles using a yardstick to measure distance from an independently agreed-upon datum. Datums in Earth history come from magnetic reversals, biostratigraphic, or paleochemical events. Elapsed time can be measured to the errors in the mean periods of the orbital series. Reasonable estimates of the uncertainties in the latter approach lie in the range of 5–20 ky. Examples of both numerical dating and the 'yardstick' modes of orbital tuning will follow below. One should note, however, that the presumed stability of the 405-ky eccentricity cycle over great stretches of geological time may allow geologists to develop a numerical time frame based on the Earth's orbit, and therefore accurate to a fraction of the 405-ky period, for perhaps the past 100 My.

Methods of Orbital Tuning

No single approach to orbital tuning is appropriate to all stratigraphic problems. The ideal case, producing an age–depth model tied precisely to an 'absolute' timescale, can only be achieved in cases where the stratigrapher has a global orbitally forced signal that can be correlated to a precise astronomical reference frame. This situation obtains from late Pleistocene (c. 4 Ma) onward, where significant changes in global ice volume altered the entire ocean inventory of oxygen isotopes on orbital timescales. Despite the general decline in the oxygen isotope signal in older marine strata, orbital tuners can find the imprint of orbital forcing in other aspects of sedimentation, such as variables like calcium carbonate content, redox state, and microfossil content. Such variations, because they originate from regional changes in the input of dust, biological productivity, and bottom water circulation, cannot provide a globally synchronous signal equal to that of isotopic measures. They can, however, be correlated with confidence to the numerical timescale up to limit of the accuracy of orbital solutions, or about 30 Ma. In older strata, orbital tuning generally produces 'floating' timescales.

One can choose among several approaches to matching sedimentary signals to an orbital forcing template. One can maximize the match of a sediment series to an orbital model by constructing an age–depth model that gives the largest correlation coefficient between presumed forcing and climatic response. An alternative is to work more selectively with the sediment record to extract the orbital components from the natural record, and to maximize the coherence of these to the corresponding orbital components. Such methods generally involve working in the frequency domain, using techniques such as band-pass filtering and complex demodulation.

The most straightforward tactic relies on visual correlation of successive peaks in a sedimentary time series to a presumed orbital signal. A new timescale emerges as peaks and troughs of a stratigraphic signal are aligned to the orbital template. The alignment may be performed visually, or by designing an objective mapping function that optimizes the fit of stratigraphic signal to orbital template. Such time-domain tuning, while it offers the highest possible age resolution, is quite sensitive to the orbital model chosen and to noise in the geological series. The possibility clearly exists to produce a tuned sedimentary series that has been forced to resemble an orbital template by overenthusiastic correlation. Several strategies exist to lessen the subjectivity and heighten the reliability of time (depth) domain tuning. The first recognizes that higher-frequency orbital cycles such as precession and obliquity have low-frequency 'envelopes' that cause their amplitude to vary systematically. The technique of complex demodulation detects such features in time series data, and the envelopes of 'tuned' stratigraphic cycles can be compared to those of the orbital cycles themselves. Erroneous tuning of individual 21- or 41-ky peaks and valleys in a stratigraphic series will be revealed by the misalignment of the envelopes of the geological cycles relative to orbital forcing.

Moving-window ('evolutive') spectral analysis also works effectively to produce smooth tunings of stratigraphic signals to time. Here the analyst divides the data into segments of a specified stratigraphic length, and searches in the frequency domain for strong signals associated with orbital components. The average sedimentation rate is deduced by optimizing the match of scaling of stratigraphic frequency (cm^{-1}) to orbital frequencies. By subdividing the data into relatively short (usually 300–500-ky length) windows, the technique reduces spectral distortions associated with changes in sedimentation rate along section, and with the nonstationarity of climate response to orbital forcing. A conceptual example of how evolutive spectral analyses can detect and tune for gradual changes in sedimentation rate is illustrated in **Figure 2**. Moving-window analyses have the virtue of producing smooth estimates of sedimentation rate, since they rely on the average spectral properties of stratigraphic signals over a depth or time range.

It is essential to remember that any 'cyclostratigraphy' carries an implicit climate model, and in many cases, an implicit sedimentation model. Our knowledge of the climate system is not adequate to predict the precise response of the Earth's climate orbital forcing over time; our correlation of paleoclimate records to orbital forcing must allow the recipe of orbital influences, which reflect changes over time in key climatic regions of the globe and in climatic feedbacks, to vary. Study of the relatively well-dated Pleistocene marine record (*see* Plio-Pleistocene Glacial Cycles and Milankovitch Variability) demonstrates that one orbital recipe may not describe the evolving Ice Ages, as the relative importance of *c.* 21-, 41-, and 100-ky ice volume cycles changes significantly over the past 2 My due to processes internal to the climate system. Orbital tuning should therefore include a healthy amount of flexibility and regard for independent checks on the quality of the result.

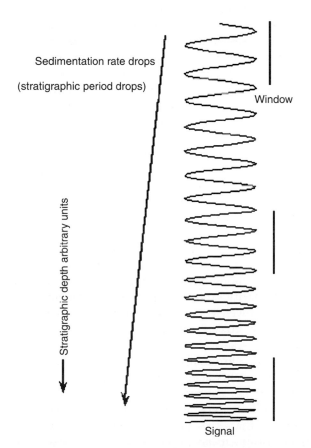

Figure 2 Changes in sedimentation rate act as frequency modulations of cyclic signals. In this synthetic example, sedimentation rate was made to decrease by a factor of four from the youngest to oldest interval of the record. Note that the repeat distance in the stratigraphic record is greatest where sediment accumulation is highest, and tapers toward the base of the sequence where accumulation rate is low. Such changes can be traced objectively by the moving window spectral method, which produces sliding estimates of the local stratigraphic frequency (the inverse of stratigraphic period) down-section. In this case, the cycle frequency moves (top–base) from low to high as the cycles become more condensed at lower sedimentation rates.

Criteria for Success

Orbital tuning is rarely applied to sediments without first considering independent age constraints from fossil events and paleomagnetic reversals. These provide a preliminary age scale and therefore a guide to approximate, time-averaged, sedimentation rates to be modified by orbital tuning. Independent age markers also provide important tests of the quality of orbital tuning. For example, one can compare the results of orbital timescales to 'known' ages of events such as magnetic reversals where high-quality radiometric ages are available. Work conducted during the last decade suggests that orbital methods yield impressively consistent estimates of the ages of biostratigraphic and magnetostratigraphic datums. For example, Wilson demonstrated that Shackleton *et al.*'s revision of the Plio-Pleistocene geomagnetic polarity timescale (GPTS) based on orbital tuning yielded smoother, and therefore more plausible, spreading rate histories on a number of mid-ocean ridge systems than did the previous timescale based on K/Ar dates of basalts.

Continuity of sedimentation is clearly another necessary condition for success of the orbital approach. The best sections have the following characteristics: they lack observable breaks in sedimentation, either erosional or depositional (e.g., turbidites), are composed of pelagic or hemipelagic facies, and do not have strong changes in overall sediment composition (e.g., major break from calcareous to detrital sedimentation) that tend to accompany strong changes in deposition rate. In the ideal case, sections will span millions of years of deposition and accumulate at a sufficient rate (empirically at $>3\,\mathrm{cm\,ky^{-1}}$) to resolve precessional variations if such exist. Paleoceanographers have also learned to drill multiple offset holes at sites of interest and to splice records across coring gaps in individual holes to create 'composite sections' that are verifiably complete.

Although orbital tuning has often been criticized as circular reasoning, it has passed a number of tests in recent years. External checks exist in the form of independent datums such as magnetic polarity reversals and biostratigraphic events. A good orbital tuning solution should produce consistent estimates of the numerical age of the datum event or the duration between datum events at a number of sites. Such concordant results have been demonstrated at least into the middle Miocene, and new studies continue to extend continuous tuning further back in time. Another powerful test of an orbital tuning model comes from time series analysis. Orbital signals have quite distinctive 'fingerprints' due to amplitude modulations. The modulations are most pronounced in the case of the eccentricity and precessional cycles, but also exist in the more sinusoidal obliquity cycle. Spectral analyses should therefore detect a hierarchy of frequencies correctly corresponding to modulation terms of the central orbital frequencies in a successful tuning. For example, one would expect to recover the 95-, 125-, and 404-ky modulating terms in a time series tuned to a presumed precessional signal.

Some Examples

A team of Dutch stratigraphers, paleontologists, and paleomagnetists has made significant progress in dating marine sedimentary successions in the Mediterranean region in the age range 0–12 Ma by astronomical tuning. The dates are particularly valuable to stratigraphers because many of the sections studied define classical substages of the Pleistocene, Pliocene, and Miocene epochs. Furthermore, many of the studied sections have good magnetic properties, permitting the Dutch team to propose an astronomical chronology for the GPTS. As magnetic polarity reversal boundaries constitute globally synchronous events recorded in many types of sediments and frozen into the ocean crust by the process of seafloor spreading, improving the GPTS has widespread implications. Hilgen and co-workers recognized orbital forcing by a grouping of sapropels (dark, organic-rich beds) into units of ~ 100 and 400 ky by eccentricity modulation of precessional climate changes. Their resulting calibration of the GPTS yielded significantly greater ages for magnetic reversal boundaries than the previously accepted dates based on K/Ar radiometric age dating. After initial controversy, the ages

proposed by Hilgen and others have largely been verified by recent advances in $^{40}Ar/^{39}Ar$ dating of volcanic ash layers at a number of magnetic reversal boundaries.

The power of orbital tuning to resolve sedimentation rates at high precision for paleoceanographic studies is nicely illustrated by Shackleton's analysis of cyclic deep sea sediments recovered from the western equatorial Atlantic by Ocean Drilling Program Leg 154. Coring was designed to aquire long records at variable water depths. Depth largely controls carbonate sedimentation rate in the region, due to a direct pressure relationship (carbonate minerals are more soluble at greater pressures) and through the indirect effect of vertical variations in water mass carbon chemistry. Carbonate sedimentation in turn largely determines overall sediment accumulation. **Figure 3** displays the results of tuning variations in magnetic susceptibility, a means of estimating the noncarbonate fraction of the sediment, to an obliquity (41 ky) pacing over an interval of early Miocene age. Sediment variations at Site 926C can be matched one-for-one to variations at drill Site 929A, approximately 750 m deeper in the water column (lower two curves, **Figure 3**). The orbital tuning also generates a sedimentation rate function at each site, displayed as the upper two curves in **Figure 3**. Higher sedimentation rates at Site 926C throughout the interval agree with expectations that carbonate dissolution should be less intense at the shallower location, while changes in the difference between the sites documents changing gradients in dissolution over time. It is important to note that conventional dating methods based on biostratigraphic or magnetic polarity stratigraphy would not have the resolving power to monitor the carbonate deep-water chemistry proxy at the resolution afforded by orbital tuning.

Orbital chronology has also played a role in studying events across one of the truly catastrophic passages of geological time at the Cretaceous/Tertiary boundary. While a number of observations suggest that the paleontological transition was very abrupt, it has proved difficult to constrain the timing of events before and after the boundary, and to differentiate rates of change across the K/T boundary quantitatively from background variability. Work by Herbert and others demonstrates that precessional cycles can be recognized at many deep sea sites that contain K/T boundary sequences. These provide a yardstick for dating events to within about 10 ky (one-half precessional wavelength) across the boundary. As one example of the information recovered, **Figure 4** displays a composite of sedimentation rates in South Atlantic pelagic sites

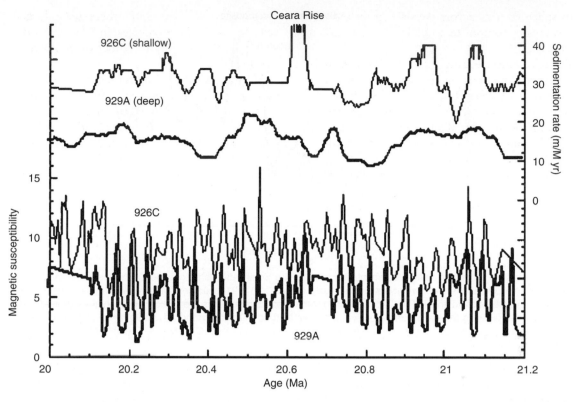

Figure 3 High-resolution orbital tuning of sediment records from different water depths allows paleoceanographers to study variations on a common accurate time base; 41-ky variations driven by obliquity forcing were identified at both shallower (ODP 926B) and deeper (ODP 929A) sites (Shackleton *et al.*, 1999). Tuning generates mapping functions of depth to time that reflect sedimentation rates at each site. Data from Shackleton NJ, Crowhurst SJ, Weedon GP, and Laskar J (1999) Astronomical calibration of Oligocene–Miocene time. *Philosophical Transactions of the Royal Society London, Series A* 357: 1907–1929, courtesy of N. J. Shackleton.

across the K/T boundary, using precessional cycles as an accumulation rate gauge. Data at each site were normalized to a value of 1 for the mean late Cretaceous accumulation rate. The step-function drop in sedimentation rate coincident with the K/T boundary reveals not only a catastrophic drop in the flux of biogenic material to the deep sea following the extinction event, but also a prolonged period of low flux in the early Tertiary as the planktonic ecosystem slowly recovered.

Where the Field is Headed

Paleoceanographers constantly search for new tools to generate objective sediment data for spectral and stratigraphic analysis. Over the last decade, sediment proxy measurements such as color reflectance, magnetic susceptibility, gamma ray attenuation porosity evaluator, X-ray fluorescence elemental scanners, p-wave velocity, and downhole geophysical and geochemical logging have produced long, densely sampled time series from many sediment coring locations. Variations in the measured parameters all

reflect in some way changes in the sediment composition (generally the proportion of carbonate to noncarbonate sediment). The variables measured yield less insight into the mechanisms of climate change than data such as stable isotopic measurements and analyses of microfossil populations, but they have the virtue of being inexpensive, rapid, and nondestructive.

At least three important conceptual targets remain to be solved by improvements in orbital tuning. The first relates directly to questions of the climate system itself. Paleoceanographers are becoming more aware of the climatic insights to be gained by comparing leads and lags in responses of various measured parameters. These can be ascertained by measuring multiple proxies in the same core, so that phase relations determined by cross-spectral analyses are independent of the absolute timescale, and between coring sites if one has a common, synchronous, tuning variable. Benthic $\delta^{18}O$ is generally assumed to provide such a chronological tie in late Pleistocene sediments. By measuring different components in the same cores, Clemens and colleagues demonstrate

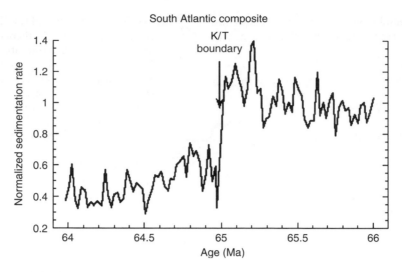

Figure 4 Composite record of pelagic sedimentation rates in the South Atlantic across the Cretaceous/Tertiary (K/T) boundary, deduced from tuning variations in sediment carbonate content and color to *c.* 21-ky precessional cycles. Note the absence of a decline in sedimentation rate prior to the K/T boundary (lack of 'precursor' events) and the prolonged interval of dramatically reduced sedimentation rate following the extinction event. Adapted from Herbert TD and D'Hondt SL (1990) Precessional climate cyclicity in late Cretaceous–early Tertiary marine sediments: A high resolution chronometer of Cretaceous–Tertiary boundary events. *Earth and Planetary Science Letters* 99: 263–275.

that the relative timing of different aspects of the monsoon system in the Indian Ocean clearly evolves over the past 6 Ma. While each component senses orbital forcing, the linkages between winds, ice volume, and sea surface temperature change have varied over time.

As long, orbitally tuned records accumulate, we can anticipate that more instances of significant and unusual biotic, climatic, and geochemical events in Earth history coincide with unusual combinations of orbital forcing. We think of orbital pacing of climate acting on characteristic timescales of 10^4–10^5 years, but much longer-scale effects may arise from low-frequency modulations of individual cycles, or from complex patterns of constructive and destructive interference that may operate on a timescale of 10^6–10^7 years. As one example, the marine extinctions that define the Miocene/Oligocene boundary, and the pronounced $\delta^{18}O$ excursion that defines a major long-lived glacial period at that time, have been convincingly tied by Zachos and colleagues to an unusual coincidence of very low obliquity variations and low orbital eccentricity. There is also a persistent indication that anomalies in the carbon cycle, as exhibited by isotopes of carbon locked into marine carbonates, may follow (and amplify) long-wavelength components of orbital forcing.

Finally, one can anticipate that the improved chronology afforded by orbital tuning will offer large refinements in the geological timescale for the Cenozoic and Mesozoic. Improvements in estimating the duration of magnetic polarity chrons will allow

geophysicists to study variations in ocean spreading rate at least into the Cretaceous without the constraint of having to specify a constant-spreading-rate ridge system. Members of the radiometric dating community have proposed that an intercalibration of radiometric and orbital dating systems may be the most accurate way to choose age standards for the $^{40}Ar/^{39}Ar$ dating system. The possibility that orbital signals preserved in marine sediments will supply celestial mechanicists with constraints on the long-term evolution of the Earth–Sun–Moon system also appears on the horizon.

See also

Paleoceanography, Climate Models in. Plio-Pleistocene Glacial Cycles and Milankovitch Variability.

Further Reading

Billups K, Palike H, Channell JET, Zachos JC, and Shackleton NJ (2004) Astronomic calibration of the late Oligocene through early Miocene geomagnetic polarity time scale. *Earth and Planetary Science Letters* 224: 33–44.

Clemens SC, Murray DW, and Prell WL (1996) Nonstationary phase of the Plio-Pleistocene Asian monsoon. *Science* 274: 943–948.

Herbert TD and D'Hondt SL (1990) Precessional climate cyclicity in late Cretaceous–early Tertiary marine sediments: A high resolution chronometer of Cretaceous–Tertiary boundary events. *Earth and Planetary Science Letters* 99: 263–275.

Hilgen FJ (1991) Astronomical calibration of Gauss to Matuyama sapropels in the Mediterranean and implication for the geomagnetic polarity time scale. *Earth and Planetary Science Letters* 104: 226–244.

Husing SK, Hilgen FJ, Aziz HA, *et al.* (2007) Completing the Neogene geological time scale between 8.5 and 12.5 Ma. *Earth and Planetary Science Letters* 253: 340–358.

Huybers P and Wunsch C (2004) A depth-derived Pleistocene age model: Uncertainty estimates, sedimentation variability, and nonlinear climate change. *Paleoceanography* 19(1): PA1028.

Laskar J, Robutel P, Joutel F, Gastineau M, Correia ACM, and Levrard B (2004) A long-term numerical solution for the insolation quantities of the Earth. *Astronomy and Astrophysics* 428: 261–285.

Lisiecki LE and Raymo ME (2005) A Pliocene–Pleistocene stack of 57 globally distributed benthic delta O-18 records. *Paleoceanography* 20: PA1003.

Lourens IJ, Sluijs A, Kroon D, *et al.* (2005) Astronomical pacing of late Palaeocene to early Eocene global warming events. *Nature* 435(7045): 1083–1087.

Martinson DG, Pisias NG, Hays JD, *et al.* (1987) Age dating and the orbital theory of the ice ages: Development of a high-resolution 0 to 300 000-year stratigraphy. *Quaternary Research* 27: 1–29.

Raffi I, Backman J, Fornaciari E, *et al.* (2006) A review of calcareous nannofossil astrobiochronology encompassing the past 25 million years. *Quaternary Science Reviews* 25: 3113–3137.

Renne PR, Deino AL, Walter RC, *et al.* (1994) Intercalibration of astronomical and radioisotopic time. *Geology* 22: 783–786.

Shackleton NJ, Crowhurst SJ, Weedon GP, and Laskar J (1999) Astronomical calibration of Oligocene–Miocene time. *Philosophical Transactions of the Royal Society London, Series A* 357: 1907–1929.

Wade BS and Pälike H (2004) Oligocene climate dynamics. *Paleoceanography* 19: PA4019 (doi:10.1029/2004PA001042).

Wilson DS (1993) Confirmation of the astronomical calibration of the magnetic polarity timescale from seafloor spreading rates. *Nature* 364: 788–790.

Zachos JC, Shackleton NJ, Revenaugh JS, *et al.* (2001) Climate response to orbital forcing across the Oligocene–Miocene boundary. *Science* 292: 274–278.

PALEOCEANOGRAPHY: THE GREENHOUSE WORLD

M. Huber, Purdue University, West Lafayette, IN, USA
E. Thomas, Yale University, New Haven, CT, USA

Introduction

Understanding and modeling a world very different from that of today, for example, a much warmer world with a so-called 'greenhouse' climate, requires a thorough grasp of such processes as the carbon cycle, ocean circulation, and heat transport in the oceans – indeed it pushes the limits of our current conceptual and numerical models. A review of what we know of past greenhouse worlds reveals that our capacity to predict and understand key features of such intervals is still limited.

Greenhouse climates represent most of the past 540 My, called the Phanerozoic, the part of Earth history for which an extensive fossil record is available. One could thus argue that a climate state with temperatures much warmer than modern, without substantial ice at sea level at one or both Poles, is the 'normal' mode, and the glaciated, cool state – such as has existed over the past several million years – is unusual. Since we were born into this glaciated climate state, however, our theories of near-modern climates are well developed and powerful, whereas greenhouse climates with their different boundary conditions challenge our understanding.

The two best-documented periods with greenhouse climates are the Cretaceous ($\sim$145–65.5 Ma) and the Eocene ($\sim$55.8–33.9 Ma), on which this article will concentrate. But, other intervals, both earlier (e.g., Silurian, 443.7–416 Ma) and more recent (e.g., early–middle Miocene, 23.03–11.61 Ma, mid-Pliocene, 3.5 Ma) are characterized by climates clearly warmer than today, although less torrid than the Eocene or Cretaceous. Furthermore, within long-term greenhouse intervals, there may lurk short-term periods of an icier state.

The alien first impression one derives from looking at the paleoclimate records of past greenhouse climates is the combination of polar temperatures too high for ice formation and warm winters within continental interiors at mid-latitudes. In fact, there are three main characteristics of past greenhouse climates that are remarkable and puzzling, and hence

are the focus of most research. These three key features are:

- Global warmth: a global mean surface temperatures much warmer ($>10\,°C$) than the modern global mean temperature ($15\,°C$);
- Equable climates: reduced seasonality in continental interiors compared to modern, with winter temperatures above freezing; and
- Low temperature gradients: a significant reduction in equator-to-pole and vertical ocean temperature gradients (i.e., to $<20\,°C$). Annual mean surface air temperatures today in the Arctic Ocean are about $-15\,°C$, in the Antarctic about $-50\,°C$, whereas tropical temperatures are on average $26\,°C$, hence the modern gradient is 40–$75\,°C$.

Not all greenhouse climates are the same: some have one or two of these key features, and only a few have all the three. For example, there is ample and unequivocal evidence for global warmth and equable climates throughout large portions of the Mesozoic (251–65.5 Ma) and the Paleogene (65.5–23.03 Ma) (**Figure 1**). But, evidence for reduced meridional and vertical ocean temperature gradients is much more controversial, however, and (largely due to the nature of the rock and fossil records) only reasonably well established for the mid-Cretaceous and the early Eocene (*see* Cenozoic Climate – Oxygen Isotope Evidence and Paleoceanography).

Superimposed on these fundamental aspects of greenhouse climate modes is a pervasive variability apparently driven by orbital variations (*see* Plio-Pleistocene Glacial Cycles and Milankovitch Variability), and abrupt climate shifts (*see* Abrupt Climate Change), indicative either of crossing of thresholds or of sudden changes in climatic forcing factors, and notable changes in climate due to shifting paleogeographies, paleotopography, and ocean circulation changes (*see* Heat Transport and Climate).

A variety of paleoclimate proxies, including oxygen isotopic paleotemperature estimates (*see* Cenozoic Climate – Oxygen Isotope Evidence), as well as newer proxies such as Mg/Ca (*see* Determination of Past Sea Surface Temperatures), and ones that might still be considered experimental, such as TEX_{86} (*see* Paleoceanography), have been used to reconstruct greenhouse climates. New proxies and long proxy records of atmospheric carbon dioxide concentrations (pCO_2)

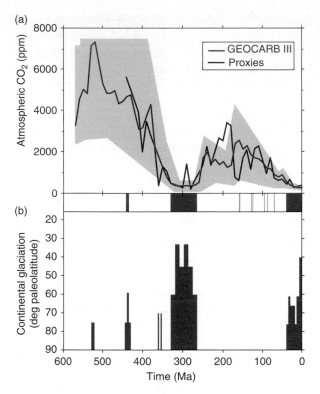

Figure 1 (a) Atmospheric carbon dioxide concentrations produced by geochemical models and proxies for the Phanerozoic. (b) Major intervals of continental glaciation and the latitude to which they extend. The major periods in Earth's history with little continental ice correspond to those periods with high greenhouse gas concentrations at this gross level of comparison. From Royer DL, Berner RA, Montanez IP, Tabor NJ, and Beerling DJ (2004) CO_2 as a primary driver of Phanerozoic climate. *GSA Today* 14: 4–12.

are becoming available and allow us to constrain the potential sensitivity of climate change to greenhouse gas forcing (**Figure 1**). The investigation of past greenhouse climates is making exciting and unprecedented progress, driven by massive innovations in multiproxy paleoclimate and paleoenvironmental reconstruction techniques in both the marine and terrestrial realms, and by significant developments in paleoclimate modeling (*see* Paleoceanography and Paleoceanography, Climate Models in).

Cretaceous

The Cretaceous as a whole was a greenhouse world (**Figure 1**): most temperature records indicate high latitude and deep ocean warmth, and pCO_2 was high. Nevertheless, substantial multiproxy evidence indicates the presence of, perhaps short-lived, below-freezing conditions at high latitudes and in continental interiors, and the apparent buildup of moderate terrestrial ice sheets (potentially in the

early Cretaceous Aptian, 125–112 Ma, and in the late Cretaceous Maastrichtian, 70.6–65.5 Ma). Peak Cretaceous warmth and the best example of a greenhouse climate lies in the mid-Cretaceous (~120–80 Ma), when tropical temperatures were between 30 and 35 °C, mean annual polar temperatures above 14 °C, and deep ocean temperatures were *c.* 12 °C. Continental interior temperatures were probably above freezing year-round, and terrestrial ice at sea level was probably absent. Global mean surface temperatures were much more than 10 °C above modern, and the equator-to-pole temperature gradient was approximately 20 °C.

It is not clear exactly when the thermal maximum occurred during this overall warm long period (e.g., in the Cenomanian, 99.6–93.5 Ma, or in the Turonian, 93.5–89.3 Ma), because regional factors such as paleogeography and ocean heat transport might substantially alter the expression of global warmth (**Figure 2**), and a global thermal maximum does not necessarily denote a global maximum everywhere at the same time.

A feature of particular interest in the Cretaceous (and to a lesser extent in the Jurassic, 199.6–145.5 Ma) oceans are the large global carbon cycle perturbations called 'oceanic anoxic events' (OAEs). These were periods of high carbon burial that led to drawdown of atmospheric carbon dioxide, lowering of bottom-water oxygen concentrations, and, in many cases, significant biological extinction. Most OAEs may have been caused by high productivity and export of organic carbon from surface waters, which has then preserved in the organic-rich sediments known as black shales. At least two Cretaceous OAEs are probably global, and indicative of ocean-wide anoxia at least at intermediate water depths, the Selli (late early Aptian, ~120 Ma) and Bonarelli events (Cenomanian–Turonian, ~93.5 Ma). During these events, global sea surface temperatures (SSTs) were extremely high, with equatorial temperatures of 32–36 °C, polar temperatures in excess of 20 °C. Multiple hypotheses exist for explaining OAEs, and different events may bear the imprint of one mechanism more than another, but one factor may have been particularly important.

Increased volcanic and hydrothermal activity, perhaps associated with large igneous provinces (LIPs), changed increased atmospheric pCO_2, and altered ocean chemistry in ways to promote upper ocean export productivity leading to oxygen depletion in the deeper ocean. These changes may have been exacerbated by changes in ocean circulation and increased thermohaline stratification which might have affected benthic oxygenation, but the direction and magnitude of this potential feedback

(a) Temperature difference

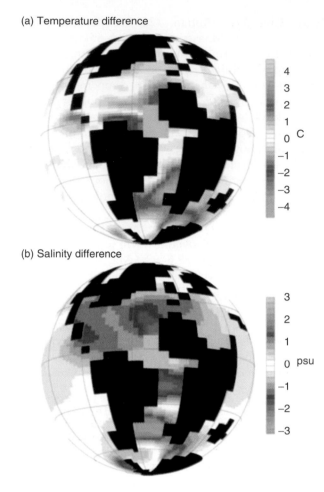

(b) Salinity difference

Figure 2 In these coupled ocean–atmospheric climate model results from Poulsen *et al.* (2003), major changes in proto-Atlantic Ocean temperature and salinity occur by deepening of the gateway between the North and South Atlantic. This study concluded that part of the pattern of warming, anoxia, and collapse of reefs in the mid-Cretaceous organic-rich sediments known as black might have been driven by ocean circulation changes engendered by rifting of the Atlantic basin.

are poorly constrained. The emplacement of LIPs was sporadic and rapid on geological timescales, but *in toto* almost 3 times as much oceanic crust was produced in the Cretaceous (in LIPs and spreading centers) as in any comparable period. The resulting fluxes of greenhouse gases and other chemical constituents, including nutrients, were unusually large compared to the remainder of the fossil record. The direct impact of this volcanism on the ocean circulation through changes in the geothermal heat flux was probably small as compared to the ocean's large-scale circulation, but perturbations to the circulation within isolated abyssal basins might have been important.

As the Cretaceous came to a close, climate fluctuated substantially (**Figure 3**), but it is impossible to know how Cretaceous climate would have

continued to evolve, because the Cretaceous ended with a bang – the bolide which hit the Yucatan Peninsula at 65.5 Ma profoundly perturbed the ocean and terrestrial ecosystems (*see* Paleoceanography). Much of the evolution of greenhouse climates is related to the carbon cycle and hence to biology and ecology, so it is difficult to know whether the processes that maintained warmth in the Cretaceous with such different biota are the same as in the Cenozoic.

Early Paleogene

After the asteroid impact and subsequent mass extinction of many groups of surface-dwelling oceanic life-forms, the world may have cooled for a few millennia, but in the Paleocene (65.5–55.8 Ma) conditions were generally much warmer than modern, although evidence for at least intervals of deep sea and polar temperatures cooler than peak Cretaceous or Eocene warmth exists. Overall, in the first 10 My after the asteroid impact, ecosystems recovered while the world followed a warming trend, reaching maximum temperatures between ~56 and ~50 Ma (latest Paleocene to early Eocene).

Within the overall greenhouse climate of the Paleocene to early Eocene lies a profoundly important, but still perplexing, abrupt climatic maximum event, the Paleocene–Eocene Thermal Maximum (PETM). The record of this time period is characterized by global negative anomalies in oxygen and carbon isotope values in surface and bottom-dwelling foraminifera and bulk carbonate. The PETM may have started in fewer than 500 years, with a recovery over ~170 ky. The negative carbon isotope excursion (CIE) was at least 2.5–3.5‰ in deep oceanic records and 5–6‰ in terrestrial and shallow marine records (**Figure 4**). These joint isotope anomalies, backed by independent temperature proxy records, indicate that rapid emission of isotopically light carbon caused severe greenhouse warming, analogous to modern anthropogenic fossil-fuel burning.

During the PETM, temperatures increased by 5–8 °C in southern high-latitude sea surface waters; *c.* 4–5 °C in the deep sea, equatorial surface waters, and the Arctic Ocean; and *c.* 5 °C on land at mid-latitudes in continental interiors. There is some indication of an increase in the vigor of the hydrologic cycle (**Figure 4**), but the true timing and spatial dependence of the change in hydrology are not clear. Diversity and distribution of surface marine and terrestrial biota shifted, with migration of thermophilic biota to high latitudes and evolutionary

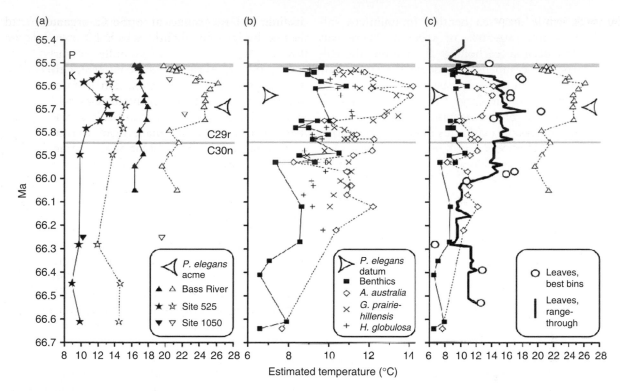

Figure 3 A summary of climate change records near the end of the Cretaceous (Maastrichtian). Paleotemperatures estimated from oxygen isotope data from benthic (filled symbols) and planktonic (open symbols) foraminiferea from middle (a) and high (b) latitudes. (c) Combined representative data (with leaf data) from (a) and (b) with terrestrial paleotemperature estimates derived from macrofloral records for North Dakota. The end of Cretaceous was clearly a time of very variable climate. From Wilf P, Johnson KR, and Huber BT (2003) Correlated terrestrial and marine evidence for global climate changes before mass extinction at the Cretaceous–Paleogene boundary. *Proceedings of the National Academy of Sciences of the United States of America* 100: 599–604.

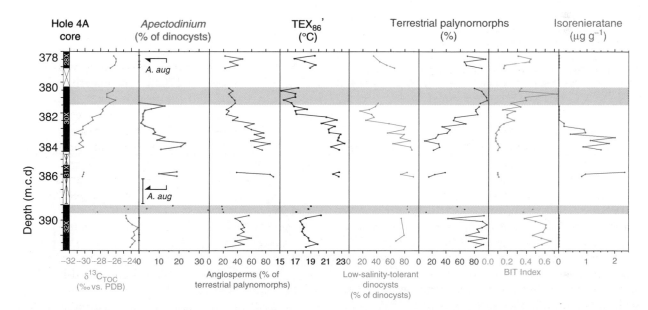

Figure 4 IODP Site 302 enabled the recovery of PETM palynological and geochemical records from the Arctic Ocean. The carbon isotope records the PETM negative excursion (far left). The dinocyst *Apectodinium* (second from left) is a warm-water species, and a nearly global indicator of the PETM. TEX_{86} is a new paleotemperature proxy (middle) which reveals extreme polar warmth even previous to the PETM, and a 5 °C warming with the event. Other indicators show an apparent decrease in salinity in the Arctic ocean. From Sluijs A, Schouten S, Pagani M, *et al.* (2006) Subtropical Arctic Ocean conditions during the Palaeocene Eocene thermal maximum. *Nature* 441: 610–613 (doi:10.1038/nature/04668).

turnover, while deep-sea benthic foraminifera suffered extinction (30–50% of species). There was widespread oceanic carbonate dissolution: the calcium carbonate compensation depth (CCD) rose by more than 2 km in the southeastern Atlantic.

One explanation for the CIE is the release of $\sim 2000–2500\,Gt$ of isotopically light ($\sim -60\permil$) carbon from methane clathrates in oceanic reservoirs (*see* Methane Hydrates and Climatic Effects). Oxidation of methane in the oceans would have stripped oxygen from the deep waters, leading to hypoxia, and the shallowing of the CCD, leading to a widespread dissolution of carbonates. Proposed triggers of gas hydrate dissociation include warming of the oceans by a change in oceanic circulation, continental slope failure, sea level lowering, explosive Caribbean volcanism, or North Atlantic basaltic volcanism.

Arguments against gas hydrate dissociation as the cause of the PETM include low estimates (500–3000 Gt C) for the size of the oceanic gas hydrate reservoir in the recent oceans, implying even smaller ones in the warm Paleocene oceans. The observed >2 km rise in the CCD is more than that estimated for a release of 2000–2500 Gt C. In addition, pre-PETM atmospheric pCO_2 levels of greater than 1000 ppm require much larger amounts than 2500 Gt C to raise global temperatures by 5 °C at estimated climate sensitivities of 1.5–4.5 °C for a doubling of CO_2. Alternate sources of carbon include a large range of options: organic matter heated by igneous intrusions in the North Atlantic, by subduction in Alaska, or by continental collision in the Himalayas; peat burning; oxidation of organic matter after desiccation of inland seas and methane release from extensive marshlands; and mantle plume-induced lithospheric gas explosions.

The interval of $\sim 52–49$ Ma is known as the early Eocene Climatic Optimum (EECO; **Figure 5**), during which temperatures reached levels unparalleled in the Cenozoic (the last 65.5 My) with the brief exception of the PETM. Crocodiles, tapir-like mammals, and palm trees flourished around an Arctic Ocean with warm, sometimes brackish surface waters. Temperatures did not reach freezing even in continental interiors at mid- to high latitudes, polar surface temperatures were $> 30\,°C$ warmer than modern, global deep water temperatures were $\sim 10–12\,°C$ warmer than today, and polar ice sheets probably did not reach sea level – if they existed at all. Interestingly, EECO tropical ocean temperatures, once thought to be at modern or even cooler values, are now considered to have been $\sim 8\,°C$ warmer than today, still much less than the extreme polar warming.

We do not know how the high latitudes were kept as warm as 15–23 °C with tropical temperatures only at $\sim 35\,°C$ (**Figure 6**). The associated small latitudinal temperature gradients make it difficult to explain either high heat transport through the atmosphere or

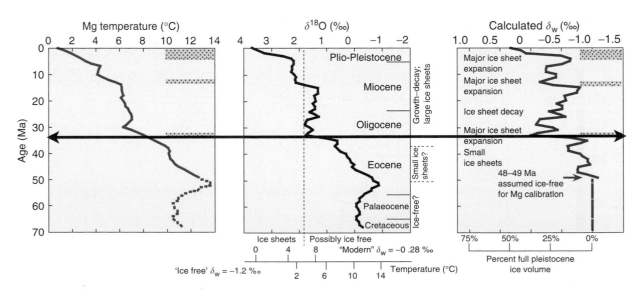

Figure 5 Lear *et al.* (2000) utilized Mg/Ca ratios from benthic foraminifera to create a paleotemperature proxy record (left). When compared against the benthic oxygen isotope record (middle) there is a general congruence in pattern which is expected, given that a large part of the oxygen isotope record also reflects temperature change. When taken in combination, the two records can be used to infer changes in the oxygen isotopic composition (right) of seawater, a proxy for terrestrial ice volume. The major ice buildup at the beginning of the Oligocene (Oi1) is apparent in the substantial change in oxygen isotopic composition in the absence of major Mg/Ca change (solid horizontal line).

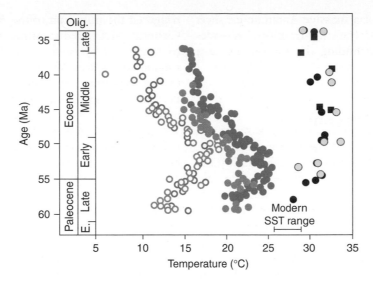

Figure 6 Pearson *et al.* (2007) paleotemperature estimates from Tanzania (red) derived from oxygen isotope ratios of planktonic foraminiferal tests and from TEX$_{86}$ (yellow). The other data points show benthic and polar surface paleotemperature estimates. The Tanzanian data reveal tropical temperatures were significantly warmer than modern but that they were stable compared to benthic and polar temperatures through the Cenozoic climate deterioration. The authors' interpretation is that previous tropical temperature estimates (filled blue and green circles) were spuriously tracking deep ocean temperatures (open blue and green circles) and temperature trends because of diagenesis. If correct, this implies that tropical temperatures were at least 5 °C warmer than previously thought in greenhouse intervals.

through the ocean given the well-proven conceptual and numerical models of climate (**Figure 7**). Despite more than two decades of work on this paradox of high polar temperatures and low heat transport, climate models consistently compute temperatures for high latitudes and mid-latitude continental winters that are lower than those indicated by biotic and chemical temperature proxies. Even this paradigm of greenhouse climate displayed profound variability. Within the EECO are alternating warm and very warm (hyperthermal) periods. These hypothermals are characterized by dissolution horizons associated with isotope anomalies and benthic foraminiferal assemblage changes and have been identified in upper Paleocene–lower Eocene sediments worldwide, reflecting events similar in nature to, but less severe than, the PETM. It remains an open question whether these hyperthermals directly reflect greenhouse gas inputs, or cumulative effects of changing ocean chemistry and circulation, perhaps driven by orbital forcing. While pCO$_2$ levels may have been high (1000–4000 ppm) to maintain the rise of the global mean temperatures of this greenhouse world, it does not simultaneously explain the small temperature gradients (as low as 15 °C), warm poles, and warm continental interiors. Furthermore, we are still far from explaining the amplitude of orbital, suborbital, and nonorbital variability exhibited by these climates. The mystery is compounded by the rapidity by which the Eocene greenhouse world came to an end with the

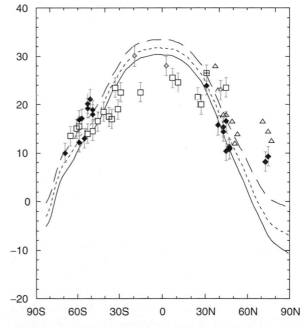

Figure 7 As described in Shellito *et al.* (2003), model-predicted zonally averaged mean annual temperatures compared with paleotemperature proxy records for the Eocene. These simulations were carried out with Eocene boundary conditions and with pCO$_2$ specified at three levels: 500 (solid), 1000 (dotted), and 2000 (dashed lines) ppm. The other symbols represent terrestrial and marine surface temperature proxy records. With increasing greenhouse gas concentrations, temperatures gradually come into better agreement with proxies at high latitudes, but the mismatch persists until extremely high values are reached, at which point tropical temperatures are too warm.

sudden emplacement of a massive Antarctic ice sheet near the epoch's end, but the transition provides critical clues to understanding the functioning of a greenhouse climate.

The End of the Greenhouse

We do not fully understand the greenhouse world, and neither do we understand the transition into the present cool world ('icehouse world') and the cause of global cooling. This much we think we do know. Beginning in the middle Eocene (at ~ 48.6 Ma), the Poles and deep oceans began to cool, and pCO_2 may have declined. The cause of this cooling and apparent carbon draw-down are unknown. As this cooling continued, the diversity of planktic and benthic oceanic life-forms declined in the late Eocene to the earliest Oligocene (~ 37.2–33.9 Ma). Antarctic ice sheets achieved significant volume and reached sea level by about 33.9 Ma, while sea ice might have covered parts of the Arctic Ocean by that time. Small, wet-based ice sheets may have existed through the late Eocene, but a rapid (~ 100 ky) increase in benthic foraminiferal $\delta^{18}O$ values in the earliest Oligocene (called the 'Oi1 event') has been interpreted as reflecting the establishment of the Antarctic ice sheet. Oxygen isotope records, however, reflect a combination of changes in ice volume and temperature (see Cenozoic Climate – Oxygen Isotope Evidence), and it is still debated whether the Oi1 event was primarily due to an increase in ice volume or to cooling.

Paleotemperature proxy data (Mg/Ca) suggest that the event was due mainly to ice volume increase, although this record might have been affected by changes in position of the CCD in the oceans at the time. Ice sheet modelers argue that the event contains both an ice volume and a cooling component, and several planktonic microfossil records suggest surface water cooling at the time. Whatever the Oi1 event was in detail, it is generally seen as the time of change from a largely unglaciated to a largely glaciated world, and the end of the last greenhouse world. There are several proposed causes of this final transition to a world with substantial glaciation, including Southern Ocean gateway opening, decreases in greenhouse gas concentrations, orbital configuration changes, and ice albedo feedbacks.

The Antarctic continent had been in a polar position for tens of millions of years before glaciation started, so why did a continental ice sheet form in ~ 100 Ky during the Oi1? A popular theory is 'thermal isolation' of the Antarctic continent: opening of the Tasman Gateway and Drake Passage triggered the initiation of the Antarctic Circumpolar Current (ACC), reducing meridional heat transport to Antarctica by isolation of the continent within a ring of cold water. We cannot constrain the validity of this hypothesis by data on the opening of Drake Passage, because even recent estimates range from middle Eocene to early Miocene (a range of 20 My). The Tasman Gateway opened several million years before Oi1 which seems to make a close connection between Southern Ocean gateway changes and glaciation unsupportable. Data on microfossil distribution indicate that there probably was no warm current flowing southward along eastern Australia, because a counterclockwise gyre in the southern Pacific prevented warm waters from reaching Antarctica. Furthermore, ocean–atmosphere climate modeling indicates that the change in meridional heat transport associated with ACC onset was insignificant at high latitudes. That does not mean that Southern Ocean gateway changes had no impact on climate. The oceanographic changes associated with opening of Drake and Tasman gateways (the so-called 'Drake Passage effect') has been reproduced by climate model investigations of this time interval, and the magnitude of the change anticipated from idealized model results has been reproduced. Opening of Drake and Tasman gateways produces a 0.6 PW decrease in southward heat transport in the southern subtropical gyre (out of a peak of ~ 1.5 PW), and shifts that heat into the Northern Hemisphere. The climatic effect of this shift is felt primarily as a small temperature change in the subtropical oceans and a nearly negligible ($< 3\,^{\circ}C$) change in the Antarctic polar ocean region.

Climate and glaciological modeling implicates a leading role of this transition to decreases in atmospheric greenhouse gas concentrations, because Antarctic climate and ice sheets have been shown to be more sensitive to pCO_2 than to other parameters. Consequently, Cenozoic cooling of Antarctica is no longer generally accepted as having been primarily caused by changes in oceanic circulation only. Instead, decreasing CO_2 levels with subsequent processes such as ice albedo and weathering feedbacks (possibly modulated by orbital variations) are seen as significant long-term climate-forcing factors.

The difficulty with this explanation is that existing pCO_2 reconstructions do not show a decrease near the Oi1 event; instead, the primary decreases were much earlier or much later. This suggests that our understanding remains incomplete. Either pCO_2 proxies or climate models could be incorrect, or, more interestingly, they may not contain key feedbacks and processes that might have transformed the long-term greenhouse gas decline into a sudden ice buildup. It is likely, both from a proxy and modeling perspective, that changes in Earth's orbital configuration played a role in setting the exact timing of major Antarctic ice accumulation, but this area remains a subject of active research.

Problems Posed by the Greenhouse Climate

A strong line of geologic evidence links higher greenhouse gas concentrations with low temperature gradients and warm global mean and deep ocean temperatures through the Cretaceous and early Paleogene. This correlation is fairly strong, but not as strong, for equable continental climates. The concentrations necessary to sufficiently warm high latitudes and continental interiors, however, are so high that keeping tropical SSTs relatively cool becomes a problem. Less obvious and more successful mechanisms have been put forward to solve this problem, such as forcing by polar stratospheric clouds and tropical cyclone-induced ocean mixing. So far, all attempts to accurately reproduce warm Eocene continental interior temperatures lead to unrealistically hot tropical temperatures in view of biotic and geochemical proxy estimates (*see* Determination of Past Sea Surface Temperatures). As summarized in **Table 1** and discussed below, many opportunities exist for continued progress to solve these problems.

Cretaceous and early Paleogene tropical SST reconstructions might be biased toward cool values, and other variables need to be constrained to more precisely define the data–model mismatch in the Tropics. For instance, if planktonic foraminifera record tropical temperatures from below the mixed-layer, that is greater than *c*. 40 m depth, rather than 6 m, the low-gradient problem is ameliorated, but such a habitat may be difficult to reconcile with evidence for the presence of photosymbionts. Alternatively, a sampling bias toward warm season values at high latitudes and cold (upwelling) season temperatures in low latitudes could partially resolve the low-gradient problem. Warm terrestrial extratropical winter temperatures tend to rule out a large high-latitude bias in planktonic foraminiferal SST estimates.

A commonly advocated conjecture might appear to resolve the mysteries of greenhouse climates, and it relies on oceanic heat transport in conjunction with high greenhouse gas concentrations. Attention has focused on the challenging problem of modeling paleo-ocean circulations and, especially, on placing bounds on the magnitude of ocean heat transport in such climates.

Climate models have been coupled to a 'slab' mixed-layer ocean model, that is, assuming that the important ocean thermal inertia is in the upper 50 m and that ocean poleward heat transport is at specified levels, to predict equator-to-pole surface temperature gradients. With appropriate (Eocene or Cretaceous) boundary conditions, near-modern pCO$_2$, and ocean heat transport specified at near-modern values, an equator-to-pole temperature gradient (and continental winter temperatures) is very close to the modern result (**Figure 7**). With higher pCO$_2$, tropical SSTs increase (to $\sim$32 °C for 1000 ppm) without changing meridional gradients substantially, and continental interiors remain well below freezing in winter. Substantial high-latitude amplification of temperature response to increases in pCO$_2$ or other forcings is generally obtained because of the nonlinearity introduced by crossing a threshold from extensive sea ice cover to little or no sea ice cover.

Proxy data imply that it was unlikely – but not impossible – that sea ice was present during the warmest of the greenhouse climates; therefore, this sensitivity is probably not representative for the true past greenhouse intervals. Nevertheless, evidence is growing of the possibility of the presence of sea ice and potentially ice sheets during some of the cooler parts of greenhouse climates. With heat transport approximately 3 times to modern in a slab ocean model, and pCO$_2$ much higher than modern ($>$2000 ppm), some of the main characteristics of greenhouse climates are reproduced, but we do not know how to reach such a high heat transport level without invoking feedbacks in the climate system that have not traditionally been considered, such as those due to tropical cyclones.

There is a range, however, of smaller-than-modern temperature gradients that may not imply an increase in ocean heat transport over modern values. Near-modern values of ocean heat transport could support an equator-to-pole surface temperature gradient of 24 °C, whereas a 15 °C gradient requires ocean heat transport roughly 2–3 times as large. In other words, warm polar temperatures of 10 °C and tropical temperatures of 34 °C can exist in equilibrium with near-modern ocean transport values, but climates with smaller gradients are paradoxical. Thus even small errors or biases in tropical or polar temperatures affect our interpretation of the magnitude of the climate mystery posed by past greenhouse climates.

In conclusion, two main hypotheses potentially operating in conjunction might explain greenhouse climates. The first is increased greenhouse gas concentration, but we need better understanding of the coupled carbon cycle and climate dynamics on short

Table 1   Greenhouse climate research: key areas to resolve the low gradient and equable climate problems

Tropical proxies may be biased to cool values	What are the true depth habits of planktonic foraminifera? Are 'mixed-layer' dwellers calcifying below the mixed layer?	Could proxies reflect seasonal biases due to upwelling? Are productivity and calcification limited to relatively cool conditions?	How much do preservation and diagenesis alter the proxy signals? Can we only trust foraminifera with 'glassy' preservation? Under what conditions do Mg/Ca produce valid temperatures? Are compound-specific isotope analyses also subject to alteration?
Polar temperature proxies might be biased to warm values	Is there sea ice? Sea ice introduces a fundamental nonlinearity into climate. When did it first form?	Are proxies recording only the warm events or warm season? Is polar warmth a fiction of aliasing?	What is the oxygen isotopic composition of polar seawater? Do hydrological cycle and ocean circulation changes alter this?
Models missing key features	Is resolution in the atmosphere and ocean models too coarse to get correct dynamics or capture the 'proxy scale'? What resolution is enough?	Is there missing physics: e.g., polar stratospheric clouds, enhanced ocean mixing; or incorrect physics: e.g., cloud parametrizations?	Are paleogeographic/ paleobathymetric/ soil/vegetation/ozone boundary conditions correct enough?
Greenhouse gases	How accurate are existing pCO_2 proxies? How can they be improved? No proxy for methane or other greenhouse gases. How did other radiatively important constituents vary?	The sensitivity of climate models to greenhouse gas forcing varies from model to model and the 'correct' values are an unknown, what can we learn from forcing models with greenhouse gases?	Carbon cycle feedbacks. What caused long-term and short-term carbon perturbations? How did climate and the carbon cycle feed back onto each other?
Circulation proxies	Although ocean circulation tracers (e.g., Nd) provide information on the sources and directions of flow, there are currently no tracers that directly reflect ocean flow rates. Can quantitative proxies for rates of deep water formation and flow be developed? Was the ocean circulation weaker or stronger?	Density-driven flows in the ocean are function of temperature and salinity distributions. Most proxies focus on temperature. Can existing techniques for estimating paleosalinity (especially surface) be refined and extended? Can new, quantitative proxies be developed? Was the ocean circulation of past greenhouse climates driven primarily by salinity gradients?	Oceanic carbon storage and oxygenation are largely controlled by ocean circulation. Can integrated carbon–oxygen-isotope-dynamical models reveal unique and important linkages between proxies and ocean dynamics?

and long timescales. The second is a physical mechanism that increases poleward heat transport strongly as the conditions warm.

See also

Abrupt Climate Change. Cenozoic Climate – Oxygen Isotope Evidence. Determination of Past Sea Surface Temperatures. Heat Transport and Climate. Methane Hydrates and Climatic Effects. Paleoceanography. Paleoceanography, Climate Models in. Plio-Pleistocene Glacial Cycles and Milankovitch Variability.

Further Reading

Barker F and Thomas E (2004) Origin, signature and palaeoclimate influence of the Antarctic Circumpolar Current. *Earth Science Reviews* 66: 143–162.

Beckmann B, Flögel S, Hofmann P, Schulz M, and Wagner T (2005) Orbital forcing of Cretaceous river discharge in tropical Africa and ocean response. *Nature* 437: 241–244 (doi:10.1038/nature03976).

Billups K and Schrag D (2003) Application of benthic foraminiferal Mg/Ca ratios to questions of Cenozoic climate change. *Earth and Planetary Science Letters* 209: 181–195.

Brinkhuis H, Schouten S, Collinson ME, et al. (2006) Episodic fresh surface waters in the early Eocene Arctic Ocean. *Nature* 441: 606–609 (doi:10.1038/nature04692).

Coxall HK, Wilson A, Palike H, Lear CH, and Backman J (2005) Rapid stepwise onset of Antarctic glaciation and deeper calcite compensation in the Pacific Ocean. *Nature* 433: 53–57.

DeConto RM and Pollard D (2003) Rapid Cenozoic glaciation of Antarctica induced by declining atmospheric CO_2. *Nature* 421: 245–249.

Donnadieu Y, Pierrehumbert R, Jacob R, and Fluteau F (2006) Modelling the primary control of paleogeography on Cretaceous climate. *Earth and Planetary Science Letters* 248: 426–437.

Greenwood DR and Wing SL (1995) Eocene continental climates and latitudinal temperature gradients. *Geology* 23: 1044–1048.

Huber BT, Macleod KG, and Wing S (1999) *Warm Climates in Earth History.* Cambridge, UK: Cambridge University Press.

Huber BT, Norris RD, and MacLeod KG (2002) Deep sea paleotemperature record of extreme warmth during the Cretaceous Period. *Geology* 30: 123–126.

Huber M, Brinkhuis H, Stickley CE, et al. (2004) Eocene circulation of the Southern Ocean: Was Antarctica kept warm by subtropical waters. *Paleoceanography* PA4026 (doi:10.1029/2004PA001014).

Jenkyns HC, Forster A, Schouten S, and Sinninge Damste JS (2004) Higher temperatures in the late Cretaceous Arctic Ocean. *Nature* 432: 888–892.

Lear CH, Elderfield H, and Wilson A (2000) Cenozoic deep-sea temperature sand global ice volumes from Mg/Ca in benthic foraminiferal calcite. *Science* 287: 269–272.

MacLeod KG, Huber BT, and Isaza-Londoño C (2005) North Atlantic warming during 'global' cooling at the end of the Cretaceous. *Geology* 33: 437–440.

MacLeod KG, Huber BT, Pletsch T, Röhl U, and Kucera M (2001) Maastrichtian foraminiferal and paleoceanographic changes on Milankovitch time scales. *Paleoceanography* 16: 133–154.

Miller KG, Wright JD, and Browning JV (2005) Visions of ice sheets in a greenhouse world. In: Paytan A and De La Rocha C (eds.) *Marine Geology, Special Issue, No. 217: Ocean Chemistry throughout the Phanerozoic,* pp. 215–231. New York: Elsevier.

Pagani M, Zachos J, Freeman KH, Bohaty S, and Tipple B (2005) Marked change in atmospheric carbon dioxide concentrations during the Oligocene. *Science* 309: 600–603.

Parrish JT (1998) *Interpreting Pre-Quaternary Climate from the Geologic Record,* 348pp. New York: Columbia University Press.

Pearson PN and Palmer MR (2000) Atmospheric carbon dioxide concentrations over the past 60 million years. *Nature* 406: 695–699.

Pearson PN, van Dongen BE, Nicholas CJ, et al. (2007) Stable tropical climate through the Eocene epoch. *Geology* 35: 211–214 (doi: 10.1130/G23175A.1).

Poulsen CJ, Barron EJ, Arthur MA, and Peterson WH (2001) Response of the mid-Cretaceous global oceanic circu-lation to tectonic and CO_2 forcings. *Paleoceanography* 16: 576–592.

Poulsen CJ, Gendaszek AS, and Jacob RL (2003) Did the rifting of the Atlantic Ocean cause the Cretaceous thermal maximum? *Geology* 31: 1115–1118.

Prahl FG, Herbert TD, Brassell SC, et al. (2000) Status of alkenone paleothermometer calibration: Report from Working Group 3: Geochemistry. *Geophysics, Geosystems* 1 (doi:10.1029/2000GC000058).

Royer DL, Berner RA, Montanez IP, Tabor NJ, and Beerling DJ (2004) CO_2 as a primary driver of Phanerozoic climate. *GSA Today* 14: 4–12.

Schouten S, Hopmans E, Forster A, van Breugel Y, Kuypers MMM, and Sinninghe Damsté JS (2003) Extremely high sea-surface temperatures at low latitudes during the middle Cretaceous as revealed by archaeal membrane lipids. *Geology* 31: 1069–1072.

Shellito C, Sloan LC, and Huber M (2003) Climate model constraints on atmospheric CO_2 levels in the early–middle Palaeogene. *Palaeogeography, Palaeoclimatology, Palaeoecology* 193: 113–123.

Sluijs A, Schouten S, Pagani M, et al. (2006) Subtropical Arctic Ocean conditions during the Palaeocene Eocene Thermal Maximum. *Nature* 441: 610–613 (doi:10.1038/nature/04668).

Tarduno JA, Brinkman DB, Renne PR, Cottrell RD, Scher H, and Castillo P (1998) Evidence for extreme climatic warmth from late Cretaceous Arctic vertebrates. *Science* 282: 2241–2244.

Thomas E, Brinkhuis H, Huber M, and Röhl U (2006) An ocean view of the early Cenozoic greenhouse world. *Oceanography* 19: 63–72.

von der Heydt A and Dijkstra HA (2006) Effect of ocean gateways on the global ocean circulation in the late Oligocene and early Miocene. *Paleoceanography* 21: 1–18 (doi:/10.1029/2005PA001149).

Wilf P, Johnson KR, and Huber BT (2003) Correlated terrestrial and marine evidence for global climate changes before mass extinction at the Cretaceous–Paleogene boundary. *Proceedings of the National Academy of Sciences of the United States of America* 100: 599–604.

Zachos JC, Röhl U, Schellenberg SA, *et al.* (2005) Rapid acidification of the ocean during the Paleocene–Eocene Thermal Maximum. *Science* 308: 1611–1615.

Zachos J, Pagani M, Sloan L, Thomas E, and Billups K (2001) Trends, rhythms, and aberrations in global climate 65 Ma to present. *Science* 292: 686–694.

HOLOCENE CLIMATE VARIABILITY

M. Maslin, C. Stickley, and V. Ettwein, University College London, London, UK

Introduction

Until a few decades ago it was generally thought that significant large-scale global and regional climate changes occurred at a gradual pace within a timescale of many centuries or millennia. Climate change was assumed to be scarcely perceptible during a human lifetime. The tendency for climate to change abruptly has been one of the most surprising outcomes of the study of Earth history. In particular, paleoceanographic records demonstrate that our present interglacial, the Holocene (the last ∼10 000 years), has not been as climatically stable as first thought. It has been suggested that Holocene climate is dominated by millennial-scale variability, with some authorities suggesting that this is a 1500 year cyclicity. These pronounced Holocene climate changes can occur extremely rapidly, within a few centuries or even within a few decades, and involve regional-scale changes in mean annual temperature of several degrees Celsius. In addition, many of these Holocene climate changes are stepwise in nature and may be due to thresholds in the climate system.

Holocene decadal-scale transitions would presumably have been quite noticeable to ancient civilizations. For instance, the emergence of crop agriculture in the Middle East corresponds very closely with a sudden warming event marking the beginning of the Holocene, and the widespread collapse of the first urban civilizations, such as the Old Kingdom in Egypt and the Akkadian Empire, coincided with a cooling event at around 4300 BP. In addition, paleo-records from the late Holocene demonstrate the possible influence of climate change on the collapse of the Mayan civilization (Classic Period), while Andean ice core records suggests that alternating wet and dry periods influenced the rise and fall of coastal and highland cultures of Ecuador and Peru.

It would be foolhardy not to bear in mind such sudden stepwise climate transitions when considering the effects that humans might have upon the present climate system, via the rapid generation of greenhouse gases for instance. Judging by what we have already learnt from Holocene records, it is not improbable that the system may gradually build up over hundreds of years to a 'breaking point' or threshold, after which some dramatic change in the system occurs over just a decade or two. At the threshold point, the climate system is in a delicate and somewhat critical state. It may take only a relatively minor 'adjustment' to trigger the transition and tip the system into abrupt change.

This article summarizes the current paleoceanographic records of Holocene climate variability and the current theories for their causes. Concentrating on records that cover a significant portion of the Holocene. The discussion is limited to centennial–millennial-scale variations. **Figure 1** illustrates the Holocene and its climate variability in context of the major global climatic changes that have occurred during the last 2.5 million years. Short-term variations such as the North Atlantic Oscillation and the El Niño-Southern Oscillation will not be discussed.

The Importance of the Oceans and Holocene Paleoceanography

Climate is created from the effects of differential latitudinal solar heating. Energy is constantly transferred from the equator (relatively hot) toward the poles (relatively cold). There are two transporters of such energy – the atmosphere and the oceans. The atmosphere responds to an internal or external change in a matter of days, months, or may be a few years. The oceans, however, have a longer response time. The surface ocean can change over months to a few years, but the deep ocean takes decades to centuries. From a physical point of view, in terms of volume, heat capacity and inertia, the deep ocean is the only viable contender for driving and sustaining long-term climate change on centennial to millennial timescales.

Since the process of oceanic heat transfer largely regulates climate change on longer time-scales and historic records are too short to provide any record of the ocean system prior to human intervention, we turn to marine sediment archives to provide information about ocean-driven climate change. Such archives can often provide a continuous record on a variety of timescales. They are the primary means for the study and reconstruction of the stability and natural variability of the ocean system prior to anthropogenic influences.

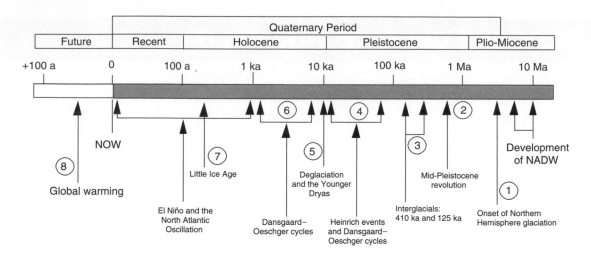

Figure 1 Log time scale cartoon, illustrating the most important climate events in the Quaternary Period. (a, ka, Ma, refer to years ago.) (1) Onset of Northern Hemisphere Glaciation (3.2–2.5 Ma), ushering in the strong glacial–interglacial cycles which are characteristic of the Quaternary Period. (2) Mid-Pleistocene Revolution when the dominant periodicity of glacial–interglacial cycles switched from 41 000 y, to every 100 000 y. The external forcing of the climate did not change; thus, the internal climate feedback's must have altered. (3) The two closest analogues to the present climate are the interglacial periods at 420 000 to 390 000 years ago (oxygen isotope stage 11) and 130 000 to 115 000 years ago (oxygen isotope stage 5e, also known as the Eemian). (4) Heinrich events and Dansgaard–Oeschger cycles (see text). (5) Deglaciation and the Younger Dryas events. (6) Holocene Dansgaard–Oeschger cycles (see text). (7) Little Ice Age (AD 1700), the most recent climate event that seems to have occurred throughout the Northern Hemisphere. (8) Anthropogenic global warming.

One advantage of marine sediments is that they can provide long, continuous records of Holocene climate at annual (sometimes intraannual) to centennial time-resolutions. However, there is commonly a trade-off between temporal and spatial resolution. Deep-ocean sediments usually represent a large spatial area, but sedimentation rates in the deep-ocean are on average between 0.002 and $0.005 \, cm \, y^{-1}$, with very productive areas producing a maximum of $0.02 \, cm \, y^{-1}$. This limits the temporal resolution to a maximum of 200 years per cm $(50 \, y \, cm^{-1}$ for productive areas). Mixing by the process of bioturbation will reduce the resolution further.

On continental shelves and in bays and other specialized sediment traps such as anoxic basins and fiords, sedimentation rates can exceed $1 \, cm \, y^{-1}$ providing temporal resolution of over $1 \, y \, cm^{-1}$. More local conditions are recorded in laminated marine sediments formed in anoxic environments, where biological activity can not disturb the sediments. For example, Pike and Kemp (1997) analysed annual and intraannual variability within the Gulf of California from laminated sediments containing a record of diatom-mat accumulation. Time series analysis highlighted a decadal-scale variability in mat-deposition associated with Pacific-wide changes in surface water circulation, suggested to be influenced by solar-cycles. In addition, anoxic sediments from the Mediterranean Ridge (ODP Site 971) reveal

seasonal-scale variability during the late Quaternary from a laminated diatom-ooze sapropel. Pearce *et al.*, (1998) inferred changes in the monsoon-related nutrient input to the Mediterranean Basin via the Nile River as the main cause of the variations in the laminated sediments, which suggests a wide influence of changes in seasonality. Other potentially extremely high-resolution studies will come from Saanich Inlet, a Canadian fiord and Prydz Bay in Antarctica, sites recently drilled by the Ocean Drilling Program.

However, the main drawback to such high-resolution locations is that they contain highly localized environmental and climate information. An additional problem associated mainly with continental margins is reworking, erosion, and redistribution of the sediment by mass density flows such as turbidities and slumps. Hence we concentrate on wider-scale records of Holocene climate change.

Holocene Climatic Variability

Initial studies of the Greenland ice core records concluded the absence of major climate variation within the Holocene. This view is being progressively eroded, particular in the light of new information being obtained from marine sediments (**Figure 2**). Long-term trends indicate an early to mid-Holocene climatic optimum with a cooling trend in the late

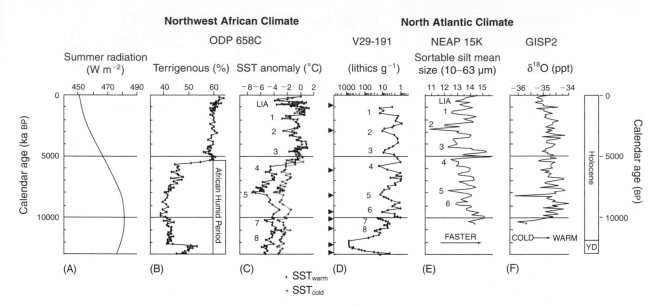

Figure 2 Comparison of summer insolation for 65°N with north-west African climate (deMenocal *et al.*, 2000) and North Atlantic climate (V29-191, Bond *et al.*, 1997; NEAP 15 K, Bianchi and McCave, 1999; GISP2, O'Brien *et al.*, 1996). Note the similarity of events labeled 1 to 8 and the Little Ice Age (LIA).

Holocene. Superimposed on this trend are several distinct oscillations or climatic cooling steps that appear to be of widespread significance (see **Figure 2**), the most dramatic of which occurred 8200, 5500, and 4400 years ago and between AD 1200 and AD 1650.

The event 8200 years ago is the most striking and abrupt, leading to widespread cool and dry conditions lasting perhaps 200 years, before a rapid return to climates warmer and generally moister than at present. This event is noticeably present in the GISP2 Greenland ice cores, from which it appears to have been about half as severe as the Younger Dryas to Holocene transition. Marine records of North African to Southern Asian climate suggest more arid conditions involving a failure of the summer monsoon rains. Cold and/or arid conditions also seem to have occurred in northernmost South America, eastern North America and parts of north-west Europe.

In the middle Holocene approximately 5500–5300 years ago there was a sudden and widespread shift in precipitation, causing many regions to become either noticeably drier or moister. The dust and sea surface temperature records off north-west Africa show that the African Humid Period, when much of subtropical West Africa was vegetated, lasted from 14 800 to 5500 years ago and was followed by a 300year transition to much drier conditions (de Menocal *et al.*, 2000). This shift also corresponds to the decline of the elm (*Ulmus*) in Europe about 5700, and of hemlock (*Tsuga*) in North America about 5300

years ago. Both vegetation changes were initially attributed to specific pathogen attacks, but it is now thought they may have been related to climate deterioration. The step to colder and drier conditions in the middle of an interglacial period is analogous to a similar change that is observed in records of the last interglacial period referred to as Marine Oxygen Isotope Stage 5e (Eemian).

There is also evidence for a strong cold and arid event occurring about 4400 years ago across the North Atlantic, northern Africa, and southern Asia. This cold, and arid event coincides with the collapse of a large number of major urban civilizations, including the Old Kingdom in Egypt, the Akkadian Empire in Mesopotamia, the Early Bronze Age societies of Anatolia, Greece, Israel, the Indus Valley civilization in India, the Hilmand civilization in Afganistan, and the Hongshan culture of China.

Little Ice Age (LIA)

The most recent Holocene cold event is the Little Ice Age (see **Figures 2** and **3**). This event really consists of two cold periods, the first of which followed the Medieval Warm Period (MWP) that ended ~1000 years ago. This first cold period is often referred to as the Medieval Cold Period (MCP) or LIAb. The MCP played a role in extinguishing Norse colonies on Greenland and caused famine and mass migration in Europe. It started gradually before AD 1200 and ended at about AD 1650. This second cold period,

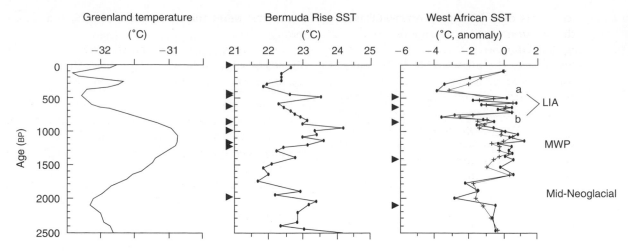

Figure 3 Comparison of Greenland temperatures, the Bermuda Rise sea surface temperatures (SST) (Keigwin, 1996), and west African and a sea surface temperature (deMenocal *et al.*, 2000) for the last 2500 years. LIALittle Ice Age; MWPMedieval Warm Period. Solid triangles indicate radiocarbon dates.

may have been the most rapid and the largest change in the North Atlantic during the Holocene, as suggested from ice-core and deep-sea sediment records. The Little Ice Age events are characterized by a drop in temperature of 0.5–1°C in Greenland and a sea surface temperature falls of 4°C off the coast of west Africa and 2°C off the Bermuda Rise (see **Figure 3**).

Holocene Dansgaard–Oeschger Cycles

The above events are now regarded as part of the millennial-scale quasiperiodic climate changes characteristic of the Holocene (see **Figure 2**) and are thought to be similar to glacial Dansgaard–Oeschger (D/O) cycles. The periodicity of these Holocene D/O cycles is a subject of much debate. Initial analysis of the GISP2 Greenland ice core and North Atlantic sediment records revealed cycles at approximately the same 1500 ($\pm$500)-year rhythm as that found within the last glacial period. Subsequent analyses have also found a strong 1000-year cycle and a 550-year cycle. These shorter cycles have also been recorded in the residual $\delta^{14}C$ data derived from dendrochronologically calibrated bidecadal tree-ring measurements spanning the last 11 500 years. In general, during the coldest point of each of the millennial-scale cycles shown in **Figure 2**, surface water temperatures of the North Atlantic were about 2–4°C cooler than during the warmest part.

One cautionary note is that Wunsch has suggested a more radical explanation for the pervasive 1500-year cycle seen in both deep-sea and ice core, glacial and interglacial records. Wunsch suggests that the extremely narrow spectral lines (less than two bandwidths) that have been found at about 1500 years in many paleo-records may be due to aliasing. The 1500-year peak appears precisely at the period predicted for a simple alias of the seasonal cycle sampled inadequately (under the Nyquist criterion) at integer multiples of the common year. When Wunsch removes this peak from the Greenland ice core data and deep-sea spectral records, the climate variability appears as expected to be a continuum process in the millennial band. This work suggests that finding a cyclicity of 1500 years in a dataset may not represent the true periodicity of the millennial-scale events. The Holocene Dansgaard–Oeschger events are quasi periodic, with different and possibly stochastic influences.

Causes of Millennial Climate Fluctuation during the Holocene

As we have already suggested, deep water circulation plays a key role in the regulation of global climate. In the North Atlantic, the north-east-trending Gulf Stream carries warm and relatively salty surface water from the Gulf of Mexico up to the Nordic seas. Upon reaching this region, the surface water has cooled sufficiently that it becomes dense enough to sink, forming the North Atlantic Deep Water (NADW). The 'pull' exerted by this dense sinking maintains the strength of the warm Gulf Stream, ensuring a current of warm tropical water into the North Atlantic that sends mild air masses across to the European continent. Formation of the NADW can be weakened by two processes. (1) The presence of huge ice sheets over North America and Europe changes the position of the atmospheric polar front, preventing the Gulf Stream from traveling so far north. This reduces the amount of cooling and the capacity of the surface water to sink. Such a reduction of formation occurred during the last glacial period. (2) The input of fresh water forms a lens of less-dense water, preventing sinking. If NADW formation is reduced, the weakening of the warm Gulf Stream causes colder conditions within the entire North Atlantic region and has a major impact on global climate. Bianchi and McCave, using deep-sea sediments from the North Atlantic, have shown that during the Holocene there have been regular reductions in the intensity of NADW (**Figure 2E**), which they link to the 1500-year D/O cycles identified by O'Brien and by Bond (1997). There are two possible causes for the millennial-scale changes observed in the intensity of the NADW: (1) instability in the North Atlantic region caused by varying

freshwater input into the surface waters; and (2) the 'bipolar seesaw'.

There are a number of possible reasons for the instability in the North Atlantic region caused by varying fresh water input into the surface waters:

- Internal instability of the Greenland ice sheet, causing increased meltwater in the Nordic Seas that reduces deep water formation.
- Cyclic changes in sea ice formation forced by solar variations.
- Increased precipitation in the Nordic Seas due to more northerly penetration of North Atlantic storm tracks.
- Changes in surface currents, allowing a larger import of fresher water from the Pacific, possibly due to reduction in sea ice in the Arctic Ocean.

The other possible cause for the millennial-scale changes is an extension of the suggested glacial intrinsic millennial-scale 'bipolar seesaw' to the

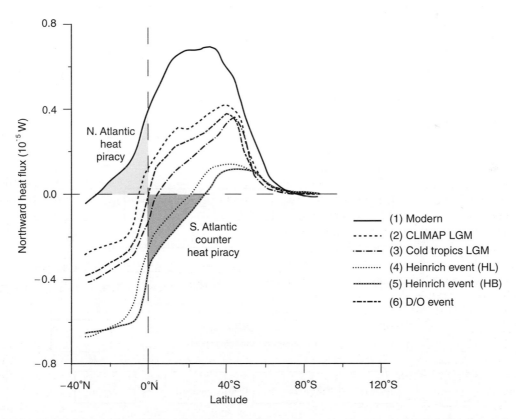

Figure 4 Atlantic Ocean poleward heat transport (positive indicates a northward movement) as given by the ocean circulation model (Seidov and Maslin, 1999) for the following scenarios: (1) present-day (warm interglacial) climate; (2) last glacial maximum (LGM) with generic CLIMAP data; (3) 'Cold tropics' LGM scenario; (4) a Heinrich-type event driven by the meltwater delivered by icebergs from decaying Laurentide ice sheet; (5) a Heinrich-type event driven by meltwater delivered by icebergs from decaying Barents Shelf ice sheet or Scandinavian ice sheet; (6) a general Holocene or glacial Dansgaard–Oeschger (D/O) meltwater confined to the Nordic Seas. Note that the total meridional heat transport can only be correctly mathematically computed in the cases of cyclic boundary conditions (as in Drake Passage for the global ocean) or between meridional boundaries, as in the Atlantic Ocean to the north of the tip of Africa. Therefore the northward heat transport in the Atlantic ocean is shown to the north of 30°S only.

Holocene. One of the most important finds in the study of glacial millennial-scale events is the apparent out-of-phase climate response of the two hemispheres seen in the ice core climate records from Greenland and Antarctica. It has been suggested that this bipolar seesaw can be explained by variations in the relative amount of deep water formation in the two hemispheres and heat piracy (**Figure 4**). This mechanism of altering dominance of the NADW and the Antarctic Bottom Water (AABW) can also be applied to the Holocene. The important difference with this theory is that the trigger for a sudden 'switching off' or a strong decrease in rate of deep water formation could occur either in the North Atlantic or in the Southern Ocean. AABW forms in a different way than NADW, in two general areas around the Antarctic continent: (1) near-shore at the shelf–ice, sea-ice interface and (2) in open ocean areas. In near-shore areas, coastal polynya are formed where katabatic winds push sea ice away from the shelf edge, creating further opportunity for sea ice formation. As ice forms, the surface water becomes saltier (owing to salt rejection by the ice) and colder (owing to loss of heat via latent heat of freezing). This density instability causes sinking of surface waters to form AABW, the coldest and

saltiest water in the world. AABW can also form in open-ocean Antarctic waters; particularly in the Weddell and Ross seas; AABW flows around Antarctica and penetrates the North Atlantic, flowing under the less dense NADW. It also flows into the Indian and Pacific Oceans, but the most significant gateway to deep ocean flow is in the south-west Pacific, where 40% of the world's deep water enters the Pacific. Interestingly, Seidov and colleagues have shown that the Southern Ocean is twice as sensitive to meltwater input as is the North Atlantic, and that the Southern Ocean can no longer be seen as a passive player in global climate change. The bipolar seesaw model may also be self sustaining, with meltwater events in either hemisphere, triggering a train of climate changes that causes a meltwater event in the opposite hemisphere, thus switching the direction of heat piracy (**Figure 5**).

Conclusion

The Holocene, or the last 10 000 years, was once thought to be climatically stable. Recent evidence, including that from marine sediments, have altered this view, showing that there are millennial-scale

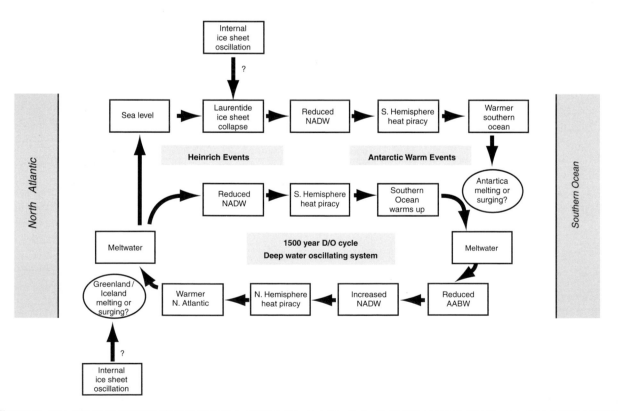

Figure 5 Possible deep water oscillatory system explaining the glacial and interglacial Dansgaard–Oeschger cycles. Additional loop demonstrates the possible link between interglacial Dansgaard–Oeschger cycles and Heinrich events.

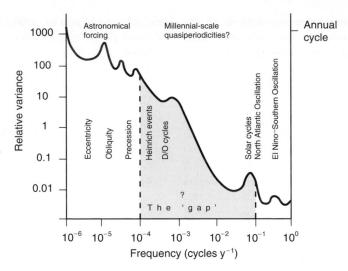

Figure 6 Spectrum of climate variance showing the climatic cycles for which we have good understanding and the 'gap' between hundreds and thousands of years for which we still do not have adequate understanding of the causes.

climate cycles throughout the Holocene. In fact we are still in a period of recovery from the last of these cycles, the Little Ice Age. It is still widely debated whether these cycles are quasiperiodic or have a regular cyclicity of 1500 years. It is also still widely debated whether these Holocene Dansgaard–Oeschger cycles are similar in time and characteristic to those observed during the last glacial period. A number of different theories have been put forward for the causes of these Holocene climate cycles, most suggesting variations in the deep water circulation system. One suggestion is that these cycles are caused by the oscillating relative dominance of North Atlantic Deep Water and Antarctic Bottom Water. Holocene climate variability still has no adequate explanation and falls in the 'gap' of our knowledge between Milankovitch forcing of ice ages and rapid variations such as El Niño and the North Atlantic Oscillation (**Figure 6**). Future research is essential for understanding these climate cycles so that we can better predict the climate response to anthropogenic 'global warming.'

See also

Antarctic Circumpolar Current. Cenozoic Oceans – Carbon Cycle Models. Millennial-Scale Climate Variability. Ocean Circulation. Paleoceanography.

Further Reading

Adams J, Maslin MA, and Thomas E (1999) Sudden climate transitions during the Quaternary. *Progress in Physical Geography* 23(1): 1–36.

Alley RB and Clark PU (1999) The deglaciation of the Northern hemisphere: A global perspective. *Annual Review of Earth and Planetary Science* 27: 149–182.

Bianchi GG and McCave IN (1999) Holocene periodicity in North Atlantic climate and deep-ocean flow south of Iceland. *Nature* 397: 515–523.

Broecker W (1998) Paleocean circulation during the last deglaciation: a bipolar seesaw? *Paleoceanography* 13: 119–121.

Bond G, Showers W, Cheseby M, *et al.* (1997) A pervasive millenial-scale cycle in North Atlantic Holocene and glacial climates. *Science* 278: 1257–1265.

Chapman MR and Shackleton NJ (2000) Evidence of 550 year and 1000 years cyclicities in North Atlantic pattern during the Holocene. *Holocene* 10: 287–291.

Cullen HM, *et al.* (2000) Climate change and the collapse of the Akkadian Empire: evidence from the deep sea. *Geology* 28: 379–382.

Dansgaard W, Johnson SJ, and Clausen HB (1993) Evidence for general instability of past climate from a 250-kyr ice-core record. *Nature* 364: 218–220.

deMenocal P, Ortiz J, Guilderson T, and Sarnthein M (2000) Coherent high- and low-latitude climate variability during the Holocene warm period. *Science* 288: 2198–2202.

Keigwin LD (1996) The Little Ice Age and Medieval warm period in the Sargasso sea. *Science* 274: 1504–1507.

Maslin MA, Seidov D, and Lowe J (2001) *Synthesis of the nature and causes of sudden climate transitions during the Quaternary. AGU Monograph: Oceans and Rapid Past and Future Climate Changes: North–South Connections.* Washington DC: American Geophysical. Union No. 119.

O'Brien SR, Mayewski A, and Meeker LD (1996) Complexity of Holocene climate as reconstructed from a Greenland ice core. *Science* 270: 1962–1964.

Pearce RB, Kemp AES, Koizumi I, Pike J, Cramp A, and Rowland SJ (1998) A lamina-scale, SEM-based study of a late quaternary diatom-ooze sapropel from the Mediterranean ridge, Site 971. In: Robertson AHF, Emeis K-C, Richter C, and Camerlenghi A (eds.) *Proceedings of the Ocean Drilling Program*, Scientific Results 160, 349–363.

Peiser BJ (1998) Comparative analysis of late Holocene environmental and social upheaval: evidence for a disaster around 4000 BP. In: Peiser BJ, Palmer T, and Bailey M (eds.) *Natural Catastrophes during Bronze Age Civilisations*, 117–139.

Pike J and Kemp AES (1997) Early Holocene decadal-scale ocean variability recorded in Gulf of California laminated sediments. *Paleoceanography* 12: 227–238.

Seidov D and Maslin M (1999) North Atlantic Deep Water circulation collapse during the Heinrich events. *Geology* 27: 23–26.

Seidov D, Barron E, Haupt BJ, and Maslin MA (2001) *Meltwater and the ocean conveyor: past, present and future of the ocean bi-polar seesaw. AGU Monograph: Oceans and Rapid Past and Future Climate Changes: North–South Connections*. Washington DC: American Geophysical. Union No. 119..

Wunsch C (2000) On sharp spectral lines in the climate record and millennial peak. *Paleoceanography* 15: 417–424.

PLANKTON AND CLIMATE

A. J. Richardson, University of Queensland,
St. Lucia, QLD, Australia

Introduction: The Global Importance of Plankton

Unlike habitats on land that are dominated by massive immobile vegetation, the bulk of the ocean environment is far from the seafloor and replete with microscopic drifting primary producers. These are the phytoplankton, and they are grazed by microscopic animals known as zooplankton. The word 'plankton' derives from the Greek *planktos* meaning 'to drift' and although many of the phytoplankton (with the aid of flagella or cilia) and zooplankton swim, none can progress against currents. Most plankton are microscopic in size, but some such as jellyfish are up to 2 m in bell diameter and can weigh up to 200 kg. Plankton communities are highly diverse, containing organisms from almost all kingdoms and phyla.

Similar to terrestrial plants, phytoplankton photosynthesize in the presence of sunlight, fixing CO_2 and producing O_2. This means that phytoplankton must live in the upper sunlit layer of the ocean and obtain sufficient nutrients in the form of nitrogen and phosphorus for growth. Each and every day, phytoplankton perform nearly half of the photosynthesis on Earth, fixing more than 100 million tons of carbon in the form of CO_2 and producing half of the oxygen we breathe as a byproduct.

Photosynthesis by phytoplankton directly and indirectly supports almost all marine life. Phytoplankton are a major food source for fish larvae, some small surface-dwelling fish such as sardine, and shoreline filter-feeders such as mussels and oysters. However, the major energy pathway to higher trophic levels is through zooplankton, the major grazers in the oceans. One zooplankton group, the copepods, is so numerous that they are the most abundant multicellular animals on Earth, outnumbering even insects by possibly 3 orders of magnitude. Zooplankton support the teeming multitudes higher up the food web: fish, seabirds, penguins, marine mammals, and turtles. Carcasses and fecal pellets of zooplankton and uneaten phytoplankton slowly yet consistently rain down on the cold dark seafloor, keeping alive the benthic (bottom-dwelling) communities of sponges, anemones, crabs, and fish.

Phytoplankton impact human health. Some species may become a problem for natural ecosystems and humans when they bloom in large numbers and produce toxins. Such blooms are known as harmful algal blooms (HABs) or red tides. Many species of zooplankton and shellfish that feed by filtering seawater to ingest phytoplankton may incorporate these toxins into their tissues during red-tide events. Fish, seabirds, and whales that consume affected zooplankton and shellfish can exhibit a variety of responses detrimental to survival. These toxins can also cause amnesic, diarrhetic, or paralytic shellfish poisoning in humans and may require the closure of aquaculture operations or even wild fisheries.

Despite their generally small size, plankton even play a major role in the pace and extent of climate change itself through their contribution to the carbon cycle. The ability of the oceans to act as a sink for CO_2 relies largely on plankton functioning as a 'biological pump'. By reducing the concentration of CO_2 at the ocean surface through photosynthetic uptake, phytoplankton allow more CO_2 to diffuse into surface waters from the atmosphere. This process continually draws CO_2 into the oceans and has helped to remove half of the CO_2 produced by humans from the atmosphere and distributed it into the oceans. Plankton play a further role in the biological pump because much of the CO_2 that is fixed by phytoplankton and then eaten by zooplankton sinks to the ocean floor in the bodies of uneaten and dead phytoplankton, and zooplankton fecal pellets. This carbon may then be locked up within sediments.

Phytoplankton also help to shape climate by changing the amount of solar radiation reflected back to space (the Earth's albedo). Some phytoplankton produce dimethylsulfonium propionate, a precursor of dimethyl sulfide (DMS). DMS evaporates from the ocean, is oxidized into sulfate in the atmosphere, and then forms cloud condensation nuclei. This leads to more clouds, increasing the Earth's albedo and cooling the climate.

Without these diverse roles performed by plankton, our oceans would be desolate, polluted, virtually lifeless, and the Earth would be far less resilient to the large quantities of CO_2 produced by humans.

Beacons of Climate Change

Plankton are ideal beacons of climate change for a host of reasons. First, plankton are ecthothermic (their body temperature varies with the surroundings), so their physiological processes such as nutrient uptake, photosynthesis, respiration, and reproductive development are highly sensitive to temperature, with their speed doubling or tripling with a $10\,°C$ temperature rise. Global warming is thus likely to directly impact the pace of life in the plankton. Second, warming of surface waters lowers its density, making the water column more stable. This increases the stratification, so that more energy is required to mix deep nutrient-rich water into surface layers. It is these nutrients that drive surface biological production in the sunlit upper layers of the ocean. Thus global warming is likely to increase the stability of the ocean and diminish nutrient enrichment and reduce primary productivity in large areas of the tropical ocean. There is no such direct link between temperature and nutrient enrichment in terrestrial systems. Third, most plankton species are short-lived, so there is tight coupling between environmental change and plankton dynamics. Phytoplankton have lifespans of days to weeks, whereas land plants have lifespans of years. Plankton systems will therefore respond rapidly, whereas it takes longer before terrestrial plants exhibit changes in abundance attributable to climate change. Fourth, plankton integrate ocean climate, the physical oceanic and atmospheric conditions that drive plankton productivity. There is a direct link between climate and plankton abundance and timing. Fifth, plankton can show dramatic changes in distribution because they are free floating and most remain so their entire life. They thus respond rapidly to changes in temperature and oceanic currents by expanding and contracting their ranges. Further, as plankton are distributed by currents and not by vectors or pollinators, their dispersal is less dependent on other species and more dependent on physical processes. By contrast, terrestrial plants are rooted to their substrate and are often dependent upon vectors or pollinators for dispersal. Sixth, unlike other marine groups such as fish and many intertidal organisms, few plankton species are commercially exploited so any long-term changes can more easily be attributed to climate change. Last, almost all marine life has a planktonic stage in their life cycle because ocean currents provide an ideal mechanism for dispersal over large distances. Evidence suggests that these mobile life stages known as meroplankton are even more sensitive to climate change than the holoplankton, their neighbors that live permanently in the plankton.

All of these attributes make plankton ideal beacons of climate change. Impacts of climate change on plankton are manifest as predictable changes in the distribution of individual species and communities, in the timing of important life cycle events or phenology, in abundance and community structure, through the impacts of ocean acidification, and through their regulation by climate indices. Because of this sensitivity and their global importance, climate impacts on plankton are felt throughout the ecosystems they support.

Changes in Distribution

Plankton have exhibited some of the fastest and largest range shifts in response to global warming of any marine or terrestrial group. The general trend, as on land, is for plants and animals to expand their ranges poleward as temperatures warm. Probably the clearest examples are from the Northeast Atlantic. Members of a warm temperate assemblage have moved more than 1000-km poleward over the last 50 years (**Figure 1**). Concurrently, species of a subarctic (cold-water) assemblage have retracted to higher latitudes. Although these translocations have been associated with warming in the region by up to $1\,°C$, they may also be a consequence of the stronger northward flowing currents on the European shelf edge. These shifts in distribution have had dramatic impacts on the food web of the North Sea. The cool water assemblage has high biomass and is dominated by large species such as *Calanus finmarchicus*. Because this cool water assemblage retracts north as waters warm, *C. finmarchicus* is replaced by *Calanus helgolandicus*, a dominant member of the warm-water assemblage. This assemblage typically has lower biomass and contains relatively small species. Despite these *Calanus* species being indistinguishable to all but the most trained eye, the two species contrast starkly in their seasonal cycles: *C. finmarchicus* peaks in spring whereas *C. helgolandicus* peaks in autumn. This is critical as cod, which are traditionally the most important fishery of the North Sea, spawn in spring. As cod eggs hatch into larvae and continue to grow, they require good food conditons, consisting of large copepods such as *C. finmarchicus*, otherwise mortality is high and recruitment is poor. In recent warm years, however, *C. finmarchicus* is rare, there is very low copepod biomass during spring, and cod recruitment has crashed.

Changes in Phenology

Phenology, or the timing of repeated seasonal activities such as migrations or flowering, is very sensitive to global warming. On land, many events in spring are happening earlier in the year, such as the arrival of swallows in the UK, emergence of butterflies in the US, or blossoming of cherry trees in Japan. Recent

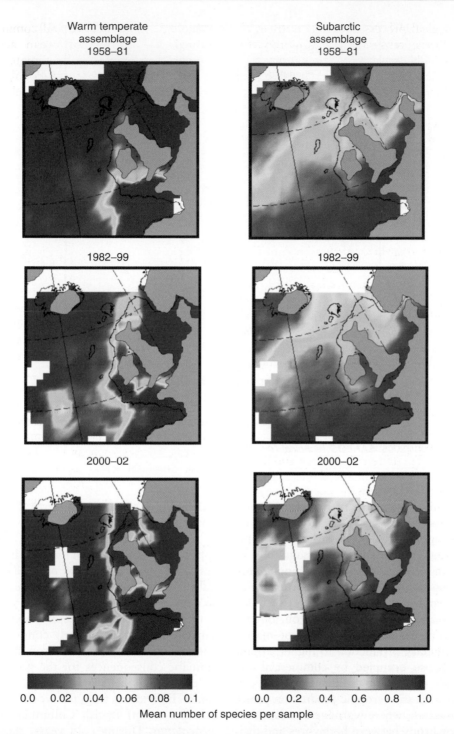

Warm temperate
assemblage
1958–81

Subarctic
assemblage
1958–81

1982–99

1982–99

2000–02

2000–02

0.0   0.02   0.04   0.06   0.08   0.1

0.0   0.2   0.4   0.6   0.8   1.0

Mean number of species per sample

Figure 1 The northerly shift of the warm temperate assemblage (including *Calanus helgolandicus*) into the North Sea and retraction of the subarctic assemblage (including *Calanus finmarchicus*) to higher latitudes. Reproduced by permission from Gregory Beaugrand.

evidence suggests that phenological changes in plankton are greater than those observed on land. Larvae of benthic echinoderms in the North Sea are now appearing in the plankton 6 weeks earlier than they did 50 years ago, and this is in response to warmer temperatures of less than 1 °C. In echinoderms,

temperature stimulates physiological developments and larval release. Other meroplankton such as larvae of fish, cirrepedes, and decapods have also responded similarly to warming (**Figure 2**).

Timing of peak abundance of plankton can have effects that resonate to higher trophic levels. In the

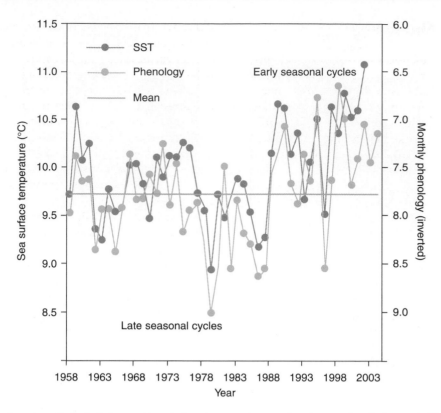

Figure 2 Monthly phenology (timing) of decapod larval abundance and sea surface temperature in the central North Sea from 1958 to 2004. Reproduced from Edwards M, Johns DG, Licandro P, John AWG, and Stevens DP (2006) Ecological status report: Results from the CPR Survey 2004/2005. *SAHFOS Technical Report* 3: 1–8.

North Sea, the timing each year of plankton blooms in summer over the last 50 years has advanced, with phytoplankton appearing 23 days earlier and copepods 10 days earlier. The different magnitude of response between phytoplankton and zooplankton may lead to a mismatch between successive trophic levels and a change in the synchrony of timing between primary and secondary production. In temperate marine systems, efficient transfer of marine primary and secondary production to higher trophic levels, such as those occupied by commercial fish species, is largely dependent on the temporal synchrony between successive trophic production peaks. This type of mismatch, where warming has disturbed the temporal synchrony between herbivores and their plant food, has been noted in other biological systems, most notably between freshwater zooplankton and diatoms, great tits and caterpillar biomass, flycatchers and caterpillar biomass, winter moth and oak bud burst, and the red admiral butterfly and stinging nettle. Such mismatches compromise herbivore survival.

Dramatic ecosystem repercussions of climate-driven changes in phenology are also evident in the subarctic North Pacific Ocean. Here a single copepod species, *Neocalanus plumchrus*, dominates the zooplankton biomass. Its vertical distribution and development are both strongly seasonal and result in an ephemeral (2-month duration) annual peak in upper ocean zooplankton biomass in late spring. The timing of this annual maximum has shifted dramatically over the last 50 years, with peak biomass about 60 days earlier in warm than cold years. The change in timing is a consequence of faster growth and enhanced survivorship of early cohorts in warm years. The timing of the zooplankton biomass peak has dramatic consequences for the growth performance of chicks of the planktivorous seabird, Cassin's auklet. Individuals from the world's largest colony of this species, off British Columbia, prey heavily on *Neocalanus*. During cold years, there is synchrony between food availability and the timing of breeding. During warm years, however, spring is early and the duration of overlap of seabird breeding and *Neocalanus* availability in surface waters is small, causing a mismatch between prey and predator populations. This compromises the reproductive performance of Cassin's auklet in warm years compared to cold years. If Cassin's auklet does not adapt to the changing food conditions, then global warming will place severe strain on its long-term survival.

Changes in Abundance

The most striking example of changes in abundance in response to long-term warming is from foraminifera in the California Current. This plankton group is valuable for long-term climate studies because it is more sensitive to hydrographic conditions than to predation from higher trophic levels. As a result, its temporal dynamics can be relatively easily linked to changes in climate. Foraminifera are also well preserved in sediments, so a consistent time series of observations can be extended back hundreds of years. Records in the California Current show increasing numbers of tropical/subtropical species throughout the twentieth century reflecting a warming trend, which is most dramatic after the 1960s (**Figure 3**). Changes in the foraminifera record echo not only increase in many other tropical and subtropical taxa in the California Current over the last few decades, but also decrease in temperate species of algae, zooplankton, fish, and seabirds.

Changes in abundance through alteration of enrichment patterns in response to enhanced stratification is often more difficult to attribute to climate change than are shifts in distribution or phenology, but may have greater ecosystem consequences. An illustration from the Northeast Atlantic highlights the role that global warming can have on stratification and thus plankton abundances. In this region, phytoplankton become more abundant when cooler regions warm, probably because warmer temperatures boost metabolic rates and enhance stratification in these often windy, cold, and well-mixed regions.

But phytoplankton become less common when already warm regions get even warmer, probably because warm water blocks nutrient-rich deep water from rising to upper layers where phytoplankton live. This regional response of phytoplankton in the North Atlantic is transmitted up the plankton food web. When phytoplankton bloom, both herbivorous and carnivorous zooplankton become more abundant, indicating that the plankton food web is controlled from the 'bottom up' by primary producers, rather than from the 'top down' by predators. This regional response to climate change suggests that the distribution of fish biomass will change in the future, as the amount of plankton in a region is likely to influence its carrying capacity of fish. Climate change will thus have regional impacts on fisheries.

There is some evidence that the frequency of HABs is increasing globally, although the causes are uncertain. The key suspect is eutrophication, particularly elevated concentrations of the nutrients nitrogen and phosphorus, which are of human origin and discharged into our oceans. However, recent evidence from the North Sea over the second half of the twentieth century suggests that global warming may also have a key role to play. Most areas of the North Sea have shown no increase in HABs, except off southern Norway where there have been more blooms. This is primarily a consequence of the enhanced stratification in the area caused by warmer temperatures and lower salinity from meltwater. In the southern North Sea, the abundance of two key HAB species over the last 45 years is positively related to warmer ocean temperatures. This work

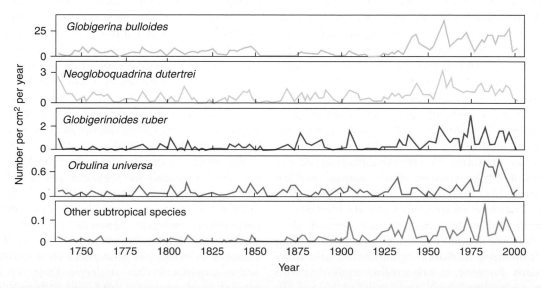

Figure 3 Fluxes of tropical/planktonic foraminifera in Santa Barbara Basin (California Current). Tropical/subtropical foraminifera showing increased abundance in the twentieth century. Reproduced from Field DB, Baumgartner TR, Charles CD, Ferreira-Bartrina V, and Ohman M (2006) Planktonic foraminifera of the California Current reflect 20th century warming. *Science* 311: 63–66.

supports the notion that the warmer temperatures and increased meltwater runoff anticipated under projected climate change scenarios are likely to increase the frequency of HABs.

Although most evidence for changes in abundance in response to climate change are from the Northern Hemisphere because this is where most (plankton) science has concentrated, there is a striking example from waters around Antarctica. Over the last 30 years, there has been a decline in the biomass of krill *Euphausia superba* in the Southern Ocean that is a consequence of warmer sea and air temperatures. In many areas, krill has been replaced by small gelatinous filter-feeding sacs known as salps, which occupy the less-productive, warmer regions of the Southern Ocean. The decline in krill is likely to be a consequence of warmer ocean temperatures impacting sea ice. It is not only that sea ice protects krill from predation, but also the algae living beneath the sea ice and photosynthesizing from the dim light seeping through are a critical food source for krill. As waters have warmed, the extent of winter sea ice and its duration have declined, and this has led to a deterioration in krill density since the 1970s. As krill are major food items for baleen whales, penguins, seabirds, fish, and seals, their declining population may have severe ramifications for the Southern Ocean food web.

Impact of Acidification

A direct consequence of enhanced CO_2 levels in the ocean is a lowering of ocean pH. This is a consequence of elevated dissolved CO_2 in seawater altering the carbonate balance in the ocean, releasing more hydrogen ions into the water and lowering pH. There has been a drop of 0.1 pH units since the Industrial Revolution, representing a 30% increase in hydrogen ions.

Impacts of ocean acidification will be greatest for plankton species with calcified (containing calcium carbonate) shells, plates, or scales. For organisms to build these structures, seawater has to be supersaturated in calcium carbonate. Acidification reduces the carbonate saturation of the seawater, making calcification by organisms more difficult and promoting dissolution of structures already formed.

Calcium carbonate structures are present in a variety of important plankton groups including coccolithophores, mollusks, echinoderms, and some crustaceans. But even among marine organisms with calcium carbonate shells, susceptibility to acidification varies depending on whether the crystalline form of their calcium carbonate is aragonite or calcite. Aragonite is more soluble under acidic conditions than calcite, making it more susceptible to dissolution. As oceans absorb more CO_2, undersaturation of aragonite and calcite in seawater will be initially most acute in the Southern Ocean and then move northward.

Winged snails known as pteropods are probably the plankton group most vulnerable to ocean acidification because of their aragonite shell. In the Southern Ocean and subarctic Pacific Ocean, pteropods are prominent components of the food web, contributing to the diet of carnivorous zooplankton, myctophids, and other fish and baleen whales, besides forming the entire diet of gymnosome mollusks. Pteropods in the Southern Ocean also account for the majority of the annual flux of both carbonate and organic carbon exported to ocean depths. Because these animals are extremely delicate and difficult to keep alive experimentally, precise pH thresholds where deleterious effects commence are not known. However, even experiments over as little as 48 h show shell deterioration in the pteropod *Clio pyrimidata* at CO_2 levels approximating those likely around 2100 under a business-as-usual emissions scenario. If pteropods cannot grow and maintain their protective shell, their populations are likely to decline and their range will contract toward lower-latitude surface waters that remain supersaturated in aragonite, if they can adapt to the warmer temperature of the waters. This would have obvious repercussions throughout the food web of the Southern Ocean.

Other plankton that produce calcite such as foraminifera (protist plankton), mollusks other than pteropods (e.g., squid and mussel larvae), coccolithophores, and some crustaceans are also vulnerable to ocean acidification, but less so than their cousins with aragonite shells. Particularly important are coccolithophorid phytoplankton, which are encased within calcite shells known as liths. Coccolithophores export substantial quantities of carbon to the seafloor when blooms decay. Calcification rates in these organisms diminish as water becomes more acidic (**Figure 4**).

A myriad of other key processes in phytoplankton are also influenced by seawater pH. For example, pH is an important determinant of phytoplankton growth, with some species being catholic in their preferences, whereas growth of other species varies considerably between pH of 7.5 and 8.5. Changes in ocean pH also affect chemical reactions within organisms that underpin their intracellular physiological processes. pH will influence nutrient uptake kinetics of phytoplankton. These effects will have repercussions for phytoplankton community

Emiliania huxleyi   *Gephyrocapsa oceanica*

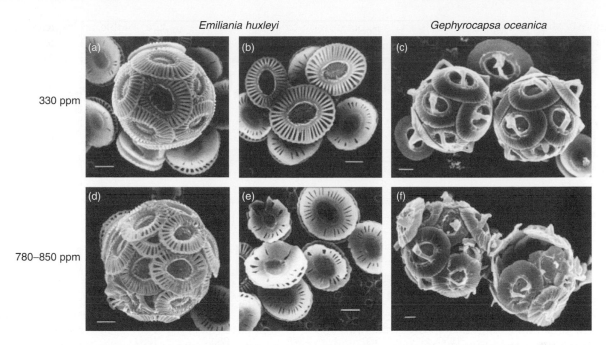

Figure 4 Scanning electron microscopy photographs of the coccolithophores *Emiliania huxleyi* and *Gephyrocapsa oceanica* collected from cultures incubated at CO_2 levels of about 300 and 780–850 ppm. Note the difference in the coccolith structure (including distinct malformations) and in the degree of calcification of cells grown at normal and elevated CO_2 levels. Scale bar = 1 mm. Reprinted by permission from Macmillan Publishers Ltd., *Nature*, Riebesell U, Zondervan I, Rost B, Tortell PD, Zeebe RE, and Morel FMM, Reduced calcification of marine plankton in response to increased atmospheric CO_2, 407: 364–376, Copyright (2000).

composition and productivity, with flow-on effects to higher trophic levels.

Climate Variability

Many impacts of climate change are likely to act through existing modes of variability in the Earth's climate system, including the well-known El Niño/Southern Oscillation (ENSO) and the North Atlantic Oscillation (NAO). Such large synoptic pressure fields alter regional winds, currents, nutrient dynamics, and water temperatures. Relationships between integrative climate indices and plankton composition, abundance, or productivity provide an insight into how climate change may affect ocean biology in the future.

ENSO is the strongest climate signal globally, and has its clearest impact on the biology of the tropical Pacific Ocean. Observations from satellite over the past decade have shown a dramatic global decline in primary productivity. This trend is caused by enhanced stratification in the low-latitude oceans in response to more frequent El Niño events. During an El Niño, upper ocean temperatures warm, thereby enhancing stratification and reducing the availability of nutrients for phytoplankton

growth. Severe El Niño events lead to alarming declines in phytoplankton, fisheries, marine birds and mammals in the tropical Pacific Ocean. Of concern is the potential transition to more frequent El Niño-like conditions predicted by some climate models. In such circumstances, enhanced stratification across vast areas of the tropical ocean may reduce primary productivity, decimating fish, mammal, and bird populations. Although it is unknown whether the recent decline in primary productivity is already a consequence of climate change, the findings and underlying understanding of climate variability are likely to provide a window to the future.

Further north in the Pacific, the Pacific Decadal Oscillation (PDO) has a strong multi-decadal signal, longer than the ENSO period of a few years. When the PDO is negative, upwelling winds strengthen over the California Current, cool ocean conditions prevail in the Northeast Pacific, copepod biomass in the region is high and is dominated by large cool-water species, and fish stocks such as coho salmon are abundant (**Figure 5**). By contrast, when the PDO is positive, upwelling diminishes and warm conditions exist, the copepod biomass declines and is dominated by small less-nutritious species, and the abundance of coho salmon plunges.

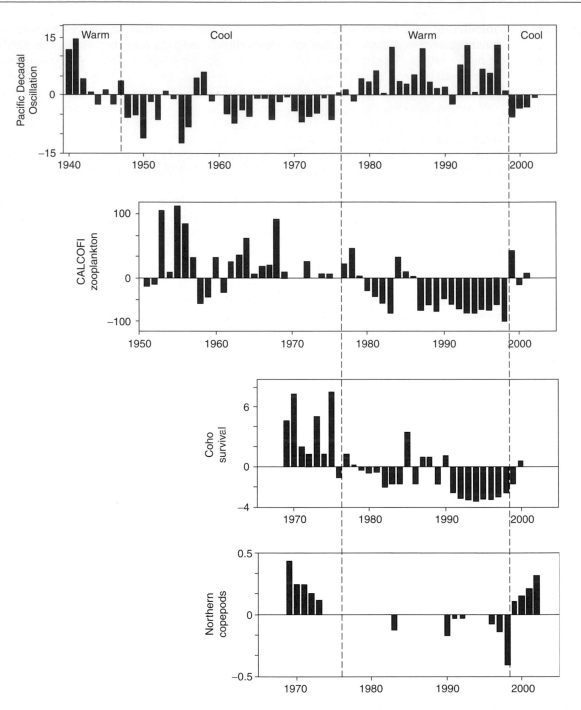

Figure 5   Annual time series in the Northeast Pacific: the PDO index from May to September; anomalies of zooplankton biomass (displacement volumes) from the California Current region (CALCOFI zooplankton); anomalies of coho salmon survival; and biomass anomalies of cold-water copepod species (northern copepods). Positive (negative) PDO index indicates warmer (cooler) than normal temperatures in coastal waters off North America. Reproduced from Peterson WT and Schwing FB (2003) A new climate regime in Northeast Pacific ecosystems. *Geophysical Research Letters* 30(17): 1896 (doi:10.1029/2003GL017528).

These transitions between alternate states have been termed regime shifts. It is possible that if climate change exceeds some critical threshold, some marine systems will switch permanently to a new state that is less favorable than present.

In Hot Water: Consequences for the Future

With plankton having relatively simple behavior, occurring in vast numbers, and amenable to experimental manipulation and automated measurements, their dynamics are far more easily studied and modeled than higher trophic levels. These attributes make it easier to model potential impacts of climate change on plankton communities. Many of our insights gained from such models confirm those already observed from field studies.

The basic dynamics of plankton communities have been captured by nutrient–phytoplankton–zooplankton (NPZ) models. Such models are based on a functional group representation of plankton communities, where species with similar ecological function are grouped into guilds to form the basic biological units in the model. Typical functional groups represented include diatoms, dinoflagellates, coccolithophores, microzooplankton, and mesozooplankton. There are many global NPZ models constructed by different research teams around the world. These are coupled to global climate models (GCMs) to provide future projections of the Earth's climate system. In this way, alternative carbon dioxide emission scenarios can be used to investigate possible future states of the ocean and the impact on plankton communities.

One of the most striking and worrisome results from these models is that they agree with fieldwork that has shown general declines in lower trophic levels globally as a result of large areas of the surface tropical ocean becoming more stratified and nutrient-poor as the oceans heat up (see sections titled 'Changes in abundance' and 'Climate variability'). One such NPZ model projects that under a middle-of-the-road emissions scenario, global primary productivity will decline by 5–10% (**Figure 6**). This trend will not be uniform, with increases in productivity by 20–30% toward the Poles, and marked declines in the warm stratified tropical ocean basins. This and other models show that warmer, more-stratified conditions in the Tropics will reduce nutrients in surface waters and lead to smaller phytoplankton cells dominating over larger diatoms. This will lengthen food webs and ultimately support fewer fish, marine mammals, and seabirds, as more trophic linkages are needed to transfer energy from small phytoplankton to higher trophic levels and 90% of the energy is lost within each trophic level through respiration. It also reduces the oceanic uptake of CO_2 by lowering the efficiency of the

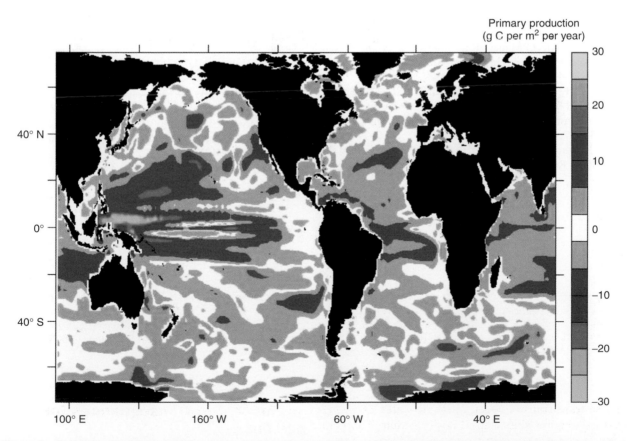

Figure 6 Change in primary productivity of phytoplankton between 2100 and 1990 estimated from an NPZ model. There is a global decline in primary productivity by 5–10%, with an increase at the Poles of 20–30%. Reproduced from Bopp L, Monfray P, Aumont O, *et al.* (2001) Potential impact of climate change on marine export production. *Global Biogeochemical Cycles* 15: 81–99.

biological pump. This could cause a positive feedback between climate change and the ocean carbon cycle: more CO_2 in the atmosphere leads to a warmer and more stratified ocean, which supports less and smaller plankton, and results in less carbon being drawn from surface ocean layers to deep waters. With less carbon removed from the surface ocean, less CO_2 would diffuse into the ocean and more CO_2 would accumulate in the atmosphere.

It is clear that plankton are beacons of climate change, being extremely sensitive barometers of physical conditions. We also know that climate impacts on plankton reverberate throughout marine ecosystems. More than any other group, they also influence the pace and extent of climate change. The impact of climate change on plankton communities will not only determine the future trajectory of marine ecosystems, but the planet.

Further Reading

Atkinson A, Siegel V, Pakhomov E, and Rothery P (2004) Long-term decline in krill stock and increase in salps within the Southern Ocean. *Nature* 432: 100–103.

Beaugrand G, Reid PC, Ibanez F, Lindley JA, and Edwards M (2002) Reorganisation of North Atlantic marine copepod biodiversity and climate. *Science* 296: 1692–1694.

Behrenfield MJ, O'Malley RT, Siegel DA, *et al.* (2006) Climate-driven trends in contemporary ocean productivity. *Nature* 444: 752–755.

Bertram DF, Mackas DL, and McKinnell SM (2001) The seasonal cycle revisited: Interannual variation and ecosystem consequences. *Progress in Oceanography* 49: 283–307.

Bopp L, Aumont O, Cadule P, Alvain S, and Gehlen M (2005) Response of diatoms distribution to global warming and potential implications: A global model study. *Geophysical Research Letters* 32: L19606 (doi:10.1029/2005GL023653).

Bopp L, Monfray P, Aumont O, *et al.* (2001) Potential impact of climate change on marine export production. *Global Biogeochemical Cycles* 15: 81–99.

Edwards M, Johns DG, Licandro P, John AWG, and Stevens DP (2006) Ecological status report: Results from the CPR Survey 2004/2005. *SAHFOS Technical Report* 3: 1–8.

Edwards M and Richardson AJ (2004) The impact of climate change on the phenology of the plankton community and trophic mismatch. *Nature* 430: 881–884.

Field DB, Baumgartner TR, Charles CD, Ferreira-Bartrina V, and Ohman M (2006) Planktonic forminifera of the California Current reflect 20th century warming. *Science* 311: 63–66.

Hays GC, Richardson AJ, and Robinson C (2005) Climate change and plankton. *Trends in Ecology and Evolution* 20: 337–344.

Peterson WT and Schwing FB (2003) A new climate regime in Northeast Pacific ecosystems. *Geophysical Research Letters* 30(17): 1896 (doi:10.1029/2003GL017528).

Raven J, Caldeira K, Elderfield II, *et al.* (2005) *Royal Society Special Report: Ocean Acidification Due to Increasing Atmospheric Carbon Dioxide*. London: The Royal Society.

Richardson AJ (2008) In hot water: Zooplankton and Climate change. *ICES Journal of Marine Science* 65: 279–295.

Richardson AJ and Schoeman DS (2004) Climate impact on plankton ecosystems in the Northeast Atlantic. *Science* 305: 1609–1612.

Riebesell U, Zondervan I, Rost B, Tortell PD, Zeebe RE, and Morel FMM (2000) Reduced calcification of marine plankton in response to increased atmospheric CO_2. *Nature* 407: 364–367.

COCCOLITHOPHORES

T. Tyrrell, National Oceanography Centre, Southampton, UK
J. R. Young, The Natural History Museum, London, UK

Introduction

Coccolithophores (**Figure 1**) are a group of marine phytoplankton belonging to the division Haptophyta. Like the other free-floating marine plants (phytoplankton), the coccolithophores are microscopic (they range in size between about 0.003 and 0.040 mm diameter) single-celled organisms which obtain their energy from sunlight. They are typically spherical in shape. They are distinguished from other phytoplankton by their construction of calcium carbonate ($CaCO_3$) plates (called coccoliths) with which they surround their cells. While not quite the only phytoplankton to use $CaCO_3$ (there are also some calcareous dinoflagellates), they are by far the most numerous; indeed they are one of the most abundant of phytoplankton groups, comprising in the order of 10% of total global phytoplankton biomass.

The first recorded observations of coccoliths were made in 1836 by Christian Gottfried Ehrenberg, a founding figure in micropaleontology. The name 'coccoliths' (Greek for 'seed-stones') was coined by Thomas Henry Huxley (famous as 'Darwin's bulldog') in 1857 as he studied marine sediment samples. Both Ehrenberg and Huxley attributed coccoliths to an inorganic origin. This was soon challenged by Henry Clifton Sorby and George Charles Wallich who inferred from the complexity of coccoliths that they must be of biological origin, and supported this with observations of groups of coccoliths aggregated into empty spheres.

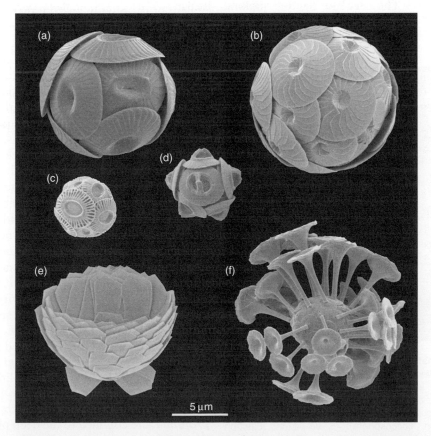

Figure 1 Electron microscope images of some major coccolithophore species: (a) *Coccolithus pelagicus*, (b) *Calcidiscus quadriperforatus*, (c) *Emiliania huxleyi*, (d) *Gephyrocapsa oceanica*, (e) *Florisphaera profunda*, (f) *Discosphaera tubifera*.

Species and Distribution

Approximately 200 species of coccolithophore have been formally described, separated into 65 genera. However, the true number of authentic modern coccolithophore species is rather unclear, for a couple of reasons. First, it is now realized that pairs of species, previously thought to be distinct and rather unrelated, are actually different life-cycle stages of the same species; coccolithophores typically have life cycles in which the haploid (single set of chromosomes, as in sex cells) and diploid (double set of chromosomes) phases can form different coccolith types: 'heterococcoliths' during the diploid life stage, and 'holococcoliths' during the haploid life stage. This type of life cycle has long been known from classic studies of laboratory cultures. It has only recently been appreciated that it is a very widespread pattern, as a result of observations of combination coccospheres representing the transition between the two life-cycle phases, that is to say possessing half a covering of heterococcoliths and half a covering of holococcoliths (**Figure 2**). Fifty of the 200 described coccolithophores are taxa known only from their holococcolith-producing phase and so may prove to be part of the life cycle of a heterococcolith-producing species. The second factor making

diversity estimates difficult is that recent research combining studies of fine-scale morphology, biogeography, and molecular genetics has suggested that many described species are actually clusters of a few closely related, but genetically distinct, species. Indeed, as a result of such studies, numerous additional morphotypes have been recognized and await formal description.

Coccolithophores occur widely throughout the world's oceans, with the exception of the very-high-latitude polar oceans. Most individual species have more restricted biogeographical ranges than the range of the group as a whole, but still typically have interoceanic distributions. Unlike diatoms (the other major group of phytoplankton that make hard mineral shields), they are absent from almost all freshwater rivers and lakes. They occur in the brackish (more saline than freshwater but less saline than seawater) Black Sea, but not in the brackish Baltic Sea. In contrast once more to diatoms, coccolithophores are almost exclusively planktonic. There are very few bottom-dwelling species, even at shallow depths experiencing adequate light levels. Most species today live in warm, nutrient-poor conditions of the subtropical oceanic gyres, where they form a prominent component of the phytoplankton; there are fewer species that inhabit coastal and temperate or subpolar waters.

Emiliania huxleyi (**Figure 1(c)**) is the best-known species, primarily because it forms intense blooms which are clearly visible in satellite images, appearing as pale turquoise swirls in the ocean (**Figure 3**). While *E. huxleyi* frequently dominates phytoplankton counts in seawater samples, at least in terms of numbers of cells, their cells (and therefore also the coccoliths that surround them) are rather small, with the cells about 5 μm across and the coccoliths about 3 μm long. No other coccolithophore species regularly forms blooms, although occasional blooms of other species, for instance *Gephyrocapsa oceanica* and *Coccolithus pelagicus*, have been recorded. Many other species, for example *Calcidiscus quadriperforatus* and *Umbilicosphaera sibogae*, are most successful in low-productivity waters but do not bloom there. Although these species are almost always much less numerous than *E. huxleyi* in water samples, they are on the other hand also significantly larger, with typical cell diameters greater than 10 μm and correspondingly larger coccoliths. Most species of coccolithophores are adapted to life in the surface mixed layer, but some species, such as *Florisphaera profunda*, are confined to the deep photic zone where they make an important contribution to the 'shade flora'.

Figure 2 A combination coccosphere, upper half (inner layer) heterococcoliths, lower half (outer layer) holococcoliths, of the species *Calcidiscus quadriperforatus* (the two stages were previously regarded as two separate species – *Calcidiscus leptoporus* and *Syracolithus quadriperforatus* – prior to discovery of this combination coccosphere). Scale = 1 μm. Electron microscope image provided by Markus Geisen (Alfred-Wegener-Institute, Germany).

15 June 1998

SeaWiFS Project NASA / GSFC ORBIMAGE

Figure 3 *SeaWiFS* satellite image from 15 June 1998 of *E. huxleyi* blooms (the turquoise patches) along the west coast of Norway and to the southwest of Iceland. The perspective is from a point over the Arctic Ocean, looking southward down the North Atlantic. The Greenland ice sheet is visible in the center foreground. Imagery provided with permission by GeoEye and NASA SeaWiFS Project.

In oligotrophic surface waters, typical concentrations of coccolithophores are in the range 5000–50 000 cells per liter. To put this in context, a teaspoonful (5 ml) of typical surface open ocean seawater will contain between 25 and 250 coccolithophore cells. Blooms of *E. huxleyi* have been defined as concentrations exceeding 1 million cells per liter; the densest bloom ever recorded, in a Norwegian fiord, had a concentration of 115 000 000 cells per liter. Blooms of *E. huxleyi* can cover large areas; the largest ever recorded bloom occurred in June 1998 (see **Figure 3**) in the North Atlantic south of Iceland and covered about 1 million km^2, 4 times the area of the United Kingdom.

Coccoliths

Coccolithophores, and the coccolith shields with which they surround themselves, are incredibly small. And yet, despite their small size, coccoliths are elegant and ornate structures, which, if the water chemistry is suitable, are produced reliably with few malformations. This efficient manufacture occurs at a miniature scale: the diameter of an *E. huxleyi* coccolith 'spoke' (**Figure 1(c)**) is of the order 50 nm, considerably smaller than the wavelength range of visible light (400–700 nm). Calcite is mostly transparent to visible light (unsurprisingly, given that coccolithophores are photosynthetic) and the small coccoliths are often at the limit of discrimination, even under high magnification. However, under cross-polarized light, coccoliths produce distinctive patterns which are closely related to their structure. As a result most coccoliths can be accurately identified by light microscopy. However, the details and beauty of coccoliths can only be properly appreciated using electron microscopy (**Figure 1**).

Coccolithophores synthesize different types of coccoliths during different life-cycle stages. Here we concentrate on the heterococcoliths associated with the diploid life stage. These heterococcoliths are formed from crystal units with complex shapes, in contrast to holococcoliths which are constructed out of smaller and simpler crystal constituents. Coccoliths are typically synthesized intracellularly (within a vesicle), probably one at a time, and subsequently extruded to the cell surface. The time taken to form a single coccolith can be less than 1 h for *E. huxleyi*. Coccoliths continue to be produced until a complete coccosphere covering (made up of maybe 20 coccoliths, depending on species) is produced.

Most coccolithophores construct only as many coccoliths as are required in order to provide a complete single layer around their cell. *Emiliania huxleyi* is unusual in that, under certain conditions, it overproduces coccoliths; many more coccoliths are built than are needed to cover the cell. In these conditions, multiple layers of coccoliths accumulate around the *E. huxleyi* cell until the excessive covering eventually becomes unstable and some of the coccoliths slough off to drift free in the water. The large number of unattached coccoliths accompanying an *E. huxleyi* bloom contributes to a great extent to the turbidity of the water and to the perturbations to optics that make the blooms so apparent from space.

Curiously, the functions of coccoliths are still uncertain. It is probable that a major function is to provide some protection from grazing by zooplankton, but many alternative hypotheses have also been advanced. For instance, the coccoliths may increase the rate of sinking of the cells through the water (and therefore also enhance the rate at which nutrient-containing water flows past the cell surface) or they may provide protection against the entry of viruses or bacteria to the cell. At one time it was thought that coccoliths might provide protection against very high light intensities, which could explain the resistance to photoinhibition apparent in *E. huxleyi*, but various experimental results make this explanation unlikely. One species, *F. profunda*, a member of the deeper 'shade flora', orients its coccoliths in such a way that they conceivably act as a light-focusing apparatus maximizing photon capture in the darker waters it inhabits (**Figure 1(e)**). Some species produce trumpet-like protrusions from each coccolith (**Figure 1(f)**), again for an unknown purpose. Currently there is a paucity of hard data with which to discriminate between the various hypotheses for coccolith function, and the diversity of coccolith morphology makes it likely that they have been adapted to perform a range of functions.

Life Cycle

Many details are still obscure, and data are only available from a limited number of species, but it appears that most coccolithophores alternate between fully armored (heterococcolith-covered) diploid life stages and less-well-armoured (either holococcolith-covered or else naked) haploid phases. Both phases are capable of indefinite asexual reproduction, which is rather unusual among protists. That sexual reproduction also occurs fairly frequently is evidenced by the observation of significant genetic diversity within coccolithophore blooms. Bloom populations do not consist of just one clone (just one genetic variant of the organism). Coccolithophore gametes (haploid stages) are radically different from those of larger (multicellular) organisms in the sense of being equipped for an independent existence: they can move about, acquire energy (photosynthesize), and divide asexually by binary fission. Naked diploid phases can be induced in cultures, but these may be mutations which are not viable in the wild. There are no confirmed identifications of resting spore or cyst stages in coccolithophores.

Coccolithophores, in common with other phytoplankton, experience only an ephemeral existence. Typical life spans of phytoplankton in nature are measured in days. Comparison of the rate at which $CaCO_3$ is being produced in open ocean waters (as measured by the rate of uptake of isotopically labeled carbon), to the amounts present (the 'standing-stock'), has led to the calculation that the average turnover (replacement) time for $CaCO_3$ averages about 3 days, ranging between a minimum of < 1 and a maximum of 7 days at different locations in the Atlantic Ocean. This implies that if a surface-dwelling coccolithophore synthesizes coccoliths on a Monday, the coccoliths are fairly unlikely to still be there on the Friday, either because they have redissolved or else because they have sunk down to deeper waters.

The genome of one species, E. huxleyi, has recently been sequenced, but at the time of writing its analysis is at an early stage.

Calcification

Calcification is the synthesis of solid calcium carbonate from dissolved substances, whether passively by spontaneous formation of crystals in a supersaturated solution (inorganic calcification) or actively through the intervention of organisms (biocalcification). The building of coccoliths by coccolithophores is a major fraction of the total biocalcification taking place in seawater. Inorganic calcification is not commonplace or quantitatively significant in the global budget, with the exception of 'whitings' that occur in just a few unusual locations in the world's oceans, such as the Persian Gulf and the Bahamas Banks. The chemical equation for calcification is

$$Ca^{2+} + 2HCO_3^- \Rightarrow CaCO_3 + H_2O + CO_2$$

Heterococcoliths are constructed out of calcite (a form of calcium carbonate; corals by contrast synthesize aragonite, which has the same chemical composition but a different lattice structure). Heterococcolith calcite typically has a very low magnesium content, making coccoliths relatively dissolution-resistant (susceptibility to dissolution increases with increasing magnesium content).

Dissolved inorganic carbon in seawater is comprised of three different components: bicarbonate ions (HCO_3^-), carbonate ions (CO_3^{2-}), and dissolved CO_2 gas ($CO_2(aq)$), of which it appears that bicarbonate or carbonate ions are taken up to provide the carbon source for $CaCO_3$ (coccoliths have a $\delta^{13}C$ isotopic composition that is very different from dissolved CO_2 gas). The exact physiological mechanisms of calcium and carbon assimilation remain to be established. Calcification (coccolith genesis) is stimulated by light but inhibited in most cases by plentiful nutrients. Separate experiments have found that increased rates of calcification in cultures can be induced by starving the cultures of phosphorus, nitrogen, and zinc. Low levels of magnesium also enhance calcification, and high levels inhibit it, but in this case probably because Mg atoms can substitute for Ca atoms in the crystalline lattice and thereby 'poison' the lattice. Calcification shows the opposite response to levels of calcium, unsurprisingly. Progressive depletion of calcium in the growth medium induces progressively less normal (smaller and more malformed) coccoliths. The calcification to photosynthesis (C:P) ratio in nutrient-replete, Ca-replete cultures is often in the vicinity of 1:1 (i.e., more or less equivalent rates of carbon uptake into the two processes). Low levels of iron appear to depress calcification and photosynthesis equally.

Measurements at sea suggest that the total amount of carbon taken up by the whole phytoplankton community to form new $CaCO_3$ is rather small compared to the total amount of carbon taken up to form new organic matter. Both calcification carbon demand and photosynthetic carbon demand have recently been measured on a long transect in the Atlantic Ocean and the ratio of the two was found to average 0.05; or, in other words, for every 20 atoms of carbon taken up by phytoplankton, only one on average was taken up into solid $CaCO_3$.

Ecological Niche

In addition to our lack of knowledge about the exact benefit of a coccosphere, we also have rather little definite knowledge as to the ecological conditions that favor coccolithophore success. There is certainly variation between species, with some being adapted to relatively eutrophic conditions (although diatoms invariably dominate the main spring blooms in temperate waters, as well as the first blooms in nutrient-rich, recently upwelled waters) and some to oligotrophic conditions. Most species are best adapted to living near to the surface, but some others to the darker conditions prevailing in the thermocline. Most species today live in warm, nutrient-poor, open ocean conditions; the highest diversity occurs in subtropical oceanic gyres, whereas lower diversity occurs in coastal and temperate waters.

Much of our knowledge of coccolithophore physiology and ecology comes from studies of *E. huxleyi*, which has attracted more scientific interest than the other coccolithophore species because of its ease of culturing and the visibility of its blooms from space. The ability to map bloom distributions from space provides unique information on the ecology of this species. Blooms of the species *E. huxleyi* occur preferentially in strongly stratified waters experiencing high light levels. Coccolithophore success may be indirectly promoted by exhaustion of silicate, due to exclusion of the more competitive diatoms. By analogy with diatoms, whose success is contingent on silicate availability for their shell building, coccolithophores might be expected to be more successful at high $CaCO_3$ saturation state $\Omega (= [CO_3^{2-}][Ca^{2+}]/K_{sp})$, because the value of Ω controls inorganic calcification and dissolution. Such a dependency would render coccolithophores vulnerable to ocean acidification, as discussed further below. It was formerly thought that *E. huxleyi* was particularly successful in phosphate-deficient waters, but a reassessment has suggested that this is not a critical factor. Many coccolithophores are restricted to the warmer parts of the oceans, although this may be coincidental rather than due to a direct temperature effect. *Emiliania huxleyi* is found to grow well at low iron concentrations, in culture experiments.

Biogeochemical Impacts

Coccolithophores assimilate carbon during photosynthesis, leading to similar biogeochemical impacts to other phytoplankton that do not possess mineral shells. They also, however, assimilate carbon into biomass.

Following death, some of the coccolith $CaCO_3$ dissolves in the surface waters inhabited by coccolithophores, with the rest of the coccolith $CaCO_3$ sinking out of the surface waters within zooplankton fecal pellets or marine snow aggregates. The exact means by which some coccoliths are dissolved in near-surface waters are unclear (dissolution within zooplankton guts may be important), but regardless of mechanisms several lines of evidence suggest that near-surface dissolution does occur. The size of

coccoliths precludes the likelihood of single coccoliths sinking at all rapidly under gravity, because of the considerable viscosity of water with respect to such small particles (Stokes' law). Stokes' law can be overcome if coccoliths become part of larger aggregates, either marine snow or zooplankton fecal pellets. Another possible fate for coccoliths is to become incorporated into the shells of tintinnid microzooplankton, which when grazing on coccolithophores make use of the coccoliths in their own shells (**Figure 4**). Regardless of their immediate fate, the coccoliths must eventually either dissolve or else sink toward the seafloor.

The construction of $CaCO_3$ coccoliths (calcification) leads to additional impacts, over and above those associated with the photosynthesis carried out by all species. The first and perhaps the most important of these is that $CaCO_3$ contains carbon and the vertical downward flux of coccoliths thereby removes carbon from the surface oceans. It might be expected that this would lead to additional removal of CO_2 from the atmosphere to the oceans, to replace that taken up into coccoliths, but in fact, because of the complex effect of calcification ($CaCO_3$ synthesis) on seawater chemistry, the production of coccoliths actually increases the partial pressure of CO_2 in surface seawater and promotes outgassing rather than ingassing. Determining the exact nature and magnitude of the overall net effect is complicated by a possible additional role of coccoliths as 'ballast' (coccoliths are denser than water and hence when

Figure 4 Tintinnid lorica (casing) with embedded coccoliths.

incorporated into aggregates of particulate fecal material may drag down extra organic carbon into the ocean interior).

Microscopic examination of seafloor sediments (if shallow enough that the $CaCO_3$ does not dissolve) and of material caught in sediment traps has revealed that much of the calcium carbonate in the samples consists of coccoliths. The flux of coccoliths probably accounts for *c.* 50% of the total vertical $CaCO_3$ flux in open ocean waters (in other words, about 50% of the inorganic carbon pump), with foraminifera shells responsible for most of the rest. It is usually not the most numerous species (*E. huxleyi*) but rather larger species (e.g., *Calcidiscus quadriperforatus* and *Coccolithus pelagicus*) that make the greatest contributions to the total coccolith flux.

Coccolithophores also impact on climate in other ways, ones that are unconnected with carbon. Coccolithophores are intense producers of a chemical called dimethylsulfoniopropionate (DMSP). The production of DMSP leads eventually (via several chemical transformations) to additional cloud condensation nuclei in the atmosphere and thereby to increased cloud cover.

Coccoliths also scatter light, polarizing it in the process. They do not reflect or block light (this would clearly be disadvantageous for the photosynthetic cell underneath), but the difference between the refractive indices of water and of calcium carbonate means that the trajectories of photons are deflected by encounters with coccoliths. A small proportion of the scattering (deflection) events are through angles greater than 90°, leading to photons being deflected into upward directions and eventually passing back out through the sea surface. Because of this light-scattering property of coccoliths, their bulk effect is to make the global oceans slightly brighter than they would otherwise be. It has been calculated that the Earth would become slightly dimmer (the albedo of the Earth would decrease by about 0.1% from its average global value of about 30%) were coccolithophores to disappear from the oceans. The effect of coccoliths in enhancing water brightness is seen in its most extreme form during coccolithophore blooms (**Figure 3**).

The Past

Coccolithophores are currently the dominant type of calcifying phytoplankton, but further back in time there were other abundant calcifying phytoplankton, for instance the nannoconids, which may or may not have been coccolithophores. The fossil calcifying phytoplankton are referred to collectively as calcareous nannoplankton.

The first calcareous nannoplankton are seen in the fossil record *c.* 225 Ma, in the late Triassic period. Abundance and biodiversity increased slowly over time, although they were at first restricted to shallow seas. During the early Cretaceous (145–100 Ma), calcareous nannoplankton also colonized the open ocean. They reached their peak, both in terms of abundance and number of different species (different morphotypes) in the late Cretaceous (100–65 Mya). 'The Chalk' was formed at this time, consisting of thick beds of calcium carbonate, predominantly coccoliths. Thick deposits of chalk are most noticeable in various striking sea cliffs, including the white cliffs of Dover in the United Kingdom, and the Isle of Rugen in the Baltic Sea. The chalk deposits were laid down in the shallow seas that were widespread and extensive at that time, because of a high sea level.

Calcareous nannoplankton, along with other biological groups, underwent long intervals of slowly but gradually increasing species richness interspersed with occasional extinction events. Their heyday in the late Cretaceous was brought to an abrupt end by the largest extinction event of all at the K/T boundary (65 Ma), at which point ~93% of all species (~85% of genera) suddenly went extinct. Although biodiversity recovered rapidly in the early Cenozoic, calcareous nannoplankton have probably never since re-attained their late Cretaceous levels.

Because the chemical and isotopic composition of coccoliths is influenced by the chemistry of the seawater that they are synthesized from, coccoliths from ancient sediments have the potential to record details of past environments. Coccoliths are therefore a widely used tool by paleooceanographers attempting to reconstruct the nature of ancient oceans. Some of the various ways in which coccoliths are put to use in interpreting past conditions are as follows: (1) elemental ratios such as Sr/Ca and Mg/Ca are used to infer past seawater chemistry, ocean productivity, and temperatures; (2) the isotopic composition ($\delta^{13}C$, $\delta^{18}O$) of the calcium carbonate is used to infer past carbon cycling, temperatures, and ice volumes; (3) the species assemblage of coccoliths (some species assemblages are characteristic of eutrophic conditions, some of oligotrophic conditions) is used to infer trophic status and productivity. Some of the organic constituents of coccolithophores are also used for paleoenvironmental reconstructions. In particular, there is a distinctive group of ketones, termed long-chain alkenones, which are specific to one family of coccolithophores and closely related haptophytes. These alkenones tend to survive degradation in sediments, and the ratio of one type of alkenones to another (the U_K^{37} index) can be used to estimate past ocean temperatures. Calcareous nannofossils are also extremely useful in determining the age of different layers in cores of ocean sediments (biostratigraphy).

The Future

The pH of the oceans is falling (they are becoming increasingly acidic), because of the invasion of fossil fuel-derived CO_2 into the oceans. Surface ocean pH has already dropped by 0.1 units and may eventually drop by as much as 0.7 units, compared to preindustrial times, depending on future CO_2 emissions. The distribution of dissolved inorganic carbon (DIC) between bicarbonate, carbonate, and dissolved CO_2 gas changes with pH in such a way that carbonate ion concentration (and therefore saturation state, Ω) is decreasing even as DIC is increasing due to the invading anthropogenic CO_2. It is predicted that, by the end of this century, carbonate ion concentration and Ω may have fallen to as little as 50% of preindustrial values. If emissions continue for decades and centuries without regulation then the surface oceans will eventually become undersaturated with respect to calcium carbonate, first with respect to the more soluble aragonite used by corals, and some time later also with respect to the calcite formed by coccolithophores.

There has been an increasing appreciation over the last few years that declining saturation states may well have significant impacts on marine life, and, in particular, on marine organisms that synthesize $CaCO_3$. Experiments on different classes of marine calcifiers ($CaCO_3$ synthesizers) have demonstrated a reduction in calcification rate in high CO_2 seawater. One such experiment showed a strong decline in coccolithophore calcification rate (and a notable increase in the numbers of malformed coccoliths) at high CO_2 (low saturation state), although some other experiments have obtained different results. If coccolithophore biocalcification is controlled by Ω then the explanation could be linked to the importance of Ω in controlling inorganic calcification, although coccolithophores calcify intracellularly and so such a link is not guaranteed. At the time of writing, further research is being undertaken to determine whether, as the oceans become more acidic, coccolithophores will continue to be able to synthesize coccoliths and subsequently maintain them against dissolution.

Our ability to predict the consequences of ocean acidification on coccolithophores is hampered by our poor understanding of the function of coccoliths (what they are for, and therefore how the cells will be affected by their absence), and also by our poor understanding of the possibilities for evolutionary adaptation to a low-pH ocean. These constraints can be overcome to an extent by examining the geological history of coccolithophores, and their (in)ability to survive previous acid ocean events in Earth history. Although coccoliths (and other calcareous nannofossils) have been widely studied by geologists, it is only recently that there has been a concerted effort to study their species turnover through events in Earth history when the oceans were more acidic than now.

Although many authors have taken the success of coccolithophores during the high-CO_2 late Cretaceous as reassuring with respect to their future prospects, the reasoning is fallacious. Levels of calcium are thought to have been higher than now during the Cretaceous, and the CCD (the depth at which $CaCO_3$ disappears from sediments due to dissolution, which is a function of deep-water Ω) was only slightly shallower than today, indicating that Cretaceous seawater conditions were not analogous to those to be expected in a future high-CO_2 world.

It turns out that coccolithophores survived the Paleocene–Eocene Thermal Maximum event (thought to more closely resemble the predicted future) fairly well, with a modest increase in extinction rates matched by a similar increase in speciation rates. On the other hand, the environmental changes at the Cretaceous–Tertiary boundary (the K/T impact event), which also appears to have induced acidification, led to a mass extinction of 93% of all coccolithophore species, as well as to extinction of many other calcifying marine organisms including ammonites. It is necessary to more accurately characterize the environmental changes that took place across such events, in order to better determine how well they correspond to the ongoing and future ocean acidification.

See also

Plankton and Climate. Benthic Foraminifera.

Further Reading

Gibbs SJ, Bown PR, Sessa JA, Bralower TJ, and Wilson PA (2007) Nannoplankton extinction and origination across the Paleocene–Eocene Thermal Maximum. *Science* 314: 1770–1773 (doi: 10.1126/science.1133902).

Holligan PM, Fernandez E, Aiken J, *et al.* (1993) A biogeochemical study of the coccolithophore *Emiliania huxleyi* in the North Atlantic. *Global Biogeochemical Cycles* 7: 879–900.

Paasche E (2002) A review of the coccolithophorid *Emiliania huxleyi* (Prymnesiophyceae), with particular reference to growth, coccolith formation, and calcification–photosynthesis interactions. *Phycologia* 40: 503–529.

Poulton AJ, Sanders R, Holligan PM, *et al.* (2006) Phytoplankton mineralization in the tropical and subtropical Atlantic Ocean. *Global Biogeochemical Cycles* 20: GB4002 (doi: 10.1029/2006GB002712).

Riebesell U, Zonderva I, Rost B, Tortell PD, Zeebe RE, and Morel FMM (2000) Reduced calcification in marine plankton in response to increased atmospheric CO_2. *Nature* 407: 634–637.

Thierstein HR and Young JR (eds.) (2004) *Coccolithophores: From Molecular Processes to Global Impact.* Berlin: Springer.

Tyrrell T, Holligan PM, and Mobley CD (1999) Optical impacts of oceanic coccolithophore blooms. *Journal of Geophysical Research, Oceans* 104: 3223–3241.

Winter A and Siesser WG (eds.) (1994) *Coccolithophores.* Cambridge, UK: Cambridge University Press.

Young JR, Geisen M, Cros L, *et al.* (2003) *Special Issue: A Guide to Extant Coccolithophore Taxonomy. Journal of Nannoplankton Research* 1–125.

Relevant Websites

http://cics.umd.edu
– Blooms of the Coccolithophorid *Emiliania huxleyi* in Global and US Coastal Waters, CICS.

http://www.ucl.ac.uk
– Calcareous Nannofossils, MIRACLE, UCL.

http://www.nanotax.org
– Calcareous Nannofossil Taxonomy.

http://www.emidas.org
– Electronic Microfossil Image Database System.

http://www.noc.soton.ac.uk
– *Emiliania huxleyi* Home Page, National Oceanography Centre, Southampton.

http://www.nhm.ac.uk
– International Nannoplankton Association page, hosted at Natural History Museum website.

BENTHIC ORGANISMS OVERVIEW

P. F. Kingston, Heriot-Watt University, Edinburgh, UK

Introduction

The term benthos is derived from the Greek word βαos (vathos, meaning depth) and refers to those organisms that live on the seabed and the bottom of rivers and lakes. In the oceans the benthos extends from the deepest oceanic trench to the intertidal spray zone. It includes those organisms that live in and on sediments, those that inhabit rocky substrata and those that make up the biodiversity of coral reefs.

The benthic environment, sometimes referred to as the benthal, may be divided up into various well defined zones that seem to be distinguished by depth (**Figure 1**).

Physical Conditions Affecting the Benthos

In most parts of the world the water level of the upper region of the benthal fluctuates, so that the animals and plants are subjected to the influence of the water only at certain times. At the highest level, only spray is involved; in the remainder of the region, the covering of water fluctuates as a result of tides and wind and other atmospheric factors. Below the

level of extreme low water, the seafloor is permanently covered in water.

Water pressure increases rapidly with depth. Pressure is directly related to depth and increases by one atmospheric (101 kPa) per 10 m. Thus at 100 m, water pressure would be 11 atmospheres and at 10 000 m, 1001 atmospheres. Water is essentially incompressible, so that the size and shape of organisms are not affected by depth, providing the species does not possess a gas space and move between depth zones.

Light is essential to plants and it is its rapid attenuation with increasing water depth that limits the distribution of benthic flora to the coastal margins. Daylight is reduced to 1% of its surface values between 10 m and 30 m in coastal waters; in addition, the spectral quality of the light changes with depth, with the longer wavelengths being absorbed first. This influences the depth zonation of plant species, which is partially based on photosynthetic pigment type.

Surface water temperatures are highest in the tropics, becoming gradually cooler toward the higher latitudes. Diurnal changes in temperature are confined to the uppermost few meters and are usually quite small (3°C); however, water temperature falls with increasing depth to between 0.5°C and 2°C in the abyssal zone.

The average salinity of the oceans is 34.7 ppt. Salinity values in the deep ocean remain very near to this value. However, surface water salinity may vary considerably. In some enclosed areas, such as the Baltic Sea, salinity may be as low as 14 ppt, while in

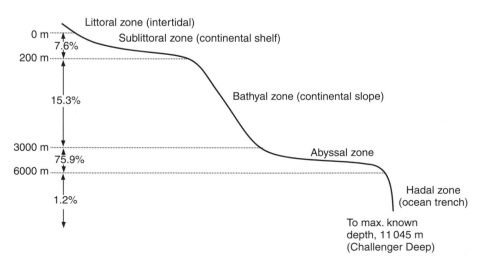

Figure 1 Classification of benthic environment with percentage representation of each depth zone.

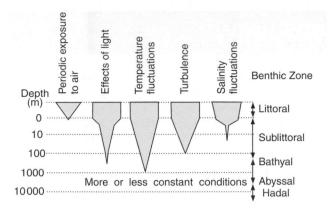

Figure 2 Diagram summarizing the influence of physical factors on the benthos. Relative influence of each factor indicated by width of shaded area.

areas of high evaporation, such as the Arabian Gulf, salinities in excess of 50 ppt may be reached. Salinity profiles may become quite complex in estuarine conditions or in coastal regions where there is a high fresh-water run off from the land (e.g., fiordic conditions) (**Figure 2**).

The composition of the benthos is profoundly affected by the nature of the substratum. Hard substrata tend to be dominated by surface dwelling forms, providing a base for the attachment of sessile animals and plants and a large variety of microhabitats for organisms of cryptic habits. In contrast, sedimentary substrata are dominated by burrowing organisms and, apart from the intertidal zone and the shallowest waters, they are devoid of plants.

Hard substrata are most common in coastal waters where there are strong tidal currents and surface turbulence. Farther offshore, and in the deep sea, the seabed is dominated by sediments, hard substrata occurring only on the slopes of seamounts, oceanic trenches, and other irregular features where the gradient of the seabed is too steep to permit significant accumulation of sedimentary material. Thus, viewed in its entirety, the seabed can be considered predominantly a level-bottom sedimentary environment (**Figure 3**).

Classification of the Benthos

It was the Danish marine scientist, Petersen working in the early part of the twentieth century, who first defined the two principal groups of benthic animals:

- the epifauna, comprising all animals living on or associated with the surface of the substratum, including sediments, rocks, shells, reefs and vegetation;
- the infauna comprising all animals living within the substratum, either moving freely through it or living in burrows and tubes or between the grains of sediments.

According to the great benthic ecologist Gunnar Thorson, the epifauna occupies less than 10% of the total area of the seabed, reaching its maximum abundance in the shallow waters and intertidal zones of tropical regions. However, the infauna, which Thorson believed occupies more than half the surface area of the planet, is most fully developed sublittorally. Nevertheless, the number of epifaunal species is far greater than the number of infaunal species. This is because the level-bottom habitat

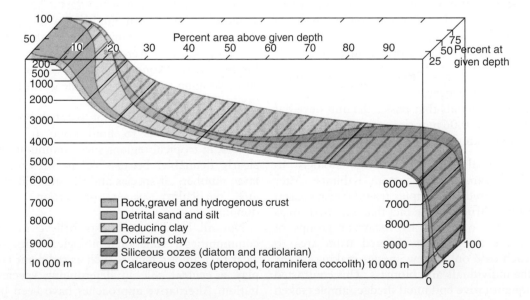

Figure 3 Distribution of sediment types in the ocean. (After Brunn, in Hedgepeth (1957).)

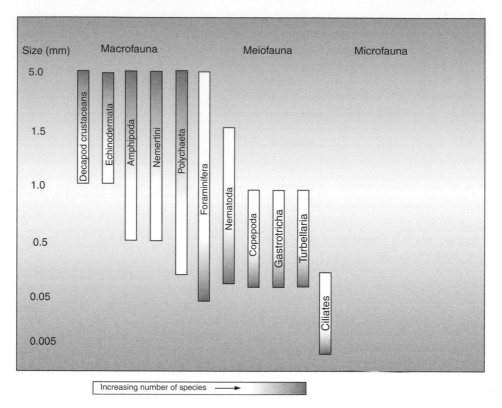

Figure 4   Principal taxa associated with each of the major categories of infauna.

provides a more uniform environment than hard substrata, with fewer potential habitat types to support a wide diversity of species. The epifauna of polar regions are not generally as well developed as those in the lower latitudes because of the effects of low temperatures in shallow waters and the effects of ice and meltwater in the intertidal region. The infauna largely escape these effects, exhibiting less latitudinal variation in number of species (**Figure 4**).

The infauna is further classified on the basis of size, the size categories broadly agreeing with major taxa that characterize the groups (**Figure 5**):

● macrofauna – animals that are retained on a 0.5 mm aperture sieve;
● meiofauna – animals that pass a 0.5 mm sieve but are retained on 0.06 mm sieve;
● microfauna – animals that pass a 0.06 mm sieve.

Petersen was also the first marine biologist to quantitatively sample soft-bottom habitats. After examining hundreds of samples from Danish coastal waters, he was struck by the fact that extensive areas of seabed were occupied by recurrent groups of species. These assemblages differed from area to area, in each case only a few species making up the bulk of the individuals and biomass. This contrasted with nonquantitative epifaunal dredge samples taken over the same range of areas, for which faunal lists

might be almost identical. Petersen proposed the infaunal assemblages that he had distinguished as communities and named them on the basis of the most visually dominant animals (**Table 1**).

Following the publication of Petersen's work in 1911–12, other marine biologists began to investigate benthic infauna quantitatively, and it began to emerge that there existed parallel bottom communities in which similar habitats around the world supported similar communities to those found by Petersen. These communities, although composed of different species, were closely similar, both ecologically and taxonomically, the characteristic species belonging to at least the same genus or a nearby taxon (**Table 2**).

Although the concept of parallel bottom communities appeared to hold good for temperate waters, in tropical regions such communities are not clearly definable, because of the presence of very large numbers of species and the small likelihood of any particular species or group of species dominating.

Not all benthic ecologists believe in Petersen's communities as functional biological units, since it is clear that abiotic factors such as sediment type must play a central role in determining species distribution. Alternative approaches have been proposed in which benthic associations are linked to

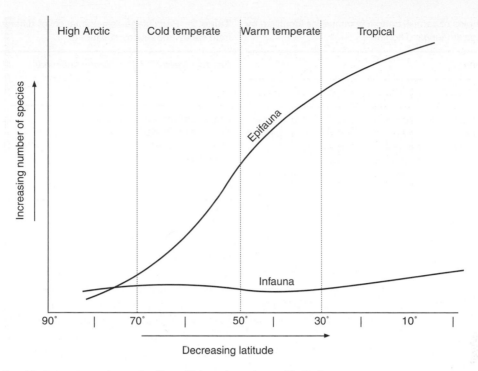

Figure 5   Relationship between numbers of epifaunal/infaunal species and latitude.

substratum (and latitude), suggesting that it is a common environmental requirement that forms the basis of the perceived community rather than an affinity of the members of the assemblage with one another.

Feeding Habits of Benthic Animals

Benthic animals, like all other animals, ultimately rely on the plant kingdom as their primary source of food. Most benthic animals are dependent on the rain of dead or partially decayed material (organic detritus) from above, and it is only in shallow coastal waters that macrophytes and phytoplankton are directly available for grazing or filtering by bottom-living forms. Much of this material consists of cellulose (dead plant cells) and chitin (crustacean exoskeletons). Few animals are able to digest these substances themselves and most rely on the action of bacteria to render them available, either as bacterial biomass or breakdown products.

Food particles are intercepted in the water before they reach the seabed, or are collected from the sediment surface after they have settled out, or are extracted from the sediment after becoming incorporated into it. These three scenarios reflect the three main types of detrivores: suspension feeders, selective deposit feeders, and direct deposit feeders.

Table 1   Petersen's benthic communities

Petersen's community	Typical species	Substratum
Macoma or Baltic community	Macoma balthica, Mya arenaria, Cerastoderma edule	Intertidal mud
Abra community	Abra alba, Macoma calcarea, Corbula gibba, Nephtys spp.	Inshore mud
Venus community	Venus striatula, Tellina fabula, Echinocardium cordatum	Offshore sand
Echninocardium–Filiformis community	E. cordatum, Amphiura filiformis, Cultellus pellucidus, Turritella communis	Offshore muddy sand
Brissopsis–Chiagei community	Brissopsis lyrifera, Amphiura chiagei, Calocaris macandreae	Offshore mud
Brissopsis–Sarsi community	B. lyrifera, Ophiura Sarsi, Abra nitida, Nucula tenuis	Deep mud
Amphilepis–Pecten community	Amphilepis norvegica, Chlamys (= Pecten) vitrea, Thyasira flexuosa	Deep mud
Haploops community	Haploops tubicola, Chlamys septemradiata, Lima loscombi	Offshore clay
Deep Venus community	Venus gallina, Spatangus purpureus, Abra prismatica	Deep sand

Table 2 Examples of parallel bottom communities identified by the Danish ecologist Gunnar Thorson

Species	Genera			
	NE Atlantic	NE Pacific	Arctic	NW Pacific
Macoma	baltica	nasuta	calcarea	incongrua
Cardium	edule	corbis	ciliatum	hungerfordi
Mya	arenaria	arenaria	truncata	
Arenicola	marina	claparedii	marina	

Table 3 Percentage representation of different feeding types from sandy and muddy sediments

Feeding types	Sandy sediment (< 15 silt/clay) (%)	Muddy sediment (> 90% silt/clay) (%)
Predators/omnivores	25.3	22.4
Suspension feeders	27.2	14.9
Selective deposit feeders	41.1	50.8
Direct deposit feeders	6.3	11.9

- Suspension feeders may be passive or active. Passive suspension feeders trap passing particles on extended appendages that are covered in sticky mucus and rely on natural water movements to bring the food to them (e.g., crinoid echinoderms). Active suspension feeders create a strong water current of their own, filtering out particles using specially modified organs (e.g., most bivalve mollusks).
- Selective deposit feeders either consume surface deposits in their immediate vicinity using unspecialized mouth parts, or, where food is less abundant, use extendable tentacles or siphons to pick up particles over a large area (e.g., terebellid polychaetes).
- Direct deposit feeders indiscriminately ingest sediment using organic matter and microbial organisms contained in the sediment as food. Polychaeta, such as the lugworm, *Arenicola*, construct L-shaped burrows and mine sediment from a horizontal gallery, reversing up the vertical shaft to defecate at the surface. Such animals play an important role in the physical turnover of the sediments.

Grazing or browsing animals are most common intertidally or in shallow waters. They are mobile consumers, cropping exposed tissues of sessile prey usually without killing the whole organism. They include animals that feed on macroalgae and those that feed on colonial cnidaria, bryozoans, and tunicates such as gastropod mollusks and echinoids. On the level bottom, the tips of tentacles and siphons of infaunal animals are grazed by demersal fish, providing a route for energy transfer from the seabed back into the pelagic system.

Predatory hunters are common among benthic epifauna and include crustaceans, asteroid echinoderms, and gastropod mollusks. These are mobile animals that seek out and consume individual prey items one at a time. Although less common, such predators are also found in the infauna, moving through the sediment, attacking their prey *in situ* (e.g., the polychaete *Glycera*).

Although, overall, deposit feeders form the largest single group of benthic infauna, the proportion of each trophic group is greatly influenced by the nature of the sediment. Thus in coarser sandy sediments, where water movement is relatively strong, the proportion of suspension feeders increases, while fine silts and muds are usually dominated by deposit feeders (see **Table 3**).

Spatial Distribution of Benthos

Competition between benthic infauna is usually for space. This is because most benthic animals are either suspension or deposit feeders and are competing for access to the same food source. In this respect, benthic communities are similar to those of terrestrial plants since, in both, competition between individuals is for an energy source that originates from above. Indeed, early approaches to the statistical analysis of benthic community structure were often rooted in principles originally developed by terrestrial botanists.

Suspension feeding may take place at more than one level. Some species, such as sabellid worms, extend their feeding organs several centimeters into the water column; some keep a lower profile, with short siphons projecting just a few millimeters above the sediment surface, while others have open burrows, drawing water into galleries below the surface. In addition, selective deposit feeders scour the sediment surface for food particles using palps or tentacles that in some species can extend up to a meter. The result is a contagious horizontal distribution of animals (i.e., neither random nor regular) that is maintained primarily by inter- and intra-specific competition for space.

Benthic infauna also show a marked vertical distribution. This is more a function of the subsurface physical conditions of the sediments and the need for the majority of species to be in communication with the surface than of competition between the animals.

Petersen chose physically large representatives to describe his benthic communities, primarily because they were easy to identify under field conditions. However, most benthic infaunal species are too small to be recognized with the naked eye, with burrows that penetrate no more than a few centimeters into the substratum. The result is that, for most level-bottom communities, some 95–99% of the animals are located within 5 cm of the sediment surface.

Reproduction in Benthic Animals

The act of reproduction offers benthic animals, the majority of which are either sessile or very restricted in their migratory powers, an opportunity to disperse and to colonize new ground. It is therefore not surprising that the majority of benthic species experience at least some sort of pelagic phase during their early development. Most invertebrates have larvae that swim for varying amounts of time before settlement and metamorphosis. The larvae, which develop freely in the surface waters of the ocean, either feed on planktonic organisms (planktotrophic larvae) or develop independently from a self-contained food supply or yolk (lecithotrophic larvae). Pelagic development in temperate waters can take several weeks, during which time developing larvae may be transported over great distances. Where it is within the interests of a particular species to ensure that its offspring are not dispersed (e.g., some intertidal habitats), a free-living larval phase may be dispensed with. In this case eggs may develop directly into miniature adults (oviparity) or may be retained within the body of the adult with the young being born fully developed (viviparity). Reproductive strategies such as these are also common in the deep-sea and polar regions where the supply of phytoplankton for feeding is unreliable or nonexistent. A good example of a latitudinal trend in this respect was demonstrated by Thorson. Analysing the developmental types of prosobranchs, he was able to show that the proportion of species with nonpelagic larvae decreases from the arctic to the tropics, while the proportion with pelagic larvae increases (**Figure 6**).

For many years deep-sea biologists believed that the energetic investment required to produce large numbers of planktotrophic larvae, and the huge distances required to be covered by such larvae in order to reach surface waters, would preclude such a reproductive strategy for deep-sea animals. However, it is now known that several species of ophiuroids living at depths of 2000–3000 m not only exhibit seasonal reproductive behavior but also produce larvae that feed in ocean surface waters.

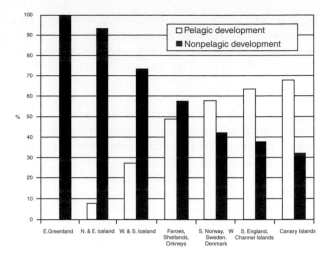

Figure 6 Percentage distribution of prosobranchs with pelagic and nonpelagic development in relation to latitude. (Adapted from Thorson (1950).)

Many benthic invertebrates are able to reproduce asexually. For example, polychaetes from the family the Syllidae are able to reproduce by budding; others, such as the cirratulid *Dodecaceria* or the ctenodrilid *Raphadrilus*, simply fragment, each fragment growing into a new individual. The ability to switch between sexual and vegetative means of propagation provides the potential for such species to rapidly colonize areas that have been disturbed. Where disturbance is accompanied by organic enrichment, for example, from sewage or paper pulp discharge, huge localized populations may result. These are the so-called opportunistic species that are sometimes used as indicators of pollution.

Although planktonic larvae are able to swim, they are very small and, for the most part, are obliged to go where ocean currents take them. The critical time arrives just before the larvae are about to settle. At one time it was thought that the process of settlement was random, with individuals that settled in unfavorable substrata perishing. Although this undoubtedly happens, most species seem to have some sort of behavioral pattern to increase their chances of finding a suitable substratum. The larvae usually pass through one or more stages of photopositive and photonegative behavior. These enable the larvae to remain near the sea surface to feed and then to drop to the bottom to seek a suitable substratum on which to settle. Depending on the species, larvae may cue on the mechanical attributes of the substratum or on its chemical nature. Chemical attraction is also important in gregarious species in which the young are attracted to settle at sites where adults of the same species are already present (e.g., oysters). Most larvae go through a period when, although able to settle permanently, they retain the ability to swim. This allows them to 'test' the substratum, rising back into the water and any prevailing currents should the nature of the ground be unsuitable. After settling, larvae may move a short distance, usually no more than a few centimeters. These early stages in the recruitment of benthic organisms are crucial in the maintenance of benthic community structure and it is now believed that it is at this time that the nature of the community is established.

It is clear that the vast majority of planktonic larvae never make it to adulthood. Mortality from predation and transport away from a suitable habitat are on a massive scale. To compensate, species with planktotrophic larvae produce huge numbers of eggs (e.g., the sea hare *Aplysia californiensis* spawns as many as 450 000 000 eggs at one time). This is possible because there is no need for a large, and energetically expensive, yolk; the larvae hatch at an early embryonic stage and rely almost entirely on plankton-derived food for their development. One consequence of this is that the recruitment varies depending on the success of the plankton production in a particular year and the vagaries of local currents. Thus, populations of benthic species that reproduce by means of planktotrophic larvae tend to fluctuate numerically from year to year, with the potential for heavy recruitment when the combination of environmental factors is favorable, or recruitment failure when they are not. Species reproducing by means of nonpelagic larvae or by direct development tend to produce fewer eggs, since there is a large yolk required to nourish the developing embryo. Although annual recruitment is relatively modest for these species, it is less variable between years, producing populations with a greater temporal stability (**Figure 7**). Because of this, these populations are

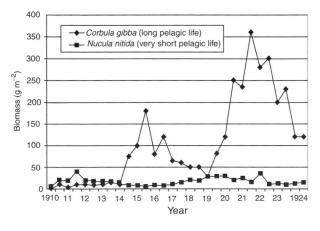

Figure 7 Example of two populations of bivalves showing the influence of type of larvae on population stability. (Adapted from Thorson (1950).)

likely to be slow to recover from major natural environmental disturbances (e.g., unusual temperature extremes or physical disturbance) or major pollution events.

Deep-sea Benthos

Because of regional differences, it is difficult to define the exact upper limit of the deep-sea benthic environment. However, it is generally regarded as beginning beyond the 200 m depth contour. At this point, where the continental shelf gives way to the continental slope, there is often a marked change in the benthic fauna. In the past there have been many attempts to produce a scheme of zonation for the deep-sea environment. There are three major regions beyond 200 m depth – these are the bathyal region, the abyssal region, and the hadal region (see **Figure 1**). The bathyal region represents the transition region between the edge of the continental shelf and the true deep sea (the continental slope). Its boundary with the abyssal region has been variously defined by workers. It is believed that the 4°C isotherm limits the depth at which the endemic abyssal organisms can survive. Since this varies in depth according to geographical and hydrographical conditions in the abyssal region, it follows that the upper depth limit of the abyssal fauna will also vary. This depth is usually between 1000 and 3000 m. The abyssal region is by far the most extensive, reaching down to 6000 m depth and accounting for over half the surface area of the planet. The hadal zone, sometimes called the ultra-abyssal zone, is largely restricted to the deep oceanic trenches. The composition of the benthos in these trenches differs from that of nearby abyssal areas. The trenches are geographically isolated from one another and the fauna exhibits a high degree of endemism.

The deep sea is aphotic and so has no primary production except in certain areas where chemosynthetic bacteria are found. Thus, the fauna of the deep sea is almost wholly reliant on organic material that has been generated in the surface layers of the oceans and has sunk to the seabed. Since the likelihood of a particle of food being consumed on its way down through the water column is related to time, the deeper the water the less food is available for the animals that live on the seafloor. This results in a relatively impoverished fauna, in which the density of organisms is low and the size of most is quite small. Paradoxically, the deep sea supports a few rare species that grow unusually large. This phenomenon is known as abyssal gigantism and is found primarily in crustacean species. The reasons for abyssal gigantism are not clear, but it is believed to be the result either of a peculiar metabolism under conditions of high pressure or of the slow growth and time taken to reach sexual maturity.

Although the number of individuals in the deep sea is quite low, there are many species. This combination of many species represented by few individuals results in high calculated diversity values and has led to the suggestion by some that the biodiversity of the deep sea is comparable to that of tropical rain forests. Accepting that diversity is high in the deep sea, there are several theories that attempt an explanation. The earliest is the stability–time hypothesis put forward by Howard Sanders of the Woods Hole Oceanographic Institution in the late 1960s. This suggests that the highly stable environmental conditions of the deep sea that have persisted over geological time might have allowed many species to evolve that are highly specialized for a particular microhabitat or food source. Another theory, the cropper or disturbance theory, suggests that, as a result of the scarcity of food in the deep sea, none of the animals is a food specialist, the animals being forced to feed indiscriminately on anything living or dead that is smaller than itself. High diversity results from intense predation, which allows a large number of species to persist, eating the same food, but never becoming abundant enough to compete with one another. More recently, Grassle and Morse-Porteus have suggested that a combination of factors might be responsible, including the patchy distribution of organically enriched areas in a background of low productivity; the occurrence of discrete, small-scale disturbances (primarily biological) in an area of otherwise great constancy; and the lack of barriers to dispersal among species distributed over a very large area.

Although most animal groups are represented in the deep sea, the fauna is often dominated by Holothuroidea (sea cucumbers) or Ophiuroidea (brittle stars). Crustacea and polychaete worms are also important members of deep-sea communities. For many years it was believed that the deep seafloor was too remote from the surface, and the physical conditions were too constant, for the organisms living there to be influenced by the seasons. Although this may be the case for many deep-sea species, long-term time-lapse photography has shown that cyclical events, such as the accumulation of organic detritus on the deep seabed, do take place. Furthermore, in temperate waters, these appear to correspond with seasonally driven processes such as the spring plankton bloom. Where they occur, these pulses of organic input inevitably influence the deep-sea communities below with the consequence that seasonal

life cycles are not uncommon in abyssal animals. Localized areas of organic enrichment can occur in the deep sea as the result of the sinking of large objects such as the carcass of a large sea mammal or fish or waterlogged tree trunk. The surprisingly quick response of deep-sea scavengers to large food-falls such as these, and the frequency with which they have been recorded, has led researchers to believe that they are important contributers to energy flow on the deep-sea floor. Hydrothermal vents are also thought to provide a significant input of energy into the benthic environment. These are a relatively recent discovery and, at first, were thought to be rare and isolated phenomena occurring only on the Galapagos Rift off Ecuador. It is now known that active vents are associated with nearly all areas of tectonic activity that have been investigated in the deep Pacific and Atlantic. These vents provide a nonphotosynthetic source of organic carbon through the medium of chemoautotrophic bacteria. Theses organisms use sulfur-containing inorganic compounds as an oxidizing substrate to synthesize organic carbon from carbon dioxide and methane without the need for sunlight. The chemicals come from hot water that originates deep within the Earth's crust. Some of the bacteria form dense white mats on the surface of the sediments similar to those of *Beggiatoa*, an anaerobic bacterium found in anoxic sediments in shallow water; others enter into symbiotic relationships with the bacteria, either hosting them on their gills (e.g., the mussel *Bathymodiolus*) or within a special internal sac, the trophosome (e.g., *Riftia*). The relationship is very complex as there has to be a compromise between the anaerobic needs of the bacteria and the aerobic needs of the animals. Nevertheless, the arrangement is very successful and the animals, which sometimes occur in huge numbers, often grow to gargantuan size. There is still much to be learned about these vent communities, which can support concentrations of biomass several orders of magnitude greater than that of the nearby seafloor. It remains a mystery how these vents become populated, since they are known to be transient and variable, probably lasting only decades or less. It has been suggested that pelagic larvae of many vent species may be able to delay settlement for months at a time so as to increase their chances of locating a suitable site.

Glossary

Abyssal region That region of the seabed from between 1000 and 3000 m depth reaching down to 6000 m.
Atmosphere Measure of pressure (1 atm = 101 kPa).

Bathyal region Region of the seabed that represents the transition region between the edge of the continental shelf and the true deep sea (the continental slope).

Benthal The benthic environment.

Benthos Those organisms that live on the seabed and the bottoms of rivers and lakes.

Continental shelf Region of the seabed from low-water mark to a depth of 200 m.

Detritivores Animals feeding on organic detritus.

Direct deposit feeders Animals indiscriminately ingesting sediment using organic matter and microbial organisms contained in the sediment as food.

Epifauna All animals living on or associated with the surface of the substratum.

Hadal zone Region of the seabed below 6000 m depth, largely restricted to the deep oceanic trenches.

Infauna All animals living within the substratum or moving freely through it.

Kilopascal (kPa) Measure of pressure (100 kPa = 1 bar = 0.987 atm).

Lecithotrophic larvae Pelagic larvae of marine animals that develop freely in the surface waters of the ocean, developing independently from a self-contained food supply or yolk.

Macrofauna Animals retained on a 0.5 mm aperture sieve.

Meiofauna Animals passing a 0.5 mm sieve but retained on a 0.06 mm sieve.

Microfauna Animals passing a 0.06 mm sieve.

Organic detritus Dead or partially decayed plant and animal material.

Oviparity Eggs laid by the adult develop directly into miniature adults.

Pelagic larvae Larvae of marine animals that swim freely in the water column.

Planktotrophic larvae Pelagic larvae of marine animals that develop freely in the surface waters of the ocean, feeding on planktonic organisms.

Selective deposit feeders Animals feeding on surface particles of organic matter or sediment particles supporting a rich bacterial flora.

Suspension feeders Animals feeding on organisms or organic detritus suspended in the water column.

Viviparity Young are born fully developed either from eggs retained within the body of the mother (oviparity) or after internal embryonic development.

See also

Benthic Foraminifera.

Further Reading

Cushing DH and Walsh JJ (eds.) (1976) *The Ecology of the Seas.* Oxford: Blackwell Scientific Publications.

Friedrich H (1969) *Marine Biology: An Introduction to Its Problems and Results.* London: Sidgwick and Jackson.

Gage JD and Tyler PA (1991) *Deep-sea Biology: A Natural History of Organisms at the Deep-Sea Floor.* Cambridge: Cambridge University Press.

Hedgepeth JW (ed.) (1957) *Treatise on Marine Ecology and Paleoecology,* Vol. 1, *Ecology,* The Geological Society of America, Memoir 67. Washington, DC: Geological Society of America.

Jøgensen CB (1990) *Bivalve Filter-Feeding: Hydrodynamics, Bioenergetics, Physiology and Ecology.* Fredensborg, Denmark: Olsen and Olsen.

Levington JS (1995) *Marine Biology: Function, Biodiversity, Ecology.* Oxford: Oxford University Press.

Nybakken JW (1993) *Marine Biology.* New York: Harper-Collins College Publishers.

Thorson G (1950) Reproduction and larval ecology of marine bottom invertebrates. *Biological Reviews* 25: 1–25.

Webber HH and Thurman HV (1991) *Marine Biology.* New York: Harper Collins.

BENTHIC FORAMINIFERA

A. J. Gooday, Southampton Oceanography Centre, Southampton, UK

Introduction

Foraminifera are enormously successful organisms and a dominant deep-sea life form. These amoeboid protists are characterized by a netlike (granuloreticulate) system of pseudopodia and a life cycle that is often complex but typically involves an alternation of sexual and asexual generations. The most obvious characteristic of foraminifera is the presence of a shell or 'test' that largely encloses the cytoplasmic body and is composed of one or more chambers. In some groups, the test is constructed from foreign particles (e.g., mineral grains, sponge spicules, shells of other foraminifera) stuck together ('agglutinated') by an organic or calcareous/organic cement. In others, it is composed of calcium carbonate (usually calcite, occasionally aragonite) or organic material secreted by the organism itself.

Although the test forms the basis of foraminiferal classification, and is the only structure to survive fossilization, the cell body is equally remarkable and important. It gives rise to the complex, highly mobile, and pervasive network of granuloreticulose pseudopodia. These versatile organelles perform a variety of functions (locomotion, food gathering, test construction, and respiration) that are probably fundamental to the ecological success of foraminifera in marine environments.

As well as being an important component of modern deep-sea communities, foraminifera have an outstandingly good fossil record and are studied intensively by geologists. Much of their research uses knowledge of modern faunas to interpret fossil assemblages. The study of deep-sea benthic foraminifera, therefore, lies at the interface between biology and geology. This articles addresses both these facets.

History of Study

Benthic foraminifera attracted the attention of some pioneer deep-sea biologists in the late 1860s. The monograph of H.B. Brady, published in 1884 and based on material collected in the *Challenger* round-the-world expedition of 1872–76, still underpins our knowledge of the group. Later biological expeditions added to this knowledge. For much of the 1900s, however, the study of deep-sea foraminifera was conducted largely by geologists, notably J.A. Cushman, F.B. Phleger, and their students, who amassed an extensive literature dealing with the taxonomy and distribution of calcareous and other hard-shelled taxa. In recent decades, the emphasis has shifted toward the use of benthic species in paleoceanographic reconstructions. Interest in deep-sea foraminifera has also increased among biologists since the 1970s, stimulated in part by the description of the Komokiacea, a superfamily of delicate, soft-shelled foraminifera, by O.S. Tendal and R.R. Hessler. This exclusively deep-sea taxon is a dominant component of the macrofauna in some abyssal regions.

Morphological and Taxonomic Diversity

Foraminifera are relatively large protists. Their tests range from simple agglutinated spheres a few tens of micrometers in diameter to those of giant tubular species that reach lengths of 10 cm or more. However, most are a few hundred micrometers in size. They exhibit an extraordinary range of morphologies (**Figures 1** and **2**), including spheres, flasks, various types of branched or unbranched tubes, and chambers arranged in linear, biserial, triserial, or coiled (spiral) patterns. In most species, the test has an aperture that assumes a variety of forms and is sometimes associated with a toothlike structure. The komokiaceans display morphologies not traditionally associated with the foraminifera. The test forms a treelike, bushlike, spherical, or lumpish body that consists of a complex system of fine, branching tubules (**Figure 2A–C**).

The foraminifera (variously regarded as a subphylum, class, or order) are highly diverse with around 900 living genera and an estimated 10 000 described living species, in addition to large numbers of fossil taxa. Foraminiferal taxonomy is based very largely on test characteristics. Organic, agglutinated, and different kinds of calcareous wall structure serve to distinguish the main groupings (orders or suborders). At lower taxonomic levels, the nature and position of the aperture and the number, shape, and arrangement of the chambers are important.

Figure 1 Scanning electron micrographs of selected deep-sea foraminifera (maximum dimensions are given in parentheses). (A) *Epistominella exigua*; 4850 m water depth, Porcupine Abyssal Plain, NE Atlantic (190 μm). (B) *Nonionella iridea*; 1345 m depth, Porcupine Seabight, NE Atlantic (110 μm). (C) *Nonionella stella*; 550 m depth, Santa Barbara Basin, California Borderland (220 μm). (D) *Brizalina tumida*; 550 m depth, Santa Barbara Basin, California Borderland (680 μm). (E) *Melonis barleaanum*; 1345 m depth, Porcupine Seabight, NE Atlantic (450 μm). (F) *Hormosina* sp., 4495 m depth, Porcupine Abyssal Plain (1.5 mm). (G) *Pyrgoella* sp.; 4550 m depth, foothills of Mid-Atlantic Ridge (620 μm). (H) *Technitella legumen*; 997–1037 m depth, NW African margin (8 mm). (A)–(E) and (G) have calcareous tests, (F) and (H) have agglutinated tests. (C) and (D), photographs courtesy of Joan Bernhard.

Methodology

Qualitative deep-sea samples for foraminiferal studies are collected using nets (e.g., trawls) that are dragged across the seafloor. Much of the *Challenger* material studied by Brady was collected in this way. Modern quantitative studies, however, require the use of coring devices. The two most popular corers used in the deep sea are the box corer, which obtains a large (e.g., 0.25 m²) sample, and the multiple corer, which collects simultaneously a battery of up to 12 smaller cores. The main advantage of the multiple corer is that it obtains the sediment–water interface in a virtually undisturbed condition.

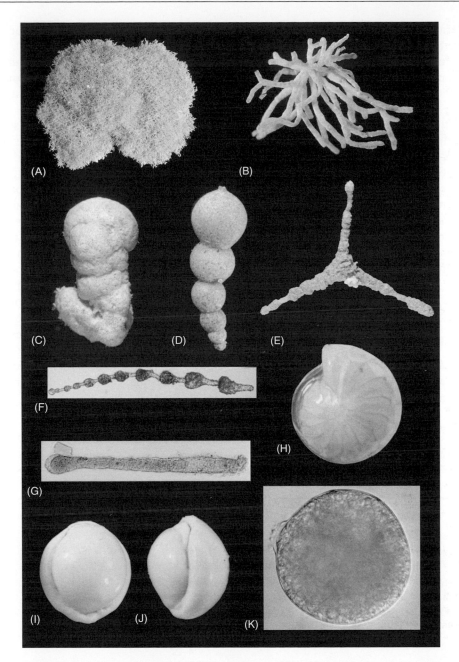

Figure 2 Light micrographs of deep-sea foraminifera (maximum dimensions are given in parentheses). (A) Species of *Lana* in which pad-like test consists of tightly meshed system of fine tubules; 5432 m water depth, Great Meteor East region, NE Atlantic (7.4 mm). (B) *Septuma* sp.; same locality (2 mm). (C) *Edgertonia* mudball; same locality (3.8 mm). (D) *Hormosina globulifera*; 4004 m depth, NW African margin (6.4 mm). (E) *Rhabdammina parabyssorum*; 3392 m depth, Oman margin, NW Arabian Sea (18 mm). (F) *Leptohalysis* sp.; 3400 m depth, Oman margin, NW Arabian Sea (520 μm). (G) Minute species of *Hyperammina*; 3400 m depth, Oman margin, NW Arabian Sea (400 μm). (H) *Lenticularia* sp.; 997–1037 m depth, NW African margin (2.5 mm). (I, J) *Biloculinella* sp.; 4004 m depth, NW African margin (3 mm). (K) Spherical allogromiid; 3400 m depth, Oman margin, NW Arabian Sea (105 μm). Specimens illustrated in (A)–(G) have agglutinated tests, in (H)–(J) calcareous tests and in (K) an organic test. (A)–(C) belong to the superfamily Komokiacea.

Foraminifera are extracted from sieved sediment residues. Studies are often based on dried residues and concern 'total' assemblages (i.e. including both live and dead individuals). To distinguish individuals that were living at the time of collection from dead tests, it is necessary to preserve sediment samples in either alcohol or formalin and then stain them with rose Bengal solution. This colors the cytoplasm red and is most obvious when residues are examined in water. Stained assemblages provide a snapshot of the foraminifera that were living when the samples were collected. Since the live assemblage varies in both time and space, it is also instructive to examine the dead assemblage that provides an averaged view of the foraminiferal fauna. Deep-sea foraminiferal assemblages are typically very diverse and therefore faunal data are often condensed mathematically by using multivariate approaches such as principal components or factor analysis.

The mesh size of the sieve strongly influences the species composition of the foraminiferal assemblage retained. Most deep-sea studies have been based on >63 μm, 125 μm, 150 μm, 250 μm, or even 500 μm meshes. In recent years, the use of a fine 63 μm mesh has become more prevalent with the realization that some small but important species are not adequately retained by coarser sieves. However, the additional information gained by examining fine fractions must be weighed against the considerable time and effort required to sort foraminifera from them.

Ecology

Abundance and Diversity

Foraminifera typically make up >50% of the soft-bottom, deep-sea meiofauna (**Table 1**). They are also often a major component of the macrofauna.

In the central North Pacific, for example, foraminifera (mainly komokiaceans) outnumber all metazoans combined by at least an order of magnitude. A few species are large enough to be easily visible to the unaided eye and constitute part of the megafauna. These include the tubular species *Bathysiphon filiformis*, which is sometimes abundant on continental slopes (**Figure 3**). Some xenophyophores, agglutinated protists that are probably closely related to the foraminifera, are even larger (up to 24 cm maximum dimension!). These giant protists may dominate the megafauna in regions of sloped topography (e.g., seamounts) or high surface productivity. In well-oxygenated areas of the deep-seafloor, foraminiferal assemblages are very species rich, with well over 100 species occurring in relatively small volumes of surface sediment (**Figure 4**). Many are undescribed delicate, soft-shelled forms. There is an urgent need to describe at least some of these species as a step toward estimating global levels of deep-sea species diversity. The common species are often widely distributed, particularly at abyssal depths, although endemic species undoubtedly also occur.

Table 1   The percentage contribution of foraminifera to the deep-sea meiofauna at sites where bottom water is well oxygenated

Area	Depth (m)	Percentage of foraminifera	Number of samples
NW Atlantic			
Off North Carolina	500–2500	11.0–90.4	14
Off North Carolina	400–4000	7.6–85.9	28
Off Martha's Vineyard	146–567	3.4–10.6	4
NE Atlantic			
Porcupine Seabight	1345	47.0–59.2	8
Porcupine Abyssal Plain[a]	4850	61.8–76.3	3
Madeira Abyssal Plain[a]	4950	61.4–76.1	3
Cape Verde Abyssal Plain[a]	4550	70.2	1
Off Mauretania	250–4250	4–27	26
46°N, 16–17°W	4000–4800	0.5–8.3	9
Indian Ocean			
NW Arabian Sea[b]	3350	54.4	1
Pacific			
Western Pacific	2000–6000	36.0–69.3	11
Central North Pacific	5821–5874	49.5	2
Arctic	1000–2600	14.5–84.1	74
Southern Ocean	1661–1680	2.2–23.7	2

[a] Data from Gooday AJ (1996) Epifaunal and shallow infaunal foraminiferal communities at three abyssal NE Atlantic sites subject to differing phytodetritus input regimes. *Deep-Sea Research I* 43: 1395–1421.
[b] Data from Gooday AJ, Bernhard JM, Levin LA and Suhr SB (2000) Foraminifera in the Arabian Sea oxygen minimum zone and other oxygen-deficient settings: taxonomic composition, diversity, and relation to metazoan faunas. *Deep-Sea Research II* 47: 25–54.
Based on Gooday AJ (1986) Meiofaunal foraminiferans from the bathyal Porcupine Seabight (northeast Atlantic): size structure, standing stock, taxonomic composition, species diversity and vertical distribution in the sediment. *Deep-Sea Research* 35: 1345–1373; with permission from Elsevier Science.

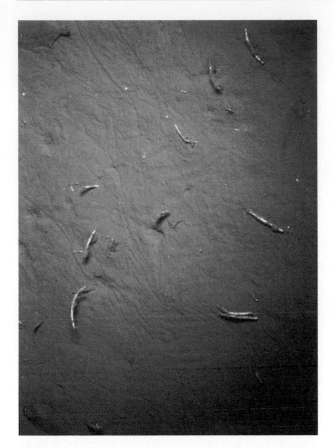

Figure 3 *Bathysiphon filiformis*, a large tubular agglutinated foraminifer, photographed from the Johnson Sealink submersible on the North Carolina continental slope (850 m water depth). The tubes reach a maximum length of about 10 cm. (Photograph courtesy of Lisa Levin.)

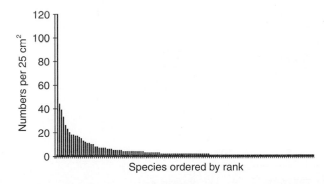

Figure 4 Deep-sea foraminiferal diversity: all species from a single multiple corer sample collected at the Porcupine Abyssal Plain, NE Atlantic (4850 m water depth), ranked by abundance. Each bar represents one 'live' (rose Bengal-stained) species. The sample was 25.5 cm² surface area, 0–1 cm depth, and sieved on a 63 μm mesh sieve. It contained 705 'live' specimens and 130 species.

Foraminifera are also a dominant constituent of deep-sea hard-substrate communities. Dense populations encrust the surfaces of manganese nodules as well as experimental settlement plates deployed on

the sea floor for periods of months. They include various undescribed matlike taxa and branched tubular forms, as well as a variety of small coiled agglutinated species (many in the superfamily Trochamminacea), and calcareous forms.

Role in Benthic Communities

The abundance of foraminifera suggests that they play an important ecological role in deep-sea communities, although many aspects of this role remain poorly understood. One of the defining features of these protists, their highly mobile and pervasive pseudopodial net, enables them to gather food particles very efficiently. As a group, foraminifera exhibit a wide variety of trophic mechanisms (e.g., suspension feeding, deposit feeding, parasitism, symbiosis) and diets (herbivory, carnivory, detritus feeding, use of dissolved organic matter). Many deep-sea species appear to feed at a low trophic level on organic detritus, sediment particles, and bacteria. Foraminifera are prey, in turn, for specialist deep-sea predators (scaphopod mollusks and certain asellote isopods), and also ingested (probably incidentally) in large numbers by surface deposit feeders such as holothurians. They may therefore provide a link between lower and higher levels of deep-sea food webs.

Some deep-sea foraminifera exhibit opportunistic characteristics – rapid reproduction and population growth responses to episodic food inputs. Well-known examples are *Epistominella exigua*, *Alabaminella weddellensis* and *Eponides pusillus*. These small (generally <200 μm), calcareous species feed on fresh algal detritus ('phytodetritus') that sinks through the water column to the deep-ocean floor after the spring bloom (a seasonal burst of phytoplankton primary production that occurs most strongly in temperate latitudes). Utilizing energy from this labile food source, they reproduce rapidly to build up large populations that then decline when their ephemeral food source has been consumed. Moreover, certain large foraminifera can reduce their metabolism or consume cytoplasmic reserves when food is scarce, and then rapidly increase their metabolic rate when food again becomes available. These characteristics, together with the sheer abundance of foraminifera, suggest that their role in the cycling of organic carbon on the deep-seafloor is very significant.

The tests of large foraminifera are an important source of environmental heterogeneity in the deep sea, providing habitats and attachment substrates for other foraminifera and metazoans. Mobile infaunal species bioturbate the sediment as they move through it. Conversely, the pseudopodial systems of foraminifera may help to bind together and stabilize deep-sea sediments, although this has not yet been clearly demonstrated.

Microhabitats and Temporal Variability

Like many smaller organisms, foraminifera reside above, on and within deep-sea sediments. Various factors influence their overall distribution pattern within the sediment profile, but food availability and geochemical (redox) gradients are probably the most important. In oligotrophic regions, the flux of organic matter (food) to the seafloor is low and most foraminifera live on or near the sediment surface where food is concentrated. At the other extreme, in eutrophic regions, the high organic-matter flux causes pore water oxygen concentrations to decrease rapidly with depth into the sediment, restricting access to the deeper layers to those species that can tolerate low oxygen levels. Foraminifera penetrate most deeply into the sediment where organic inputs are of intermediate intensity and the availability of food and oxygen within the sediment is well balanced.

Underlying these patterns are the distributions of individual species. Foraminifera occupy more or less distinct zones or microenvironments ('microhabitats'). For descriptive purposes, it is useful to recognize a number of different microhabitats: epifaunal and shallow infaunal for species living close to the sediment surface (upper 2 cm); intermediate infaunal for species living between about 1 cm and 4 cm (**Figure 5**); and deep infaunal for species that occur at depths

down to 10 cm or more (**Figure 6**). A few deep-water foraminifera, including the well-known calcareous species *Cibicidoides wuellerstorfi*, occur on hard substrates (e.g., stones) that are raised above the sediment–water interface (elevated epifaunal microhabitat). There is a general relation between test morphotypes and microhabitat preferences. Epifaunal and shallow infaunal species are often trochospiral with large pores opening on the spiral side of the test; infaunal species tend to be planispiral, spherical, or ovate with small, evenly distributed pores. It is important to appreciate that foraminiferal microhabitats are by no means fixed. They may vary between sites and over time and are modified by the burrowing activities of macrofauna. Foraminiferal microhabitats should therefore be regarded as dynamic rather than static. This tendency is most pronounced in shallow-water settings where environmental conditions are more changeable and macrofaunal activity is more intense than in the deep sea.

The microhabitats occupied by species reflect the same factors that constrain the overall distribution patterns of foraminifera within the sediment. Epifaunal and shallow infaunal species cannot tolerate low oxygen concentrations and also require a diet of relatively fresh organic matter. Deep infaunal foraminifera are less opportunistic but are more tolerant of oxygen depletion than are species living close to the sediment–water interface (**Figure 6**). It has been suggested that species of genera such as *Globobulimina* may consume either sulfate-reducing bacteria or labile organic matter released by the metabolic

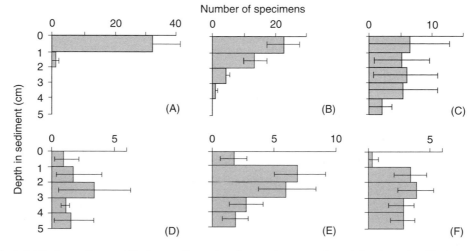

Figure 5 Vertical distribution patterns within the top 5 cm of sediment of common foraminiferal species ('live', rose Bengal-stained specimens) in the Porcupine Seabight, NW Atlantic (51°36′N, 13°00′W; 1345 m water depth). Based on >63 μm sieve fraction. (A) *Ovammina* sp. (mean of 20 samples). (B) *Nonionella iridea* (20 samples). (C) *Leptohalysis* aff. *catenata* (7 samples). (D) *Melonis barleeanum* (9 samples). (E) *Haplophragmoides bradyi* (19 samples). (F) '*Turritellella*' *laevigata* (21 samples). (Amended and reprinted from Gooday AJ (1986) Meiofaunal foraminiferans from the bathyal Porcupine Seabight (northeast Atlantic): size structure, standing stock, taxonomic composition, species diversity and vertical distribution in the sediment. *Deep-Sea Research* 35: 1345–1373; permission from Elsevier Science.)

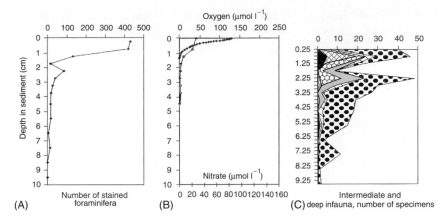

Figure 6 Vertical distribution of (A) total 'live' (rose Bengal-stained) foraminifera, (B) pore water oxygen and nitrate concentrations, and (C) intermediate and deep infaunal foraminiferal species within the top 10 cm of sediment on the north-west African margin (21°28.8′N, 17°57.2′W, 1195 m). All foraminiferal counts based on >150 μm sieve fraction, standardized to a 34 cm³ volume. Species are indicated as follows: *Pullenia salisburyi* (black), *Melonis barleeanum* (crossed pattern), *Chilostomella oolina* (honeycomb pattern), *Fursenkoina mexicana* (grey), *Globobulimina pyrula* (diagonal lines), *Bulimina marginata* (large dotted pattern). (Adopted and reprinted from Jorissen FJ, Wittling I, Peypouquet JP, Rabouille C and Relexans JC (1998) Live benthic foraminiferal faunas off Cape Blanc, northwest Africa: community structure and microhabitats. *Deep-Sea Research I* 45: 2157–2158; with permission from Elsevier Science.)

activities of these bacteria. These species move closer to the sediment surface as redox zones shift upward in the sediment under conditions of extreme oxygen depletion. Although deep-infaunal foraminifera must endure a harsh microenvironment, they are exposed to less pressure from predators and competitors than those occupying the more densely populated surface sediments.

Deep-sea foraminifera may undergo temporal fluctuations that reflect cycles of food and oxygen availability. Changes over seasonal timescales in the abundance of species and entire assemblages have been described in continental slope settings (**Figure 7**). These changes are related to fluctuations in pore water oxygen concentrations resulting from episodic (seasonal) organic matter inputs to the seafloor. In some cases, the foraminifera migrate up and down in the sediment, tracking critical oxygen levels or redox fronts. Population fluctuations also occur in abyssal settings where food is a limiting ecological factor. In these cases, foraminiferal population dynamics reflect the seasonal availability of phytodetritus ('food'). As a result of these temporal processes, living foraminifera sampled during one season often provide an incomplete view of the live fauna as a whole.

Environmental Controls on Foraminiferal Distributions

Our understanding of the factors that control the distribution of foraminifera on the deep-ocean floor is very incomplete, yet lack of knowledge has not prevented the development of ideas. It is likely that foraminiferal distribution patterns reflect a combination of influences. The most important first-order factor is calcium carbonate dissolution. Above the carbonate compensation depth (CCD), faunas include calcareous, agglutinated, and allogromiid taxa. Below the CCD, calcareous species are almost entirely absent. At oceanwide or basinwide scales, the organic carbon flux to the seafloor (and its seasonality) and bottom-water hydrography appear to be particularly important, both above and below the CCD.

Studies conducted in the 1950s and 1960s emphasized bathymetry (water depth) as an important controlling factor. However, it soon became apparent that the bathymetric distribution of foraminiferal species beyond the shelf break is not consistent geographically. Analyses of modern assemblages in the North Atlantic, carried out in the 1970s, revealed a much closer correlation between the distribution of foraminiferal species and bottom-water masses. For example, *Cibicidoides wuellerstorfi* was linked to North Atlantic Deep Water (NADW) and *Nuttallides umbonifera* to Antarctic Bottom Water (AABW). At this time, it was difficult to explain how slight physical and chemical differences between water masses could influence foraminiferal distributions. However, recent work in the south-east Atlantic, where hydrographic contrasts are strongly developed, suggests that the distributions of certain foraminiferal species are controlled in part by the lateral advection of water masses. In the case of *N. umboniferus* there is good evidence that the main

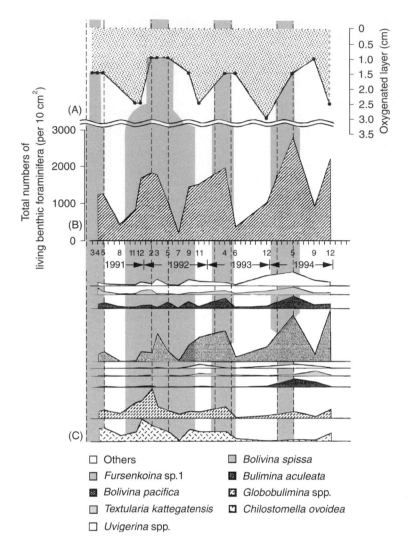

Figure 7 Seasonal changes over a 4-year period (March 1991 to December 1994) in (A) the thickness of the oxygenated layer, (B) the total population density of live benthic foraminifera, and (C) the abundances of the most common species at a 1450 m deep site in Sagami Bay, Japan. (Reprinted from Ohga T and Kitazato H (1997) Seasonal changes in bathyal foraminiferal populations in response to the flux of organic matter (Sagami Bay, Japan). *Terra Nova* 9: 33–37; with permission from Blackwell Science Ltd.)

factor is the degree of undersaturation of the bottom water in calcium carbonate. This abyssal species is found typically in the carbonate-corrosive (and highly oligotrophic) environment between the calcite lysocline and the CCD, a zone that may coincide approximately with AABW. Where water masses are more poorly delineated, as in the Indian and Pacific Oceans, links with faunal distributions are less clear.

During the past 15 years, attention has focused on the impact on foraminiferal ecology of organic matter fluxes to the seafloor. The abundance of dead foraminiferal shells >150 μm in size correlates well with flux values. There is also compelling evidence that the distributions of species and species associations are linked to flux intensity. Infaunal species, such as *Melonis barleeanum*, *Uvigerina peregrina*,

Chilostomella ovoidea and *Globobulimina affinis*, predominate in organically enriched areas, e.g. beneath upwelling zones. Epifaunal species such as *Cibicidoides wuellerstorfi* and *Nuttallides umbonifera* are common in oligotrophic areas, e.g. the central oceanic abyss. In addition to flux intensity, the degree of seasonality of the food supply (i.e., whether it is pulsed or continuous) is a significant factor. *Epistominella exigua*, one of the opportunists that exploit phytodetritus, occurs in relatively oligotrophic areas where phytodetritus is deposited seasonally.

Recent analysis of a large dataset relating the relative abundance of 'live' (stained) foraminiferal assemblages in the north-east Atlantic and Arctic Oceans to flux rates to the seafloor has provided a quantitative framework for these observations. Although species are associated with a wide flux range, this range diminishes as a species become relatively more abundant and conditions become increasingly optimum for it. When dominant occurrences (i.e., where species represent a high percentages of the fauna) are plotted against flux and water depth, species fall into fields bounded by particular flux and depth values (**Figure 8**). Despite a good deal of overlap, it is possible to distinguish a series of dominant species that succeed each other bathymetrically on relatively eutrophic continental slopes and other species that dominate on the more oligotrophic abyssal plains.

Other environmental attributes undoubtedly modify the species composition of foraminiferal assemblages in the deep sea. Agglutinated species with tubular or spherical tests are found in areas where the seafloor is periodically disturbed by strong currents capable of eroding sediments. Forms projecting into the water column may be abundant where steady flow rates convey a continuous supply of suspended food particles. Other species associations may be linked to sedimentary characteristics.

Low-oxygen Environments

Oxygen availability is a particularly important ecological parameter. Since oxygen is consumed during the degradation of organic matter, concentrations of oxygen in bottom water and sediment pore water are inversely related to the organic flux derived from surface production. In the deep sea, persistent oxygen depletion ($O_2 < 1$ ml l^{-1}) occurs at bathyal depths (<1000 m) in basins (e.g., on the California Borderland) where circulation is restricted by a sill and in areas where high primary productivity resulting from the upwelling of nutrient-rich water leads to the development of an oxygen minimum zone (OMZ; e.g., north-west Arabian Sea and the Peru margin). Subsurface sediments also represent an oxygen-limited setting, although oxygen penetration is generally greater in oligotrophic deep-sea sediments than in fine-grained sediments on continental shelves.

On the whole, foraminifera exhibit greater tolerance of oxygen deficiency than most metazoan taxa, although the degree of tolerance varies among species. Oxygen probably only becomes an important limiting factor for foraminifera at concentrations well below 1 ml l^{-1}. Some species are abundant at levels of 0.1 ml l^{-1} or less. A few apparently live in permanently anoxic sediments, although anoxia sooner or later results in death when accompanied by high concentrations of hydrogen sulfide. Oxygen-deficient areas are characterized by high foraminiferal densities but low, sometimes very low (<10), species numbers. This assemblage structure (high dominance, low species richness) arises because (i) low oxygen

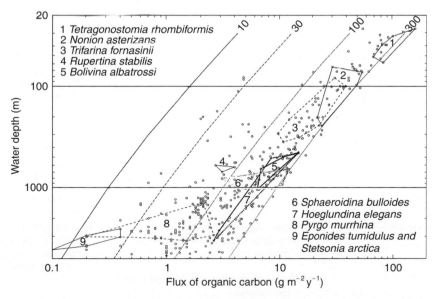

Figure 8 Dominant 'live' (rose Bengal-stained) occurrences of foraminiferal species in relation to water depth and flux or organic carbon to seafloor in the North Atlantic from the Guinea Basin to the Arctic Ocean. Each open circle corresponds to a data point. The polygonal areas indicate the combination of water depth and flux conditions under which nine different species are a dominant faunal component. The diagonal lines indicate levels of primary production (10, 30, 100, 300 g m^{-2} y^{-1}) that result in observed flux rates. Based on $>250 \, \mu m$ sieve fraction plus 63–250 μm fraction from Guinea Basin and Arctic Ocean. (Reprinted from Altenbach AV, Pflaumann U, Schiebel R *et al.* (1999) Scaling percentages and distribution patterns of benthic foraminifera with flux rates of organic carbon. *Journal of Foraminiferal Research* 29: 173–185; with permission from The Cushman Foundation.)

concentration acts as a filter that excludes non-tolerant species and (ii) the tolerant species that do survive are able to flourish because food is abundant and predation is reduced. Utrastructural studies of some species have revealed features, e.g., bacterial symbionts and unusually high abundances of peroxisomes, that may be adaptations to extreme oxygen depletion. In addition, mitachondria-laden pseudopodia have the potential to extend into overlying sediment layers where some oxygen may be present.

Many low-oxygen-tolerant foraminifera belong to the Orders Rotaliida and Buliminida. They often have thin-walled, calcareous tests with either flattened, elongate biserial or triserial morphologies (e.g., *Bolivina, Bulimina, Globobulimina, Fursenkoina, Loxotomum, Uvigerina*) or planispiral/lenticular morphologies (e.g., *Cassidulina, Chilostomella, Epistominella, Loxotomum, Nonion, Nonionella*). Some agglutinated foraminifera, e.g.,

Textularia, Trochammina (both multilocular), *Bathysiphon*, and *Psammosphaera* (both unilocular), are also abundant. However, miliolids, allogromiids, and other soft-shelled foraminifera are generally rare in low-oxygen environments. It is important to note that no foraminiferal taxon is currently known to be confined entirely to oxygen-depleted environments.

Deep-sea Foraminifera in Paleo-oceanography

Geologists require proxy indicators of important environmental variables in order to reconstruct ancient oceans. Benthic foraminifera provide good proxies for seafloor parameters because they are widely distributed, highly sensitive to environmental conditions, and abundant in Cenozoic and Cretaceous deep-sea sediments (note that deep-sea

Table 2 Benthic foraminiferal proxies or indicators (both faunal and chemical) useful in paleo-oceanographic reconstruction

Environmental parameter/property	Proxy or indicator	Remarks
Water depth	Bathymetric ranges of abundant species in modern oceans	Depth zonation largely local although broad distinction between shelf, slope and abyssal depth zones possible
Distribution of bottom water masses	Characteristic associations of epifaunal species	Relations between species and water masses may reflect lateral advection
Carbonate corrosiveness of bottom water	Abundance of *Nuttallides umbonifera*	Corrosive bottom water often broadly corresponds to Antarctic Bottom Water
Deep-ocean thermohaline circulation	Cd/Ca ratios and $\delta^{13}C$ values for calcareous tests	Proxies reflect 'age' of bottom watermasses; i.e., period of time elapsed since formation at ocean surface
Oxygen-deficient bottom-water and pore water	Characteristic species associations; high-dominance, low-diversity assemblages	Species not consistently associated with particular range of oxygen concentrations and also found in high-productivity areas
Primary productivity	Abundance of foraminiferal tests > 150 μm	Transfer function links productivity to test abundance (corrected for differences in sedimentation rates between sites) in oxygenated sediments
Organic matter flux to seafloor	(i) Assemblages of high productivity taxa (e.g. *Globobulimina, Melonis barleeanum*) (ii) Ratio between infaunal and epifaunal morphotypes (iii) Ratio between planktonic and benthic tests	Assemblages indicate high organic matter flux to seafloor, with or without corresponding decrease in oxygen concentrations
Seasonality in organic matter flux	Relative abundance of 'phytodetritus species'	Reflects seasonally pulsed inputs of labile organic matter to seafloor
Methane release	Large decrease (2–3‰) in $\delta^{13}C$ values of benthic and planktonic tests	Inferred sudden release of ^{12}C enriched methane from clathrate deposits following temperature rise

sediments older than the middle Jurassic age have been destroyed by subduction, except where preserved in ophiolite complexes).

Foraminiferal faunas, and the chemical tracers preserved in the tests of calcitic species, can be used to reconstruct a variety of paleoenvironmental parameters and attributes. The main emphasis has been on organic matter fluxes and bottom-water/pore water oxygen concentrations (inversely related parameters), the distribution of bottom-water masses, and the development of thermohaline circulation (**Table 2**). Modern deep-sea faunas became established during the Middle Miocene (10–15 million years ago), and these assemblages can often be interpreted in terms of modern analogues. This approach is difficult or impossible to apply to sediments from the Cretaceous and earlier Cenozoic, which contain many foraminiferal species that are now extinct. In these cases, it can be useful to work with test morphotypes (e.g., trochospiral, cylindrical, biserial/triserial) rather than species. The relative abundance of infaunal morphotypes, for example, has been used as an index of bottom-water oxygenation or relative intensities of organic matter inputs. The trace element (e.g., cadmium) content and stable isotope ($\delta^{13}C$; i.e., the deviation from a standard $^{12}C:^{12}C$ ratio) chemistry of the calcium carbonate shells of benthic foraminifera provide powerful tools for making paleo-oceanographic reconstructions, particularly during the climatically unstable Quaternary period.

The cadmium/calcium ratio is a proxy for the nutrient (phosphate) content of sea water that reflects abyssal circulation patterns. Carbon isotope ratios also reflect deep-ocean circulation and the strength of organic matter fluxes to the seafloor.

It is important to appreciate that the accuracy with which fossil foraminifera can be used to reconstruct ancient deep-sea environments is often limited. These limitations reflect the complexities of deep-sea foraminiferal biology, many aspects of which remain poorly understood. Moreover, simple relationships between the composition of foraminiferal assemblages and environmental variables are elusive, and it is often difficult to identify faunal characteristics that can be used as precise proxies for paleo-oceanographic parameters. For example, geologists often wish to establish paleobathymetry. However, the bathymetric distributions of foraminiferal species are inconsistent and depend largely on the organic flux to the seafloor, which decreases with increasing depth (**Figure 8**) and is strongly influenced by surface productivity. Thus, foraminifera can be used only to discriminate in a general way between shelf, slope, and abyssal faunas, but not to estimate precise paleodepths. Oxygen concentrations and organic matter inputs are particularly problematic. Certain species and morphotypes dominate in low-oxygen habitats that also are usually characterized by high organic loadings. However, the same foraminifera may occur in organically enriched settings where oxygen levels are

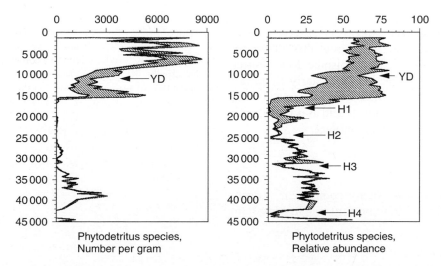

Figure 9 (A) Absolute (specimens per gram of dry sediment) and (B) relative (percentage) abundances of *Alabaminella weddellensis* and *Epistominella exigua* (>63 μm fraction) in a long-sediment core from the North Atlantic (50°41.3′N, 21°51.9′W, 3547 m water depth). In modern oceans, these two species respond to pulsed inputs of organic matter ('phytodetritus') derived from surface primary production. Note that they increased in abundance around 15 000 years ago, corresponding to the main Northern Hemisphere deglaciation and the retreat of the Polar Front. Short period climatic fluctuations (YD = Younger Dryas; H1–4 = Heinrich events, periods of very high meltwater production) are also evident in the record of these two species. (Reprinted from Thomas E, Booth L, Maslin M and Shackleton NJ (1995). Northeast Atlantic benthic foraminifera during the last 45 000 years: change in productivity seen from the bottom up. *Paleoceanography*. 10: 545–562; with permission from the American Geophysical Union.)

not severely depressed, making it difficult for paleo-oceanographers to disentangle the influence of these two variables. Finally, biological factors such as microhabitat preferences and the exploitation of phytodetrital aggregates ('floc') influence the stable isotope chemistry of foraminiferal tests.

There are many examples of the use of benthic foraminiferal faunas to interpret the geological history of the oceans. Only one is given here. Cores collected at 50°41′N, 21°52′W (3547 m water depth) and 58°37′N, 19°26′W (1756 m water depth) were used by E. Thomas and colleagues to study changes in the North Atlantic over the past 45 000 years. The cores yielded fossil specimens of two foraminiferal species, *Epistominella exigua* and *Alabaminella weddellensis*, both of which are associated with seasonal inputs of organic matter (phytodetritus) in modern oceans. In the core from 51°N, these 'phytodetritus species' were uncommon during the last glacial maximum but increased sharply in absolute and relative abundance during the period of deglaciation 15 000–16 000 years ago (**Figure 9**). At the same time there was a decrease in the abundance of *Neogloboquadrina pachyderma*, a planktonic foraminifer found in polar regions, and an increase in the abundance of *Globigerina bulloides*, a planktonic species characteristic of warmer water. These changes were interpreted as follows. Surface primary productivity was low at high latitudes in the glacial North Atlantic, but was much higher to the south of the Polar Front. At the end of the glacial period, the ice sheet shrank and the Polar Front retreated northwards. The 51°N site was now overlain by more productive surface water characterized by a strong spring bloom and a seasonal flux of phytodetritus to the seafloor. This episodic food source favored opportunistic species, particularly *E. exigua* and *A. weddellensis*, which became much more abundant both in absolute terms and as a proportion of the entire foraminiferal assemblage.

Conclusions

Benthic foraminifera are a major component of deep-sea communities, play an important role in ecosystem functioning and biogeochemical cycling, and are enormously diverse in terms of species numbers and test morphology. These testate (shell-bearing) protists are also the most abundant benthic organisms preserved in the deep-sea fossil record and provide powerful tools for making paleo-oceanographic reconstructions. Our understanding of their biology has advanced considerably during the last two decades, although much remains to be learnt.

See also

Abrupt Climate Change. Benthic Foraminifera. Benthic Organisms Overview. Cenozoic Oceans – Carbon Cycle Models. Ocean Carbon System, Modeling of. Radiocarbon. Stable Carbon Isotope Variations in the Ocean.

Further Reading

Fischer G and Wefer G (1999) *Use of Proxies in Paleoceanography: Examples from the South Atlantic*. Berlin: Springer-Verlag.

Gooday AJ, Levin LA, Linke P, and Heeger T (1992) The role of benthic foraminifera in deep-sea food webs and carbon cycling. In: Rowe GT and Pariente V (eds.) *Deep-Sea Food Chains and the Global Carbon Cycle*, pp. 63–91. Dordrecht: Kluwer Academic.

Jones RW (1994) *The Challenger Foraminifera*. Oxford: Oxford University Press.

Loeblich AR and Tappan H (1987) *Foraminiferal Genera and their Classification*, vols 1, 2. New York: Van Nostrand Reinhold.

Murray JW (1991) *Ecology and Palaeoecology of Benthic Foraminifera*. New York: Wiley; Harlow: Longman Scientific and Technical.

SenGupta BK (ed.) (1999) *Modern Foraminifera*. Dordrecht: Kluwer Academic.

Tendal OS and Hessler RR (1977) An introduction to the biology and systematics of Komokiacea. *Galathea Report* 14: 165–194, plates 9–26.

Van der Zwan GJ, Duijnstee IAP, den Dulk M, *et al.* (1999) Benthic foraminifers:: proxies or problems? A review of paleoecological concepts. *Earth Sciences Reviews* 46: 213–236.

PAST CLIMATE FROM CORALS

A. G. Grottoli, University of Pennsylvania,
Philadelphia, PA, USA

Introduction

The influence of the tropics on global climate is well recognized. Our ability to understand, model and predict the interannual, decadal and long-term variability in tropical climate depends on our knowledge of past climate. However, our understanding of the natural variability in tropical climate is limited because long-term instrumental records prior to 1950 are sparse or nonexistent in many tropical regions. Continuous satellite monitoring did not begin until the 1970s and *in situ* equatorial Pacific Ocean monitoring has only existed since the 1980s. Therefore, we depend on proxy records to provide information about past climate. Proxy records are indirect measurements of the physical and chemical structure of past environmental conditions chronicled in natural archives such as ice cores, sediment cores, coral cores and tree rings.

In the tropical oceans, the isotopic, trace and minor elemental signatures of coral skeletons can vary as a result of environmental conditions such as temperature, salinity, cloud cover and upwelling. As such, coral cores offer a suite of proxy records with potential for reconstructing tropical paleoclimate on intraannual-to-centennial timescales. Massive, symbiotic stony corals are good tropical climate proxy recorders because: (1) they are widely distributed throughout the tropics; (2) their unperturbed annual skeletal banding pattern offers excellent chronological control; (3) they incorporate a variety of climate tracers from which paleo sea surface temperature (SST), sea surface salinity (SSS), cloud cover, upwelling, ocean circulation, ocean mixing patterns, and other climatic and oceanic features can be reconstructed; (4) their proxy records can be almost as good as instrumental records; (5) their records can span several centuries; and (6) their high skeletal growth rate (usually in the range of $5-25\,mm\ y^{-1}$) permits subseasonal sampling resolution. Thus proxy records in corals provide the best means of obtaining long seasonal-to-centennial timescale paleoclimate information in the tropics.

Coral-based paleoclimate research has grown tremendously in the last few decades. Most of the published records come from living corals and report the reconstructed SST, SSS, rainfall, water circulation pattern, or some combination of these variables. Records from fossil and deep-sea corals are also increasing. The goal of this chapter is to provide an overview of the current state of coral-based paleoclimate research and a list of further readings for those interested in more detailed information.

Coral Biology

General

Corals are animals in the cnidarian family, order Scleractinia (class Anthozoa). Their basic body plan consists of a polyp containing unicellular endosymbiotic algae known as zooxanthellae overlaying a calcium carbonate exoskeleton (**Figure 1**). Most coral species are colonial and include many polyps interconnected by a lateral layer of tissue. The polyp, consisting of tentacles (used to capture prey), an oral opening and a gastrovascular cavity, has three tissue layers: the epidermis, mesoglea, and gastrodermis. The symbiotic zooxanthellae are located in the gastrodermis. Corals deposit a calcium carbonate ($CaCO_3$) aragonite skeleton below the basal epidermis. Corals reproduce sexually during mass-spawning events by releasing egg and/or sperm, or egg–sperm bundles into the water column. Mass spawning events typically occur a few times a year for each species, and are triggered by the lunar cycle. Corals can also reproduce asexually by fragmentation.

Animal–Zooxanthellae Symbiosis

Corals acquire the greater part of their food energy by two mechanisms: photosynthesis and heterotrophy (direct ingestion of zooplankton and other

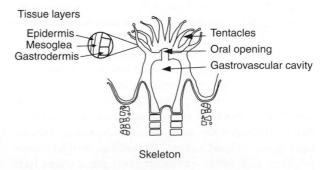

Figure 1   Cross-section of coral polyp and skeleton.

organic particles in the water column by the coral animal). Photosynthesis is carried out by the endo-symbiotic zooxanthellae. The bulk of photo-synthetically fixed carbon is translocated directly to the coral host. In some cases, the coral animal can obtain all of its daily energy requirements via photosynthesis alone. In general, as light intensity increases, photosynthesis increases.

Skeleton

Corals deposit skeleton below the basal epidermis. Typically, corals deposit one high-density and one low-density band of skeleton each year. The high-density band has thicker skeletal elements than the low-density band. Each band is often composed of several finer bands called dissepiments, deposited directly at the base of the coral tissue. At discrete intervals, the polyp presumably detaches from the dissepiment, and begins to lay down a new skeletal dissepiment. Some evidence suggests that dissepi-ments may form on a lunar cycle.

High- and low-density bands are deposited sea-sonally. Overall, high-density bands form during suboptimal temperature conditions and low-density bands form during optimal temperature conditions. At higher latitudes (i.e., Hawaii, Florida) optimal growth temperature occurs in summer. At lower latitudes (i.e., Galápagos, equatorial Pacific regions, Australian Great Barrier Reef), optimal growth temperatures occur in the cooler months. The width and density of growth bands also vary with en-vironmental variables such as light, sedimentation, season length, and salinity. In general, as light levels decrease due to increased cloud cover, increased sedimentation or due to increasing depth, maximum linear skeletal extension decreases, calcification de-creases, and skeletal density increases.

The Interpretation of Isotopes, Trace Elements, and Minor Elements in Corals

Several environmental variables can be reconstructed by measuring changes in the skeletal isotope ratios, trace and minor elemental composition, and growth rate records in coral cores (**Table 1**). The width, density, and chemical composition of each band are generally thought to reflect the average environ-mental conditions that prevailed during the time over which that portion of the skeleton was calcified. Reconstructions of seawater temperature, salinity, light levels (cloud cover), upwelling, nutrient com-position and other environmental parameters have been obtained from coral records (**Table 1**).

Table 1 Environmental variable(s) that can be reconstructed from coral skeletal isotopes, trace and minor elements, and growth records

Proxy	Environmental variable
Isotopes	
$\delta^{13}C$	Light (seasonal cloud cover), nutrients/zooplankton levels
$\delta^{18}O$	Sea surface temperature, sea surface salinity
$\Delta^{14}C$	Ocean ventilation, water mass circulation
Trace and minor elements	
Sr/Ca	Sea surface temperature
Mg/Ca	Sea surface temperature
U/Ca	Sea surface temperature
Mn/Ca	Wind anomalies, upwelling
Cd/Ca	Upwelling
$\delta^{11}B$	pH
F	Sea surface temperature
Ba/Ca	Upwelling, river outflow, sea surface temperature
Skeleton	
Skeletal growth bands	Light (seasonal changes), stress, water motion, sedimentation, sea surface temperature
Fluorescence	River outflow

Method

Continuous records of past tropical climate con-ditions can be obtained by extracting a core from an individual massive coral head along its major axis of growth. Typically, this involves placing a coring de-vice on the top and center of the coral head (**Figure 2A**). The extracted core is cut longitudinally into slabs ranging in thickness from 0.7 to 1 cm that are then X-rayed. X-ray-positive prints reveal the banding pattern of the slab and are used: (1) as a guide for sample drilling and (2) to establish a chronology for the entire coral record when the banding pattern is clear (**Figure 2B**). Samples are drilled out along the major axis of growth by grinding the skeletal material with a diamond-tipped dental drill. For high-resolution climate re-constructions, samples are extracted every millimeter or less down the entire length of the core. Since corals grow about 5–15 mm per year, this sampling method can yield approximately bimonthly-to-monthly resolution. Much higher resolution sam-pling is possible, yielding approximately weekly samples, but this is not commonly performed. In most cases, the $\delta^{13}C$ (the per mil deviation of the ratio of $^{13}C/^{12}C$ relative to the Peedee Belemnite (PDB) Limestone Standard) and $\delta^{18}O$ (ratio of $^{18}O/^{16}O$ relative to PDB) values of each sample are

Porites sp.

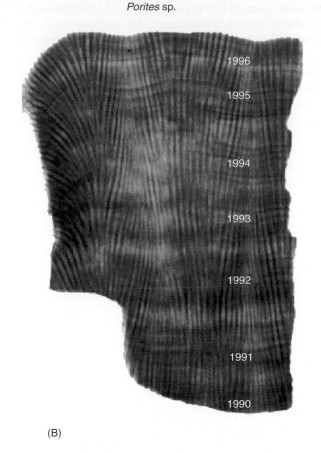

Figure 2 Collecting coral cores. (A) Coral core being extracted from top and center of an individual massive coral head using a pneumatic coring device. (Photo courtesy of M. Kazmers/Shark Song Tax ID #374-50-5314.) (B) X-ray positive print reveals the banding pattern of the slab and is used to help establish a chronology for the entire coral record.

measured. Since the $\delta^{13}C$ and/or $\delta^{18}O$ compositions of corals usually have a strong seasonal component, they are often used to establish an accurate chronology and/or to confirm the chronology established from the X-rays.

Temperature and Salinity Reconstructions

Coral skeletal $\delta^{18}O$ reflects a combination of the local SST and SSS. In the many regions of the tropical ocean where the natural variation in salinity is small, changes in coral skeletal $\delta^{18}O$ primarily reflect changes in SST. The $\delta^{18}O$ of coral skeleton responds to changes in temperature usually according to the standard paleotemperature relationship for carbonates. Based on empirical studies, a 1°C increase in water temperature corresponds to a decrease of about 0.22‰ (parts per thousand) in $\delta^{18}O$. Precipitation has a low $\delta^{18}O$ value relative to that of sea water. Therefore, in regions with pronounced variability in rainfall and/or river runoff, coral $\delta^{18}O$ values reflect changes in SSS. Thus, depending on the nature of the coral collection site, the $\delta^{18}O$ record is

used to reconstruct the SST and/or SSS. Additional studies show that other proxy indicators of temperature include the ratios of strontium/calcium (Sr/Ca), magnesium/calcium (Mg/Ca), and uranium/calcium (U/Ca), fluorine levels (F) and skeletal band thickness (**Table 1**). The ratios of Sr/Ca, Mg/Ca, and U/Ca incorporated into the skeleton is largely determined by the temperature-dependent distribution coefficient of Sr/Ca, Mg/Ca, and U/Ca between aragonite and sea water. As temperatures increase, the Sr/Ca and U/Ca ratios decrease and the Mg/Ca ratio increases.

Cloud Cover and Upwelling

$\delta^{13}C$ seems to indicate seasonal changes in cloud cover and upwelling. Thus far, only a small number of studies have used $\delta^{13}C$ records to confirm seasonal rainfall patterns established using the $\delta^{18}O$ signature. Only one study has directly linked a $\delta^{13}C$ record with seasonal upwelling. $\delta^{13}C$ in coral skeletons has been difficult to use as a paleoclimate tracer because it is heavily influenced by metabolic processes, namely photosynthesis and heterotrophy. Firstly, as light levels decrease due to cloud cover, the rate of photosynthesis by the coral's symbiotic zooxanthellae decreases, and skeletal $\delta^{13}C$ decreases. The reverse occurs when light levels increase. Secondly, zooplankton have a low $\delta^{13}C$ value relative to coral. During upwelling events in the Red Sea, nutrient and zooplankton level increases have been linked to decreases in coral skeletal $\delta^{13}C$ values. Other upwelling tracers include cadmium (Cd) and barium (Ba) concentrations, and $\Delta^{14}C$. Cadmium and barium are trace elements whose concentrations are greater in deep water than in surface water. During upwelling events, deep water is driven to the surface and cadmium/calcium (Cd/Ca) and barium/ calcium (Ba/Ca) ratios in the surface water, and consequently in the coral skeleton, increase (Table 1). Although SST also influences Ba/Ca ratios, most of the variation in Ba/Ca ratios in corals is due to nutrient fluxes and upwelling. $\Delta^{14}C$ is also an excellent tracer for detecting upwelling and changes in seawater circulation ($\Delta^{14}C$ is the per mil deviation of the ratio of $^{14}C/^{12}C$ relative to a nineteenth century wood standard). For example, in the eastern equatorial Pacific Ocean, the $\Delta^{14}C$ value of deep water tends to be very low relative to the $\Delta^{14}C$ of surface water. Here, increased upwelling or increases in the proportion of deep water contributing to surface water results in a decrease in the $\Delta^{14}C$ of the coral skeleton. Manganese (Mn) is a trace element whose concentration is highest in surface waters and decreases with depth. Therefore, during upwelling events, Mn/Ca ratios decrease. The ratio of Mn/Ca can also record prolonged and sustained changes in winds. In at least one case, Mn/Ca ratios from a Tarawa Atoll coral increased during El Niño events as a result of strong and prolonged wind reversals that had remobilized manganese from the lagoon sediments.

Other Proxy Indicators

Other environmental parameters that can be inferred from coral skeleton structure and composition are river outflow (fluorescence bands) and pH (boron isotope levels) (Table 1). Large pulses in river outflow can result in an ultraviolet-sensitive fluorescent band in the coral record. New evidence strongly suggests that the fluorescent patterns in coral skeletal records are due to changes in skeletal density, not terrestrially derived humics as previously thought. Variations in salinity associated with fresh water discharge pulses from rivers appear to cause changes in coral skeletal growth density which can be observed in the skeletal fluorescence pattern. In the case of boron, $\delta^{11}B$ levels in sea water increase as pH increases. Changes in the pH at the site of coral calcification seem to reflect changes in productivity of the symbiotic zooxanthellae. As photosynthesis increases, pH increases, and $\delta^{11}B$ levels in the coral skeleton increase.

Coral Records: What has been Learned About Climate From Corals?

To date, there are over 100 sites where coral cores have been recovered and analyzed (Figure 3). Of these, at least 22 have records that exceed 120 years in length. In most cases, $\delta^{13}C$ and $\delta^{18}O$ have been measured at annual-to-subannual resolution. $\delta^{18}O$ as a SST and/or SSS proxy is the best understood and most widely reported of all the coral proxy measurements. In a few cores, other isotopic, trace, and minor elements, or skeletal density and growth measurements have also been made.

Typically, coral-derived paleoclimate records are studied on three timescales: seasonal, interannual-to-decadal, and long-term trends. The seasonal variation refers to one warm and one cool phase each year. An abrupt shift in the proxy's long-term mean often indicates a decadal modulation in the data. Long-term trends are usually associated with a gradual increase or decrease in the measured proxy over the course of several decades or centuries. The following sections explore some of the seasonal, decadal and long-term trends in coral-derived paleoclimate records and some of the limitations associated with interpreting coral proxy records.

Seasonal Variation in Coral Climate Records

Seasonal variation accounts for the single largest percentage of the variance in most coral isotope, trace, and minor element records. In regions such as Japan and the Galápagos with distinct SST seasonality, the annual periodicity in $\delta^{18}O$, Sr/Ca, Mg/Ca, and U/Ca is pronounced. In regions heavily affected by monsoonal rains such as Tarawa Atoll, the seasonal variation in $\delta^{13}C$ is regular and pronounced. $\Delta^{14}C$ also has an annual periodicity in coral from regions with a strong seasonal upwelling regime such as is seen in the Galápagos (see next section). The strength and duration of the upwelling season is reflected in the length and degree of $\Delta^{14}C$ decrease in the coral record.

Interannual-to-decadal Variation in Climate and El Niño Southern Oscillation (ENSO)

The second largest component of the variance in Pacific coral isotope records is associated with the

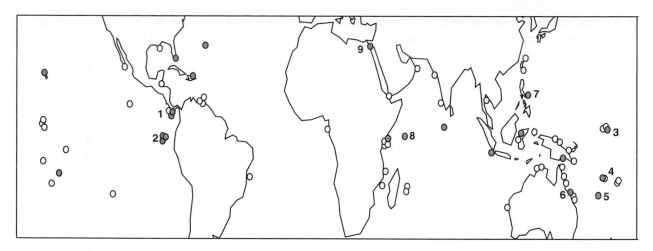

Figure 3 Map indicating the approximate locations of current paleoclimate research. The coral sites involve the work of many investigators and may be incomplete. The $\delta^{18}O$ records from the numbered sites are shown in **Figure 5** and are discussed in the text. ● , sites with records longer than 120 years (most are published); ○, sites where cores have been recovered and data collection is underway. (Reproduced from Gagan *et al.* (2000), with permission from Elsevier Science.)

interannual-to-decadal variation in the El Niño Southern Oscillation (ENSO)[1]. Several of the longer Pacific $\delta^{18}O$ coral records reveal that the frequency of ENSO has changed on decadal timescales over the past few centuries. Over the last 300 years, the dominant mode of ENSO recorded by a Galápagos coral has been at 4.6 years. However, during that time period there have been shifts in that mode from 4.6 to 7 years during the 1600s, 3–4.6 years from 1700–1750, and 3.5 years from 1800–1850. These major shifts in ENSO frequency may indicate major reorganizations in Pacific climate at various intervals over time. A 101-year long $\delta^{18}O$ record from Clipperton Atoll reveals a pronounced period of reduced ENSO frequency from ~1925 to 1940 suggesting a reduced coupling between the eastern and western Pacific. At Clipperton, decadal timescale variability represents the largest percentage of the variance in $\delta^{18}O$ and appears to be related to the processes influencing the Pacific Decadal Oscillation phenomenon (PDO)[2].

Another component of ENSO variability recovered from coral $\delta^{18}O$ records is the shift in rainfall patterns during El Niños associated with: (1) the migration of the Indonesian Low pressure cell to the region of the date line and the equator in the western Pacific, and (2) the northern migration of the intertropical convergence zone (ITCZ) in the eastern Pacific. Eastward migration of the Indonesian Low results in decreased precipitation in the Indian Ocean and increased precipitation in the western and central Pacific. These phenomena are reflected in the $\delta^{18}O$ record of Seychelles and Tarawa Atoll corals, respectively. Decadal variability in the Seychelles

---

[1] ENSO refers to the full range of variability observed in the Southern Oscillation, including both El Niño and La Niña events in the Pacific. The Southern Oscillation Index (SOI) is a measure of the normalized difference in the surface air pressure between Tahiti, French Polynesia and Darwin, Australia. Most of the year, under normal seasonal Southern Oscillation cool phase conditions, easterly trade winds induce upwelling in the eastern equatorial Pacific and westward near-equatorial surface flow. The westward flowing water warms and piles up in the western Pacific creating a warm pool and elevating sea level. The wind-driven-transport of this water from the eastern Pacific leads to an upward tilt of the thermocline and increases the efficiency of the local trade-wind-driven equatorial upwelling to cool the surface resulting in an SST cold tongue that extends from the coast of South America to near the international date line. Normal seasonal Southern Oscillation warm-phase conditions are marked by a relaxation of the zonal component of trade winds, reduced upwelling, and a weakening or reversal of the westward flowing current coupled with a deepening of the thermocline in the eastern equatorial Pacific Ocean, and increased SST in the central and eastern equatorial Pacific. This oscillation between cool and warm phases normally occurs annually. Exaggerated and/or prolonged warm-phase conditions are called El Niño events. They usually last 6–18 months, occur irregularly at intervals of 2–7 years, and average about once every 3–4 years. The SOI is low during El Niño events. Exaggerated and/or prolonged ENSO cool phase conditions are called La El Niña events. They often follow El Niño events (but not necessarily). La El Niña events are marked by unusually low surface temperatures in the eastern and central equatorial Pacific and a high SOI. For a detailed description of ENSO, *see* El Niño Southern Oscillation (ENSO).

---

[2] The PDO appears to be a robust, recurring two-to-three decade pattern of ocean–atmosphere climate variability in the North Pacific. A positive PDO index is characterized by cooler than average SST in the central North Pacific and warmer than average SST in the Gulf of Alaska and along the Pacific Coast of North America and corresponds to warm phases of ENSO. The reverse is true with a negative PDO.

record suggests that regional rainfall variability may originate from the ocean. Decadal variability in a 280-year $\delta^{18}O$ record from a Panamanian coral indicates decadal periods in the strength and position of the ITCZ.

Changes in the decadal variability of coral skeletal $\Delta^{14}C$ reveals information about the natural variability in ocean circulation, water mass movement and ventilation rates in surface water. Biennial-to-decadal shifts in $\Delta^{14}C$ between 1880 and 1955 in a Bermuda coral indicates that rapid pulses of increased mixing between surface and subsurface waters occurred in the North Atlantic Ocean during the past century and that these pulses appeared to correlate with fluctuations in the North Atlantic Oscillation. In a post-bomb Galápagos coral record, abrupt increases in monthly $\Delta^{14}C$ values during the upwelling season after 1976 suggest a decadal time-scale shift in the vertical thermal structure of the eastern tropical Pacific (**Figure 4**). The decadal variability in $\Delta^{14}C$ in the Bermuda and Galápagos records are testimony to the power of coral proxy records to provide information about ocean circulation patterns. Additional $\Delta^{14}C$ records from Nauru, Fanning Island, Great Barrier Reef, Florida, Belize, Guam, Brazil, Cape Verde, French Frigate Shoals, Tahiti, Fiji, Hawaii and a few other locations are either published or in progress. As the number of coral $\Delta^{14}C$ records increases, our understanding of

the relationship between climate and ocean circulation patterns will also increase.

Some decadal-to-centennial trends in climate are consistent among many of the longer $\delta^{18}O$ coral records. For example in **Figure 5**, all six records longer than 200 years show a cooler/dryer period from AD 1800 to 1840. Cooling may be related to enhanced volcanism during this period. Following this cooler/dryer interval, four of the six records show shifts towards warmer/wetter conditions around 1840–1860 and five of the six show another warming around 1925–1940. These abrupt shifts towards warmer/wetter conditions detected in corals from a variety of tropical locations suggest that corals may be responding to global climate forcing.

Long-term Trends in Climate

There are three major long-term trends observed in several coral records: (1) a prolonged cool phase prior to 1900 generally consistent with the Little Ice Age; (2) a gradual warming/freshening trend over the past century; and (3) evidence of increased burning of fossil fuels. First, in three of the four longest $\delta^{18}O$ coral records the cool/dry period of the Little Ice Age is observed from the beginning of their respective records, up to the mid to late 1800s (**Figure 5**). However, the lack of this cool/dry period in the Galápagos coral indicates that the Little Ice Age

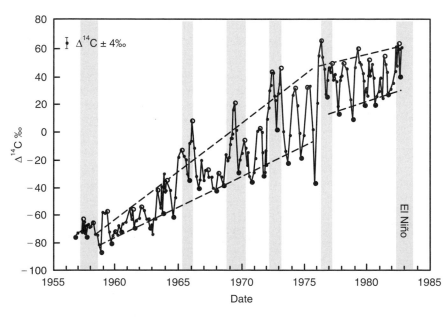

Figure 4   Galápagos coral $\Delta^{14}C\Delta$ record from 1957 to 1983. El Niños are indicated by the shaded bars. ●, upwelling maxima; ○, nonupwelling season. Dashed lines indicated linear trend in the upwelling and nonupwelling seasons. The seasonal variation in $\Delta^{14}C$ is pronounced with high $\Delta^{14}C$ during the nonupwelling season and low $\Delta^{14}C$ values during the upwelling season. A shift in $\Delta^{14}C$ baselines began in 1976. (Reproduced from Guilderson TP and Schrag DP (1998) Abrupt shift in subsurface temperatures in the tropical Pacific associated with changes in El Niño. *Science* 281: 240–243; with permission from the American Association for the Advancement of Science.)

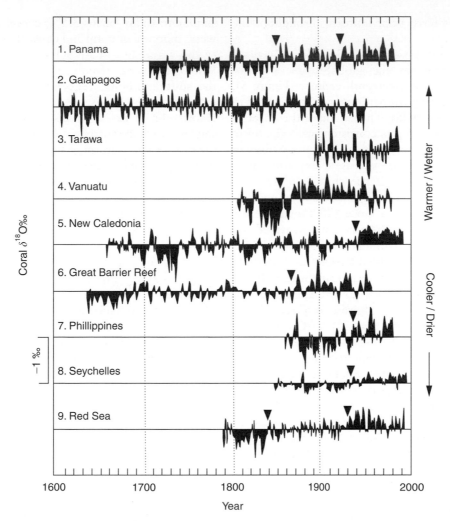

Figure 5 Annual mean coral $\delta^{18}O$ records in the Pacific and Indian Ocean region extending back at least for 100 years (locations of cores are indicated in **Figure 3**). Mean $\delta^{18}O$ values for each site indicated with a horizontal line. Abrupt shifts in $\delta^{18}O$ towards warmer/ wetter conditions indicated by black triangles. Data are from the World Data Center-A for Paleoclimatology, NOAA/NGDC Paleoclimatology Program, Boulder, Colorado, USA (http://www.ngdc.noaa.gov/paleo/corals.html) and the original references. Core details list locality, species name, record length, and original reference: 1, Gulf of Chiriqui, Panama, *Porites lobata* 1708–1984 Linsley BK, Dunbar RB, Wellington GM, Mucciarone DA (1994) A coral-based reconstruction of intertropical convergence zone variability over Central America since 1707. *Journal of Geophysical Research* 99: 9977–9994); 2, Urvina Bay, Galápagos, *Pavona clavus* and *Pavona gigantea*, 1607–1981 Dunbar RG, Wellington GM, Colgan MW, Glynn PW (1994) Eastern Pacific sea surface temperature since 1600 A.D.: the $\delta^{18}O$ record of climate variability in Galápagos corals. *Paleoceanography* 9: 291–315; 3, Tarawa Atoll, Republic of Kiribati, *Porites* spp., 1893–1989 Cole JE, Fairbanks RG, Shen GT (1993) Recent variability in the Southern Oscillation: isotopic results from Tarawa Atoll coral. *Science* 260: 1790–1793; 4, Espiritu Santo, Vanuatu, *Platygyra lamellina*, 1806–1979 Quinn TM, Taylor FW, Crowley TJ (1993) A 173 year stable isotope record from a tropical south Pacific coral. *Quaternary Science Review* 12: 407–418; 5, Amedee Lighthouse, New Caledonia, *Porites lutea*, 1657–1992 Quinn TM, Crowley TJ, Taylor FW, Henin C, Joannot P, Join Y (1998) A multicentury stable isotope record from a New Caledonia coral: Interannual and decadal sea surface temperature variability in the southwest Pacific since 1657 A.D. *Paleoceanography* 13: 412–426; 6, Abraham Reef, Great Barrier Reef, Australia, *Porites australiensis*, 1635–1957 Druffel ERM, Griffin S (1993) Large variations of surface ocean radiocarbon: evidence of circulation changes in the southwestern Pacific. *Journal of Geophysical Research* 98: 20 249–22 259; 7, Cebu, Philippines, *Porites lobata*, 1859–1980 Pätzold J (1986) Temperature and CO_2 changes in the tropical surface waters of the Philippines during the past 120 years: record in the stable isotopes of hermatypic corals. Berichte Reports, Gol.-Paläont, Inst. Univ. Kiel, 12; 8, Mahe Island, Seychelles, *Porites lutea*, 1846–1995 Charles CD, Hunter ED, Fairbanks RG (1997) Interaction between the ENSO and the Asian monsoon in a coral record of tropical climate. *Science* 277: 925–928; 9, Aqaba, Red Sea, *Porites* sp., 1788–1992 Heiss GA (1996) Annual band width variation in *Porites* sp. from Aquaba, Gulf of Aquaba, Red Sea. *Bulletin of Marine Science* 59: 393–403; (Reproduced from Gagan (2000), with permission from Elsevier Science.)

effects may not have been uniform throughout the tropical oceans.

Second, this cool phase was followed by a general warming/freshening of the global tropical ocean beginning during the nineteenth century (**Figure 5**). This overall warming/freshening trend is observed in seven of the nine records. The timing of the onset of this warming/freshening is consistent with the onset of industrialization and the consequent increases in greenhouse gases due to increased emissions from fossil fuel consumption. If the shift in $\delta^{18}O$ were solely due to increases in SST, it would be equivalent to an increase of 0.3–2.0°C since 1800. Instrumental data indicate that the tropics only warmed by ~ 0.5°C since 1850. The influence of SSS on $\delta^{18}O$ is probably responsible for the difference and needs to be taken into account when interpreting $\delta^{18}O$ records. Although Sr/Ca ratios are thought to be un-affected by SSS, only a few shorter coral records are currently published. Until recently, Sr/Ca measure-ments were very time-consuming. With recently developed technology, the use of Sr/Ca as a paleo-thermometer proxy should increase. Two main limitations exist with the correct interpretation of decadal and long-term $\delta^{18}O$ trends: (1) an inter-decadal cycle of unknown origin is commonly identified in long coral $\delta^{18}O$ records; and (2) long-term trends of increasing $\delta^{18}O$ are observed in some coral while other coral $\delta^{18}O$ records show a de-creasing trend. Whether these trends are due to biological processes or are the result of gradual environmental changes (i.e., global warming) is unclear.

Finally, evidence of increased fossil fuel emission into the atmosphere can be seen in the general de-crease in $\Delta^{14}C$ from 1850 to 1955 in shallow corals from the Atlantic and Pacific Oceans. This phe-nomenon, referred to as the Suess Effect, is mainly the result of ^{14}C-free CO_2 produced from combusted fossil fuel entering the atmosphere, the oceans and eventually, the coral skeleton (post-1950, coral $\Delta^{14}C$ values skyrocketed as a result of ^{14}C produced by thermonuclear bombs effectively swamping out the Suess effect).

Fossil Corals

Fossil corals provide windows into past climate. Records covering several decades to centuries offer the opportunity to compare the same three com-ponents (seasonal, interannual-to-decadal, and long term) in climate in the distant past to the present. A 3.0 million-year-old south-western Florida coral re-veals a seasonal $\delta^{18}O$ derived temperature pattern

similar to today but ~ 3.5°C cooler. A North Sula-wesi, Indonesian coral indicates that 124 000 years BP the variability in ENSO was similar to modern ENSO frequency from 1856 to 1976. However, the shift in ENSO frequency observed in modern records after 1976 is not found in the fossil coral record nor in pre-1976 instrumental records. This suggests that the current state of ENSO frequency is outside of the natural range of ENSO variability. Perhaps anthropogenic effects are having an effect on ENSO frequency. Finally, long-term changes in climate can also be reconstructed from fossil coral records. A series of coral records from Vanuatu in-dicate that $\sim 10\,300$ years BP the south-western tropical Pacific was 6.5°C cooler than today fol-lowed by a rapid rise in temperature over the sub-sequent 15 000 years. This rapid rise in temperature lags the post-Younger Dryas warming of the Atlantic by ~ 3000 years suggesting that the mechanism for deglacial climate change may not have been globally uniform. How seasonal, decadal (ENSO) and long-term climate changes varied in the distant past throughout the tropics can be addressed using fossil coral records and can offer us a better idea of the natural variability in tropical climate over geologic time.

Deep-sea Corals

Deep sea corals do deposit calcium carbonate exo-skeleton but do not contain endosymbiotic zoox-anthellae and are not colonial. Their isotopic and trace mineral composition reflects variation in am-bient conditions on the seafloor. Although this does not directly reflect changes in climate on the surface, ocean circulation patterns are tightly coupled with atmospheric climatic conditions. Understanding the history of deep and intermediate water circulation lends itself to a better understanding of climate. For example, the origin of the Younger Dryas cooling event (13 000 to 11 700 years BP) has recently been attributed to a cessation or slowing of North Atlantic deep water formation and subsequent reduction in heat flux. Isotopic evidence from deep-sea corals suggests that profound changes in intermediate-water circulation also occurred during the Younger Dryas. Other studies of deep-sea corals show rapid changes in deep ocean circulation on decadal-to-centennial timescales at other intervals during the last deglaciation. Reconstructing intermediate and deep ocean circulation patterns and their relationship to climate using isotopic, trace element and minor element records in deep sea coral promises to be an expanding line of paleoclimate research.

Discussion

The geochemical composition of coral skeletons currently offers the only means of recovering multi-century records of seasonal-to-centennial timescale variation in tropical climate. $\delta^{18}O$-derived SST and SSS records are the workhorse of coral-based paleoclimate reconstructions to date. Improved methodologies are now making high resolution, multicentury Sr/Ca records feasible. Since Sr/Ca is potentially a less ambiguous SST recorder, coupling Sr/Ca with $\delta^{18}O$ records could yield more reliable SST and SSS reconstructions. $\delta^{13}C$ as a paleorecorder of seasonal variation in cloud cover and upwelling is also gaining credibility. However, more experimental research needs to be done before $\delta^{13}C$ records can be used more widely for paleoclimate reconstructions. Coral $\Delta^{14}C$ records are highly valued as an ocean circulation/ventilation proxy. Increasing numbers of high-resolution $\Delta^{14}C$ coral records are being published shedding invaluable new light on links between climate and ocean circulation processes. Coral trace and minor element records are also becoming more common and can add critical information about past upwelling regimes, wind patterns, pH, river discharge patterns, and SST.

The growing number of multicentury coral oxygen isotope records is yielding new information on the natural variability in tropical climate. Eastern equatorial Pacific corals track ENSO-related changes in SST and upwelling. Further west, coral records track ENSO-related changes in SST and SSS related to the displacement of rainfall associated with the Indonesian Low. Decadal timescale changes in ENSO frequency and in ocean circulation and water mass movement detected in $\delta^{18}O$ and $\Delta^{14}C$ records, respectively, indicate a major re-organization in Pacific climate at various intervals over time. Long-term trends in coral oxygen isotope records point to a gradual warming/freshening of the oceans over the past century suggesting that the tropics are responding to global forcings.

Although the coral-based paleoclimate records reconstructed to date are impressive, much work remains to be done. It is necessary to develop multiple tracer records from each coral record in order to establish a more comprehensive reconstruction of several concurrent climatic features. In addition, replication of long isotopic and elemental records from multiple sites is invaluable for establishing better signal precision and reproducibility. Coupled with fossil and deep-sea coral records, coral proxy records offer a comprehensive and effective means of reconstructing tropical paleoclimates.

Acknowledgments

I thank B Linsley, E Druffel, T Guilderson, J Adkins and an anonymous reviewer for their comments on the manuscript. I thank the Henry and Camille Dreyfus Foundation for financial support.

See also

El Niño Southern Oscillation (ENSO). Pacific Ocean Equatorial Currents.

Further Reading

Beck JW, Recy J, Taylor F, Edwards RL, and Cabioch G (1997) Abrupt changes in early Holocene tropical sea surface temperature derived from coral records. *Nature* 385: 705–707.

Druffel ERM (1997) Geochemistry of corals: proxies of past ocean chemistry, ocean circulation, and climate. *Proceedings of the National Academy of Sciences of the USA* 94: 8354–8361.

Druffel ERM, Dunbar RB, Wellington GM, and Minnis SS (1990) Reef-building corals and identification of ENSO warming episodes. In: Glynn PW (ed.) *Global Ecological Consequences of the 1982–83 El Niño – Southern Oscillation pp. 233–253* Elsevier Oceanography, Series 52. New York: Elsevier.

Dunbar RB and Cole JE (1993) Coral records of ocean-atmosphere variability. *NOAA Climate and Global Change Program Special Report* No. 10, Boulder, CO: UCAR.

Dunbar RB and Cole JE (1999) *Annual Records of Tropical Systems (ARTS)*. Kauai ARTS Workshop, September 1996. Pages workshop report series 99–1.

Fairbanks RG, Evans MN, Rubenstone JL *et al.* (1997) Evaluating climate indices and their geochemical proxies measured in corals. *Coral Reefs* 16 suppl.: s93–s100.

Felis T, Pätzold J, Loya Y, and Wefer G (1998) Vertical water mass mixing and plankton blooms recorded in skeletal stable carbon isotopes of a Red Sea coral. *Journal of Geophysical Research* 103: 30731–30739.

Gagan MK, Ayliffe LK, Beck JW, *et al.* (2000) New views of tropical paleoclimates from corals. *Quaternary Science Reviews* 19: 45–64.

Grottoli AG (2000) Stable carbon isotopes ($\delta^{13}C$) in coral skeletons. *Oceanography* 13: 93–97.

Linsley BK, Ren L, Dunbar RB, and Howe SS (2000) El Niño Southern Oscillation (ENSO) and decadal-scale climate variability at 10°N in the eastern Pacific from 1893 to 1994: a coral-based reconstruction from Clipperton Atoll. *Paleoceanography* 15: 322–335. http://pangea. stanford. edu/Oceans/ARTS/arts_report/arts_report_ home. html

Hudson JH, Shinn EA, Halley RB, and Lidz B (1981) Sclerochronology: a tool for interpreting past environments. *Geology* 4: 361–364.

NOAA/NGCD Paleoclimatology Program http://www.ngdc.noaa.gov/paleo/corals.thml

Shen GT (1993) Reconstruction of El Niño history from reef corals. *Bull. Inst. fr. études andines* 22(1): 125–158.

Smith JE, Risk MJ, Schwarcz HP, and McConnaughey TA (1997) Rapid climate change in the North Atlantic during the Younger Dryas recorded by deep-sea corals. *Nature* 386: 818–820.

Swart PK (1983) Carbon and oxygen isotope fractionation in scleractinian corals: a review. *Earth-Science Reviews* 19: 51–80.

Weil SM, Buddemeier RW, Smith SV, and Kroopnick PM (1981) The stable isotopic composition of coral skeletons: control by environmental variables. *Geochimica et Cosmochimica Acta* 45: 1147–1153.

Wellington GM, Dunbar RB, and Merlen G (1996) Calibration of stable oxygen isotope signatures in Galapagos corals. *Paleoceanography* 11: 467–480.

CARBON SYSTEM

CARBON CYCLE

C. A. Carlson, University of California, Santa Barbara, CA, USA

N. R. Bates, Bermuda Biological Station for Research, St George's, Bermuda, USA

D. A. Hansell, University of Miami, Miami FL, USA

D. K. Steinberg, College of William and Mary, Gloucester Pt, VA, USA

Introduction

Why is carbon an important element? Carbon has several unique properties that make it an important component of life, energy flow, and climate regulation. It is present on the Earth in many different inorganic and organic forms. Importantly, it has the ability to form complex, stable carbon compounds, such as proteins and carbohydrates, which are the fundamental building blocks of life. Photosynthesis provides marine plants (phytoplankton) with an ability to transform energy from sunlight, and inorganic carbon and nutrients dissolved in sea water, into complex organic carbon materials. All organisms, including autotrophs and heterotrophs, then catabolize these organic compounds to their inorganic constituents via respiration, yielding energy for their metabolic requirements. Production, consumption, and transformation of these organic materials provide the energy to be transferred between all the trophic states of the ocean ecosystem.

In its inorganic gaseous phases (carbon dioxide, CO_2; methane, CH_4; carbon monoxide, CO), carbon has important greenhouse properties that can influence climate. Greenhouse gases in the atmosphere act to trap long-wave radiation escaping from Earth to space. As a result, the Earth's surface warms, an effect necessary to maintain liquid water and life on Earth. Human activities have led to a rapid increase in greenhouse gas concentrations, potentially impacting the world's climate through the effects of global warming. Because of the importance of carbon for life and climate, much research effort has been focused on understanding the global carbon cycle and, in particular, the functioning of the ocean carbon cycle. Biological and chemical processes in the marine environment respond to and influence climate by helping to regulate the concentration of CO_2 in the atmosphere. We will discuss (1) the importance of the ocean to the global carbon cycle; (2) the mechanisms of carbon exchange between the ocean and atmosphere; (3) how carbon is redistributed throughout the ocean by ocean circulation; and (4) the roles of the 'solubility', 'biological,' and 'carbonate' pumps in the ocean carbon cycle.

Global Carbon Cycle

The global carbon cycle describes the complex transformations and fluxes of carbon between the major components of the Earth system. Carbon is stored in four major Earth reservoirs, including the atmosphere, lithosphere, biosphere, and hydrosphere. Each reservoir contains a variety of organic and inorganic carbon compounds ranging in amounts. In addition, the exchange and storage times for each carbon reservoir can vary from a few years to millions of years. For example, the lithosphere contains the largest amount of carbon (10^{23} g C), buried in sedimentary rocks in the form of carbonate minerals ($CaCO_3$, $CaMgCO_3$, and $FeCO_3$) and organic compounds such as oil, natural gas, and coal (fossil fuels). Carbon in the lithosphere is redistributed to other carbon reservoirs on timescales of millions of years by slow geological processes such as chemical weathering and sedimentation. Thus, the lithosphere is considered to be a relatively inactive component of the global carbon cycle (though the fossil fuels are now being added to the biologically active reservoirs at unnaturally high rates). The Earth's active carbon reservoirs contain approximately 43×10^{18} g of carbon, which is partitioned between the atmosphere (750×10^{15} g C), the terrestrial biosphere (2190×10^{15} g C), and the ocean ($39\,973 \times 10^{15}$ g C; **Figure 1**). While the absolute sum of carbon found in the active reservoirs is maintained in near steady state by slow geological processes, more rapid biogeochemical processes drive the redistribution of carbon among the active reservoirs.

Human activities, such as use of fossil fuels and deforestation, have significantly altered the amount of carbon stored in the atmosphere and perturbed the fluxes of carbon between the atmosphere, the terrestrial biosphere, and the ocean. Since the emergence of the industrial age 200 years ago, the release of CO_2 from fossil fuel use, cement manufacture, and deforestation has increased the partial pressure of atmospheric CO_2 from 280 ppm to present day values of 360 ppm; an increase of 25% in the last century (**Figure 2**). Currently, as a result of human activities, approximately 5.5×10^{15} g of

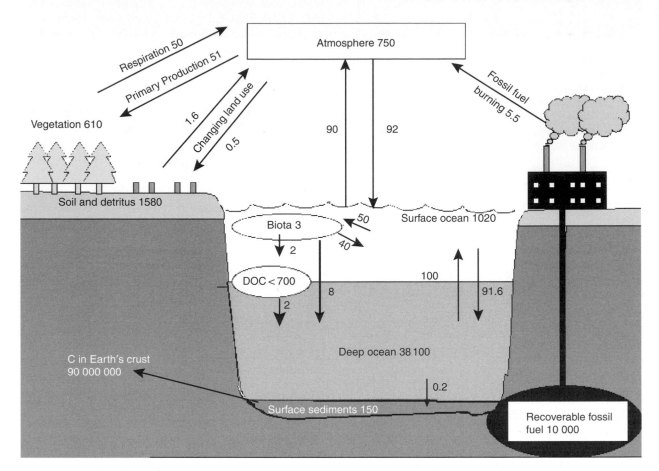

Figure 1   The global carbon cycle. Arrows indicate fluxes of carbon between the various reservoirs of the atmosphere, lithosphere, terrestrial biosphere, and the ocean. All stocks are expressed as 10^{15} g C. All fluxes are decadal means and expressed as 10^{15} g C y^{-1}. (Adapted with permission from Sigenthaler and Sarmiento, 1993), copyright 1993, Macmillan Magazines Ltd.). Data used to construct this figure came from Sigenthaler and Sarmiento (1993), Hansell and Carlson (1998), and Sarmiento and Wofsy (1999).

'anthropogenic' carbon is added to the atmosphere every year. About half of the anthropogenic CO_2 is retained in the atmosphere, while the remaining carbon is transferred to and stored in the ocean and the terrestrial biosphere. Carbon reservoirs that remove and sequester CO_2 from the atmosphere are referred to as carbon 'sinks'. The partitioning of anthropogenic carbon between oceanic and terrestrial sinks is not well known. Quantifying controls on the partitioning is necessary for understanding the dynamics of the global carbon cycle. The terrestrial biosphere may be a significant sink for anthropogenic carbon, but scientific understanding of the causative processes is hindered by the complexity of terrestrial ecosystems.

Global ocean research programs such as Geochemical Ocean Sections (GEOSEC), the Joint Global Ocean Flux Study (JGOFS), and the JGOFS/ World Ocean Circulation Experiment (WOCE) Ocean CO_2 Survey have resulted in improvements in our understanding of physical circulation and biological processes of the ocean. These studies have also allowed oceanographers to better constrain the role of the ocean in CO_2 sequestration compared to terrestrial systems. Based on numerical models of ocean circulation and ecosystem processes, oceanographers estimate that 70% (2×10^{15} g C) of the anthropogenic CO_2 is absorbed by the ocean each year. The fate of the remaining 30% (0.75×10^{15} g) of anthropogenic CO_2 is unknown. Determining the magnitude of the oceanic sink of anthropogenic CO_2 is dependent on understanding the interplay of various chemical, physical, and biological factors.

Oceanic Carbon Cycle

The ocean is the largest reservoir of the Earth's active carbon, containing $39\,973 \times 10^{15}$ g C. Oceanic carbon occurs as a variety of inorganic and organic

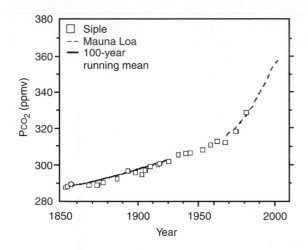

Figure 2 Atmospheric CO_2 concentrations from 1850 to 1996. These data illustrate an increase in atmospheric CO_2 concentration from pre-industrial concentration of 280 ppmv to present-day concentrations of 360 ppmv. Human activities of fossil fuel burning and deforestation have caused this observed increase in atmospheric CO_2. (Adapted from Houghton *et al.* (1996) with permission from Intergovernmental Panel on Climate Change (IPCC). The original figure was constructed from Siple ice core data and (from 1958) data collected at the Mauna Loa sampling site.)

forms, including dissolved CO_2, bicarbonate (HCO_3^-), carbonate (CO_3^{2-}) and organic compounds. CO_2 is one of the most soluble of the major gases in sea water and the ocean has an enormous capacity to buffer changes in the atmospheric CO_2 content.

The concentration of dissolved CO_2 in sea water is relatively small because CO_2 reacts with water to form a weak acid, carbonic acid (H_2CO_3), which rapidly dissociates (within milliseconds) to form HCO_3^- and CO_3^{2-} (eqn [I]).

$$CO_2(gas) + H_2O \rightleftharpoons H_2CO_3(aq) \rightleftharpoons H^+(aq) \quad \text{[I]}$$
$$+ HCO_3^-(aq) \rightleftharpoons 2H^+(aq) + CO_3^{2-}(aq)$$

For every 20 molecules of CO_2 absorbed by the ocean, 19 molecules are rapidly converted to HCO_3^- and CO_3^{2-}; at the typical range of pH in sea water (7.8–8.2; see below), most inorganic carbon is found in the form of HCO_3^-. These reactions (eqn [I]) provide a chemical buffer, maintain the pH of the ocean within a small range, and constrain the amount of atmospheric CO_2 that can be taken up by the ocean.

The amount of dissolved CO_2 in sea water cannot be determined analytically but can be calculated after measuring other inorganic carbon species. Dissolved inorganic carbon (DIC) refers to the total amount of CO_2, HCO_3^- plus CO_3^{2-} in sea water, while the partial pressure of CO_2 (P_{CO_2}) measures the contribution of CO_2 to total gas pressure. The alkalinity of sea water (A) is a measure of the bases present in sea

water, consisting mainly of HCO_3^- and CO_3^{2-} ($A[HCO_3^-] + 2[CO_3^{2-}]$) and minor constituents such as borate (BO_4) and hydrogen ions (H^+). Changes in DIC concentration and alkalinity affect the solubility of CO_2 in sea water (i.e., the ability of sea water to absorb CO_2) (see below).

The concentrations of inorganic carbon species in sea water are controlled not only by the chemical reactions outlined above (i.e., eqn [I]) but also by various physical and biological processes, including the exchange of CO_2 between ocean and atmosphere; the solubility of CO_2; photosynthesis and respiration; and the formation and dissolution of calcium carbonate ($CaCO_3$).

Typical surface sea water ranges from pH of 7.8 to 8.2. On addition of acid (i.e., H^+), the chemical reactions shift toward a higher concentration of CO_2 in sea water (eqn [IIa]) and pH decreases from 8.0 to 7.8 and then pH will rise from 8.0 to 8.2.

$$H^+ + HCO_3^- \rightarrow H_2CO_3 \rightarrow CO_2(aq) + H_2O \quad \text{[IIa]}$$

If base is added to sea water (eqn [IIb]), then pH will rise.

$$H_2CO_3 \rightarrow H^+ + HCO_3 \quad \text{[IIb]}$$

Solubility and Exchange of CO_2 between the Ocean and Atmosphere

The solubility of CO_2 in sea water is an important factor in controlling the exchange of carbon between the ocean and atmosphere. Henry's law (eqn [1]) describes the relationship between solubility and sea water properties, where S equals the solubility of gas in liquid, k is the solubility constant (k is a function mainly of temperature) and P is the overlying pressure of the gas in the atmosphere.

$$S = kP \quad \text{[1]}$$

Sea water properties such as temperature, salinity, and partial pressure of CO_2 determine the solubility of CO_2. For example, at 0°C in sea water, the solubility of CO_2 is double that in sea water at 20°C; thus colder water will tend to absorb more CO_2 than warmer water.

Henry's law also describes the relationship between the partial pressure of CO_2 in solution (P_{CO_2}) and its concentration (i.e., $[CO_2]$). Colder waters tend to have lower P_{CO_2} than warmer waters: for every 1°C temperature increase, sea water P_{CO_2} increases by $\sim4\%$. Sea water P_{CO_2} is also influenced by complicated thermodynamic relationships

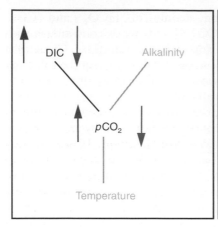 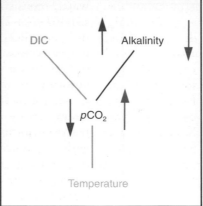 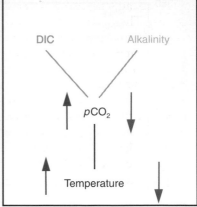

Figure 3 The response of P_{CO_2} to changes in the sea water properties of (A) DIC concentration, (B) alkalinity and (C) temperature. Each panel describes how P_{CO_2} will respond to the changes in the relevant sea water property. The blue arrows illustrate the response of P_{CO_2} to an increase in the sea water property and the red arrows illustrate the response to a decrease in the sea water property. For example, as DIC or temperature increases, P_{CO_2} increases; whereas an increase in alkalinity results in a decrease in P_{CO_2}.

between the different carbon species. For example, a decrease in sea water DIC or temperature acts to decrease P_{CO_2}, while a decrease in alkalinity acts to increase P_{CO_2} (**Figure 3**).

Carbon dioxide is transferred across the air–sea interface by molecular diffusion and turbulence at the ocean surface. The flux (F) of CO_2 between the atmosphere and ocean is driven by the concentration difference between the reservoirs (eqn [2]).

$$F = \Delta P_{CO_2} K_W \qquad [2]$$

In eqn [2] ΔP_{CO_2} is the difference in P_{CO_2} between the ocean and atmosphere and K_W is the transfer coefficient across the air–sea interface, termed the piston velocity. In cold waters, sea water P_{CO_2} tends to be lower than atmospheric P_{CO_2}, thus driving the direction of CO_2 gas exchange from atmosphere to ocean (**Figure 3**). In warmer waters, sea water P_{CO_2} is greater than atmospheric P_{CO_2}, and CO_2 gas exchange occurs in the opposite direction, from the ocean to the atmosphere. The rate at which CO_2 is transferred between the ocean and the atmosphere depends not only on the P_{CO_2} difference but on turbulence at the ocean surface. The piston velocity of CO_2 is related to solubility and the strength of the wind blowing on the sea surface. As wind speed increases, the rate of air–sea CO_2 exchange also increase. Turbulence caused by breaking waves also influences gas exchange because air bubbles may dissolve following entrainment into the ocean mixed layer.

Ocean Structure

Physically, the ocean can be thought of as two concentric spheres, the surface ocean and the deep

ocean, separated by a density discontinuity called the pycnocline. The surface ocean occupies the upper few hundred meters of the water column and contains approximately 1020×10^{15} g C of DIC (**Figure 1**). The absorption of CO_2 by the ocean through gas exchange takes place in the mixed layer, the upper portion of the surface ocean that makes direct contact with the atmosphere. The surface ocean reaches equilibrium with the atmosphere within one year. The partial pressure of CO_2 in the surface ocean is slightly less than or greater than that of the atmosphere, depending on the controlling variables as described above, and varies temporally and spatially with changing environmental conditions. The deeper ocean represents the remainder of the ocean volume and is supersaturated with CO_2, with a DIC stock of $38\ 100 \times 10^{15}$ g C (**Figure 1**), or 50 times the DIC contained in the atmosphere.

CO_2 absorbed by the ocean through gas exchange has a variety of fates. Physical and biological mechanisms can return the CO_2 back to the atmosphere or transfer carbon from the surface ocean to the deep ocean and ocean sediments through several transport processes termed the 'solubility', 'biological', and 'carbonate' pumps.

The Solubility Pump, Oceanic Circulation, and Carbon Redistribution

The 'solubility pump' is defined as the exchange of carbon between the atmosphere and the ocean as mediated by physical processes such as heat flux, advection and diffusion, and ocean circulation. It assists in the transfer of atmospheric CO_2 to the deep ocean. This transfer is controlled by circulation patterns of the surface ocean (wind-driven circulation) and the deep ocean (thermohaline circulation). These circulation patterns assist in the transfer of atmospheric CO_2 to the deep ocean and help to maintain the vertical gradient of DIC found in the ocean (**Figure 4**). The ability of the ocean to take up anthropogenic CO_2 via the solubility pump is limited by the physical structure of the ocean, the distribution of oceanic DIC, ocean circulation patterns, and the exchange between the surface and deep ocean layers. To be an effective sink for anthropogenic carbon, CO_2 must be transferred to the deep ocean by mixing and biological processes (see below).

Wind-driven circulation occurs as a consequence of friction and turbulence imparted by wind blowing over the sea surface. This circulation pattern is primarily horizontal in movement and is responsible for transporting warm water from lower latitudes (warm) to higher latitudes (cold). Surface currents move water and carbon great distances within ocean

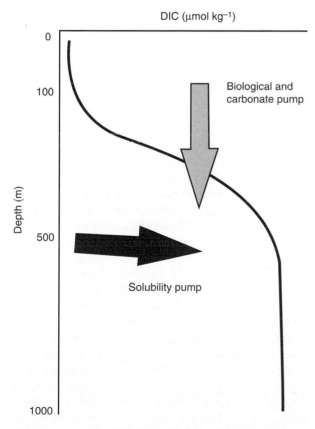

Figure 4   Illustration of the vertical gradient of DIC in the ocean. The uptake of DIC by phytoplankton and conversion into sinking organic matter ('biological pump'; gray arrow) and sinking calcium carbonate skeletal matter ('carbonate pump'; gray arrow) contributes to the maintenance of the vertical gradient. Introduction of DIC to the deep waters via the 'solubility pump' at high latitudes and subsequent deep water formation also helps maintain this vertical gradient (black arrow; see **Figure 5**).

basins on timescales of months to years. As surface sea water moves from low latitudes to high latitudes, the increasing solubility of CO_2 in the sea water (due to sea surface cooling) allows atmospheric CO_2 to invade the surface mixed layer (**Figure 5** and **6A**).

Exchange of surface waters with the deep ocean through wind-driven mixing is limited because of strong density stratification of the water column over the majority of the world's oceans. However, thermohaline (overturning) circulation at high latitudes provides a mechanism for surface waters to exchange with the deep ocean. Passage of cold and dry air masses over high-latitude regions, such as the Greenland and Labrador Seas in the North Atlantic or the Weddell Sea in the Southern Ocean, forms cold and very dense sea water ('deep water' formation). Once formed, these dense water masses sink vertically until they reach a depth at which water is of similar density (i.e., 2000–4000 m deep). Following sinking, the dense waters are transported slowly throughout all of the deep ocean basins by advection and diffusion, displacing other deep water that eventually is brought back to the surface by upwelling (**Figure 5**).

Because of the smaller volume and faster circulation, the residence time of the surface ocean is only one decade compared to 600–1000 years for the deep ocean. The process of deep water formation transfers CO_2, absorbed from the atmosphere by the solubility pump, into the deep ocean. The effect is that DIC concentration increases with depth in all ocean basins (**Figure 4, 5** and **6A,**). As a result of the long residence time of the deep ocean, carbon, once removed from the surface ocean to the deep ocean through the effects of solubility and deep water formation, is stored without contact with the atmosphere for hundreds to thousands of years. At present, deep water formed at the surface that is in equilibrium with the atmosphere (sea water P_{CO_2} of ~360 ppm), carries more CO_2 to depth than deep water formed prior to the industrial age (e.g., ~280 ppm). Furthermore, P_{CO_2} of upwelled deep water is less than that in the recently formed deep water, indicating that the deep water formation and the 'solubility pump' allow the ocean to be a net sink for anthropogenic CO_2. The vertical gradient in DIC (**Figure 4**) and the ability of the ocean to take up atmospheric CO_2 is augmented by biological processes known as the 'biological pump'.

The Biological Pump

Although the standing stock of marine biota in the ocean is relatively small (3×10^{15} g C), the activity

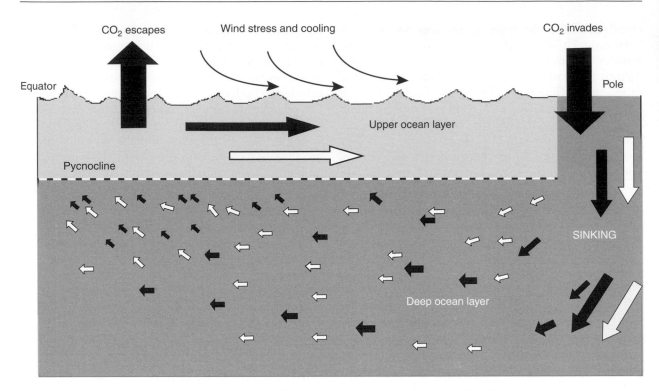

Figure 5 Conceptual model of the 'solubility pump'. White arrows represent movement of water; black arrows represent movement of CO_2 within, and into and out of, the ocean. Cooling increases the solubility of CO_2 and results in a flux of CO_2 from the atmosphere to the surface ocean. At subpolar latitudes the water density increases and the CO_2-enriched water sinks rapidly. At depth, the CO_2 enriched water moves slowly as is it is dispersed throughout the deep ocean. The sinking water displaces water that is returned to the surface ocean in upwelling regions. As the water warms, P_{CO_2} increases, resulting in escape of CO_2 from the surface water to the atmosphere.

associated with the biota is extremely important to the cycling of carbon between the atmosphere and the ocean. The largest and most rapid fluxes in the global carbon cycle are those that link atmospheric CO_2 to photosynthetic production (primary production) on the land and in the ocean. Globally, marine phytoplankton are responsible for more than one-third of the total gross photosynthetic production (50×10^{15} g C y^{-1}). In the sea, photosynthesis is limited to the euphotic zone, the upper 100–150 m of the water column where light can penetrate. Photosynthetic organisms use light energy to reduce CO_2 to high-energy organic compounds. In turn, a portion of these synthesized organic compounds are utilized by heterotrophic organisms as an energy source, being remineralized to CO_2 via respiration. Eqn [III] represents the overall reactions of photosynthesis and respiration.

$$CO_2(gas) + H_2O \xrightleftharpoons[\text{Metabolic energy (respiration)}]{\text{Light energy (photosynthesis)}} (CH_2O)_n \quad [III]$$
$$+ O_2(gas)$$

In the sea, net primary production (primary production in excess of respiration) converts CO_2 to

organic matter that is stored as particulate organic carbon (POC; in living and detrital particles) and as dissolved organic carbon (DOC). In stratified regions of the ocean (lower latitudes), net primary production results in a drawdown of DIC and an accumulation of organic matter as POC and DOC (**Figure 6B**). However, it is the portion of organic carbon production that can be exported from the surface ocean and remineralized in the deep ocean that is important in the exchange of CO_2 between the atmosphere and the ocean. The biological pump refers to the processes that convert CO_2 (thereby drawing down DIC) to organic matter by photosynthesis, and remove the organic carbon to depth (where it is respired) via sinking, mixing, and active transport mechanisms (**Figure 7**). Once at great depth, it is effectively removed from exchange with the atmosphere. As living biomass is produced, some particles becomes senescent and form sinking aggregates, while other particles are consumed by herbivores and sinking fecal pellets (POC) are formed. These sinking aggregates and pellets remove carbon from the surface to be remineralized at depth via decomposition by bacteria or consumption by zooplankton and fish (**Figure 7**). In addition, DOC produced by phytoplankton or by animal excretion

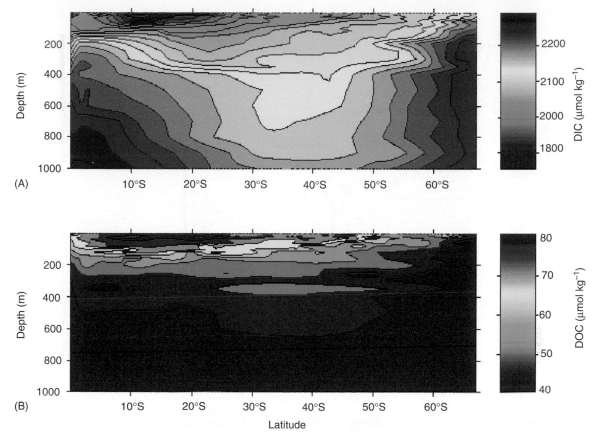

Figure 6 Contour plot of (A) DIC and (B) DOC along a transect line in the South Pacific between the equator (0°; 170° W) and the Antarctic Polar Front (66° S; 170° W). Note that in the low-latitude stratified waters DIC concentrations are depleted in surface water relative to deep water, as a result of net primary production and air–sea exchange. DOC concentrations are elevated relative to deep water. In high-latitude regions, DIC concentration are elevated in the surface water as a result of increased solubility of cooler surface waters.

in surface waters can also be transported downward by subduction or convective mixing of surface waters (**Figure 7**). Finally, vertically migrating zooplankton that feed in the surface waters at night and return to deep waters during the day actively transport dissolved and particulate material to depth, where a portion is metabolized (**Figure 7**).

Production via photosynthesis can occur only in the surface ocean, whereas remineralization can occur throughout the water column. The biological pump serves to spatially separate the net photosynthetic from net respiratory processes. Thus, the conversion of DIC to exportable organic matter acts to reduce the DIC concentration in the surface water and its subsequent remineralization increases DIC concentration in the deep ocean (**Figure 6**). The biological pump is important to the maintenance of a vertical DIC profile of undersaturation in the surface and supersaturation at depth (**Figure 4** and **5A**). Undersaturation of DIC in the surface mixed layer, created by the biological pump, allows for the influx

of CO_2 from the atmosphere (see Henry's law above; **Figure 7**).

Gross export of organic matter out of the surface waters is approximately 10×10^{15} g C y^{-1} (**Figure 1**). Less than 1% of the organic matter exported from the surface waters is stored in the abyssal sediment. In fact, most of the exported organic matter is remineralized to DIC in the upper 500 m of the water column. It is released back to the atmosphere on timescales of months to years via upwelling, mixing, or ventilation of high-density water at high latitudes. It is that fraction of exported organic matter that actually reaches the deep ocean (>1000 m) that is important for long-term atmospheric CO_2 regulation. Once in the deep ocean, the organic matter either remains as long-lived DOC or is remineralized to DIC and is removed from interaction with the atmosphere on timescales of centuries to millennia. Thus, even though less than 1% of the exported carbon is stored in marine sediments, the activities of the biological pump are very important in mediating the air–sea

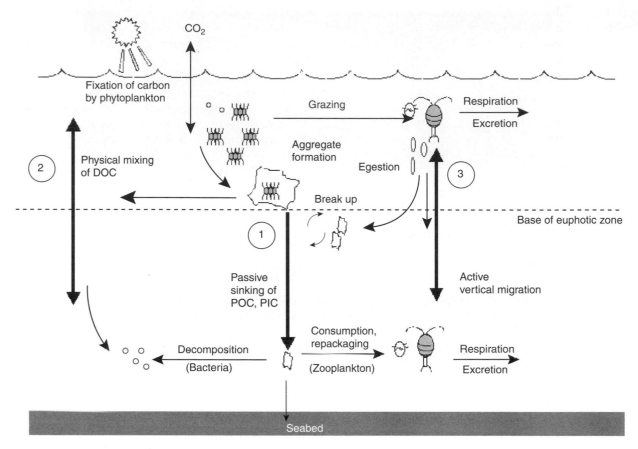

Figure 7 Conceptual diagram depicting components of the 'biological pump'. CO_2 is taken up by phytoplankton and organic matter is produced. As this organic matter is processed through the marine food web, fecal pellets or aggregates are produced, a portion of which sink from the surface waters to depth (1). As organic matter is processed through the food web, DOC is also produced. DOC is removed from the surface waters to depth via physical mixing of the water by convective overturn (2). DOC and DIC are also actively transported to depth by vertically migrating organisms such as copepods that feed in surface waters and excrete and respire the consumed organic carbon at depth (3).

transfer of CO_2. Without this pump in action, atmospheric CO_2 concentration might be as high as 500 to 1000 ppm versus the 360 ppm observed today.

Contribution of POC Versus DOC in the Biological Pump

Historically, sinking particles were thought of as the dominant export mechanism of the biological pump and the primary driver of respiration in the ocean's interior. However, downward mixing of surface water can also transport large quantities of DOC trapped within the sinking water mass. In order for DOC to be an important contributor to the biological pump two sets of conditions must exist. First, the producer–consumer dynamics in the surface waters must yield DOC of a quality that is resistant to rapid remineralization by bacteria and lead to net DOC production. Second, the physical system must undergo periods of deep convective mixing or subduction in order to remove surface waters and DOC

to depth. Although approximately 80% of the globally exported carbon is in the form of POC, DOC can represent 30–50% of carbon export in the upper 500 m of the water column at specific ocean sites. The biological/physical controls on DOC export are complex and are currently being assessed for various regions of the worlds' ocean.

Factors that affect the Efficiency of the Biological Pump

An efficient biological pump means that a large fraction of the system's net production is removed from the surface waters via export mechanisms. Factors that affect the efficiency of the biological pump are numerous and include nutrient supply and plankton community structure.

Nutrient supply Does an increased partial pressure of atmospheric CO_2 lead to a more efficient biological pump? Not necessarily, since net primary production is limited by the availability of other inorganic nutrients such as nitrogen, phosphorus, silicon and iron. Because these inorganic nutrients are continuously being removed from the surface waters with vertical export of organic particles, their concentrations are often below detection limits in highly stratified water columns. As a result, primary production becomes limited by the rate at which these nutrients can be re-supplied to the surface ocean by mixing, by atmospheric deposition, or by heterotrophic recycling. Primary production supported by the recycling of nutrients in the surface ocean is referred to as 'regenerative' production and contributes little to the biological pump. Primary production supported by the introduction of new nutrients from outside the system, via mixing from below or by atmospheric deposition (e.g. dust), is referred to as 'new' production. New nutrients enhance the amount of net production that can be exported (new production). Because CO_2 is not considered to be a limiting nutrient in marine systems, the increase in atmospheric CO_2 is not likely to stimulate net production for most of the world's ocean unless it indirectly affects the introduction of new nutrients as well.

Community structure Food web structure also plays an important role in determining the size distribution of the organic particles produced and whether the organic carbon and associated nutrients are exported from or recycled within the surface waters. The production of large, rapidly settling cells will make a greater contribution to the biological pump than the production of small, suspended particles. Factors such as the number of trophic links and the size of the primary producers help determine the overall contribution of sinking particles. The number of trophic steps is inversely related to the magnitude of the export flux. For example, in systems where picoplankton are the dominant primary producers there may be 4–5 steps before reaching a trophic level capable of producing sinking particles. With each trophic transfer, a

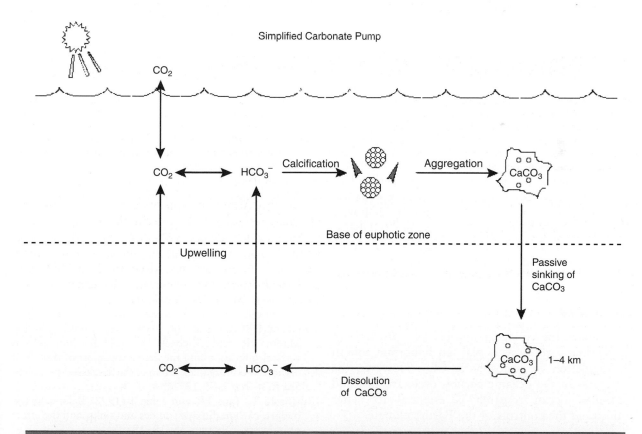

Figure 8 Conceptual diagram of a simplified 'carbonate pump'. Some marine organisms form calcareous skeletal material, a portion of which sinks as calcium carbonate aggregates. These aggregates are preserved in shallow ocean sediments or dissolve at greater depths (3000–5000 m), thus increasing DIC concentrations in the deep ocean. The calcium and bicarbonate are returned to the surface ocean through upwelling.

percentage (50–70%) of the organic carbon is respired, so only a small fraction of the original primary production forms sinking particles. Although picoplankton may dominate primary production in oceanic systems, their production is considered 'regenerative' and contributes little to the production of sinking material. Alternatively, production by larger phytoplankton such as diatoms ($>20\,\mu m$ in size) may represent a smaller fraction of primary production, but their contribution to the biological pump is larger because fewer trophic steps are taken to produce sinking particles.

The Carbonate Pump

A process considered part of the biological pump (depending how it is defined) is the formation and sinking of calcareous skeletal material by some marine phytoplankton (e.g., coccolithophores) and animals (e.g., pteropods and foraminifera). Calcification is the process by which marine organisms combine calcium with carbonate ions to form hard body parts. The resulting calcium carbonate ($CaCO_3$) is dense and sinks out of the surface water with export production (**Figure 8**). The global mean ratio for carbon sinking from the surface ocean as $CaCO_3$ or organic carbon is 1:4. However, unlike organic matter, $CaCO_3$ is not remineralized as it sinks; it only begins to dissolve in intermediate and deep waters, waters undersaturated with respect to $CaCO_3$. Complete dissolution of $CaCO_3$ skeletons typically occurs at depths of 1–4 km (in the north Pacific Ocean) to 5 km (in the North Atlantic). This depth zone is known as the carbonate compensation depth. $CaCO_3$ is only found in sediments shallower than the carbonate compensation depth. Globally, the CO_2 sink in sedimentary rock is four times greater than the sink in organic sediments.

Summary

In summary, the biological and physical processes of the oceanic carbon cycle play an important role in the regulation of atmospheric CO_2. However, the intricacies of the oceanic carbon cycle are vast and continued ocean research is essential to better understand the controls of the Earth's climate.

See also

Carbon Dioxide (CO_2) Cycle.

Further Reading

Bates NR, Michaels AF, and Knap AH (1996) Seasonal and interannual variability of oceanic carbon dioxide species at the U.S. JGOFS Bermuda Atlantic Time-series Study (BATS) site. *Deep-Sea Research II* 43: 347–383.

Bolin B (ed.) (1983) *The Major Biogeochemical Cycles and Their Interactions: SCOPE 21*. New York: Wiley.

Carlson CA, Ducklow HW, and Michaels AF (1994) Annual flux of dissolved organic carbon from the euphotic zone in the northwestern Sargasso Sea. *Nature* 371: 405–408.

Denman K, Hofman H, and Marchant H (1996) Marine biotic responses to environmental change and feedbacks to climate. In: Houghton JT, Meira Filho LG, and Callander BA, *et al.* (eds.) *Climate Change 1995: The Science of Climate Change*. New York: Cambridge University Press.

Follows MJ, Williams RG, and Marshall JC (1996) The solubility pump of carbon in the subtropical gyre of the North Atlantic. *Journal of Marine Research* 54: 605–630.

Hansell DA and Carlson CA (1998) Net community production of dissolved organic carbon. *Global Biogeochemical Cycles* 12: 443–453.

Holmén K (1992) The global carbon cycle. In: Butcher SS, Charlson RJ, Orians GH, and Wolfe GV (eds.) *Global Biogeochemical Cycles*, pp. 239–262. New York: Academic Press.

Houghton JT, Meira Filho LG, and Callander BA, *et al.* (eds.) (1996) *Climate Change 1995: The Science of Climate Change*. New York: Cambridge University Press.

Michaels AF and Silver MW (1988) Primary producers, sinking fluxes and the microbial food web. *Deep-Sea Research* 35: 473–490.

Sarmiento JL and Wofsy (eds.) (1999) *A U.S. Carbon Cycle Science Plan*. Washington, DC: U.S. Global Change Research Program.

Sarmiento JL, Hughes TMC, Stouffer RJ, and Manabe S (1998) Simulated response of the ocean carbon cycle to anthropogenic climate warming. *Nature* 393: 245–249.

Schlesinger WH (1997) *Biogeochemistry: An Analysis of Global Change*. New York: Academic Press.

Siegenthaler U and Sarmiento JL (1993) Atmospheric carbon dioxide and the ocean. *Nature* 365: 119–125.

Steinberg DK, Carlson CA, Bates NR, Goldthwait SA, Madin LP, and Michaels AF (2000) Zooplankton vertical migration and the active transport of dissolved organic and inorganic carbon in the Sargasso Sea. *Deep-Sea Research I* 47: 137–158.

Takahashi T, Tans PP, and Fung I (1992) Balancing the budget: carbon dioxide sources and sinks, and the effect of industry. *Oceanus* 35: 18–28.

Varney M (1996) The marine carbonate system. In: Summerhayes CP and Thorpe SA (eds.) *Oceanography an Illustrated Guide*, pp. 182–194. London: Manson Publishing.

CARBON DIOXIDE (CO₂) CYCLE

CARBON DIOXIDE (CO$_2$) CYCLE

T. Takahashi, Lamont Doherty Earth Observatory, Columbia University, Palisades, NY, USA

Introduction

The oceans, the terrestrial biosphere, and the atmosphere are the three major dynamic reservoirs for carbon on the earth. Through the exchange of CO$_2$ between them, the atmospheric concentration of CO$_2$ that affects the heat balance of the earth, and hence the climate, is regulated. Since carbon is one of the fundamental constituents of living matter, how it cycles through these natural reservoirs has been one of the fundamental questions in environmental sciences. The oceans contain about 50 times as much carbon (about 40 000 Pg-C or 10^{15} g as carbon) as the atmosphere (about 750Pg-C). The terrestrial biosphere contains about three times as much carbon (610 Pg-C in living vegetation and 1580 Pg-C in soil organic matter) as the atmosphere. The air–sea exchange of CO$_2$ occurs via gas exchange processes across the sea surface; the natural air-to-sea and sea-to-air fluxes have been estimated to be about 90 Pg-C y^{-1} each. The unperturbed uptake flux of CO$_2$ by global terrestrial photosynthesis is roughly balanced with the release flux by respiration, and both have been estimated to be about 60 Pg-C y^{-1}. Accordingly, atmospheric CO$_2$ is cycled through the ocean and terrestrial biosphere with a time scale of about 7 years.

The lithosphere contains a huge amount of carbon (about 100 000 000 Pg-C) in the form of limestones ((Ca, Mg) CO$_3$), coal, petroleum, and other forms of organic matter, and exchanges carbon slowly with the other carbon reservoirs via such natural processes as chemical weathering and burial of carbonate and organic carbon. The rate of removal of atmospheric CO$_2$ by chemical weathering has been estimated to be of the order of 1 Pg-Cy^{-1}. Since the industrial revolution in the nineteenth century, the combustion of fossil fuels and the manufacturing of cement have transferred the lithospheric carbon into the atmosphere at rates comparable to the natural CO$_2$ exchange fluxes between the major carbon reservoirs, and thus have perturbed the natural balance significantly (6 Pg-Cy^{-1} is about an order of magnitude less than the natural exchanges with the oceans (90 Pg-C y^{-1}) and land (60 Pg-C y^{-1})). The industrial carbon emission rate has been about 6 Pg-C y^{-1} for the 1990s, and the cumulative industrial emissions since the nineteenth century to the end of the twentieth century have been estimated to be about 250 Pg-C. Presently, the atmospheric CO$_2$ content is increasing at a rate of about 3.5 Pg-C y^{-1} (equivalent to about 50% of the annual emission) and the remainder of the CO$_2$ emitted into the atmosphere is absorbed by the oceans and terrestrial biosphere in approximately equal proportions. These industrial CO$_2$ emissions have caused the atmospheric CO$_2$ concentration to increase by as much as 30% from about 280 ppm (parts per million mole fraction in dry air) in the pre-industrial year 1850 to about 362 ppm in the year 2000. The atmospheric CO$_2$ concentration may reach 580 ppm, double the pre-industrial value, by the mid-twenty first century. This represents a significant change that is wholly attributable to human activities on the Earth.

It is well known that the oceans play an important role in regulating our living environment by providing water vapor into the atmosphere and transporting heat from the tropics to high latitude areas. In addition to these physical influences, the oceans partially ameliorate the potential CO$_2$-induced climate changes by absorbing industrial CO$_2$ in the atmosphere.

Therefore, it is important to understand how the oceans take up CO$_2$ from the atmosphere and how they store CO$_2$ in circulating ocean water. Furthermore, in order to predict the future course of the atmospheric CO$_2$ changes, we need to understand how the capacity of the ocean carbon reservoir might be changed in response to the Earth's climate changes, that may, in turn, alter the circulation of ocean water. Since the capacity of the ocean carbon reservoir is governed by complex interactions of physical, biological, and chemical processes, it is presently not possible to identify and predict reliably various climate feedback mechanisms that affect the ocean CO$_2$ storage capacity.

Units

In scientific and technical literature, the amount of carbon has often been expressed in three different units: giga tons of carbon (Gt-C), petagrams of carbon (Pg-C) and moles of carbon or CO$_2$. Their relationships are: 1Gt-C = 1 Pg-C = 1×10^{15} g of carbon = 1000 million metric tonnes of carbon = $(1/12) \times 10^{15}$ moles of carbon. The equivalent quantity as CO$_2$ may be obtained by multiplying the above numbers by 3.67 (= 44/12 = the molecular weight of CO$_2$ divided by the atomic weight of carbon).

459

The magnitude of CO_2 disequilibrium between the atmosphere and ocean water is expressed by the difference between the partial pressure of CO_2 of ocean water, $(pCO_2)sw$, and that in the overlying air, $(pCO_2)air$. This difference represents the thermodynamic driving potential for CO_2 gas transfer across the sea surface. The pCO_2 in the air may be estimated using the concentration of CO_2 in air, that is commonly expressed in terms of ppm (parts per million) in mole fraction of CO_2 in dry air, in the relationship:

$$p(CO_2)air = (CO_2\ conc.)air \times (Pb - pH_2O) \quad [1]$$

where Pb is the barometric pressure and pH_2O is the vapor pressure of water at the sea water temperature. The partial pressure of CO_2 in sea water, $(pCO_2)sw$, may be measured by equilibration methods or computed using thermodynamic relationships. The unit of microatmospheres (μatm) or 10^{-6} atm is commonly used in the oceanographic literature.

History

The air–sea exchange of CO_2 was first investigated in the 1910s through the 1930s by a group of scientists including K. Buch, H. Wattenberg, and G.F.R. Deacon. Buch and his collaborators determined in land-based laboratories CO_2 solubility, the dissociation constants for carbonic and boric acids in sea water, and their dependence on temperature and chlorinity (the chloride ion concentration in sea water). Based upon these dissociation constants along with the shipboard measurements of pH and titration alkalinity, they computed the partial pressure of CO_2 in surface ocean waters. The Atlantic Ocean was investigated from the Arctic to Antarctic regions during the period 1917–1935, especially during the METEOR Expedition 1925–27, in the North and South Atlantic. They discovered that temperate and cold oceans had lower pCO_2 than air (hence the sea water was a sink for atmospheric CO_2), especially during spring and summer seasons, due to the assimilation of CO_2 by plants. They also observed that the upwelling areas of deep water (such as African coastal areas) had greater pCO_2 than the air (hence the sea water was a CO_2 source) due to the presence of respired CO_2 in deep waters.

With the advent of the high-precision infrared CO_2 gas analyzer, a new method for shipboard measurements of pCO_2 in sea water and in air was introduced during the International Geophysical Year, 1956–59. The precision of measurements was improved by more than an order of magnitude. The global oceans were investigated by this new method, which rapidly yielded high precision data. The

equatorial Pacific was identified as a major CO_2 source area. The GEOSECS Program of the International Decade of Ocean Exploration, 1970–80, produced a global data set that began to show systematic patterns for the distribution of CO_2 sink and source areas over the global oceans.

Methods

The net flux of CO_2 across these a surface, Fs-a, may be estimated by:

$$\begin{aligned} Fs\text{-}a &= E \times [(pCO_2)sw - (pCO_2)air] \\ &= k \times \alpha \times [(pCO_2)sw - (pCO_2)air] \end{aligned} \quad [2]$$

where E is the CO_2 gas transfer coefficient expressed commonly in (moles $CO_2/m^2/y/uatm$); k is the gas transfer piston velocity (e.g. in (cmh^{-1})) and α is the solubility of CO_2 in sea water at a given temperature and salinity (e.g. (moles $CO_2\ kg\text{-}sw^{-1}\ atm^{-1}$)). If $(pCO_2)sw < (pCO_2)$ air, the net flux of CO_2 is from the sea to the air and the ocean is a source of CO_2; if $(pCO_2)sw < (pCO_2)$ air, the ocean water is a sink for atmospheric CO_2. The sea–air pCO_2 difference may be measured at sea and α has been determined experimentally as a function of temperature and salinity. However, the values of E and k that depend on the magnitude of turbulence near the air–water interface cannot be simply characterized over complex ocean surface conditions. Nevertheless, these two variables have been commonly parameterized in terms of wind speed over the ocean. A number of experiments have been performed to determine the wind speed dependence under various wind tunnel conditions as well as ocean and lake environments using different nonreactive tracer gases such as SF_6 and ^{222}Rn. However, the published results differ by as much as 50% over the wind speed range of oceanographic interests.

Since ^{14}C is in the form of CO_2 in the atmosphere and enters into the surface ocean water as CO_2 in a timescale of decades, its partition between the atmosphere and the oceans yields a reliable estimate for the mean CO_2 gas transfer rate over the global oceans. This yields a CO_2 gas exchange rate of 20 ± 3 mol $CO_2\ m^{-2}\ y^{-1}$ that corresponds to a sea–air CO_2 transfer coefficient of 0.067 mol $CO_2\ m^{-2}\ y^{-1}\ uatm^{-1}$. Wanninkhof in 1992 presented an expression that satisfies the mean global CO_2 transfer coefficient based on ^{14}C and takes other field and wind tunnel results into consideration. His equation for variable wind speed conditions is:

$$k\left(cm\ h^{-1}\right) = 0.39 \times (u_{av})^2 \times (Sc/660)^{-0.5} \quad [3]$$

where u_{av} is the average wind speed in ms^{-1} corrected to $10\,m$ above sea surface; Sc(dimensionless) is the Schmidt number (kinematic viscosity ofwater)/ (diffusion coefficient of CO_2 gas inwater); and 660 represents the Schmidt number for CO_2 in seawater at $20°C$.

In view of the difficulties in determining gas transfer coefficients accurately, direct methods for CO_2 flux measurements aboard the ship are desirable. Sea–air CO_2 flux was measured directly by means of the shipboard eddy-covariance method over the North Atlantic Ocean by Wanninkhof and McGillis in 1999. The net flux of CO_2 across the sea surface was determined by a covariance analysis of the tri-axial motion of air with CO_2 concentrations in the moving air measured in short time intervals ($\sim ms$) as a ship moved over the ocean. The results obtained over awind speed range of 2–$13.5\,m$ s^{-1} are consistent with eqn [3] within about $\pm 20\%$. If the data obtainedin wind speeds up to $15\,m\,s^{-1}$ are taken into consideration, they indicate that the gas transfer piston velocity tends to increase as a cubeof wind speed. However, because of a large scatter ($\pm 35\%$) ofthe flux values at high wind speeds, further work is needed to confirm the cubic dependence.

In addition to the uncertainties in the gas transfer coefficient (or piston velocity), the CO_2 fluxestimated with eqn [2] is subject to errors in (pCO₂)sw caused by the difference between the bulk water temperature and the temperature of the thin skin of ocean water at the sea–air interface. Ordinarily the (pCO₂)sw is obtained at the bulk seawater temperature, whereas the relevant value for the flux calculation is (pCO₂)sw at the 'skin'temperature, that depends on the rate of evaporation, the incoming solar radiation, the wind speed, and the degree of turbulence near the interface. The 'skin' temperature is often cooler than the bulk water temperatureby as much as $0.5°C$ if the water evaporates rapidly to a dry air mass, but is not always so if a warm humid air mass covers over the ocean. Presently, the time–space distribution of the 'skin' temperature is not well known. This, therefore, could introduce errors in (pCO₂)sw up to about $6\,\mu atm$ or 2%.

CO₂ Sink/Source areas of the Global Ocean

The oceanic sink and source areas for atmospheric CO_2 and the magnitude of the sea–air CO_2 flux over the global ocean vary seasonally and annually as well as geographically. These changes are the manifestation of changes in the partial pressure of sea water,

(pCO₂)sw, which are caused primarily by changes in the water temperature, in the biological utilization of CO_2, and in the lateral/vertical circulation of ocean waters including the upwelling of deep water rich in CO_2. Over the global oceans, sea water temperatures change from the pole to the equator by about $32°C$. Since the pCO₂ in sea water doubles with each $16°C$ of warming, temperature changes should cause a factor of 4 change in pCO₂. Biological utilization of CO_2 over the global oceans is about $200\,\mu mol$ $CO_2\,kg^{-1}$, which should reduce pCO₂ in sea water by a factor of 3. If this is accompanied with growths of $CaCO_3$-secreting organisms, the reduction of pCO₂ could be somewhat smaller. While these effects are similar in magnitude, they tend to counteract each other seasonally, since the biological utilization tends to be large when waters are warm. In subpolar and polar areas, winter cooling of surface waters induces deep convective mixing that brings high pCO₂ deep waters to the surface. The lowering effect on CO_2 by winter cooling is often compensated for or some times over compensated for by the increasing effect of the upwelling of high CO_2 deep waters. Thus, in high latitude oceans, surface waters may become a source for atmospheric CO_2 during the winter time when the water is coldest.

In **Figure 1**, the global distribution map of the sea–air pCO₂ differences for February and August 1995, are shown. These maps were constructed on the basis of about a half million pairs of atmospheric and seawater pCO₂ measurements made at sea over the 40-year period, 1958–98, by many investigators. Since the measurements were made in different years, during which the atmospheric pCO₂ was increasing, they were corrected to a single reference year (arbitrarily chosen to be 1995) on the basis of the following observations. Warm surface waters in subtropical gyres communicate slowly with the underlying subsurface waters due to the presence of a strong stratification at the base of the mixed layer. This allows a long time for the surface mixed-layer-waters ($\sim 75\,m$ thick) to exchange CO_2 with the atmosphere. Therefore, their CO_2 chemistry tends to follow the atmospheric CO_2 increase. Accordingly, the pCO₂ in the warm water follows the increasing trend of atmospheric CO_2, and the sea–air pCO₂ difference tends to be independent of the year of measurements. On the other hand, since surface waters in high latitude regions are replaced partially with subsurface waters by deep convection during the winter, the effect of increased atmospheric CO_2 is diluted to undetectable levels and their CO_2 properties tend to remain unchanged from year to year. Accordingly, the sea–air pCO₂ difference measured in a given year increases as the atmospheric CO_2

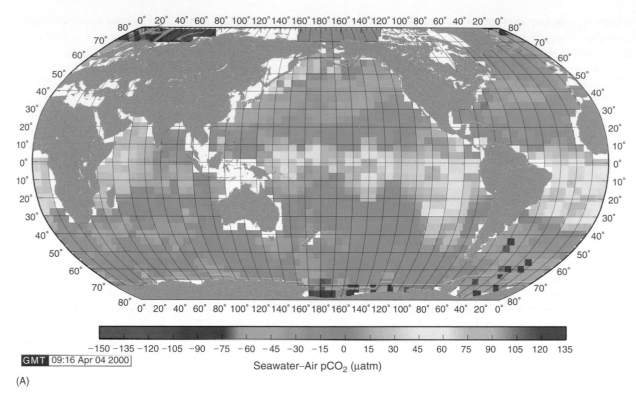

(A)

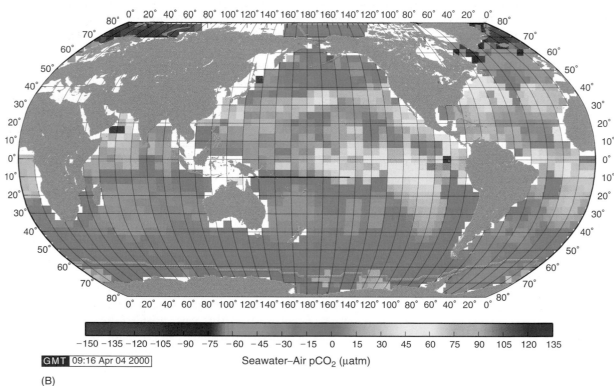

(B)

Figure 1   The sea–air pCO₂ difference in μatm (ΔpCO₂) for (A) February and (B) August for the reference year 1995. The purple-blue areas indicate that the ocean is a sink for atmospheric CO₂, and the red-yellow areas indicate that the ocean is source. The pink lines in the polar regions indicate the edges of ice fields.

concentration increases with time. This effect was corrected to the reference year using the observed increase in the atmospheric CO_2 concentration. During El Niño periods, sea–air pCO_2 differences over the equatorial belt of the Pacific Ocean, which are large in normal years, are reduced significantly and observations are scarce. Therefore, observations made between 10°N and 10°S in the equatorial Pacific for these periods were excluded from the maps. Accordingly, these maps represent the climatological means for non-El Niño period oceans for the past 40 years. The purple-blue areas indicate that the ocean is a sink for atmospheric CO_2, and the red-yellow areas indicate that the ocean is a source.

Strong CO_2 sinks (blue and purple areas) are present during the winter months in the Northern (**Figure 1A**) and Southern (**Figure 1B**) Hemispheres along the poleward edges of the subtropical gyres, where major warm currents are located. The Gulf Stream in the North Atlantic and the Kuroshio Current in the North Pacific are both major CO_2 sinks (**Figure 1A**) due primarily to cooling as they flow from warm tropical oceans to subpolar zones.

Similarly, in the Southern Hemisphere, CO_2 sink areas are formed by the cooling of poleward-flowing currents such as the Brazil Current located along eastern South America, the Agulhus Current located south of South Africa, and the East Australian Current located along south-eastern Australia. These warm water currents meet with cold currents flowing equator ward from the Antarctic zone along the northern border of the Southern (or Antarctic) Ocean. As the sub Antarctic waters rich in nutrients flow northward to more sunlit regions, CO_2 is drawn down by photosynthesis, thus creating strong CO_2 sink conditions, as exemplified by the Falkland Current in the western South Atlantic (**Figure 1A**). Confluence of subtropical waters with polar waters forms broad and strong CO_2 sink zones as a result of the juxta position of the lowering effects on pCO_2 of the cooling of warm waters and the photosynthetic drawdown of CO_2 in nutrient-rich subpolar waters. This feature is clearly depicted in azone between 40°S and 60°S in **Figure 1A** representing the austral summer, and between 20°S and 40°S in **Figure 1B** representing the austral winter.

During the summer months, the high latitude areas of the North Atlantic Ocean (**Figure 1A**) and the Weddell and Ross Seas, Antarctica(**Figure 1B**), are intense sink areas for CO_2. This is attributed to the intense biological utilization of CO_2 within the strongly stratified surface layer caused by solar warming and ice melting during the summer. The winter convective mixing of deep waters rich in CO_2 and nutrient seliminates the strong CO_2 sink and

replenishes the depleted nutrients in the surface waters.

The Pacific equatorial belt is a strong CO_2 source which is caused by the warming of upwelled deep waters along the coast of South America as well as by the upward entrainment of the equatorial under current water. The source strengths are most intense in the eastern equatorial Pacific due to the strong upwelling, and decrease to the west as a result of the biological utilization of CO_2 and nutrients during the westward flow of the surface water.

Small but strong source areas in the north-western subArctic Pacific Ocean are due to the winter convective mixing of deep waters (**Figure 1A**). The lowering effect on pCO_2 of cooling in the winter is surpassed by the increasing effect of high CO_2 concentration in the upwelled deep waters. During the summer (**Figure 1B**), however, these source areas become a sink for atmospheric CO_2 due to the intense biological utilization that overwhelms the increasing effect on pCO_2 of warming. A similar area is found in the Arabian Sea, where upwelling of deep-waters is induced by the south-west monsoon during July–August(**Figure 1B**), causing the area tobecome a strong CO_2 source. This source area is eliminated by the photosynthetic utilization of CO_2 following the end of the upwelling period (**Figure 1A**).

As illustrated in **Figure 1A** and **B**, the distribution of oceanic sink and source areas for atmospheric CO_2 varies over a wide range in space and time. Surface ocean waters are out of equilibrium with respect to atmospheric CO_2 by as much as $\pm 200\,\mu atm$ (or by $\pm 60\%$).The large magnitudes of CO_2 disequilibrium between the sea and theair is in contrast with the behavior of oxygen, another biologically mediated gas, that shows only up to $\pm 10\%$ sea–air disequilibrium. The large CO_2 disequilibrium may be attributed to the fact that the internal ocean processes that control pCO_2 in sea water, such as the temperature of water, the photosynthesis, and the upwelling of deep waters,occur at much faster rates than the sea–air CO_2 transfer rates. The slow rate of CO_2 transfer across the sea surface is due to the slow hydration rates of CO_2 as well as to the large solubility of CO_2 in sea water attributable to the formation of bicarbonate and carbonate ions. The latter effect does not exist at all for oxygen.

Net CO_2 Flux Across the Sea Surface

The net sea–air CO_2 flux over the global oceans may be computed using eqns [2] and [3]. **Figure 2** shows the climatological mean distribution of the annual sea–air CO_2 flux for the reference year 1995 using the following set of information. (1) The monthly mean ΔpCO_2 values in $4° \times 5°$ pixel areas for the reference year 1995 (**Figure 1A** and **B** for all other months); (2) the Wanninkhof formulation, eqn [3], for the effect of wind speed on the CO_2 gas transfer coefficient; and (3) the climatological mean wind speeds for each month compiled by Esbensen and Kushnir in 1981. This set yields a mean global gas transfer rate of $0.063\,\text{mole}\ CO_2\,\text{m}^{-2}\ \mu\text{atm}^{-1}\,\text{y}^{-1}$, that is consistent with $20\,\text{moles}\ CO_2\ \text{m}^{-2}\text{y}^{-1}$ estimated on the basis of carbon-14 distribution in the atmosphere and the oceans.

Figure 2 shows that the equatorial Pacific is a strong CO_2 source. On the other hand, the areas along the poleward edges of the temperate gyres in both hemispheres are strong sinks for atmospheric CO_2. This feature is particularly prominent in the southern Indian and Atlantic Oceans between 40°S and 60°S, and is attributable to the combined effects of negative sea–air pCO_2 differences with strong winds ('the roaring 40 s') that accelerate sea–air gas transfer rates. Similarly strong sink zones are formed in the North Pacific and North Atlantic between 45°N and 60°N. In the high latitude Atlantic, strong

sink areas extend into the Norwegian and Greenland Seas. Over the high latitude Southern Ocean areas, the sea–air gas transfer is impeded by the field of ice that covers the sea surface for $\geqslant 6$ months in a year.

The net sea–air CO_2 fluxes computed for each ocean basin for the reference year of 1995, representing non-El Niño conditions, are summarized in **Table 1**. The annual net CO_2 uptake by the global ocean is estimated to be about 2.0 Pg-C y^{-1}. This is consistent with estimates obtained on the basis of a number of different ocean–atmosphere models including multi-box diffusion advection models and three-dimensional general circulation models.

The uptake flux for the Northern Hemisphere ocean (north of 14°N) is 1.2 Pg-C y^{-1}, whereas that for the Southern Hemisphere ocean (south of 14°S) is 1.7 Pg-C y^{-1}. Thus, the Southern Hemisphere ocean is astronger CO_2 sink by about 0.5 Pg-C y^{-1}. This is due partially to the much greater oceanic areas in the Southern Hemisphere. In addition, the Southern Ocean south of 50°S is an efficient CO_2 sink, for it takes up about 26% of the global ocean CO_2 uptake, while it has only 10% of the global ocean area. Cold temperature and moderate photosynthesis are both

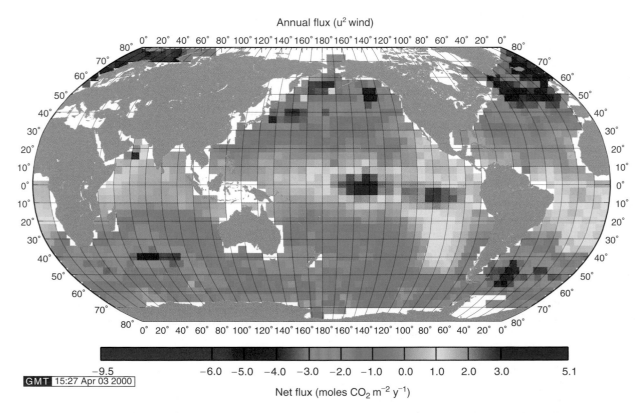

Figure 2 The mean annual sea–air flux of CO_2 in the reference year 1995. The red-yellow areas indicate that the flux is from sea to air, whereas blue-purple areas indicate that the flux is from air to sea. The flux is given in moles of $CO_2\ \text{m}^{-2}\text{y}^{-1}$. The map gives a total annual air-to-sea flux of 2.0 Pg-C y^{-1}.

Table 1 The net sea–air flux of CO_2 estimated for a reference year of 1995 using the effect of wind speed on the CO_2 gas transfer coefficient, eqn [3], of Wanninkhof and the monthly wind field of Esbensen and Kushnir

Latitudes	Pacific Ocean	Atlantic Ocean	Indian Ocean	Southern Ocean	Global Oceans
	Sea–air flux in 10^{15}g Carbon y^{-1}				
North of 50°N	− 0.02	− 0.44	—	—	− 0.47
50°N–14°N	− 0.47	− 0.27	+ 0.03	—	− 0.73
14°N–14°S	+ 0.64	+ 0.13	+ 0.09	—	+ 0.86
14°S–50°S	− 0.37	− 0.20	− 0.60	—	− 1.17
South of 50°S	—	—	—	− 0.52	− 0.52
Total	− 0.23	− 0.78	− 0.47	− 0.52	− 2.00
%Uptake	11%	39%	24%	26%	100%
Area (10^6 km²)	151.6	72.7	53.2	31.7	309.1
Area (%)	49.0%	23.5%	17.2%	10.2%	100%

Positive values indicate sea-to-air fluxes, and negative values indicate air-to-sea fluxes.

responsible for the large uptake by the Southern Ocean.

The Atlantic Ocean is the largest net sink for atmospheric CO_2 (39%); the Southern Ocean (26%) and the Indian Ocean (24%) are next; and the Pacific Ocean (11%) is the smallest. The intense biological drawdown of CO_2 in the high latitude areas of the North Atlantic and Arctic seasduring the summer months is responsible for the Atlantic being a major sink. This is also due to the fact that the upwelling deep waters in the North Atlantic contain low CO_2 concentrations, which are in turn caused primarily by the short residence time (~ 80y) of the North Atlantic Deep Waters. The small uptake flux of the Pacific can be attributed to the fact that the combined sink flux of the northern and southern subtropical gyres is roughly balanced by the source flux from the equatorial Pacific during non-El Niño periods. On the other hand, the equatorial Pacific CO_2 source flux is significantly reduced or eliminated during El Niño events. As a result the equatorial zone is covered with the eastward spreading of the warm, low pCO_2 western Pacific waters in response to the relaxation of the trade wind. Although the effects of El Niño and Southern Ocean Oscillation may be far reaching beyond the equatorial zone as far as to the polar areas, the El Niño effects on the equatorial Pacific alone could reduce the equatorial CO_2 source. Hence, this could increase the global ocean uptake flux by up to 0.6 Pg-Cy^{-1} during an El Niño year.

The sea–air CO_2 flux estimated above is subject to three sources of error: (1) biases in sea–air ΔpCO_2 values interpolated from relatively sparse observations, (2) the 'skin' temperature effect, and(3) uncertainties in the gas transfer coefficient estimated on the basis of the wind speed dependence. Possible biases in ΔpCO_2 differences have been tested using sea surface temperatures (SST) as a proxy. The systematic error in the global sea–air CO_2 flux resulting from sampling and interpolation has been estimated to be about $\pm 30\%$ or ± 0.6 Pg-Cy^{-1}. The'skin' temperature of ocean water may affect ΔpCO_2 by as much as ± 6 μatm depending upon time and place, as discussed earlier.

Although the distribution of the 'skin' temperature over the global ocean is not known, it may be cooler than the bulk water temperature by a few tenths of a degree on the global average. This may result in an under estimation of the ocean uptake by 0.4 Pg-Cy^{-1}. The estimated global sea–air flux depends on the wind speed data used. Since the gas transfer rate increases nonlinearly with wind speed, the estimated CO_2 fluxes tend to be smaller when mean monthly wind speeds are used instead of high frequency wind data.

Furthermore, the wind speed dependence on the CO_2 gas transfer coefficient in high wind speed regimes is still questionable. If the gas transfer rate is taken to be a cubic function of wind speed instead of the square dependence as shown above, the global ocean uptake would be increased by about 1 Pg-C y^{-1}. The effect is particularly significant over the high latitude oceans where the winds are strong. Considering various uncertainties discussed above, the global ocean CO_2 uptake presented in **Table 1** is uncertain by about 1 Pg-C y^{-1}.

See also

Carbon Cycle. Ocean Carbon System, Modeling of. Ocean Circulation. Radiocarbon. Stable Carbon Isotope Variations in the Ocean.

Further Reading

Broecker WS and Peng TH (1982) *Tracers in the Sea.* Palisades, NY: Eldigio Press.

Broecker WS, Ledwell JR, Takahashi, *et al.* (1986) Isotopic versus micrometeorologic ocean CO$_2$ fluxes a: serious conflict. *Journal of Geophysical Research* 91: 10517–10527.

Keeling R, Piper SC, and Heinmann M (1996) Global and hemispheric CO$_2$ sinks deduced from changes in atmospheric O$_2$ concentration. *Nature* 381: 218–221.

Sarmiento JL, ʻ ʻırnanʻ ˉ and Le Quere C (1995) Air–sea CO$_2$ transter and ɯ.ɯ carbon budget of the North Atlantic. *Philosophical Transactions of the Royal Society of London, series B* 343: 211–219.

Sundquist ET (1985) Geological perspectives on carbon dioxide and carbon cycle. In: Sundquist ET and Broecker WS (eds.) *The Carbon Cycle and Atmospheric CO$_2$ N:atural Variations, Archean to Present, Geophysical Monograph 32*, pp. 5–59. Washington, DC: American Geophysical Union.

Takashahi T, Olafsson J, Goddard J, Chipman DW, and Sutherland SC (1993) Seasonal variation of CO$_2$ and nutrients in the high-latitude surface oceans a: comparative study. *Global Biogeochemical Cycles* 7: 843–878.

Takahashi T, Feely RA, Weiss R, *et al.* (1997) Global air–sea flux of CO$_2$ a:n estimate based on measurements of sea–air pCO$_2$ difference. *Proceedings of the National Academy of science USA* 94: 8292–8299.

Tans PP, Fung IY, and Takahashi T (1990) Observational constraints on the global atmospheric CO$_2$ budget. *Sciece* 247: 1431–1438.

Wanninkhof R (1992) Relationship between wind speed and gas exchange. *Journal of Geophysical Research* 97: 7373–7382.

Wanninkhof R and McGillis WM (1999) A cubic relationship between gas transfer and wind speed. *Geophysical Research Letters* 26: 1889–1893.

OCEAN CARBON SYSTEM, MODELING OF

S. C. Doney and D. M. Glover, Woods Hole
Oceanographic Institution, Woods Hole, MA, USA

Introduction

Chemical species such as radiocarbon, chloro-fluorocarbons, and tritium–^{3}He are important tools for ocean carbon cycle research because they can be used to trace circulation pathways, estimate time-scales, and determine absolute rates. Such species, often termed chemical tracers, typically have rather simple water-column geochemistry (e.g., conservative or exponential radioactive decay), and reasonably well-known time histories in the atmosphere or surface ocean. Large-scale ocean gradients of nutrients, oxygen, and dissolved inorganic carbon reflect a combination of circulation, mixing, and the production, transport, and oxidation (or remineralization) of organic matter. Tracers provide additional, often independent, information useful in separating these biogeochemical and physical processes. Biogeochemical and tracer observations are often framed in terms of ocean circulation models, ranging from simple, idealized models to full three-dimensional (3-D) simulations. Model advection and diffusion rates are typically calibrated or evaluated against transient tracer data. Idealized models are straightforward to construct and computationally inexpensive and are thus conducive to hypothesis testing and extensive exploration of parameter space. More complete and sophisticated dynamics can be incorporated into three-dimensional models, which are also more amenable for direct comparisons with field data. Models of both classes are used commonly to examine specific biogeochemical process, quantify the uptake of anthropogenic carbon, and study the carbon cycle responses to climate change. All models have potential drawbacks, however, and part of the art of numerical modeling is deciding on the appropriate model(s) for the particular question at hand.

Carbon plays a unique role in the Earth's environment, bridging the physical and biogeochemical systems. Carbon dioxide (CO_2), a minor constituent in the atmosphere, is a so-called greenhouse gas that helps to modulate the planet's climate and temperature. Given sunlight and nutrients, plants and some microorganisms convert CO_2 via photosynthesis into organic carbon, serving as the building blocks and energy source for most of the world's biota. The concentration of CO_2 in the atmosphere is affected by the net balance of photosynthesis and the reverse reaction respiration on land and in the ocean. Changes in ocean circulation and temperature can also change CO_2 levels because carbon dioxide is quite soluble in sea water. In fact, the total amount of dissolved inorganic carbon (DIC) in the ocean is about 50 times larger than the atmospheric inventory. The air–sea exchange of carbon is governed by the gas transfer velocity and the surface water partial pressure of CO_2 (pCO_2), which increases with warmer temperatures, higher DIC, and lower alkalinity levels. The natural carbon cycle has undergone large fluctuations in the past, the most striking during glacial periods when atmospheric CO_2 levels were about 30% lower than preindustrial values. The ocean must have been involved in such a large redistribution of carbon, but the exact mechanism is still not agreed upon.

Human activities, including fossil-fuel burning and land-use practices such as deforestation and biomass burning, are altering the natural carbon cycle. Currently about 7.5 Pg C yr^{-1} (1 Pg $= 10^{15}$g) are emitted into the atmosphere, and direct measurements show that the atmospheric CO_2 concentration is indeed growing rapidly with time. Elevated atmospheric CO_2 levels are projected to heat the Earth's surface, and the evidence for climate warming is mounting. Only about 40% of the released anthropogenic carbon remains in the atmosphere, the remainder is taken up in about equal portions (or 2 Pg C yr^{-1}) by land and ocean sinks (**Figure 1**). The future magnitude of these sinks is not well known, however, and is one of the major uncertainties in climate simulations.

Solving this problem is complicated because human impacts appear as relatively small perturbations on a large natural background. In the ocean, the reservoir of organic carbon locked up as living organisms, mostly plankton, is only about 3 Pg C. The marine biota in the sunlit surface ocean are quite productive though, producing roughly 50 Pg of new organic carbon per year. Most of this material is recycled near the ocean surface by zooplankton grazing or microbial consumption. A small fraction, something like 10–20% on average, is exported to the deep ocean as sinking particles or as dissolved organic matter moving with the ocean circulation. Bacteria and other organisms in the deep ocean feed on this source of organic matter from above, releasing DIC and associated nutrients back into the

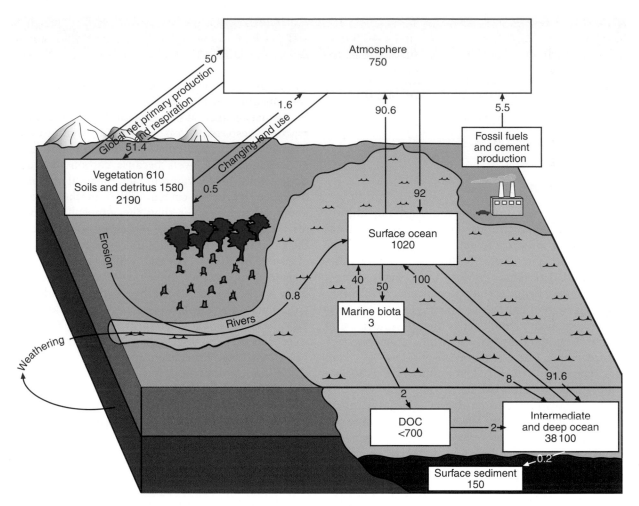

Figure 1 Schematic of global carbon cycle for the 1980s including natural background and human perturbations. Carbon inventories are in Pg C (1 Pg = 10^{15} g) and fluxes are in Pg C yr^{-1}. DOC, dissolved organic carbon. Adapted with permission from Schimel D, Enting IG, Heimann M, *et al.* (1995) CO$_2$ and the carbon cycle. In: Houghton JT, Meira Filho LG, Bruce J, *et al.* (eds.) *Climate Change 1994, Intergovernmental Panel on Climate Change*, pp. 39–71. Cambridge, UK: Cambridge University Press.

water, a process termed respiration or remineralization. The export flux from the surface ocean is a key factor driving the marine biogeochemical cycles of carbon, oxygen, nitrogen, phosphorus, silicon, and trace metals such as iron.

The surface export and subsurface remineralization of organic matter are difficult to measure directly. Biogeochemical rates, therefore, are often inferred based on the large-scale distributions of DIC, alkalinity, inorganic nutrients (nitrate, phosphate, and silicate), and dissolved oxygen. The elemental stoichiometry of marine organic matter, referred to as the Redfield ratio, is with some interesting exceptions relatively constant in the ocean, simplifying the problem of interrelating the various biogeochemical fields. Geochemical distributions have the advantage that they integrate over much of the localized time/space variability in the ocean and can be used to extrapolate to region and basin scales. Property fields, though,

reflect a combination of the net biogeochemical uptake and release as well as physical circulation and turbulent mixing. Additional information is required to separate these signals and can come from a mix of dynamical constraints, numerical models, and ocean process tracers.

The latter two approaches are related because natural and artificial tracers are used to calibrate or evaluate ocean models. A key aspect of these tracers is that they provide independent information on timescale, either because they decay or are produced at some known rate, for example, due to radioactivity, or because they are released into the ocean with a known time history. The different chemical tracers can be roughly divided into two classes. Circulation tracers such as radiocarbon, tritium–^{3}He, and the chlorofluorocarbons are not strongly impacted by biogeochemical cycling and are used primarily to quantify physical advection and mixing rates. These tracers are the major focus here. The distribution of other tracer species is more closely governed by biology and chemistry, for example, the thorium isotope series, which is used to study export production, particle scavenging, vertical transport, and remineralization rates.

Ocean Tracers and Dynamics: A One-dimensional (1-D) Example

Natural radiocarbon (^{14}C), a radioactive isotope of carbon, is a prototypical example of a (mostly) passive ocean circulation tracer. Radiocarbon is produced by cosmic rays in the upper atmosphere and enters the surface ocean as radiolabeled carbon dioxide ($^{14}CO_2$) via air–sea gas exchange. The ^{14}C DIC concentrations in the ocean decrease away from the surface, reflecting the passage of time since the water was last exposed to the atmosphere. Some radiolabeled carbon is transported to the deep ocean in sinking particulate organic matter, which can be largely corrected for in the analysis. The ^{14}C deficits relative to the surface water can be converted into age estimates for ocean deep waters using the radioactive decay half-life (5730 years). Natural radiocarbon is most effective for describing the slow thermohaline overturning circulation of the deep ocean, which has timescales of roughly a few hundred to a thousand years.

The main thermocline of the ocean, from the surface down to about 1 km or so, has more rapid ventilation timescales, from a few years to a few decades. Tracers useful in this regard are chlorofluorocarbons, tritium and its decay product ^3He, and bomb radiocarbon, which along with tritium was released into the atmosphere in large quantities in the 1950s and 1960s by atmospheric nuclear weapons testing.

When properly formulated, the combination of ocean process tracers and numerical models provides powerful tools for studying ocean biogeochemistry. At their most basic level, models are simply a mathematical statement quantifying the rates of the essential physical and biogeochemical processes. For example, advection–diffusion models are structured around coupled sets of differential equations:

$$\frac{\partial C}{\partial t} = -\nabla \cdot (\vec{u}C) + \nabla \cdot (K\nabla C) + J \qquad [1]$$

describing the time rate of change of a generic species C. The first and second terms on the right-hand side of eqn [1] stand for the local divergence due to physical advection and turbulent mixing, respectively. All of the details of the biogeochemistry

are hidden in the net source/sink term J, which for radiocarbon would include net input from particle remineralization (R) and radioactive loss ($-\lambda^{14}C$).

One of the first applications of ocean radiocarbon data was as a constraint on the vertical diffusivity, upwelling, and oxygen consumption rates in the deep waters below the main thermocline. As illustrated in **Figure 2**, the oxygen and radiocarbon concentrations in the North Pacific show a minimum at mid-depth and then increase toward the ocean seabed. This reflects particle remineralization in the water column and the inflow and gradual upwelling of more recently ventilated bottom waters from the Southern Ocean. Mathematically, the vertical profiles for radiocarbon, oxygen (O_2), and a conservative tracer salinity (S) can be posed as steady-state, 1-D balances:

$$0 = K_z \frac{d^2 S}{dz^2} - w \frac{dS}{dz} \qquad [2]$$

$$0 = K_z \frac{d^2 O_2}{dz^2} - w \frac{dO_2}{dz} + R_{O_2} \qquad [3]$$

$$0 = K_z \frac{d^2\,^{14}C}{dz^2} - w \frac{d^{14}C}{dz} + (^{14}C:O_2)R_{O_2} - \lambda^{14}C \qquad [4]$$

K_z and w are the vertical diffusivity and upwelling rates, and $^{14}C:O_2$ is a conversion factor.

Looking carefully at eqn [2], one sees that the solution depends on the ratio K_z/w but not K_z or w separately. Similarly the equation for oxygen gives us information on the relative rates of upwelling and remineralization. It is only by the inclusion of radiocarbon, with its independent clock due to radioactive decay, that we can solve for the absolute physical and biological rates. The solutions to eqns [2]–[4] can be derived analytically, and as shown in **Figure 2** parameter values of $w = 2.3 \times 10^{-5}\,\mathrm{cm\,s^{-1}}$, $K_z = 1.3\,\mathrm{cm^2\,s^{-2}}$, and $R_{O_2} = 0.13 \times 10^{-6}\,\mathrm{mol\,kg^{-1}\,yr^{-1}}$ fit the data reasonably well. The 1-D model-derived vertical diffusivity is about an order of magnitude larger than estimates from deliberate tracer release experiments and microscale turbulence measurements in the upper thermocline. However, they may be consistent with recent observations of enhanced deep-water vertical mixing over regions of rough bottom topography.

Ocean Circulation and Biogeochemical Models

The 1-D example shows the basic principles behind the application of tracer data to the ocean carbon

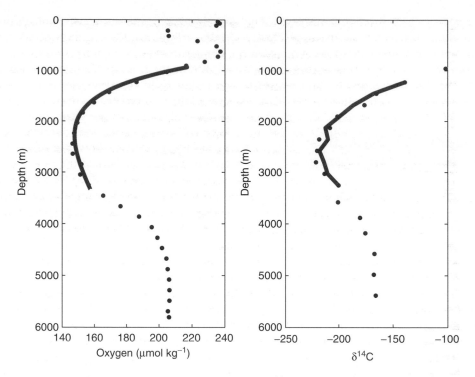

Figure 2   Observed vertical profiles of oxygen (O_2) and radiocarbon ($\delta^{14}C$) in the North Pacific. The solid curves are the model solution of a 1-D advection–diffusion equation.

cycle, but the complexity (if not always the sophistication) of the models and analysis has grown with time. Ocean carbon models can be roughly divided into idealized models (multi-box, 1-D and 2-D advection–diffusion models) and 3-D general circulation models (GCMs). Although the distinction can be blurry at times, idealized models are characterized typically by reduced dimensionality and/or kinematic rather than dynamic physics. That is, the circulation and mixing are specified rather than computed by the model and are often adjusted to best match the transient tracer data.

Global Box Models

An example of a simple, high-latitude outcrop box model is shown in **Figures 3(a)** and **3(b)**. The boxes represent the atmosphere, the high- and low-latitude surface ocean, and the deep interior. In the model, the ocean thermohaline circulation is represented by one-way flow with high-latitude sinking, low-latitude upwelling, and poleward surface return flow. Horizontal and vertical mixing is included by two-way exchange of water between each pair of boxes. The physical parameters are constrained so that model natural radiocarbon values roughly match observations. Note that the ^{14}C concentration in the deep water is significantly depleted relative to the surface

boxes and amounts to a mean deep-water ventilation age of about 1150 years.

The model circulation also transports phosphate, inorganic carbon, and alkalinity. Biological production, particle export, and remineralization are simulated by the uptake of these species in the surface boxes and release in the deep box. The model allows for air–sea fluxes of CO_2 between the surface boxes and the atmosphere. The low-latitude nutrient concentrations are set near zero as observed. The surface nutrients in the high-latitude box are allowed to vary and are never completely depleted in the simulation of modern conditions. Similar regions of 'high-nutrient, low-chlorophyll' concentrations are observed in the subpolar North Pacific and Southern Oceans and are thought to be maintained by a combination of light and iron limitation as well as zooplankton grazing. The nutrient and DIC concentrations in the deep box are higher than either of the surface boxes, reflecting the remineralization of sinking organic particles.

The three-box ocean outcrop model predicts that atmospheric CO_2 is controlled primarily by the degree of nutrient utilization in high-latitude surface regions. Marine production and remineralization occur with approximately fixed carbon-to-nutrient ratios, the elevated nutrients in the deep box are associated with an equivalent increase in DIC and $p\mathrm{CO}_2$. Large adjustments in the partitioning of

(a)

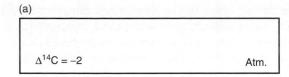

$\Delta^{14}C = -2$ Atm.

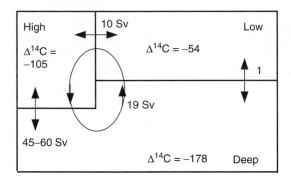

High   10 Sv   Low

$\Delta^{14}C = -105$

$\Delta^{14}C = -54$

1

19 Sv

45–60 Sv

$\Delta^{14}C = -178$ Deep

(b)

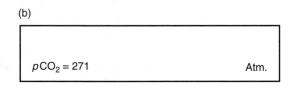

$pCO_2 = 271$ Atm.

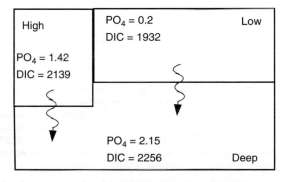

High $PO_4 = 0.2$ Low
       $DIC = 1932$

$PO_4 = 1.42$
$DIC = 2139$

$PO_4 = 2.15$
$DIC = 2256$ Deep

Figure 3 Results from a simple three-box ocean carbon cycle model. (a) The physical circulation and modeled radiocarbon ($\Delta^{14}C$) values. (b) The model biogeochemical fields, ocean DIC, and phosphate (PO_4) and atmospheric pCO_2. From Toggweiler JR and Sarmiento JL, Glacial to inter-glacial changes in atmospheric carbon dioxide: The critical role of ocean surface waters in high latitudes, *The Carbon Cycle and Atmospheric CO₂: Natural Variations Archean to Present*, Sundquist ET and Broecker WS (eds.), pp. 163–184, 1985, Copyright [1985]. American Geophysical Union. Adapted by permission of American Geophysical Union.

carbon between the ocean and atmosphere can occur only where this close coupling of the carbon and nutrient cycles breaks down. When subsurface water is brought to the surface at low latitude, production draws the nutrients down to near zero and removes to first order all of the excess seawater DIC and pCO_2. Modifications in the upward nutrient flux to the low latitudes have relatively little impact on the

model atmospheric CO_2 as long as the surface nutrient concentrations stay near zero.

At high latitudes, however, the nutrients and excess DIC are only partially utilized, resulting in higher surface water pCO_2 and, over decades to centuries, higher atmospheric CO_2 concentrations. Depending on the polar biological efficiency, the model atmosphere effectively sees more or less of the high DIC concentrations (and pCO_2 levels) of the deep ocean. Thus changes in ocean biology and physics can have a correspondingly large impact on atmospheric CO_2. On longer timescales (approaching a few millennia), these variations are damped to some extent by adjustments of the marine calcium carbonate cycle and ocean alkalinity.

The three-box outcrop model is a rather crude representation of the ocean, and a series of geographical refinements have been pursued. Additional boxes can be added to differentiate the individual ocean basins (e.g., Atlantic, Pacific, and Indian), regions (e.g., Tropics and subtropics), and depths (e.g., thermocline, intermediate, deep, and bottom waters), leading to a class of models with a half-dozen to a few dozen boxes. The larger number of unknown advective flows and turbulent exchange parameters, however, complicates the tuning procedure. Other model designs take advantage of the vertical structure in the tracer and biogeochemical profile data. The deep box (es) is discretized in the vertical, essentially creating a continuous interior akin to a 1-D advection–diffusion model. This type of model was often used in the 1970s and 1980s for the initial anthropogenic CO_2 uptake calculations, where it is important to differentiate between the decadal ventilation timescales of the thermocline and the centennial timescales of the deep water.

Intermediate Complexity and Inverse Models

In terms of global models, the next step up in sophistication from box models is intermediate complexity models. These models typically have higher resolution and/or include more physical dynamics but fall well short of being full GCMs. Perhaps the most common examples for ocean carbon cycle research are zonal average basin models. The dynamical equations are similar to a GCM but are integrated in 2-D rather than in 3-D, the third east–west dimension removed by averaging zonally across the basin. In some versions, multiple basins are connected by an east–west Southern Ocean channel. The zonal average models often have a fair representation of the shallow wind-driven Ekman and deep thermohaline overturning circulations but obviously lack western boundary currents and gyre circulations. Tracer data remain an important element in tuning some of the mixing coefficients and surface boundary conditions and in evaluating the model solutions.

Based on resolution, many inverse models can also be categorized as intermediate complexity, but their mode of operation differs considerably from the models considered so far. In an inverse model, the circulation field and biogeochemical net source/sink (the J terms in the notation above) are solved for using the observed large-scale hydrographic and tracer distributions as constraints. Additional dynamic information may also be incorporated such as the geostrophic velocity field, general water mass properties, or float and mooring velocities.

Inverse calculations are typically posed as a large set of simultaneous linear equations, which are then solved using standard linear algebra methods. The inverse techniques are most commonly applied to steady-state tracers, though some exploration of transient tracers has been carried out. The beauty of the inverse approach is that it tries to produce dynamically consistent physical and/or biogeochemical solutions that match the data within some assigned error. The solutions are often underdetermined in practice, however, which indicates the existence of a range of possible solutions. From a biogeochemical perspective, the inverse circulation models provide estimates of the net source/sink patterns, which can then be related to potential mechanisms.

Thermocline Models

Ocean process tracers and idealized models have also been used extensively to study the ventilation of the main thermocline. The main thermocline includes the upper 1 km of the tropical to subpolar ocean where the temperature and potential density vertical gradients are particularly steep. Thermocline ventilation refers to the downward transport of surface water recently exposed to the atmosphere, replenishing the oxygen and other properties of the subsurface interior. Based on the vertical profiles of tritium and ^{3}He as well as simple 1-D and box models, researchers in the early 1980s showed that ventilation of the main thermocline in the subtropical gyres occurs predominately as a horizontal process along surfaces of constant density rather than by local vertical mixing. Later work on basin-scale bomb-tritium distributions confirmed this result and suggested the total magnitude of subtropical ventilation is large, comparable to the total wind-driven gyre circulation.

Two dimensional gyre-scale tracer models have been fruitfully applied to observed isopycnal tritium–^{3}He and chlorofluorocarbon patterns. As shown in **Figure 4**, recently ventilated water (near-zero tracer age) enters the thermocline on the poleward side of the gyre and is swept around the clockwise circulation of the gyre (for a Northern Hemisphere case). Comparisons of model tracer patterns and property–property relationships constrain the absolute ventilation rate and the relative effects of isopycnal advection versus turbulent mixing by depth and region.

Thermocline tracer observations are also used to estimate water parcel age, from which biogeochemical rate can be derived. For example, remineralization produces an apparent oxygen deficit relative to atmospheric solubility. Combining the oxygen deficit with an age estimate, one can compute the rate of oxygen utilization. Similar geochemical approaches have been or can be applied to a host of problems: nitrogen fixation, denitrification, dissolved organic matter remineralization, and nutrient resupply to the upper ocean. The biogeochemical application of 2-D gyre models has not been pursued in as much detail.

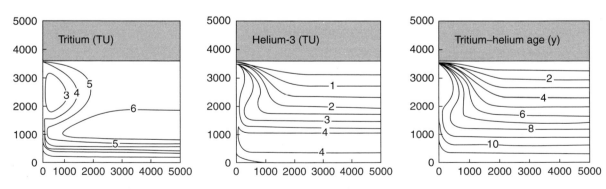

Figure 4 Tracer results from a two-dimensional gyre model. The model represents the circulation on a constant density surface (isopycnal) in the main thermocline that outcrops along the northern boundary (shaded region). Thermocline ventilation is indicated by the gradual increase of tritium–^{3}He ages around the clockwise flowing gyre circulation. TU, tritium unit (1 TU = 1 ^{3}H atom/10^{18} H atoms). Reproduced with permission from Musgrave DL (1990) Numerical studies of tritium and helium-3 in the thermocline. *Journal of Physical Oceanography* 20: 344–373.

This form of modeling, however, is particularly useful for describing regional patterns and in the areas where simple tracer age approaches breakdown.

Three-dimensional (3-D) Biogeochemical Simulations

By their very nature, idealized models neglect many important aspects of ocean dynamics. The alternative is full 3-D ocean GCMs, which incorporate more realistic spatial and temporal geometry and a much fuller suite of physics. There are several different families of ocean physical models characterized by the underlying governing equations and the vertical discretization schemes. Within model families, individual simulations will differ in important factors of surface forcing and choice of physical parametrizations. Often these parametrizations account for complex processes, such as turbulent mixing, that occur on small space scales that are not directly resolved or computed by the model but which can have important impacts on the larger-scale ocean circulation; often an exact description of specific events is not required and a statistical representation of these subgrid scale processes is sufficient. For example, turbulent mixing is commonly treated using equations analogous to those for Fickian molecular diffusion. Ocean GCMs, particularly coarse-resolution global versions, are sensitive to the subgrid scale parametrizations used to account for unresolved processes such as mesoscale eddy mixing, surface and bottom boundary layer dynamics, and air–sea and ocean–ice interactions. Different models will be better or worse for different biogeochemical problems.

Considerable progress has been achieved over the last two decades on the incorporation of chemical tracers, biogeochemistry, and ecosystem dynamics into both regional and global 3-D models. Modeling groups now routinely simulate the more commonly measured tracers (e.g., radiocarbon, chlorofluoro-carbons, and tritium–^{3}He). Most of these tracers have surface sources and, therefore, provide clear indications of the pathways and timescales of subsurface ventilation in the model simulations. The bottom panel of **Figure 5** shows from a simulation of natural radiocarbon the layered structure of the intermediate and deep water flows in the Atlantic basin, with Antarctic Intermediate and Bottom Waters penetrating from the south and southward-flowing North Atlantic Deep Water sandwiched in between.

In a similar fashion, biogeochemical modules have been incorporated to simulate the 3-D fields of nutrients, oxygen, DIC, etc. Just as with the physics, the degree of biological complexity can vary considerably

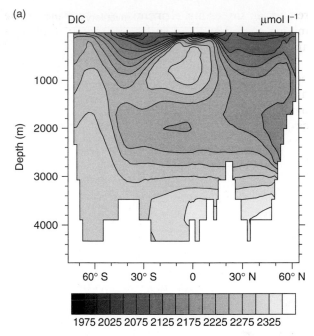

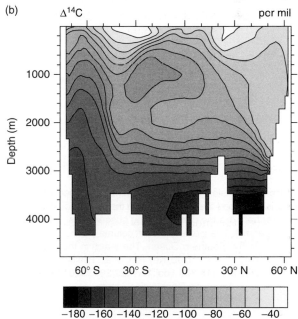

Figure 5 Simulated radiocarbon and carbon cycle results from a three-dimensional global ocean biogeochemical model. Depth–latitude sections are shown for (a) DIC concentration (μmol l^{-1}) and (b) natural (preindustrial) radiocarbon (Δ^{14}C, per mil) along the prime meridian in the Atlantic Ocean.

across ocean carbon models. At one end are diagnostic models where the production of organic matter in the surface ocean is prescribed based on satellite ocean color data or computed implicitly by forcing the simulated surface nutrient field to match observations (assuming that any net reduction in nutrients is driven by organic matter production). At the other extreme are simulations that couple prognostic

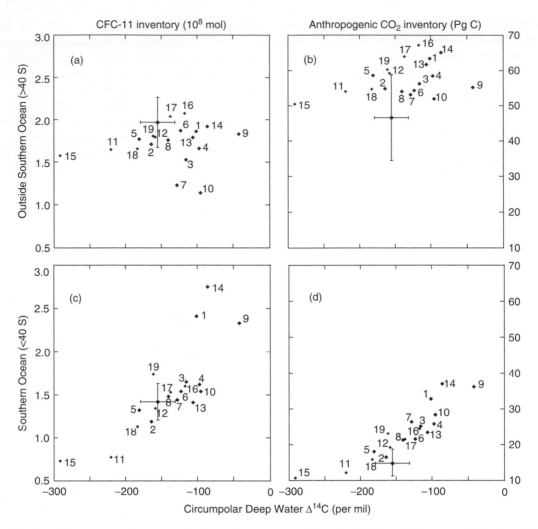

Figure 6 Transient tracer constraints on numerical model ocean carbon dynamics. Each panel shows results for the suite of global ocean simulations that participated in the international Ocean Carbon-Cycle Model Intercomparison Project (OCMIP). The points with the error bars are the corresponding observations. The x-axis in all panels is the average Circumpolar Deep Water radiocarbon level ($\Delta^{14}C$, per mil). The y-axis in the left column is the chlorofluorocarbon (CFC-11) inventory (10^8 mol) outside of the Southern Ocean (top row) and in the Southern Ocean. The y-axis in the right column is the corresponding anthropogenic carbon inventory (Pg C). From Matsumoto K, Sarmiento JL, Key RM, *et al.*, Evaluation of ocean carbon cycle models with data-based metrics, *Geophysical Research Letters*, Vol. 31, L07303 (doi:10.1029/2003GL018970), 2004. Copyright [2004] American Geophysical Union. Reproduced by permission of American Geophysical Union.

treatments of multiple phytoplankton, zooplankton, and bacteria species to the biogeochemical tracers. Transient tracers enter the picture because the biogeochemical distributions are also strongly influenced circulation. This is nicely illustrated by comparing the simulated radiocarbon and DIC sections in **Figure 5**. To first order, older waters marked by low radiocarbon exhibit high DIC levels due to the accumulation of carbon released from particle remineralization. Tracers can be used to guide the improvement of the physical circulation and highlight areas where the circulation is poor and thus the biogeochemical tracers may be suspect. They can also be used more quantitatively to create model skill

metrics; this approach can be applied across multiple model simulations to provide the best overall constraint on quantities like the ocean uptake of anthropogenic CO_2 or export production (**Figure 6**).

Models as Research Tools

Designed around a particular question or hypothesis, conceptual models attempt to capture the basic elements of the problem while remaining amenable to straightforward analysis and interpretation. They are easy to construct and computationally inexpensive, requiring only a desktop PC rather than a supercomputer. When well-formulated, idealized models provide a practical method to analyze ocean physical and biogeochemical dynamics and in some cases to quantitatively constrain specific rates. Their application is closely tied to ocean tracer observations, which are generally required for physical calibration and evaluation. Idealized models remain a valuable tool for estimating the oceanic uptake of anthropogenic carbon and the long time-scale responses (centuries to millennia) of the natural carbon cycle. Also, some of the more memorable and lasting advances in tracer oceanography are directly linked to simple conceptual models. Examples include constraints on the deep-water large-scale vertical diffusivity and demonstration of the dominance of lateral over vertical ventilation of the subtropical main thermocline.

Three-dimensional models offer more realism, at least apparently, but with the cost of greater complexity, a more limited number of simulations, and a higher probability of crucial regional errors in the base solutions, which may compromise direct, quantitative model-data comparisons. Ocean GCM solutions, however, should be exploited to address exactly those problems that are intractable for simpler conceptual and reduced dimensional models. For example, two key assumptions of the 1-D advection–diffusion model presented in **Figure 2** are that the upwelling occurs uniformly in the horizontal and vertical and that mid-depth horizontal advection is not significant. Ocean GCMs and tracer data, by contrast, show a rich three-dimensional circulation pattern in the deep Pacific.

The behavior of idealized models and GCMs can diverge, and it is not always clear that complexity necessarily leads to more accurate results. In the end, the choice of which model to use depends on the scientific problem and the judgment of the researcher. Probably the best advice is to explore solutions from a hierarchy of models and to thoroughly evaluate the skill of the models against a range of tracers and other dynamical measures. Just because a model does a good job reproducing the distribution of one tracer field does not mean that it can be applied indiscriminately to another variable, especially if the underlying dynamics or timescales differ.

Models can be quite alluring in the sense that they provide concrete answers to questions that are often difficult or nearly impossible to address from sparsely sampled field data. However, one should not forget that numerical models are simply a set of tools for doing science. They are no better than the foundations upon which they are built and should not be carried out in isolation from observations of the real ocean. For ocean carbon cycle models, the two key elements are the ocean physical circulation and the biogeochemical processes. Even the best biogeochemical model will perform poorly in an ill-constructed physical model. Conversely, if the underlying biogeochemical mechanisms are poorly known, a model may be able to correctly reproduce the distributions of biogeochemical tracers but for the wrong reasons. Mechanistic-based models are critical in order to understand and predict natural variability and the response of ocean biogeochemistry to perturbations such as climate change.

Glossary

Anthropogenic carbon   The additional carbon that has been released to the environment over the last several centuries by human activities including fossil-fuel combustion, agriculture, forestry, and biomass burning.

Excess ^{3}He   Computed as the ^{3}He in excess of gas solubility equilibrium with the atmosphere.

Export production   That part of the organic matter formed in the surface layer by photosynthesis that is transported out of the surface layer and into the interior of the ocean by particle sinking, mixing, and circulation, or active transport by organisms.

Radiocarbon ^{14}C   Either δ^{14}C or Δ^{14}C where δ^{14}C $= [(^{14}C/^{12}C)_{sample}/(^{14}C/^{12}C)_{standard} - 1] \times 1000$ (in parts per thousand or 'per mil'); Δ^{14}C is similar but corrects the sample ^{14}C for biological fractionation using the sample ^{13}C/^{12}C ratio.

Transient tracers   Chemical tracers that contain time information either because they are radioactive or because their source, usually anthropogenic, has evolved with time.

Tritium–^{3}He age   An age is computed assuming that all of the excess ^{3}He in a sample is due to the radioactive decay of tritium, age $= \ln[(^3H + ^3He)/^3H]/\lambda$, where λ is the decay constant for tritium.

Ventilation   The physical process by which surface properties are transported into the ocean interior.

See also

Carbon Cycle. Carbon Dioxide (CO₂) Cycle. Ocean Circulation. Ocean Circulation: Meridional Overturning Circulation. Radiocarbon.

Further Reading

Broecker WS and Peng T-H (1982) *Tracers in the Sea*. Palisades, NY: Lamont-Doherty Geological Observatory, Columbia University.

Charnock H, Lovelock JE, Liss P, and Whitfield M (eds.) (1988) *Tracers in the Ocean.* London: The Royal Society.

Doney SC, Lindsay K, Caldeira K, *et al.* (2004) Evaluating global ocean carbon models: The importance of realistic physics. *Global Biogeochemical Cycles* 18: GB3017 (doi:10.1029/2003GB002150).

England MH and Maier-Reimer E (2001) Using chemical tracers in ocean models. *Reviews in Geophysics* 39: 29–70.

Fasham M (ed.) (2003) *Ocean Biogeochemistry.* New York: Springer.

Jenkins WJ (1980) Tritium and ^{3}He in the Sargasso Sea. *Journal of Marine Research* 38: 533–569.

Kasibhatla P, Heimann M, and Rayner P (eds.) (2000) *Inverse Methods in Global Biogeochemical Cycles.* Washington, DC: American Geophysical Union.

Matsumoto K, Sarmiento JL, Key RM, *et al.* (2004) Evaluation of ocean carbon cycle models with data-based metrics. *Geophysical Research Letters* 31: L07303 (doi:10.1029/2003GL018970).

Munk WH (1966) Abyssal recipes. *Deep-Sea Research* 13: 707–730.

Musgrave DL (1990) Numerical studies of tritium and helium-3 in the thermocline. *Journal of Physical Oceanography* 20: 344–373.

Sarmiento JL and Gruber N (2006) *Ocean Biogeochemical Dynamics.* Princeton, NJ: Princeton University Press.

Schimel D, Enting IG, Heimann M, *et al.* (1995) CO_2 and the carbon cycle. In: Houghton JT, Meira Filho LG, and Bruce J, *et al.* (eds.) *Climate Change 1994, Intergovernmental Panel on Climate Change,* pp. 39–71. Cambridge, UK: Cambridge University Press.

Siedler G, Church J, and Gould J (eds.) (2001) *Ocean Circulation and Climate.* New York: Academic Press.

Siegenthaler U and Oeschger H (1978) Predicting future atmospheric carbon dioxide levels. *Science* 199: 388–395.

Toggweiler JR and Sarmiento JL (1985) Glacial to inter-glacial changes in atmospheric carbon dioxide: The critical role of ocean surface waters in high latitudes. In: Sundquist ET and Broecker WS (eds.) *The Carbon Cycle and Atmospheric CO_2: Natural Variations Archean to Present,* pp. 163–184. Washington, DC: American Geophysical Union.

Relevant Websites

http://cdiac.ornl.gov
 – Global Ocean Data Analysis Project, Carbon Dioxide Information Analysis Center.

http://w3eos.whoi.edu
 – Modeling, Data Analysis, and Numerical Techniques for Geochemistry (resource page for MIT/WHOI course number 12.747), at Woods Hole Oceanographic Institution.

http://us-osb.org
 – Ocean Carbon and Biogeochemistry (OCB) Program

http://www.ipsl.jussieu.fr
 – Ocean Carbon-Cycle Model Intercomparison Project (OCMIP), Institut Pierre Simon Laplace.

RADIOCARBON

R. M. Key, Princeton University, Princeton, NJ, USA

Introduction

In 1934 F.N.D. Kurie at Yale University obtained the first evidence for existence of radiocarbon (carbon-14, ^{14}C). Over the next 20 years most of the details for measuring ^{14}C and for its application to dating were worked out by W.F. Libby and co-workers. Libby received the 1960 Nobel Prize in chemistry for this research.

The primary application of ^{14}C is to date objects or to determine various environmental process rates. The ^{14}C method is based on the assumption of a constant atmospheric formation rate. Once produced, atmospheric ^{14}C reacts to form $^{14}CO_2$, which participates in the global carbon cycle processes of photosynthesis and respiration as well as the physical processes of dissolution, particulate deposition, evaporation, precipitation, transport, etc. Atmospheric radiocarbon is transferred to the ocean primarily by air–sea gas exchange of $^{14}CO_2$. Once in the ocean, $^{14}CO_2$ is subject to the same physical, chemical, and biological processes that affect CO_2. While alive, biota establish an equilibrium concentration of radiocarbon with their surroundings; that is, ^{14}C lost by decay is replaced by uptake from the environment. Once the tissue dies or is removed from an environment that contains ^{14}C, the decay is no longer compensated. The loss of ^{14}C by decay can then be used to determine the time of death or removal from the original ^{14}C source. After death or removal of the organism, it is generally assumed that no exchange occurs between the tissue and its surroundings; that is, the system is assumed to be closed. As a result of the ^{14}C decay rate, the various reservoir sizes involved in the carbon cycle, and exchange rates between the reservoirs, the ocean contains approximately 50 times as much natural radiocarbon as does the atmosphere.

Carbon-14 is one of three naturally occurring carbon isotopes; ^{14}C is radioactive, has a half-life of 5730 years and decays by emitting a β-particle with an energy of about 156 keV. On the surface of the earth, the abundance of natural ^{14}C relative to the two stable naturally occurring carbon isotopes is $^{12}C:^{13}C:^{14}C = 98.9\%:1.1\%:1.2 \times 10^{-10}$ %. Natural radiocarbon is produced in the atmosphere, primarily by the collision of cosmic ray produced neutrons with nitrogen according to the reaction [I].

$$_0^1 n + {}_7^{14} N \Rightarrow {}_6^{14} C + {}_1^1 H \qquad [I]$$

where n is a neutron and H is the proton emitted by the product nucleus. Similarly, the decay of ^{14}C takes place by emission of a β-particle and leads to stable nitrogen according to reaction (II),

$$_6^{14} C \Rightarrow {}_7^{14} N + \beta^- + \bar{\nu} + Q \qquad [II]$$

where $\bar{\nu}$ is an antineutrino and Q is the decay energy.

The atmospheric production rate varies somewhat and is influenced by changes in the solar wind and in the earth's geomagnetic field intensity. A mean of 1.57 atom $cm^{-2} s^{-1}$ is estimated based on the long-term record preserved in tree rings and a carbon reservoir model. This long-term production rate yields a global natural ^{14}C inventory of approximately 50 t ($1t = 10^6$ g). Production estimates based on the more recent record of neutron flux measurements tend to be higher, with values approaching 2 atom $cm^{-2} s^{-1}$. **Figure 1** shows the atmospheric history of ^{14}C from AD 1511 to AD 1954 measured by Minze Stuiver (University of Washington) using tree growth rings. The strong decrease that occurs after about AD 1880 is due to dilution by anthropogenic addition of CO_2 during the industrial revolution by the burning of fossil fuels (coal, gas, oil). This dilution has come to be known as the Suess effect (after Hans E. Suess).

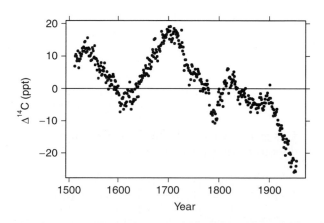

Figure 1 Atmospheric history of Δ^{14}C measured by M. Stuiver in tree rings covering AD 1511 to AD 1954. Most of the decrease over the last hundred years is due to the addition of anthropogenic CO_2 to the atmosphere during the industrial revolution by the burning of fossil fuels.

Prior to 16 July 1945 all radiocarbon on the surface of the earth was produced naturally. On that date, US scientists carried out the first atmospheric atomic bomb test, known as the Trinity Test. Between 1945 and 1963, when the Partial Test Ban Treaty was signed and atmospheric nuclear testing was banned, approximately 500 atmospheric nuclear explosions were carried out by the United States (215), the former Soviet Union (219), the United Kingdom (21) and France (50). After the signing, a few additional atmospheric tests were carried out by China (23) and other countries not participating in the treaty. The net effect of the testing was to significantly increase ^{14}C levels in the atmosphere and subsequently in the ocean. Anthropogenic ^{14}C has also been added to the environment from some nuclear power plants, but this input is generally only detectable near the reactor.

It is unusual to think of any type of atmospheric contamination – especially by a radioactive species – as beneficial; however, bomb-produced radiocarbon (and tritium) has proven to be extremely valuable to oceanographers. The majority of the atmospheric testing, in terms of number of tests and ^{14}C production, occurred over a short time interval, between 1958 and 1963, relative to many ocean circulation processes. This time history, coupled with the level of contamination and the fact that ^{14}C becomes intimately involved in the oceanic carbon cycle, allows bomb-produced radiocarbon to be valuable as a tracer for several ocean processes including biological activity, air–sea gas exchange, thermocline ventilation, upper ocean circulation, and upwelling.

Oceanographic radiocarbon results are generally reported as $\Delta^{14}C$, the activity ratio relative to a standard (NBS oxalic acid, 13.56 dpm per g of carbon) with a correction applied for dilution of the radiocarbon by anthropogenic CO_2 with age corrections of the standard material to AD1950. $\Delta^{14}C$ is defined by eqn [1].

$$\Delta^{14}C = \delta^{14}C - 2(\delta^{13}C + 25)\left(1 + \frac{\delta^{14}C}{1000}\right) \quad [1]$$

$\delta^{14}C$ is given by eqn [2] and the definition of $\delta^{13}C$ is analogous to that for $\delta^{14}C$.

$$\delta^{14}C = \left[\frac{^{14}C/C\big|_{smp} - ^{14}C/C\big|_{std}}{^{14}C/C\big|_{std}}\right] \times 1000 \quad [2]$$

The first part of the second term in the right side of eqn [1] $2(\delta^{13}C + 25)$, corrects for fractionation effects. The factor of 2 accounts for the fact that ^{14}C fractionation is expected to be twice as much as for

^{13}C and the additive constant 25 is a normalization factor conventionally applied to all samples and based on the mean value of terrestrial wood. The details of ^{14}C calculations can be significantly more involved than expressed in the above equations; however, there is a general consensus that the calculations and reporting of results be done as described by Minze Stuiver and Henry Polach in a paper specifically written to eliminate differences that existed previously. $\Delta^{14}C$ has units of parts per thousand (ppt). That is, 1 ppt means that $^{14}C/^{12}C$ for the sample is greater than $^{14}C/^{12}C$ for the standard by 0.001. In these units the radioactive decay rate of ^{14}C is approximately 1 ppt per 8.1 years.

The number of surface ocean measurements made before any bomb-derived contamination are insufficient to provide the global distribution before input from explosions. It is now possible to measure $\Delta^{14}C$ values in the annual growth rings of corals. By establishment of the exact year associated with each ring, reconstruction of the surface ocean $\Delta^{14}C$ history is possible. Applying the same procedure to long-lived mollusk shells extends the method to higher latitudes than is possible with corals. Whether corals or shells are used, it must be demonstrated that the coral or shell incorporates ^{14}C in the same ratio as the water in which it grew or at least that the fractionation is known. This method works only over the depth range at which the animal lived. **Figure 2** shows the $\Delta^{14}C$ record from two Pacific coral reefs measured by Ellen Druffel. Vertical lines indicate the period of atmospheric nuclear tests (1945–1963). The relatively small variability over the first ~ 300 years of the record includes variations due to weather events, climate change, ocean circulation, atmospheric production, etc. The last 50 years of the sequence records the

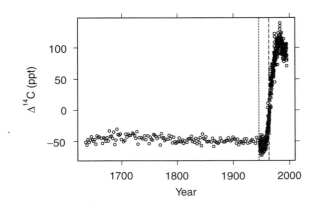

Figure 2 Long-term history of $\Delta^{14}C$ in the surface Pacific Ocean measured by E. Druffel in two coral reefs. The vertical lines surround the period of atmospheric nuclear weapons testing. The oceanic response to bomb contamination is delayed relative to the atmosphere because of the relatively long equilibration time between the ocean and atmosphere for $^{14}CO_2$.

invasion of bomb-produced $\Delta^{14}C$. Worth noting is the fact that the coral record of the bomb signal is lagged. That is, the coral values did not start to increase immediately testing began nor did they cease to increase when atmospheric testing ended. The lag is due to the time for the northern and southern hemisphere atmospheres to mix ($\sim$1 year) and to the relatively long time required for the surface ocean to equilibrate with the atmosphere with respect to $\Delta^{14}C$ ($\sim$10 years). Because of the slow equilibration, the surface ocean is frequently not at equilibrium with the atmosphere. This disequilibrium is one of the reasons why pre-bomb surface ocean results, when expressed as ages rather than ppt units, are generally 'old' rather than 'zero' as might be expected.

Figure 3A shows measured atmospheric $\Delta^{14}C$ levels from 1955 to the present in New Zealand (data from T.A. Rafter, M.A. Manning, and co-workers) and Germany (data from K.O. Munnich and co-workers) as well as older estimates based on tree ring measurements (data from M. Stuiver). The beginning of the significant increase in the mid-1950s marks the atmospheric testing of hydrogen bombs. Atmospheric levels increased rapidly from that point until the mid-1960s. Soon after the ban on atmospheric testing, levels began a decrease that continues up to the present. The rate of decrease in the atmosphere is about 0.055 y^{-1}. Also clearly evident in the figure is that the German measurements were significantly higher than those from New Zealand between approximately 1962 and 1970. The difference reflects the facts that most of the atmospheric tests were carried out in the Northern Hemisphere and that approximately 1 year is required for atmospheric mixing across the Equator. During that interval some of the atmospheric $^{14}CO_2$ is removed. Once atmospheric testing ceased, the two hemispheres equilibrated to the same radiocarbon level.

Figure 3B shows detailed Pacific Ocean $\Delta^{14}C$ coral ring data (J.R. Toggwelier and E. Druffel). This surface ocean record shows an increase during the 1960s; however, the peak occurs somewhat later than in the atmosphere and is significantly less pronounced. Careful investigation of coral data also demonstrates the north–south difference evidenced in the atmospheric record.

Sampling and Measurement Techniques

The radiocarbon measurement technique has existed for only 50 years. The first ^{14}C measurement was made in W.F. Libby's Chicago laboratory in 1949

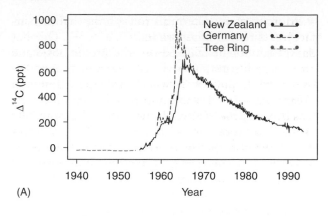

(A)

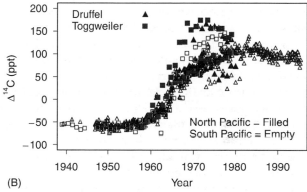

(B)

Figure 3 (A) Detailed atmospheric $\Delta^{14}C$ history as recorded in tree rings for times prior to 1955 and in atmospheric gas samples from both New Zealand and Germany subsequently. The large increase in the late 1950s and 1960s was due to the atmospheric testing of nuclear weapons (primarily fusion devices). The hemispheric difference during the 1960s is because most atmospheric bomb tests were carried out north of the Equator and there is a resistance to atmospheric mixing across the equator. Atmospheric levels began to decline shortly after the ban on atmospheric bomb testing. (B) $\Delta^{14}C$ in the surface Pacific Ocean as recorded in the annual growth rings of corals. The same general trend seen in the atmosphere is present. The bomb contamination peak is broadened and time-lagged relative to the atmosphere due to both mixing and to the time required for transfer from the atmosphere to the ocean.

and the first list of ages was published in 1951. A necessary prerequisite to the age determination was accurate measurement of the radiocarbon half-life. This was done in 1949 in Antonia Engelkeimer's laboratory at the Argonne National Laboratory. Between 1952 and 1955 several additional radiocarbon dating laboratories opened. By the early 1960s several important advances had occurred including the following.

- Significantly improved counting efficiency and lower counting backgrounds, resulting in much greater measurement precision and longer time-scale over which the technique was applicable.
- Development of the extraction and concentration technique for sea water samples.

- More precise determination of the half-life by three different laboratories.
- Recognition by Hans Suess, while at the USGS and Scripps Institution of Oceanography, that radio-carbon in modern samples (since the beginning of the industrial revolution) was being diluted by anthropogenic CO_2 addition to the atmosphere and biosphere.
- Recognition that atmospheric and oceanic $\Delta^{14}C$ levels were increasing as a result of atmospheric testing of nuclear weapons.

During the 1970s and 1980s incremental changes in technique and equipment further increased the precision and lowered the counting background. With respect to the ocean, this was a period of sample collection, analysis, and interpretation. The next significant change occurred during the 1990s with application of the accelerator mass spectrometry (AMS) technique to oceanic samples. This technique counts ^{14}C atoms rather than detecting the energy released when a ^{14}C atom decays. The AMS technique allowed reduction of the sample size required for oceanic $\Delta^{14}C$ determination from approximately 250 liters of water to 250 milliliters! By 1995 the AMS technique was yielding results that were as good as the best prior techniques using large samples and decay counting. This size reduction and concurrent automation procedures had a profound effect on sea water $\Delta^{14}C$ determination. Many of the AMS techniques were developed and most of the oceanographic AMS $\Delta^{14}C$ measurements have been made at the National Ocean Sciences AMS facility in Woods Hole, Massachusetts, by Ann McNichol, Robert Schneider, and Karl von Reden under the initial direction of Glenn Jones and more recently John Hayes.

The natural concentration of ^{14}C in sea water is extremely low ($\sim 1 \times > 10^9$ atoms kg^{-1}). Prior to AMS, the only available technique to measure this low concentration was radioactive counting using either gas proportional or liquid scintillation detectors. Large sample were needed to obtain high precision and to keep counting times reasonable. Between about 1960 and 1995 most subsurface open-ocean radiocarbon water samples were collected using a Gerard–Ewing sampler commonly known as a Gerard barrel. The final design of the Gerard barrel consisted of a stainless steel cylinder with a volume of approximately 270 liters. An external scoop and an internal divider running the length of the cylinder resulted in efficient flushing while the barrel was lowered through the water on wire rope. When the barrel was returned to the ship deck, the water was transferred to a gas-tight container and acidified to convert carbonate species to CO_2. The CO_2 was swept from the water with a stream of inert gas and absorbed in a solution of sodium hydroxide. The solution was returned to shore where the CO_2 was extracted, purified, and counted. When carefully executed, the procedure produced results which were accurate to 2–4 ppt based on counting errors alone. Because of the expense, time, and difficulty, samples for replicate analyses were almost never collected.

With the AMS technique only 0.25 liter of sea water is required. Generally a 0.5 liter water sample is collected at sea and poisoned with $HgCl_2$ to halt all biological activity. The water is returned to the laboratory and acidified, and the CO_2 is extracted and purified. An aliquot of the CO_2 is analyzed to determine $\delta^{13}C$ and the remainder is converted to carbide and counted by AMS. Counting error for the AMS technique can be <2 ppt, however, replicate analysis shows the total sample error to be approximately 4.5 ppt.

Sampling History

Soon after the radiocarbon dating method was developed, it was applied to oceanic and atmospheric samples. During the 1950s and 1960s most of the oceanographic samples were limited to the shallow waters owing to the difficulty of deep water sampling combined with the limited analytical precision. The majority of the early samples were collected in the Atlantic Ocean and the South-west Pacific Ocean. Early sample coverage was insufficient to give a good description of the global surface ocean radiocarbon content prior to the onset of atmospheric testing of thermonuclear weapons; however, repeated sampling at the same location was sufficient to record the surface water increase due to bomb-produced fallout. A very good history of radiocarbon activity, including the increase due to bomb tests and subsequent decrease, exists primarily as a result of the work of R. Nydal and co-workers (Trondheim) and K. Munnich and co-workers (Heidelberg).

The primary application of early radiocarbon results was to estimate the flux of CO_2 between the atmosphere and ocean and the average residence time in the ocean. Sufficient subsurface ocean measurements were made, primarily by W. Broecker (Lamont–Doherty Earth Observatory LDEO) and H. Craig (Scripps Institution of Oceanography SIO), to recognize that radiocarbon had the potential to be an important tracer of deep ocean circulation and mixing rates.

During the 1970s the Geochemical Ocean Sections (GEOSECS) program provided the first full water column global survey of the oceanic radiocarbon distribution. The GEOSECS cruise tracks were approximately meridional through the center of the major ocean basins. Radiocarbon was sampled with a station spacing of approximately 500 km and an average of 20 samples per station. All of the GEOSECS $\Delta^{14}C$ measurements were made by G. Östlund (University of Miami) and M. Stuiver (University of Washington) using traditional β counting of large-volume water samples with a counting accuracy of ~ 4 ppt. GEOSECS results revolutionized what was

known about the oceanic $\Delta^{14}C$ distribution and the applications for which radiocarbon is used.

During the early 1980s the Atlantic Ocean was again surveyed for radiocarbon as part of the Transient Tracers in the Ocean (TTO) North Atlantic Study (NAS) and Tropical Atlantic Study (TAS) programs and the South Atlantic Ventilation Experiment (SAVE). Sampling for these programs was designed to enable mapping of property distributions on constant pressure or density surfaces with reasonable gridding uncertainty. The radiocarbon portion of these

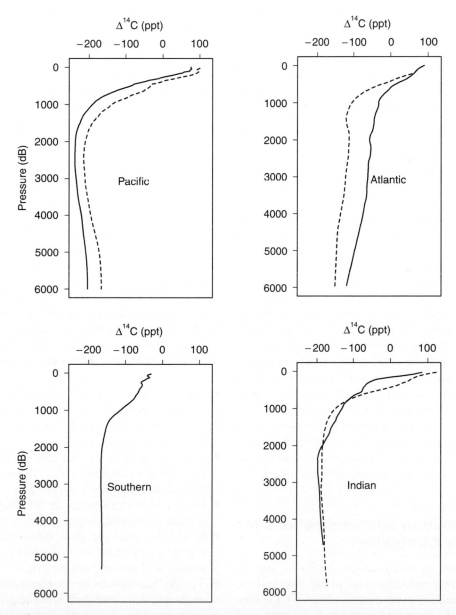

Figure 4 Average vertical $\Delta^{14}C$ profiles for the major ocean basins. Except for the Southern Ocean the dotted line is for the Southern Hemisphere and the solid line for the Northern Hemisphere. The Pacific and Southern Ocean profiles were compiled from WOCE data; the Atlantic profiles from TTO and SAVE data; and the Indian Ocean profiles from GEOSECS data. In approximately the upper 1000 m(= 1000 dB) of each profile, the natural $\Delta^{14}C$ is contaminated with bomb-produced radiocarbon.

programs was directed by W. Broecker. Östlund made the $\Delta^{14}C$ measurements with $\delta^{13}C$ provided by Stuiver using the GEOSECS procedures. Comparison of TTO results to GEOSECS gave the first clear evidence of the penetration of the bomb-produced radiocarbon signal into the subsurface North Atlantic waters. The French carried out a smaller scale (INDIGO) ^{14}C program in the Indian Ocean during this time with Östlund and P. Quay (University of Washington) collaborating. These data also quantified upper ocean changes since GEOSECS and relied on the same techniques.

The most recent oceanic survey was carried out during the 1990s as part of the World Ocean Circulation Experiment (WOCE). This program was a multinational effort. The US ^{14}C sampling effort was heavily focused on the Pacific (1991–1993) and Indian oceans (1995–1996) since TTO and SAVE had provided reasonable Atlantic coverage. R. Key (Princeton University) directed the US radiocarbon effort with collaboration from P. Schlosser (LDEO) and Quay. In the deep Pacific where gradients were known to be small, most radiocarbon sampling was by the proven large-volume β technique. The Pacific thermocline, however, was sampled using the AMS technique. Shifting techniques allowed thermocline waters to be sampled at approximately 2–3 times the horizontal density used for large volume sampling. Östlund and Stuiver again measured the large-volume samples while the AMS samples were measured at the National Ocean Sciences AMS facility (NOSAMS) at Woods Hole Oceanographic Institution. By 1994 the analytical precision at NOSAMS had improved to the point that all US Indian Ocean WOCE ^{14}C sampling used this technique. WOCE sampling increased the total number of ^{14}C results for the Pacific and Indian Oceans by approximately an order of magnitude. Analysis of the Pacific Ocean samples was completed in 1998. US WOCE ^{14}C sampling in the Atlantic was restricted to two zonal sections in the north-west basin using the AMS technique. Analysis of the Atlantic and Indian Ocean samples is expected to be finished during 2000–2001.

$\Delta^{14}C$ Distribution and Implications for Large-scale Circulation

The distribution of radiocarbon in the ocean is controlled by the production rate in the atmosphere, the spatial variability and magnitude of $^{14}CO_2$ flux across the air–sea interface, oceanic circulation and mixing, and the carbon cycle in the ocean. **Figure 4** shows average vertical radiocarbon profiles for the Pacific, Atlantic, Southern, and Indian oceans with

the dotted line being southern basin and solid line northern basin. All of the profiles have higher $\Delta^{14}C$ in shallow waters, reflecting proximity to the atmospheric source. The different collection times combined with the penetration of the bomb-produced signal into the upper thermocline negate the possibility of detailed comparison for the upper 600–800 dB (deeper for the North Atlantic). Detailed comparison is justified for deeper levels. The strongest signal in deep and bottom waters is that the North Atlantic is significantly younger (higher $\Delta^{14}C$) than the South Atlantic, while the opposite holds for the Pacific. Second, the average age of deep water increases ($\Delta^{14}C$ decreases) from Atlantic to Indian to Pacific. Third, the Southern Ocean $\Delta^{14}C$ is very uniform below approximately 1800 dB at a level (~ -160 ppt). This is similar to the near bottom water values for all three southern ocean basins. All three differences are directly attributable to the large-scale thermohaline circulation.

Figure 5 shows meridional sections for the Atlantic, Indian and Pacific oceans using subsets of the data from **Figure 4**. As with **Figure 4**, the $\Delta^{14}C$ values in the upper water column have been increased by invasion of the bomb signal. The pattern of these contours, however, is generally representative of the natural $\Delta^{14}C$ signal. The $\Delta^{14}C = -100‰$ contour can be taken as the approximate demarcation between the bomb-contaminated waters and those having only natural radiocarbon.

Comparison of the major features in each section shows that the meridional $\Delta^{14}C$ distributions in the Pacific and Indian Oceans are quite similar. The greatest difference between these two is that the Indian Ocean deep water (1500–3500 m) is significantly younger than Pacific deep waters. In both oceans:

- The near bottom water has higher $\Delta^{14}C$ than the overlying deep water.
- The deep and bottom waters have higher $\Delta^{14}C$ at the south than the north.
- The lowest $\Delta^{14}C$ values are found as a tongue extending southward from the north end of the section at a depth of ~ 2500 m.
- Deep and bottom water at the south end of each section is relatively uniform with $\Delta^{14}C \sim -160$ ppt.
- The $\Delta^{14}C$ gradient with latitude from south to north is approximately the same for both deep waters and for bottom waters.
- The $\Delta^{14}C$ contours in the thermocline shoal both at the equator and high latitudes. (This feature is suppressed in the North Indian Ocean owing to the limited geographic extent and the influence of flows through the Indonesian Seas region and from the Arabian Sea.)

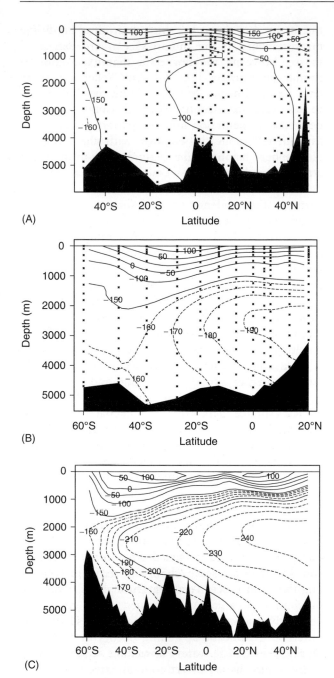

Figure 5 Typical meridional sections for each ocean compiled from a subset of the data used for **Figure 4**. The deep water contour patterns are primarily due to the large-scale thermohaline circulation. The highest deep water $\Delta^{14}C$ values are found in the North Atlantic and the lowest in the North Pacific. The natural $\Delta^{14}C$ in the upper ocean is contaminated by the influx of bomb-produced radiocarbon.

In the Atlantic Ocean the pattern in the shallow water down through the upper thermocline is similar to that in the other oceans. The $\Delta^{14}C$ distribution in the deep and bottom waters of the Atlantic is, however, radically different. The only similarities to the other oceans are (1) the $\Delta^{14}C$ value for deep and bottom water at the southern end of the section, (2) a southward-pointing tongue in deep water, and (3) the apparent northward flow indicated by the near-bottom tongue-shaped contour. Atlantic deep water has higher $\Delta^{14}C$ than the bottom water, and the deep and bottom waters at the north end of the section have higher rather that lower $\Delta^{14}C$ as found in the Indian and Pacific. Additionally, the far North Atlantic deep and bottom waters have relatively uniform values rather than a strong vertical gradient.

The reversal of the Atlantic deep and bottom water $\Delta^{14}C$ gradients with latitude relative to those in the Indian and Pacific is due to the fact that only the Atlantic has the conditions of temperature and salinity at the surface (in the Greenland–Norwegian Sea and Labrador Sea areas) that allow formation of a deep water mass (commonly referred to as North Atlantic Deep Water, NADW). Newly formed NADW flows down slope from the formation region until it reaches a level of neutral buoyancy. Flow is then southward, primarily as a deep western boundary current constrained by the topography of the North American slope. In its southward journey, NADW encounters and overrides northward-flowing denser waters of circumpolar origin. This general circulation pattern can be very clearly demonstrated by comparing the invasion of the bomb-produced tritium and radiocarbon signals obtained during GEOSECS to those from the TTO programs. This large circulation pattern leads to the observed $\Delta^{14}C$ distribution in the deep Atlantic.

Since neither the Pacific nor the Indian Ocean has a northern hemisphere source of deep water, the large-scale circulation is simpler. The densest Pacific waters originate in the Southern Ocean and flow northward along the sea floor (Circumpolar Deep Water, CDW). In the Southern Ocean, CDW is partially ventilated, either by direct contact with the atmosphere or by mixing with waters that have contacted the atmosphere, resulting in somewhat elevated $\Delta^{14}C$. As CDW flows northward, it ages, warms, mixes with overlying water, and slowly upwells. This upwelling, combined with mixing with overlying lower thermocline waters, results in the water mass commonly known as Pacific Deep Water (PDW). PDW has the lowest $\Delta^{14}C$ values found anywhere in the oceans. The long-term mean flow pattern for PDW is somewhat controversial; however, the radiocarbon distribution supports a southward flow with the core of the flow centered around 2500 m. WOCE results further imply that if there is a mean southward flow of PDW, it may be concentrated toward the eastward and westward boundaries rather than uniformly distributed zonally. **Figure 6** shows a zonal Pacific WOCE $\Delta^{14}C$ section at 32°S contoured at the same intervals as the previous sections. PDW is identified by the minimum layer between 2000 and 3000 m. The PDW core appears segregated into two channels, one against the South American slope and the other over the Kermadec Trench. The actual minimum values in the latter were found at ~170°W, essentially abutting the western wall of the trench. The northward-flowing CDW is also clearly indicated in this section by the relatively high $\Delta^{14}C$ values near the bottom between 140°W and the Date Line.

Little has been said about the natural $\Delta^{14}C$ values found in the upper ocean where bomb-produced radiocarbon is prevalent. GEOSECS samples were collected only ~10 years after the maximum in atmospheric $\Delta^{14}C$. GEOSECS surface water measurements almost always had the highest $\Delta^{14}C$ values. Twenty years later during WOCE, the maximum $\Delta^{14}C$ was generally below the surface.

Broecker and Peng (1982, p. 415, Figures 8–19) assembled the few surface ocean $\Delta^{14}C$ measurements made prior to bomb contamination for comparison to the GEOSECS surface ocean data. For the Atlantic and Pacific Oceans, their plot of $\Delta^{14}C$ versus latitude shows a characteristic 'M' shape with maximum $\Delta^{14}C$ values of approximately − 50 ppt centered in the main ocean gyres between latitudes 20° and 40°. Each ocean had a relative minimum $\Delta^{14}C$ value of approximately − 70 ppt in the equatorial latitudes, 20° S to 20° N and minima at high latitudes ranging from − 70 ppt for the far North Atlantic to

− 150 ppt for the other high latitudes. Pre-bomb measurements in the Indian Ocean are extremely sparse; however, the few data that exist imply a similar distribution. The GEOSECS surface ocean data had the same 'M' shape; however, all of the values were significantly elevated owing to bomb-derived contamination and the pattern was slightly asymmetric about the equator with the Northern Hemisphere having higher values since most of the atmospheric bomb tests were carried out there. The 'M' shape of $\Delta^{14}C$ with latitude is due to circulation patterns, the residence time of surface water in an ocean region, and air–sea gas exchange rates. At mid-latitudes the water column is relatively stable and surface waters reside sufficiently long to absorb a significant amount of ^{14}C from the atmosphere. In the equatorial zone, upwelling of deeper (and therefore lower $\Delta^{14}C$) waters lowers the surface ocean value. At high latitudes, particularly in the Southern Ocean, the near-surface water is relatively unstable, resulting in a short residence time. In these regions $\Delta^{14}C$ acquired from the atmosphere is more than compensated by upwelling, mixing, and convection.

Figure 7 shows a comparison for GEOSECS and WOCE surface data from the Pacific Ocean. The GEOSECS $\Delta^{14}C$ values are higher than WOCE everywhere except for the Equator. The difference is due to two factors. First, GEOSECS sampling occurred shortly after the atmospheric maximum. At that time the air–sea $\Delta^{14}C$ gradient was large and the surface ocean $\Delta^{14}C$ values were dominated by air–sea gas exchange processes. Second, by the 1990s, atmospheric $\Delta^{14}C$ levels had declined significantly and sufficient time had occurred for ocean mixing to compete with air–sea exchange in terms of controlling the surface ocean values. During the 1990s, the maximum oceanic $\Delta^{14}C$ values were frequently below the surface. Near the Equator the situation is different. Significant upwelling occurs in this zone. During GEOSECS, waters upwelling at low latitude in the Pacific were not yet contaminated with bomb radiocarbon. Twenty years later, the upwelling waters had acquired a bomb radiocarbon component.

While surface ocean $\Delta^{14}C$ generally decreased between GEOSECS and WOCE, values throughout the upper kilometer of the water column generally increased as mixing and advection carried bomb-produced radiocarbon into the upper thermocline. The result of these processes on the bomb-produced $\Delta^{14}C$ signal can be visualized by comparing GEOSECS and WOCE depth distributions. **Figure 8** shows such a comparison. To produce this figure the WOCE data from section P16 (152°W) were gridded (center panel). GEOSECS data collected east of the data line were then gridded to the same grid (top

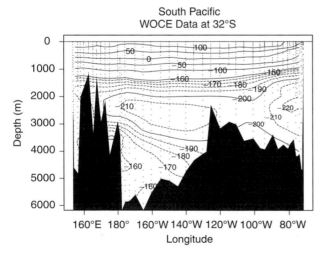

Figure 6 Zonal section of $\Delta^{14}C$ in the South Pacific collected during the WOCE program. The two minima at 2000–2500 m depth are thought to be the core of southward-flowing North Pacific Deep Water. Northward-flowing Circumpolar Deep Water is identified by the relatively high values in the Kermadec Trench area at the bottom between 140°W and the Date Line.

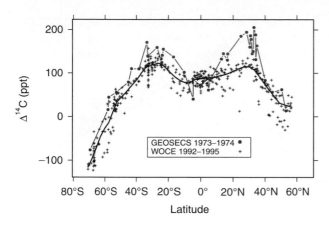

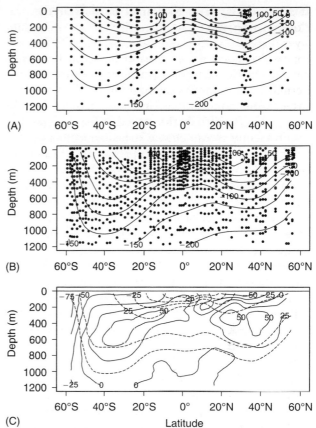

Figure 7 Distribution of Δ^{14}C in the surface Pacific Ocean as recorded by the GEOSECS program in the early 1970s and the WOCE program in the early 1990s. From **Figure 3B** it follows that GEOSECS recorded the maximum bomb-contamination. Over the 20 years separating the programs, mixing and advection dispersed the signal. By the time of WOCE the maximum contamination level was found below the surface at many locations. The asymmetry about the Equator in the GEOSECS data is a result of most atmospheric bomb tests being executed in the Northern Hemisphere.

Figure 8 Panel (C) shows the change in the meridional eastern Pacific thermocline distribution of Δ^{14}C between the GEOSECS (1973–1974, (A)) and WOCE (1991–1994, (B)) surveys. The change was computed by gridding each section then finding the difference. The dashed lines in (C) indicate constant potential density surfaces. Negative near-surface values indicate maximum concentration surfaces moving down into the thermocline after GEOSECS. The region of greatest increase in the southern hemisphere is ventilated in the Southern Ocean.

panel). Once prepared, the two sections were simply subtracted grid box by grid box (bottom panel). One feature of **Figure 8** is the asymmetry about the Equator. The difference at the surface in **Figure 8** reflects the same information (and data) as in **Figure 7**. The greatest increase (up to 60 ppt) along the section is in the Southern Hemisphere mid-latitude thermocline at a depth of 300–800 m. This concentration change decreases in both depth and magnitude toward the Equator. All of the potential density isolines that pass through this region of significant increase (dashed lines in the bottom panel) outcrop in the Southern Ocean. These outcrops (especially during austral winter) provide the primary pathway by which radiocarbon is entering the South Pacific thermocline. In the North Pacific the surface ocean decrease extends as a blob well into the water column (>200 m). This large change is due to the extremely high surface concentrations measured during GEOSECS and to subsurface mixing and ventilation processes that have diluted or dispersed the peak signal. The values contoured in the bottom panel represent the change in Δ^{14}C between the two surveys, not the total bomb Δ^{14}C.

WOCE results from the Indian Ocean are not yet available. Once they are, changes since GEOSECS in the South Indian Ocean should be quite similar to those in the South Pacific because the circulation and ventilation pathways are similar. Changes in the North Indian Ocean are difficult to predict owing to water inputs from the Red Sea and the Indonesian

throughflow region and to the changing monsoonal circulation patterns.

Göte Östlund and Claes Rooth described radiocarbon changes in the North Atlantic Ocean using data from GEOSECS (1972) and the TTO North Atlantic Study (1981–1983). The pattern of change they noted is different from that in the Pacific because of the difference in thermohaline circulation mentioned previously. Prior to sinking, the formation waters for NADW are at the ocean surface long enough to pick up significant amounts of bomb radiocarbon from the atmosphere. The circulation pattern coupled with the timing of GEOSECS and TTO sampling resulted in increased Δ^{14}C levels during the latter program. The significant changes were mostly limited to the deep water region north of 40°N latitude. When the WOCE Atlantic samples are analyzed, we expect to see changes extending farther southward.

Separating the Natural and Bomb Components

Up to this point the discussion has been limited to changes in radiocarbon distribution due to oceanic uptake of bomb-produced radiocarbon. Many radiocarbon applications, however, require not the change but the distribution of either bomb or natural radiocarbon. Ocean water measurements give the total of natural plus bomb-produced $\Delta^{14}C$. Since these two are chemically and physically identical, no analytical procedure can differentiate one from the other. Far too few $\Delta^{14}C$ measurements were made in the upper ocean prior to contamination by the bomb component for us to know what the upper ocean natural $\Delta^{14}C$ distribution was.

One separation approach derived by Broecker and co-workers at LDEO uses the fact that $\Delta^{14}C$ is linearly anticorrelated with silicate in waters below the depth of bomb-^{14}C penetration. By assuming the same correlation extends to shallow waters, the natural $\Delta^{14}C$ can be estimated for upper thermocline and near surface water. Pre-bomb values for the ocean surface were approximated from the few pre-bomb surface ocean measurements. The silicate method is limited to temperate and low latitude waters since the correlation fails at high latitudes, especially for waters of high silicate concentration. More recent work by S. Rubin and R. Key indicates that potential alkalinity (alkalinity + nitrate normalized to salinity of 35) may be a better co-variable than silicate and can be used at all latitudes. **Figure 9** illustrates the silicate and PALK correlations using the GEOSECS data set. Regardless of the co-variable, the correlation is used to estimate pre-bomb $\Delta^{14}C$ in contaminated regions. The difference between the measured and estimated natural $\Delta^{14}C$ is the bomb-produced $\Delta^{14}C$.

In **Figure 10** the silicate and potential alkalinity (PALK) methods are illustrated and compared. The upper panel (A) shows the measured $\Delta^{14}C$ and estimates of the natural $\Delta^{14}C$ using both methods. The bomb $\Delta^{14}C$ is then just the difference between the measured value and the estimate of the natural value (B). For this example, taken from the mid-latitude Pacific, the two estimates are quite close; however this is not always true.

In **Figure 11A** the upper 1000 m of the Pacific WOCE $\Delta^{14}C$ section shown in **Figure 5C** is reproduced. **Figure 11B** shows the estimated natural $\Delta^{14}C$ using the potential alkalinity method. The shape of the two contour sets is quite similar; however, the contour values and vertical gradients are very different, illustrating the strong influence of bomb-produced radiocarbon on the upper ocean. The

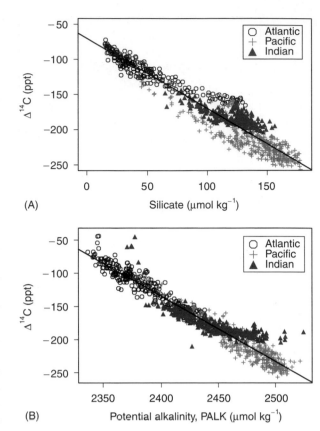

(A)

(B)

Figure 9 Comparison of the correlation of natural $\Delta^{14}C$ with silicate (A) and potential alkalinity (PALK = [alkalinity + nitrate] × 35/salinity) (B) using the GEOSECS global data. Samples from high southern latitudes are excluded from the silicate relation. The presence of tritium was used to surmise the presence of bomb-$\Delta^{14}C$. The somewhat anomalous high PALK values from the Indian Ocean are from upwelling–high productivity zones and may be influenced by nitrogen fixation and/or particle flux.

integrated difference between these two sections would yield an estimate of the bomb-produced $\Delta^{14}C$ inventory for the section.

Oceanographic Applications

As illustrated, the $\Delta^{14}C$ distribution can be used to infer general large-scale circulation patterns. The most valuable applications for radiocarbon derive from the fact that it is radioactive and has a half-life appropriate to the study of deep ocean processes and that the bomb component is transient and is useful as a tracer for upper ocean processes. A few of the more common uses are described below.

Deep Ocean Mixing and Ventilation Rate, and Residence Time

Since the first subsurface measurements of radiocarbon, one of the primary applications has been the

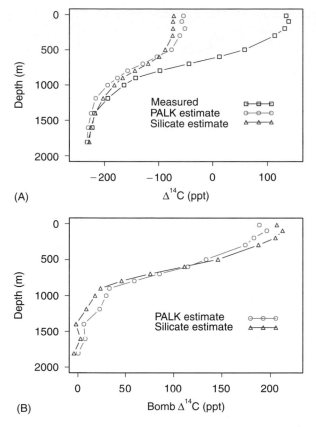

Figure 10 Panel (A) compares measured $\Delta^{14}C$ from a mid-latitude Pacific WOCE station with natural $\Delta^{14}C$ estimated using the silicate and potential alkalinity methods. Bomb-$\Delta^{14}C$, the difference between measured and natural $\Delta^{14}C$, estimated with both methods is compared in (B). Integration of estimated bomb-$\Delta^{14}C$ from the surface down to the depth where the estimate approaches zero yields an estimate of the bomb-$\Delta^{14}C$ inventory. Inventory is generally expressed in units of atoms per unit area.

determination of deep ocean ventilation rates. Most of these calculations have used a box model to approximate the ocean system. The first such estimates yielded mean residence times for the various deep and abyssal ocean basins of 350–900 years. Solution of these models generally assumes a steady-state circulation, identifiable source water regions with known $\Delta^{14}C$, no mixing between water masses, and no significant biological sources or sinks. Another early approach assumed that the vertical distribution of radiocarbon in the deep and abyssal ocean could be described by a vertical advection–diffusion equation. This type of calculation leads to estimates of the effect of biological particle flux and dissolution and to the vertical upwelling and diffusion rates. The 1D vertical advection–diffusion approach has been abandoned for 2D and 3D calculations as the available data and our knowledge of oceanic processes have increased.

When the GEOSECS data became available, box models were again used to estimate residence times

and mass fluxes for the abyssal ocean. In this case the model had only four boxes, one for the deep region (>1500 m) of each ocean. New bottom water formation (NADW and Antarctic Bottom Water, AABW) were included as inputs to the Atlantic and Circumpolar boxes. Upwelling was allowed in the Atlantic, Pacific, and Indian boxes and exchange was considered between the Circumpolar box and each of the other three ocean boxes. Results from this calculation gave mean replacement times of 510, 250, 275, and 85 years for the deep Pacific, Indian, Atlantic, and Southern Ocean, respectively, and 500 years for the deep waters of the entire world. Upwelling rates were estimated at $4–5\,\mathrm{m\,y}^{-1}$ and mass transports generally agreed with contemporary geostrophic calculations. Applying the same model to more recent data sets would yield the same results.

Oxygen Utilization Rate

Radiocarbon can be used to determine the rate of biological or geochemical processes such as the rate at which oxygen is consumed in deep ocean water. The simplest example of this would be the case of a water mass moving away from a source region at a steady rate, undergoing constant biological oxygen uptake and not subject to mixing. In such a situation the oxygen utilization rate could be obtained from the slope of oxygen versus ^{14}C in appropriate units. The closest approximation to this situation is the northward transport of CDW in the abyssal Pacific, although the mixing requirement is only approximate. **Figure 12** shows such a plot for WOCE Pacific Ocean samples from deeper than 4000 m and north of 40°S. In this case, apparent oxygen utilization (saturated oxygen concentration at equilibration temperature − measured oxygen concentration) rather than oxygen concentration is plotted, to remove the effect of temperature on oxygen solubility. The least-squares slope of $0.83\,\mu\mathrm{mol\,kg}^{-1}$ per ppt converts to $0.1\,\mu\mathrm{mol\,kg}^{-1}\,\mathrm{y}^{-1}$ for an oxygen utilization rate. Generally, mixing with other water masses must be accounted for prior to evaluating the gradient. With varied or additional approximations, very similar calculations have been used to estimate the mean formation rates of various deep water masses.

Ocean General Circulation Model Calibration

Oceanographic data are seldom of value for prediction. Additionally, the effect of a changing

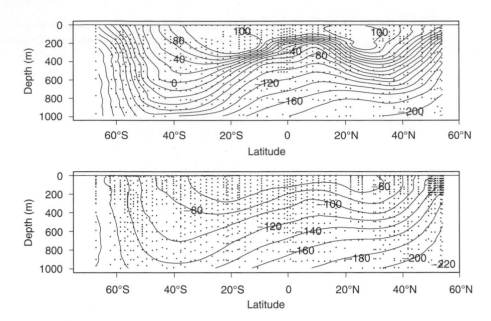

Figure 11 Upper thermocline meridional sections along 152° W in the central Pacific. (A) The same measured data as in **Figure 5C** (B) An estimate of thermocline Δ¹⁴C values prior to the invasion of bomb-produced radiocarbon.

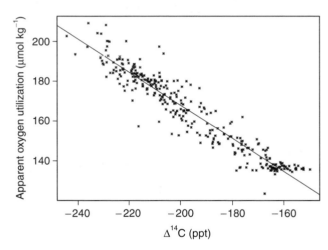

Figure 12 Apparent oxygen utilization plotted against measured $\Delta^{14}C$ for WOCE Pacific Ocean samples taken at depths greater than 4000 m and north of 40°S. The slope of the line (-0.831 ± 0.015) can be used to estimate an approximate oxygen utilization rate of 0.1 µmol kg⁻¹ y⁻¹ if steady state and no mixing with other water masses is assumed.

oceanographic parameter on another parameter can be difficult to discern directly from data. These research questions are better investigated with numerical ocean models. Before an ocean model result can be taken seriously, however, the model must demonstrate reasonable ability to simulate current conditions. This generally requires that various model inputs or variables be 'tuned' or calibrated to match measured distributions and rates. Radiocarbon is the only common measurement that can be used to calibrate the various rates of abyssal processes in general circulation models. M. Fiadeiro carried out the first numerical simulation for the abyssal Pacific and used the GEOSECS ¹⁴C data to calibrate the model. J.R. Toggweiler extended this study using a global model.

Both the Fiadeiro and Toggweiler models, and all subsequent models that include the deep water $\Delta^{14}C$, are of coarse resolution owing to current computer limitations. As the much larger WOCE ¹⁴C data set becomes available, the failure of these models, especially in detail, becomes more evident. Toggweiler's model, for example, has advective mixing in the Southern Ocean that is significantly greater than supported by data. Additionally, the coarse resolution of the model prevents the formation of, or at least retards the importance of, deep western boundary currents. Significant model deficiencies appear when the bomb-¹⁴C distribution and integrals at the time of GEOSECS and WOCE are compared with data.

During the last 10 years the number and variety of numerical ocean models has expanded greatly, in large part because of the availability and speed of modern computers. The Ocean Carbon Model Intercomparison Project (OCMIP) brought ocean modelers together with data experts in the first organized effort to compare model results with data, with the long-term goals of understanding the processes that cause model differences and of improving the prediction capabilities of the models. The unique aspect of this study was that each participating group

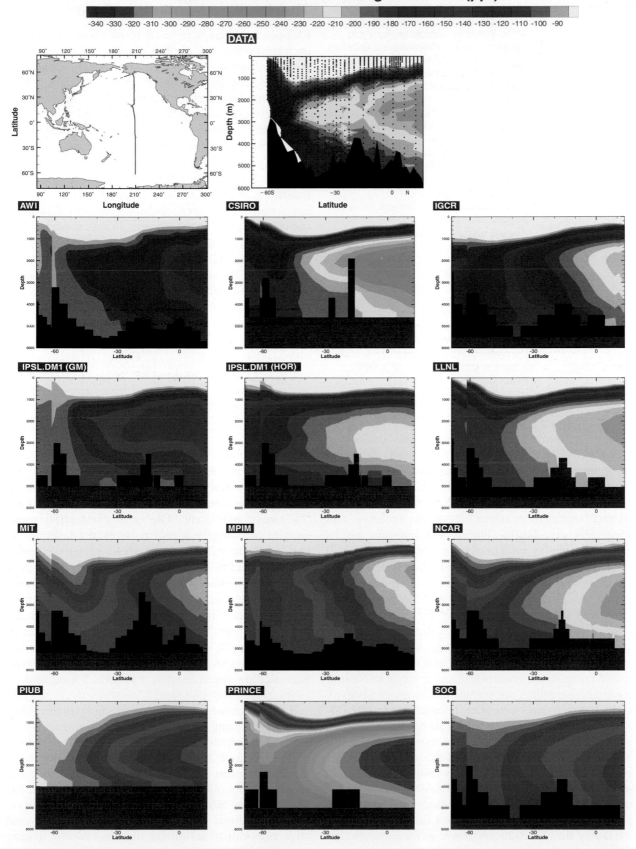

Figure 13 (Right) Global ocean circulation model results from 12 different coarse-resolution models participating in OCMIP-2 compared to WOCE data for natural $\Delta^{14}C$ on a meridional Pacific section. The model groups are identified in each subpanel and in **Table 1** All of the models used the same chemistry and boundary conditions.

Table 1 OCMIP-2 participants

	Model groups
AWI	Alfred Wegener Institute for Polar and Marine Research, Bremerhaven, Germany
CSIRO	Commonwealth Science and Industrial Research Organization, Hobart, Australia
IGCR/CCSR	Institute for Global Change Research, Tokyo, Japan
IPSL	Institut Pierre Simon Laplace, Paris, France
LLNL	Lawrence Livermore National Laboratory, Livermore, CA, USA
MIT	Massachusetts Institute of Technology, Cambridge, MA, USA
MPIM	Max Planck Institut fur Meteorologie, Hamburg, Germany
NCAR	National Center for Atmospheric Research, Boulder, CO, USA
PIUB	Physics Institute, University of Bern, Switzerland
PRINCETON	Princeton University AOS, OTL/GFDL, Princeton, NJ, USA
SOC	Southampton Oceanography Centre/ SUDO/Hadley Center, UK Met. Office
	Data groups
PMEL	Pacific Marine Environmental Laboratory, NOAA, Seattle, WA, USA
PSU	Pennsylvania State University, PA, USA
PRINCETON	Princeton University AOS, OTL/GFDL, Princeton, NJ, USA

essentially 'froze' development of the underlying physics in their model and then used the same boundary conditions and forcing in order to eliminate as many potential variables as possible. Radiocarbon, both bomb-derived and natural, were used as tracers in each model to examine air–sea gas exchange and long-term circulation. **Figure 13** compares results from 12 global ocean circulation models with WOCE data from section P16. The tag in the top left corner of each panel identifies the institution of the modeling group. All of the model results and the data are colored and scaled identically and the portion of the section containing bomb radiocarbon has been masked. While all of the models get the general shape of the contours, the concentrations vary widely. Detailed comparison is currently under way, but cursory examination points out significant discrepancies in all model results and remarkable model-to-model differences. Similar comparisons can be made focusing on the bomb component. Discussion of model differences is beyond the scope of this work. For information, see publications by the various groups having results in **Figure 13** (listed in **Table 1**). These radiocarbon results are not yet published, but an overview of the OCMIP-2 program can be found in the work of Dutay on chlorofluorocarbon in the same models (see Further Reading).

Air–Sea Gas Exchange and Thermocline Ventilation Rate

Radiocarbon has been used to estimate air–sea gas exchange rates for almost as long as it has been measured in the atmosphere and ocean. Generally, these calculations are based on box models, which have both included and excluded the influence of bomb contamination. W. Broecker and T.-H. Peng summarized efforts to estimate air–sea transfer rates up to 1974 and gave examples based on GEOSECS results using both natural and bomb-^{14}C and a stagnant film model. In this, the rate-limiting step for transfer is assumed to be molecular diffusion of the gas across a thin layer separating the mixed layer of the ocean from the atmosphere. In this model, if one assumes steady state for the ^{14}C and ^{12}C distribution and uniform ^{14}C/^{12}C for the atmosphere and surface ocean then the amount of ^{14}C entering the ocean must be balanced by decay. For this model the solution is given by eqn [3].

$$\frac{D}{z} = \frac{\sum[CO_2]|_{ocean}}{\sum[CO_2]|_{mix}} \frac{V}{A} \frac{\left[\frac{^{14}C/C|_{ocean}}{^{14}C/C|_{mix}}\right]\frac{\alpha_{14_{CO_2}}}{\alpha_{CO_2}}}{1 - \left[\frac{^{14}C/C|_{mix}}{^{14}C/C|_{atm}}\right]\frac{\alpha_{14_{CO_2}}}{\alpha_{CO_2}}} \lambda \quad [3]$$

Here D is the molecular diffusivity of CO_2, z is the film thickness, α_i is the solubility of i, V and A are the volume and surface area of the ocean, and λ is the ^{14}C decay coefficient. Use of pre-industrial mean concentrations gave a global boundary layer thickness of $30\,\mu m$ ($D/z \sim 1800\,m\,y^{-1}$ = piston velocity). The film thickness is then used to estimate gas residence times either in the atmosphere or in the mixed layer of the ocean. For CO_2 special consideration must be made for the chemical speciation in the ocean, and for $^{14}CO_2$ further modification is necessary for isotopic effects. The equilibration times for CO_2 with respect to gas exchange, chemistry, and isotopics are approximately 1 month, 1 year, and 10 years, respectively.

Radiocarbon has been used to study thermocline ventilation using tools ranging from simple 3-box models to full 3D ocean circulation models. Many of the 1D and 2D models are based on work by W. Jenkins using tritium in the North Atlantic. In a recent example, R. Sonnerup and co-workers at the University of Washington used chlorofluorocarbon data to calibrate a 1D (meridional) along-isopycnal advection–diffusion model in the North Pacific with WOCE data. Equation [4] is the basic equation

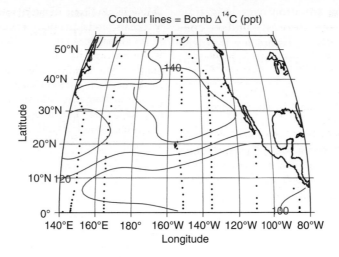

Contour lines = Bomb Δ¹⁴C (ppt)

Figure 14 Bomb-Δ^{14}C on the potential density surface $\sigma_\theta = 26.1$ in the North Pacific. The blue line is the wintertime outcrop of the surface based on long-term climatology. The Sea of Okhotsk is a known region of thermocline ventilation for the North Pacific.

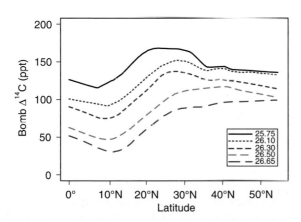

Figure 15 Meridional distribution of bomb-Δ^{14}C on potential density surfaces in the North Pacific thermocline.

for the model.

$$\frac{dC}{dt} = -v\frac{dC}{dx} + K\frac{d^2C}{dx^2} \qquad [4]$$

In eqn [4] C is concentration, K is along-isopycnal eddy diffusivity, $-v$ is the southward component of along isopycnal velocity, t is time, and x is the meridional distance. Upper-level isopycnal surfaces outcrop at the surface. Once the model is calibrated, the resulting values are used to investigate the distribution of other parameters. The original work and the references cited there should be read for details, but **Figure 14** shows an objective map of the bomb-^{14}C distribution on the potential density surface 26.1 for the North Pacific and **Figure 15** summarizes the bomb-^{14}C distribution as a function of latitude. These figures illustrate the type of data that

would be input considerations to an investigation of thermocline ventilation.

Conclusions

Since the very earliest measurements, radiocarbon has proven to be an extremely powerful tracer, and sometimes the only available tracer, for the study of many oceanographic processes. Perhaps the most important of these today are large-scale deep ocean mixing and ventilation processes and the calibration of numerical ocean models. The first global survey of the radiocarbon distribution collected on the GEOSECS program resulted in radical changes in the way the abyssal ocean is viewed. The newer and much denser WOCE survey will certainly add significant detail and precision to what is known and will probably result in other, if not so many, totally new discoveries. Progress with this tracer today is due largely to the decrease in required sample size from ~ 250 liters to ~ 250 milliliters and to the availability and application of fast, inexpensive computers.

Glossary

dpm   Disintegrations per minute: a measure of the activity of a radioactivesubstance frequently used rather than concentration.

$t_{1/2}$ Half-life: time required for one half of the atoms of a radio active species to decay.

λ Decay constant for a radioactive species $= \ln(2)/t_{1/2}$

Mean life, λ^{-1}   Average time expected for a given radioactive atom to decay.

Abyssal   Very deep ocean, often near bottom.

Steady state   Unchanging situation over long time interval relative to the process under consideration; frequently assumed state for the deep and abyssal ocean with respect to many parameters.

See also

Carbon Cycle. Carbon Dioxide (CO$_2$) Cycle. Ocean Carbon System, Modeling of. Stable Carbon Isotope Variations in the Ocean. Ocean Circulation: Meridional Overturning Circulation.

Further Reading

Broecker WS, Gerard R, Ewing M, and Heezen BC (1960) Natural radiocarbon in the Atlantic Ocean. *Journal of Geophysical Research* 65(a): 2903–2931.

Broecker WS and Peng T-H (1974) Gas exchange rates between air and sea. *Tellus* 26: 21–34.

Broecker WS and Peng T-H (1982) *Tracers in the Sea.* Columbia University, Palisades, NY: Lamont-Doherty Geological Observatory.

Broecker WS, Sutherland S, Smethie W, Peng T-H, and Östlund G (1995) Oceanic radiocarbon: separation of natural and bomb components. *Global Biogeochemical Cycles* 9(2): 263–288.

Craig H (1969) Abyssal carbon radiocarbon in the Pacific. *Journal of Geophysical Research* 74(23): 5491–5506.

Druffel ERM and Griffin S (1999) Variability of surface ocean radiocarbon and stable isotopes in the southwestern Pacific. *Journal of Geophysical Research* 104(C10): 23607–23614.

Dutay JC, Bullister JL, and Doney SC (2001) Evaluation of ocean model ventilation with CFC-11: comparison of 13 global ocean models. *Global Biogeochemical Cycles.* (in press).

Fiadeiro ME (1982) Three-dimensional modeling of tracers in the deep Pacific Ocean, II. Radiocarbon and circulation. *Journal of Marine Research* 40: 537–550.

Key RM, Quay PD, Jones GA, *et al.* (1996) WOCE AMS radiocarbon I: Pacific Ocean results (P6, P16 and P17). *Radiocarbon* 38(3): 425–518.

Libby WF (1955) *Radiocarbon Dating.* Chicago: University of Chicago Press.

Östlund HG and Rooth CGH (1990) The North Atlantic tritium and radiocarbon transients 1972–1983. *Journal of Geophysical Research* 95(C11): 20147–20165.

Schlosser P, Bönisch G, and Kromer B (1995) Mid-1980s distribution of tritium, ^{3}He, ^{14}C and ^{39}Ar in the Greenland/Norwegian Seas and the Nansen Basin of the Arctic Ocean. *Progress in Oceanography* 35: 1–28.

Sonnerup RE, Quay PD, and Bullister JL (1999) Thermocline ventilation and oxygen utilization rates in the subtropical North Pacific based on CFC distributions during WOCE. *Deep-Sea Research I* 46: 777–805.

Stuiver M and Polach HA (1977) Discussion: Reporting of ^{14}C data. *Radiocarbon* 19(3): 355–363.

Stuiver M and Quay P (1983) Abyssal water carbon-14 distribution and the age of the World Ocean. *Science* 219: 849–851.

Taylor RE, Long A, and Kra RS (eds.) (1992) *Radiocarbon After Four Decades, An Interdisciplinary Perspective.* New York: Springer.

Toggweiler JR, Dixon K, and Bryan K (1989) Simulations of radiocarbon in a coarse-resolution World Ocean model. 1. Steady state prebomb distributions. *Journal of Geophysical Research* 94(C6): 8217–8242.

STABLE CARBON ISOTOPE VARIATIONS IN THE OCEAN

K. K. Turekian, Yale University, New Haven, CT, USA

The two stable isotopes of carbon, ^{12}C and ^{13}C, vary in proportions in different reservoirs on earth. The ratio of ^{13}C to ^{12}C is commonly given relative to a standard (a belemnite from the Peedee formation in South Carolina and therefore called PDB). On the basis of this standard $\delta^{13}C$ is defined as:

$$\left[\frac{^{13}C/^{12}C_{sample}}{^{13}C/^{12}C_{standard}} - 1 \right] \times 1000$$

The values for some major carbon reservoirs are: marine limestones, $\delta^{13}C = 0$; C-3 plants, $\delta^{13}C = -25$; air CO_2, $\delta^{13}C = -7$. The inorganic carbon in the surface ocean is in isotopic equilibrium with atmospheric CO_2 and has a value of about 2. Organic matter in the shallow ocean ranges from -19 at high latitudes to -28 at low latitudes. The midlatitude value is around -21. The transport of organic matter to depth and subsequent metabolism adds inorganic carbon to the water. The isotopic composition of dissolved inorganic carbon then reflects the amount of addition of this metabolic carbon. **Figure 1** is a profile of $\delta^{13}C$ for the North Pacific. It is typical of other profiles in the oceans.

Carbon isotope measurements in all the oceans were made on the GEOSECS expedition. These values are given in the *GEOSECS Atlas* (1987).

See also

Carbon Cycle. Carbon Dioxide (CO_2) Cycle.

Further Reading

Chesselet R, Fontagne M, Buat-Menard P, Ezat U, and Lambert CE (1981) The origin of particulate organic matter in the marine atmosphere as indicated by its stable carbon isotopic composition. *Geophysical Research Letters* 8: 345–348.

GEOSECS Atlantic, Pacific, and Indian Ocean Expeditions, vol 7: *Shorebased Data and Graphics*. National Science Foundation. 200 pp. (1987).

Kroopnick P, Deuser WG, and Craig H (1970) Carbon-13 measurements on dissolved inorganic carbon at the North Pacific (1969) GEOSECS station. *Journal of Geophysical Research* 75: 7668–7671.

Sackett WM (1964) The depositional history and isotopic organic composition of marine sediments. *Marine Geology* 2: 173–185.

Figure 1 Variation of $\delta^{13}C$ in dissolved inorganic carbon with depth in the Pacific Ocean at GEOSECS Station 346 (28°N, 121°W) (Kroopnick, Deuser and Craig, 1970).

CENOZOIC OCEANS – CARBON CYCLE MODELS

L. François and Y. Goddéris, University of Liège, Liège, Belgium

Introduction

The story of the Cenozoic is essentially a story of global cooling. The last 65 million years of the Earth's history mark the transition from the Cretaceous 'greenhouse' climate toward the present-day 'icehouse' conditions. Particularly, the cooling by about 8–10°C of deep ocean waters since the Cretaceous was linked to a reorganization of the oceanic circulation triggered by tectonic plate movements. These oceanic circulation changes were coeval with continental climatic change, as demonstrated by abundant evidence for global cooling (pollen, faunal assemblages, development of glaciers, etc.). For instance, most of western Europe and the western United States had a subtropical climate during the Eocene, despite the fact that they were located at the same latitude as today. Another striking feature of the changes that have occurred during Cenozoic times is the decrease of the partial pressure of atmospheric CO_2 (P_{CO_2}). Since CO_2 is a greenhouse gas, there might be a causal relationship between the decrease in P_{CO_2} and the general cooling trend of Cenozoic climate. The global cooling might be the result of the changes in oceanic circulation and atmospheric CO_2, both probably influencing each other and possibly initiated by tectonic processes.

Indicators of Atmospheric CO_2 Change

It should be kept in mind that there are no direct proxies of ancient levels of CO_2 in the atmosphere. Methods rely on three indirect indicators.

1. The $\delta^{13}C$ measured in ancient soil carbonates can be directly linked to the atmospheric P_{CO_2}. This method reveals declining atmospheric P_{CO_2} over the last 65 million years, from about 650 ppm by volume (ppmv) in the Paleocene (**Figure 1**).
2. The biological isotopic fractionation occurring during assimilation of carbon by the marine biosphere (ε_p) depends on the partial pressure of CO_2 dissolved in sea water, itself directly related to the atmospheric P_{CO_2}. The estimation of ε_p for ancient organic sediments indicates high P_{CO_2} values in the Eocene (620 ppmv), followed by a constant decline toward the present-day pressure through the Cenozoic (**Figure 1**).
3. The measured boron isotopic composition of marine carbonates gives insight into the pH of ancient sea water. Assuming a plausible history for the ocean alkalinity, lower or higher pH values can be respectively related to higher or lower P_{CO_2}. This method has been applied to the last 60 million years, showing values as high as 3500 ppmv CO_2 during the Paleocene. The decline in P_{CO_2} was then roughly linear through time until the late Eocene. During the last 25 million years of the Earth's history, P_{CO_2} was relatively constant, possibly displaying lower values than present-day ones during Miocene. No data are available for the Oligocene (**Figure 1**).

Despite some disagreements between the three reconstructions, they all indicate a major reduction of the atmospheric CO_2 partial pressure during the Cenozoic, which might potentially play an important role in the coeval global cooling. Any exploration of the cause of the decline in P_{CO_2} requires the identification of the sources and sinks of oceanic and atmospheric carbon, and some knowledge of their relative changes during the Cenozoic.

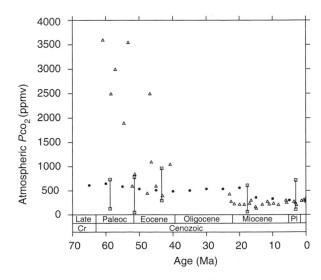

Figure 1 The reconstructed partial pressure of atmospheric CO_2. Δ, from boron isotopes (Pearson and Palmer (2000)) ; ●, from ε_p (Kump and Arthur, 1997); □, from paleosoils analysis (compilation by Berner (1998)). Timescale according to Harland *et al.* (1990) (CR = Cretaceous; PALEOC = Paleocene; PL = Pleistocene.)

Carbon Cycle Changes and Processes Involved

Long-term Regulation of Atmospheric CO₂

On the geological timescale, and neglecting at this point the possible impact of sedimentary organic carbon cycling, the sources of carbon for the ocean–atmosphere system are the degassing of the mantle and metamorphic processes. Carbon is injected as CO_2 into the ocean–atmosphere system today through the degassing of fresh basalts along mid-oceanic ridges (MOR) (1.5–2.2×10^{12} mol y^{-1}), and through plume events and arc volcanism (1.5–5.5×10^{12} mol y^{-1}). These various sources account for the total degassing flux F_{VOL}. Once released, this carbon is rapidly (within 10^3 years) redistributed between the atmosphere and the ocean, reaching a steady-state repartition after a negligible time compared to the geological timescale. Carbon can leave the system mainly through the deposition of carbonate minerals on the seafloor. The rate of carbon removal through carbonate deposition is controlled by the saturation state of the ocean, and thus by the rate of supply of alkalinity by the chemical weathering of continental minerals. Carbonate and silicate minerals exposed at the continental surface weather under the corrosive action of atmospheric CO_2 dissolved in rain water as carbonic acid, and eventually concentrated by the microbial respiration in soils. Regarding the chemical weathering of carbonate minerals, the net budget of the dissolution reaction can be written as follows eqn [I].

$$CaCO_3(rock) + H_2CO_3(atmosphere)$$
$$\rightarrow Ca^{2+}(rivers) + 2HCO_3^-(rivers) \qquad [I]$$

In this reaction, there is a net transfer of Ca^{2+} (or Mg^{2+} in the case of magnesium carbonates) from the continental crust to the ocean, a creation of two equivalents of alkalinity and a transfer to the river system of two moles of carbon per mole of Ca^{2+}, one coming from the crust, the other one from the atmosphere. Once the weathering products reach the ocean, they will increase the saturation state of surface waters with respect to calcite and induce rapidly (within 10^3 years) the biologically driven precipitation of one mole of $CaCO_3$ followed by its deposition on the seafloor. The precipitation–deposition reaction is the reverse of reaction [1]. The net carbon budget of the weathering of carbonate minerals followed by deposition of sedimentary carbonate is thus equal to zero.

The chemical weathering of continental silicate rocks is fundamentally different, since silicate rocks do not contain carbon. The budget can be written as

$$CaSiO_3(rock) + 2H_2CO_3(atmosphere) + H_2O$$
$$\rightarrow Ca^{2+}(rivers) + 2HCO_3^-(rivers) + H_4SiO_4(rivers)$$
$$[II]$$

Here $CaSiO_3$ stands for a 'generic' silicate mineral. The weathering reaction of more realistic Ca- (or Mg-) silicate minerals, if more complex, displays the same budget in terms of alkalinity versus carbon fluxes. Again, once reaching the ocean, the excess Ca^{2+} will precipitate as $CaCO_3$, thus removing one mole of carbon from the ocean per mole of weathered silicates. The net budget of this reaction, after sedimentary carbonate precipitation, is the transfer of exospheric carbon to the crust. Chemical weathering of continental silicate minerals thus acts as the main sink of carbon on the geological timescale. Today, about 6×10^{12} mol y^{-1} of Ca^{2+} and Mg^{2+} are released from silicate weathering.

The size of the exospheric carbon pool (ocean + atmosphere) is about 3.2×10^{18} mol today. As mentioned above, the fluxes entering and leaving this reservoir are of the order of 10^{12}–10^{13} mol y^{-1}. A relatively small imbalance between the input and output of carbon of 10^{12} mol y^{-1}, the output being higher, but persisting for several million years, will result in a drastic reduction in the exospheric content. Three million years will be sufficient to remove all the carbon from the exospheric system, thus forcing the atmospheric $P\text{CO}_2$ to zero. There is no lithological, fossil, or geochemical record of such a dramatic event during the Cenozoic, or event during the complete Phanerozoic. To avoid the occurrence of such events for which there is no evidence, the perturbations of the carbon cycle had to be limited in time and amplitude, and thus the past exospheric carbon cycle was not strictly at, but always close to, steady state. The same considerations apply to the alkalinity budget. These steady-state conditions require that the amount of carbon removed from the atmosphere–ocean system by continental silicate weathering must always closely track the amount of carbon released by degassing. Mathematically, these conditions translate into eqn [1].

$$F_{SW} = F_{VOL} \qquad [1]$$

The question is now how to physically force F_{SW} to follow the degassing. The answer lies in the fact that the chemical weathering of continental silicates appears to be dependent on air temperature, the dissolution being enhanced during warmer climates. This dependence provides a negative feedback that not only allows equilibration of the carbon and alkalinity budgets on the geological timescale, but also stabilizes the Earth's climate. When the degassing increases suddenly, for instance as a result of a higher spreading rate of the oceanic floor, the amount of carbon in the ocean and atmosphere will first increase, increasing the atmospheric P_{CO_2}. Because CO_2 is a greenhouse gas, the climate will become warmer, and this will enhance the weathering of silicate rocks. As a result, any increase in the input of carbon will be counterbalanced by an increase of the output through silicate weathering, thus stabilizing the system through a negative feedback loop. The P_{CO_2} will stabilize at a somewhat higher level than before the perturbation. Similarly, the decline in P_{CO_2} through the Cenozoic could be due to a decreasing degassing rate, which acts as the driving force of changes. This simple process is the basis of all existing long-term geochemical cycle models. It was first identified in 1981 by Walker, Hays, and Kasting. Breaking this feedback loop would result in fluctuations in calculated P_{CO_2} and thus presumably in climate, that are not reflected in the geological record.

Himalayan Uplift, $^{87}Sr/^{86}Sr$ Record, and Possible Implications for Weathering History

M.E. Raymo in 1991 put forward another explanation of the global Cenozoic P_{CO_2} decline. Instead of a decreased degassing rate, she suggested that continental silicate weathering rates increased drastically over the last 40 million years, although degassing conditions remained more or less constant. This assertion was originally based on the Cenozoic carbonate record of $^{87}Sr/^{86}Sr$. The isotope ^{86}Sr is stable, whereas ^{87}Sr is produced by the radioactive β-decay of ^{87}Rb. Strontium ions easily replace calcium ions in mineral lattices, since their ionic diameters are comparable. The present-day sea water $^{87}Sr/^{86}Sr$ equals 0.709. Two main processes impinge on this ratio: the chemical weathering of continents, delivering strontium with a mean $^{87}Sr/^{86}Sr$ of 0.712, and the exchanges between seafloor basalts and sea water, resulting in the release of mantle strontium into the ocean ($^{87}Sr/^{86}Sr = 0.703$). In other words, chemical weathering of continental rocks tends to increase the strontium isotopic ratio of sea water, while exchanges with seafloor basalts at low or high temperature tend to decrease it.

The sea water $^{87}Sr/^{86}Sr$ recorded over the last 65 million years displays a major increase, starting about 37–38 million years ago (**Figure 2**). An event approximately coeval with the sea water $^{87}Sr/^{86}Sr$ upward shift is the Himalayan uplift, which was

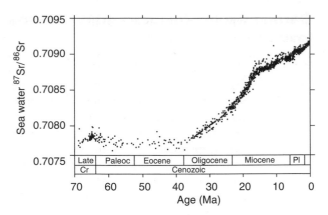

Figure 2 Sea water $^{87}Sr/^{86}Sr$ recorded in ancient carbonate sediments (Ottawa-Bochum database: http://www.science. uottawa.ca/geology/isotope_data/). Timescale abbreviations as **Figure 1**.

initiated by the India–Asia collision some 50 million years ago. Raymo has proposed that, in an uplifted area, the mechanical breakdown of rocks increases owing to the cooling and development of glaciers, to the development of steep slopes, and to temperatures oscillating below and above the freezing point at high altitudes. Furthermore, the development of the monsoon regime, about 10 million years ago, resulted in increased runoff, and thus an enhanced water availability for weathering, over at least the southern side of the Himalayan range. All these uplift-related changes might result in an enhanced chemical dissolution of minerals, since the surface in contact with the corrosive solutions increases when rocks fragment. The consequence of the Himalayan uplift might thus be an increase in the consumption of exospheric carbon by enhanced weathering on the continents, a process recorded in the sea water $^{87}Sr/^{86}Sr$ rise. The system depicted in this hypothesis is new, compared to the hypothesis described in the previous section. Here, tectonic processes result in uplift, followed by enhanced weathering, itself consuming atmospheric CO_2, thus cooling the climate. This cooling favors the development of glaciers not only in the uplifted area, but also globally, producing a global increase in mechanical and subsequent chemical weathering, a positive feedback that further cools the Earth. In Raymo's hypothesis, the negative feedback proposed by Walker *et al.*, stabilizing P_{CO_2} no longer exists. Chemical weathering is mainly controlled by tectonic processes with high rates in a cool world (Raymo's world), while it was controlled by climate and P_{CO_2} with high rates in a warm world in the Walker hypothesis (Walker's world). However, as mentioned above, negative feedbacks are needed to stabilize P_{CO_2}, especially since the degassing remained more or less constant over the period of interest. Raymo's world has the ability to exhaust atmospheric CO_2 within a few million years.

In an attempt to reconcile the two approaches, François and Walker proposed in 1992 the addition of a new CO_2 consumption flux to the carbon cycle, identified as the precipitation of abiotic carbonates within the oceanic crust, subsequent to its alteration at low temperature. This flux is directly dependent on deep water temperature, which has decreased by $\sim 8°C$ over the Cenozoic. An increase in the continental weathering rate might be compatible with a constant degassing rate, since the sink of carbon through low-temperature alteration of the oceanic crust is decreasing. The balance between input and output is thus still in place. However, this additional sink of carbon is poorly constrained. The present-day consumption of carbon is estimated to be about 1.4×10^{12} mol y^{-1}, but the kinetics of the process is essentially unknown. This attractive hypothesis still needs experimental verification.

Finally, it should be noted that Raymo's hypothesis interprets the increase in the sea water $^{87}Sr/^{86}Sr$ in terms of an increase in the weathering fluxes. However, silicate minerals exposed in the Himalayan area, particularly in the High Himalayan Crystalline Series, display unusually high isotopic ratios (reaching 0.740). Sediments of Proterozoic age with a $^{87}Sr/^{86}Sr$ reaching 0.8 are also exposed in the Lesser Himalaya area. For this reason, rivers draining the Himalayan area (Ganges, Brahmapoutra, etc.) display an isotopic ratio (0.725 for the Ganges) higher than the mean global value (0.712). At least part of the Cenozoic increase in the sea water $^{87}Sr/^{86}Sr$ might thus be due to changes in the isotopic composition of source rocks.

Lysocline and Carbonate Accumulation Changes

Other indicators of a possible increase in the continental weathering rate over the course of Cenozoic exist. For instance, the global mean Calcite Compensation Depth (CCD) sank by about 1 km over the last 40 million years (**Figure 3**), a change possibly linked to an increased supply of alkalinity from rivers caused by the Himalayan uplift. Paradoxically, there is no evidence of major changes in the carbonate accumulation flux during the Cenozoic (**Figure 4**). The deepening of the CCD might thus be linked, at least partially, to the global Cenozoic marine regression, reducing the area of shallow epicontinental seas and thus the area available for the accumulation of coral reefs. In that case, carbon and alkalinity will be preferentially removed from the ocean through enhanced formation of calcitic shells in open waters, leading to the deepening of the CCD. This process might have been favored by the coeval

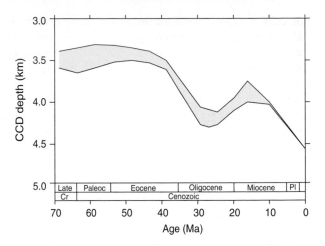

Figure 3 Reconstructed Carbonate Compensation Depth (CCD) through the Cenozoic (Van Andel, 1975; Broecker and Peng, 1982). Timescale as **Figure 1**.

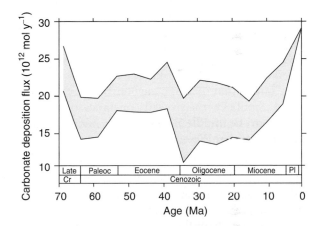

Figure 4 Total carbonate accumulation flux reconstructed from paleodata (Opdyke and Wilkinson, 1988). Timescale as **Figure 1**.

development of new foraminiferal species. The cause of the CCD deepening thus remains unresolved.

Organic Carbon Subcycle

The Cenozoic history of sea water $\delta^{13}C$ recorded in marine limestones (**Figure 5**) is marked by an ample fluctuation in the Paleocene and early Eocene, a roughly constant background value with superimposed high-frequency variations from the middle Eocene to the middle Miocene, and a sharp decrease from the middle Miocene to the present. Since organic matter is enriched in the lighter ^{12}C isotope with respect to sea water (owing to photosynthetic fractionation), this $\delta^{13}C$ record can be used to constrain the temporal changes in the organic fluxes of the carbon cycle. The burial of organic matter on the sea floor preferentially removes ^{12}C from the ocean and hence tends to increase seawater $\delta^{13}C$. Conversely, the oxidation of old organic carbon

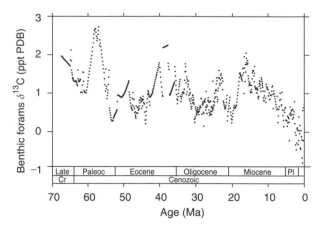

Figure 5 Sea water $\delta^{13}C$ recorded in ancient carbonate sediments (Ottawa-Bochum database: http://www.science.uottawa.ca/geology/isotope_data/). Timescale as **Figure 1**. ppt PDB = parts per thousand PeeDee Belemnite.

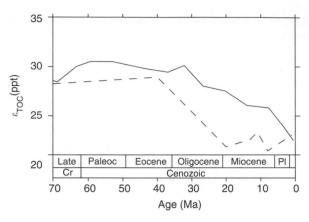

Figure 6 Average carbon isotopic fractionation (ε_{TOC}) between total organic carbon and sedimentary carbonate (solid line, Hayes *et al.* (1999); dashed line, Freeman and Hayes (1992)). Timescale as **Figure 1**.

(kerogen) contained in weathered sedimentary rocks is a source of isotopically light carbon for the ocean, tending to decrease its $\delta^{13}C$. The recent decrease in $\delta^{13}C$ since mid-Miocene times might thus be interpreted as the result of kerogen carbon oxidation being larger than organic carbon burial during that period. Similarly, the overall constancy of sea water $\delta^{13}C$ from the middle Eocene to the middle Miocene may suggest that the organic subcycle was essentially balanced at that time. However, the average carbon isotopic fractionation (ε_{TOC}) between total organic carbon and sedimentary carbonate (which is close to coeval sea water) has decreased from the Eocene to the present (**Figure 6**). With a balanced organic subcycle, this change in ε_{TOC} would imply a decrease of the sea water $\delta^{13}C$ over time, as it forces the $\delta^{13}C$ of organic deposits to become closer to the sea water value than it is for kerogen carbon. For the isotopic composition of the ocean to remain constant from the middle Eocene to the middle Miocene, the trend associated with ε_{TOC} variations must be compensated for by an imbalance in the organic subcycle in which the burial of organic carbon exceeds kerogen oxidation. This imbalance was progressively reduced after mid-Miocene times, but may have persisted until very recently.

The late Cenozoic was therefore a time of unusually high organic carbon deposition rates, leading to an increase in the size of the sedimentary organic carbon reservoir. The organic subcycle thus acted as a carbon sink over the course of the Himalayan uplift. There are two possible causes of this evolution.

1. The increase in chemical weathering rates in the Himalayan region during the uplift (Raymo's world) leads to enhanced delivery of nutrients to the ocean, forcing the oceanic primary productivity to increase. This might result in an increased burial of organic matter.

2. Enhanced mechanical weathering in the Himalayan region increased the sedimentation rate on the ocean floor, so that organic carbon was more easily preserved. This hypothesis does not require any increase in the chemical weathering rate in the Himalayan region. This facilitated burial might have significantly contributed to the Cenozoic P_{CO_2} decrease, since carbon is stored in a sedimentary reservoir. C. France-Lanord and L. Derry argued in 1997 that the consumption of CO_2 through organic carbon burial might be three times more important today than the amount of CO_2 consumed by silicate weathering within the orogen. Even if this hypothesis still links the climatic cooling with the Himalayan uplift, the origin of the CO_2 sink is quite different from that hypothesized in Raymo's world.

Observational data argue toward the second hypothesis, indicating that the Cenozoic increase in the sea water $^{87}Sr/^{86}Sr$ might be of isotopic origin. Calcium silicates are indeed not the most common mineral exposed in the Himalayan orogen, and thus cannot contribute widely to the CO_2 consumption. Furthermore, reverse weathering reactions take place in the Bengal Fan, releasing CO_2 and thus reducing the impact of the Himalayan silicate weathering on P_{CO_2}.

It has been suggested that the emission at some time in the past of large amounts of methane from gas hydrates may have influenced the $\delta^{13}C$ of the ocean. This may invalidate the interpretation of the carbon isotopic record if the gas hydrate reservoir has had long-term as well as shorter-term effects.

Organic carbon deposition on the seafloor is linked to ocean biological productivity, itself

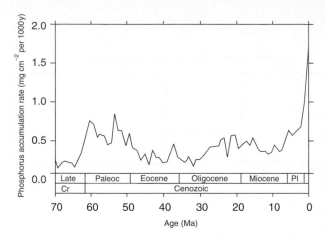

Figure 7 Average accumulation of phosphorus in sediments through the Cenozoic (Föllmi, 1995). Timescale as **Figure 1**.

depending on the availability of nutrients, among which phosphorus is thought to play a key role. The global phosphorus accumulation into sediments increased by a factor of about 4 over the last 10 million years (**Figure 7**), interpreted as the record of a global continental weathering enhanced by the onset of large ice sheets and glaciers at the end of Cenozoic. However, the question whether this increase is related to an increase in continental chemical weathering, or in mechanical weathering alone, is not clear.

Modelling: An Attempt to Integrate the Records into a Unified Framework

The Concept of Box Models

Biogeochemical cycles are usually described with box models. Such models provide a simple mathematical framework appropriate for calculating the geochemical evolution of the Earth through geological times. The Earth system is split into a relatively small number of components or reservoirs assumed to be homogeneous, such as the atmosphere, the ocean, the biosphere, the continental or oceanic crust, and the (upper) mantle. These reservoirs are connected by a series of 'arrows' representing the flows of material between them. The biogeochemical cycle of each element is thus represented as a set of interconnected reservoirs and, at any time, its state is characterized by the reservoir sizes or contents q_i (amount of the element in reservoir 'I', units: mol or kg) and the fluxes F_{ij} (amount of the element transferred per unit time from reservoir 'I' to reservoir 'j', units: mol y^{-1} or kg y^{-1}).

The temporal evolution of the system can be calculated by making a budget of input and output fluxes for each reservoir (eqn [2]).

$$\frac{\mathrm{d}q_i}{\mathrm{d}t} = \sum_{j=1j\neq i}^{N} F_{ji} - \sum_{j=1j\neq i}^{N} F_{ij}(i = 1, ..., N) \quad [2]$$

To solve this system of differential equations, the values of the fluxes must be provided at each time step. Kinetic rate laws describing the dependence of the fluxes F_{ij} on the reservoir contents q_i, time t, or some external forcing are thus needed. Defining such kinetic rate laws is the most critical task of modeling. The reliability of the solution and hence the usefulness of the results depend strongly on the adopted rate laws. The challenge is clearly to get at least a first-order estimate of the fluxes from a very broad knowledge of the system, i.e., from the values of its state variables $q_1, ..., q_N$.

A useful concept in box modeling is that of turnover time. The turnover time of an element in a given reservoir is defined as the ratio between its reservoir content and its total output flux (eqn [3]).

$$\tau_i = \frac{q_i}{\sum_{j=1j\neq i}^{N} F_{ij}} \quad [3]$$

The turnover time can be seen as the time needed to empty the reservoir if the input happened to stop suddenly and the current output flux were held constant. It provides a first-order idea of the evolution timescale of a reservoir. At steady state (i.e., when input and output fluxes balance each other), the turnover time is equal to the residence time, which is the average time spent by individual atoms of the element in the reservoir. Finally, the response time of a reservoir characterizes the time needed for the reservoir to adjust to a new equilibrium after a perturbation.

Models of the Carbon Cycle

Figure 8 illustrates the present state of the long-term carbon cycle from a recent (unpublished) box model simulation of the authors. The reservoirs and fluxes that have been included in this figure are those that are important to describe the evolution of atmospheric CO_2 at the geological timescale. The values of reservoir sizes and fluxes are consistent with current knowledge of the system. Crustal reservoirs include continental (5000×10^{18} mol C) and pelagic (150×10^{18} mol C) carbonates, as well organic carbon (1250×10^{18} mol C) from the sedimentary cover. The atmosphere and ocean have been lumped into one single reservoir containing 3.2×10^{18} mol C, since the time necessary for the atmosphere to reach equilibrium with the ocean is much shorter than ~1 My, the timescale of geological processes. Indeed,

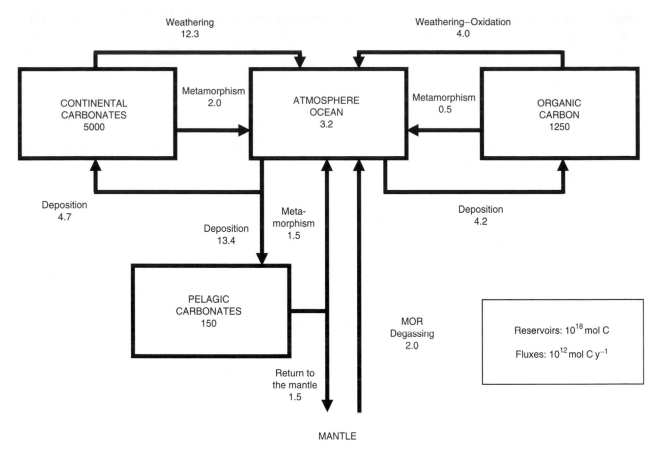

Figure 8   Present-day state of the long-term carbon cycle. Numbers represent 10^{18} mol C for reservoirs (boxes) and 10^{12} mol C y^{-1} for fluxes (arrows).

with a modern atmosphere-ocean exchange flux of 7.5×10^{15} mol C y^{-1} (i.e., 90 Gt C y^{-1}) and an atmospheric content of 62.5×10^{15} mol C (i.e., 750 Gt C), the turnover time of carbon in the atmosphere can be calculated to be only 8.33 years. Similarly, the terrestrial biosphere has not been included, since its size is small compared to other reservoirs and it can be assumed in equilibrium with the atmosphere–ocean system. The fluxes involved in the cycle are MOR/metamorphic CO_2 release ('volcanism'), weathering fluxes, and deposition of carbonates or organic carbon on the seafloor. The reported values of these fluxes are long-term averages, that is, they should be thought of as averages over several glacial–interglacial oscillations of the Pleistocene, although such averages cannot always be estimated from presently available data. The turnover time of carbon in the atmosphere–ocean reservoir in this 'geological' system can be calculated to be 143 000 y. Owing to this relatively short turnover time with respect to the timescale of long-term geological changes, the atmosphere–ocean system is essentially at equilibrium. By contrast, crustal reservoirs that exhibit much larger turnover times are not at equilibrium. This is

clearly the case of continental ($\tau_i = 350$ My) and pelagic ($\tau_i = 50$ My) carbonate reservoirs, as a result of the Cenozoic deepening of the ocean lysocline and the associated transfer of carbonate deposition from the shelf to the pelagic environment.

To distribute the carbon content of the atmosphere–ocean system among its two components, and hence to derive the atmospheric P_{CO_2} value (and its effect on the climate), it is necessary to know the alkalinity content of the ocean. For this reason, the evolution of the ocean alkalinity (A_T) and ocean–atmosphere carbon (C_T) content are always calculated in parallel. Writing eqn [2] for these two variables yields eqns [4a] and [4b].

$$\frac{dA_T}{dt} = 2F_{SW} + 2F_{CW} - 2F_{CD} \qquad [4a]$$

$$\frac{dC_T}{dt} = F_{VOL} + F_{CW} - F_{CD} + F_{OW} - F_{OD} \qquad [4b]$$

F_{VOL} represents the total CO_2 release flux from volcanic origin (i.e., the sum of all metamorphic and MOR fluxes in **Figure 8**), F_{CW} and F_{SW} are the weathering fluxes from respectively carbonate and silicate rocks expressed in moles of divalent ions (Ca^{2+} or Mg^{2+}) per unit of time (i.e., the rates of reactions [I] and [II]), F_{CD} is the carbonate deposition flux, F_{OW} is the carbon input flux from weathering-oxidation of crustal organic carbon, and F_{OD} is the organic carbon deposition flux. Note that the silicate weathering flux does not appear in the carbon budget, eqn [4b], since silicate weathering (reaction [II]) transfers carbon from the atmosphere to the ocean but does not remove it from the atmosphere–ocean system. The factor of 2 in eqn [4a] results from the fact that two equivalents of alkalinity are transferred to the ocean when one Ca^{2+} or Mg^{2+} ion is delivered to the ocean by rivers (reactions [I] and [II]). The same factor of 2 holds for carbonate deposition, which is the reverse of reaction [I]. As already mentioned, the atmosphere–ocean system must be close to equilibrium, so that the derivatives on the left-hand side of eqn [4a] and [4b] can be set to zero. This assumption transforms the differential equation system into a set of two algebraic equations, which can be solved to yield eqn [5].

$$F_{VOL} - F_{SW} = F_{OD} - F_{OW} \qquad [5]$$

This equation leads to eqn [1] if the effect of the organic subcycle is neglected (i.e., when this subcycle is set to equilibrium). Hence, eqn [5] is a generalization of the Walker, Hays, and Kasting budget. It states that the disequilibrium of the inorganic part of the carbon cycle must be compensated for by a disequilibrium of opposite sign in the organic subcycle.

Use of Isotopic Data (Inverse Modeling)

To solve eqns [4] or [5], some kinetic laws must be provided for the fluxes, that is, the relations between these fluxes, time t, and the reservoir contents, or atmospheric P_{CO_2}, must be known. Such kinetic laws are, however, poorly known, so it may be preferable, at least for some fluxes, to use forcing functions in the calculation of these fluxes. For example, volcanic fluxes are often made proportional to the seafloor spreading rate and weathering fluxes to land area, for which past reconstructions are available. Ocean isotopic records, such as those presented earlier, can also be used to force the model. Budget equations similar to [2] are then written for the relevant isotopes and transformed into equations containing isotopic ratios r (or δ, the relative departure of the isotopic ratio from a standard). The sea water $^{87}Sr/^{86}Sr$ ratio has been used in this way to estimate the silicate weathering flux F_{SW}, but as discussed earlier the results are strongly dependent on the hypothesis of constancy for the isotopic ratios of

weathered products. The ^{13}C isotopic history of the ocean has been used in many models, since the beginning of the 1980s, to constrain the organic carbon subcycle. The ^{13}C isotopic budget for the ocean can be written as eqn [6].

$$C_T \frac{d\delta_{OC}}{dt} = (\delta_{VOL} - \delta_{OC})F_{VOL} + (\delta_{CW} - \delta_{OC})F_{CW}$$
$$+ (\delta_{OW} - \delta_{OC})F_{OW} - (\delta_{OD} - \delta_{OC})F_{OD} \qquad [6]$$

δ_{oc} here is the $\delta^{13}C$ of the ocean (more precisely, this should be the $\delta^{13}C$ of the atmosphere–ocean system); δ_{VOL}, δ_{CW}, and δ_{OW} are the $\delta^{13}C$ of the carbon inputs from respectively volcanic, carbonate weathering, and crustal organic carbon weathering–oxidation fluxes. It is assumed that no fractionation occurs with respect to average oceanic carbon during carbonate precipitation, so that this flux does not appear in the equation. $\delta_{OD} = \delta_{oc} - \Delta$ is the $\delta^{13}C$ of the organic carbon deposited on the seafloor, with θ being the average fractionation of photosynthesis with respect to oceanic carbon (this includes both terrestrial and marine photosynthesis). The past values of δ_{oc} are known from the ^{13}C isotopic history of sea water (**Figure 5**). Equation [6] can then be solved with respect to F_{OD} and the resulting expression for F_{OD} is then used in eqn [4b] or [5]. The isotopic composition of the input fluxes must, however, be known or derived from similar isotopic budgets for the crustal reservoirs. This procedure is actually an inverse method, since it derives model parameters (fluxes) from an observed signal (ocean isotopic composition) linked to the model parameters through a mathematical operator (the isotopic budget equation). Y. Goddéris and L.M. François in 1996, and L.R. Kump and M.A. Arthur in 1997, published two separate models inverting the oceanic $\delta^{13}C$ signal over the Cenozoic, making use of an isotopic fractionation Δ variable with age and derived from paleodata. The Cenozoic histories of silicate weathering from these models are compared in **Figure 9**. The predicted trend of the carbonate deposition flux is broadly consistent with an available reconstruction based on carbonate accumulation data (**Figure 10**).

A classical example of a box model using ^{13}C isotopic data to constrain the organic carbon subcycle is the BLAG model of Lasaga, Berner, and Garrels published in 1985. R.A. Berner in 1990 also used such an isotopic budget in GEOCARB to calculate the history of atmospheric CO_2 over the Phanerozoic. The results show a decreasing trend of atmospheric CO_2 over the Cenozoic. The trend is consistent with the overall trend reconstructed with other models (e.g., François and Walker, 1992) or

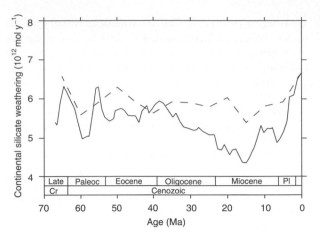

Figure 9 Model silicate weathering flux (solid line, Goddéris and François (1996); dashed line, Kump and Arthur (1997)). Timescale as **Figure 1**.

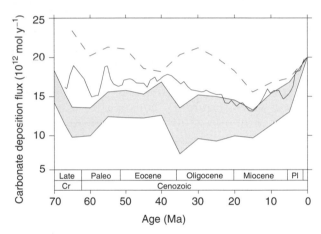

Figure 10 Model carbonate accumulation flux compared to the reconstruction of **Figure 4** normalized to the present day value of 20×10^{12} mol y^{-1} (solid line: Goddéris and François (1996); dashed line, Kump and Arthur (1997); shaded area, reconstruction from **Figure 4**).

from various paleoindicators (**Figure 1**). This does not mean, however, that we understand the carbon cycle (and climate) trends of the Cenozoic, since different models can produce similar trends from completely different underlying mechanisms. To be reliable, models should not rest only on a limited set of data but should be able to explain a wide range of geochemical records.

Conclusions

Proxy records indicate that the Earth's climate cooled gradually over the Cenozoic. This cooling trend was accompanied by a decrease of atmospheric P_{CO_2}. Other striking features of the Cenozoic are the sharp increase of the $^{87}Sr/^{86}Sr$ ratio of sea water and

the overall deepening of the lysocline from late Eocene time to the present, and after the mid-Miocene a marked decrease of ocean $\delta^{13}C$ together with an increase in total carbonate accumulation and possibly phosphorus deposition. Are these environmental changes related? The role of models is to synthesize and provide a coherent explanation of such records, and then reconstruct the history of other key variables not directly accessible from paleodata. Today, we are still far from this goal. It is fundamental that models use multiple proxy data both as forcings and for validation, implying that other biogeochemical cycles for which proxies are available are modeled together with the carbon cycle. A coupling to other major biogeochemical cycles is also essential because of the interactions with the carbon cycle and the feedbacks involved.

See also

Carbon Cycle. Cenozoic Climate – Oxygen Isotope Evidence. Ocean Carbon System, Modeling of. Paleoceanography, Climate Models in. Past Climate from Corals. Stable Carbon Isotope Variations in the Ocean.

Further Reading

Berner RA (1990) Atmospheric carbon dioxide levels over Phanerozoic time. *Science* 249: 1382–1386.

Berner RA (1998) The carbon cycle and CO_2 over Phanerozoic time: the role of land plants. *Philosophical Transactions of the Royal Society of London B* 353: 75–82.

Broecker WS and Peng TH (1982) *Tracers in the Sea*. Palisades: Eldigio Press.

Butcher SS, Charlson RJ, Orians GH, and Wolfe GV (eds.) (1992) *Global Biogeochemical Cycles*. London: Academic Press.

Chameides WL and Perdue EM (1997) *Biogeochemical Cycles: A Computer-Interactive Study of Earth System Science and Global Change*. Oxford: Oxford University Press.

Föllmi KB (1995) 160 My record of marine sedimentary phosphorus burial: coupling of climate and continental weathering under greenhouse and icehouse conditions. *Geology* 23: 859–862.

France-Lanord C and Derry LA (1997) Organic carbon burial forcing of the carbon cycle from Himalayan erosion. *Nature* 390: 65–67.

François LM and Walker JCG (1992) Modelling the Phanerozoic carbon cycle and climate: constraints from the $^{87}Sr/^{86}Sr$ isotopic ratio of sea water. *American Journal of Science* 292: 81–135.

Freeman KH and Hayes JM (1992) Fractionation of carbon isotopes by phytoplankton and estimates of ancient CO_2 levels. *Global Biogeochemical Cycles* 6: 185–198.

Goddéris Y and François LM (1996) Balancing the Cenozoic carbon and alkalinity cycles: constraints from isotopic records. *Geophysical Research Letters* 23: 3743–3746.

Harland WB, Armstrong RL, Cox AV, *et al.* (1990) *A Geologic Time Scale 1989*. Cambridge: Cambridge University Press.

Hayes JM, Strauss H, and Kaufman AJ (1999) The abundance of ^{13}C in marine organic matter and isotopic fractionation in the global biogeochemical cycle of carbon during the past 800 Ma. *Chemical Geology* 161: 103–125.

Kump LR and Arthur MA (1997) Global chemical erosion during the Cenozoic: weatherability balances the budgets. In: Ruddiman WF (ed.) *Tectonic Uplift and Climate Change*. New York: Plenum Press.

Kump LR, Kasting JF, and Crane RG (1999) *The Earth System*. New Jersey: Prentice Hall.

Lasaga AC, Berner RA, and Garrels RM (1985) An improved geochemical model of atmospheric CO_2 fluctuations over the past 100 million years. In: Sundquist E and Broecker WS (eds.) *The Carbon Cycle and Atmospheric CO_2: Natural Variations*

Archean to Present Geophysical Monograph, vol. 32, pp. 397–411. Washington, DC: American Geophysical Union.

Opdyke BN and Wilkinson BH (1988) Sea surface area control of shallow cratonic to deep marine carbonate accumulation. *Paleoceanography* 3: 685–703.

Pearson PN and Palmer MR (2000) Atmospheric carbon dioxide concentrations over the past 60 million years. *Nature* 406: 695–699.

Raymo ME (1991) Geochemical evidence supporting T.C. Chamberlin's theory of glaciation. *Geology* 19: 344–347.

Ruddiman WF (ed.) (1997) *Tectonic Uplift and Climate Change*. New York: Plenum Press.

Van Andel TH (1975) Mesozoic-Cenozoic calcite compensation depth and the global distribution of calcareous sediments. *Earth and Planetary Science Letters* 26: 187–194.

Van Andel TH (1994) *New Views on an Old Planet: a History of Global Change*, 2nd edn. Cambridge: Cambridge Universitys Press.

Walker JCG, Hays PB, and Kasting JF (1981) A negative feedback mechanism for the long-term stabilization of Earth's surface temperature. *Journal of Geophysical Research* 86: 9776–9782.

EFFECTS & REMEDIATION

SEA LEVEL CHANGE

J. A. Church, Antarctic CRC and CSIRO Marine Research, TAS, Australia
J. M. Gregory, Hadley Centre, Berkshire, UK

Introduction

Sea-level changes on a wide range of time and space scales. Here we consider changes in mean sea level, that is, sea level averaged over a sufficient period of time to remove fluctuations associated with surface waves, tides, and individual storm surge events. We focus principally on changes in sea level over the last hundred years or so and on how it might change over the next one hundred years. However, to understand these changes we need to consider what has happened since the last glacial maximum 20 000 years ago. We also consider the longer-term implications of changes in the earth's climate arising from changes in atmospheric greenhouse gas concentrations.

Changes in mean sea level can be measured with respect to the nearby land (relative sea level) or a fixed reference frame. Relative sea level, which changes as either the height of the ocean surface or the height of the land changes, can be measured by a coastal tide gauge.

The world ocean, which has an average depth of about 3800 m, contains over 97% of the earth's water. The Antarctic ice sheet, the Greenland ice sheet, and the hundred thousand nonpolar glaciers/ ice caps, presently contain water sufficient to raise sea level by 61 m, 7 m, and 0.5 m respectively if they were entirely melted. Ground water stored shallower than 4000 m depth is equivalent to about 25 m (12 m stored shallower than 750 m) of sea-level change. Lakes and rivers hold the equivalent of less than 1 m, while the atmosphere accounts for only about 0.04 m.

On the time-scales of millions of years, continental drift and sedimentation change the volume of the ocean basins, and hence affect sea level. A major influence is the volume of mid-ocean ridges, which is related to the arrangement of the continental plates and the rate of sea floor spreading.

Sea level also changes when mass is exchanged between any of the terrestrial, ice, or atmospheric reservoirs and the ocean. During glacial times (ice ages), water is removed from the ocean and stored in large ice sheets in high-latitude regions. Variations in the surface loading of the earth's crust by water and ice change the shape of the earth as a result of the elastic response of the lithosphere and viscous flow of material in the earth's mantle and thus change the level of the land and relative sea level. These changes in the distribution of mass alter the gravitational field of the earth, thus changing sea level. Relative sea level can also be affected by local tectonic activities as well as by the land sinking when ground water is extracted or sedimentation increases. Sea water density is a function of temperature. As a result, sea level will change if the ocean's temperature varies (as a result of thermal expansion) without any change in mass.

Sea-Level Changes Since the Last Glacial Maximum

On timescales of thousands to hundreds of thousands of years, the most important processes affecting sea-level are those associated with the growth and decay of the ice sheets through glacial–interglacial cycles. These are also relevant to current and future sea level rise because they are the cause of ongoing land movements (as a result of changing surface loads and the resultant small changes in the shape of the earth – postglacial rebound) and ongoing changes in the ice sheets.

Sea-level variations during a glacial cycle exceed 100 m in amplitude, with rates of up to tens of millimetres per year during periods of rapid decay of the ice sheets (**Figure 1**). At the last glacial maximum (about 21 000 years ago), sea level was more than 120 m below current levels. The largest contribution to this sea-level lowering was the additional ice that formed the North American (Laurentide) and European (Fennoscandian) ice sheets. In addition, the Antarctic ice sheet was larger than at present and there were smaller ice sheets in presently ice-free areas.

Observed Recent Sea-Level Change

Long-term relative sea-level changes have been inferred from the geological records, such as radio-carbon dates of shorelines displaced from present day sea level, and information from corals and sediment cores. Today, the most common method of measuring sea level relative to a local datum is by tide gauges at coastal and island sites. A global data set is maintained by the Permanent Service for Mean Sea Level (PSMSL). During the 1990s, sea level has been measured globally with satellites.

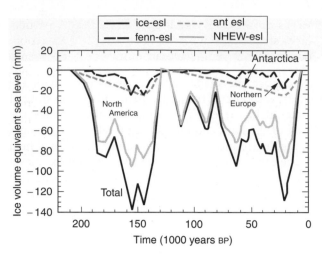

Figure 1 Change in ice sheet volume over the last 200 000 years. Fenn, Fennoscandian; ant, Antarctic; NHEW; North America; esl, equivalent sea level. (Reproduced from Lambeck, 1998.)

Tide-gauge Observations

Unfortunately, determination of global-averaged sea-level rise is severely limited by the small number of gauges (mostly in Europe and North America) with long records (up to several hundred years, **Figure 2**). To correct for vertical land motions, some sea-level change estimates have used geological data, whereas others have used rates of present-day vertical land movement calculated from models of postglacial rebound.

A widely accepted estimate of the current rate of global-average sea-level rise is about $1.8 \, \mathrm{mm \, y^{-1}}$. This estimate is based on a set of 24 long tide-gauge records, corrected for land movements resulting from deglaciation. However, other analyses produce different results. For example, recent analyses suggest that sea-level change in the British Isles, the North Sea region and Fennoscandia has been about $1 \, \mathrm{mm \, y^{-1}}$ during the past century. The various assessments of the global-average rate of sea-level change over the past century are not all consistent within stated uncertainties, indicating further sources of error. The treatment of vertical land movements remains a source of potential inconsistency, perhaps amounting to $0.5 \, \mathrm{mm \, y^{-1}}$. Other sources of error include variability over periods of years and longer and any spatial distribution in regional sea level rise (perhaps several tenths of a millimeter per year).

Comparison of the rates of sea-level rise over the last 100 years ($1.0–2.0 \, \mathrm{mm \, y^{-1}}$) and over the last two millennia ($0.1–0.0 \, \mathrm{mm \, y^{-1}}$) suggests the rate has accelerated fairly recently. From the few very long tide-gauge records (**Figure 2**), it appears that an acceleration of about $0.3–0.9 \, \mathrm{mm \, y^{-1}}$ per century occurred over the nineteenth and twentieth century. However, there is little indication that sea-level rise accelerated during the twentieth century.

Altimeter Observations

Following the advent of high-quality satellite radar altimeter missions in the 1990s, near-global and homogeneous measurement of sea level is possible, thereby overcoming the inhomogeneous spatial sampling from coastal and island tide gauges. However, clarifying rates of global sea-level change requires continuous satellite operations over many

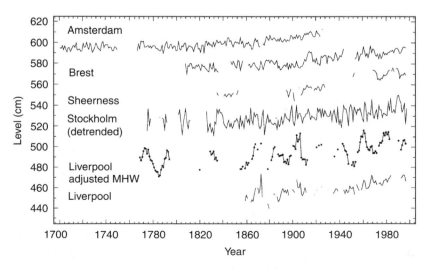

Figure 2 Time series of relative sea level over the last 300 years from several European coastal locations. For the Stockholm record, the trend over the period 1774 to 1873 has been removed from the entire data set. For Liverpool two series are given. These are mean sea level and, for a longer period, the mean high water (MHW) level. (Reproduced with permission from Woodworth, 1999.)

years and careful control of biases within and between missions.

To date, the TOPEX/POSEIDON satellite-altimeter mission, with its (near) global coverage from 66°N to 66°S (almost all of the ice-free oceans) from late 1992 to the present, has proved to be of most value in producing direct estimates of sea-level change. The present data allow global-average sea level to be estimated to a precision of several millimeters every 10 days, with the absolute accuracy limited by systematic errors. The most recent estimates of global-average sea level rise based on the short (since 1992) TOPEX/POSEIDON time series range from $2.1 \, \text{mm y}^{-1}$ to $3.1 \, \text{mm y}^{-1}$.

The alimeter record for the 1990s indicates a rate of sea-level rise above the average for the twentieth century. It is not yet clear if this is a result of an increase in the rate of sea-level rise, systematic differences between the tide-gauge and altimeter data sets or the shortness of the record.

Processes Determining Present Rates of Sea-Level Change

The major factors determining sea-level change during the twentieth and twenty-first century are ocean thermal expansion, the melting of nonpolar glaciers and ice caps, variation in the mass of the Antarctic and Greenland ice sheets, and changes in terrestrial storage.

Projections of climate change caused by human activity rely principally on detailed computer models referred to as atmosphere–ocean general circulation models (AOGCMs). These simulate the global three-dimensional behavior of the ocean and atmosphere by numerical solution of equations representing the underlying physics. For simulations of the next hundred years, future atmospheric concentrations of gases that may affect the climate (especially carbon dioxide from combustion of fossil fuels) are estimated on the basis of assumptions about future population growth, economic growth, and technological change. AOGCM experiments indicate that the global-average temperature may rise by 1.4–5.8°C between 1990 and 2100, but there is a great deal of regional and seasonal variation in the predicted changes in temperature, sea level, precipitation, winds, and other parameters.

Ocean Thermal Expansion

The broad pattern of sea level is maintained by surface winds, air–sea fluxes of heat and fresh water (precipitation, evaporation, and fresh water runoff from the land), and internal ocean dynamics. Mean sea level varies on seasonal and longer timescales. A particularly striking example of local sea-level variations occurs in the Pacific Ocean during El Niño events. When the trade winds abate, warm water moves eastward along the equator, rapidly raising sea level in the east and lowering it in the west by about 20 cm.

As the ocean warms, its density decreases. Thus, even at constant mass, the volume of the ocean increases. This thermal expansion is larger at higher temperatures and is one of the main contributors to recent and future sea-level change. Salinity changes within the ocean also have a significant impact on the local density, and thus on local sea level, but have little effect on the global-average sea level.

The rate of global temperature rise depends strongly on the rate at which heat is moved from the ocean surface layers into the deep ocean; if the ocean absorbs heat more readily, climate change is retarded but sea level rises more rapidly. Therefore, time-dependent climate change simulation requires a model that represents the sequestration of heat in the ocean and the evolution of temperature as a function of depth. The large heat capacity of the ocean means that there will be considerable delay before the full effects of surface warming are felt throughout the depth of the ocean. As a result, the ocean will not be in equilibrium and global-average sea level will continue to rise for centuries after atmospheric greenhouse gas concentrations have stabilized. The geographical distribution of sea-level change may take many decades to arrive at its final state.

While the evidence is still somewhat fragmentary, and in some cases contradictory, observations indicate ocean warming and thus thermal expansion, particularly in the subtropical gyres, at rates resulting in sea-level rise of order $1 \, \text{mm y}^{-1}$. The observations are mostly over the last few decades, but some observations date back to early in the twentieth century. The evidence is most convincing for the subtropical gyre of the North Atlantic, for which the longest temperature records (up to 73 years) and most complete oceanographic data sets exist. However, the pattern also extends into the South Atlantic and the Pacific and Indian oceans. The only areas of substantial ocean cooling are the subpolar gyres of the North Atlantic and perhaps the North Pacific. To date, the only estimate of a global average rate of sea-level rise from thermal expansion is $0.55 \, \text{mm y}^{-1}$.

The warming in the Pacific and Indian Oceans is confined to the main thermocline (mostly the upper 1 km) of the subtropical gyres. This contrasts with the North Atlantic, where the warming is also seen at greater depths.

AOGCM simulations of sea level suggest that during the twentieth century the average rate

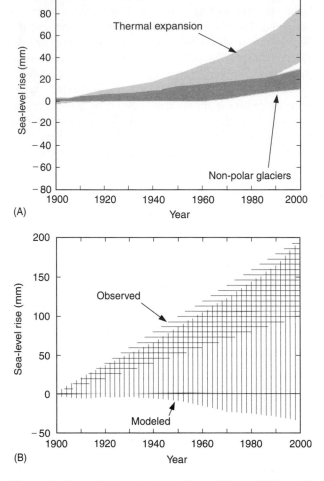

Figure 3 Computed sea-level rise from 1900 to 2000 AD. (A) The estimated thermal expansion is shown by the light stippling; the estimated nonpolar glacial contribution is shown by the medium-density stippling. (B) The computed total sea level change during the twentieth century is shown by the vertical hatching and the observed sea level change is shown by the horizontal hatching.

of change due to thermal expansion was of the order of $0.3–0.8\,\mathrm{mm\,y^{-1}}$ (**Figure 3**). The rate rises to $0.6–1.1\,\mathrm{mm\,y^{-1}}$ in recent decades, similar to the observational estimates of ocean thermal expansion.

Nonpolar Glaciers and Ice Caps

Nonpolar glaciers and ice caps are rather sensitive to climate change, and rapid changes in their mass contribute significantly to sea-level change. Glaciers gain mass by accumulating snow, and lose mass (ablation) by melting at the surface or base. Net accumulation occurs at higher altitude, net ablation at lower altitude. Ice may also be removed by discharge into a floating ice shelf and/or by direct calving of icebergs into the sea.

In the past decade, estimates of the regional totals of the area and volume of glaciers have been improved. However, there are continuous mass balance records longer than 20 years for only about 40 glaciers worldwide. Owing to the paucity of measurements, the changes in mass balance are estimated as a function of climate.

On the global average, increased precipitation during the twenty-first century is estimated to offset only 5% of the increased ablation resulting from warmer temperatures, although it might be significant in particular localities. (For instance, while glaciers in most parts of the world have had negative mass balance in the past 20 years, southern Scandinavian glaciers have been advancing, largely because of increases in precipitation.) A detailed computation of transient response also requires allowance for the contracting area of glaciers.

Recent estimates of glacier mass balance, based on both observations and model studies, indicate a contribution to global-average sea level of 0.2 to $0.4\,\mathrm{mm\,y^{-1}}$ during the twentieth century. The model results shown in **Figure 3** indicate an average rate of 0.1 to $0.3\,\mathrm{mm\,y^{-1}}$.

Greenland and Antarctic Ice Sheets

A small fractional change in the volume of the Greenland and Antarctic ice sheets would have a significant effect on sea level. The average annual solid precipitation falling onto the ice sheets is equivalent to 6.5 mm of sea level, but this input is approximately balanced by loss from melting and iceberg calving. In the Antarctic, temperatures are so low that surface melting is negligible, and the ice sheet loses mass mainly by ice discharge into floating ice shelves, which melt at their underside and eventually break up to form icebergs. In Greenland, summer temperatures are high enough to cause widespread surface melting, which accounts for about half of the ice loss, the remainder being discharged as icebergs or into small ice shelves.

The surface mass balance plays the dominant role in sea-level changes on a century timescale, because changes in ice discharge generally involve response times of the order of 10^2 to 10^4 years. In view of these long timescales, it is unlikely that the ice sheets have completely adjusted to the transition from the previous glacial conditions. Their present contribution to sea-level change may therefore include a term related to this ongoing adjustment, in addition to the effects of climate change over the last hundred years. The current rate of change of volume of the polar ice sheets can be assessed by estimating the individual mass balance terms or by monitoring surface elevation changes directly (such as by airborne and satellite altimetry during the 1990s). However, these techniques give results with large uncertainties. Indirect methods (including numerical modeling of ice-sheets, observed sea-level changes over the last few millennia, and changes in the earth's rotation parameters) give narrower bounds, suggesting that the present contribution of the ice sheets to sea level is a few tenths of a millimeter per year at most.

Calculations suggest that, over the next hundred years, surface melting is likely to remain negligible in Antarctica. However, projected increases in precipitation would result in a net negative sea-level contribution from the Antarctic ice sheet. On the other hand, in Greenland, surface melting is projected to increase at a rate more than enough to offset changes in precipitation, resulting in a positive contribution to sea-level rise.

Changes in Terrestrial Storage

Changes in terrestrial storage include reductions in the volumes of some of the world's lakes (e.g., the Caspian and Aral seas), ground water extraction in excess of natural recharge, more water being impounded in reservoirs (with some seeping into aquifers), and possibly changes in surface runoff. Order-of-magnitude evaluations of these terms are uncertain but suggest that each of the contributions could be several tenths of millimeter per year, with a small net effect (**Figure 3**). If dam building continues at the same rate as in the last 50 years of the twentieth century, there may be a tendency to reduce sea-level rise. Changes in volumes of lakes and rivers will make only a negligible contribution.

Permafrost currently occupies about 25% of land area in the northern hemisphere. Climate warming leads to some thawing of permafrost, with partial runoff into the ocean. The contribution to sea level in the twentieth century is probably less than 5 mm.

Projected Sea-Level Changes for the Twenty-first Century

Detailed projections of changes in sea level derived from AOCGM results are given in material listed as Further Reading. The major components are thermal expansion of the ocean (a few tens of centimeters), melting of nonpolar glaciers (about 10–20 cm), melting of Greenland ice sheet (several centimeters), and increased storage in the Antarctic (several centimeters).

After allowance for the continuing changes in the ice sheets since the last glacial maximum and the

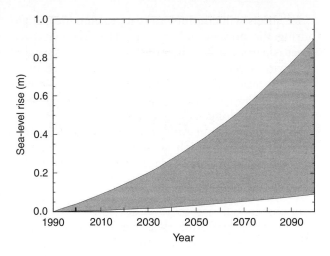

Figure 4 The estimated range of future global-average sea-level rise from 1990 to 2100 AD for a range of plausible projections in atmospheric greenhouse gas concentrations.

melting of permafrost (but not including changes in terrestrial storage), total projected sea-level rise during the twenty-first century is currently estimated to be between about 9 and 88 cm (**Figure 4**).

Regional Sea-Level Change

Estimates of the regional distribution of sea-level rise are available from several AOGCMs. Our confidence in these distributions is low because there is little similarity between model results. However, models agree on the qualitative conclusion that the range of regional variation is substantial compared with the global-average sea-level rise. One common feature is that nearly all models predict less than average sea-level rise in the Southern Ocean.

The most serious impacts of sea-level change on coastal communities and ecosystems will occur during the exceptionally high water levels known as storm surges produced by low air pressure or driving winds. As well as changing mean sea level, climate change could also affect the frequency and severity of these meteorological conditions, making storm surges more or less severe at particular locations.

Longer-term Changes

Even if greenhouse gas concentrations were to be stabilized, sea level would continue to rise for several hundred years. After 500 years, sea-level rise from thermal expansion could be about 0.3–2 m but may be only half of its eventual level. Glaciers presently contain the equivalent of 0.5 m of sea level. If the CO_2 levels projected for 2100 AD were sustained, there would be further reductions in glacier mass.

Ice sheets will continue to react to climatic change during the present millennium. The Greenland ice sheet is particularly vulnerable. Some models suggest that with a warming of several degrees no ice sheet could be sustained on Greenland. Complete melting of the ice sheet would take at least a thousand years and probably longer.

Most of the ice in Antarctica forms the East Antarctic Ice Sheet, which would disintegrate only if extreme warming took place, beyond what is currently thought possible. The West Antarctic Ice Sheet (WAIS) has attracted special attention because it contains enough ice to raise sea level by 6 m and because of suggestions that instabilities associated with its being grounded below sea level may result in rapid ice discharge when the surrounding ice shelves are weakened. However, there is now general agreement that major loss of grounded ice, and accelerated sea-level rise, is very unlikely during the twenty-first century. The contribution of this ice sheet to sea level change will probably not exceed 3 m over the next millennium.

Summary

On timescales of decades to tens of thousands of years, sea-level change results from exchanges of mass between the polar ice sheets, the nonpolar glaciers, terrestrial water storage, and the ocean. Sea level also changes if the density of the ocean changes (as a result of changing temperature) even though there is no change in mass. During the last century, sea level is estimated to have risen by 10–20 cm, as a result of combination of thermal expansion of the ocean as its temperature rose and increased mass of the ocean from melting glaciers and ice sheets. Over the twenty-first century, sea level is expected to rise as a result of anthropogenic climate change. The main contributors to this rise are expected to be thermal expansion of the ocean and the partial melting of nonpolar glaciers and the Greenland ice sheet. Increased precipitation in Antarctica is expected to offset some of the rise from other contributions. Changes in terrestrial storage are uncertain but may also partially offset rises from other contributions. After allowance for the continuing changes in the ice sheets since the last glacial maximum, the total projected sea-level rise over the twenty-first century is currently estimated to be between about 9 and 88 cm.

See also

Abrupt Climate Change. El Niño Southern Oscillation (ENSO). El Niño Southern Oscillation (ENSO) Models. Glacial Crustal Rebound, Sea Levels, and Shorelines. Ice-shelf Stability. Ocean Circulation. Past Climate from Corals. Sea Level Variations Over Geologic Time.

Further Reading

Church JA, Gregory JM, Huybrechts P, et al. (2001) Changes in sea level. In: Houghton JT (ed.) Climate Change 2001; The Scientific Basis. Cambridge: Cambridge University Press.

Douglas BC, Keaney M, and Leatherman SP (eds.) (2000) Sea Level Rise: History and Consequences, 232 pp. San Diego: Academic Press.

Fleming K, Johnston P, Zwartz D, et al. (1998) Refining the eustatic sea-level curve since the Last Glacial Maximum using far- and intermediate-field sites. Earth and Planetary Science Letters 163: 327–342.

Lambeck K (1998) Northern European Stage 3 ice sheet and shoreline reconstructions: Preliminary results. News 5, Stage 3 Project, Godwin Institute for Quaternary Research, 9 pp.

Peltier WR (1998) Postglacial variations in the level of the sea: implications for climate dynamics and solid-earth geophysics. Review of Geophysics 36: 603–689.

Summerfield MA (1991) Global Geomorphology. Harlowe: Longman.

Warrick RA, Barrow EM, and Wigley TML (1993) Climate and Sea Level Change: Observations, Projections and Implications. Cambridge: Cambridge University Press.

WWW pages of the Permanent Service for Mean Sea Level, http://www.pol.ac.uk/psmsl/

SEA LEVEL VARIATIONS OVER GEOLOGICAL TIME

M. A. Kominz, Western Michigan University,
Kalamazoo, MI, USA

Introduction

Sea level changes have occurred throughout Earth history. The magnitudes and timing of sea level changes are extremely variable. They provide considerable insight into the tectonic and climatic history of the Earth, but remain difficult to determine with accuracy.

Sea level, where the world oceans intersect the continents, is hardly fixed, as anyone who has stood on the shore for 6 hours or more can attest. But the ever-changing tidal flows are small compared with longer-term fluctuations that have occurred in Earth history. How much has sea level changed? How long did it take? How do we know? What does it tell us about the history of the Earth?

In order to answer these questions, we need to consider a basic question: what causes sea level to change? Locally, sea level may change if tectonic forces cause the land to move up or down. However, this article will focus on global changes in sea level. Thus, the variations in sea level must be due to one of two possibilities: (1) changes in the volume of water in the oceans or (2) changes in the volume of the ocean basins.

Sea Level Change due to Volume of Water in the Ocean Basin

The two main reservoirs of water on Earth are the oceans (currently about 97% of all water) and glaciers (currently about 2.7%). Not surprisingly, for at least the last three billion years, the main variable controlling the volume of water filling the ocean basins has been the amount of water present in glaciers on the continents. For example, about 20 000 years ago, great ice sheets covered northern North America and Europe. The volume of ice in these glaciers removed enough water from the oceans to expose most continental shelves. Since then there has been a sea level rise (actually from about 20 000 to about 11 000 years ago) of about 120 m (**Figure 1A**).

A number of methods have been used to establish the magnitude and timing of this sea level change. Dredging on the continental shelves reveals human activity near the present shelf-slope boundary. These data suggest that sea level was much lower a relatively short time ago. Study of ancient corals shows that coral species which today live only in very shallow water are now over 100 m deep. The carbonate skeletons of the coral, which once thrived in the shallow waters of the tropics, yield a detailed picture of the timing of sea level rise, and, thus, the melting of the glaciers. Carbon-14, a radioactive isotope formed by carbon-12 interacting with high-energy solar radiation in Earth's atmosphere allows us to determine the age of Earth materials, which are about 30 thousand years old.

This is just the most recent of many, large changes in sea level caused by glaciers, (**Figure 1B**). These variations in climate and subsequent sea level changes have been tied to quasi-periodic variations in the Earth's orbit and the tilt of the Earth's spin axis. The record of sea level change can be estimated by observing the stable isotope, oxygen-18 in the tests (shells) of dead organisms. When marine microorganisms build their tests from the calcium, carbon, and oxygen present in sea water they incorporate both the abundant oxygen-16 and the rare oxygen-18 isotopes. Water in the atmosphere generally has a lower oxygen-18 to oxygen-16 ratio because evaporation of the lighter isotope requires less energy. As a result, the snow that eventually forms the glaciers is depleted in oxygen-18, leaving the ocean proportionately oxygen-18-enriched. When the microorganisms die, their tests sink to the seafloor to become part of the deep marine sedimentary record. The oxygen-18 to oxygen-16 ratio present in the fossil tests has been calibrated to the sea level change, which occurred from 20 000 to 11 000 years ago, allowing the magnitude of sea level change from older times to be estimated. This technique does have uncertainties. Unfortunately, the amount of oxygen-18 which organisms incorporate in their tests is affected not only by the amount of oxygen-18 present but also by the temperature and salinity of the water. For example, the organisms take up less oxygen-18 in warmer waters. Thus, during glacial times, the tests are even more enriched in oxygen-18, and any oxygen isotope record reveals a combined record of changing local temperature and salinity in addition to the record of global glaciation.

Moving back in time through the Cenozoic (zero to 65 Ma), paleoceanographic data remain excellent due to relatively continuous sedimentation on the ocean floor (as compared with shallow marine and

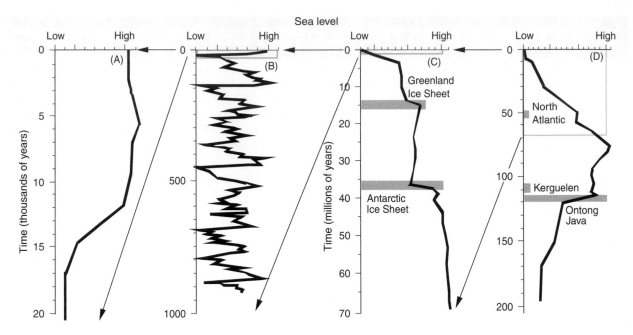

Figure 1   (A) Estimates of sea level change over the last 20 000 years. Amplitude is about 120 m. (B) Northern Hemisphere glaciers over the last million years or so generated major sea level fluctuations, with amplitudes as high as 125 m. (C) The long-term oxygen isotope record reveals rapid growth of the Antarctic and Greenland ice sheets (indicated by gray bars) as Earth's climate cooled. (D) Long-term sea level change as indicated from variations in deep-ocean volume. Dominant effects include spreading rates and lengths of mid-ocean ridges, emplacement of large igneous provinces (the largest, marine LIPs are indicated by gray bars), breakup of supercontinents, and subduction of the Indian continent. The Berggren *et al.* (1995) chronostratigraphic timescale was used in (C) and (D).

terrestrial sedimentation). Oxygen-18 in the fossil shells suggests a general cooling for about the last 50 million years. Two rapid increases in the oxygen-18 to oxygen-16 ratio about 12.5 Ma and about 28 Ma are observed (**Figure 1C**). The formation of the Greenland Ice Sheet and the Antarctic Ice Sheet are assumed to be the cause of these rapid isotope shifts. Where oxygen-18 data have been collected with a resolution finer than 20 000 years, high-frequency variations are seen which are presumed to correspond to a combination of temperature change and glacial growth and decay. We hypothesize that the magnitudes of these high-frequency sea level changes were considerably less in the earlier part of the Cenozoic than those observed over the last million years. This is because considerably less ice was involved.

Although large continental glaciers are not common in Earth history they are known to have been present during a number of extended periods ('ice house' climate, in contrast to 'greenhouse' or warm climate conditions). Ample evidence of glaciation is found in the continental sedimentary record. In particular, there is evidence of glaciation from about 2.7 to 2.1 billion years ago. Additionally, a long period of glaciation occurred shortly before the first fossils of multicellular organisms, from about 1

billion to 540 million years ago. Some scientists now believe that during this glaciation, the entire Earth froze over, generating a 'snowball earth'. Such conditions would have caused a large sea level fall. Evidence of large continental glaciers are also seen in Ordovician to Silurian rocks (~420 to 450 Ma), in Devonian rocks (~380 to 390 Ma), and in Carboniferous to Permian rocks (~350 to 270 Ma).

If these glaciations were caused by similar mechanisms to those envisioned for the Plio-Pleistocene (**Figure 1B**), then predictable, high-frequency, periodic growth and retreat of the glaciers should be observed in strata which form the geologic record. This is certainly the case for the Carboniferous through Permian glaciation. In the central United States, UK, and Europe, the sedimentary rocks have a distinctly cyclic character. They cycle in repetitive vertical successions of marine deposits, near-shore deposits, often including coals, into fluvial sedimentary rocks. The deposition of marine rocks over large areas, which had only recently been nonmarine, suggests very large-scale sea level changes. When the duration of the entire record is taken into account, periodicities of about 100 and 400 thousand years are suggested for these large sea level changes. This is consistent with an origin due to a response to changes in the eccentricity of the Earth's orbit. Higher-frequency cyclicity associated with the tilt of the spin axis and precession of the equinox is more difficult to prove, but has been suggested by detailed observations.

It is fair to say that large-scale (10 to >100 m), relatively high-frequency (20 000–400 000 years; often termed 'Milankovitch scale') variations in sea level occurred during intervals of time when continental glaciers were present on Earth (ice house climate). This indicates that the variations of Earth's orbit and the tilt of its spin axis played a major role in controlling the climate. During the rest of Earth history, when glaciation was not a dominant climatic force (greenhouse climate), sea level changes corresponding to Earth's orbit did occur. In this case, the mechanism for changing the volume of water in the ocean basins is much less clear.

There is no geological record of continental ice sheets in many portions of Earth history. These time periods are generally called 'greenhouse' climates. However, there is ample evidence of Milankovitch scale variations during these periods. In shallow marine sediments, evidence of orbitally driven sea level changes has been observed in Cambrian and Cretaceous age sediments. The magnitudes of sea level change required (perhaps 5–20 m) are far less than have been observed during glacial climates. A possible source for these variations could be variations in average ocean-water temperature. Water expands as it is heated. If ocean bottom-water sources were equatorial rather than polar, as they are today, bottom-water temperatures of about 2°C today might have been about 16°C in the past. This would generate a sea level change of about 11 m. Other causes of sea level change during greenhouse periods have been postulated to be a result of variations in the magnitude of water trapped in inland lakes and seas, and variations in volumes of alpine glaciers. Deep marine sediments of Cretaceous age also show fluctuations between oxygenated and anoxic conditions. It is possible that these variations were generated when global sea level change restricted flow from the rest of the world's ocean to a young ocean basin. In a more recent example, tectonics caused a restriction at the Straits of Gibraltar. In that case, evaporation generated extreme sea level changes and restricted their entrance into the Mediterranean region.

Sea Level Change due to Changing Volume of the Ocean Basin

Tectonics is thought to be the main driving force of long-term (≥ 50 million years) sea level change. Plate tectonics changes the shape and/or the areal extent of the ocean basins.

Plate tectonics is constantly reshaping surface features of the Earth while the amount of water present has been stable for about the last four billion years. The reshaping changes the total area taken up by oceans over time. When a supercontinent forms, subduction of one continent beneath another decreases Earth's ratio of continental to oceanic area, generating a sea level fall. In a current example, the continental plate including India is diving under Asia to generate the Tibetan Plateau and the Himalayan Mountains. This has probably generated a sea level fall of about 70 m over the last 50 million years. The process of continental breakup has the opposite effect. The continents are stretched, generating passive margins and increasing the ratio of continental to oceanic area on a global scale (**Figure 2A**). This results in a sea level rise. Increments of sea level rise resulting from continental breakup over the last 200 million years amount to about 100 m of sea level rise.

Some bathymetric features within the oceans are large enough to generate significant changes in sea level as they change size and shape. The largest physiographic feature on Earth is the mid-ocean ridge system, with a length of about 60 000 km and a width of 500–2000 km. New ocean crust and lithosphere are generated along rifts in the center of these ridges. The ocean crust is increasingly old, cold, and dense away from the rift. It is the heat of ocean lithosphere formation that actually generates this feature. Thus, rifting of continents forms new ridges, increasing the proportionate area of young, shallow, ocean floor to older, deeper ocean floor (**Figure 2B**). Additionally, the width of the ridge is a function of the rates at which the plates are moving apart. Fast spreading ridges (e.g. the East Pacific Rise) are very broad while slow spreading ridges (e.g. the North Atlantic Ridge) are quite narrow. If the average spreading rates for all ridges decreases, the average volume taken up by ocean ridges would decrease. In this case, the volume of the ocean basin available for water would increase and a sea level fall would occur. Finally, entire ridges may be removed in the process of subduction, generating fairly rapid sea level fall.

Scientists have made quantitative estimates of sea level change due to changing ocean ridge volumes. Since ridge volume is dependent on the age of the ocean floor, where the age of the ocean floor is known, ridge volumes can be estimated. Seafloor magnetic anomalies are used to estimate the age of the ocean floor, and thus, spreading histories of the oceans 256. The oldest ocean crust is about 200 million years old. Older oceanic crust has been subducted. Thus, it is not surprising that quantitative estimates of sea level change due to ridge volumes are increasingly uncertain and cannot be calculated

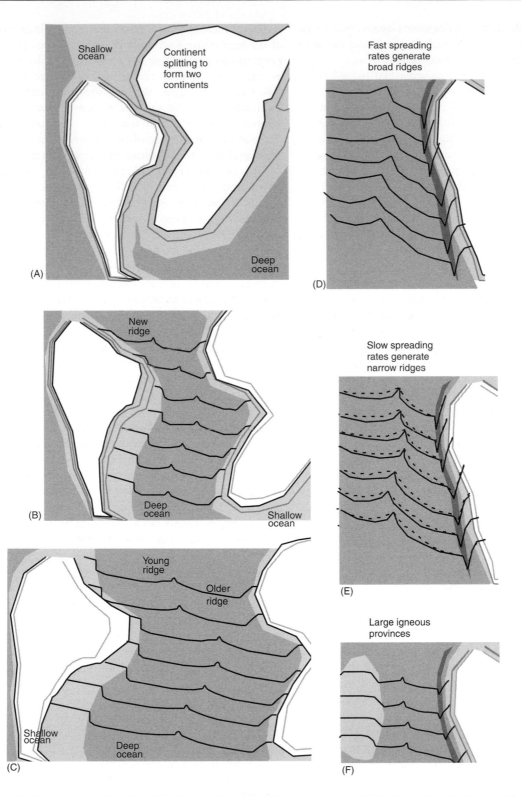

Figure 2   Diagrams showing a few of the factors which affect the ocean volume. (A) Early breakup of a large continent increases the area of continental crust by generating passive margins, causing sea level to rise. (B) Shortly after breakup a new ocean is formed with very young ocean crust. This young crust must be replacing relatively old crust via subduction, generating additional sea level rise. (C) The average age of the ocean between the continents becomes older so that young, shallow ocean crust is replaced with older, deeper crust so that sea level falls. (D) Fast spreading rates are associated with relatively high sea level. (E) Relatively slow spreading ridges (solid lines in ocean) take up less volume in the oceans than high spreading rate ridges (dashed lines in ocean), resulting in relatively low sea level. (F) Emplacement of large igneous provinces generates oceanic plateaus, displaces ocean water, and causes a sea level rise.

before about 90 million years. Sea level is estimated to have fallen about 230 m ($\pm$120 m) due to ridge volume changes in the last 80 million years.

Large igneous provinces (LIPs) are occasionally intruded into the oceans, forming large oceanic plateaus (see Igneous Provinces). The volcanism associated with LIPs tends to occur over a relatively short period of time, causing a rapid sea level rise. However, these features subside slowly as the lithosphere cools, generating a slow increase in ocean volume, and a long-term sea level fall. The largest marine LIP, the Ontong Java Plateau, was emplaced in the Pacific Ocean between about 120 and 115 Ma (**Figure 1D**). Over that interval it may have generated a sea level rise of around 50 m.

In summary, over the last 200 million years, long-term sea level change (**Figure 1D**) can be largely attributed to tectonics. Continental crust expanded by extension as the supercontinents Gondwana and Laurasia split to form the continents we see today. This process began about 200 Ma when North America separated from Africa and continues with the East African Rift system and the formation of the Red Sea. The generation of large oceans occurred early in this period and there was an overall rise in sea level from about 200 to about 90 million years. New continental crust, new mid-ocean ridges, and very fast spreading rates were responsible for the long-term rise (**Figure 1D**). Subsequently, a significant decrease in spreading rates, a reduction in the total length of mid-ocean ridges, and continent–continent collision coupled with an increase in glacial ice (**Figure 1C**) have resulted in a large-scale sea level fall (**Figure 1D**). Late Cretaceous volcanism associated with the Ontong Java Plateau, a large igneous province (see Igneous Provinces), generated a significant sea level rise, while subsequent cooling has enhanced the 90 million year sea level fall. Estimates of sea level change from changing ocean shape remain quite uncertain. Magnitudes and timing of stretching associated with continental breakup, estimates of shortening during continental assembly, volumes of large igneous provinces, and volumes of mid-ocean ridges improve as data are gathered. However, the exact configuration of past continents and oceans can only be a mystery due to the recycling character of plate tectonics.

Sea Level Change Estimated from Observations on the Continents

Long-term Sea Level Change

Estimates of sea level change are also made from sedimentary strata deposited on the continents. This is actually an excellent place to obtain observations of sea level change not only because past sea level has been much higher than it is now, but also because in many places the continents have subsequently uplifted. That is, in the past they were below sea level, but now they are well above it. For example, studies of 500–400 million year old sedimentary rocks which are now uplifted in the Rocky Mountains and the Appalachian Mountains indicate that there was a rise and fall of sea level with an estimated magnitude of 200–400 m. This example also exemplifies the main problem with using the continental sedimentary record to estimate sea level change. The continents are not fixed and move vertically in response to tectonic driving forces. Thus, any indicator of sea level change on the continents is an indicator of relative sea level change. Obtaining a global signature from these observations remains extremely problematic. Additionally, the continental sedimentary record contains long periods of non-deposition, which results in a spotty record of Earth history. Nonetheless a great deal of information about sea level change has been obtained and is summarized here.

The most straightforward source of information about past sea level change is the location of the strand line (the beach) on a stable continental craton (a part of the continent, which was not involved in local tectonics). Ideally, its present height is that of sea level at the time of deposition. There are two problems encountered with this approach. Unfortunately, the nature of land–ocean interaction at their point of contact is such that those sediments are rarely preserved. Where they can be observed, there is considerable controversy over which elements have moved, the continents or sea level. However, data from the past 100 million years tend to be consistent with calculations derived from estimates of ocean volume change. This is not saying a lot since uncertainties are very large (see above).

Continental hypsography (cumulative area versus height) coupled with the areal extent of preserved marine sediments has been used to estimate past sea level. In this case only an average result can be obtained, because marine sediments spanning a time interval (generally 5–10 million years) have been used. Again, uncertainties are large, but results are consistent with calculations derived from estimates of ocean volume change.

Backstripping is an analytical tool, which has been used to estimate sea level change. In this technique, the vertical succession of sedimentary layers is progressively decompacted and unloaded (**Figure 3A**). The resulting hole is a combination of the subsidence generated by tectonics and by sea level change

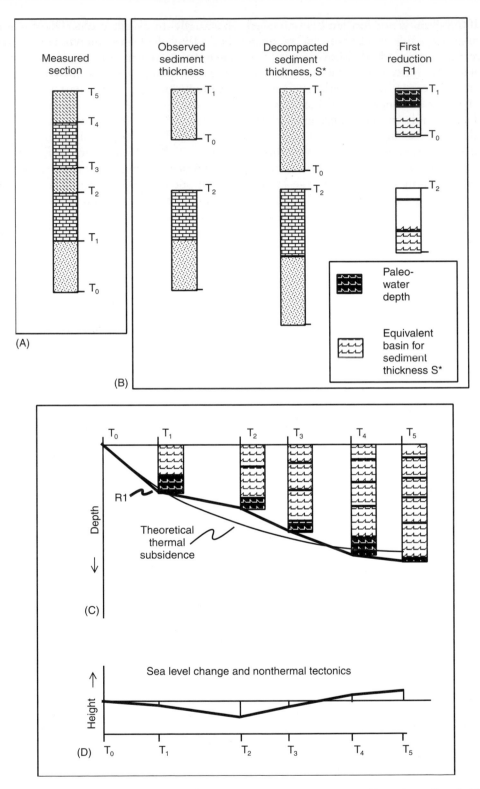

Figure 3 Diagrams depicting the backstripping method for obtaining sea level estimates in a thermally subsiding basin. (A) A stratigraphic section is measured either from exposed sedimentary rocks or from drilling. These data include lithologies, ages, and porosity. Note that the oldest strata are always at the base of the section (T0). (B) Porosity data are used to estimate the thickness that each sediment section would have had at the time of deposition (S*). They are also used to obtain sediment density so that the sediments can be unloaded to determine how deep the basin would have been in the absence of the sediment load (R1). This calculation also requires an estimate of the paleo-water depth (the water depth at the time of deposition). (C) A plot of R1 versus time is compared (by least-squares fit) to theoretical tectonic subsidence in a thermal setting. (D) The difference between R1 and thermal subsidence yields a quantitative estimate of sea level change if other, nonthermal tectonics, did not occur at this location.

(Figure 3B). If the tectonic portion can be established then an estimate of sea level change can be determined (Figure 3C). This method is generally used in basins generated by the cooling of a thermal anomaly (e.g. passive margins). In these basins, the tectonic signature is predictable (exponential decay) and can be calibrated to the well-known subsidence of the mid-ocean ridge.

The backstripping method has been applied to sedimentary strata drilled from passive continental margins of both the east coast of North America and the west coast of Africa. Again, estimates of sea level suggest a rise of about 100–300 m from about 200–110 Ma followed by a fall to the present level (Figure 1D). Young interior basins, such as the Paris

Basin, yield similar results. Older, thermally driven basins have also been analyzed. This was the method used to determine the (approximately 200 m) sea level rise and fall associated with the breakup of a Pre-Cambrian supercontinent in earliest Phanerozoic time.

Million Year Scale Sea Level Change

In addition to long-term changes in sea level there is evidence of fluctuations that are considerably shorter than the 50–100 million year variations discussed above, but longer than those caused by orbital variations (≤ 0.4 million years). These variations appear to be dominated by durations which last either tens

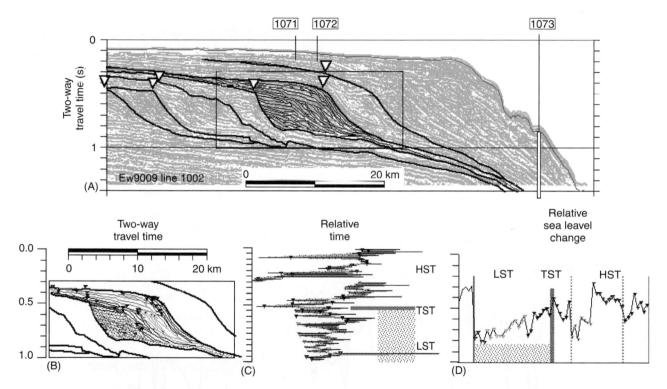

Figure 4 Example of the sequence stratigraphic approach to estimates of sea level change. (A) Multichannel seismic data (gray) from the Baltimore Canyon Trough, offshore New Jersey, USA (Miller *et al.*, 1998). Black lines are interpretations traced on the seismic data. Thick dark lines indicate third-order Miocene-aged (5–23 Ma) sequence boundaries. They are identified by truncation of the finer black lines. Upside-down deltas indicate a significant break in slope associated with each identified sequence boundary. Labeled vertical lines (1071–1073) show the locations of Ocean Drilling Project wells, used to help date the sequences. The rectangle in the center is analyzed in greater detail. (B) Detailed interpretation of a single third-order sequence from (A). Upside-down deltas indicate a significant break in slope associated with each of the detailed sediment packages. Stippled fill indicates the low stand systems tract (LST) associated with this sequence. The gray packages are the transgressive systems tract (TST), and the overlying sediments are the high stand systems tract (HST). (C) Relationships between detailed sediment packages (in B) are used to establish a chronostratigraphy (time framework). Youngest sediment is at the top. Each observed seismic reflection is interpreted as a time horizon, and each is assigned equal duration. Horizontal distance is the same as in (A) and (B). A change in sediment type is indicated at the break in slope from coarser near-shore sediments (stippled pattern) to finer, offshore sediments (parallel, sloping lines). Sedimentation may be present offshore but at very low rates. LST, TST, and HST as in (B). (D) Relative sea level change is obtained by assuming a consistent depth relation at the change in slope indicated in (B). Age control is from the chronostratigraphy indicated in (C). Time gets younger to the right. The vertical scale is in two-way travel time, and would require conversion to depth for a final estimate of the magnitude of sea level change. LST, TST, and HST as in (B). Note that in (B), (C), and (D), higher frequency cycles (probably fourth-order) are present within this (third-order) sequence. Tracing and interpretations are from the author's graduate level quantitative stratigraphy class project (1998, Western Michigan University).

of millions of years or a half to three million years. These sea level variations are sometimes termed second- and third-order sea level change, respectively. There is considerable debate concerning the source of these sea level fluctuations. They have been attributed to tectonics and changing ocean basin volumes, to the growth and decay of glaciers, or to continental uplift and subsidence, which is independent of global sea level change. As noted above, the tectonic record of subsidence and uplift is intertwined with the stratigraphic record of global sea level change on the continents. Synchronicity of observations of sea level change on a global scale would lead most geoscientists to suggest that these signals were caused by global sea level change. However, at present, it is nearly impossible to globally determine the age equivalency of events which occur during intervals as short as a half to two million years. These data limitations are the main reason for the heated controversy over third-order sea level.

Quantitative estimates of second-order sea level variations are equally difficult to obtain. Although the debate is not as heated, these somewhat longer-term variations are not much larger than the third-order variations so that the interference of the two signals makes definition of the beginning, ending and/or magnitude of second-order sea level change

equally problematic. Recognizing that our understanding of second- and third-order (million year scale) sea level fluctuations is limited, a brief review of that limited knowledge follows.

Sequence stratigraphy is an analytical method of interpreting sedimentary strata that has been used to investigate second- and third-order relative sea level changes. This paradigm requires a vertical succession of sedimentary strata which is analyzed in at least a two-dimensional, basinal setting. Packages of sedimentary strata, separated by unconformities, are observed and interpreted mainly in terms of their internal geometries (e.g. **Figure 4**). The unconformities are assumed to have been generated by relative sea level fall, and thus, reflect either global sea level or local or regional tectonics. This method of stratigraphic analysis has been instrumental in hydrocarbon exploration since its introduction in the late 1970s. One of the bulwarks of this approach is the 'global sea level curve' most recently published by Haq *et al.* (1987). This curve is a compilation of relative sea level curves generated from sequence stratigraphic analysis in basins around the world. While sequence stratigraphy is capable of estimating relative heights of relative sea level, it does not estimate absolute magnitudes. Absolute dating requires isotope data or correlation via fossil data into the

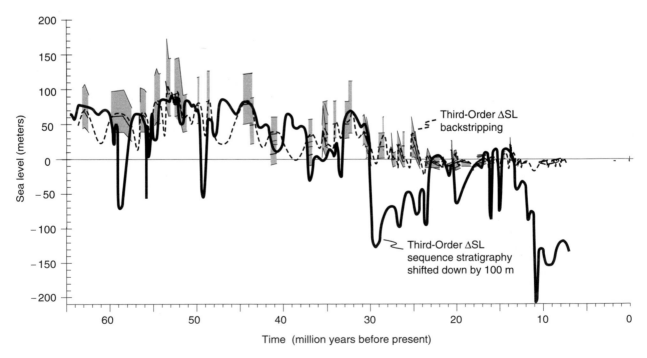

Figure 5 Million year scale sea level fluctuations. Estimates from sequence stratigraphy (Haq *et al.*, 1987; solid curve) have been shifted down by 100 m to allow comparison with estimates of sea level from backstripping (Kominz *et al.*, 1998; dashed curve). Where sediments are present, the backstripping results, with uncertainty ranges, are indicated by gray fill. Between backstrip observations, lack of preserved sediment is presumed to have been a result of sea level fall. The Berggren *et al.* (1995) biostratigraphic timescale was used.

chronostratigraphic timescale. However, the two-dimensional nature of the data allows for good to excellent relative age control.

Backstripping has been used, on a considerably more limited basis, in an attempt to determine million year scale sea level change. This approach is rarely applied because it requires very detailed, quantitative, estimates of sediment ages, paleo-environments and compaction in a thermal tectonic setting. A promising area of research is the application of this method to coastal plain boreholes from the mid-Atlantic coast of North America. Here an intensive Ocean Drilling Project survey is underway which is providing sufficiently detailed data for this type of analysis. Initial results suggest that magnitudes of million year scale sea level change are roughly one-half to one-third that reported by Haq *et al*. However, in glacial times, the timing of the cycles was quite consistent with those of this 'global sea level curve' derived by application of sequence stratigraphy (**Figure 5**). Thus, it seems reasonable to conclude that, at least during glacial times, global, third-order sea level changes did occur.

Summary

Sea level changes are either a response to changing ocean volume or to changes in the volume of water contained in the ocean. The timing of sea level change ranges from tens of thousands of years to over 100 million years. Magnitudes also vary significantly but may have been as great as 200 m or more. Estimates of sea level change currently suffer from significant ranges of uncertainty, both in magnitude and in timing. However, scientists are converging on consistent estimates of sea level changes by using very different data and analytical approaches.

Further Reading

Allen PA and Allen JR (1990) *Basin Analysis: Principless, Applications*. Oxford: Blackwell Scientific Publications.

Berggren WA, Kent DV, Swisher CC, and Aubry MP (1995) A revised Cenozoic geochronology and chronostratigraphy. In: Berggren WA, Kent DV, and Hardenbol J (eds.) *Geochronology, Time Scales and Global Stratigraphic Correlations: A Unified Temporal Framework for an Historical Geology,* SEPM Special Publication No. 54, 131–212.

Bond GC (1979) Evidence of some uplifts of large magnitude in continental platforms. *Tectonophysics* 61: 285–305.

Coffin MF and Eldholm O (1994) Large igneous provinces: crustal structure, dimensions, and external consequences. *Reviews of Geophysics* 32: 1–36.

Crowley TJ and North GR (1991) *Paleoclimatology.* Oxford Monographs on Geology and Geophysics, no. 18

Fairbanks RG (1989) A 17,000-year glacio-eustatic sea level record: influence of glacial melting rates on the Younger Dryas event and deep-ocean circulation. *Nature* 6250: 637–642.

Hallam A (1992) *Phanerozoic Sea-Level Changes*. NY: Columbia University Press.

Haq BU, Hardenbol J, and Vail PR (1987) Chronology of fluctuating sea levels since the Triassic (250 million years ago to present). *Science* 235: 1156–1167.

Harrison CGA (1990) Long term eustasy and epeirogeny in continents. In: Sea-Level Change, pp. 141–158. Washington, DC: US National Research Council Geophysics Study Committee.

Hauffman PF and Schrag DP (2000) Snowball Earth. *Scientific American* 282: 68–75.

Kominz MA, Miller KG, and Browning JV (1998) Long-term and short term global Cenozoic sea-level estimates. *Geology* 26: 311–314.

Miall AD (1997) *The Geology of Stratigraphic Sequences.* Berlin: Springer-Verlag.

Miller KG, Fairbanks RG, and Mountain GS (1987) Tertiary oxygen isotope synthesis, sea level history, and continental margin erosion. *Paleoceanography* 2: 1–19.

Miller KG, Mountain GS, Browning J, *et al*. (1998) Cenozoic global sea level, sequences, and the New Jersey transect; results from coastal plain and continental slope drilling. *Reviews of Geophysics* 36: 569–601.

Sahagian DL (1988) Ocean temperature-induced change in lithospheric thermal structure: a mechanism for long-term eustatic sea level change. *Journal of Geology* 96: 254–261.

Wilgus CK, Hastings BS, Kendall CG St C *et al*. (1988) *Sea Level Changes: An Integrated Approach*. Special Publication no. 42. Society of Economic Paleontologists and Mineralogists.

GLACIAL CRUSTAL REBOUND, SEA LEVELS, AND SHORELINES

K. Lambeck, Australian National University, Canberra, ACT, Australia

Introduction

Geological, geomorphological, and instrumental records point to a complex and changing relation between land and sea surfaces. Elevated coral reefs or wave-cut rock platforms indicate that in some localities sea levels have been higher in the past, while observations elsewhere of submerged forests or flooded sites of human occupation attest to levels having been lower. Such observations are indicators of the relative change in the land and sea levels: raised shorelines are indicative of land having been uplifted or of the ocean volume having decreased, while submerged shorelines are a consequence of land subsidence or of an increase in ocean volume. A major scientific goal of sea-level studies is to separate out these two effects.

A number of factors contribute to the instability of the land surfaces, including the tectonic upheavals of the crust emanating from the Earth's interior and the planet's inability to support large surface loads of ice or sediments without undergoing deformation. Factors contributing to the ocean volume changes include the removal or addition of water to the oceans as ice sheets wax and wane, as well as addition of water into the oceans from the Earth's interior through volcanic activity. These various processes operate over a range of timescales and sea level fluctuations can be expected to fluctuate over geological time and are recorded as doing so.

The study of such fluctuations is more than a scientific curiosity because its outcome impacts on a number of areas of research. Modern sea level change, for example, must be seen against this background of geologically–climatologically driven change before contributions arising from the actions of man can be securely evaluated. In geophysics, one outcome of the sea level analyses is an estimate of the viscosity of the Earth, a physical property that is essential in any quantification of internal convection and thermal processes. Glaciological modeling of the behavior of large ice sheets during the last cold

period is critically dependent on independent constraints on ice volume, and this can be extracted from the sea level information. Finally, as sea level rises and falls, so the shorelines advance and retreat. As major sea level changes have occurred during critical periods of human development, reconstructions of coastal regions are an important part in assessing the impact of changing sea levels on human movements and settlement.

Tectonics and Sea Level Change

Major causes of land movements are the tectonic processes that have shaped the planet's surface over geological time. Convection within the high-temperature viscous interior of the Earth results in stresses being generated in the upper, cold, and relatively rigid zone known as the lithosphere, a layer some 50–100 km thick that includes the crust. This convection drives plate tectonics — the movement of large parts of the lithosphere over the Earth's surface — mountain building, volcanism, and earthquakes, all with concomitant vertical displacements of the crust and hence relative sea level changes. The geological record indicates that these processes have been occurring throughout much of the planet's history. In the Andes, for example, Charles Darwin identified fossil seashells and petrified pine trees trapped in marine sediments at 4000 m elevation. In Papua New Guinea, 120 000-year-old coral reefs occur at elevations of up to 400 m above present sea level (**Figure 1**).

One of the consequences of the global tectonic events is that the ocean basins are being continually reshaped as mid-ocean ridges form or as ocean floor collides with continents. The associated sea level changes are global but their timescale is long, of the order 10^7 to 10^8 years, and the rates are small, less than 0.01 mm per year. **Figure 2** illustrates the global sea level curve inferred for the past 600 million years from sediment records on continental margins. The long-term trends of rising and falling sea levels on timescales of 50–100 million years are attributed to these major changes in the ocean basin configurations. Superimposed on this are smaller-amplitude and shorter-period oscillations that reflect more regional processes such as large-scale volcanism in an ocean environment or the collision of continents. More locally, land is pushed up or down episodically

Figure 1 Raised coral reefs from the Huon Peninsula, Papua New Guinea. In this section the highest reef indicated (point 1) is about 340 m above sea level and is dated at about 125 000 years old. Elsewhere this reef attains more than 400 m elevation. The top of the present sea cliffs (point 2) is about 7000 years old and lies at about 20 m above sea level. The intermediate reef tops formed at times when the rate of tectonic uplift was about equal to the rate of sea level rise, so that prolonged periods of reef growth were possible. Photograph by Y. Ota.

in response to the deeper processes. The associated vertical crustal displacements are rapid, resulting in sea level rises or falls that may attain a few meters in amplitude, but which are followed by much longer periods of inactivity or even a relaxation of the original displacements. The raised reefs illustrated in **Figure 1**, for example, are the result of a large number of episodic uplift events each of typically a meter amplitude. Such displacements are mostly local phenomena, the Papua New Guinea example extending only for some 100–150 km of coastline.

The episodic but local tectonic causes of the changing position between land and sea can usually be identified as such because of the associated seismic activity and other tell-tale geological signatures. The development of geophysical models to describe these local vertical movements is still in a state of infancy

and, in any discussion of global changes in sea level, information from such localities is best set aside in favor of observations from more stable environments.

Glacial Cycles and Sea Level Change

More important for understanding sea level change on human timescales than the tectonic contributions – important in terms of rates of change and in terms of their globality – is the change in ocean volume driven by cyclic global changes in climate from glacial to interglacial conditions. In Quaternary time, about the last two million years, glacial and interglacial conditions have followed each other on timescales of the order of 10^4–10^5 years. During interglacials, climate conditions were similar to those of today and sea levels were within a few meters of their present day position. During the major glacials, such as 20 000 years ago, large ice sheets formed in the northern hemisphere and the Antarctic ice sheet expanded, taking enough water out of the oceans to lower sea levels by between 100 and 150 m. **Figure 3** illustrates the changes in global sea level over the last 130 000 years, from the last interglacial, the last time that conditions were similar to those of today, through the Last Glacial Maximum and to the present. At the end of the last interglacial, at 120 000 years ago, climate began to fluctuate; increasingly colder conditions were reached, ice sheets over North America and Europe became more or less permanent features of the landscape, sea levels reached progressively lower values, and large parts of today's coastal shelves were exposed. Soon after the culmination of maximum glaciation, the ice sheets again disappeared, within a period of about

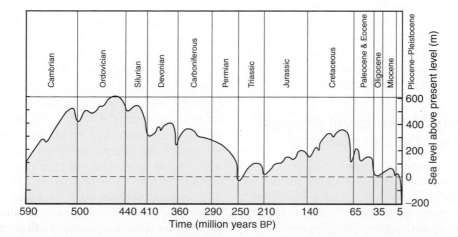

Figure 2 Global sea level variations through the last 600 million years estimated from seismic stratigraphic studies of sediments deposited at continental margins. Redrawn with permission from Hallam A (1984) Pre-Quaternary sea-level changes. *Annual Review of Earth and Planetary Science* 12: 205–243.

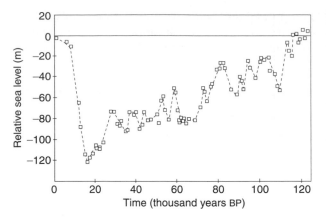

Figure 3 Global sea level variations (relative to present) since the time of the last interglacial 120 000–130 000 years ago when climate and environmental conditions were last similar to those of the last few thousand years. Redrawn with permission from Chapell J *et al.* (1996) Reconciliation of Late Quaternary sea level changes derived from coral terraces at Huon Peninsula with deep sea oxygen isotope records. *Earth and Planetary Science Letters* 141b: 227–236.

10 000 years, and climate returned to interglacial conditions.

The global changes illustrated in **Figure 3** are only one part of the sea level signal because the actual ice–water mass exchange does not give rise to a spatially uniform response. Under a growing ice sheet, the Earth is stressed; the load stresses are transmitted through the lithosphere to the viscous underlying mantle, which begins to flow away from the stressed area and the crust subsides beneath the ice load. When the ice melts, the crust rebounds. Also, the meltwater added to the oceans loads the seafloor and the additional stresses are transmitted to the mantle, where they tend to dissipate by driving flow to unstressed regions below the continents. Hence the seafloor subsides while the interiors of the continents rise, causing a tilting of the continental margins. The combined adjustments to the changing ice and water loads are called the glacio- and hydro-isostatic effects and together they result in a complex pattern of spatial sea level change each time ice sheets wax and wane.

Observations of Sea Level Change Since the Last Glacial Maximum

Evidence for the positions of past shorelines occurs in many forms. Submerged freshwater peats and tree stumps, tidal-dwelling mollusks, and archaeological sites would all point to a rise in relative sea level since the time of growth, deposition, or construction. Raised coral reefs, such as in **Figure 1**, whale bones cemented in beach deposits, wave-cut rock platforms and notches, or peats formed from saline-loving

plants would all be indicative of a falling sea level since the time of formation. To obtain useful sea level measurements from these data requires several steps: an understanding of the relationship between the feature's elevation and mean sea level at the time of growth or deposition, a measurement of the height or depth with respect to present sea level, and a measurement of the age. All aspects of the observation present their own peculiar problems but over recent years a substantial body of observational evidence has been built up for sea level change since the last glaciation that ended at about 20 000 years ago. Some of this evidence is illustrated in **Figure 4**, which also indicates the very significant spatial variability that may occur even when, as is the case here, the evidence is from sites that are believed to be tectonically stable.

In areas of former major glaciation, raised shorelines occur with ages that are progressively greater with increasing elevation and with the oldest shorelines corresponding to the time when the region first became ice-free. Two examples, from northern Sweden and Hudson Bay in Canada, respectively, are illustrated in **Figure 4A**. In both cases sea level has been falling from the time the area became ice-free and open to the sea. This occurred at about 9000 years ago in the former case and about 7000 years later in the second case. Sea level curves from localities just within the margins of the former ice sheet are illustrated in **Figure 4B**, from western Norway and Scotland, respectively. Here, immediately after the area became ice-free, the sea level fell but a time was reached when it again rose. A local maximum was reached at about 6000 years ago, after which the fall continued up until the present. Farther away from the former ice margins the observed sea level pattern changes dramatically, as is illustrated in **Figure 4C**. Here the sea level initially rose rapidly but the rate decreased for the last 7000 years up to the present. The two examples illustrated, from southern England and the Atlantic coast of the United States, are representative of tectonically stable localities that lie within a few thousand kilometers from the former centers of glaciation. Much farther away again from the former ice sheets the sea level signal undergoes a further small but significant change in that the present level was first reached at about 6000 years ago and then exceeded by a small amount before returning to its present value. The two examples illustrated in **Figure 4D** are from nearby localities in northern Australia. At both sites present sea level was reached about 6000 years ago after a prolonged period of rapid rise with resulting highstands that are small in amplitude but geographically variable.

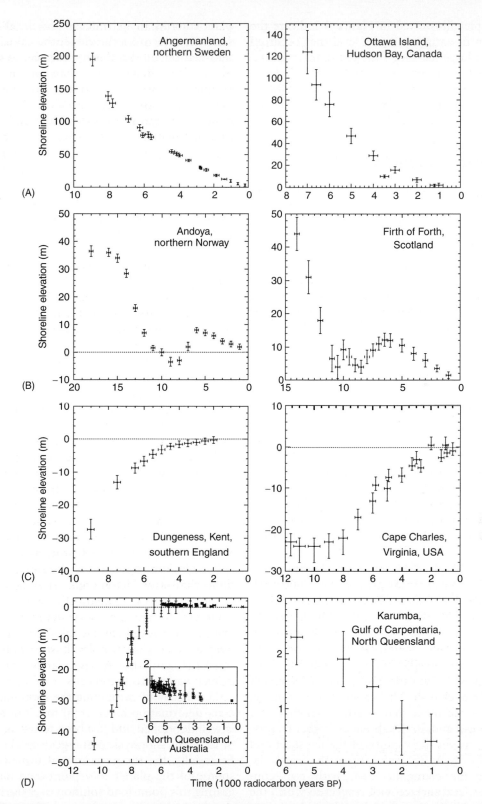

Figure 4 Characteristic sea level curves observed in different localities (note the different scales used). (A) From Ångermanland, Gulf of Bothnia, Sweden, and from Hudson Bay, Canada. Both sites lie close to centers of former ice sheet, over northern Europe and North America respectively. (B) From the Atlantic coast of Norway and the west coast of Scotland. These sites are close to former ice-sheet margins at the time of the last glaciation. (C) From southern England and Virginia, USA. These sites lie outside the areas of former glaciation and at a distance from the margins where the crust is subsiding in response to the melting of the nearby icesheet. (D) Two sites in northern Australia, one for the Coral Sea coast of Northern Queensland and the second from the Gulf of Carpentaria some 200 km away but on the other side of the Cape York Peninsula. At both localities sea level rose until about 6000 years ago before peaking above present level.

The examples illustrated in **Figure 4** indicate the rich spectrum of variation in sea level that occurred when the last large ice sheets melted. Earlier glacial cycles resulted in a similar spatial variability, but much of the record before the Last Glacial Maximum has been overwritten by the effects of the last deglaciation.

Glacio-Hydro-Isostatic Models

If, during the decay of the ice sheets, the meltwater volume was distributed uniformly over the oceans, then the sea level change at time t would be

$$\Delta\zeta_e(t) = (\text{change in ice volume/ocean surface area})$$
$$\times \rho_i/\rho_w \qquad [1a]$$

where ρ_i and ρ_w are the densities of ice and water, respectively. (See end of article – symbols used.) This term is the ice-volume equivalent sea level and it provides a measure of the change in ice volume through time. Because the ocean area changes with time a more precise definition is

$$\Delta\zeta_e(t) = \frac{dV_i}{dt} dt \frac{\rho_i}{\rho_0} \int_t \frac{\rho_i}{A_0(t)} \qquad [1b]$$

where A_0 is the area of the ocean surface, excluding areas covered by grounded ice. The sea level curve illustrated in **Figure 3** is essentially this function. However, it represents only a zero-order approximation of the actual sea level change because of the changing gravitational field and deformation of the Earth.

In the absence of winds or ocean currents, the ocean surface is of constant gravitational potential. A planet of a defined mass distribution has a family of such surfaces outside it, one of which – the geoid – corresponds to mean sea level. If the mass·distribution on the surface (the ice and water) or in the interior (the load-forced mass redistribution in the mantle) changes, so will the gravity field, the geoid, and the sea level change. The ice sheet, for example, represents a large mass that exerts a gravitational pull on the ocean and, in the absence of other factors, sea level will rise in the vicinity of the ice sheet and fall farther away. At the same time, the Earth deforms under the changing load, with two consequences: the land surface with respect to which the level of the sea is measured is time-dependent, as is the gravitational attraction of the solid Earth and hence the geoid.

The calculation of the change of sea level resulting from the growth or decay of ice sheets therefore involves three steps: the calculation of the amount of water entering into the ocean and the distribution of this meltwater over the globe; the calculation of the deformation of the Earth's surface; and the calculation of the change in the shape of the gravitational equipotential surfaces. In the absence of vertical tectonic motions of the Earth's surface, the relative sea level change $\Delta\zeta(\varphi, t)$ at a site φ and time t can be written schematically as

$$\Delta\zeta(\varphi, t) = \Delta\zeta_e(t) + \Delta\zeta_I(\varphi, t) \qquad [2]$$

where $\Delta\zeta_I(\varphi, t)$ is the combined perturbation from the uniform sea level rise term [1]. This is referred to as the isostatic contribution to relative sea level change.

In a first approximation, the Earth's response to a global force is that of an elastic outer spherical layer (the lithosphere) overlying a viscous or viscoelastic mantle that itself contains a fluid core. When subjected to an external force (e.g., gravity) or a surface load (e.g., an ice cap), the planet experiences an instantaneous elastic deformation followed by a time-dependent or viscous response with a characteristic timescale(s) that is a function of the viscosity. Such behavior of the Earth is well documented by other geophysical observations: the gravitational attraction of the Sun and Moon raises tides in the solid Earth; ocean tides load the seafloor with a time-dependent water load to which the Earth's surface responds by further deformation; atmospheric pressure fluctuations over the continents induce deformations in the solid Earth. The displacements, measured with precision scientific instruments, have both an elastic and a viscous component, with the latter becoming increasingly important as the duration of the load or force increases. These loads are much smaller than the ice and water loads associated with the major deglaciation, the half-daily ocean tide amplitudes being only 1% of the glacial sea level change, and they indicate that the Earth will respond to even small changes in the ice sheets and to small additions of meltwater into the oceans.

The theory underpinning the formulation of planetary deformation by external forces or surface loads is well developed and has been tested against a range of different geophysical and geological observations. Essentially, the theory is one of formulating the response of the planet to a point load and then integrating this point-load solution over the load through time, calculating at each epoch the surface deformation and the shape of the equipotential surfaces, making sure that the meltwater is appropriately distributed into the oceans and that the total ice–water mass is preserved. Physical inputs into the formulation are the time–space history of the ice sheets, a description of the ocean basins from which the water is extracted or into which it is added, and a description of the rheology, or response parameters, of the Earth. For the last requirement, the elastic properties of the Earth, as well as the density distribution with depth, have been determined from seismological and geodetic studies. Less well determined are the viscous properties of the mantle, and the usual procedure is to adopt a simple parametrization of the viscosity structure and to estimate the relevant parameters from analyses of the sea level change itself.

The formulation is conveniently separated into two parts for schematic reasons: the glacio-isostatic effect representing the crustal and geoid displacements due to the ice load, and the hydro-isostatic effect due to the water load. Thus

$$\Delta\zeta(\varphi,t) = \Delta\zeta_e(t) + \Delta\zeta_i(\varphi,t) + \Delta\zeta_w(\varphi,t) \qquad [3]$$

where $\Delta\zeta_i(\varphi,t)$ is the glacio-isostatic and $\Delta\zeta_w(\varphi,t)$ the hydro-isostatic contribution to sea level change. (In reality the two are coupled and this is included in the formulation). If $\Delta\zeta_i(\varphi,t)$ is evaluated everywhere, the past water depths and land elevations $H(\varphi,t)$ measured with respect to coeval sea level are given by

$$H(\varphi,t) H_0(\varphi) - \Delta\zeta(\varphi,t) \qquad [4]$$

where $H(\phi)$ is the present water depth or land elevation at location φ.

A frequently encountered concept is eustatic sea level, which is the globally averaged sea level at any time t. Because of the deformation of the seafloor during and after the deglaciation, the isostatic term $\Delta\zeta(\varphi,t)$ is not zero when averaged over the ocean at any time t, so that the eustatic sea level change is

$$\Delta\zeta_{eus}(t) = \Delta\zeta_e(t) + \langle\Delta\zeta_I\{\varphi,t\}\rangle_0$$

where the second term on the right-hand side denotes the spatially averaged isostatic term. Note that $\Delta\zeta_e(t)$ relates directly to the ice volume, and not $\Delta\zeta_{eus}(t)$.

The Anatomy of the Sea Level Function

The relative importance of the two isostatic terms in eqn. [3] determines the spatial variability in the sea level signal. Consider an ice sheet of radius that is much larger than the lithospheric thickness: the limiting crustal deflection beneath the center of the load is $I\rho_i/I\rho_m$ where I is the maximum ice thickness and $I\rho_m$ is the upper mantle density. This is the local isostatic approximation and it provides a reasonable approximation of the crustal deflection if the loading time is long compared with the relaxation time of the

mantle. Thus for a 3 km thick ice sheet the maximum deflection of the crust can reach 1 km, compared with a typical ice volume for a large ice sheet that raises sea level globally by 50–100 m. Near the centers of the formerly glaciated regions it is the crustal rebound that dominates and sea level falls with respect to the land.

This is indeed observed, as illustrated in **Figure 4A**, and the sea level curve here consists of essentially the sum of two terms, the major glacio-isostatic term and the minor ice-volume equivalent sea level term $\Delta\zeta_e(t)$ (**Figure 5A**). Of note is that the rebound continues long after all ice has disappeared, and this is evidence that the mantle response includes a memory of the earlier load. It is the decay time of this part of the curve that determines the mantle viscosity. As the ice margin is approached, the local ice thickness becomes less and the crustal rebound is reduced and at some stage is equal, but of opposite sign, to $\Delta\zeta_e(t)$. Hence sea level is constant for a period (**Figure 5B**) before rising again when the rebound becomes the minor term. After global melting has ceased, the dominant signal is the late stage of the crustal rebound and levels fall up to the present. Thus the oscillating sea level curves observed in areas such as Norway and Scotland (**Figure 4B**) are essentially the sum of two effects of similar magnitude but opposite sign. The early part of the observation contains information on earth rheology as well as on the local ice thickness and the globally averaged rate of addition of meltwater into the oceans. Furthermore, the secondary maximum is indicative of the timing of the end of global glaciation and the latter part of the record is indicative mainly of the mantle response.

At the sites beyond the ice margins it is the meltwater term $\Delta\zeta_e(t)$ that is important, but it is not the sole factor. When the ice sheet builds up, mantle material flows away from the stressed mantle region and, because flow is confined, the crust around the periphery of the ice sheet is uplifted. When the ice sheet melts, subsidence of the crust occurs in a broad zone peripheral to the original ice sheet and at these locations the isostatic effect is one of an apparent subsidence of the crust. This is illustrated in **Figure 5C**. Thus, when the ocean volumes have stabilized, sea level continues to rise, further indicating that the planet responds viscously to the changing surface loads. The early part of the observational record (e.g., **Figure 4C**) is mostly indicative of the rate at which meltwater is added into the ocean, whereas the latter part is more indicative of mantle viscosity.

In all of the examples considered so far it is the glacial-rebound that dominates the total isostatic adjustment, the hydro-isostatic term being present but comparatively small. Consider an addition of

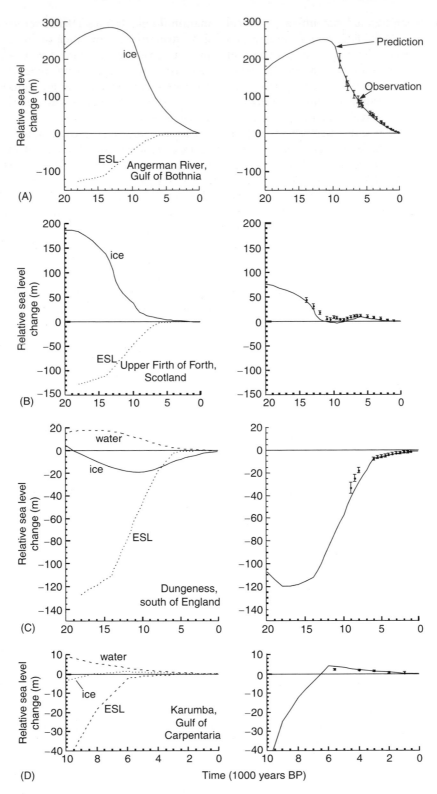

Figure 5 Schematic representation of the sea level curves in terms of the ice-volume equivalent function $\Delta\zeta_e(t)$ (denoted by ESL) and the glacio- and hydro-isostatic contributions $\Delta\zeta_i(\phi, t)$, $\Delta\zeta_w(\phi, t)$ (denoted by ice and water, respectively). The panels on the left indicate the predicted individual components and the panels on the right indicate the total predicted change compared with the observed values (data points). (A) For the sites at the ice center where $\Delta\zeta_i(\phi, t) \gg \Delta\zeta_e(t) \gg \Delta\zeta_w(\phi, t)$. (B) For sites near but within the ice margin where $|\Delta\zeta_i(\phi, t)| \sim |\Delta\zeta_e(t)| \gg |\Delta\zeta_w(\phi, t)|$, but the first two terms are of opposite sign. (C) For sites beyond the ice margin where $|\Delta\zeta_e(t)| > |\Delta\zeta_i(\phi, t)| \gg |\Delta\zeta_w(\phi, t)|$. (D) For sites at continental margins far from the former ice sheets where $|\Delta\zeta_w(\phi, t)| > |\Delta\zeta_i(t)|$. Adapted with permission from Lambeck K and Johnston P (1988). The viscosity of the mantle: evidence from analyses of glacial rebound phenomena. In: Jackson ISN (ed.) *The Earth's Mantle – Composition, Structure and Evolution*, pp. 461–502. Cambridge: Cambridge University Press.

water that raises sea level by an amount D. The local isostatic response to this load is $D\rho_w/\rho_m$, where ρ_w is the density of ocean water. This gives an upper limit to the amount of subsidence of the sea floor of about 30 m for a 100 m sea level rise. This is for the middle of large ocean basins and at the margins the response is about half as great. Thus the hydro-isostatic effect is significant and is the dominant perturbing term at margins far from the former ice sheets. This occurs at the Australian margin, for example, where the sea level signal is essentially determined by $\Delta\zeta_e(t)$ and the water-load response (**Figure 4D**). Up to the end of melting, sea level is dominated by $\Delta\zeta_e(t)$ but thereafter it is determined largely by the water-load term $\Delta\zeta_w(\varphi, t)$, such that small highstands develop at 6000 years. The amplitudes of these highstands turn out to be strongly dependent on the geometry of the water-load distribution around the site: for narrow gulfs, for example, their amplitude increases with distance from the coast and from the water load at rates that are particularly sensitive to the mantle viscosity.

While the examples in **Figure 5** explain the general characteristics of the global spatial variability of the sea level signal, they also indicate how observations of such variability are used to estimate the physical quantities that enter into the schematic model (3). Thus, observations near the center of the ice sheet partially constrain the mantle viscosity and central ice thickness. Observations from the ice sheet margin partially constrain both the viscosity and local ice thickness and establish the time of termination of global melting. Observations far from the ice margins determine the total volumes of ice that melted into the oceans as well as providing further constraints on the mantle response. By selecting data from different localities and time intervals, it is possible to estimate the various parameters that underpin the sea level eqn. [3] and to use these models to predict sea level and shoreline change for unobserved areas. Analyses of sea-level change from different regions of the world lead to estimates for the lithospheric thickness of between 60–100 km, average upper mantle viscosity (from the base of the lithosphere to a depth of 670 km) of about $(1–5)10^{20}$ Pa s and an average lower mantle viscosity of about $(1–5)10^{22}$ Pa s. Some evidence exists that these parameters vary spatially; lithosphere thickness and upper mantle viscosity being lower beneath oceans than beneath continents.

Sea Level Change and Shoreline Migration: Some Examples

Figure 6 illustrates the ice-volume equivalent sea level function $\Delta\zeta_e(t)$ since the time of the last

maximum glaciation. This curve is based on sea level indicators from a number of localities, all far from the former ice sheets, with corrections applied for the isostatic effects. The right-hand axis indicates the corresponding change in ice volume (from the relation [1b]). Much of this ice came from the ice sheets over North America and northern Europe but a not insubstantial part also originated from Antarctica. Much of the melting of these ice sheets occurred within 10 000 years, at times the rise in sea level exceeding 30 mm per year, and by 6000 years ago most of the deglaciation was completed. With this sea level function, individual ice sheet models, the formulation for the isostatic factors, and a knowledge of the topography and bathymetry of the world, it becomes possible to reconstruct the paleo shorelines using the relation [4].

Scandinavia is a well-studied area for glacial rebound and sea level change since the time the ice retreat began about 18 000 years ago. The observational evidence is quite plentiful and a good record of ice margin retreat exists. **Figure 7** illustrates examples for two epochs. The first (**Figure 7A**), at 16 000 years ago, corresponds to a time after onset

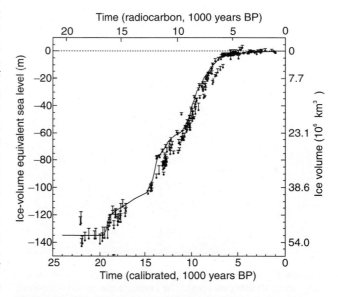

Figure 6 The ice-volume equivalent function $\Delta\zeta_e(t)$ and ice volumes since the time of the last glacial maximum inferred from corals from Barbados, from sediment facies from north-western Australia, and from other sources for the last 7000 years. The actual sea level function lies at the upper limit defined by these observations (continuous line). The upper time scale corresponds to the radiocarbon timescale and the lower one is calibrated to calendar years. Adapted with permission from Fleming K *et al.* (1998) Refining the eustatic sea-level curve since the LGM using the far- and intermediate-field sites. *Earth and Planetary Science Letters* 163: 327–342; and Yokoyama Y *et al.* (2000) Timing of the Last Glacial Maximum from observed sea-level minimum. *Nature* 406: 713–716.

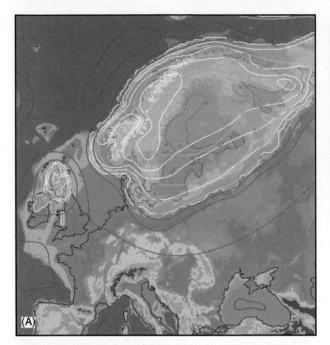

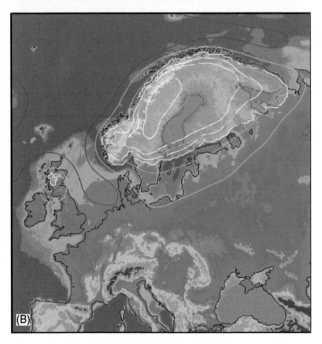

Figure 7 Shoreline reconstructions for Europe (A) at 16 000 and (B) at 10 500 years ago. The contours are at 400 m intervals for the ice thickness (white) and 25 m for the water depths less than 100 m. The orange, yellow, and red contours are the predicted lines of equal sea level change from the specified epoch to the present and indicate where shorelines of these epochs could be expected if conditions permitted their formation and preservation. The zero contour is in yellow; orange contours, at intervals of 100 m, are above present, and red contours, at 50 m intervals, are below present. At 10 500 years ago the Baltic is isolated from the Atlantic and its level lies about 25 m above that of the latter.

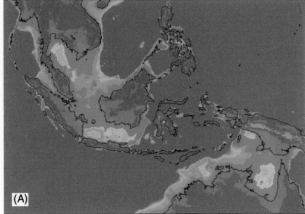

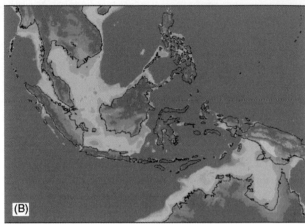

Figure 8 Shoreline reconstructions for South East Asia and northern Australia at the time of the last glacial maximum, (A) at about 18 000 years ago and (B) at 12 000 years ago. The water depth contours are the same as in **Figure 7**. The water depths in the inland lakes at 18 000 years ago are contoured at 25 m intervals relative to their maximum levels that can be attained without overflowing.

of deglaciation. A large ice sheet existed over Scandinavia with a smaller one over the British Isles. Globally sea level was about 110 m lower than now and large parts of the present shallow seas were exposed, for example, the North Sea and the English Channel but also the coastal shelf farther south such as the northern Adriatic Sea. The red and orange contours indicate the sea level change between this period and the present. Beneath the ice these rebound contours are positive, indicating that if shorelines could form here they would be above sea level today. Immediately beyond the ice margin, a broad but shallow bulge develops in the topography, which will subside as the ice sheet retreats. At the second epoch selected (**Figure 7B**) 10 500 years ago, the ice has retreated and reached a temporary halt as the climate briefly returned to colder conditions; the Younger Dryas time of Europe. Much of the Baltic was then ice-free and a freshwater lake developed at some

25–30 m above coeval sea level. The flooding of the North Sea had begun in earnest. By 9000 years ago, most of the ice was gone and shorelines began to approach their present configuration.

The sea level change around the Australian margin has also been examined in some detail and it has been possible to make detailed reconstructions of the shoreline evolution there. **Figure 8** illustrates the reconstructions for northern Australia, the Indonesian islands, and the Malay–IndoChina peninsula. At the time of the Last Glacial Maximum much of the shallow shelves were exposed and deeper depressions within them, such as in the Gulf of Carpentaria or the Gulf of Thailand, would have been isolated from the open sea. Sediments in these depressions will sometimes retain signatures of these pre-marine conditions and such data provide important constraints on the models of sea level change. Part of the information illustrated in **Figure 6**, for example, comes from the shallow depression on the Northwest Shelf of Australia. By 12 000 years ago the sea has begun its encroachment of the shelves and the inland lakes were replaced by shallow seas.

Symbols used

$\Delta\zeta_e(t)$ Ice-volume equivalent sea level. Uniform change in sea level produced by an ice volume that is distributed uniformly over the ocean surface.

$\Delta\zeta_I(t)$ Perturbation in sea level due to glacio-isostatic $\Delta\zeta_i(t)$ and hydro-isostatic $\Delta\zeta_w(t)$ effects.

$\Delta\zeta_{eus}(t)$ Eustatic sea level change. The globally averaged sea level at time t.

See also

Past Climate from Corals. Sea Level Change.

Further Reading

Lambeck K (1988) *Geophysical Geodesy: The Slow Deformations of the Earth*. Oxford: Oxford University Press.

Lambeck K and Johnston P (1999) The viscosity of the mantle: evidence from analysis of glacial-rebound phenomena. In: Jackson ISN (ed.) *The Earth's Mantle – Composition, Structure and Evolution*, pp. 461–502. Cambridge: Cambridge University Press.

Lambeck K, Smither C, and Johnston P (1998) Sea-level change, glacial rebound and mantle viscosity for northern Europe. *Geophysical Journal International* 134: 102–144.

Peltier WR (1998) Postglacial variations in the level of the sea: implications for climate dynamics and solid-earth geophysics. *Reviews in Geophysics* 36: 603–689.

Pirazzoli PA (1991) *World Atlas of Holocene Sea-Level Changes*. Amsterdam: Elsevier.

van de Plassche O (ed.) (1986) *Sea-Level Research: A Manual for the Collection and Evaluation of Data*. Norwich: Geo Books.

Sabadini R, Lambeck K, and Boschi E (1991) *Glacial Isostasy, Sea Level and Mantle Rheology*. Dordrecht: Kluwer.

ECONOMICS OF SEA LEVEL RISE

R. S. J. Tol, Economic and Social Research Institute, Dublin, Republic of Ireland

Introduction

The economics of sea level rise is part of the larger area of the economics of climate change. Climate economics is concerned with five broad areas:

- What are the economic implications of climate change, and how would this be affected by policies to limit climate change?
- How would and should people, companies, and government adapt and at what cost?
- What are the economic implications of greenhouse gas emission reduction?
- How much should emissions be reduced?
- How can and should emissions be reduced?

The economics of sea level rise is limited to the first two questions, which can be rephrased as follows:

- What are the costs of sea level rise?
- What are the costs of adaptation to sea level rise?
- How would and should societies adapt to sea level rise?

These three questions are discussed below.

There are a number of processes that cause the level of the sea to rise. Thermal expansion is the dominant cause for the twenty-first century. Although a few degrees of global warming would expand seawater by only a fraction, the ocean is deep. A 0.01% expansion of a column of 5 km of water would give a sea level rise of 50 cm, which is somewhere in the middle of the projections for 2100. Melting of sea ice does not lead to sea level rise – as the ice currently displaces water. Melting of mountain glaciers does contribute to sea level, but these glaciers are too small to have much of an effect. The same is true for more rapid runoff of surface and groundwater. The large ice sheets on Greenland and Antarctica could contribute to sea level rise, but the science is not yet settled. A complete melting of Greenland and West Antarctica would lead to a sea level rise of some 12 m, and East Antarctica holds some 100 m of sea level. Climate change is unlikely to warm East Antarctica above freezing point, and in fact additional snowfall is likely to store more ice on East Antarctica. Greenland and West Antarctica are much warmer, and climate change may

well imply that these ice caps melt in the next 1000 years or so. It may also be, however, that these ice caps are destabilized – which would mean that sea level would rise by 10–12 m in a matter of centuries rather than millennia.

Concern about sea level rise is only one of many reasons to reduce greenhouse gas emissions. In fact, it is only a minor reason, as sea level rise responds only with a great delay to changes in emissions. Although the costs of sea level rise may be substantial, the benefits of reduced sea level rise are limited – for the simple fact that sea level rise can be slowed only marginally. However, avoiding a collapse of the Greenland and West Antarctic ice sheets would justify emission reduction – but for the fact that it is not known by how much or even whether emission abatement would reduce the probability of such a collapse. Therefore, the focus here is on the costs of sea level rise and the only policy considered is adaptation to sea level rise.

What are the Costs of Sea Level Rise?

The impacts of sea level rise are manifold, the most prominent being erosion, episodic flooding, permanent flooding, and saltwater intrusion. These impacts occur onshore as well as near shore/on coastal wetlands, and affect natural and human systems. Most of these impacts can be mitigated with adaptation, but some would get worse with adaptation elsewhere. For instance, dikes protect onshore cities, but prevent wetlands from inland migration, leading to greater losses.

The direct costs of erosion and permanent flooding equal the amount of land lost times the value of that land. Ocean front property is often highly valuable, but one should not forget that sea level rise will not result in the loss of the ocean front – the ocean front simply moves inland. Therefore, the average land value may be a better approximation of the true cost than the beach property value.

The amount of land lost depends primarily on the type of coast. Steep and rocky coasts would see little impact, while soft cliffs may retreat up to a few hundred meters. Deltas and alluvial plains are at a much greater risk. For the world as a whole, land loss for 1 m of sea level rise is a fraction of a percent, even without additional coastal protection. However, the distribution of land loss is very skewed. Some islands, particularly atolls, would disappear altogether – and this may lead to the disappearance of entire nations and cultures. The number of people involved is small, though. This is different for deltas, which tend to be densely populated and heavily used because of superb soils and excellent transport. Without additional protection, the deltas of the Ganges–Brahmaputra, Mekong, and Nile could lose a quarter of their area even for relative modest sea level rise. This would force tens of millions of people to migrate, and ruin the economies of the respective countries.

Besides permanent inundation, sea level rise would also cause more frequent and more intense episodic flooding. The costs of this are more difficult to estimate. Conceptually, one would use the difference in the expected annual flood damage. In practice, however, floods are infrequent and the stock-at-risk changes rapidly. This makes estimates particularly uncertain. Furthermore, floods are caused by storms, and climate models cannot yet predict the effect of the enhanced greenhouse effect on storms. That said, tropical cyclones kill only about 10 000 people per year worldwide, and economic damages similarly are only a minor fraction of gross national product (GDP). Even a 10-fold increase because of sea level rise, which is unlikely even without additional protection, would not be a major impact at the global scale. Here, as above, the global average is likely to hide substantial regional differences, but there has been too little research to put numbers on this.

Sea level rise would cause salt water to intrude in surface and groundwater near the coast. Saltwater intrusion would require desalination of drinking water, or moving the water inlet upstream. The latter is not an option on small islands, which may lose all freshwater resources long before they are submerged by sea level rise. The economic costs of desalination are relatively small, particularly near the coast, but desalination is energy-intensive and produces brine, which may cause local ecological problems. Because of saltwater intrusion, coastal agriculture would suffer a loss of productivity, and may become impossible. Halophytes (salt-tolerant plants) are a lucrative nice market for vegetables, but not one that is constrained by a lack of brackish water. Halophytes' defense mechanisms against salt work at the expense of overall plant growth. Saltwater intrusion could induce a shift from agriculture to aquaculture, which may be more profitable. Few studies are available for saltwater intrusion.

Sea level rise would erode coastal wetlands, particularly if hard structures protect human occupations. Current estimates suggest that a third of all coastal wetlands could be lost for less than 40-cm sea level rise, and up to half for 70 cm. Coastal wetlands provide many services. They are habitats for fish (also as nurseries), shellfish, and birds (including migratory birds). Wetlands purify water, and protect coast against storms. Wetlands also provide food and recreation. If wetlands get lost, so do these functions – unless scarce resources are reallocated to provide these services. Estimates of the value of wetlands vary between $100 and $10 000 per hectare, depending on the type of wetland, the services it provides, population density, and per capita income. At present, there are about 70 million ha of coastal wetlands, so the economic loss of sea level rise would be measured in billions of dollars per year.

Besides the direct economic costs, there are also higher-order effects. A loss of agricultural land would restrict production and drive up the price of food. Some estimates have that a 25-cm sea level rise would increase the price of food by 0.5%. These effects would not be limited to the affected regions, but would be spread from international trade. Australia, for instance, has few direct effects from sea level rise – and may therefore benefit from increased exports to make up for the reduced production elsewhere. Other markets would be affected too, as farmers would buy more fertilizer to produce more on the remaining land, and as workers elsewhere would demand higher salaries to compensate for the higher cost of living. Such spillovers are small in developed economies, because agriculture is only a small part of the economy. They are much larger in developing countries.

What are the Costs of Adaptation to Sea Level Rise?

There are three basic ways to adapt to sea level rise, although in reality mixes and variations will dominate. The two extremes are protection and retreat. In between lies accommodation. Protection entails such things as building dikes and nourishing beaches – essentially, measures are taken to prevent the impact of sea level rise. Retreat implies giving up land and moving people and infrastructure further inland – essentially, the sea is given free range, but people and their things are moved out of harm's way. Accommodation means coping with the consequences of sea level rise. Examples include placing houses on stilts and purchasing flood insurance.

Besides adaptation, there is also failure to (properly) adapt. In most years, this will imply that adaptation costs (see below) are less. In some years, a storm will come to kill people and damage property. The next section has a more extensive discussion on adaptation.

Estimating the costs of protection is straightforward. Dike building and beach nourishment are routine operations practised by many engineering companies around the world. The same holds for accommodation. Estimates have that the total annual cost of coastal protection is less than 0.01% of global GDP. Again, the average hides the extremes. In small island nations, the protection bill may be over 1% of GDP per year and even reach 10%. Note that a sacrifice of 10% is still better than 100% loss without protection.

There may be an issue of scale. Gradual sea level rise would imply a modest extension of the 'wet engineering' sector, if that exists, or a limited expense for hiring foreign consultants. However, rapid sea level rise would imply (local) shortages of qualified engineers – and this would drive up the costs and drive down the effectiveness of protection and accommodation. If materials are not available locally, they will need to be transported to the coast. This may be a problem in densely populated deltas. Few studies have looked into this, but those that did suggest that a sea level rise of less than 2 m per century would not cause logistical problems. Scale is therefore only an issue in case of ice sheet collapse.

Costing retreat is more difficult. The obvious impact is land loss, but this may be partly offset by wetland gains (see above). Retreat also implies relocation. There are no solid estimates of the costs of forced migration. Again, the issue is scale. People and companies move all the time. A well-planned move by a handful of people would barely register. A hasty retreat by a large number would be noticed. If coasts are well protected, the number of forced migrants would be limited to less than 10 000 a year. If coastal protection fails, or is not attempted, over 100 000 people could be displaced by sea level rise each year.

Like impacts, adaptation would have economy-wide implications. Dike building would stimulate the construction sector and crowd out other investments and consumption. In most countries, these effects would be hard to notice, but in small island economies this may lead to stagnation of economic growth.

How Would and Should Societies Adapt to Sea Level Rise?

Decision analysis of coastal protection goes back to Von Dantzig, one of the founding fathers of operations research. This has also been applied to additional protection for sea level rise. Some studies simply compare the best guess of the costs of dike building against the best guess of the value of land lost if no dikes were built, and decide to protect or not on the basis of a simple cost–benefit ratio. Other studies use more advanced methods that, however, conceptually boil down to cost–benefit analysis as well. These

studies invariably conclude that it is economically optimal to protect most of the populated coast against sea level rise. Estimates go up to 85% protection, and 15% abandonment. The reason for these high protection levels are that coastal protection is relatively cheap, and that low-lying deltas are often very densely populated so that the value per hectare is high even if people are poor.

However, coastal protection has rarely been based on cost–benefit analysis, and there is no reason to assume that adaptation to sea level rise will be different. One reason is that coastal hazards manifest themselves irregularly and with unpredictable force. Society tends to downplay most risks, and overemphasize a few. The selection of risks shifts over time, in response to events that may or may not be related to the hazards. As a result, coastal protection tends to be neglected for long periods, interspersed with short periods of frantic activity. Decisions are not always rational. There are always 'solutions' waiting for a problem, and under pressure from a public that demands rapid and visible action, politicians may select a 'solution' that is in fact more appropriate for a different problem. The result is a fairly haphazard coastal protection policy.

Some policies make matters worse. Successful past protection creates a sense of safety and hence neglect, and attracts more people and business to the areas deemed safe. Subsidized flood insurance has the same effect. Local authorities may inflate their safety record, relax their building standards and land zoning, and relocate their budget to attract more people.

A fundamental problem is that coastal protection is partly a public good. It is much cheaper to build a dike around a 'community' than around every single property within that community. However, that does require that there is an authority at the appropriate level. Sometimes, there is no such authority and homeowners are left to fend for themselves. In other cases, the authority for coastal protection sits at provincial or national level, with civil servants who are occupied with other matters.

A number of detailed studies have been published on decision making on adaptation to coastal and other hazards. The unfortunate lesson from this work is that every case is unique – extrapolation is not possible except at the bland, conceptual level.

Conclusion

The economics of climate change are still at a formative stage, and so are the economics of sea level rise. First estimates have been developed of the order of magnitude of the problem. Sea level rise is not a substantial economic problem at the global or even the continental scale. It is, however, a substantial problem for a number of countries, and a dominant issue for some local economies. The main issue, therefore, is the distribution of the impacts, rather than the impacts themselves.

Future estimates are unlikely to narrow down the current uncertainties. The uncertainties are not so much due to a paucity of data and studies, but are intrinsic to the problem. The impacts of sea level rise are local, complex, and in a distant future. The priority for economic research should be in developing dynamic models of economies and their interaction with the coast, to replace the current, static assessments.

See also

Abrupt Climate Change. Monsoons, History of. Paleoceanography: the Greenhouse World. Sea Level Change.

Further Reading

Burton I, Kates RW, and White GF (1993) *The Environment as Hazard*, 2nd edn. New York: The Guilford Press.

Nicholls RJ, Wong PP, Burkett VR, *et al.* (2007) Coastal systems and low-lying areas. In: Parry ML, Canziani O, Palutikof J, van der Linden P, and Hanson C (eds.) *Climate Change 2007: Impacts, Adaptation and Vulnerability – Contribution of Working Group II to the Fourth Assessment Report of the Intergovernmental Panel on Climate Change*, pp. 315–356. Cambridge, UK: Cambridge University Press.

LAND–SEA GLOBAL TRANSFERS

F. T. Mackenzie and L. M. Ver, University of Hawaii, Honolulu, HI, USA

Introduction

The interface between the land and the sea is an important boundary connecting processes operating on land with those in the ocean. It is a site of rapid population growth, industrial and agricultural practices, and urban development. Large river drainage basins connect the vast interiors of continents with the coastal zone through river and groundwater discharges. The atmosphere is a medium of transport of substances from the land to the sea surface and from that surface back to the land. During the past several centuries, the activities of humankind have significantly modified the exchange of materials between the land and sea on a global scale – humans have become, along with natural processes, agents of environmental change. For example, because of the combustion of the fossil fuels coal, oil, and gas and changes in land-use practices including deforestation, shifting cultivation, and urbanization, the direction of net atmospheric transport of carbon (C) and sulfur (S) gases between the land and sea has been reversed. The global ocean prior to these human activities was a source of the gas carbon dioxide (CO_2) to the atmosphere and hence to the continents. The ocean now absorbs more CO_2 than it releases. In pre-industrial time, the flux of reduced sulfur gases to the atmosphere and their subsequent oxidation and deposition on the continental surface exceeded the transport of oxidized sulfur to the ocean via the atmosphere. The situation is now reversed because of the emissions of sulfur to the atmosphere from the burning of fossil fuels and biomass on land. In addition, river and groundwater fluxes of the bioessential elements carbon, nitrogen (N), and phosphorus (P), and certain trace elements have increased because of human activities on land. For example, the increased global riverine (and atmospheric) transport of lead (Pb) corresponds to its increased industrial use. Also, recent changes in the concentration of lead in coastal sediments appear to be directly related to changes in the use of leaded gasoline in internal combustion engines. Synthetic substances manufactured by modern society, such as pesticides and pharmaceutical products, are now appearing in river

and groundwater flows, thus moving toward the sea in a greater degree than before.

The Changing Picture of Land–Sea Global Transfers

Although the exchange through the atmosphere of certain trace metals and gases between the land and the sea surface is important, rivers are the main purveyors of materials to the ocean. The total water discharge of the major rivers of the world to the ocean is $36\,000\,km^3\,yr^{-1}$. At any time, the world's rivers contain about 0.000 1% of the total water volume of $1459 \times 10^6\,km^3$ near the surface of the Earth and have a total dissolved and suspended solid concentration of c. 110 and $540\,ppm\,l^{-1}$, respectively. The residence time of water in the world's rivers calculated with respect to total net precipitation on land is only 18 days. Thus the water in the world's rivers is replaced every 18 days by precipitation. The global annual direct discharge to the ocean of groundwater is about 10% of the surface flow, with a recent estimate of $2400\,km^3\,yr^{-1}$. The dissolved constituent content of groundwater is poorly known, but one recent estimate of the dissolved salt groundwater flux is $1300 \times 10^6\,t\,yr^{-1}$.

The chemical composition of average river water is shown in **Table 1**. Note that the major anion in river water is bicarbonate, HCO_3^-; the major cation is calcium, Ca^{2+}, and that even the major constituent concentrations of river water on a global scale are influenced by human activities. The dissolved load of the major constituents of the world's rivers is derived from the following sources: about 7% from beds of salt and disseminated salt in sedimentary rocks, 10% from gypsum beds and sulfate salts disseminated in rocks, 38% from carbonates, and 45% from the weathering of silicate minerals. Two-thirds of the HCO_3^- in river waters are derived from atmospheric CO_2 via the respiration and decomposition of organic matter and subsequent conversion to HCO_3^- through the chemical weathering of silicate (~30% of total) or carbonate (~70% of total) minerals. The other third of the river HCO_3^- comes directly from carbonate minerals undergoing weathering.

It is estimated that only about 20% of the world's drainage basins have pristine water quality. The organic productivity of coastal aquatic environments has been heavily impacted by changes in the

Table 1 Chemical composition of average river water

By continent	River water concentration[a] ($mg\,l^{-1}$)												Water discharge ($10^3\,km^3\,yr^{-1}$)	Runoff ratio[c]
	Ca^{2+}	Mg^{2+}	Na^+	K^+	Cl^-	SO_4^{2-}	HCO_3^-	SO_2	TDS[b]	TOC[b]	TDN[b]	TDP[b]		
Africa														
Actual	5.7	2.2	4.4	1.4	4.1	4.2	36.9	12.0	60.5				3.41	0.28
Natural	5.3	2.2	3.8	1.4	3.4	3.2	26.7	12.0	57.8					
Asia														
Actual	17.8	4.6	8.7	1.7	10.0	13.3	67.1	11.0	134.6				12.47	0.54
Natural	16.6	4.3	6.6	1.6	7.6	9.7	66.2	11.0	123.5					
S. America														
Actual	6.3	1.4	3.3	1.0	4.1	3.8	24.4	10.3	54.6				11.04	0.41
Natural	6.3	1.4	3.3	1.0	4.1	3.5	24.4	10.3	54.3					
N. America														
Actual	21.2	4.9	8.4	1.5	9.2	18.0	72.3	7.2	142.6				5.53	0.38
Natural	20.1	4.9	6.5	1.5	7.0	14.9	71.4	7.2	133.5					
Europe														
Actual	31.7	6.7	16.5	1.8	20.0	35.5	86.0	6.8	212.8				2.56	0.42
Natural	24.2	5.2	3.2	1.1	4.7	15.1	80.1	6.8	140.3					
Oceania														
Actual	15.2	3.8	7.6	1.1	6.8	7.7	65.6	16.3	125.3				2.40	
Natural	15.0	3.8	7.0	1.1	5.9	6.5	65.1	16.3	120.3					
World average														
Actual	14.7	3.7	7.2	1.4	8.3	11.5	53.0	10.4	110.1	12.57	0.574	0.053	37.40	0.46
Natural (unpolluted)	13.4	3.4	5.2	1.3	5.8	5.3	52.0	10.4	99.6	9.89	0.386	0.027	37.40	0.46
Pollution	1.3	0.3	2.0	0.1	2.5	6.2	1.0	0.0	10.5	2.67	0.187	0.027		
World % pollutive	9%	8%	28%	7%	30%	54%	2%	0%	10%	21%	33%	50%		

[a] Actual concentrations include pollution. Natural concentrations are corrected for pollution.
[b] TDS, total dissolved solids; TOC, total organic carbon; TDN, total dissolved nitrogen; TDP, total dissolved phosphorus.
[c] Runoff ratio = average runoff per unit area/average rainfall.
Revised after Meybeck M (1979) Concentrations des eaux fluviales en elements majeurs et apports en solution aux oceans. *Revue de Geologie Dynamique et de Geographie Physique* 21: 215–246; Meybeck M (1982) Carbon, nitrogen, and phosphorus transport by world rivers. *American Journal of Science* 282: 401–450; Meybeck M (1983) C, N, P and S in rivers: From sources to global inputs. In: Wollast R, Mackenzie FT, and Chou L (eds.) *Interactions of C, N, P and S Biogeochemical Cycles and Global Change*, pp. 163–193. Berlin: Springer.

dissolved and particulate river fluxes of three of the major bioessential elements found in organic matter, C, N, and P (the other three are S, hydrogen (H), and oxygen (O)). Although these elements are considered minor constituents of river water, their fluxes may have doubled over their pristine values on a global scale because of human activities. Excessive river-borne nutrients and the cultural eutrophication of freshwater and coastal marine ecosystems go hand in hand. In turn, these fluxes have become sensitive indicators of the broader global change issues of population growth and land-use change (including water resources engineering works) in the coastal zone and upland drainage basins, climatic change, and sea level rise.

In contrast to the situation for the major elements, delivery of some trace elements from land to the oceans via the atmosphere can rival riverine inputs. The strength of the atmospheric sources strongly depends on geography and meteorology. Hence the North Atlantic, western North Pacific, and Indian Oceans, and their inland seas, are subjected to large atmospheric inputs because of their proximity to both deserts and industrial sources. Crustal dust is the primary terrestrial source of these atmospheric inputs to the ocean. Because of the low solubility of dust in both atmospheric precipitation and seawater and the overwhelming inputs from river sources, dissolved sources of the elements are generally less important. However, because the oceans contain only trace amounts of iron (Fe), aluminum (Al), and manganese (Mn) (concentrations are in the ppb level), even the small amount of dissolution in seawater ($\sim$10% of the element in the solid phase) results in eolian dust being the primary source for the dissolved transport of these elements toremote areas of the ocean. Atmospheric transport of the major nutrients N, silicon (Si), and Fe to the ocean has been hypothesized to affect and perhaps limit primary productivity in certain regions of the ocean at certain times. Modern processes of fossil fuel combustion and biomass burning have significantly modified the atmospheric transport from land to the ocean of trace metals like Pb, copper (Cu), and zinc (Zn), C in elemental and organic forms, and nutrient N.

As an example of global land–sea transfers involving gases and the effect of human activities on the exchange, consider the behavior of CO_2 gas. Prior to human influence on the system, there was a net flux of CO_2 out of the ocean owing to organic metabolism (net heterotrophy). This flux was mainly supported by the decay of organic matter produced by phytoplankton in the oceans and part of that transported by rivers to the oceans. An example

overall reaction is:

$$C_{106}H_{263}O_{110}N_{16}S_2P + 141O_2 \Rightarrow 106CO_2$$
$$+ 6HNO_3 + 2H_2SO_4 + H_3PO_4 + 120H_2O \quad [1]$$

Carbon dioxide was also released to the atmosphere due to the precipitation of carbonate minerals in the oceans. The reaction is:

$$Ca^{2+} + 2HCO_3^- \Rightarrow CaCO_3 + CO_2 + H_2O \quad [2]$$

The CO_2 in both reactions initially entered the dissolved inorganic carbon (DIC) pool of seawater and was subsequently released to the atmosphere at an annual rate of about 0.2×10^9 t of carbon as CO_2 gas. It should be recognized that this is a small number compared with the 200×10^9 t of carbon that exchanges between the ocean and atmosphere each year because of primary production of organic matter and its subsequent respiration.

Despite the maintenance of the net heterotrophic status of the ocean and the continued release of CO_2 to the ocean–atmosphere owing to the formation of calcium carbonate in the ocean, the modern ocean and the atmosphere have become net sinks of anthropogenic CO_2 from the burning of fossil fuels and the practice of deforestation. Over the past 200 years, as CO_2 has accumulated in the atmosphere, the gradient of CO_2 concentration across the atmosphere–ocean interface has changed, favoring uptake of anthropogenic CO_2 into the ocean. The average oceanic carbon uptake for the decade of the 1990s was c. 2×10^9 t annually. The waters of the ocean have accumulated about 130×10^9 t of anthropogenic CO_2 over the past 300 years.

The Coastal Zone and Land–Sea Exchange Fluxes

The global coastal zone environment is an important depositional and recycling site for terrigenous and oceanic biogenic materials. The past three centuries have been the time of well-documented human activities that have become an important geological factor affecting the continental and oceanic surface environment. In particular, historical increases in the global population in the areas of the major river drainage basins and close to oceanic coastlines have been responsible for increasing changes in land-use practices and discharges of various substances into oceanic coastal waters. As a consequence, the global C cycle and the cycles of N and P that closely interact with the carbon cycle have been greatly affected. Several major perturbations of the past three centuries of the industrial age have affected the processes of transport from land, deposition of terrigenous materials, and *in situ* production of organic matter in coastal zone environments. In addition, potential future changes in oceanic circulation may have significant effects on the biogeochemistry and CO_2 exchange in the coastal zone.

The coastal zone is that environment of continental shelves, including bays, lagoons, estuaries, and near-shore banks that occupy 7% of the surface area of the ocean ($36 \times 10^{12}\,m^2$) or *c.* 9% of the volume of the surface mixed layer of the ocean ($3 \times 10^{15}\,m^3$). The continental shelves average 75 km in width, with a bottom slope of $1.7\,m\,km^{-1}$. They are generally viewed as divisible into the interior or proximal shelf, and the exterior or distal shelf. The mean depth of the global continental shelf is usually taken as the depth of the break between the continental shelf and slope at *c.* 200 m, although this depth varies considerably throughout the world's oceans. In the Atlantic, the median depth of the shelf-slope break is at 120 m, with a range from 80 to 180 m. The depths of the continental shelf are near 200 m in the European section of the Atlantic, but they are close to 100 m on the African and North American coasts. Coastal zone environments that have high sedimentation rates, as great as $30–60\,cm\,ky^{-1}$ in active depositional areas, act as traps and filters of natural and human-generated materials transported from continents to the oceans via river and groundwater flows and through the atmosphere. At present a large fraction ($\sim$ 80%) of the land-derived organic and inorganic materials that are transported to the oceans is trapped on the proximal continental shelves. The coastal zone also accounts for 30–50% of total carbonate and 80% of organic carbon accumulation in the ocean. Coastal zone environments are also regions of higher biological production relative to that of average oceanic surface waters, making them an important factor in the global carbon cycle. The higher primary production is variably attributable to the nutrient inflows from land as well as from coastal upwelling of deeper ocean waters.

Fluvial and atmospheric transport links the coastal zone to the land; gas exchange and deposition are its links with the atmosphere; net advective transport of water, dissolved solids and particles, and coastal upwelling connect it with the open ocean. In addition, coastal marine sediments are repositories for much of the material delivered to the coastal zone. In the last several centuries, human activities on land have become a geologically important agent affecting the land–sea exchange of materials. In particular, river and groundwater flows and atmospheric transport of materials to the coastal zone have been substantially altered.

Bioessential Elements

Continuous increase in the global population and its industrial and agricultural activities have created four major perturbations on the coupled system of the biogeochemical cycles of the bioessential elements C, N, P, and S. These changes have led to major alterations in the exchanges of these elements between the land and sea. The perturbations are: (1) emissions of C, N, and S to the atmosphere from fossil-fuel burning and subsequent partitioning of anthropogenic C and deposition of N and S; (2) changes in land-use practices that affect the recycling of C, N, P, and S on land, their uptake by plants and release from decaying organic matter, and the rates of land surface denudation; (3) additions of N and P in chemical fertilizers to cultivated land area; and (4) releases of organic wastes containing highly reactive C, N, and P that ultimately enter the coastal zone. A fifth major perturbation is a climatic one: (5) the rise in mean global temperature of the lower atmosphere of about 1 °C in the past 300 years, with a projected increase of about 1.4–5.8 °C relative to 1990 by the year 2100. **Figure 1** shows how the fluxes associated with these activities have changed during the past three centuries with projections to the year 2040.

Partially as a result of these activities on land, the fluxes of materials to the coastal zone have changed historically. **Figure 2** shows the historical and projected future changes in the river fluxes of dissolved inorganic and organic carbon (DIC, DOC), nitrogen (DIN, DON), and phosphorus (DIP, DOP), and fluxes associated with the atmospheric deposition and denitrification of N, and accumulation of C in organic matter in coastal marine sediments. It can be seen in **Figure 2** that the riverine fluxes of C, N, and P all increase in the dissolved inorganic and organic phases from about 1850 projected to 2040. For example, for carbon, the total flux (organic + inorganic) increases by about 35% during this period. These increased fluxes are mainly due to changes in land-use practices, including deforestation, conversion of forest to grassland, pastureland, and urban centers, and regrowth of forests, and application of fertilizers to croplands and the subsequent leaching of N and P into aquatic systems.

Inputs of nutrient N and P to the coastal zone which support new primary production are from the land by riverine and groundwater flows, from the open ocean by coastal upwelling and onwelling, and to a lesser extent by atmospheric deposition of nitrogen. New primary production depends on the

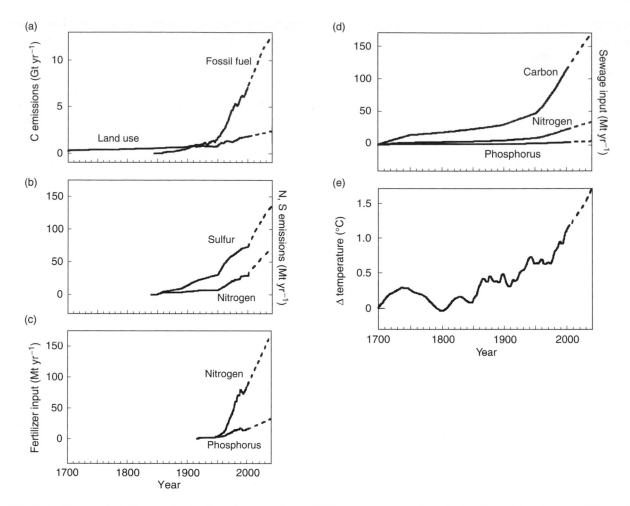

Figure 1 Major perturbations on the Earth system over the past 300 years and projections for the future: (a) emissions of CO_2 and (b) gaseous N and S from fossil-fuel burning and land-use activities; (c) application of inorganic N and P in chemical fertilizers to cultivated land; (d) loading of highly reactive C, N, and P into rivers and the coastal ocean from municipal sewage and wastewater disposal; and (e) rise in mean global temperature of the lower atmosphere relative to 1700. Revised after Ver LM, Mackenzie FT, and Lerman A (1999) Biogeochemical responses of the carbon cycle to natural and human perturbations: Past, present, and future. *American Journal of Science* 299: 762–801.

availability of nutrients from these external inputs, without consideration of internal recycling of nutrients. Thus any changes in the supply of nutrients to the coastal zone owing to changes in the magnitude of these source fluxes are likely to affect the cycling pathways and balances of the nutrient elements. In particular, input of nutrients from the open ocean by coastal upwelling is quantitatively greater than the combined inputs from land and the atmosphere. This makes it likely that there could be significant effects on coastal primary production because of changes in ocean circulation. For example, because of global warming, the oceans could become more strongly stratified owing to freshening of polar oceanic waters and warming of the ocean in the tropical zone. This could lead to a reduction in the intensity of the oceanic thermohaline circulation (oceanic circulation owing to differences in density of water masses, also

popularly known as the 'conveyor belt') and hence the rate at which nutrient-rich waters upwell into coastal environments.

Another potential consequence of the reduction in the rate of nutrient inputs to the coastal zone by upwelling is the change in the CO_2 balance of coastal waters: reduction in the input of DIC to the coastal zone from the deeper ocean means less dissolved CO_2, HCO_3^-, and CO_3^{2-} coming from that source. With increasing accumulation of anthropogenic CO_2 in the atmosphere, the increased dissolution of atmospheric CO_2 in coastal water is favored. The combined result of a decrease in the upwelling flux of DIC and an enhancement in the transfer of atmospheric CO_2 across the air–sea interface of coastal waters is a lower saturation state for coastal waters with respect to the carbonate minerals calcite, aragonite, and a variety of magnesian calcites. The

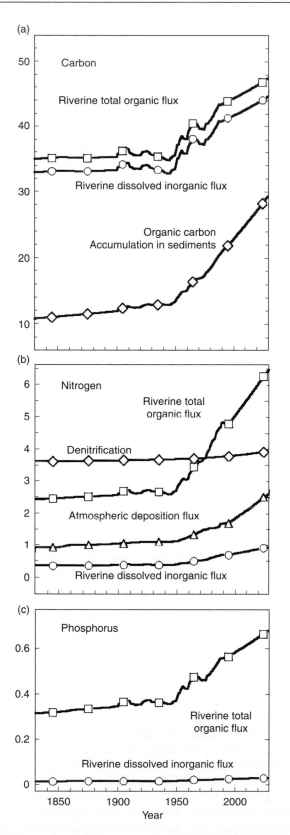

Figure 2 Past, present, and predicted fluxes of carbon, nitrogen, and phosphorus into or out of the global coastal margin, in 10^{12} mol yr^{-1}.

lower saturation state in turn leads to the likelihood of lower rates of inorganic and biological precipitation of carbonate and hence deposition and accumulation of sedimentary carbonate.

In addition, the present-day burial rate of organic carbon in the ocean may be about double that of the late Holocene flux, supported by increased fluxes of organic carbon to the ocean via rivers and groundwater flows and increased *in situ* new primary production supported by increased inputs of inorganic N and P from land and of N deposited from the atmosphere. The organic carbon flux into sediments may constitute a sink of anthropogenic CO_2 and a minor negative feedback on accumulation of CO_2 in the atmosphere.

The increased flux of land-derived organic carbon delivered to the ocean by rivers may accumulate there or be respired, with subsequent emission of CO_2 back to the atmosphere. This release flux of CO_2 may be great enough to offset the increased burial flux of organic carbon to the seafloor due to enhanced fertilization of the ocean by nutrients derived from human activities. The magnitude of the CO_2 exchange is a poorly constrained flux today. One area for which there is a substantial lack of knowledge is the Asian Pacific region. This is an area of several large seas, a region of important river inputs to the ocean of N, P, organic carbon, and sediments from land, and a region of important CO_2 exchange between the ocean and the atmosphere.

Anticipated Response to Global Warming

From 1850 to modern times, the direction of the net flux of CO_2 between coastal zone waters and the atmosphere due to organic metabolism and calcium carbonate accumulation in coastal marine sediments was from the coastal surface ocean to the atmosphere (negative flux, **Figure 3**). This flux in 1850 was on the order of -0.2×10^9 t yr^{-1}. In a condition not disturbed by changes in the stratification and thermohaline circulation of the ocean brought about by a global warming of the Earth, the direction of this flux is projected to remain negative (flux out of the coastal ocean to the atmosphere) until early in the twenty-first century. The increasing partial pressure of CO_2 in the atmosphere because of emissions from anthropogenic sources leads to a reversal in the gradient of CO_2 across the air–sea interface of coastal zone waters and, hence, invasion of CO_2 into the coastal waters. From that time on the coastal ocean will begin to operate as a net sink (positive flux) of atmospheric CO_2 (**Figure 3**). The role of the open ocean as a sink for anthropogenic CO_2 is slightly reduced while that of the coastal oceans

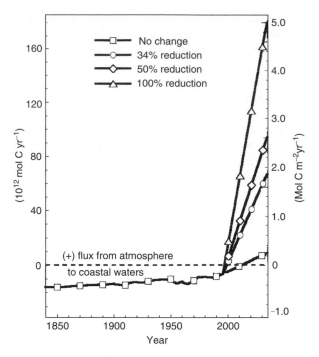

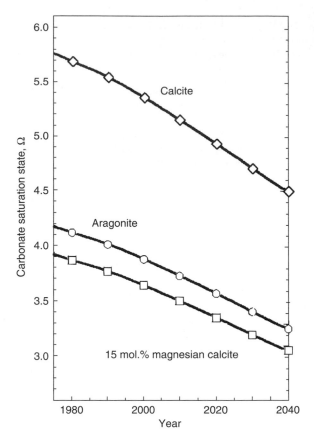

Figure 3 The net flux of CO_2 between coastal zone waters and the atmosphere due to organic metabolism and calcium carbonate accumulation in coastal marine sediments, under three scenarios of changing thermohaline circulation rate compared to a business-as-usual scenario, in units of 10^{12} mol C yr^{-1}.

Figure 4 Changes in saturation state with respect to carbonate minerals of surface waters of the coastal ocean projected from 1999 to 2040. Calculations are for a temperature of 25 °C.

increases. The net result is the maintenance of the role of the global oceans as a net sink for anthropogenic CO_2.

The saturation index (Ω) for calcite or aragonite (both $CaCO_3$) is the ratio of the ion activity product IAP in coastal waters to the respective equilibrium constant K at the *in situ* temperature. For aqueous species, IAP $= a$Ca$^{2+} \times a$CO$_3^{2-}$ (where a is the activity; note that for 15 mol.% magnesian calcite, the IAP also includes the activity of the magnesium cation, aMg^{2+}). Most coastal waters and open-ocean surface waters currently are supersaturated with respect to aragonite, calcite, and magnesian calcite containing 15 mol.% Mg, that is, $\Omega_{calcite}$, $\Omega_{aragonite}$, and $\Omega_{15\%magnesian\ calcite}$ are >1. Because of global warming and the increasing land-to-atmosphere-to-seawater transport of CO_2 (due to the continuing combustion of fossil fuels and burning of biomass), the concentration of the aqueous CO_2 species in seawater increases and the pH of the water decreases slightly. This results in a decrease in the concentration of the carbonate ion, CO_3^{2-}, resulting in a decrease in the degree of supersaturation of coastal zone waters. **Figure 4** shows how the degree of saturation might change into the next century because of rising atmospheric CO_2 concentrations. The overall reduction in the saturation state of coastal

zone waters with respect to aragonite from 1997 projected to 2040 is about 16%, from 3.89 to 3.26.

Modern carbonate sediments deposited in shoal-water ('shallow-water') marine environments (including shelves, banks, lagoons, and coral reef tracts) are predominantly biogenic in origin derived from the skeletons and tests of benthic and pelagic organisms, such as corals, foraminifera, echinoids, mollusks, algae, and sponges. One exception to this statement is some aragonitic muds that may, at least in part, result from the abiotic precipitation of aragonite from seawater in the form of whitings. Another exception is the sand-sized, carbonate oöids composed of either aragonite with a laminated internal structure or magnesian calcite with a radial internal structure. In addition, early diagenetic carbonate cements found in shoal-water marine sediments and in reefs are principally aragonite or magnesian calcite. Thus carbonate production and accumulation in shoal-water environments are dominated by a range of metastable carbonate minerals associated with skeletogenesis and abiotic processes, including calcite, aragonite, and a variety of magnesian calcite compositions.

With little doubt, as has been documented in a number of observational and experimental studies, a reduction in the saturation state of ocean waters will lead to a reduction in the rate of precipitation of both inorganic and skeletal calcium carbonate. Conversely, increases in the degree of supersaturation and temperature will increase the precipitation rates of calcite and aragonite from seawater. During global warming, rising sea surface temperatures and declining carbonate saturation states due to the absorption of anthropogenic CO_2 by surface ocean waters result in opposing effects. However, experimental evidence suggests that within the range of temperature change predicted for the next century due to global warming, the effect of changes in saturation state will be the predominant factor affecting precipitation rate. Thus decreases in precipitation rates should lead to a decrease in the production and accumulation of shallow-water carbonate sediments and perhaps changes in the types and distribution of calcifying biotic species found in shallow-water environments.

Anticipated Response to Heightened Human Perturbation: The Asian Scenario

In the preceding sections it was shown that the fluxes of C, N, and P from land to ocean have increased because of human activities (refer to **Table 1** for data comparing the actual and natural concentrations of C, N, P, and other elements in average river water). During the industrial era, these fluxes mainly had their origin in the present industrialized and developed countries. This is changing as the industrializing and developing countries move into the twenty-first century. A case in point is the countries of Asia.

Asia is a continent of potentially increasing contributions to the loading of the environment owing to a combination of such factors as its increasing population, increasing industrialization dependent on fossil fuels, concentration of its population along the major river drainage basins and in coastal urban centers, and expansion of land-use practices. It is anticipated that Asia will experience similar, possibly even greater, loss of storage of C and nutrient N and P on land and increased storage in coastal marine sediments per unit area than was shown by the developed countries during their period of industrialization. The relatively rapid growth of Asia's population along the oceanic coastal zone indicates that higher inputs of both dissolved and particulate organic nutrients may be expected to enter coastal waters.

A similar trend of increasing population concentration in agricultural areas inland, within the drainage basins of the main rivers flowing into the ocean, is also expected to result in increased dissolved and particulate organic nutrient loads that may eventually reach the ocean. Inputs from inland regions to the ocean would be relatively more important if no entrapment or depositional storage occurred en route, such as in the dammed sections of rivers or in alluvial plains. In the case of many of China's rivers, the decline in sediment discharge from large rivers such as the Yangtze and the Yellow Rivers is expected to continue due to the increased construction of dams. The average decadal sediment discharge from the Yellow River, for example, has decreased by 50% from the 1950s to the 1980s. If the evidence proposed for the continental United States applies to Asia, the damming of major rivers would not effectively reduce the suspended material flow to the ocean because of the changes in the erosional patterns on land that accompany river damming and more intensive land-use practices. These flows on land and into coastal ocean waters are contributing factors to the relative importance of autotrophic and heterotrophic processes, competition between the two, and the consequences for carbon exchange between the atmosphere and land, and the atmosphere and ocean water. The change from the practices of land fertilization by manure to the more recent usage of chemical fertilizers in Asia suggests a shift away from solid organic nutrients and therefore a reduced flow of materials that might promote heterotrophy in coastal environments.

Sulfur is an excellent example of how parts of Asia can play an important role in changing land–sea transfers of materials. Prior to extensive human interference in the global cycle of sulfur, biogenically produced sulfur was emitted from the sea surface mainly in the form of the reduced gas dimethyl sulfide (DMS). DMS was the major global natural source of sulfur for the atmosphere, excluding sulfur in sea salt and soil dust. Some of this gas traveled far from its source of origin. During transport the reduced gas was oxidized to micrometer-size sulfate aerosol particles and rained out of the atmosphere onto the sea and continental surface. The global sulfur cycle has been dramatically perturbed by the industrial and biomass burning activities of human society. The flux of gaseous sulfur dioxide to the atmosphere from the combustion of fossil fuels in some regions of the world and its conversion to sulfate aerosol greatly exceeds natural fluxes of sulfur gases from the land surface. It is estimated that this flux for the year 1990 was equivalent to 73×10^6 $t\,yr^{-1}$, nearly 4 times the natural DMS flux from the ocean. This has led to a net transport of sulfur from the land to the ocean via the atmosphere, completely reversing the flow direction in preindustrial times. In addition, the sulfate aerosol content of the atmosphere derived from human activities has increased. Sulfate aerosols affect global climate directly as particles that scatter incoming solar radiation and indirectly as cloud condensation nuclei (CCNs), which lead to an increased number of cloud droplets and an increase in the solar reflectance of clouds. Both effects cause the cooling of the planetary surface. As can be seen in **Figure 5** the eastern Asian region is an important regional source of sulfate aerosol because of the combustion of fossil fuels,

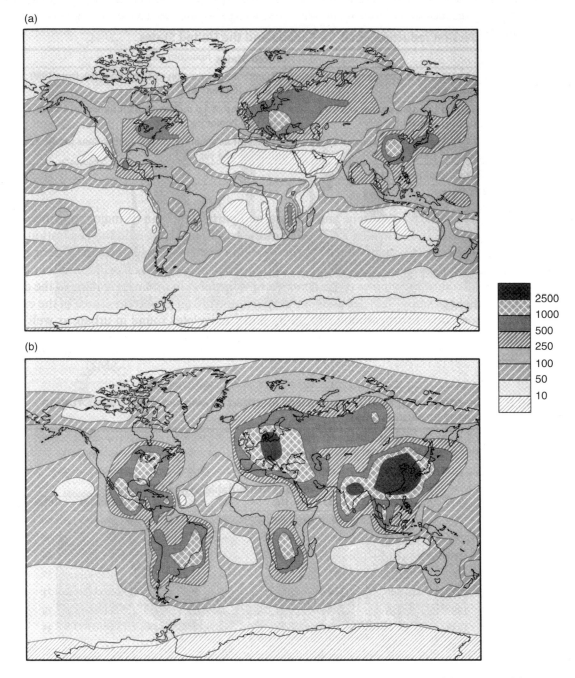

Figure 5 Comparison of the magnitude of atmospheric sulfur deposition for the years 1990 (a) and 2050 (b). Note the large increases in both spatial extent and intensity of sulfur deposition in both hemispheres and the increase in importance of Asia, Africa, and South America as sites of sulfur deposition between 1990 and 2050. The values on the diagrams are in units of kg $S m^{-2} yr^{-1}$. Revised after Mackenzie FT (1998) *Our Changing Planet*: *An Introduction to Earth System Science and Global Environmental Change*. Upper Saddle River, NJ: Prentice Hall; Rodhe H, Langner J, Gallardo L, and Kjellström E (1995) Global transport of acidifying pollutants. *Water, Air and Soil Pollution* 85: 37–50.

particularly coal. This source is predicted to grow in strength during the early- to mid-twenty-first century (**Figure 5**).

Conclusion

Land–sea exchange processes and fluxes of the bioessential elements are critical to life. In several cases documented above, these exchanges have been substantially modified by human activities. These modifications have led to a number of environmental issues including global warming, acid deposition, excess atmospheric nitrogen deposition, and production of photochemical smog. All these issues have consequences for the biosphere – some well known, others not so well known. It is likely that the developing world, with increasing population pressure and industrial development and with no major changes in agricultural technology and energy consumption rates, will become a more important source of airborne gases and aerosols and materials for river and groundwater systems in the future. This will lead to further modification of land–sea global transfers. The region of southern and eastern Asia is particularly well poised to influence significantly these global transfers.

See also

Carbon Cycle. Carbon Dioxide (CO₂) Cycle. Ocean Carbon System, Modeling of. Ocean Circulation: Meridional Overturning Circulation. Past Climate from Corals.

Further Reading

Berner EA and Berner RA (1996) *Global Environment: Water, Air and Geochemical Cycles*. Upper Saddle River, NJ: Prentice Hall.

Galloway JN and Melillo JM (eds.) (1998) *Asian Change in the Context of Global Change*. Cambridge, MA: Cambridge University Press.

Mackenzie FT (1998) *Our Changing Planet: An Introduction to Earth System Science and Global Environmental Change*. Upper Saddle River, NJ: Prentice Hall.

Mackenzie FT and Lerman A (2006) *Carbon in the Geobiosphere – Earth's Outer Shell*. Dordrecht: Springer.

Meybeck M (1979) Concentrations des eaux fluviales en elements majeurs et apports en solution aux oceans. *Revue de Geologie Dynamique et de Geographie Physique* 21: 215–246.

Meybeck M (1982) Carbon, nitrogen, and phosphorus transport by world rivers. *American Journal of Science* 282: 401–450.

Meybeck M (1983) C, N, P and S in rivers: From sources to global inputs. In: Wollast R, Mackenzie FT, and Chou L (eds.) *Interactions of C, N, P and S Biogeochemical Cycles and Global Change*, pp. 163–193. Berlin: Springer.

Rodhe H, Langner J, Gallardo L, and Kjellström E (1995) Global transport of acidifying pollutants. *Water, Air and Soil Pollution* 85: 37–50.

Schlesinger WH (1997) *Biogeochemistry: An Analysis of Global Change*. San Diego, CA: Academic Press.

Smith SV and Mackenzie FT (1987) The ocean as a net heterotrophic system: Implications from the carbon biogeochemical cycle. *Global Biogeochemical Cycles* 1: 187–198.

Ver LM, Mackenzie FT, and Lerman A (1999) Biogeochemical responses of the carbon cycle to natural and human perturbations: Past, present, and future. *American Journal of Science* 299: 762–801.

Vitousek PM, Aber JD, and Howarth RW (1997) Human alteration of the global nitrogen cycle: Sources and consequences. *Ecological Applications* 7(3): 737–750.

Wollast R and Mackenzie FT (1989) Global biogeochemical cycles and climate. In: Berger A, Schneider S, and Duplessy JC (eds.) *Climate and Geo-Sciences*, pp. 453–473. Dordrecht: Kluwer Academic Publishers.

EFFECTS OF CLIMATE CHANGE ON MARINE MAMMALS

I. Boyd and N. Hanson, University of St. Andrews, St. Andrews, UK

Introduction

The eff'ects of climate change on marine mammals will be caused by changes in the interactions between the physiological state of these animals and the physical changes in their environment caused by climate change. In this article, climate change is defined as a long-term (millennial) trend in the physical climate. This distinguishes it from short-term, regional fluctuations in the physical climate. Marine mammals are warm-blooded vertebrates living in a highly conductive medium often with a steep temperature gradient across the body surface. They also have complex behavioral repertoires that adapt rapidly to changes in the conditions of the external environment. In general, we would expect the changes in the physical environment at the scales envisaged under climate change scenarios to be well within the homeostatic capacity of these species. Effects of climate upon the prey species normally eaten by marine mammals, most of which do not have the same level of homeostatic control to stresses in their physical environment, may be the most likely mechanism of interaction between marine mammals and climate change. However, we should not assume that effects of climate change on marine mammals should necessarily be negative.

Responses to Normal Environmental Variation

Marine mammals normally experience variation in their environment that is very large compared with most variance predicted due to climate change. Examples include the temperature gradients that many marine mammals experience while diving through the water column and the extreme patchiness of the prey resources for marine mammals. Marine mammals in the Pacific have life-histories adapted to transient climatic phenomena such as El Niño, which oscillate every 4 years or so. Consequently, the morphologies, physiologies, behaviors, and life histories of marine mammals will have evolved to cope with this high level of variance. However, it is generally accepted that climate change is occurring too rapidly for the life histories of marine mammals to adapt to longer periods of adverse conditions than are experienced in examples like El Niño.

Marine mammals appear to cope with other longer wavelength oscillations including the North Pacific and North Atlantic Oscillations and the Antarctic Circumpolar Wave and it is possible that their life histories have evolved to cope with this type of long wavelength variation. Nevertheless, nonoscillatory climate change could result in nonlinear processes of change in some of the physical and biological features of the environment that are important to some marine mammal species. Although speculative, obvious changes such as the extent of Arctic and Antarctic seasonal ice cover could affect the presence of essential physical habitat for marine mammals as well as food resources, and there may be other changes in the structure of marine mammal habitats that are less obvious and difficult to both identify and quantify. Changes in the trophic structure of the oceans in icebound regions, where the ecology is very reliant on sea ice, may lead the trophic pyramid that supports these top predators to alter substantially. The polar bear is particularly an obvious example of this type of effect where both loss of hunting habitat in the form of sea ice and the potential effects on prey abundance are already having measurable effects upon populations. Similarly, changes in coastal habitats resulting from changes in sea level, changes in run-off and salinity, and changes in nutrient and sediment loads are likely to have important effects on some species of small cetaceans with localized distributions. Many sirenians rely upon seagrass communities and anything that affects the sustainability of this food source is likely to have a negative effect on these species.

Many marine mammal species have already experienced range retraction and population depletion because of direct interaction with man. Monk seals appear to be particularly vulnerable because they rely upon small pockets of beach habitat, many of which are threatened directly by man and also by rising sea level.

Marine mammals are known to be vulnerable to the effects of toxic algal blooms. Toxins may lead to sublethal effects, such as reduced rates of reproduction as well as direct mortality. Several mortality events including coastal whale and dolphin species as well as seals have been attributed to these effects.

Climate change could result in increased frequency of the conditions that lead to such effects, perhaps as a result of interactions between temperature and eutrophication of coastal habitats.

Classifying Effects

A common approach to the assessment of the effects of climate change is to divide these into 'direct' and 'indirect' effects. In this case, 'direct' effects are those associated with changes in the physical environment, such as those that affect the availability of suitable habitat. 'Indirect' effects are those that operate through the agency of food availability because of changes in ecology, susceptibility to disease, changed exposure to pollution, or changes in competitive interactions. Würsig et al. in 2002 added a third level of effect which was the result of human activities occurring in response to climate change that tend to increase conflicts between man and marine mammals. This division has little utility in terms of rationalizing the effects of climate change because, in simple terms, the effects will operate ultimately through the availability of suitable habitat. Assessing the effects of climate change rests upon an assessment of whether there is a functional relationship between the availability of suitable habitat and climate and the form of these functional relationships, which will differ between species, has not been determined.

The expansion and contraction of suitable habitat can be affected by a broad range of factors and some of these can operate on their own but others are often closely related and synergistic, such as the combined effect of retraction of sea ice upon the availability of breeding habitat for seals and also for the food chains that support these predators.

Evidence for the Effects of Climate Change on Marine Mammals

There is no strong evidence that current climate change scenarios are affecting marine mammals although there are studies that suggest some typical effects of climate change could affect marine mammal distribution and abundance. There is an increasing body of literature that links apparent variability in marine mammal abundance, productivity, or behavior with climate change processes. However, with the probable exception of those documenting the changes occurring to the extent of breeding habitat for ringed seals within some sections of the Arctic, and the consequences of this also for polar bears, most of these studies simply reflect a trend toward the interpretation of responses of marine mammals to large-scale regional variability in the physical environment, as has already been well documented in the Pacific for El Niño, in terms of climate trends. Long-term trends in the underlying regional ecosystem structure are sometimes extrapolated as evidence of climate change. In few, if any, of these cases is there strong evidence that the physical environmental variability being observed is derived from irreversible trends in climate. Some of the current literature confounds understanding of the responses of marine mammals to regional variability with that of climate change, albeit that an understanding of one may be useful in the interpretation and prediction of the effects of the other.

Based upon records of species from strandings, MacLeod et al. in 2005 have suggested that the species diversity of cetaceans around the UK has increased recently and that this may be evidence of range expansion in some species. However, the sample sizes involved are small and there are difficulties in these types of studies accounting for observer effort. This is a common story for marine mammals, and many other marine predators including seabirds, in that, there is a great deal of theory about what the effects of climate change might be but little convincing evidence that backs up these suggestions. Even process studies, involving research on the mechanisms underlying how climate change could affect marine mammals, when considered in detail make a tenuous linkage between the physical variables and the biological response of the marine mammals.

Is Climate Change Research on Marine Mammals Scientific?

Although it is beyond dispute that marine mammals respond to physical changes in habitat suitability, the relationship between a particular effect and the response from the marine mammal is seldom clear. Where data from time series are analyzed, as in the case of Forcada et al., they are used to test post hoc for relationships between climate and biological variables. There is a tendency in these circumstances to test for all possible relationships using a range of physical and biological variables. Such post hoc testing is fraught with pitfalls because invariably the final apparently statistically significant relationships are not downweighted in their significance by all the other nonsignificant relationships that were investigated alongside those that proved to be statistically significant. Of course, there may be a priori reasons for accepting that a particular relationship is true, but the approach to examining time series rarely provides an analysis of the relationships that were not statistically significant or the a priori reasons there might be for rejection of these. Consequently, current suggestions from the literature about the potential effects of climate change may be exaggerated because of the strong possibility of the presence of type I and type II statistical error in the assessment process. Moreover, in the great majority of examples, it will be almost impossible to clearly demonstrate effects of climate change, as has been the case with partitioning the variance between a range of causes of the decline of the Steller sea lion (*Eumetopias jubatus*) in the North Pacific and Bering Sea.

Identifying Situations in Which Climate Change is Likely to Have a Negative Effect on Marine Mammals: Future Work

To date, little has been done to build predictive frameworks for assessing the effects of climate change of marine mammals. There have been broad assessments and focused ecological studies but these are a fragile foundation for guiding policy and management, and for identifying populations that are at greatest risk. The resilience of marine mammal populations to climate change will reflect resilience to any other change in habitat quality, that is, it will depend upon the extent of suitable habitat, the degree to which populations currently fill that habitat, the dispersal capacity of the species, and the structure of the current population, including its capacity for increase and demographics. Clearly, populations that are already in a depleted state, or that are dependent upon habitat that is diminishing for reasons other than climate change, will be more vulnerable to the effects of climate change. There are also some, as yet unconvincing, suggestions that habitat degradation may occur through effects of climate upon pollutant burdens.

The general demographic characteristics of marine mammal populations are relatively well known so there are simple ways of assessing the risk to populations under different scenarios of demographic stochasticity, population size, and isolation. An analysis of this type could only provide a very broad guide to the types of effects that could be expected but, whereas no such analysis has been carried out to date, this should be seen as a first step in the risk-assessment process.

The metapopulation structure of many marine mammal populations will affect resilience to climate change and will be reflected in the dispersal capacity of the population. Again, this type of effect could be included within an analysis of the sensitivity of marine mammal populations to climate change under different metapopulation structures. A feature of climate change is that it is likely to have global as well as local effects and the sensitivity to the relative contribution from these would be an important feature of such an analysis.

Further Reading

Atkinson A, Siegel V, Pakhamov E, and Rothery P (2004) Long-term decline in krill stocks and increase in salps within the Southern Ocean. *Nature* 432: 100–103.

Cavalieri DJ, Parkinson CL, and Vinnikov KY (2003) 30-year satellite record reveals contrasting Arctic and Antarctic decadal sea ice variability. *Geophysical Research Letters* 30: 1970 (doi:10.1029/2003GL018031).

Derocher E, Lunn N, and Stirling I (2004) Polar bears in a warming climate. *Integrative and Comparative Biology* 44: 163–176.

Ferguson S, Stirling I, and McLoughlin P (2005) Climate change and ringed seal (*Phoca hispida*) recruitment in western Hudson Bay. *Marine Mammal Science* 21: 121–135.

Forcada J, Trathan P, Reid K, and Murray E (2005) The effects of global climate variability in pup production of Antarctic fur seals. *Ecology* 86: 2408–2417.

Grebmeier J, Overland J, Moore S, et al. (2006) A major ecosystem shift in the northern Bering Sea. *Science* 311: 1461–1464.

Green C and Pershing A (2004) Climate and the conservation biology of North Atlantic right whales: The right whale at the wrong time? *Frontiers in Ecology and the Environment* 2: 29–34.

Heide-Jorgensen MP and Lairde KL (2004) Declining extent of open water refugia for top predators in Baffin Bay and adjacent waters. *Ambio* 33: 487–494.

Hunt G, Stabeno P, Walters G, et al. (2002) Climate change and control of southeastern Bering Sea pelagic ecosystem. *Deep Sea Research II* 49: 5821–5853.

Laidre K and Heide-Jorgensen M (2005) Artic sea ice trends and narwhal vulnerability. *Biological Conservation* 121: 509–517.

Leaper R, Cooke J, Trathan P, Reid K, Rowntree V, and Payne R (2005) Global climate drives southern right whale (*Eubaena australis*) population dynamics. *Biology Letters* 2 (doi:10.1098/rsbl.2005.0431).

Lusseau RW, Wilson B, Grellier K, Barton TR, Hammond PS, and Thompson PM (2004) Parallel influence of climate on the behaviour of Pacific killer whales and Atlantic bottlenose dolphins. *Ecology Letters* 7: 1068–1076.

MacDonald R, Harner T, and Fyfe J (2005) Recent climate change in the Artic and its impact on contaminant pathways and interpretation of temporal trend data. *Science of the Total Environment* 342: 5–86.

MacLeod C, Bannon S, Pierce G, et al. (2005) Climate change and the cetacean community of north-west Scotland. *Biological Conservation* 124: 477–483.

McMahon C and Burton C (2005) Climate change and seal survival: Evidence for environmentally mediated changes in elephant seal *Mirounga leonina* pup survival. *Proceedings of the Royal Society B* 272: 923–928.

Robinson R, Learmouth J, Hutson A, *et al.* (2005) Climate change and migratory species. *BTO Research Report 414*. London: Defra. http://www.bto.org/research/reports/researchrpt_abstracts/2005/RR414%20_summary_report.pdf (accessed Mar. 2008).

Sun L, Liu X, Yin X, Zhu R, Xie Z, and Wang Y (2004) A 1,500-year record of Antarctic seal populations in response to climate change. *Polar Biology* 27: 495–501.

Trillmich F, Ono KA, Costa DP, *et al.* (1991) The effects of El Niño on pinniped populations in the eastern Pacific. In: Trillmich F and Ono KA (eds.) *Pinnipeds and El Niño: Responses to Environmental Stress*, pp. 247–270. Berlin: Springer.

Trites A, Miller A, Maschner H, *et al.* (2006) Bottom up forcing and decline of Stellar Sea Lions in Alaska: Assessing the ocean climate hypothesis. *Fisheries Oceanography* 16: 46–67.

Walther G, Post E, Convey P, *et al.* (2002) Ecological responses to recent climate change. *Nature* 416: 389–395.

Würsig B, Reeves RR, and Ortega-Ortiz JG (2002) Global climate change and marine mammals. In: Evans PGH and Raga JA (eds.) *Marine Mammals – Biology and Conservation*, pp. 589–608. New York: Kluwer Academic/Plenum Publishers.

FISHERIES AND CLIMATE

K. M. Brander, DTU Aqua, Charlottenlund, Denmark

Introduction

Concern over the effects of climate has increased in recent years as concentrations of greenhouse gases have risen in the atmosphere. We all observe changes which are attributed, with more or less justification, to the influence of climate. Anglers and fishermen in many parts of the world have noticed a steady increase in warm-water species. The varieties of locally caught fish which are sold in markets and shops are changing. For example, fish shops in Scotland now sell locally caught bass (*Dicentrarchus labrax*) and red mullet (*Mullus surmuletus*) – species which only occurred in commercial quantities south of the British Isles (600 km to the south) until the turn of the millennium. The northward spread of two non-commercial, subtropical species is shown in **Figure 1**.

They were first recorded off Portugal in the 1960s and were then progressively found further north, until by the mid-1990s they occurred over 1000 km

north of the Iberian Peninsula. *Zenopsis conchifer* was recorded at Iceland for the first time in 2002.

In addition to distribution changes, the growth, reproduction, migration, and seasonality of fish are affected by climate. The productivity and composition of the ecosystems on which fish depend are altered, as is the incidence of pathogens. The changes are not only due to temperature; but winds, ocean currents, vertical mixing, sea ice, freshwater runoff, cloud cover, oxygen, salinity, and pH are also part of climate change, with effects on fish. The processes by which these act will be explored in the examples cited later.

The effect of climate on fisheries is not a new phenomenon or a new area of scientific investigation. Like the effects of climate on agriculture or on hunted and harvested animal populations, it has probably been systematically observed and studied since humans began fishing. However, the anthropogenic component of climate change is a new phenomenon, which is gradually pushing the ranges of atmospheric and oceanic conditions outside the envelope experienced during human history. This article begins by describing some of the past effects of climate on fisheries and then goes on to review expected impacts of anthropogenic climate change.

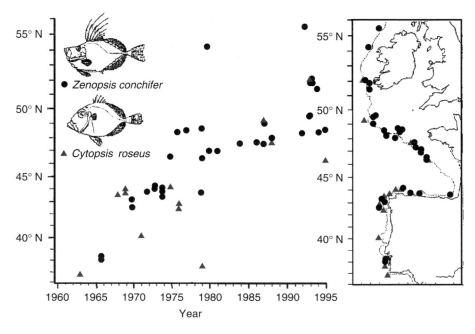

Figure 1 First records of two subtropical fish species (silvery John Dory and rosy Dory) along the NW European continental shelf. From Quéro JC, Buit HD, and Vayne JJ (1998) Les observations de poissons tropicaux et le réchauffement des eaux dans l'Atlantique européen. *Oceanologica Acta* 21: 345–351.

Climate Timescales and Terminology

A wide range of timescales of change in the physical and chemical environment may be included in the term 'climate'. In this article, 'climate variability' refers to changes in temperature, wind fields, hydrological cycles, etc., at annual to decadal timescales and 'climate change' denotes longer-term shifts in the mean values. There are both natural and anthropogenic causes of climate variability and it is not always easy to distinguish the underlying causes of a particular observed effect. Changes in the physical and chemical environment occur naturally on daily, seasonal, and longer-term (e.g., 18.6-year nodal tide) cycles, which can be related to planetary motion. Natural variability in the environment overlies these cycles, so that one can, for example, speak of a windy month or a wet year. Underlying such statements is the idea of a 'normal' month or year, which is generally defined in relation to a climatology, that is, by using a long-term mean and distribution of the variable in question. Volcanic activity and solar fluctuations are other nonanthropogenic factors which affect climate.

History

The Norwegian spring spawning herring (*Clupea harengus* L.) has been a major part of the livelihood of coastal communities in western and northern Norway for over 1000 years.

> It comes up to the shore here from the great fish pond which is the Icelandic Sea, towards the winter when the great part of other fish have left the land. And the herring does not seek the shore along the whole, but at special points which God in his Good Grace has found fitting, and here in my days there have been two large and wonderful herring fisheries at different places in Norway. The first was between Stavanger and Bergen and much further north, and this fishery did begin to diminish and fall away in the year 1560. And I do not believe there is any man to know how far the herring travelled. For the Norwegian Books of Law show that the herring fishery in most of the northern part of Norway has continued for many hundreds of years, although it may well be that in punishment for the unthankfulness of men it has moved from place to place, or has been taken away for a long period. (Clergyman Peder Claussøn Friis (1545–1614))

The changes in distribution of herring which Clergyman Friis wrote about in the sixteenth century have occurred many times since. We now know that herring, which spawn along the west coast of Norway in spring, migrate out to feed in the Norwegian Sea in summer, mainly on the copepod *Calanus finmarchicus*. The distribution shifts in response to decadal changes in oceanic conditions.

Beginning in the 1920s much of the North Atlantic became warmer, the polar front moved north, and the summer feeding migration of this herring stock expanded along the north coast of Iceland. For the next four decades, the resulting fishery provided economic prosperity for northern Iceland and for the country as a whole. When the polar front, which separates cold, nutrient-poor polar water from warmer, nutrient-rich Atlantic water, shifted south and east again, in 1964–65, the herring stopped migrating along the north coast of Iceland. At the same time, the stock collapsed due to overfishing and the herring processing business in northern Iceland died out (**Figure 2**).

There are many similar examples of fisheries on pelagic and demersal fish stocks, which have changed their distribution or declined and recovered as oceanic conditions switched between favorable and unfavorable periods. The Japanese Far Eastern sardine has undergone a series of boom and bust cycles lasting several decades as has the Californian sardine and anchovy. Along the west coast of South America the great Peruvian anchoveta fishery is subject to great fluctuations in abundance, which are not only driven by El Niño/Southern Oscillation (ENSO), but also by decadal-scale variability in the Pacific circulation. In the North Atlantic, the economy of fishing communities in Newfoundland, Greenland, and Faroe has been severely affected by climate-induced fluctuations in the cod stocks (*Gadus morhua* L.) on which they depend.

A number of lessons can be learned from history:

- Fish stocks have always been subject to climate-induced changes.
- Stocks can recover and recolonize areas from which they had disappeared, but such recoveries may take a long time.
- The risk of collapse increases when unfavorable environmental changes overwhelm the resilience of heavily fished stocks.

The present situation is different from the past in at least two important respects: (1) the current rate of climate change is very rapid; and (2) fish stocks are currently subjected to extra mortality and stress due to overfishing and other anthropogenic impacts. History may therefore not be a reliable guide to the future changes in fisheries.

Effects on Individuals, Populations, and Ecosystems

Climatic factors act directly on the growth, survival, reproduction, and movement of individuals and

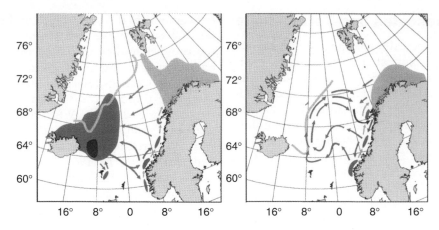

Figure 2 Left panel shows the distribution of Norwegian spring spawning herring between the 1920s and 1965, with areas of spawning (dark blue), nursery (light blue), feeding (green), and overwintering (red). Right panel is the distribution after 1990, when the stock had recovered, but had not reoccupied the area north of Iceland. The polar front separating warm, nutrient-rich Atlantic water from cooler, nutrient-depleted polar water is shown as a light blue line. From Vilhjalmsson H (1997) Climatic variations and some examples of their effects on the marine ecology of Icelandic and Greenland waters, in particular during the present century. *Rit Fiskideildar* XV(1): 9–27.

some of these processes can be studied experimentally, as well as by sampling in the ocean. The integrated effects of the individual processes, are observed at the level of populations, communities, or ccosystems.

In order to convincingly attribute a specific change to climate, one needs to be able to identify and describe the processes involved. Even if the processes are known, and have been studied experimentally, such as the direct effects of temperature on growth rate, the outcome at the population level, over periods of years, can be complex and uncertain. Temperature affects frequency of feeding, meal size, rate of digestion, and rate of absorption of food. Large-scale experiments in which a range of sizes of fish were fed to satiation (Atlantic cod, *G. morhua* L. in this case; see **Figure 3**) show that there is an optimum temperature for growth, but that this depends on the size of the fish, with small fish showing a higher optimum. The optimum temperature for growth also depends on how much food is available, since the energy required for basic maintenance metabolism increases at higher temperature. This means that if food is in short supply then fish will grow faster at cooler temperature, but if food is plentiful then they will grow faster at higher temperatures. A further complication is that temperature typically has a seasonal cycle so the growth rate may decline due to higher summer temperature, but increase due to higher winter temperature.

Recruitment of young fish to an exploited population is variable from year to year for a number of reasons, including interannual and long-term climatic variability. The effects of environmental

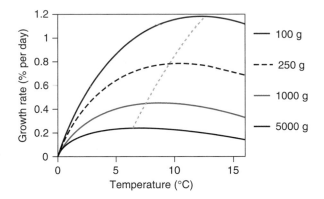

Figure 3 Growth rate of four sizes of Atlantic cod (*Gadus morhua* L.) in rearing experiments at different temperatures in which they were provided unlimited food. The steep dashed line intersects the growth curves at their maximum values, to show how the temperature for maximum growth rate declines as fish get bigger.

variability on survival during early life, when mortality rates in the plankton are very high, is thought to be critical in determining the number of recruits. When recruitment is compared across a number of cod populations, a consistent domed pattern emerges (**Figure 4**).

The relationship between temperature and recruitment for cod has an ascending limb from $c.$ 0 to $4\,°C$ and a descending limb above $c.$ $7\,°C$.

In addition to temperature, the distribution of fish depends on salinity, oxygen, and depth, which is only affected by climate at very long timescales. A 'bioclimate envelope', defining the limits of the range, can be constructed by taking all climate-related variables together. Such 'envelopes' are in reality

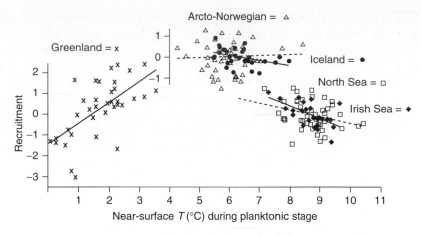

Figure 4 Composite pattern of recruitment for five of the North Atlantic cod stocks to illustrate the effect of temperature during the planktonic stage of early life on the number of recruiting fish. The scales are $\log_e$ (number of 1-year-old fish) with the means adjusted to zero. The axes for the Arcto-Norwegian and Iceland stocks have been displaced vertically.

quite complex because the tolerance and response of fish may vary with size (as shown for growth in **Figure 3**); there may be interactions between the factors (e.g., tolerance of low oxygen level depends on salinity) and different subpopulations may have different tolerances because they have developed local adaptations (e.g., cold-adapted populations may produce antifreeze molecules).

Distribution and abundance of a population are the outcome of their rates of growth, reproductive output, survival to maturity, mortality (due to fishing and natural causes), migration, and location of spawning. Unfavorable changes in any of these, due to climate and other factors, will cause changes in distribution and abundance, with the result that a species may no longer be able to maintain a population in areas which are affected. Favorable changes allow a species to increase its population or to colonize previously unsuitable areas.

Climate and fishing can both be considered as additional stresses on fish populations. In order to manage fisheries in a sustainable way the effects of both (and other factors, such as pollution or loss of essential habitat) need to be recognized, as well as the interactions between them. When the age structure and geographic extent of fish populations are reduced by fishing pressure, then they become less resilient to climate change. Conversely, if climate change reduces the surplus production of a population (by altering growth, survival, or reproductive success), then that population will decline at a level of fishing which had previously been sustainable. Because of these interactions it is not possible to deal with adaption to climate change and fisheries management as separate issues. The most effective strategy to assist fish stocks in adapting to climate change is to reduce the mortality due to fishing. Sustainable fisheries require continuous monitoring of the consequences of climate change.

Regional Effects of Climate

Tropical Pacific

The tuna of the Pacific provide one of the very few examples in which the consequences of climate change have been modeled to include geographic and trophic detail all the way through from primary and secondary production to top fish predators. The tuna species skipjack (*Katsuwonus pelamis*) and yellowfin (*Thunnus albacares*) are among the top predators of the tropical pelagic ecosystem and produced a catch of 3.6 million t in 2003, which represents *c.* 5.5% of total world capture fisheries in weight and a great deal more in value. Their forage species include the Japanese sardine *Engraulis japonicus*, which itself provided a catch of over 2 million t in 2003. The catches and distribution of these species and other tuna species (e.g., albacore *Thunnus alalunga*) are governed by variability in primary and secondary production and location of suitable habitat for spawning and for adults, which in turn are linked to varying regimes of the principal climate indices, such as the El Niño–La Niña Southern Oscillation index (SOI) and the related Pacific Decadal Oscillation (PDOI). Statistical and coupled biogeochemical models have been developed to explore the causes of regional variability in catches and their connection with climate. The model area includes the Pacific from 40° S to 60° N and timescales range from short-term to decadal regime shifts. The model captures the slowdown of Pacific meridional overturning circulation and decrease of equatorial upwelling, which has caused primary production and biomass to decrease by about 10% since 1976–77 in the equatorial Pacific. Further climate change will affect the distribution and production of tuna fisheries in rather complex ways. Warmer surface waters and lower primary production in the central and eastern Pacific may result in a redistribution of tuna to higher latitudes (such as Japan) and toward the western equatorial Pacific.

North Pacific

Investigations into the effects of climate change in the North Pacific have focused strongly on regime shifts. The physical characteristics of these regime shifts and the biological consequences differ between the major regions within the North Pacific.

The PDO tracks the dominant spatial pattern of sea surface temperature (SST). The alternate phases of the PDO represent cooling/warming in the central subarctic Pacific and warming/cooling along the North American continental shelf. This 'classic' pattern represents change along an east–west axis, but since 1989 a north–south pattern has also emerged. Other commonly used indices track the intensity of the winter Aleutian low-pressure system and the sea level pressure over the Arctic. North Pacific regime shifts are reported to have occurred in 1925, 1947, 1977, 1989, and 1998, and paleo-ecological records show many earlier ones. The duration of these regimes appears to have shortened from 50–100 years to c. 10 years for the two most recent regimes, although whether this apparent shortening of regimes is real and whether it is related to other aspects of climate change is a matter of current debate and concern. The SOI also has a large impact on the North Pacific, adding an episodic overlay with a duration of 1 or 2 years to the dec-adal-scale regime behavior.

Regime shifts, such as the one in 1998, have well-documented effects on ocean climate and biological systems. Sea surface height (SSH) in the central North Pacific increased, indicating a gain in thickness of the upper mixed layer, while at the same time SSH on the eastern and northern boundaries of the North Pacific dropped. The position of the transition-zone chlorophyll front, which separates subarctic from subtropical waters and is a major migration and forage habitat for large pelagic species, such as albacore tuna, shifted northward. In addition to its effects on pelagic fish species, shifts in the winter position of the chlorophyll front affect other species, such as Hawaiian monk seals, whose pup survival rate is lower when the front, with its associated production, is far north of the islands. Spiny lobsters (*Panulirus*

marginatus) recruitment in the Northwestern Hawaiian Islands is also affected.

In the California Current System (CCS), zoo-plankton species characteristic of shelf waters have since 1999 replaced the southerly, oceanic species which had been abundant since 1989 and northern fish species (Pacific salmon, cod, and rockfish species) have increased, while the southern migratory pelagics such as Pacific sardines, have declined. The distribution of Pacific hake (*Merluccius productus*), which range from Baja California to the Gulf of Alaska, is closely linked to hydrographic conditions. During the 1990s, the species occurred as far north as the Gulf of Alaska, but following a contraction of range by several hundred kilometers in 2000 and 2001, its northern limit has reverted to northern Vancouver Island, a return to the distribution observed in the 1980s.

The biological response to the 1998 regime shift was weaker in the Gulf of Alaska and the Bering Sea than in the central North Pacific and CCS. The northern regions of the western North Pacific ressembled the southern regions of the eastern North Pacific in showing an increase in biological production. Zooplankton biomass increased in the Sea of Okhotsk and the previously dominant Japanese sardine was replaced by herring, capelin, and Japanese anchovy.

North Atlantic

Some of the consequences of the warming of the North Atlantic from the 1920s to the 1960s have already been described. There are numerous excellent publications on the subject, dating back to the 1930s and 1940s, which show how much scientific interest there was in the effects of climate change 65 years ago.

The history of cod stocks at Greenland since the early 1900s is particularly well documented, showing how rapidly a species can extend its range (at a rate of 50 km yr^{-1}) and then decline again. This rate of range extension is matched by the examples shown in **Figure 1** and also by the plankton species which have been collected systematically since the 1930s by ships of opportunity towing continuous plankton recorder on many routes in the North Atlantic (*see*). This and other evidence indicate that the rates of change in distribution in response to climate are much more rapid in the sea than they are on land. There are fewer physical barriers in the sea; many marine species have dispersive planktonic life stages; the diverse, immobile habitats, which are created on land by large, long-lived plants, do not

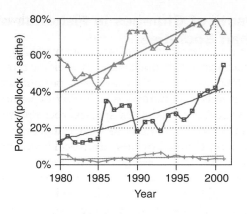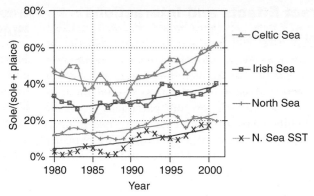

Figure 5 Increasing abundance of warm-adapted species (pollock: Pollachius pollachius and sole: *Solea solea*) relative to similar cold-adapted species (saithe: *Pollachius virens* and plaice: *Pleuronectes platessa*) as shown by catch ratios in the Celtic Sea, Irish Sea, and North Sea. The SST for the North Sea is also shown.

occur in the sea, with the exception of kelp forests in coastal areas. Coral reefs are another exception.

One of the difficulties in ascribing observed changes in fish stocks to climate is that fishing now has such a pervasive effect and exerts such high mortalities on most species. Where similar warm and cold-adapted species co-occur in a particular fishing area, the ratio of their catches gives an indication of which of them is being favored by climate trends (see **Figure 5**).

Baltic Sea

The Baltic Sea is the largest brackish water sea in the world, with much lower species diversity than the adjacent North Sea. The fisheries depend on just three marine species (cod, herring, and sprat), which are at the extreme limits of their tolerance of low salinity and oxygen. Over time, the Baltic has become fresher and less oxygenated, but there are periodic inflows of high-salinity, oxygen-rich water from the Skagerrak (North Sea) which flush through the deep basins and restore more favorable, saline, oxygenated conditions for the fish to reproduce. A specific, short-term weather pattern is required to generate an inflow: a period of easterly wind lowers the sea level in the Baltic, then several weeks of westerly wind force water from the Skagerrak through the shallow Danish Straits and into the deeper areas to the east.

On average, there has been approximately one major inflow per year since 1897 and the benefit to the fish stocks lasts for more than 1 year, but the frequency with which the required short-term weather pattern occurs depends on longer-term climatic factors. Inflows have been much less frequent over the past two decades, the last two major inflows happening in 1993 and 2003. The changes in

windfields associated with climate change may continue to reduce the frequency of inflows to the Baltic.

The reproductive potential of cod in particular has been badly affected by the reduced volume of suitable water for development of its eggs. Cod spawn in the deep basins and if the salinity and density are too low, then their eggs sink into the anoxic layers near the bottom, where they die. This is one of the very few examples in which laboratory studies on individual fish can be applied almost directly to make inferences about the effects of climate. The buoyancy of cod eggs can be measured in a density gradient and their tolerance of low oxygen and salinity studied in incubation chambers. With this information, it is possible to determine what depth they will sink to under the conditions of temperature and salinity found in the deep basins and hence whether they remain in sufficiently oxygenated water for survival.

Coral Reef Fisheries

Coral reefs have suffered an increasing frequency of bleaching events due to loss of symbiotic algae. This occurs when SST remains ~1 °C above the current seasonal maxima. Mortality of corals occurs when the increment in SST is >2 °C. Extensive and extreme bleaching occurred in 1998, associated with the strong El Niño and the hottest year on record.

The mass coral bleaching in the Indian Ocean in 1998 apparently did not result in major short-term impacts on coastal reef fisheries. However, in the longer term, the loss of coral communities and reduced structural complexity of the reefs are expected to have serious consequences for fisheries production with reduced fish species richness, local extinctions, and loss of species within key functional groups of reef fish.

Indirect Effects and Interactions

Spread of Pathogens

Pathogens have been implicated in mass mortalities of many aquatic species, including plants, fish, corals, and mammals, but lack of standard epidemiological data and information on pathogens generally make it difficult to attribute causes. An exception is the northward spread of two protozoan parasites (*Perkinsus marinus* and *Haplosporidium nelsoni*) from the Gulf of Mexico to Delaware Bay and further north, where they have caused mass mortalities of Eastern oysters (*Crassostrea virginica*). Winter temperatures consistently lower than 3 °C limit the development of the MSX disease caused by *Perkinsus* and the poleward spread of this and other pathogens can be expected to continue as such winter temperatures become rarer. This example also illustrates the relevance of seasonal information when considering the effects of climate change, since in this case it is winter temperature which controls the spread of the pathogen.

Effects of Changing Primary Production

Changes in primary production and in food-chain processes due to climate will probably be the major cause of future changes in fisheries production. Reduced nutrient supply to the upper ocean due to slower vertical mixing and changes in the balance between primary production and respiration at higher temperature will result in less energy passing to higher trophic levels. Many other processes will also have an effect, which makes prediction uncertain. For example, nutrient-depleted conditions favor small phytoplankton and longer food chains at the expense of diatoms and short food chains. Altered seasonality of primary production will cause mismatches with zooplankton whose phenology is adapted to the timing of the spring bloom. As a result, a greater proportion of the pelagic production may settle out of the water column to the benefit of benthic production.

There is evidence from satellite observations and from *in situ* studies that primary production in some parts of the ocean has begun to decline, but the results of modeling studies suggest that primary production will change very little up to 2050. Within this global result, there are large regional differences. For example, the highly productive sea–ice margin in the Arctic is retreating, but as a result a greater area of ocean is exposed to direct light and therefore becomes more productive.

Impacts of Changes in Fisheries on Human Societies

Fluctuations in fish stocks have had major economic consequences for human societies throughout history. Fishing communities which were dependent on local resources of just a few species have always been vulnerable to fluctuations in stocks, whether due to overfishing, climate, or other causes. The increase in distant water fleets during the last century reduced the dependence of that sector of the fishing industry on a particular area or species, but the resulting increase in rates of exploitation also reduced stock levels and increased their variability.

Many examples can be cited to show the effects of fish-stock fluctuations. The history of herring in European waters over the past 1000 years influenced the economic fortunes of the Hanseatic League and had a major impact on the economy of northern Europe. Climate-dependent fluctuations in the Far Eastern sardine population influenced their fisheries and human societies dependent on them. Variability in cod stocks at Newfoundland, Greenland, and the Faroe Islands due to a combination of climate-related effects and overfishing had major impacts on the economies and societies, resulting in changes in migration and human demography. The investigation of economic effects of climate change on fisheries is a rapidly developing field, which can be expected to help considerably when planning strategies for adaptation or, in some cases, mitigation of future impacts.

Given the uncertainties over future marine production and consequences for fish stocks, it is not surprising that projections of impacts on human societies and economies are also uncertain. Global aquaculture production increased by *c.* 50% between 1997 and 2003, while capture production decreased by *c.* 5% and the likelihood that these trends will continue also affects the way in which climate change will affect fisheries production.

Some areas, such as Greenland, which are strongly affected by climate variability and which have been undergoing a relatively cold period since the 1960s, can be expected to benefit from warmer oceanic conditions and changes in the marine ecosystem are occuring there quite rapidly. In other areas, such as Iceland, the positive and negative impacts are more finely balanced.

It is very difficult to judge at a global level who the main losers and winners will be from changes in fisheries as a result of climate change, aside from the obvious advantages of being well informed, well capitalized, and able to shift to alternative areas or kinds of fishing activity (or other nonfishery activities) as circumstances change. Some of the most vulnerable systems may be in the mega-deltas of rivers in Asia, such as the Mekong, where 60 million people are active in fisheries in some way or other. These are mainly seasonal floodplain fisheries, which, in addition to overfishing, are increasingly threatened by changes in the hydrological cycle and in land use, damming, irrigation, and channel alteration. Thus the impact of climate change is just one of a number of pressures which require integrated international solutions if the fisheries are to be maintained.

Further Reading

ACIA (2005) *Arctic Climate Impact Assessment Scientific Report*. Cambridge, UK: Cambridge University Press.

Cushing DH (1982) *Climate and Fisheries*. London: Academic Press.

Drinkwater KF, Loeng H, Megrey BA, Bailey N, and Cook RM (eds.) (2005) The influence of climate change on North Atlantic fish stocks. *ICES Journal of Marine Science* 62(7): 1203–1542.

German Advisory Council on Global Change (2006) *The Future Oceans – Warming up, Rising High, Turning Sour*. Special Report. Berlin: German Advisory Council on Global Change (WBGU). http://www.wbgu.de/wbgu_sn2006_en.pdf (accessed Mar. 2008).

King JR (ed.) (2005) *Report of the Study Group on Fisheries and Ecosystem Responses to Recent Regime Shifts*. PICES Scientific Report 28, 162pp.

Lehodey P, Chai F, and Hampton J (2003) Modelling climate-related variability of tuna populations from a coupled ocean biogeochemical-populations dynamics model. *Fisheries Oceanography* 12: 483–494.

Quéro JC, Buit HD, and Vayne JJ (1998) Les observations de poissons tropicaux et le réchauffement des eaux dans l'Atlantique européen. *Oceanologica Acta* 21: 345–351.

Stenseth Nils C, Ottersen G, Hurrell JW, and Belgrano A (eds.) (2004) *Marine Ecosystems and Climate Variation: The North Atlantic. A Comparative Perspective*. Oxford, UK: Oxford University Press.

Vilhjalmsson H (1997) Climatic variations and some examples of their effects on the marine ecology of Icelandic and Greenland waters, in particular during the present century. *Rit Fiskideildar* XV(1): 9–27.

Wood CM and McDonald DG (eds.) (1997) *Global Warming: Implications for Freshwater and Marine Fish*. Cambridge, UK: Cambridge University Press.

Relevant Websites

http://www.ipcc.ch
– IPCC Fourth Assessment Report.

SEABIRD RESPONSES TO CLIMATE CHANGE

David G. Ainley, H.T. Harvey and Associates, San
Jose, CA, USA
G. J. Divoky, University of Alaska, Fairbanks, AK,
USA

Introduction

This article reviews examples showing how seabirds
have responded to changes in atmospheric and
marine climate. Direct and indirect responses take
the form of expansions or contractions of range;
increases or decreases in populations or densities
within existing ranges; and changes in annual cycle,
i.e., timing of reproduction. Direct responses are
those related to environmental factors that affect the
physical suitability of a habitat, e.g., warmer or
colder temperatures exceeding the physiological tolerances of a given species. Other factors that can
affect seabirds directly include: presence/absence of
sea ice, temperature, rain and snowfall rates, wind,
and sea level. Indirect responses are those mediated
through the availability or abundance of resources
such as food or nest sites, both of which are also
affected by climate change.

Seabird response to climate change may be most
apparent in polar regions and eastern boundary
currents, where cooler waters exist in the place of the
warm waters that otherwise would be present. In
analyses of terrestrial systems, where data are in
much greater supply than marine systems, it has been
found that range expansion to higher (cooler but
warming) latitudes has been far more common than
retraction from lower latitudes, leading to speculation that cool margins might be more immediately
responsive to thermal variation than warm margins.
This pattern is evident among sea birds, too. During
periods of changing climate, alteration of air temperatures is most immediate and rapid at high latitudes due to patterns of atmospheric circulation.
Additionally, the seasonal ice and snow cover characteristic of polar regions responds with great sensitivity to changes in air temperatures. Changes in
atmospheric circulation also affect eastern boundary
currents because such currents exist only because of
wind-induced upwelling.

Seabird response to climate change, especially in
eastern boundary currents but true elsewhere,
appears to be mediated often by El Niño or La Niña.
In other words, change is expressed stepwise, each
step coinciding with one of these major, short-term
climatic perturbations. Intensive studies of seabird
populations have been conducted, with a few exceptions, only since the 1970s; and studies of seabird
responses to El Niño and La Niña, although having a
longer history in the Peruvian Current upwelling
system, have become commonplace elsewhere only
since the 1980s. Therefore, our knowledge of seabird responses to climate change, a long-term process, is in its infancy. The problem is exacerbated by
the long generation time of seabirds, which is 15–70
years depending on species.

Evidence of Sea-bird Response to Prehistoric Climate Change

Reviewed here are well-documented cases in which
currently extant seabird species have responded to
climate change during the Pleistocene and Holocene
(last 3 million years, i.e., the period during which
humans have existed).

Southern Ocean

Presently, 98% of Antarctica is ice covered, and only
5% of the coastline is ice free. During the Last
Glacial Maximum (LGM: 19 000 BP), marking the
end of the Pleistocene and beginning of the Holocene, even less ice-free terrain existed as the ice sheets
grew outward to the continental shelf break
and their mass pushed the continent downward.
Most likely, land-nesting penguins (Antarctic genus
Pygoscelis) could not have nested on the Antarctic
continent, or at best at just a few localities (e.g.,
Cape Adare, northernmost Victoria Land). With
warming, loss of mass and subsequent retreat of the
ice, the continent emerged.

The marine-based West Antarctic Ice Sheet (WAIS)
may have begun to retreat first, followed by the land-
based East Antarctic Ice Sheet (EAIS). Many Adélie
penguin colonies now exist on the raised beaches
remaining from successive periods of rapid ice retreat. Carbon-dated bones from the oldest beaches
indicate that Adélie penguins colonized sites soon
after they were exposed. In the Ross Sea, the WAIS
receded south-eastward from the shelf break to its
present position near Ross Island approximately
6200 BP. Penguin remains from Cape Bird, Ross
Island (southwestern Ross Sea), date back to

7070 ± 180 BP; those from the adjacent southern Victoria Land coast (Terra Nova Bay) date to 7505 ± 230 BP. Adélie penguin remains at capes Royds and Barne (Ross Island), which are closest to the ice-sheet front, date back to 500 BP and 375 BP, respectively. The near-coast Windmill Islands, Indian Ocean sector of Antarctica, were covered by the EAIS during the LGM. The first islands were deglaciated about 8000 BP, and the last about 5500 BP. Penguin material from the latter was dated back to 4280–4530 BP, with evidence for occupation 500–1000 years earlier. Therefore, as in Victoria Land, soon after the sea and land were free from glaciers, Adélie penguins established colonies.

The study of raised beaches at Terra Nova Bay also investigated colony extinction. In that area several colonies were occupied 3905–4930 BP, but not since. The period of occupancy, called 'the penguin optimum' by geologists, corresponds to one of a warmer climate than at present. Currently, this section of Victoria Land is mostly sea-ice bound and penguins nest only at Terra Nova Bay owing to a small, persistent polynya (open-water area in the sea ice).

A study that investigated four extinct colonies of chinstrap penguin in the northern part of the Antarctic Peninsula confirmed the rapidity with which colonies can be founded or deserted due to fluctuations in environmental conditions. The colonies were dated at about 240–440 BP. The chinstrap penguin, an open-water loving species, occupied these former colonies during infrequent warmer periods indicated in glacial ice cores from the region. Sea ice is now too extensive for this species offshore of these colonies. Likewise, abandoned Adélie penguin nesting areas in this region were occupied during the Little Ice Age (AD 1500–1850), but since have been abandoned as sea ice has dissipated in recent years (see below).

South-east Atlantic

A well-documented avifaunal change from the Pleistocene to Recent times is based on bone deposits at Ascension and St Helena Islands. During glacial maxima, winds were stronger and upwelling of cold water was more pronounced in the region. This pushed the 23°C surface isotherm north of St Helena, thus accounting for the cool-water seabird avifauna that was present then. Included were some extinct species, as well as the still extant sooty shearwater and white-throated storm petrel. As the glacial period passed, the waters around St Helena became warmer, thereby encouraging a warm-water avifauna similar to that which exists today at Ascension Island; the cool-water group died out or decreased

substantially in representation. Now, a tropical avifauna resides at St Helena including boobies, a frigatebird not present earlier, and Audubon's shearwater. Most recently these have been extirpated by introduced mammals.

North-west Atlantic/Gulf of Mexico

Another well-documented change in the marine bird fauna from Plio-Pleistocene to Recent times is available for Florida. The region experienced several major fluctuations in sea level during glacial and interglacial periods. When sea level decreased markedly to expose the Isthmus of Panama, thus, changing circulation, there was a cessation of upwelling and cool, productive conditions. As a result, a resident cool-water avifauna became extinct. Subsequently, during periods of glacial advance and cooler conditions, more northerly species visited the area; and, conversely, during warmer, interglacial periods, these species disappeared.

Direct Responses to Recent Climate Change

A general warming, especially obvious after the mid-1970s, has occurred in ocean temperatures especially west of the American continents. Reviewed here are sea-bird responses to this change.

Chukchi Sea

The Arctic lacks the extensive water–ice boundaries of the Antarctic and as a result fewer seabird species will be directly affected by the climate-related changes in ice edges. Reconstructions of northern Alaska climatology based on tree rings show that temperatures in northern Alaska are now the warmest within the last 400 years with the last century seeing the most rapid rise in temperatures following the end of the Little Ice Age (AD 1500–1850). Decreases in ice cover in the western Arctic in the last 40 years have been documented, but the recent beginnings of regional ornithological research precludes examining the response of birds to these changes.

Changes in distribution and abundance related to snow cover have been found for certain cavity-nesting members of the auk family (Alcidae). Black guillemots and horned puffins require snow-free cavities for a minimum of 80 and 90 days, respectively, to successfully pair, lay and incubate eggs, and raise their chicks(s). Until the mid-1960s the snow-free period in the northern Chukchi Sea was usually shorter than 80 days but with increasing spring air temperatures, the annual date of spring snow-melt has advanced more than 5 days per decade over the last five decades (**Figure 1**). The annual snow-free period now regularly exceeds 90 days, which reduces the likelihood of chicks being trapped in their nest sites before fledging. This has allowed black guillemots and horned puffins to establish colonies (range expansion) and increase population size in the northern Chukchi.

California Current

Avifaunal changes in this region are particularly telling because the central portion of the California Current marks a transitional area where subtropical and subarctic marine faunas meet, and where north–south faunal shifts have been documented at a variety of temporal scales. Of interest here is the invasion of brown pelicans northward from their 'usual' wintering grounds in central California to the Columbia River mouth and Puget Sound, and the invasion of various terns and black skimmers northward from Mexico into California. The pelican and terns are tropical and subtropical species.

During the last 30 years, air and sea temperatures have increased noticeably in the California Current region. The response of seabirds may be mediated by thermoregulation as evidenced by differences in the amount of subcutaneous fat held by polar and tropical seabird species, and by the behavioral responses of seabirds to inclement air temperatures.

The brown pelican story is particularly well documented and may offer clues to the mechanism by which similar invasions come about. Only during the very intense El Niño of 1982–83 did an unusual

number of pelicans move northward to the Columbia River mouth. They had done this prior to 1900, but then came a several-decade long period of cooler temperatures. Initially, the recent invasion involved juveniles. In subsequent years, these same birds returned to the area, and young-of-the-year birds followed. Most recently, large numbers of adult pelicans have become a usual feature feeding on anchovies that have been present all along. This is an example of how tradition, or the lack thereof, may facilitate the establishment (or demise) of expanded range, in this case, compatible with climate change.

The ranges of skimmers and terns have also expanded in pulses coinciding with El Niño. This pattern is especially clear in the case of the black skimmer, a species whose summer range on the east coast of North America retracts southward in winter. On the west coast, almost every step in a northward expansion of range from Mexico has coincided with ocean warming and, in most cases, El Niño: first California record, 1962 (not connected to ocean warming); first invasion *en masse*, 1968 (El Niño); first nesting at Salton Sea, 1972 (El Niño); first nesting on coast (San Diego), 1976 (El Niño); first nesting farther north at Newport Bay and inland, 1986 (El Niño). Thereafter, for Southern California as a whole, any tie to El Niño became obscure, as (1) average sea-surface temperatures off California became equivalent to those reached during the intense 1982–83 El Niño, and (2) population increase became propelled not just by birds dispersing north from Mexico, but also by recruits raised locally. By 1995, breeding had expanded north to Central California. In California, with warm temperatures year round, skimmers winter near where they breed.

The invasion northward by tropical/subtropical terns also relates to El Niño or other warm-water incursions. The first US colony (and second colony in the world) of elegant tern, a species usually present off California as post-breeders (July through October), was established at San Diego in 1959, during the strongest El Niño event in modern times. A third colony, farther north, was established in 1987 (warm-water year). The colony grew rapidly, and in 1992–93 (El Niño) numbers increased 300% (to 3000 pairs). The tie to El Niño for elegant terns is confused by the strong correlation, initially, between numbers breeding in San Diego and the biomass of certain prey (anchovies), which had also increased. Recently, however, anchovies have decreased. During the intense 1997–98 El Niño, hundreds of elegant terns were observed in courtship during spring even farther north (central California). No colony formed.

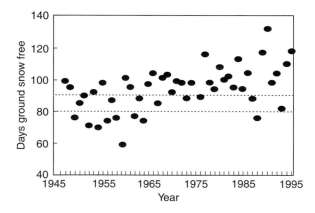

Figure 1 Changes in the length of the annual snow-free period at Barrow, Alaska, 1947–1995. Dashed lines show the number of days that black guillemots and horned puffins require a snow-free cavity (80 and 90 days, respectively). Black guillemots first bred near Barrow in 1966 and horned puffins in 1986. (Redrawn from Divoky, 1998.)

Climate change, and El Niño as well, may be involved in the invasion of Laysan albatross to breed on Isla de Guadalupe, in the California Current off northern Mexico. No historical precedent exists for the breeding by this species anywhere near this region. First nesting occurred in 1983 (El Niño) and by 1988, 35–40 adults were present, including 12 pairs. Ocean temperatures off northern Mexico are now similar to those in the Hawaiian Islands, where nesting of this species was confined until recently. In the California Current, sea temperatures are the warmest during the autumn and winter, which, formerly, were the only seasons when these albatross occurred there. With rising temperatures in the California Current, more and more Laysan albatross have been remaining longer into the summer each year. Related, too, may be the strengthening of winds off the North American west coast to rival more closely the trade winds that buffet the Hawaiian Islands. Albatross depend on persistent winds for efficient flight, and such winds may limit where albatrosses occur, at least at sea.

Several other warm-water species have become more prevalent in the California Current. During recent years, dark-rumped petrel, a species unknown in the California Current region previously, has occurred regularly, and other tropical species, such as Parkinson's petrel and swallow-tailed gull have been observed for the first time in the region.

In response to warmer temperatures coincident with these northward invasions of species from tropical and subtropical regions, a northward retraction of subarctic species appears to be underway, perhaps related indirectly to effects of prey availability. Nowadays, there are markedly fewer black-footed albatross and sooty and pink-footed shearwaters present in the California Current system than occurred just 20 years ago (**Figure 2**). Cassin's auklet is becoming much less common at sea in central California, and its breeding populations have also been declining. Similarly, the southern edge of the breeding range of common murres has retreated north. The species no longer breeds in Southern California (Channel Islands) and numbers have become greatly reduced in Central California (**Figure 3**). Moreover, California Current breeding populations have lost much of their capacity, demonstrated amply as late as the 1970s, to recover

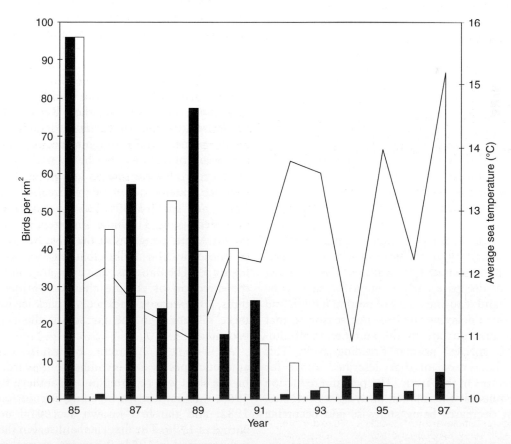

Figure 2   The density (■) (plus 3-point moving average, □) of a cool-water species, the sooty shearwater, in the central portion of the California Current, in conjunction with changes in marine climate (sea surface temperature, (—)), 1985–1997.

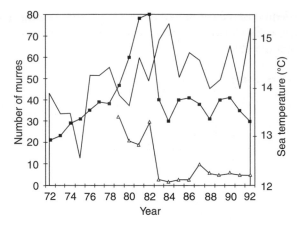

Figure 3 Changes in the number of breeding common murres in central (California, ■) and northern (Washington, △) portions of the California Current during recent decades. Sea surface temperature (—) from central California shown for comparison. During the 1970s, populations had the capacity to recover from reductions due to anthropogenic factors (e.g., oil spills). Since the 1982–83 El Niño event and continued higher temperatures, however, the species' capacity for population recovery has been lost (cf. **Figure 5**)

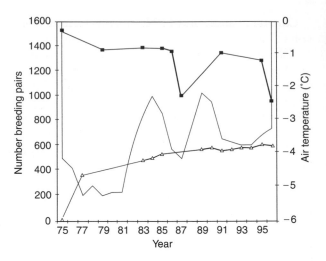

Figure 4 Changes in the number of breeding pairs of two species of penguins at Arthur Harbor, Anvers Island, Antarctica (64°S, 64°W), 1975–1996. The zoogeographic range of Adélie penguins (■) is centered well to the south of this site; the range of chinstrap penguins (△) is centered well to the north. Arthur Harbor is located within a narrow zone of overlap (200 km) between the two species and is at the northern periphery of the Adélie penguins' range.

from catastrophic losses. The latter changes may or may not be more involved with alterations of the food web (see below).

Northern Bellingshausen Sea

Ocean and air temperatures have been increasing and the extent of pack ice has been decreasing for the past few decades in waters west of the Antarctic peninsula. In response, populations of the Adélie penguin, a pack-ice species, have been declining, while those of its congener, the open-water dwelling chinstrap penguin have been increasing (**Figure 4**). The pattern contrasts markedly with the stability of populations on the east side of the Antarctic Peninsula, which is much colder and sea-ice extent has changed little.

The reduction in Adélie penguin populations has been exacerbated by an increase in snowfall coincident with the increased temperatures. Deeper snow drifts do not dissipate early enough for eggs to be laid on dry land (causing a loss of nesting habitat and eggs), thus also delaying the breeding season so that fledging occurs too late in the summer to accommodate this species' normal breeding cycle. This pattern is the reverse of that described above for black guillemots in the Arctic. The penguin reduction has affected mostly the smaller, outlying subcolonies, with major decreases being stepwise and occurring during El Niño.

Similar to the chinstrap penguin, some other species, more typical of the Subantarctic, have been

expanding southward along the west side of the Antarctic Peninsula. These include the brown skua, blue-eyed shag, southern elephant seal and Antarctic fur seal.

Ross Sea

A large portion (32%) of the world's Adélie penguin population occurs in the Ross Sea (South Pacific sector), the southernmost incursion of ocean on the planet (to 78°S). This species is an obligate inhabitant of pack ice, but heavy pack ice negatively affects reproductive success and population growth. Pack-ice extent decreased noticeably in the late 1970s and early 1980s and air temperatures have also been rising. The increasing trends in population size of Adélie penguins in the Ross Sea are opposite to those in the Bellingshausen Sea (see above; **Figure 5**). The patterns, however, are consistent with the direction of climate change: warmer temperatures, less extensive pack ice. As pack ice has become more dispersed in the far south, the penguin has benefited.

As with the Antarctic Peninsula region, subantarctic species are invading southward. The first brown skua was reported in the southern Ross Sea in 1966; the first known breeding attempt occurred in 1982; and the first known successful nesting occurred in 1996. The first elephant seal in the Ross Sea was reported in 1974; at present, several individuals occur there every year.

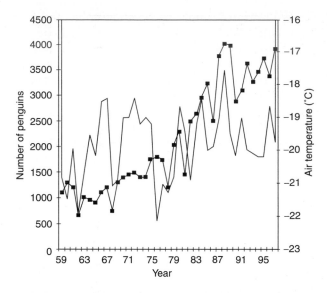

Figure 5 Changes in numbers of breeding pairs of Adélie penguins (■) at Cape Royds, Ross Island, Antarctica (77°S, 166°E), during the past four decades. This is the southernmost breeding site for any penguin species. Although changes in sea ice (less extensive now) is the major direct factor involved, average air temperatures (—) (indirect factor) of the region are shown.

Indirect Responses to Recent Climate Change

California Current

The volume of zooplankton has declined over the past few decades, coincident with reduced upwelling and increased sea-surface temperatures. In response, numbers of sooty shearwaters, formerly, the most abundant avian species in the California Current avifauna, have declined 90% since the 1980s (**Figure 2**). The shearwater feeds heavily on euphausiids in the California Current during spring. The decline, however, has occurred in a stepwise fashion with each El Niño or warm-water event. Sooty shearwaters are now ignoring the Peru and California currents as wintering areas, and favoring instead those waters of the central North Pacific transition zone, which have been cooling and increasing in productivity (see below).

The appearance of the elegant and royal terns as nesting species in California (see above) may in part be linked to the surge in abundance in northern anchovy, which replaced the sardine in the 1960s–1980s. More recently, the sardine has rebounded and the anchovy has declined, but the tern populations continue to grow (see above). Similarly, the former breeding by the brown pelican as far north as Central California was linked to the former presence of sardines. However, the pelicans recently invaded

northward (see above) long before the sardine resurgence began. Farthest north the pelicans feed on anchovies.

Central Pacific

In the central North Pacific gyre, the standing crop of chlorophyll-containing organisms increased gradually between 1965 and 1985, especially after the mid-1970s. This was related to an increase in storminess (winds), which in turn caused deeper mixing and the infusion of nutrients into surface waters. The phenomenon reached a maximum during the early 1980s, after which the algal standing crop subsided. As ocean production increased, so did the reproductive success of red-billed tropicbirds and red-footed boobies nesting in the Leeward Hawaiian Islands (southern part of the gyre). When production subsided, so did the breeding success of these and other species (lobsters, seals) in the region. Allowing for lags of several years as dictated by demographic characteristics, the increased breeding success presumably led to larger populations of these seabird species.

Significant changes in the species composition of seabirds in the central Pacific (south of Hawaii) occurred between the mid-1960s and late 1980s. Densities of Juan Fernandez and black-winged petrels and short-tailed shearwaters were lower in the 1980s, but densities of Stejneger's and mottled petrels and sooty shearwaters were higher. In the case of the latter, the apparent shift in migration route (and possibly destination) is consistent with the decrease in sooty shearwaters in the California Current (see above).

Peru Current

The Peruvian guano birds – Peruvian pelican, piquero, and guanay – provide the best-documented example of changes in seabird populations due to changes in prey availability. Since the time of the Incas, the numbers of guano birds have been strongly correlated with biomass of the seabirds' primary prey, the anchoveta. El Niño 1957 (and earlier episodes) caused crashes in anchoveta and guano bird populations, but these were followed by full recovery. Then, with the disappearance of the anchoveta beginning in the 1960s (due to over-fishing and other factors), each subsequent El Niño (1965, 1972, etc.) brought weaker recovery of the seabird populations.

Apparently, the carrying capacity of the guano birds' marine habitat had changed, but population decreases occurred stepwise, coinciding with mortality caused by each El Niño. However, more than just fishing caused anchoveta to decrease; without fishing pressure, the anchoveta recovered quickly (to its lower level) following El Niño 1982–83, and

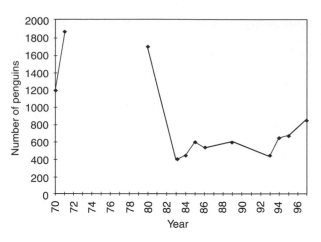

Figure 6 Changes in numbers of breeding Galápagos penguins, 1970–1997. With the 1982–83 El Niño event, the species lost the capacity to recover from periodic events leading to increased mortality. Compare to **Figure 3**, which represents another cool-water species having a similar life-history strategy, but which resides in the other eastern Pacific boundary current (i.e., the two species are ecological complements in the Northern and Southern Hemispheres).

trends in the sardine were contrary to those of the anchoveta beginning in the late 1960s. The seabirds that remain have shifted their breeding sites southward to southern Peru and northern Chile in response to the southward, compensatory increase in sardines. A coincident shift has occurred in the zooplankton and mesopelagic fish fauna. All may be related to an atmospherically driven change in ocean circulation, bringing more subtropical or oceanic water onshore. It is not just breeding seabird species that have been affected, nonbreeding species of the region, such as sooty shearwater, have been wintering elsewhere than the Peru Current (see above).

Trends in penguin populations on the Galapagos confirm that a system-wide change has occurred off western South America (**Figure 6**). Galapagos penguins respond positively to cool water and negatively to warm-water conditions; until recently they recovered after each El Niño, just like the Peruvian guano birds. Then came El Niño 1982–83. The population declined threefold, followed by just a slight recovery. Apparently, the carrying capacity of the habitat of this seabird, too, is much different now than a few decades ago. Like the diving, cool-water species of the California Current (see above), due to climate change, the penguin has lost its capacity for population growth and recovery.

Gulf of Alaska and Bering Sea

A major 'regime' shift involving the physical and biological make-up of the Gulf of Alaska and Bering

Sea is illustrated amply by oceanographic and fisheries data. Widespread changes and switches in populations of ecologically competing fish and invertebrate populations have been underway since the mid-1970s. Ironically, seabird populations in the region show few geographically consistent patterns that can be linked to the biological oceanographic trends. There have been no range expansions or retractions, no doubt because this region, in spite of its great size, does not constitute a faunal transition; and from within the region, species to species, some colonies have shown increases, others decreases, and others stability. Unfortunately, the picture has been muddled by the introduction of exotic terrestrial mammals to many seabird nesting islands. Such introductions have caused disappearance and serious declines in sea-bird numbers. In turn, the subsequent eradication of mammals has allowed recolonization and increases in seabird numbers.

In the Bering Sea, changes in the population biology of seabirds have been linked to decadal shifts in the intensity and location of the Aleutian Low Pressure System (the Pacific Decadal Oscillation), which affects sea surface temperatures among other things. For periods of 15–30 years the pressure (North Pacific Index) shifts from values that are above (high pressure state) or below (low pressure state) the long-term average. Kittiwakes in the central Bering Sea (but not necessarily the Gulf of Alaska) do better with warmer sea temperatures; in addition, the relationship of kittiwake productivity to sea surface temperature changes sign with switches from the high to low pressure state. Similarly, the dominance among congeneric species of murres at various sympatric breeding localities may flip-flop depending on pressure state. Although these links to climate have been identified, cause–effect relationships remain obscure. At the Pribilof Islands in the Bering Sea, declines in seabird numbers, particularly of kittiwakes, coincided with the regime shift that began in 1976. Accompanying these declines has been a shift in diets, in which lipid-poor juvenile walleye pollock have been substituted for lipid-rich forage fishes such as sand lance and capelin. Thus, the regime shifts may have altered trophic pathways to forage fishes and in turn the availability of these fish to seabirds. Analogous patterns are seen among the seabirds of Prince William Sound.

Chukchi Sea

Decrease of pack ice extent in response to recent global temperature increases has been more pronounced in the Arctic than the Antarctic. This decrease has resulted in changes in the availability of under-ice fauna, fish, and zooplankton that are important to certain arctic seabirds. The decline of black guillemot populations in northern Alaska in the last decade may be associated with this pack ice decrease.

North Atlantic

The North Atlantic is geographically confined and has been subject to intense human fishery pressure for centuries. Nevertheless, patterns linked to climate change have emerged. In the North Sea, between 1955 and 1987, direct, positive links exist between the frequency of westerly winds and such biological variables as zooplankton volumes, stock size of young herring (a seabird prey), and the following parameters of black-legged kittiwake biology: date of laying, number of eggs laid per nest, and breeding success. As westerly winds subsided through to about 1980, zooplankton and herring decreased, kittiwake laying date retarded and breeding declined. Then, with a switch to increased westerly winds, the biological parameters reversed.

In the western Atlantic, changes in the fish and seabird fauna correlate to warming sea surface temperatures near Newfoundland. Since the late 1800s, mackerel have moved in, as has one of their predators, the Atlantic gannet. Consequently, the latter has been establishing breeding colonies farther and farther north along the coast.

South-east Atlantic, Benguela Current

A record of changes in sea-bird populations relative to the abundance and distribution of prey in the Benguela Current is equivalent to that of the Peruvian upwelling system. As in other eastern boundary currents, Benguela stocks of anchovy and sardine have flip-flopped on 20–30 year cycles for as long as records exist (to the early 1900s). Like other eastern boundary currents, the region of concentration of prey fish, anchovy versus sardine, changes with sardines being relatively more abundant poleward in the region compared to anchovies. Thus, similar to the Peruvian situation, seabird populations have shifted, but patterns also are apparent at smaller timescales depending on interannual changes in spawning areas of the fish. As with all eastern boundary currents, the role of climate in changing pelagic fish populations is being intensively debated.

See also

El Niño Southern Oscillation (ENSO). Sea Ice: Overview. Sea Level Change.

Further Reading

Aebischer NJ, Coulson JC, and Colebrook JM (1990) Parallel long term trends across four marine trophic levels and weather. *Nature* 347: 753–755.

Crawford JM and Shelton PA (1978) Pelagic fish and seabird interrelationships off the coasts of South West and South Africa. *Biological Conservation* 14: 85–109.

Decker MB, Hunt GL, and Byrd GV (1996) The relationship between sea surface temperature, the abundance of juvenile walleye pollock (*Theragra chalcogramma*), and the reproductive performance and diets of seabirds at the Pribilof islands, southeastern Bering Sea. *Canadian Journal of Fish and Aquatic Science* 121: 425–437.

Divoky GJ (1998) *Factors Affecting Growth of a Black Guillemot Colony in Northern Alaska*. PhD Dissertation, University of Alaska, Fairbanks.

Emslie SD (1998) *Avian Community, Climate, and Sea-level Changes in the Plio-Pleistocene of the Florida Peninsula*. Ornithological Monograph No. 50. Washington, DC: American Ornithological Union.

Furness RW and Greenwood JJD (eds.) (1992) *Birds as Monitors of Environmental Change*. London, New York: Chapman and Hall.

Olson SL (1975) *Paleornithology of St Helena Island, South Atlantic Ocean*. Smithsonian Contributions to Paleobiology, No. 23. Washington, Dc..

Smith RC, Ainley D, Baker K, et al. (1999) Marine ecosystem sensitivity to climate change. *BioScience* 49(5): 393–404.

Springer AM (1998) Is it all climate change? Why marine bird and mammal populations fluctuate in the North Pacific. In: Holloway G, Muller P, and Henderson D (eds.) *Biotic Impacts of Extratropical Climate Variability in the Pacific*. Honolulu: University of Hawaii: SOEST Special Publication.

Stuiver M, Denton GH, Hughes T, and Fastook JL (1981) History of the marine ice sheet in West Antarctica during the last glaciation: a working hypothesis. In: Denton GH and Hughes T (eds.) *The Last Great Ice Sheets*. New York: Wiley.

CARBON SEQUESTRATION VIA DIRECT INJECTION INTO THE OCEAN

E. E. Adams, Massachusetts Institute of Technology, Cambridge, MA, USA
K. Caldeira, Stanford University, Stanford, CA, USA

Introduction

Global climate change, triggered by a buildup of greenhouse gases, is emerging as perhaps the most serious environmental challenge in the twenty-first century. The primary greenhouse gas is CO_2, whose concentration in the atmosphere has climbed from its preindustrial level of *c.* 280 to >380 ppm. Stabilization at no more than 500–550 ppm is a target frequently discussed to avoid major climatic impact.

The primary source of CO_2 is the burning of fossil fuels – specifically gas, oil, and coal – so stabilization of atmospheric CO_2 concentration will clearly require substantial reductions in CO_2 emissions from these sources. For example, one commonly discussed scenario to stabilize at 500 ppm by the mid-twenty-first century suggests that about 640 Gt CO_2 (*c.* 175 Gt C) would need to be avoided over 50 years, with further emission reductions beyond 50 years. As references, a 1000 MW pulverized coal plant produces 6–8 Mt CO_2 (*c.* 2 Mt C) per year, while an oil-fired single-cycle plant produces about two-thirds this amount and a natural gas combined cycle plant produces about half this amount. Thus the above scenario would require that the atmospheric emissions from the equivalent of 2000–4000 large power plants be avoided by approximately the year 2050.

Such changes will require a dramatic reduction in our current dependence on fossil fuels through increased conservation and improved efficiency, as well as the introduction of nonfossil energy sources like solar, wind, and nuclear. While these strategies will slow the buildup of atmospheric CO_2, it is probable that they will not reduce emissions to the required level. In other words, fossil fuels, which currently supply over 85% of the world's energy needs, are likely to remain our primary energy source for the foreseeable future. This has led to increased interest in a new strategy termed carbon capture and storage, or sequestration. The importance of this option for mitigating climate change is highlighted by the recent

Special Report on Carbon Dioxide Capture and Storage published by the Intergovernmental Panel on Climate Change, to which the reader is referred for more information.

Carbon sequestration is often associated with the planting of trees. As they mature, the trees remove carbon from the atmosphere. As long as the forest remains in place, the carbon is effectively sequestered. Another type of sequestration involves capturing CO_2 from large, stationary sources, such as a power plant or chemical factory, and storing the CO_2 in underground reservoirs or the deep ocean, the latter being the focus of this article. There has been much attention paid recently to underground storage with several large-scale field sites in operation or being planned. Conversely, while there have been many studies regarding use of the deep ocean as a sink for atmospheric carbon, there have been only a few small-scale field studies.

Why is the ocean of interest as a sink for anthropogenic CO_2? The ocean already contains an estimated 40 000 Gt C compared with about 800 Gt C in the atmosphere and 2200 Gt C in the land biosphere. As a result, the amount of carbon that would cause a doubling of the atmospheric concentration would only change the ocean concentration by about 2%. In addition, natural chemical equilibration between the atmosphere and ocean would result in about 80% of present-day emissions ultimately residing in the ocean. Discharging CO_2 directly to the ocean would accelerate this slow, natural process, thus reducing both peak atmospheric CO_2 concentrations and their rate of increase. It is noted that a related strategy for sequestration – not discussed here – would be to enhance the biological sink using nutrients such as iron to fertilize portions of the world's oceans, thus stimulating phytoplankton growth. The phytoplankton would increase the rate of biological uptake of CO_2, and a portion of the CO_2 would be transported to ocean depths when the plankton die.

The indirect flux of CO_2 to the ocean from the atmosphere is already quite apparent: since preindustrial times, the pH of the surface ocean has been reduced by about 0.1 units, from an initial surface pH of about 8.2. **Figure 1** illustrates what could happen to ocean pH under conditions of continued atmospheric release of CO_2. Under the conditions simulated, the pH of the surface would drop by over 0.7 units. Conversely, by injecting some of the CO_2

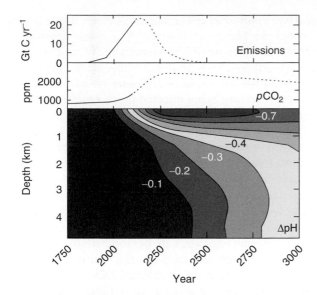

Figure 1 Model simulations of long-term ocean pH changes, averaged horizontally, as a result of atmospheric CO_2 emissions shown in the top panel. Reprinted from Caldeira K and Wickett ME (2003) Anthropogenic carbon and ocean pH. *Nature* 425: 365.

to the deep ocean, the change in pH could be more uniformly distributed.

Ocean sequestration of CO_2 by direct injection assumes that a relatively pure CO_2 stream has been generated at a power plant or chemical factory and transported to an injection point. To better understand the role the ocean can play, we address the capacity of the ocean to sequester CO_2, its effectiveness at reducing atmospheric CO_2 levels, how to inject the CO_2, and possible environmental consequences and issues of public perception.

Capacity

How much carbon can the ocean sequester? At over 70% of the Earth's surface and an average depth of 3800 m, the ocean has enormous storage capacity; based on physical chemistry, the amount of CO_2 that could be dissolved in the deep ocean far exceeds the estimated available fossil energy resources of 5000–10 000 Gt C. However, a more realistic criterion needs to be based on an understanding of ocean biogeochemistry and expected environmental impact.

CO_2 exists in seawater in various forms as part of the carbonate system:

$$CO_2(aq) + H_2O \leftrightarrow H_2CO_3(aq)$$
$$\leftrightarrow H^+ + HCO_3^- \leftrightarrow 2H^+ + CO_3^{2-} \qquad [1]$$

Dissolving additional CO_2 increases the hydrogen ion concentration (lowering the pH), but the change

is buffered by the fact that total alkalinity is conserved, which results in carbonate ion being converted into bicarbonate. Thus, the principal reactions occurring when CO_2 is dissolved in seawater are

$$CO_2 + H_2O + CO_3^{2-} \rightarrow 2HCO_3^- \qquad [2]$$

$$CO_2 + H_2O + H^+ \rightarrow HCO_3^- \qquad [3]$$

which result in a decrease in pH and carbonate ion, and an increase in bicarbonate ion.

Reduced pH is one of the principal environmental impacts threatening marine organisms, the other being the concentration of CO_2 itself. At short travel times from the injection point, the changes in pH and CO_2 concentration will be greatest, which suggests that injection schemes should achieve the maximum dilution possible to minimize potential acute impacts in the vicinity of injection. See further discussion below.

At longer travel times, injected carbon would be distributed widely in the oceans and any far-field impact of the injected CO_2 on the oceans would be similar to the impact of anthropogenic CO_2 absorbed from the atmosphere. As indicated above, such changes are already taking place within the surface ocean, where the pH has been reduced by about 0.1 unit. Adding about 2000 Gt CO_2 to the ocean would reduce the average ocean pH by about 0.1 unit, while adding about 5600 Gt CO_2 (about 200 years of current emissions) would decrease the average ocean pH by about 0.3 units. (It should be noted that with stabilization of atmospheric CO_2 at 550 ppm, natural chemical equilibration between the atmosphere and ocean will result in eventual storage of over 6000 Gt CO_2 in the ocean.)

The impacts of such changes are poorly understood. The deep-ocean environment has been relatively stable and it is unknown to what extent changes in dissolved carbon or pH would affect these ecosystems. However, one can examine measured spatial and temporal variation in ocean pH to understand how much change might be tolerated. The spatial variability within given zoogeographic regions and bathymetric ranges (where similar ecosystems might be expected), and the temporal variability at a particular site, have both been found to vary by about 0.1 pH unit. If it is assumed that a change of 0.1 unit is a threshold tolerance, and that CO_2 should be stored in the bottom half of the ocean's volume (to maximize retention), nearly 1000 Gt CO_2 might be stored, which exceeds the 640 Gt CO_2 over 50 years estimated above. It is important to recognize that the long-term changes in ocean pH would ultimately be much the same whether the CO_2 is released into the atmosphere or the deep ocean. However, in the shorter term, releasing the CO_2 in the deep ocean will diminish the pH change in the near-surface ocean, where marine biota are most plentiful. Thus, direct injection of CO_2 into the deep ocean could reduce adverse impacts presently occurring in the surface ocean. In the long run, however, a sustainable solution to the problem of climate change must ultimately entail a drastic reduction of total CO_2 emissions.

Effectiveness

Carbon dioxide is constantly exchanged between the ocean and atmosphere. Each year the ocean and atmosphere exchange about 350 Gt CO_2, with a net ocean uptake currently of about 8 Gt CO_2. Because of this exchange, questions arise as to how effective ocean sequestration will be at keeping the CO_2 out of the atmosphere. Specifically, is the sequestration permanent, and if not, how fast does the CO_2 leak back to the atmosphere. Because there has been no long-term CO_2 direct-injection experiment in the ocean, the long-term effectiveness of direct CO_2 injection must be predicted based on observations of other oceanic tracers (e.g., radiocarbon) and on computer models of ocean circulation and chemistry.

As implied earlier, because the atmosphere and ocean are currently out of equilibrium, most CO_2 emitted to either media will ultimately enter the ocean. The percentage that is permanently sequestered depends on the atmospheric CO_2 concentration, through the effect of atmospheric CO_2 on surface ocean chemistry (see **Table 1**). At today's concentration of c. 380 ppm, nearly 80% of any

Table 1 Percent of injected CO_2 permanently sequestered from the atmosphere as a function of atmospheric CO_2 stabilization concentration

Atmospheric carbon dioxide concentration (ppm)	Percentage of carbon dioxide permanently sequestered
350	80
450	77
550	74
650	72
750	70
1000	66

Based on data in IPCC (2005) *Special Report on Carbon Dioxide Capture and Storage*. Prepared by Working Group III of the Intergovernmental Panel on Climate Change. Cambridge, UK: Cambridge University Press. http://arch.rivm.nl/env/int/ipcc/pages_media/SRCCS-final/IPCCSpecialReportonCarbondioxideCaptureandStorage.htm (accessed Mar. 2008) and references therein.

carbon emitted to either the atmosphere or the ocean would be permanent, while at a concentration of 550 ppm, 74% would be permanent. Of course, even at equilibrium, CO_2 would continue to be exchanged between the atmosphere and oceans, so the carbon that is currently being injected is not exactly the same carbon that will reside in equilibrium.

For CO_2 injected to the ocean today, the net quantity retained in the ocean ranges from 100% (now) to about 80% as equilibrium between the atmosphere and oceans is approached. (A somewhat greater percentage will ultimately be retained as CO_2 reacts with ocean sediments over a timescale of thousands of years.) The nomenclature surrounding ocean carbon storage can be somewhat confusing. The percentage retained in the ocean shown in **Figure 2** is the fraction of injected CO_2 that has never interacted with the atmosphere. **Table 1** shows the fraction of CO_2 that contacts the atmosphere that remains permanently in the ocean. So, for example, for a 550 ppm atmosphere, even as the 'retained fraction' approaches zero (**Figure 2**), the amount permanently stored in the ocean approaches 74% (see **Table 1**). The exact time course depends on the location and depth of the injection.

Several computer modeling studies have studied the issue of retention. The most comprehensive summary is the Global Ocean Storage of Anthropogenic Carbon (GOSAC) intercomparison study of several ocean general circulation models (OGCMs). In this study a number of OGCMs simulated the fate of CO_2, injected over a period of 100 years at seven locations and three depths, for a period of 500 years. The CO_2 retained as a function of time, averaged over the seven sites, is shown in **Figure 2**. While there is variability among models, they all show that retention increases with injection depth, with most simulations predicting over 70% retention after 500 years for an injection depth of 3000 m.

The time required for injected carbon to mix from the deep ocean to the atmosphere is roughly equal to the time required for carbon to mix from the atmosphere to the deep ocean. This can be estimated through observations of radiocarbon (carbon-14) in the ocean. Correcting for mixing of ocean waters from different sources, the age of North Pacific deep water is in the range of 700–1000 years, while other basins, such as the North Atlantic, have overturning times of 300 years or more. These estimates are consistent with output from OGCMs and, collectively, suggest that outgassing of the 20% of injected carbon would occur on a timescale of 300–1000 years.

It is important to stress that leakage to the atmosphere would take place gradually and over large areas of the ocean. Thus, unlike geological sequestration, it

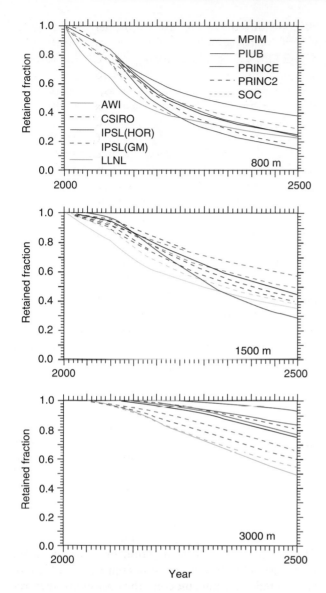

Figure 2 Model-intercomparison study reported by Orr in 2004 showing fraction of CO_2, injected from 2000 through 2100, that remains isolated from the atmosphere as a function of time and injection depth. Results are averaged over seven injection locations. Most of the CO_2 that does interact with the atmosphere remains in the ocean (see **Table 1**), so the amount of CO_2 remaining in the ocean is much greater than shown here. Reprinted with permission from IEA Greenhouse Gas R&D Programme.

would not be possible to produce a sudden release that could lead to harmful CO_2 concentrations at the ocean or land surface.

Injection Methods

The first injection concept was proposed by the Italian physicist Cesare Marchetti, who thought to introduce CO_2 into the outflow of the Mediterranean

Sea, where the relatively dense seawater would cause the CO_2 to sink as it entered the Atlantic Ocean. As illustrated in **Figure 3**, a number of options have been considered since then.

Understanding these methods requires some background information on the CO_2–seawater system. Referring to **Figure 4**, at typical ocean pressures and temperatures, pure CO_2 would be a gas above a depth of 400–500 m and a liquid below that depth. Liquid CO_2 is more compressible than seawater, and would be positively buoyant (i.e., it will rise) down to about 3000 m, but negatively buoyant (i.e., it will sink) below that depth. At about 3700 m, the liquid becomes negatively buoyant compared to seawater saturated with CO_2. In seawater–CO_2 systems, CO_2 hydrate ($CO_2 \cdot nH_2O$, $n \sim 5.75$) can form below $c.$ 400 m depth depending on the relative compositions of CO_2 and H_2O. CO_2 hydrate is a solid with a density about 10% greater than that of seawater.

The rising droplet plume has been the most studied and is probably the easiest scheme to implement. It would rely on commercially available technology to inject the CO_2 as a stream of buoyant droplets from a bottom manifold. Effective sequestration can be achieved by locating the manifold below the thermocline, and dilution can be increased by increasing the manifold length. Even better dilution can be achieved by releasing the CO_2 droplets from a moving ship whose motion provides additional dispersal. Although the means of delivery are different, the plumes resulting from these two options would be similar, each creating a vertical band of CO_2-enriched seawater over a prescribed horizontal region.

Another promising option is to inject liquid CO_2 into a reactor where it can react at a controlled rate with seawater to form hydrates. While it is difficult to achieve 100% reaction efficiency, laboratory and field experiments indicate that negative buoyancy, and hence sinking, can be achieved with as little as about 25% reaction efficiency. The hydrate reactor could be towed from a moving ship to encourage dilution, or attached to a fixed platform, where the large concentration of dense particles, and the increased seawater density caused by hydrate dissolution, would induce a negatively buoyant plume.

The concept of a CO_2 lake is based on a desire to minimize leakage to the atmosphere and exposure to biota. This would require more advanced technology and perhaps higher costs, as the depth of the lake should be at least 3000 m, which exceeds the depths at which the offshore oil industry currently works. The CO_2 in the lake would be partly in the form of solid hydrates. This would limit the CO_2 dissolution into the water column, further slowing leakage to the atmosphere from that shown in **Figure 2**, which

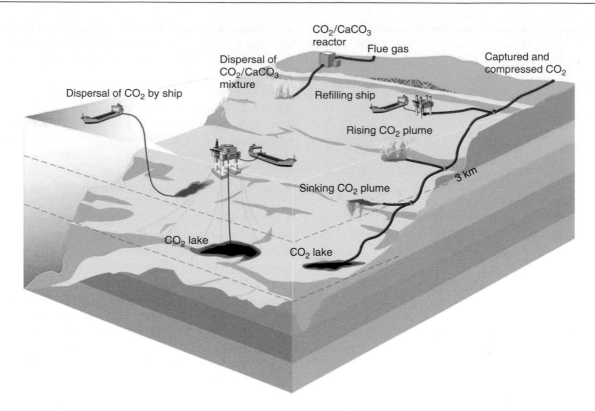

Figure 3 Different strategies for ocean carbon sequestration. Reprinted with permission from IPCC (2005) *Special Report on Carbon Dioxide Capture and Storage*, figure TS-9 (printed as *Special Report on Safeguarding the Ozone Layer and the Global Climate System*, Figure 6.1). Prepared by Working Group III of the Intergovernmental Panel on Climate Change. Cambridge, UK: Cambridge University Press. http://arch.rivm.nl/env/int/ipcc/pages_media/SRCCS-final/IPCCSpecialReportonCarbondioxideCaptureandStorage.htm, with permission from the Intergovernmental Panel on Climate Change.

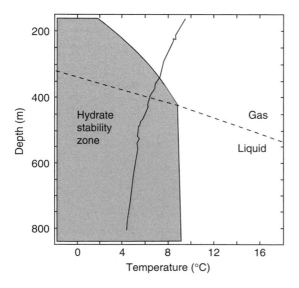

Figure 4 Phase diagram for CO_2 including typical ocean temperature profile (solid line). Reprinted from Brewer PG, Peltzer E, Aya I, *et al.* (2004) Small scale field study of an ocean CO_2 plume. *Journal of Oceanography* 60(4): 751.

assumes that CO_2 is injected into the water column. It is also possible that various approaches could be engineered to physically contain CO_2 on the seafloor and isolate the CO_2 from the overlying water column (and perhaps the sediments); however, this would entail an additional cost.

Another method that has received attention is injecting a dense CO_2–seawater mixture at a depth of 500–1000 m, forming a sinking bottom gravity current. CO_2-enriched seawater is less than 1% heavier than seawater, but this is sufficient to promote a sinking density current, especially if the current were formed along a submarine canyon. However, the environmental impacts would be greater with this option due to the concentrated nature of the plume, and its contact with the seafloor.

As discussed earlier, the deep ocean equilibrates with the surface ocean on the scale of 300–1000 years, and by injecting anthropogenic CO_2 into the deep ocean, the surface-to-deep mixing timescale is effectively bypassed. Anthropogenic CO_2 also equilibrates with carbonate sediments, but over a much longer time, about 6000 years. Technical means could also be used to bypass this timescale, thereby increasing the effectiveness and diminishing the environmental impacts of intentional storage of carbon dioxide in the ocean. For example, CO_2 reacts with carbonate sediments to form bicarbonate ions (HCO_3^-) as indicated by eqn [2]. Power plant CO_2 could be dissolved in seawater, then reacted with crushed limestone, either at the power plant or at the point of release, thus minimizing changes in plume pH. Or an emulsion of liquid CO_2-in-water could be stabilized by fine particles of pulverized limestone; the emulsion would be sufficiently dense to form a sinking plume, whose pH change would be buffered by the limestone. Drawbacks of these approaches include the cost to mine and transport large quantities of carbonate minerals.

Local Environmental Impacts and Public Perception

Environmental impacts may be the most significant factor determining the acceptability of ocean storage, since the strategy is predicated on the notion that impacts to the ocean will be significantly less than the avoided impacts of continued emission to the atmosphere. Earlier, environmental impacts were discussed from the global viewpoint. Here, we examine the environmental impacts near the injection point.

A number of studies have summarized potential impacts to different types of organisms, including adult fish, developmental fish, zooplankton, and benthic fauna. While earlier studies focused mainly on lethal impacts to coastal fauna exposed to strong acids, recent data have focused on deep-water organisms exposed to CO_2, and have included sublethal effects. Impacts include respiratory stress (reduced pH limits oxygen binding and transport of respiratory proteins), acidosis (reduced pH disrupts an organism's acid/basis balance), and metabolic depression (elevated CO_2 causes some animals to reach a state of torpor).

Data generally show that CO_2 causes greater stress than an equivalent change in pH caused by a different acid, that there are strong differences in tolerance among different species and among different life stages of the same species, and that the duration of stress, as well as the level of stress, are important. While some studies imply that deep organisms would be less tolerant than surface organisms, other studies have found the opposite. Likewise, some animals are able to avoid regions of high CO_2 concentration, while others appear less able to. Results generally suggest that lethal effects can be avoided by achieving high near-field dilution. However, more research is needed to resolve impacts, especially at the community level (e.g., reduced lifespan and reproduction effects).

The viability of ocean storage as a greenhouse gas mitigation option hinges on social, political, and regulatory considerations. In view of public precaution toward the ocean, the strategy will require that all parties (private, public, nongovernmental organizations) be included in ongoing research and debate. But the difficulty in this approach is highlighted by the recent experience of an international research team working on ocean carbon sequestration research. A major part of their collaboration was to have included a field experiment involving release of 5 t of CO_2 off the coast of Norway. Researchers would have monitored the physical, chemical, and biological effects of the injected CO_2 over a period of about a week. However, lobbying from environmental groups caused the Norwegian Minister of Environment to rescind the group's permit that had previously been granted. Such actions are unfortunate, because field experiments of this type are what is needed to produce data that would help policymakers decide if full-scale implementation would be prudent.

See also

Abrupt Climate Change. Carbon Cycle. Carbon Dioxide (CO_2) Cycle. Ocean Carbon System, Modeling of. Paleoceanography: the Greenhouse World.

Further Reading

Alendal G and Drange H (2001) Two-phase, near field modeling of purposefully released CO_2 in the ocean. *Journal of Geophysical Research* 106(C1): 1085–1096.

Brewer PG, Peltzer E, Aya I, *et al.* (2004) Small scale field study of an ocean CO_2 plume. *Journal of Oceanography* 60(4): 751–758.

Caldeira K and Rau GH (2000) Accelerating carbonate dissolution to sequester carbon dioxide in the ocean: Geochemical implications. *Geophysical Research Letters* 27(2): 225–228.

Caldeira K and Wickett ME (2003) Anthropogenic carbon and ocean pH. *Nature* 425: 365.

Giles J (2002) Norway sinks ocean carbon study. *Nature* 419: 6.

Golomb D, Pennell S, Ryan D, Barry E, and Swett P (2007) Ocean sequestration of carbon dioxide: Modeling the deep ocean release of a dense emulsion of liquid CO_2-in-water stabilized by pulverized limestone particles. *Environmental Science and Technology* 41(13): 4698–4704.

Haugan H and Drange H (1992) Sequestration of CO_2 in the deep ocean by shallow injection. *Nature* 357(28): 1065–1072.

IPCC (2005) *Special Report on Carbon Dioxide Capture and Storage*. Prepared by Working Group III of the Intergovernmental Panel on Climate Change. Cambridge, UK: Cambridge University Press. http://arch.rivm.nl/env/int/ipcc/pages_media/SRCCS-final/IPCCSpecialReporton CarbondioxideCaptureandStorage.htm (accessed Mar. 2008).

Ishimatsu A, Kikkawa T, Hayashi M, and Lee KS (2004) Effects of CO_2 on marine fish: Larvae and adults. *Journal of Oceanography* 60: 731–741.

Israelsson P and Adams E (2007) Evaluation of the Acute Biological Impacts of Ocean Carbon Sequestration. *Final Report for US Dept. of Energy, under grant DE-FG26-98FT40334*. Cambridge, MA: Massachusetts Institute of Technology.

Kikkawa T, Ishimatsu A, and Kita J (2003) Acute CO_2 tolerance during the early developmental stages of four marine teleosts. *Environmental Toxicology* 18(6): 375–382.

Ohsumi T (1995) CO_2 storage options in the deep-sea. *Marine Technology Society Journal* 29(3): 58–66.

Orr JC (2004) *Modeling of Ocean Storage of CO_2 – The GOSAC Study*, Report PH4/37, 96pp. Paris: Greenhouse Gas R&D Programme, International Energy Agency.

Ozaki M, Minamiura J, Kitajima Y, Mizokami S, Takeuchi K, and Hatakenka K (2001) CO_2 ocean sequestration by moving ships. *Journal of Marine Science and Technology* 6: 51–58.

Pörtner HO, Reipschläger A, and Heisler N (2004) Biological impact of elevated ocean CO_2 concentrations: Lessons from animal physiology and Earth history. *Journal of Oceanography* 60(4): 705–718.

Riestenberg D, Tsouris C, Brewer P, *et al.* (2005) Field studies on the formation of sinking CO_2 particles for ocean carbon sequestration: Effects of injector geometry on particle density and dissolution rate and model simulation of plume behavior. *Environmental Science and Technology* 39: 7287–7293.

Sato T and Sato K (2002) Numerical prediction of the dilution process and its biological impacts in CO_2 ocean sequestration. *Journal of Marine Science and Technology* 6(4): 169–180.

Vetter EW and Smith CR (2005) Insights into the ecological effects of deep-ocean CO_2 enrichment: The impacts of natural CO_2 venting at Loihi seamount on deep sea scavengers. *Journal of Geophysical Research* 110: C09S13 (doi:10.1029/2004JC002617).

Wannamaker E and Adams E (2006) Modeling descending carbon dioxide injections in the ocean. *Journal of Hydraulic Research* 44(3): 324–337.

Watanabe Y, Yamaguchi A, Ishida H, *et al.* (2006) Lethality of increasing CO_2 levels on deep-sea copepods in the western North Pacific. *Journal of Oceanography* 62: 185–196.

IRON FERTILIZATION

K. H. Coale, Moss Landing Marine Laboratories, CA, USA

Introduction

The trace element iron has been shown to play a critical role in nutrient utilization and phytoplankton growth and therefore in the uptake of carbon dioxide from the surface waters of the global ocean. Carbon fixation in the surface waters, via phytoplankton growth, shifts the ocean–atmosphere exchange equilibrium for carbon dioxide. As a result, levels of atmospheric carbon dioxide (a greenhouse gas) and iron flux to the oceans have been linked to climate change (glacial to interglacial transitions). These recent findings have led some to suggest that large-scale iron fertilization of the world's oceans might therefore be a feasible strategy for controlling climate. Others speculate that such a strategy could deleteriously alter the ocean ecosystem, and still others have calculated that such a strategy would be ineffective in removing sufficient carbon dioxide to produce a sizable and rapid result. This article focuses on carbon and the major plant nutrients, nitrate, phosphate, and silicate, and describes how our recent discovery of the role of iron in the oceans has increased our understanding of phytoplankton growth, nutrient cycling, and the flux of carbon from the atmosphere to the deep sea.

Major Nutrients

Phytoplankton growth in the oceans requires many physical, chemical, and biological factors that are distributed inhomogenously in space and time. Because carbon, primarily in the form of the bicarbonate ion, and sulfur, as sulfate, are abundant throughout the water column, the major plant nutrients in the ocean commonly thought to be critical for phytoplankton growth are those that exist at the micromolar level such as nitrate, phosphate, and silicate. These, together with carbon and sulfur, form the major building blocks for biomass in the sea. As fundamental cellular constituents, they are generally thought to be taken up and remineralized in constant ratio to one another. This is known as the Redfield

ratio (Redfield, 1934, 1958) and can be expressed on a molar basis relative to carbon as 106C:16N:1P.

Significant local variations in this uptake/regeneration relationship can be found and are a function of the phytoplankton community and growth conditions, yet this ratio can serve as a conceptual model for nutrient uptake and export.

The vertical distribution of the major nutrients typically shows surface water depletion and increasing concentrations with depth. The schematic profile in **Figure 1** reflects the processes of phytoplankton uptake within the euphotic zone and remineralization of sinking planktonic debris via microbial degradation, leading to increased concentrations in the deep sea. Given favorable growth conditions, the nutrients at the surface may be depleted to zero. The rate of phytoplankton production of new biomass, and therefore the rate of carbon uptake, is controlled by the resupply of nutrients to the surface waters, usually via the upwelling of deep waters. Upwelling occurs over the entire ocean basin at the rate of approximately 4 m per year but increases in coastal and

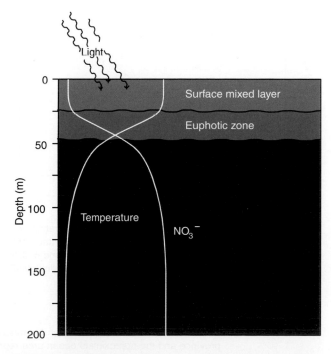

Figure 1 A schematic profile indicating the regions of the upper water column where phytoplankton grow. The surface mixed layer is that region that is actively mixed by wind and wave energy, which is typically depleted in major nutrients. Below this mixed layer temperatures decrease and nutrients increase as material sinking from the mixed layer is regenerated by microbial decomposition.

regions of divergent surface water flow, reaching average values of 15 to 30 or greater. Thus, those regions of high nutrient supply or persistent high nutrient concentrations are thought to be most important in terms of carbon removal.

Nitrogen versus Phosphorus Limitation

Although both nitrogen and phosphorus are required at nearly constant ratios characteristic of deep water, nitrogen has generally been thought to be the limiting nutrient in sea water rather than phosphorus. This idea has been based on two observations: selective enrichment experiments and surface water distributions. When ammonia and phosphate are added to sea water in grow-out experiments, phytoplankton growth increases with the ammonia addition and not with the phosphate addition, thus indicating that reduced nitrogen and not phosphorus is limiting. Also, when surface water concentration of nitrate and phosphate are plotted together (Figure 2), it appears that there is still residual phosphate after the nitrate has gone to zero.

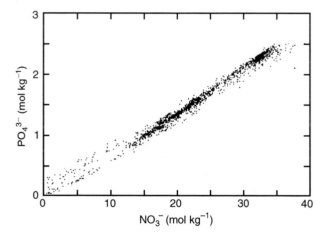

Figure 2 A plot of the global surface water concentrations of phosphate versus nitrate indicating a general positive intercept for phosphorus when nitrate has gone to zero. This is one of the imperical observations favoring the notion of nitrate limitation over phosphate limitation.

The notion of nitrogen limitation seems counterintuitive when one considers the abundant supply of dinitrogen (N_2) in the atmosphere. Yet this nitrogen gas is kinetically unavailable to most phytoplankton because of the large amount of energy required to break the triple bond that binds the dinitrogen molecule. Only those organisms capable of nitrogen fixation can take advantage of this form of nitrogen and reduce atmospheric N_2 to biologically available nitrogen in the form of urea and ammonia. This is, energetically, a very expensive process requiring specialized enzymes (nitrogenase), an anaerobic microenvironment, and large amounts of reducing power in the form of electrons generated by photosynthesis. Although there is currently the suggestion that nitrogen fixation may have been underestimated as an important geochemical process, the major mode of nitrogen assimilation, giving rise to new plant production in surface waters, is thought to be nitrate uptake.

The uptake of nitrate and subsequent conversion to reduced nitrogen in cells requires a change of five in the oxidation state and proceeds in a stepwise fashion. The initial reduction takes place via the nitrate/nitrite reductase enzyme present in phytoplankton and requires large amounts of the reduced nicotinamide–adenine dinucleotide phosphate (NADPH) and of adenosine triphosphate (ATP) and thus of harvested light energy from photosystem II. Both the nitrogenase enzyme and the nitrate reductase enzyme require iron as a cofactor and are thus sensitive to iron availability.

Ocean Regions

From a nutrient and biotic perspective, the oceans can be generally divided into biogeochemical provinces that reflect differences in the abundance of macronutrients and the standing stocks of phytoplankton. These are the high-nitrate, high-chlorophyll (HNHC); high-nitrate, low-chlorophyll (HNLC); low-nitrate, high-chlorophyll (LNHC); and low-nitrate, low-chlorophyll (LNLC) regimes (**Table 1**). Only the HNLC and LNLC regimes are relatively stable, because the high phytoplankton

Table 1 The relationship between biomass and nitrate as a function of biogeochemical province and the approximate ocean area represented by these regimes

	High-chlorophyll	Low-chlorophyll
High-nitrate	Unstable/coastal (5%)	Stable/Subarctic/Antarctic/equatorial Pacific (20%)
Low-nitrate	Unstable/coastal (5%)	Oligotrophic gyres (70%)

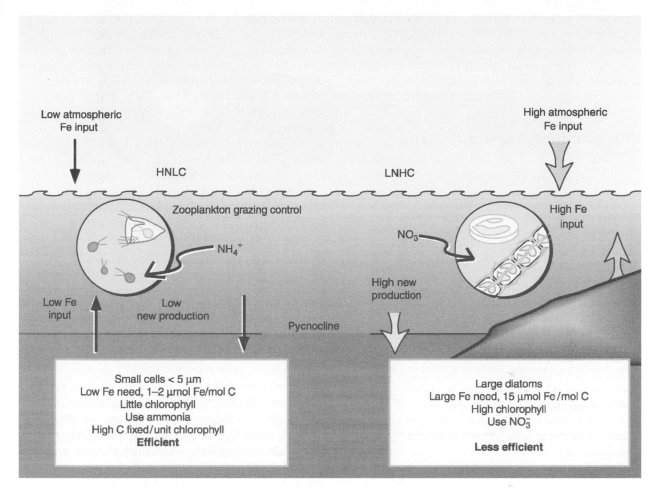

Figure 3 A schematic representation of the 'iron theory' as it functions in offshore HNLC regions and coastal transient LNHC regions. It has been suggested that iron added to the HNLC regions would induce them to function as LNHC regions and promote carbon export.

growth rates in the other two systems will deplete any residual nitrate and sink out of the system. The processes that give rise to these regimes have been the subject of some debate over the last few years and are of fundamental importance relative to carbon export (**Figure 3**).

High-nitrate, Low-chlorophyll Regions

The HNLC regions are thought to represent about 20% of the areal extent of the world's oceans. These are generally regions characterized by more than $2 \, \mu mol \, l^{-1}$ nitrate and less than $0.5 \, \mu g \, l^{-1}$ chlorophyll-a, a proxy for plant biomass. The major HNLC regions are shown in **Figure 4** and represent the Subarctic Pacific, large regions of the eastern equatorial Pacific and the Southern Ocean. These HNLC regions persist in areas that have high macronutrient concentrations, adequate light, and physical characteristics required for phytoplankton growth but have very low plant biomass. Two explanations have been

given to describe the persistence of this condition. (1) The rates of zooplankton grazing of the phytoplankton community may balance or exceed phytoplankton growth rates in these areas, thus cropping plant biomass to very low levels and recycling reduced nitrogen from the plant community, thereby decreasing the uptake of nitrate. (2) Some other micronutrient (possibly iron) physiologically limits the rate of phytoplankton growth. These are known as top-down and bottom-up control, respectively.

Several studies of zooplankton grazing and phytoplankton growth in these HNLC regions, particularly the Subarctic Pacific, confirm the hypothesis that grazers control production in these waters. Recent physiological studies, however, indicate that phytoplankton growth rates in these regions are suboptimal, as is the efficiency with which phytoplankton harvest light energy. These observations indicate that phytoplankton growth may be limited by something other than (or in addition to) grazing. Specifically, these studies implicate the lack of

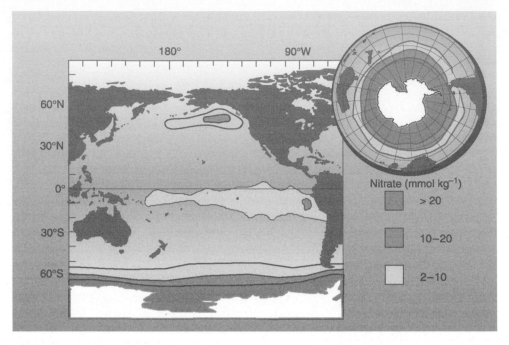

Figure 4 Current HNLC regions of the world's oceans covering an extimated 20% of the ocean surface. These regions include the Subarctic Pacific, equatorial Pacific and Southern Ocean.

sufficient electron transport proteins and the cell's ability to transfer reducing power from the photocenter. These have been shown to be symptomatic of iron deficiency.

The Role of Iron

Iron is a required micronutrient for all living systems. Because of its d-electron configuration, iron readily undergoes redox transitions between Fe(II) and Fe(III) at physiological redox potentials. For this reason, iron is particularly well suited to many enzyme and electron carrier proteins. The genetic sequences coding for many iron-containing electron carriers and enzymes are highly conserved, indicating iron and iron-containing proteins were key features of early biosynthesis. When life evolved, the atmosphere and waters of the planet were reducing and iron was abundant in the form of soluble Fe(II). Readily available and at high concentration, iron was not likely to have been limiting in the primordial biosphere. As photosynthesis evolved, oxygen was produced as a by-product. As the biosphere became more oxidizing, iron precipitated from aquatic systems in vast quantities, leaving phytoplankton and other aquatic life forms in a vastly changed and newly deficient chemical milieu. Evidence of this mass Fe(III) precipitation event is captured in the ancient banded iron formations in many parts of the world. Many primitive aquatic and terrestrial

organisms have subsequently evolved the ability to sequester iron through the elaboration of specific Fe(II)-binding ligands, known as siderophores. Evidence for siderophore production has been found in several marine dinoflagellates and bacteria and some researchers have detected similar compounds in sea water.

Today, iron exists in sea water at vanishingly small concentrations. Owing to both inorganic precipitation and biological uptake, typical surface water values are on the order of 20 pmol l^{-1}, perhaps a billion times less than during the prehistoric past. Iron concentrations in the oceans increase with depth, in much the same manner as the major plant nutrients (**Figure 5**).

The discovery that iron concentrations in surface waters is so low and shows a nutrient-like profile led some to speculate that iron availability limits plant growth in the oceans. This notion has been tested in bottle enrichment experiments throughout the major HNLC regions of the world's oceans. These experiments have demonstrated dramatic phytoplankton growth and nutrient uptake upon the addition of iron relative to control experiments in which no iron was added.

Criticism that such small-scale, enclosed experiments may not accurately reflect the response of the HNLC system at the level of the community has led to several large-scale iron fertilization experiments in the equatorial Pacific and Southern Ocean. These have been some of the most dramatic

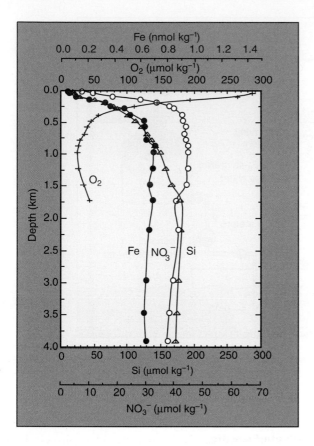

Figure 5 The vertical distributions of iron, nitrate, silicate, and oxygen in sea water. This figure shows how iron is depleted to picomolar levels in surface waters and has a profile that mimics other plant nutrients.

oceanographic experiments of our times and have led to a profound and new understanding of ocean systems.

Open Ocean Iron Enrichment

The question of iron limitation was brought into sharp scientific focus with a series of public lectures, reports by the US National Research Council, papers, special publications, and popular articles between 1988 and 1991. What was resolved was the need to perform an open ocean enrichment experiment in order to definitively test the hypothesis that iron limits phytoplankton growth and nutrient and carbon dioxide uptake in HNLC regions. Such an experiment posed severe logistical challenges and had never been conducted.

Experimental Strategy

The mechanics of producing an iron-enriched experimental patch and following it over time was

developed in four release experiments in the equatorial Pacific (IronEx I and II) and more recently in the Southern Ocean (SOIREE). At this writing, a similar strategy is being employed in the Caruso experiments now underway in the Atlantic sector of the Southern Ocean. All of these strategies were developed to address certain scientific questions and were not designed as preliminary to any geoengineering effort.

Form of Iron

All experiments to date have involved the injection of an iron sulfate solution into the ship's wake to achieve rapid dilution and dispersion throughout the mixed layer (**Figure 6**). The rationale for using ferrous sulfate involved the following considerations: (1) ferrous sulfate is the most likely form of iron to enter the oceans via atmospheric deposition; (2) it is readily soluble (initially); (3) it is available in a relatively pure form so as to reduce the introduction of other potentially bioactive trace metals; and (4) its counterion (sulfate) is ubiquitous in sea water and not likely to produce confounding effects. Although mixing models indicate that Fe(II) carbonate may reach insoluble levels in the ship's wake, rapid dilution reduces this possibility.

New forms of iron are now being considered by those who would seek to reduce the need for subsequent infusions. Such forms could include iron lignosite, which would increase the solubility and residence time of iron in the surface waters. Since this is a chelated form of iron, problems of rapid precipitation are reduced. In addition, iron lignosulfonate is about 15% Fe by weight, making it a space-efficient form of iron to transport. As yet untested is the extent to which such a compound would reduce the need for re-infusion.

Although solid forms of iron have been proposed (slow-release iron pellets; finely milled magnetite or iron ores), the ability to trace the enriched area with an inert tracer has required that the form of iron added and the tracer both be in the dissolved form.

Inert Tracer

Concurrent with the injection of iron is the injection of the inert chemical tracer sulfur hexafluoride (SF_6). By presaturating a tank of sea water with SF_6 and employing an expandable displacement bladder, a constant molar injection ratio of Fe : SF_6 can be achieved (**Figure 6**). In this way, both conservative and nonconservative removal of iron can be quantified. Sulfur hexafluoride traces the physical properties of the enriched patch; the relatively rapid shipboard detection of SF_6 can be used to track and

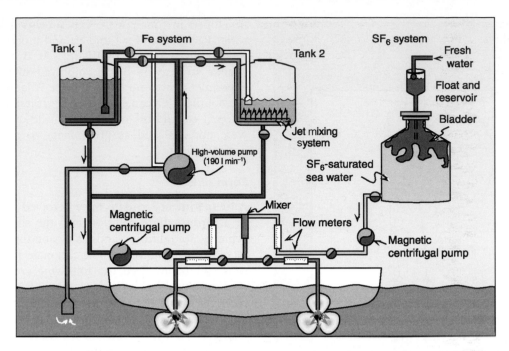

Figure 6 The iron injection system used during the IronEx experiments utilized two polyethylene tanks that could be sequentially filled with sea water and iron sulfate solution while the other was being injected behind the ship's propellers. A steel tank of sea water saturated with 40 g of sulfur hexafluoride (SF_6) was simultaneously mixed with the iron sulfate solution to provide a conservative tracer of mixing.

map the enriched area. The addition of helium-3 to the injected tracer can provide useful information regarding gas transfer.

Fluorometry

The biophysical response of the phytoplankton is rapid and readily detectable. Thus shipboard measurement of relative fluorescence (F_v/F_m) using fast repetition rate fluorometry has been shown to be a useful tactical tool and gives nearly instantaneous mapping and tracking feedback.

Shipboard Iron Analysis

Because iron is rapidly lost from the system (at least initially), the shipboard determination of iron is necessary to determine the timing and amount of subsequent infusions. Several shipboard methods, using both chemiluminescent and catalytic colorimetric detection have proven useful in this regard.

Lagrangian Drifters

A Lagrangian point of reference has proven to be very useful in every experiment to date. Depending upon the advective regime, this is the only practical way to achieve rapid and precise navigation and mapping about the enriched area.

Remote Sensing

A variety of airborne and satellite-borne active and passive optical packages provide rapid, large-scale mapping and tracking of the enriched area. Although SeaWiffs was not operational during IronEx I and II, AVHRR was able to detect the IronEx II bloom and airborne optical LIDAR was very useful during IronEx I. SOIREE has made very good use of the more recent SeaWiffs images, which have markedly extended the observational period and led to new hypotheses regarding iron cycling in polar systems.

Experimental Measurements

In addition to the tactical measurements and remote sensing techniques required to track and ascertain the development of the physical dynamics of the enriched patch, a number of measurements have been made to track the biogeochemical development of the experiment. These have typically involved a series of underway measurements made using the ship's flowing sea water system or towed fish. In addition, discrete measurements are made in the vertical dimension at every station occupied both inside and outside of the fertilized area. These measurements include temperature salinity, fluorescence (a measure of plant biomass), transmissivity (a measure of suspended particles), oxygen, nitrate, phosphate, silicate, carbon dioxide partial pressure, pH, alkalinity, total carbon dioxide, iron-binding ligands, $^{234}Th : ^{238}U$ radioisotopic disequilibria (a proxy for particle removal), relative fluorescence (indicator of photosynthetic competence), primary production, phytoplankton and zooplankton enumeration, grazing rates, nitrate uptake, and particulate and dissolved organic carbon and nitrogen. These parameters allow for the general characterization of both the biological and geochemical response to added iron. From the results of the equatorial enrichment experiments (IronEx I and II) and the Southern Ocean Iron Enrichment Experiment (SOIREE), several general features have been identified.

Findings to Date

Biophysical Response

The experiments to date have focused on the high-nitrate, low-chlorophyll (HNLC) areas of the world's oceans, primarily in the Subarctic, equatorial Pacific and Southern Ocean. In general, when light is abundant many researchers find that HNLC systems are iron-limited. The nature of this limitation is similar between regions but manifests itself at different levels of the trophic structure in some characteristic ways. In general, all members of the HNLC photosynthetic community are physiologically limited by iron availability. This observation is based primarily on the examination of the efficiency of photosystem II, the light-harvesting reaction centers. At ambient levels of iron, light harvesting proceeds at suboptimal rates. This has been attributed to the lack of iron-dependent electron carrier proteins at low iron concentrations. When iron concentrations are increased by subnanomolar amounts, the efficiency of light harvesting rapidly increases to maximum levels. Using fast repetition rate fluorometry and non-heme iron proteins, researchers have described these observations in detail. What is notable about these results is that iron limitation seems to affect the photosynthetic energy conversion efficiency of even the smallest of phytoplankton. This has been a unique finding that stands in contrast to the hypothesis that, because of diffusion, smaller cells are not iron limited but larger cells are.

Nitrate Uptake

As discussed above, iron is also required for the reduction (assimilation) of nitrate. In fact, a change of oxidation state of five is required between nitrate and the reduced forms of nitrogen found in amino acids and proteins. Such a large and energetically unfavorable redox process is only made possible by substantial reducing power (in the form of NADPH) made available through photosynthesis and active nitrate reductase, an iron-requiring enzyme. Without iron, plants cannot take up nitrate efficiently. This provided original evidence implicating iron deficiency as the cause of the HNLC condition. When phytoplankton communities are relieved from iron deficiency, specific rates of nitrate uptake increase. This has been observed in both the equatorial Pacific and the Southern Ocean using isotopic tracers of nitrate uptake and conversion. In addition, the accelerated uptake of nitrate has been observed in both the mesoscale iron enrichment experiments to date, IronEx and SOIREE.

Growth Response

When iron is present, phytoplankton growth rates increase dramatically. Experiments over widely differing oceanographic regimes have demonstrated that, when light and temperature are favorable, phytoplankton growth rates in HNLC environments increase to their maximum at dissolved iron concentrations generally below 0.5 nmol l^{-1}. This observation is significant in that it indicates that phytoplankton are adapted to very low levels of iron and they do not grow faster if given iron at more than 0.5 nmol l^{-1}. Given that there is still some disagreement within the scientific community about the validity of some iron measurements, this phytoplankton response provides a natural, environmental, and biogeochemical benchmark against which to compare results.

The iron-induced transient imbalance between phytoplankton growth and grazing in the equatorial Pacific during IronEx II resulted in a 30-fold increase in plant biomass (**Figure 7**). Similarly, a 6-fold increase was observed during the SOIREE experiment in the Southern Ocean. These are perhaps the most dramatic demonstrations of iron limitation of nutrient cycling, and phytoplankton growth to date and has fortified the notion that iron fertilization may be a useful strategy to sequester carbon in the oceans.

Heterotrophic Community

As the primary trophic levels increase in biomass, growth in the small microflagellate and heterotrophic bacterial communities increase in kind. It appears that these consumers of recently fixed carbon (both particulate and dissolved) respond to the food source and not necessarily the iron (although some have been found to be iron-limited). Because their division rates are fast, heterotrophic bacteria, ciliates, and flagellates can rapidly divide and

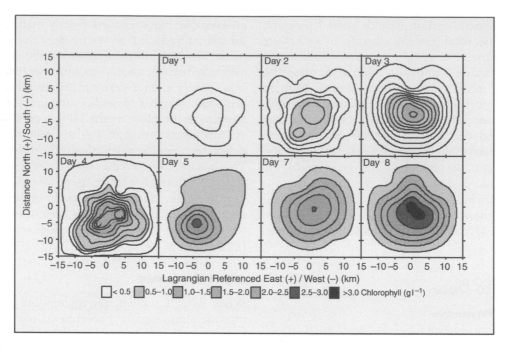

Figure 7 Chlorophyll concentrations during IronEx II were mapped daily. This figure shows the progression of the phytoplankton bloom that reached over 30 times the background concentrations.

respond to increasing food availability to the point where the growth rates of the smaller phytoplankton can be overwhelmed by grazing. Thus there is a much more rapid turnover of fixed carbon and nitrogen in iron replete systems. M. Landry and coworkers have documented this in dilution experiments conducted during IronEx II. These results appear to be consistent with the recent SOIREE experiments as well.

Nutrient Uptake Ratios

An imbalance in production and consumption, however, can arise at the larger trophic levels. Because the reproduction rates of the larger micro- and mesozooplankton are long with respect to diatom division rates, iron-replete diatoms can escape the pressures of grazing on short timescales (weeks). This is thought to be the reason why, in every iron enrichment experiment, diatoms ultimately dominate in biomass. This result is important for a variety of reasons. It suggests that transient additions of iron would be most effective in producing net carbon uptake and it implicates an important role of silicate in carbon flux. The role of iron in silicate uptake has been studied extensively by Franck and colleagues. The results, together with those of Takeda and coworkers, show that iron alters the uptake ratio of nitrate and silicate at very low levels (**Figure 8**). This is thought to be brought about by the increase in nitrate uptake rates relative to silica.

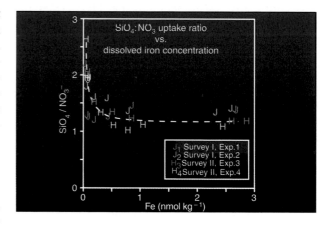

Figure 8 Bottle enrichment experiments show that the silicate : nitrate uptake ratio changes as a function of the iron added. This is thought to be due to the increased rate of iron uptake relative to silicate in these experimental treatments.

Organic Ligands

Consistent with the role of iron as a limiting nutrient in HNLC systems is the notion that organisms may have evolved competitive mechanisms to increase iron solubility and uptake. In terrestrial systems this is accomplished using extracellularly excreted or membrane-bound siderophores. Similar compounds have been shown to exist in sea water where the competition for iron may be as fierce as it is on land. In open ocean systems where it has been measured, iron-binding ligand production increases with the addition of iron. Whether this is a competitive response to added iron or a function of phytoplankton biomass and grazing is not yet well understood. However, this is an important natural mechanism for reducing the inorganic scavenging of iron from the surface waters and increasing iron availability to phytoplankton. More recent studies have considerably advanced our understanding of these ligands, their distribution and their role in ocean ecosystems.

Carbon Flux

It is the imbalance in the community structure that gives rise to the geochemical signal. Whereas iron stimulation of the smaller members of the community may result in chemical signatures such as an increased production of beta-dimethylsulfoniopropionate (DMSP), it is the stimulation of the larger producers that decouples the large cell producers from grazing and results in a net uptake and export of nitrate, carbon dioxide, and silicate.

The extent to which this imbalance results in carbon flux, however, has yet to be adequately described. The inability to quantify carbon export has primarily been a problem of experimental scale. Even though mesoscale experiments have, for the first time, given us the ability to address the effect of iron on communities, the products of surface water processes and the effects on the midwater column have been difficult to track. For instance, in the IronEx II experiment, a time-series of the enriched patch was diluted by 40% per day. The dilution was primarily in a lateral (horizontal/isopycnal) dimension. Although some correction for lateral dilution can be made, our ability to quantify carbon export is dependent upon the measurement of a signal in waters below the mixed layer or from an uneroded enriched patch. Current data from the equatorial Pacific showed that the IronEx II experiment advected over six patch diameters per day. This means that at no time during the experiment were the products of increased export reflected in the waters below the enriched area. A transect through the IronEx II patch is shown in **Figure 9**. This figure indicates the massive production of plant biomass with a concomitant decrease in both nitrate and carbon dioxide.

The results from the equatorial Pacific, when corrected for dilution, suggest that about 2500 t of carbon were exported from the mixed layer over a 7-day period. These results are preliminary and subject to more rigorous estimates of dilution and export production, but they do agree favorably with estimates based upon both carbon and nitrogen budgets. Similarly, thorium export was observed in this experiment, confirming some particle removal.

The results of the SOIREE experiment were similar in many ways but were not as definitive with respect to carbon flux. In this experiment biomass increased 6-fold, nitrate was depleted by $2\,\mu mol\,l^{-1}$ and carbon dioxide by 35–40 microatmospheres (3.5–4.0 Pa). This was a greatly attenuated signal relative to IronEx II. Colder water temperatures likely led to slower rates of production and bloom evolution and there was no observable carbon flux.

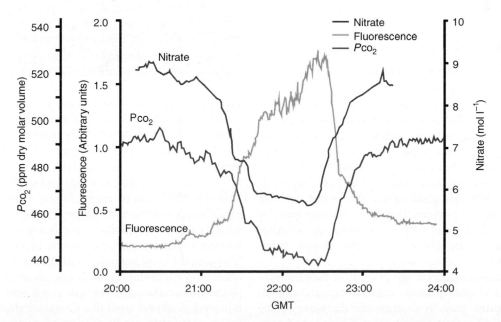

Figure 9 A transect through the IronEx II patch. The *x*-axis shows GMT as the ship steams from east to west through the center of the patch. Simultaneously plotted are the iron-induced production of chlorophyll, the drawdown of carbon dioxide, and the uptake of nitrate in this bloom.

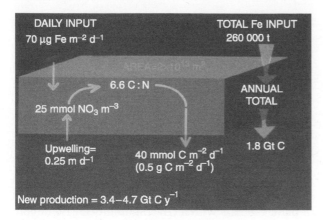

Figure 10 Simple calculations of the potential for carbon export for the Southern Ocean. These calculations are based on the necessary amount of iron required to efficiently utilize the annual upwelled nitrate and the subsequent incorporation into sinking organic matter. An estimated 1.8×10^9 t (Gt) of carbon export could be realized in this simple model.

Original estimates of carbon export in the Southern Ocean based on the iron-induced efficient utilization of nitrate suggest that as much as 1.8×10^9 t of carbon could be removed annually (**Figure 10**). These estimates of carbon sequestration have been challenged by some modelers yet all models lack important experimental parameters which will be measured in upcoming experiments.

Remaining Questions

A multitude of questions remain regarding the role of iron in shaping the nature of the pelagic community. The most pressing question is whether iron enrichment accelerates the downward transport of carbon from the surface waters to the deep sea? More specifically, how does iron affect the cycling of carbon in HNLC, LNLC, and coastal systems? Recent studies indicate that coastal systems may be iron-limited and the iron requirement for nitrogenase activity is quite large, suggesting that iron may limit nitrogen fixation, but there have been limited studies to test the former and none to test the latter. If iron does stimulate carbon uptake, what are the spatial scales over which this fixed carbon may be remineralized? This is crucial to predicting whether fertilization is an effective carbon sequestration mechanism.

Given these considerations, the most feasible way to understand and quantify carbon export from an enriched water mass is to increase the scale of the experiment such that both lateral dilution and sub-mixed-layer relative advection are small with respect to the size of the enriched patch. For areas such as the equatorial Pacific, this would be very large (hundreds of kilometers on a side). For other areas, it could be much smaller.

The focus of the IronEx and SOIREE experiments has been from the scientific perspective, but this focus is shifting toward the application of iron enrichment as a carbon sequestration strategy. We have come about rapidly from the perspective of trying to understand how the world works to one of trying to make the world work for us. Several basic questions remain regarding the role of natural or anthropogenic iron fertilization on carbon export. Some of the most pressing questions are: What are the best proxies for carbon export? How can carbon export best be verified? What are the long-term ecological consequences of iron enrichment on surface water community structure, midwater processes, and benthic processes? Even with answers to these, there are others that need to be addressed prior to any serious consideration of iron fertilization as an ocean carbon sequestration option.

Simple technology is sufficient to produce a massive bloom. The technology required either for a large-scale enrichment experiment or for purposeful attempts to sequester carbon is readily available. Ships, aircraft (tankers and research platforms), tracer technology, a broad range of new Autonomous Underwater Vehicles (AUVs) and instrument packages, Lagrangian buoy tracking systems, together with aircraft and satellite remote sensing systems and a new suite of chemical sensors/*in situ* detection technologies are all available, or are being developed. Industrial bulk handling equipment is available for large-scale implementation. The big questions, however, are larger than the technology.

With a slow start, the notion of both scientific experimentation through manipulative experiments, as well as the use of iron to purposefully sequester carbon, is gaining momentum. There are now national, international, industrial, and scientific concerns willing to support larger-scale experiments. The materials required for such an experiment are inexpensive and readily available, even as industrial by-products (of paper, mining, and steel processing).

Given the concern over climate change and the rapid modernization of large developing countries such as China and India, there is a pressing need to address the increased emission of greenhouse gases. Through the implementation of the Kyoto accords or other international agreements to curb emissions (Rio), financial incentives will reach into the multibillion dollar level annually. Certainly there will soon be an overwhelming fiscal incentive to investigate, if not implement, purposeful open ocean carbon sequestration trials.

A Societal Challenge

The question is not whether we have the capability of embarking upon such an engineering strategy but whether we have the collective wisdom to responsibly negotiate such a course of action. Posing the question another way: If we do not have the social, political and economic tools or motivation to control our own population and greenhouse gas emissions, what gives us the confidence that we have the wisdom and ability to responsibly manipulate and control large ocean ecosystems without propagating yet another massive environmental calamity? Have we as an international community first tackled the difficult but obvious problem of overpopulation and implemented alternative energy technologies for transportation, industry, and domestic use?

Other social questions arise as well. Is it appropriate to use the ocean commons for such a purpose? What individuals, companies, or countries would derive monetary compensation for such an effort and how would this be decided?

It is clear that there are major scientific investigations and findings that can only benefit from large-scale open ocean enrichment experiments, but certainly a large-scale carbon sequestration effort should not proceed without a clear understanding of both the science and the answers to the questions above.

Glossary

ATP	Adenosine triphosphate
AVHRR	Advanced Very High Resolution Radiometer
HNHC	High-nitrate high-chlorophyll
HNLC	High-nitrate low-chlorophyll
IronEx	Iron Enrichment Experiment
LIDAR	Light detection and ranging
LNHC	Low-nitrate high-chlorophyll
LNLC	Low-nitrate low-chlorophyll
NADPH	Reduced form of nicotinamide–adenine dinucleotide phosphate
SOIREE	Southern Ocean Iron Enrichment Experiment

Further Reading

Abraham ER, Law CS, Boyd PW, et al. (2000) Importance of stirring in the development of an iron-fertilized phytoplankton bloom. Nature 407: 727–730.

Barbeau K, Moffett JW, Caron DA, Croot PL, and Erdner DL (1996) Role of protozoan grazing in relieving iron limitation of phytoplankton. Nature 380: 61–64.

Behrenfeld MJ, Bale AJ, Kobler ZS, Aiken J, and Falkowski PG (1996) Confirmation of iron limitation of phytoplankton photosynthesis in Equatorial Pacific Ocean. Nature 383: 508–511.

Boyd PW, Watson AJ, Law CS, et al. (2000) A mesoscale phytoplankton bloom in the polar Southern Ocean stimulated by iron fertilization. Nature 407: 695–702.

Cavender-Bares KK, Mann EL, Chishom SW, Ondrusek ME, and Bidigare RR (1999) Differential response of equatorial phytoplankton to iron fertilization. Limnology and Oceanography 44: 237–246.

Coale KH, Johnson KS, Fitzwater SE, et al. (1996) A massive phytoplankton bloom induced by an ecosystem-scale iron fertilization experiment in the equatorial Pacific Ocean. Nature 383: 495–501.

Coale KH, Johnson KS, Fitzwater SE, et al. (1998) IronEx-I, an in situ iron-enrichment experiment: experimental design, implementation and results. Deep-Sea Research Part II 45: 919–945.

Elrod VA, Johnson KS, and Coale KH (1991) Determination of subnanomolar levels of iron (II) and total dissolved iron in seawater by flow injection analysis with chemiluminescence dection. Analytical Chemistry 63: 893–898.

Fitzwater SE, Coale KH, Gordon RM, Johnson KS, and Ondrusek ME (1996) Iron deficiency and phytoplankton growth in the equatorial Pacific. Deep-Sea Research Part II 43: 995–1015.

Greene RM, Geider RJ, and Falkowski PG (1991) Effect of iron lititation on photosynthesis in a marine diatom. Limnology Oceanogrography 36: 1772–1782.

Hoge EF, Wright CW, Swift RN, et al. (1998) Fluorescence signatures of an iron-enriched phytoplankton community in the eastern equatorial Pacific Ocean. Deep-Sea Research Part II 45: 1073–1082.

Johnson KS, Coale KH, Elrod VA, and Tinsdale NW (1994) Iron photochemistry in seawater from the Equatorial Pacific. Marine Chemistry 46: 319–334.

Kolber ZS, Barber RT, Coale KH, et al. (1994) Iron limitation of phytoplankton photosynthesis in the Equatorial Pacific Ocean. Nature 371: 145–149.

Landry MR, Ondrusek ME, Tanner SJ, et al. (2000) Biological response to iron fertilization in the eastern equtorial Pacific (Ironex II). I. Microplankton community abundances and biomass. Marine Ecology Progress Series 201: 27–42.

LaRoche J, Boyd PW, McKay RML, and Geider RJ (1996) Flavodoxin as an in situ marker for iron stress in phytoplankton. Nature 382: 802–805.

Law CS, Watson AJ, Liddicoat MI, and Stanton T (1998) Sulfer hexafloride as a tracer of biogeochemical and physical processes in an open-ocean iron fertilization experiment. Deep-Sea Research Part II 45: 977–994.

Martin JH, Coale KH, Johnson KS, et al. (1994) Testing the iron hypothesis in ecosystems of the equatorial Pacific Ocean. Nature 371: 123–129.

Nightingale PD, Liss PS, and Schlosser P (2000) Measurements of air–gas transfer during an open ocean algal bloom. *Geophysical Research Letters* 27: 2117–2121.

Obata H, Karatani H, and Nakayama E (1993) Automated determination of iron in seawater by chelating resin concentration and chemiluminescence detection. *Analytical Chemistry* 65: 1524–1528.

Redfield AC (1934) On the proportions of organic derivatives in sea water and their relation to the composition of plankton. *James Johnstone Memorial Volume*, pp. 177–192. Liverpool: Liverpool University Press.

Redfield AC (1958) The biological control of chemical factors in the environment. *American Journal of Science* 46: 205–221.

Rue EL and Bruland KW (1997) The role of organic complexation on ambient iron chemistry in the equatorial Pacific Ocean and the response of a mesoscale iron addition experiment. *Limnology and Oceanography* 42: 901–910.

Smith SV (1984) Phosphorus versus nitrogen limitation in the marine environment. *Limnology and Oceanography* 29: 1149–1160.

Stanton TP, Law CS, and Watson AJ (1998) Physical evolutation of the IronEx I open ocean tracer patch. *Deep-Sea Research Part II* 45: 947–975.

Takeda S and Obata H (1995) Response of equatorial phytoplankton to subnanomolar Fe enrichment. *Marine Chemistry* 50: 219–227.

Trick CG and Wilhelm SW (1995) Physiological changes in coastal marine cyanobacterium *Synechococcus* sp. PCC 7002 exposed to low ferric ion levels. *Marine Chemistry* 50: 207–217.

Turner SM, Nightingale PD, Spokes LJ, Liddicoat MI, and Liss PS (1996) Increased dimethyl sulfide concentrations in seawater from *in situ* iron enrichment. *Nature* 383: 513–517.

Upstill-Goddard RC, Watson AJ, Wood J, and Liddicoat MI (1991) Sulfur hexafloride and helium-3 as sea-water tracers: deployment techniques and continuous underway analysis for sulphur hexafloride. *Analytica Chimica Acta* 249: 555–562.

Van den Berg CMG (1995) Evidence for organic complesation of iron in seawater. *Marine Chemistry* 50: 139–157.

Watson AJ, Liss PS, and Duce R (1991) Design of a small-scale *in situ* iron fertilization experiment. *Limnology and Oceanography* 36: 1960–1965.

METHANE HYDRATES AND CLIMATIC EFFECTS

B. U. Haq, Vendome Court, Bethesda, MD, USA

Introduction

Natural gas hydrates are crystalline solids that occur widely in marine sediment of the world's continental margins. They are composed largely of methane and water, frozen in place in the sediment under the dual conditions of high pressure and frigid temperatures at the sediment–water interface (**Figure 1**). When the breakdown of the gas hydrate (also known as clathrate) occurs in response to reduced hydrostatic pressure (e.g., sea level fall during glacial periods), or an increase in bottom-water temperature, it causes dissociation of the solid hydrate at its base, creating a zone of reduced sediment strength that is prone to structural faulting and sediment slumping. Such sedimentary failure at hydrate depths could inject large quantities of methane (a potent greenhouse gas) in the water column, and eventually into the atmosphere, leading to enhanced greenhouse warming. Ice core records of the recent geological past show that climatic warming occurs in tandem with rapid increase in atmospheric methane. This suggests that catastrophic release of methane into the atmosphere during periods of lowered sea level may have been a causal factor for abrupt climate change. Massive injection of methane in sea water following hydrate

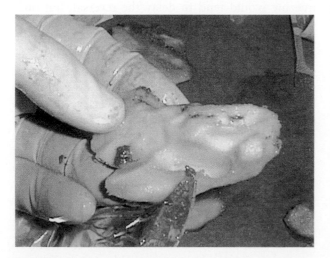

Figure 1 A piece of natural gas hydrate from the Gulf of Mexico. (Photograph courtesy of I. MacDonald, Texas A & M University.)

dissociation during periods of warm bottom temperatures are also suspected to be responsible for major shifts in carbon-isotopic ratios of sea water and associated changes in benthic assemblages and hydrographic conditions.

Hydrate Stability and Detection

Gas hydrate stability requires high hydrostatic pressure (>5 bars) and low bottom-water temperature (<7°C) on the seafloor. These requirements dictate that hydrates occur mostly on the continental slope and rise, below 530 m water depth in the low latitudes, and below 250 m depth in the high latitudes. Hydrated sediments may extend from these depths to c. 1100 m sub-seafloor. In higher latitudes, hydrates also occur on land, in association with the permafrost.

Rapidly deposited sediments with high biogenic content are amenable to the genesis of large quantities of methane by bacterial alteration of the organic matter. Direct drilling of hydrates on the Blake Ridge, a structural high feature off the US east coast, indicated that the clathrate is only rarely locally concentrated in the otherwise widespread field of thinly dispersed hydrated sediments. The volume of the solid hydrate based on direct measurements on Blake Ridge suggested that it occupies between 0 and 9% of the sediment pore space within the hydrate stability zone (190–450 m sub-seafloor). It has been estimated that a relatively large amount, c. 35 Gt (Gigaton = 10^{15} g), of methane carbon was tied up on Blake Ridge, which is equal to carbon from about 7% of the total terrestrial biota.

Hydrates can be detected remotely through the presence of acoustic reflectors, known as bottom simulating reflectors (BSR), that mimic the seafloor and are caused by acoustic velocity contrast between the solid hydrate above and the free gas below. However, significant quantities of free gas need to be present below the hydrate to provide the velocity contrast for the presence of a BSR. Thus, hydrate may be present at the theoretical hydrate stability depths even when no BSR is observed. Presence of gas hydrate can also be inferred through the sudden reduction in pore-water chlorinity (salinity) of the hydrated sediments during drilling, as well as through gas escape features on land and on the seafloor.

Global estimates of methane trapped in gas hydrate reservoirs (both in the hydrate stability zone and as free gas beneath it) vary widely. For example, the Arctic permafrost is estimated to hold anywhere between 7.6 and 18 000 Gt of methane carbon, while marine sediments are extrapolated to hold between 1700 and 4 100 000 Gt of methane carbon globally. Obtaining more accurate global estimates of methane sequestered in the clathrate reservoirs remains one of the more significant challenges in gas hydrate research.

Another major unknown, especially for climatic implications, is the mode of expulsion of methane from the hydrate. How and how much of the gas escapes from the hydrate zone and how much of it is dissolved in the water column versus escaping into the atmosphere? In a steady state much of the methane diffusing from marine sediments is believed to be oxidized in the surficial sediment and the water column above. However, it is not clear what happens to significant volumes of gas that might be catastrophically released from the hydrates when they disintegrate. How much of the gas makes it to the atmosphere (in the rapid climate change scenarios it is assumed that much of it does), or is dissolved in the water column?

Hydrate dissociation and methane release into the atmosphere from continental margin and permafrost sources and the ensuing accelerated greenhouse heating also have important implications for the models of global warming over the next century. The results of at least one modeling study play down the role of methane release from hydrate sources. When heat transfer and methane destabilization process in oceanic sediments was modeled in a coupled atmosphere–ocean model with various input assumptions and anthropogenic emission scenarios, it was found that the hydrate dissociation effects were smaller than the effects of increased carbon dioxide emissions by human activity. In a worst case scenario global warming increased by 10–25% more with clathrate destabilization than without. However, these models did not take into account the associated free gas beneath the hydrate zone that may play an additional and significant role as well. It is obvious from drilling results on Blake Ridge that large volumes of free methane are readily available for transfer without requiring dissociation.

Hydrate Breakdown and Rapid Climate Change

The temperature and pressure dependency for the stability of hydrates implies that any major change in either of these controlling factors will also modify the zone of hydrate stability. A notable drop in sea level, for example, will reduce the hydrostatic pressure on the slope and rise, altering the temperature–pressure regime and leading to destabilization of the gas hydrates. It has been suggested that a sea level drop of nearly 120 m during the last glacial maximum (c. 30 000 to 18 000 years BP) reduced the hydrostatic pressure sufficiently to raise the lower limit of gas hydrate stability by about 20 m in the low latitudes. When a hydrate dissociates, its consistency changes from a solid to a mixture of sediment, water, and free gas. Experiments on the mechanical strength behavior of hydrates has shown that the hydrated sediment is markedly stronger than water ice (10 times stronger than ice at 260 K). Thus, such conversion would create a zone of weakness where sedimentary failure could take place, encouraging low-angle faulting and slumping on the continental margins. The common occurrence of Pleistocene slumps on the seafloor have been ascribed to this catastrophic mechanism and major slumps have been identified in sediments of this age in widely separated margins of the world.

When slumping occurs it would be accompanied by the liberation of a significant amount of methane trapped below the level of the slump, in addition to the gas emitted from the dissociated hydrate itself. These emissions are envisaged to increase in the low latitudes, along with the frequency of slumps, as glaciation progresses, eventually triggering a negative feedback to advancing glaciation, encouraging the termination of the glacial cycle. If such a scenario is true then there may be a built-in terminator to glaciation, via the gas hydrate connection.

In this scenario, the negative feedback to glaciation, can initially function effectively only in the lower latitudes. At higher latitudes glacially induced freezing would tend to delay the reversal, but once deglaciation begins, even a relatively small increase in atmospheric temperature of the higher latitudes could cause additional release of methane from near-surface sources, leading to further warming. One scenario suggests that a small triggering event and liberation of one or more Arctic gas pools could initiate massive release of methane frozen in the permafrost, leading to accelerated warming. The abrupt nature of the termination of the Younger Dryas glaciation (some 10 000 years ago) has been ascribed to such an event. Modeling results of the effect of a pulse of 'realistic' amount of methane release at the glacial termination as constrained by ice core records indicate that the direct radiative effects of such an emission event may be too small to account for deglaciation alone. However, with certain combinations of methane, CO_2, and heat transport changes, it may be possible to simulate changes of the same magnitude as those indicated by empirical data.

The Climate Feedback Loop

The paleoclimatic records of the recent past, e.g., Vostock ice core records of the past 420 000 years from Antarctica, show the relatively gradual decrease in atmospheric carbon dioxide and methane at the onset of glaciations. Deglaciations, on the other hand, tend to be relatively abrupt and are associated with equally rapid increases in carbon dioxide and methane. Glaciations are thought to be initiated by Milankovitch forcing (a combination of variations in the Earth's orbital eccentricity, obliquity and precession), a mechanism that also can explain the broad variations in glacial cycles, but not the relatively abrupt terminations. Degassing of carbon dioxide from the ocean surface alone cannot explain the relatively rapid switch from glacials to interglacials.

The delayed response to glacially induced sea level fall in the high latitudes (as compared with low latitudes) is a part of a feedback loop that could be an effective mechanism for explaining the rapid warmings at the end of glacial cycles (also known as the Dansgaard-Oeschger events) in the late Quaternary. These transitions often occur only on decadal to centennial time scales. In this scenario it is envisioned that the low-stand-induced slumping and methane emissions in lower latitudes lead to greenhouse warming and trigger a negative feedback to glaciation. This also leads to an increase in carbon dioxide degassing for the ocean. Once the higher latitudes are warmed by these effects, further release of methane from near-surface sources could provide a positive feedback to warming. The former (methane emissions in the low latitudes) would help force a reversal of the glacial cooling, and the latter (additional release of methane from higher latitudes) could reinforce the trend, resulting in apparent rapid warming observed at the end of the glacial cycles (see **Figure 2**).

The record of stable isotopes of carbon from Santa Barbara Basin, off California, has revealed rapid warmings in the late Quaternary that are synchronous with warmings associated with Dansgaard-Oeschger (D-O) events in the ice record from Greenland. The energy needed for these rapid warmings could have come from methane hydrate dissociation. Relatively large excursions of $\delta^{13}C$ (up to 5 ppm) in benthic foraminifera are associated with the D-O events. However, during several brief intervals the planktonics also show large negative shifts in $\delta^{13}C$ (up to 2.5 ppm), implying that the

entire water column may have experienced rapid ^{12}C enrichment. One plausible mechanism for these changes may be the release of methane from the clathrates during the interstadials. Thus, abrupt warmings at the onset of D-O events may have been forced by dissociation of gas hydrates modulated by temperature changes in overlying intermediate waters.

For the optimal functioning of the negative–positive feedback model discussed above, methane would have to be constantly replenished from new and larger sources during the switchover. Although as a greenhouse gas methane is nearly 10 times as effective as carbon dioxide, its residence time in the atmosphere is relatively short (on the order of a decade and a half), after which it reacts with the hydroxyl radical and oxidizes to carbon dioxide and water. The atmospheric retention of carbon dioxide is somewhat more complex than methane because it is readily transferred to other reservoirs, such as the oceans and the biota, from which it can re-enter the atmosphere. Carbon dioxide accounts for up to 80% of the contribution to greenhouse warming in the atmosphere. An effective residence time of about 230 years has been estimated for carbon dioxide. These retention times are short enough that for cumulative impact of methane and carbon dioxide through the negative–positive feedback loop to be effective methane levels would have to be continuously sustained from gas hydrate and permafrost sources. The feedback loop would close when a threshold is reached where sea level is once again high enough that it can stabilize the residual clathrates and encourage the genesis of new ones.

Several unresolved problems remain with the gas hydrate climate feedback model. The negative–positive feedback loop assumes a certain amount of time lag between events as they shift from lower to higher latitudes, but the duration of the lag remains unresolved, although a short duration (on decadal to centennial time scales) is implied by the ice core records. Also, it is not clear whether hydrate dissociation leads to initial warming, or warming caused by other factors leads to increased methane emissions from hydrates. Data gathered imply a time lag of c. 50 (± 10) years between abrupt warming and the peak in methane values at the Blling transition (around 14 500 years Bp), although an increase in methane emissions seems to have begun almost simultaneously with the warming trend (± 5 years). However, this does not detract from the notion that there may be a built-in feedback between increased methane emissions from gas hydrate sources and accelerated warming. If smaller quantities of methane released from hydrated sediments are

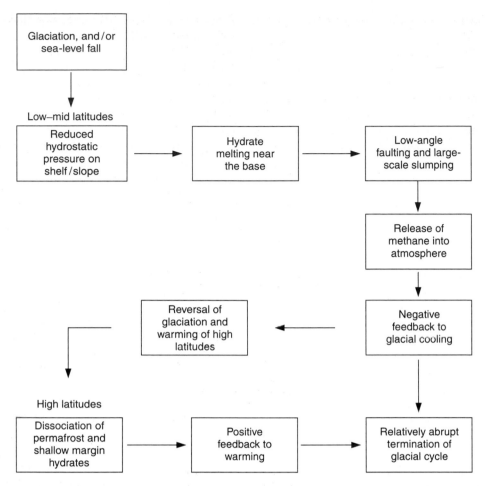

Figure 2 The negative–positive feedback loop model of sea level fall, hydrate decomposition, and climate change (reversal of glaciation and rapid warming) through methane release in the low and high latitudes. (Adapted with permission from Haq, 1993.)

oxidized in the water column, initial releases of methane from dissociated hydrates may not produce a significant positive shift in the atmospheric content of methane. However, as the frequency of catastrophic releases from this source increases, more methane is expected to make it to the atmosphere. And, although the atmospheric residence time of methane itself is relatively short, when oxidized, it adds to the greenhouse forcing of carbon dioxide. This may explain the more gradual increase in methane, and is not inconsistent with the short temporal difference between the initiation of the warming trend and methane increase, as well as the time lag between the height of warming trend and the peak in methane values.

Although there is still no evidence to suggest that the main forcing for the initiation of deglaciation is to be found in hydrate dissociation, once begun, a positive feedback of methane emissions from hydrate sources (and its by-product, carbon dioxide) can only help accelerate the warming trend.

Gas Hydrates and the Long-term Record of Climate Change

Are there any clues in the longer term geological record where cause and/or effect can be ascribed to gas hydrates? One potential clue for the release of significant volumes of methane into the ocean waters is the changes in $\delta^{13}C$ composition of the carbon reservoir. The $\delta^{13}C$ of methane in hydrates averages $c.$ -60 ppm; perhaps the lightest (most enriched in ^{12}C) carbon anywhere in the Earth system. It has been argued that massive methane release from gas hydrate sources is the most likely mechanism for the pronounced input of carbon greatly enriched in ^{12}C during a period of rapid bottom-water warming. The dissolution of methane (and its oxidative by-product, CO_2) in the sea water should also coincide with increased dissolution of carbonate on the seafloor. Thus, a major negative shift in $\delta^{13}C$ that occurs together with an increase in benthic temperature (bottom-water warming) or a sea level fall event (reducing hydrostatic pressure) may provide clues to past behavior of gas hydrates.

A prominent excursion in global carbonate and organic matter $\delta^{13}C$ during the latest Paleocene peak warming has been explained as a consequence of such hydrate breakdown due to rapid warming of the bottom waters. The late Paleocene–early Eocene was a period of peak warming, and overall the warmest interval in the Cenozoic when latitudinal thermal gradients were greatly reduced. In the latest Paleocene bottom-water temperature also increased rapidly by as much as 4°C, with a coincident excursion of about −2 to −3 ppm in $\delta^{13}C$ of all carbon reservoirs in the global carbon cycle. A high resolution study of a sediment core straddling the Paleocene–Eocene boundary concluded that much of this carbon-isotopic shift occurred within no more than a few thousand years and was synchronous in oceans and on land, indicating a catastrophic release of carbon, probably from a methane source. The late Paleocene thermal maximum was also coincident with a major benthic foraminiferal mass extinction and widespread carbonate dissolution and low oxygen conditions on the seafloor. This rapid excursion cannot be explained by conventional mechanisms (increased volcanic emissions of carbon dioxide, changes in oceanic circulation and/or terrestrial and marine productivity, etc.). A rapid warming of bottom waters from 11 to 15°C could abruptly alter the sediment thermal gradients leading to methane release from gas hydrates. Increased flux of methane into the ocean–atmosphere system and its subsequent oxidation is considered sufficient to explain the −2.5 ppm excursion in $\delta^{13}C$ in the inorganic carbon reservoir. Explosive volcanism and rapid release of carbon dioxide and changes in the sources of bottom water during this time are considered to be plausible triggering mechanisms for the peak warming leading to hydrate dissociation. Another recent high-resolution study supports the methane hydrate connection to latest Paleocene abrupt climate change. Stable isotopic evidence from two widely separated sites from low- and southern high-latitude Atlantic Ocean indicates multiple injections of methane with global consequences during the relatively short interval at the end of Paleocene. Modeling results, as well as wide empirical data, suggest warm and wet climatic conditions with less vigorous atmospheric circulation during the late Paleocene thermal optimum.

The eustatic record of the late Paleocene–early Eocene could offer further clues for the behavior of the gas hydrates and their contribution to the overall peak warm period of this interval. The longer term trend shows a rising sea level through the latest Paleocene and early Eocene, but there are several shorter term sea level drops throughout this period and one prominent drop straddling the Paleocene–Eocene boundary (which could be an additional forcing component to hydrate dissociation for the terminal Paleocene event). Early Eocene is particularly rich in high-frequency sea level drops of several tens of meters. Could these events have contributed to the instability of gas hydrates, adding significant quantities of methane to the atmosphere and maintaining the general warming of the period? These ideas seem testable if detailed faunal and isotopic data for the interval in question were available with at least the same kind of resolution as that obtained for the latest Paleocene interval.

Timing of the Gas Hydrate Development

When did the gas hydrates first develop in the geological past? The specific low temperature–high pressure requirement for the stability of gas hydrates suggests that they may have existed at least since the latest Eocene, the timing of the first development of the oceanic psychrosphere and cold bottom waters. Theoretically clathrates could exist on the slope and rise when bottom-water temperatures approach those estimated for late Cretaceous and Paleogene (c. 7–15°C), although they would occur deeper within the sedimentary column and the stability zone would be relatively slimmer. A depth of c. 900 m below sea level has been estimated for the hydrate stability zone in the late Paleocene. If the bottom waters were to warm up to 22°C only then would most margins of the world be free of gas hydrate accumulation. The implied thinner stability zone during warm bottom-water regimes, however, does not necessarily mean an overall reduced methane reservoir, since it also follows that the sub-hydrate free gas zone could be larger, making up to the hydrate deficiency.

Prior to late Eocene there is little evidence of large polar ice caps, and the mechanism for short-term sea level changes remains uncertain. And yet, the Mesozoic–Early Cenozoic eustatic history is replete with major sea level falls of 100 m or more that are comparable in magnitude, if not in frequency, to glacially induced eustatic changes of the late Neogene. If gas hydrates existed in the pre-glacial times, major sea level falls would imply that hydrate dissociation may have contributed significantly to climate change and shallow-seated tectonics along continental margins. However, such massive methane emissions should also be accompanied by prominent $\delta^{13}C$ excursions, as exemplified by the terminal Paleocene climatic optimum.

The role of gas hydrate as a significant source of greenhouse emissions in global change scenarios and as a major contributor of carbon in global carbon cycle remains controversial. It can only be resolved with more detailed studies of hydrated intervals, in conjunction with high-resolution studies of the ice cores, preferably with decadal time resolution. A better understanding of gas hydrates may well show their considerable role in controlling continental margin stratigraphy and shallow structure, as well as in global climatic change, and through it, as agents of biotic evolution.

See also

Paleoceanography, Climate Models in.

Further Reading

Dickens GR, O'Neil JR, Rea DK, and Owen RM (1995) Dissociation of oceanic methane hydrate as a cause of the carbon isotope excursion at the end of the Paleocene. *Paleoceanography* 10: 965–971.

Dillon WP Paul CK (1983) Marine gas hydrates II. Geophysical evidence. In: Cox JS (ed.) *Natural Gas Hyrate*. London: Butterswarth, pp. 73–90.

Haq BU (1998) Gas hydrates: Greenhouse nightmare? Energy panacea or pipe dream? *GSA Today, Geological Society of America* 8(11): 1–6.

Henriet J-P and Mienert J (eds.) (1998) *Gas Hydrates: Relevance to World Margin Stability and Climate Change*, vol. 137. London: Geological Society Special Publications.

Kennett JP, Cannariato KG, Hendy IL, and Behl RJ (2000) Carbon isotopic evidence for methane hydrate instability during Quaternary interstadials. *Science* 288: 128–133.

Kvenvolden KA (1998) A primer on the geological occurrence of gas hydrates. In: Henriet J-P and Mienert J (eds.) *Gas Hydrates: Relevance to World Margin Stability and Climate Change*, vol. 137, pp. 9–30. London: Geological Society, Special Publications.

Max MD (ed.) (2000) Natural Gas Hydrates: In: *Oceanic and Permafrost Environments*. Dordrecht: Kluwer Academic Press.

Nisbet EG (1990) The end of ice age. *Canadian Journal of Earth Sciences* 27: 148–157.

Paull CK, Ussler W, and Dillon WP (1991) Is the extent of glaciation limited by marine gas hydrates. *Geophysical Research Letters* 18(3): 432–434.

Sloan ED Jr (1998) *Clathrate Hydrates of Natural Gases*. New York: Marcel Dekker.

Thorpe RB, Pyle JA, and Nisbet EG (1998) What does the ice-core record imply concerning the maximum climatic impact of possible gas hydrate release at Termination 1A? In: Henriet J-P and Mienert J (eds.) *Gas Hydrates: Relevance to World Margin Stability and Climate Change*, vol. 137, pp. 319–326. London: Geological Society, Special Publications.

APPENDICES

APPENDIX 1. SI UNITS AND SOME EQUIVALENCES

Wherever possible the units used are those of the International System of Units (SI). Other "conventional" units (such as the liter or calorie) are frequently used, especially in reporting data from earlier work. Recommendations on standardized scientific terminology and units are published periodically by international committees, but adherence to these remains poor in practice. Conversion between units often requires great care.

The base SI units

Quantity	Unit	Symbol
Length	meter	m
Mass	kilogram	kg
Time	second	s
Electric current	ampere	A
Thermodynamic temperature	kelvin	K
Amount of substance	mole	mol
Luminous intensity	candela	cd

Some SI derived and supplementary units

Quantity	Unit	Symbol	Unit expressed in base or other derived units
Frequency	hertz	Hz	s^{-1}
Force	newton	N	$kg\,m\,s^{-2}$
Pressure, stress	pascal	Pa	$N\,m^{-2}$
Energy, work, quantity of heat	joule	J	$N\,m$
Power	watt	W	$J\,s^{-1}$
Electric charge, quantity of electricity	coulomb	C	$A\,s$
Electric potential, potential difference, electromotive force	volt	V	$J\,C^{-1}$
Electric capacitance	farad	F	$C\,V^{-1}$
Electric resistance	ohm	ohm (Ω)	$V\,A^{-1}$
Electric conductance	Siemens	S	Ω^{-1}
Magnetic flux	weber	Wb	$V\,s$
Magnetic flux density	tesla	T	$Wb\,m^{-2}$
Inductance	henry	H	$Wb\,A^{-1}$
Luminous flux	lumen	lm	$cd\,sr$
Illuminance	lux	lx	$lm\,m^{-2}$
Activity (of a radionuclide)	becquerel	Bq	s^{-1}
Absorbed dose, specific energy	gray	Gy	$J\,kg^{-1}$
Dose equivalent	sievert	Sv*	$J\,kg^{-1}$
Plane angle	radian	rad	
Solid angle	steradian	sr	

*Not to be confused with Sverdrup conventionally used in oceanography: see SI Equivalences of Other Units.

SI base units and derived units may be used with multiplying prefixes (with the exception of kg, though prefixes may be applied to gram $= 10^{-3}$kg; for example, 1 Mg $= 10^6$ g $= 10^6$kg)

Prefixes used with SI units

Prefix	Symbol	Factor
yotta	Y	10^{24}
zetta	Z	10^{21}
exa	E	10^{18}
peta	P	10^{15}
tera	T	10^{12}
giga	G	10^{9}
mega	M	10^{6}
kilo	k	10^{3}
hecto	h	10^{2}
deca	da	10
deci	d	10^{-1}
centi	c	10^{-2}
milli	m	10^{-3}
micro	μ	10^{-6}
nano	n	10^{-9}
pico	p	10^{-12}
femto	f	10^{-15}
atto	a	10^{-18}
zepto	z	10^{-21}
yocto	y	10^{-24}

SI Equivalences of Other Units

Physical quantity	Unit	Equivalent	Reciprocal
Length	nautical mile (nm)	1.85318 km	km = 0.5396 nm
Mass	tonne (t)	10^3 kg = 1 Mg	
Time	min	60 s	
	h	3600 s	
	day or d	86 400 s	s $= 1.1574 \times 10^{-5}$ day
	y	3.1558×10^7 s	s $= 3.1688 \times 10^{-8}$ y
Temperature	°C	°C = K − 273.15	
Velocity	knot (1 nm h^{-1})	0.51477 m s^{-1}	m s^{-1} = 1.9426 knot
		44.5 km d^{-1}	
		16 234 km y^{-1}	
Density	gm cm^{-3}	tonne m^{-3} = 10^3 kg m^{-3}	
Force	dyn	10^{-5} N	
Pressure	dyn cm^{-2}	10^{-1} N m^{-2} = 10^{-1} Pa	
	bar	10^5 N m^{-2} = 10^5 Pa	
	atm (standard atmosphere)	101 325 N m^{-2} = 101.325 kPa	
Energy	erg	10^{-7} J	
	cal (I.T.)	4.1868 J	
	cal (15°C)	4.1855 J	
	cal (thermochemical)	4.184 J	J = 0.239 cal

(*Note*: The last value is the one used for subsequent conversions involving calories.)

Energy flux	langley (ly) min^{-1} = cal cm^{-2} min^{-1}	697 W m^{-2}	W m^{-2} = 1.434 × 10^{-3} ly min^{-1}
	ly h^{-1}	11.6 W m^{-2}	W m^{-2} = 0.0860 ly h^{-1}
	ly d^{-1}	0.484 W m^{-2}	W m^{-2} = 2.065 ly d^{-1}
	kcal cm^{-2} y^{-1}	1.326 W m^{-2}	W m^{-2} = 0.754 kly y^{-1}
Volume flux	Sverdrup	10^6 m^3 s^{-1}	
		3.6 km^3 h^{-1}	
Latent heat	cal g^{-1}	4184 J kg^{-1}	J kg^{-1} = 2.39 × 10^{-4} cal g^{-1}
Irradiance	Einstein m^{-2} s^{-1} (mol photons m^{-2} s^{-1})		

*Most values are taken from or derived from *The Royal Society Conference of Editors Metrication in Scientific Journals*, 1968, The Royal Society, London.

The SI units for pressure is the pascal (1 Pa = 1 N m^{-2}). Although the bar (1 bar = 10^5 Pa) is also retained for the time being, it does not belong to the SI system. Various texts and scientific papers still refer to gas pressure in units of the torr (symbol: Torr), the bar, the conventional millimetre of mercury (symbol: mmHg), atmospheres (symbol: atm), and pounds per square inch (symbol: psi) – although these units will gradually disappear (see Conversions between Pressure Units).

Irradiance is also measured in W m^{-2}. Note: 1 mol photons − 6.02 × 10^{23} photons.

The SI unit used for the amount of substance is the mole (symbol: mol), and for volume the SI unit is the cubic metre (symbol: m^3). It is technically correct, therefore, to refer to concentration in units of mol m^3. However, because of the volumetric change that sea water experiences with depth, marine chemists prefer to express sea water concentrations in molal units, mol kg^{-1}.

Conversions between Pressure Units

	Pa	kPa	bar	atm	Torr	psi
1 Pa =	1	10^{-3}	10^{-5}	9.869 23 × 10^{-6}	7.500 62 × 10^{-3}	1.450 38 × 10^{-4}
1 kPa =	10^3	1	10^{-2}	9.869 23 × 10^{-3}	7.500 62	0.145 038
1 bar =	10^5	10^2	1	0.986 923	750.062	145.038
1 atm =	101 325	101.325	1.013 25	1	760	14.6959
1 Torr =	133.322	0.133 322	1.333 22 × 10^{-3}	1.315 79 × 10^{-3}	1	1.933 67 × 10^{-2}
1 psi	6894.76	6.894 76	6.894 76 × 10^{-2}	6.804 60 × 10^{-2}	51.715 07	1

psi = pounds force per square inch.
1 mmHg = 1 Torr to better than 2 × 10^{-7} Torr.

APPENDIX 2. USEFUL VALUES

Molecular mass of dry air, $m_a = 28.966$

Molecular mass of water, $m_w = 18.016$

Universal gas constant, $R = 8.31436 \, \mathrm{J \, mol^{-1} K^{-1}}$

Gas constant for dry air, $R_a = R/m_a = 287.04 \, \mathrm{J \, kg^{-1} K^{-1}}$

Gas constant for water vapor, $R_v = R/m_w = 461.50 \, \mathrm{J \, kg^{-1} K^{-1}}$

Molecular weight ratio $\varepsilon \equiv m_w/m_a = R_a/R_v = 0.62197$

Stefan's constant $\sigma = 5.67 \times 10^{-8} \, \mathrm{W \, m^{-2} K^{-4}}$

Acceleration due to gravity, $g \, (\mathrm{m \, s^{-2}})$ as a function of latitude φ and height $z \, (\mathrm{m})$

$$g = (9.78032 + 0.005172 \sin^2 \varphi - 0.00006 \sin^2 2\varphi)(1 + z/a)^{-2}$$

Mean surface value, $\bar{g} = \int_0^{\pi/2} g \cos \varphi \, \mathrm{d}\varphi = 9.7976$

Radius of sphere having the same volume as the Earth, $a = 6371 \, \mathrm{km}$ (equatorial radius $= 6378 \, \mathrm{km}$, polar radius $= 6357 \, \mathrm{km}$)

Rotation rate of earth, $\Omega = 7.292 \times 10^{-5} \, \mathrm{s^{-1}}$

Mass of earth $= 5.977 \times 10^{24} \, \mathrm{kg}$

Mass of atmosphere $= 5.3 \times 10^{18} \, \mathrm{kg}$

Mass of ocean $= 1400 \times 10^{18} \, \mathrm{kg}$

Mass of ground water $= 15.3 \times 10^{18} \, \mathrm{kg}$

Mass of ice caps and glaciers $= 43.4 \times 10^{18} \, \mathrm{kg}$

Mass of water in lakes and rivers $- 0.1267 \times 10^{18} \, \mathrm{kg}$

Mass of water vapor in atmosphere $= 0.0155 \times 10^{18} \, \mathrm{kg}$

Area of earth $= 5.10 \times 10^{14} \, \mathrm{m^2}$

Area of ocean $= 3.61 \times 10^{14} \, \mathrm{m^2}$

Area of land $= 1.49 \times 10^{14} \, \mathrm{m^2}$

Area of ice sheets and glaciers $= 1.62 \times 10^{13} \, \mathrm{m^2}$

Area of sea ice $= 1.9 \times 10^{13} \, \mathrm{m^2}$ in March and $2.9 \times 10^{13} \, \mathrm{m^2}$ in September (averaged between 1979 and 1987)

APPENDIX 3. PERIODIC TABLE OF THE ELEMENTS

Atomic number
Element symbol
Atomic mass

1	2	3	4	5	6	7	8	9	10	11	12	13	14	15	16	17	18
1 H 1.00794																	2 He 4.00260
3 Li 6.941	4 Be 9.01218											5 B 10.811	6 C 12.011	7 N 14.0067	8 O 15.9994	9 F 18.9984	10 Ne 20.1797
11 Na 22.9898	12 Mg 24.3050											13 Al 26.9815	14 Si 28.0855	15 P 30.9738	16 S 32.066	17 Cl 35.4527	18 Ar 39.948
19 K 39.0983	20 Ca 40.078	21 Sc 44.9559	22 Ti 47.88	23 V 50.9415	24 Cr 51.9961	25 Mn 54.9380	26 Fe 55.847	27 Co 58.9332	28 Ni 58.69	29 Cu 63.546	30 Zn 65.39	31 Ga 69.723	32 Ge 72.61	33 As 74.9216	34 Se 78.96	35 Br 79.904	36 Kr 83.80
37 Rb 85.4678	38 Sr 87.62	39 Y 88.9059	40 Zr 91.224	41 Nb 92.9064	42 Mo 95.94	43 Tc (98)	44 Ru 101.07	45 Rh 102.906	46 Pd 106.42	47 Ag 107.868	48 Cd 112.411	49 In 114.82	50 Sn 118.710	51 Sb 121.75	52 Te 127.60	53 I 126.905	54 Xe 131.29
55 Cs 132.905	56 Ba 137.327	57 La 138.906 ★	72 Hf 178.49	73 Ta 180.948	74 W 183.85	75 Re 186.207	76 Os 190.2	77 Ir 192.22	78 Pt 195.08	79 Au 196.967	80 Hg 200.59	81 Tl 204.383	82 Pb 207.2	83 Bi 208.980	84 Po (209)	85 At (210)	86 Rn (222)
87 Fr (223)	88 Ra 226.025	89 Ac 227.028 ◀	104 (261)	105 (262)	106 (263)	107 (262)	108 (265)	109 (267)									

★ Lanthanides

58 Ce 140.115	59 Pr 140.908	60 Nd 144.24	61 Pm (145)	62 Sm 150.36	63 Eu 151.965	64 Gd 157.25	65 Tb 158.925	66 Dy 162.50	67 Ho 164.930	68 Er 167.26	69 Tm 168.934	70 Yb 173.04	71 Lu 174.967

◀ Actinides

90 Th 232.038	91 Pa 231.036	92 U 238.029	93 Np 237.048	94 Pu (244)	95 Am (243)	96 Cm (247)	97 Bk (247)	98 Cf (251)	99 Es (252)	100 Fm (257)	101 Md (258)	102 No (259)	103 Lr (260)

APPENDIX 4. THE GEOLOGIC TIME SCALE

Eon	Era	Period	Epoch	Millions of Years Ago
Phanerozoic	Cenozoic	(Quaternary)	Holocene	
				0.011
			Pleistocene	
				1.82
		(Tertiary)	Pliocene	
				5.32
			Miocene	
				23
			Oligocene	
				33.7
			Eocene	
				55
			Paleocene	
				65
	Mesozoic	Cretaceous		
				144
		Jurassic		
				200
		Triassic		
				250
	Paleozoic	Permian		
				295
		Carboniferous Pennsylvanian		
				320
		Mississippian		
				355
		Devonian		
				410
		Silurian		
				440
		Ordovician		
				500
		Cambrian		
				540
Proterozoic				
				2500
Archean		Oldest Rock		4400
		Age of the Solar System		4550

APPENDIX 5. PROPERTIES OF SEAWATER

A5.1 The Equation of State

It is necessary to know the equation of state for the ocean very accurately to determine stability properties, particularly in the deep ocean. The equation of state defined by the Joint Panel on Oceanographic Tables and Standards fits available measurements with a standard error of 3.5 ppm for pressure up to 1000 bar, for temperatures between freezing and 40 °C, and for salinities between 0 and 42. The density ρ (kg m^{-3}) is expressed in terms of pressure p (bar), temperature t (°C), and practical salinity S. The last quantity is defined in such a way that its value (in practical salinity units or psu) is very close to the old value expressed in parts per thousand (‰ or ppt). Its relation to previously defined measures of salinity is given by Lewis and Perkin.

The equation for ρ is obtained in a sequence of steps. First, the density ρ_w of pure water ($S=0$) is given by

$$\rho_w = 999.842\ 594 + 6.793\ 952 \times 10^{-2}t$$
$$- 9.095\ 290 \times 10^{-3}t^2 + 1.001\ 685 \times 10^{-4}t^3$$
$$- 1.120\ 083 \times 10^{-6}t^4$$
$$+ 6.536\ 332 \times 10^{-9}t^5 \qquad \text{[A5.1]}$$

Second, the density at one standard atmosphere (effectively $p=0$) is given by

$$\rho(S,t,0) = \rho_w + S(0.824\ 493 - 4.089\ 9 \times 10^{-3}t$$
$$+ 7.643\ 8 \times 10^{-5}t^2 - 8.246\ 7 \times 10^{-7}t^3$$
$$+ 5.387\ 5 \times 10^{-9}t^4) + S^{3/2}(-5.724\ 66$$
$$\times 10^{-3} + 1.022\ 7 \times 10^{-4}t - 1.654\ 6$$
$$\times 10^{-6}t^2) + 4.831\ 4 \times 10^{-4}S^2 \qquad \text{[A5.2]}$$

Finally, the density at pressure p is given by

$$\rho(S,t,p) = \frac{\rho(S,t,0)}{1 - p/K(S,t,p)} \qquad \text{[A5.3]}$$

where K is the secant bulk modulus. The pure water value K_w is given by

$$K_w = 19\ 652.21 + 148.420\ 6t - 2.327\ 105t^2$$
$$+ 1.360\ 477 \times 10^{-2}t^3 - 5.155\ 288$$
$$\times 10^{-5}t^4 \qquad \text{[A5.4]}$$

The value at one standard atmosphere ($p=0$) is given by

$$K(S,t,0) = K_w + S(54.674\ 6 - 0.603\ 459t$$
$$+ 1.099\ 87 \times 10^{-2}t^2 - 6.167\ 0$$
$$\times 10^{-5}t^3) + S^{3/2}(7.944 \times 10^{-2}$$
$$+ 1.648\ 3 \times 10^{-2}t$$
$$- 5.300\ 9 \times 10^{-4}t^2) \qquad \text{[A5.5]}$$

and the value at pressure p by

$$K(S,t,p) = K(S,t,0) + p(3.239\ 908 + 1.437\ 13$$
$$\times 10^{-3}t + 1.160\ 92 \times 10^{-4}t^2$$
$$- 5.779\ 05 \times 10^{-7}t^3)$$
$$+ pS(2.283\ 8 \times 10^{-3} - 1.098\ 1$$
$$\times 10^{-5}t - 1.607\ 8 \times 10^{-6}t^2)$$
$$+ 1.910\ 75 \times 10^{-4}pS^{3/2}$$
$$+ p^2(8.509\ 35 \times 10^{-5} - 6.122\ 93$$
$$\times 10^{-6}t + 5.278\ 7 \times 10^{-8}t^2)$$
$$+ p^2S(-9.934\ 8 \times 10^{-7} + 2.081\ 6$$
$$\times 10^{-8}t + 9.169\ 7 \times 10^{-10}t^2) \qquad \text{[A5.6]}$$

Values for checking the formula are $\rho(0,5,0) = 999.966\ 75$, $\rho(35, 5, 0) = 1027.675\ 47$, and $\rho(35, 25, 1000) = 1062.538\ 17$.

Since ρ is always close to 1000 kg m^{-3}, values quoted are usually those of the difference $(\rho - 1000)$ in kg m^{-3} as is done in **Table A5.1**. The table is constructed so that values can be calculated for 98% of the ocean (see **Figure A5.1**). The maximum errors in density on straight linear interpolation are 0.013 kg m^{-3} for both temperature and pressure interpolation and only 0.006 for salinity interpolation in the range of salinities between 30 and 40. The error when combining all types of interpolation for the 98% range of values is less than 0.03 kg m^{-3}.

A5.2 Other Quantities Related to Density

Older versions of the equation of state usually gave formulas not for calculating the absolute density ρ, but for the 'specific gravity' ρ/ρ_m, where ρ_m is the maximum density of pure water. Since this is always

Table A5.1

p (bar)	S	t (°C)	$\rho - 1000$ (kg m^{-3})	$\partial\rho/\partial S$	α (10^{-7} K^{-1})	$\partial\alpha/\partial S$	c_p (J kg^{-1} K^{-1})	$\partial c_p/\partial S$	θ (10^{-3} °C)	$\partial\theta/\partial S$	c_s (m s^{-1})	$\partial c_s/\partial S$
0	35	-2	28.187	0.814	254	33	3989	-6.2	-2000	0	1439.7	1.37
0	35	-0	28.106	0.808	526	31	3987	-6.1	0	0	1449.1	1.34
0	35	2	27.972	0.801	781	28	3985	-5.9	2000	0	1458.1	1.31
0	35	4	27.786	0.796	1021	26	3985	-5.8	4000	0	1466.6	1.29
0	35	7	27.419	0.788	1357	23	3985	-5.6	7000	0	1478.7	1.25
0	35	10	26.952	0.781	1668	20	3986	-5.5	10000	0	1489.8	1.22
0	35	13	26.394	0.775	1958	17	3988	-5.3	13000	0	1500.2	1.19
0	35	16	25.748	0.769	2230	15	3991	-5.2	16000	0	1509.8	1.16
0	35	19	25.022	0.764	2489	14	3993	-5.1	19000	0	1518.7	1.13
0	35	22	24.219	0.760	2734	12	3996	-4.9	22000	0	1526.8	1.10
0	35	25	23.343	0.756	2970	11	3998	-4.9	25000	0	1534.4	1.08
0	35	28	22.397	0.752	3196	9	4000	-4.8	28000	0	1541.3	1.06
0	35	31	21.384	0.749	3413	8	4002	-4.7	31000	0	1547.6	1.03
100	35	-2	32.958	0.805	552	31	3953	-5.8	-2029	-2	1456.1	1.38
100	35	-0	32.818	0.799	799	28	3953	-5.7	-45	-2	1465.5	1.35
100	35	2	32.629	0.793	1031	26	3954	-5.6	1939	-2	1474.5	1.33
100	35	4	32.393	0.788	1251	24	3955	-5.5	3923	-2	1483.1	1.30
100	35	7	31.958	0.781	1559	21	3957	-5.3	6901	-1	1495.1	1.26
100	35	10	31.431	0.774	1844	18	3960	-5.2	9879	-1	1506.3	1.22
100	35	13	30.818	0.769	2111	16	3963	-5.1	12858	-1	1516.7	1.19
100	35	16	30.126	0.763	2363	14	3967	-5.0	15838	-1	1526.4	1.16
100	35	19	29.359	0.759	2603	13	3970	-4.9	18819	-1	1535.3	1.13
200	35	-2	37.626	0.797	834	28	3922	-5.5	-2076	-3	1472.8	1.39
200	35	0	37.429	0.791	1058	26	3923	-5.4	-107	-3	1482.3	1.36
200	35	2	37.187	0.786	1269	24	3925	-5.3	1862	-3	1491.2	1.33
200	35	4	36.903	0.781	1469	22	3927	-5.2	3832	-3	1499.8	1.30
200	35	7	36.402	0.774	1750	19	3931	-5.1	6789	-3	1511.8	1.26
300	35	-2	42.191	0.789	1101	26	3893	-5.2	-2140	-5	1489.9	1.39
300	35	-0	41.941	0.783	1303	24	3896	-5.1	-186	-5	1499.3	1.36
300	35	2	41.649	0.778	1494	22	3899	-5.0	1771	-5	1508.2	1.33
300	35	4	41.319	0.774	1676	20	3903	-5.0	3728	-5	1516.6	1.30
400	35	-2	46.658	0.781	1351	24	3867	-4.9	-2221	-7	1507.2	1.39
400	35	-0	46.356	0.776	1534	22	3871	-4.8	-279	-6	1516.5	1.36
400	35	2	46.017	0.771	1707	20	3876	-4.8	1665	-6	1525.3	1.33
400	35	4	45.643	0.767	1872	19	3880	-4.7	3610	-6	1533.7	1.30
500	35	-2	51.029	0.773	1587	22	3844	-4.7	-2316	-8	1524.8	1.38
500	35	0	50.678	0.769	1751	20	3849	-4.6	-386	-8	1534.0	1.35
500	35	2	50.293	0.764	1907	19	3854	-4.6	1546	-7	1542.7	1.32
600	35	-2	55.305	0.766	1807	20	3824	-4.4	-2426	-9	1542.6	1.37
600	35	-0	54.908	0.762	1954	18	3829	-4.4	-506	-9	1551.6	1.34
600	35	2	54.481	0.758	2094	17	3835	-4.4	1416	-9	1560.2	1.31

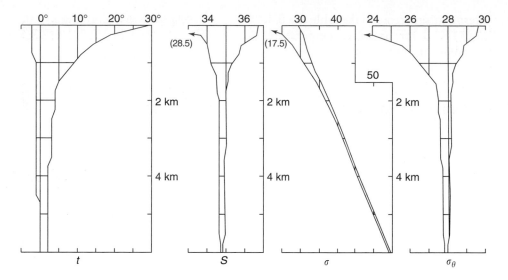

Figure A5.1 The ranges of temperature t (in °C) and salinity S for 98% of the ocean as a function of depth and the corresponding ranges of density σ and potential density σ_θ. From Bryan K and Cox MD (1972) An approximate equation of state for numerical models of ocean circulation. *Journal of Physical Oceanography* 2: 510–514.

close to unity, a quantity called σ was defined by

$$1000\left(\frac{\rho}{\rho_m} - 1\right) - \frac{1000}{\rho_m}(\rho \quad \rho_m) \quad [A5.7]$$

Since

$$\rho_m = 999.975 \text{ kg m}^{-3} \quad [A5.8]$$

it follows that σ, as defined above, is related to the $(\rho - 1000)$ values by

$$\sigma = (\rho - 1000) + 0.025 \quad [A5.9]$$

that is, 0.025 must be added to the values of $(\rho - 1000)$ on the table to obtain the old σ value. The notation σ, (sigma tau) was used for the value of σ calculated at zero pressure, and σ_θ (sigma theta) for the quantity corresponding to potential density. Another quantity commonly used in oceanography is the specific volume (or steric) 'anomaly' δ defined by

$$\delta = v_s(S, t, p) - v_s(35, 0, p) \quad [A5.10]$$

and usually reported in units of $10^{-8} \text{ m}^3 \text{ kg}^{-1}$.

A5.3 Expansion Coefficients

The thermal expansion coefficient α is given in **Table A5.1** in units of 10^{-7}K^{-1} along with its S derivative. The maximum error from pressure interpolation is 2 units, that from temperature interpolation is 3 units, and that for salinity interpolation $(30 < S < 40)$ is 2 units plus a possible round-off error of 2 units.

The salinity expansion coefficient β can be calculated by using the given values of $\partial \rho / \partial S$.

A5.4 Specific Heat

The specific heat at surface pressure is given by Millero *et al.* and can be calculated in two stages. First, the value in $\text{J kg}^{-1} - \text{K}^{-1}$ for fresh water is given by

$$c_p(0, t, 0) = 4217.4 - 3.720\ 283t + 0.141\ 285\ 5t^2$$
$$- 2.654\ 387 \times 10^{-3}t^3$$
$$+ 2.093\ 236 \times 10^{-5}t^4 \quad [A5.11]$$

Second,

$$c_p(S, t, 0) = c_p(0, t, 0) + S(-7.644\ 4$$
$$+ 0.107\ 276t - 1.383\ 9 \times 10^{-3}t^2)$$
$$+ S^{3/2}(0.177\ 09 - 4.077\ 2 \times 10^{-3}t$$
$$+ 5.353\ 9 \times 10^{-5}t^2) \quad [A5.12]$$

The formula can be checked against the result $c_p(40, 40, 0) = 3981.050$. The standard deviation of the algorithm fit is 0.074. Values at nonzero pressures can be calculated by using eqn [A5.13] and the equation of state:

$$\left(\frac{\partial c_p}{\partial p}\right)_T = -T\left(\frac{\partial^2 v_s}{\partial T^2}\right)_p \quad [A5.13]$$

The values in **Table A5.1** are based on the above formula and a polynomial fit for higher pressures derived from the equation of state by N.P. Fofonoff.

The intrinsic interpolation errors in the table are 0.4, 0.1, and $0.3\,\mathrm{J\,kg^{-1}\,K^{-1}}$ for pressure, temperature, and salinity interpolation, respectively, and there are additional obvious round-off errors.

A5.5 Potential Temperature

The 'adiabatic lapse rate' Γ is given by

$$\Gamma = \frac{g\alpha T}{c_p} \qquad [\text{A5.14}]$$

and therefore can be calculated from the above formulas. The definition of potential temperature

$$\frac{\theta}{T} = \left(\frac{p_r}{p}\right)^{\kappa} \qquad [\text{A5.15}]$$

where p_r is a reference pressure level (usually 1 bar) and $\kappa = (\gamma - 1)/\gamma$, where γ is the ratio of specific heats at constant pressure and at constant volume, can then be used to obtain θ. The following algorithm, however, was derived by Bryden, using experimental compressibility data, to give $\theta\,(^{\circ}\mathrm{C})$ as a function of salinity S, temperature $t\,(^{\circ}\mathrm{C})$, and pressure p (bar) for $30 < S < 40$, $2 < t < 30$, and $0 < p < 1000$:

$$\begin{aligned}
\theta(S, t, p) =\ & t - p(3.650\ 4 \times 10^{-4} + 8.319\ 8 \times 10^{-5}t \\
& - 5.406\ 5 \times 10^{-7}t^2 + 4.027\ 4 \times 10^{-9}t^3) \\
& - p(S - 35)(1.743\ 9 \times 10^{-5} - 2.977\ 8 \\
& \times 10^{-7}t) - p^2(8.930\ 9 \times 10^{-7} - 3.162\ 8 \\
& \times 10^{-8}t + 2.198\ 7 \times 10^{-10}t^2) + 4.105\ 7 \\
& \times 10^{-9}(S - 35)p^2 - p^3(-1.605\ 6 \times 10^{-10} \\
& + 5.048\ 4 \times 10^{-12}t)
\end{aligned} \qquad [\text{A5.16}]$$

A check value is $\theta(25, 10, 1000) = 8.467\ 851\ 6$, and the standard deviation of Bryden's polynomial fit was 0.001 K. Values in **Table A5.1** are given in millidegrees, the intrinsic interpolation errors being 2, 0.3, and 0 millidegrees for pressure, temperature, and salinity interpolation, respectively (**Figure A5.2**).

A5.6 Speed of Sound

The speed of sound c_s can be calculated from the equation of state, using eqn [A5.17]

$$c_s^2 = \left(\frac{\partial p}{\partial \rho}\right)_{\theta, S} \qquad [\text{A5.17}]$$

Values given in **Table A5.1** use algorithms derived by Chen and Millero on the basis of direct measurements. The formula applies for $0 < S < 40$,

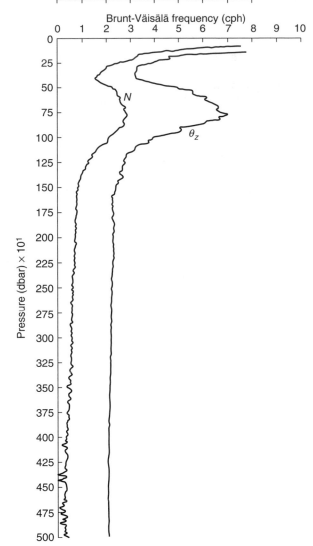

Figure A5.2 A profile of buoyancy frequency N in the ocean. From the North Atlantic near 28° N, 70° W, courtesy of Dr. R.C. Millard.

$0 < t < 40$, $0 < p < 1000$ with a standard deviation of $0.19\,\mathrm{ms^{-1}}$. Values in the table are given in meters per second, the intrinsic interpolation errors being 0.05, 0.10, and $0.04\,\mathrm{ms^{-1}}$ for pressure, temperature, and salinity interpolation, respectively.

A5.7 Freezing Point of Sea Water

The freezing point t_f of sea water $(^{\circ}\mathrm{C})$ is given by

$$\begin{aligned}
t_f(S, p) =\ & -0.057\ 5S + 1.710\ 523 \\
& \times 10^{-3}S^{3/2} - 2.154\ 996 \\
& \times 10^{-4}S^2 - 7.53 \times 10^{-3}p
\end{aligned} \qquad [\text{A5.18}]$$

The formula fits measurements to an accuracy of ± 0.004 K.

Further Reading

Bryan K and Cox MD (1972) An approximate equation of state for numerical models of ocean circulation. *Journal of Physical Oceanography* 2: 510–514.

Bryden HL (1973) New polynomials for thermal expansion, adiabatic temperature gradient and potential temperature gradient of sea water. *Deep Sea Research* 20: 401–408.

Chen C-T and Millero FJ (1977) Speed of sound in seawater at high pressures. *Journal of the Acoustical Society of America* 62: 1129–1135.

Dauphinee TM (1980) Introduction to the special issue on the Practical Salinity Scale 1978. *IEEE, Journal of Oceanic Engineering* OE 5: 1–2.

Gill AE (1982) *Atmosphere–Ocean Dynamics*, International Geophysics Series Volume 30. San Diego, CA: Academic Press.

Kraus EB (1972) *Atmosphere–Ocean Interaction*. London: Oxford University Press.

Lewis EL and Perkin RG (1981) The Practical Salinity Scale 1978: Conversion of existing data. *Deep Sea Research* 28A: 307–328.

Millero FJ (1978) Freezing point of seawater. In: *Eighth Report of the Joint Panel on Oceanographic Tables and Standards*, UNESCO Technical Papers in Marine Science No. 28, Annex 6. Paris: UNESCO.

Millero FJ, Chen C-T, Bradshaw A, and Schleicher K (1980) A new high pressure equation of state for seawater. *Deep Sea Research* 27A: 255–264.

Millero FJ and Poisson A (1981) International one-atmosphere equation of state for seawater. *Deep Sea Research* 28A: 625–629.

Millero FJ, Perron G, and Desnoyers JE (1973) Heat capacity of seawater solutions from 5 to 25 °C and 0.5 to 22% chlorinity. *Journal of Geophysical Research* 78: 4499–4507.

UNESCO (1981) *Tenth Report of the Joint Panel on Oceanographic Tables and Standards*, UNESCO Technical Papers in Marine Science No. 36. Paris: UNESCO.

APPENDIX 6. THE BEAUFORT WIND SCALE AND SEASTATE

Beaufort wind scale	Wind description	Wind speed		Seastate	Probable wave height		Description of sea
		Mean (m s⁻¹)	Range (m s⁻¹)		Mean (m)	Maximum (m)	
0	Calm	0	0–0.2	0			'Calm': Sea like a mirror
1	Light airs	0.8	0.3–1.5	1	0.1	0.1	'Calm': Ripples with the appearance of scales are formed, but without foam crests
2	Light breeze	2.4	1.6–3.3	2	0.2	0.3	'Smooth': Small wavelets, still short, but more pronounced. Crests have a glassy appearance and do not break
3	Gentle breeze	4.3	3.4–5.4	3	0.6	1.0	'Slight': Large wavelets. Crests begin to break. Foam of glassy appearance. Perhaps scattered whitecaps
4	Moderate breeze	6.7	5.5–7.9	3–4	1.0	1.5	'Slight–moderate': Small waves, becoming larger; fairly frequent whitecaps
5	Fresh breeze	9.3	8.0–10.7	4	2.0	2.5	'Moderate': Moderate waves, taking a more pronounced long form; many whitecaps formed
6	Strong breeze	12.3	10.8–13.8	5	3.0	4.0	'Rough': Large waves begin to form; the white foam crests are more extensive everywhere
7	Near gale	15.5	13.9–17.1	5–6	4.0	5.5	'Rough–very rough': Sea heaps up and white foam begins to be blown in streaks along the direction of the wind
8	Gale	18.9	17.2–20.7	6–7	5.5	7.5	'Very rough–high': Moderately high waves of greater length; edges of crests begin to break into spindrift. The foam is blown in well-marked streaks along the direction of the wind
9	Severe gale	22.6	20.8–24.4	7	7.0	10.0	'High': High waves. Dense streaks of foam along the direction ofthe wind. Crests of waves begin to topple, tumble, and roll over. Spray may affect visibility

(Continued)

Continued

Beaufort wind scale	Wind description	Wind speed		Seastate	Probable wave height		Description of sea
		Mean $(m\,s^{-1})$	Range $(m\,s^{-1})$		Mean (m)	Maximum (m)	
10	Storm	26.4	24.5–28.4	8	9.0	12.5	'Very high': Very high waves with long overhanging crests. The resulting foam, in great patches, is blown in dense white streaks along the direction of the wind. On the whole the surface of the sea takes on a white appearance. The 'tumbling' of the sea becomes heavy and shock-like. Visibility affected
11	Violent storm	30.5	28.5–32.6	8	11.5	16.0	'Very high': Exceptionally high waves (small and medium-size ships might be for a time lost to view behind the waves). The sea is completely covered with long white patches of foam lying along the direction of the wind. Everywhere the edges of the wave crests are blown into froth. Visibility affected
12	Hurricane	–	32.7+	9	–	14.0+	'Phenomenal': The air is filled with foam and spray. Sea completely white with driving spray. Visibility very severely affected

The 'Probable wave heights' refer to well-developed wind waves in the open sea. There is a lag between the wind getting up and the sea increasing, which should be borne in mind.
This table is derived from the UK Met Office table (http://www.metoffice.gov.uk/weather/marine/guide/beaufortscale.html) and the corresponding scale given in the Met Office Observers Handbook.

INDEX

Notes

Cross-reference terms in italics are general cross-references, or refer to subentry terms within the main entry (the main entry is not repeated to save space). Readers are also advised to refer to the end of each article for additional cross-references - not all of these cross-references have been included in the index cross-references.

The index is arranged in set-out style with a maximum of three levels of heading. Major discussion of a subject is indicated by bold page numbers. Page numbers suffixed by T and F refer to Tables and Figures respectively. vs. indicates a comparison.

This index is in letter-by-letter order, whereby hyphens and spaces within index headings are ignored in the alphabetization. For example, 'oceanography' is alphabetized before 'ocean optics.' Prefixes and terms in parentheses are excluded from the initial alphabetization.

Where index subentries and sub-subentries pertaining to a subject have the same page number, they have been listed to indicate the comprehensiveness of the text.

Abbreviations used in subentries

AUV - autonomous underwater vehicle
$\delta^{18}O$ - oxygen isotope ratio
DIC - dissolved inorganic carbon
DOC - dissolved organic carbon
ENSO - El Niño Southern Oscillation
MOC - meridional overturning circulation
MOR - mid-ocean ridge
NADW - North Atlantic Deep Water
POC - particulate organic carbon
SAR - synthetic aperture radar
SST - sea surface temperature

Additional abbreviations are to be found within the index.

A

AABW *see* Antarctic Bottom Water (AABW)
AAIW *see* Antarctic Intermediate Water (AAIW)
Absolute humidity, definition, 121T
Abyssal, definition, 491
Abyssal depths, 27, 29
Abyssal gigantism, 423
Abyssal zone, 419T, 422–423, 423
ACC *see* Antarctic Circumpolar Current (ACC)
Accelerator mass spectrometry (AMS), 305–307, 480
Acoustic waves, 19
Active layer, definition, 252–253

Active sensors, 98
Adaptations
  benthic organisms, 416–417
Adelie coast, 28–29
Adée penguin (*Pygoscelis adeliae*)
  response to climate change, 562, 562F, 563F
  prehistoric, 558–559
ADEOS-I *see* Advanced Earth Observing Satellite
ADEOS II (Advanced Earth Observing Satellite II), 282
Adhemar, Joseph Alphonse, 344
Advanced ATSR (AATSR), 302
Advanced Earth Observing Satellite (ADEOS), 111
Advanced Earth Observing Satellite II (ADEOS II), 282

Advanced Microwave Scanning Radiometer (AMSR), 114, 282
Advanced microwave sounding unit (AMSU-A), 115
Advanced Scatterometer (ASCAT), 111, 112T
Advanced Very High Resolution Radiometer (AVHRR), 292, 294–295
  channels, 294
  day-time contamination by reflected sunlight, 294
  evaporation, estimation of, 125
  non-linear SST and Pathfinder SST, 292–293
  operating method, 294
  post-process data, 292–293
  scan geometry, 294–295, 295F

Advection, 129
  ocean-atmosphere interactions, 73
Advection-diffusion equation, general
    circulation models, 22
Advection-diffusion models, chemical
    tracers and, 469
Advective feedback, 43
Advective fluxes, upper ocean, 129
Aerosols
  atmospheric, infrared atmospheric
    correction algorithms, SST
    measurement, 293
  cloud formation and, 543–545
Africa
  El Niño and, 57
    precipitation, 59–60
  river water, composition, 537T
African Humid Period, 391
Agassiz, Louis, 344
AGCMs see Atmospheric general
    circulation models (AGCMs)
Agulhas Current
  flow variability, 6F
Agulhas rings, 108
AIDJEX experiment, 193–194
Air
  density, 120
  temperature
    satellite remote sensing, 114–115,
    115
Air–sea gas exchange
  carbon dioxide cycle, 459
    El Niño Southern Oscillation and, 68
    see also Carbon dioxide (CO_2)
  radiocarbon, 490–491
  satellite remote sensing of SST
    application, 301–302
Air–sea heat flux
  transfer processes, 110
Air–water interface
  temperature difference vs. Ui, 185, 185F
Alabaminella weddellensis foraminifer,
    435F, 436
Alaska
  sea ice cover, 234
Alaskan Beaufort Sea, sub-sea permafrost,
    259–260
Albatross, 355
  sediment core collection, 356
Albedo, 103, 186–187
Algorithms
  infrared atmospheric correction, SST
    measurement, 293
  NASA Team, 283, 288F
Alkenone unsaturation index
  bias, 328
  sea surface temperature and, 336F
  sea surface temperature
    paleothermometry and, 331T
  error sources, 336
Along-Track Scanning Radiometer
    (ATSR), 297
  brightness temperature calibration,
    296
  infrared rotating scan mirror, 295–296,
    295F

relatively narrow swath width, 296–297
  two brightness temperature
    measurements, 295
Alps, North Atlantic Oscillation and, 37
Altimeters
  in estimation of sea level variation,
    508–509
Altimetry, 270
  satellite see Satellite altimetry
Ammonia
  profiles, 159F
AMS see Accelerator mass spectrometry
AMSR see Advanced microwave scanning
    radiometer (AMSR)
Amundsen Basin
  deep water, 162–163
  temperature and salinity profiles, 156F,
    157F, 160F
Amundsen Sea, sea ice cover, 239–240
Anistropy, 183–184
Anoxic zones
  formation, 77–78
  and sediment bioturbation, 77
Antarctic see Antarctic Ocean
Antarctica, 30
  ice shelves, distribution of, 202F
  sub-sea permafrost, 260
  see also Antarctic Ocean; Antarctic sea
    ice
Antarctic bottom water (AABW), 28F, 29,
    31, 141–142
  Drake Passage closure, 363
    Panama Passage opening and, 364,
    365F
  glacial climate change and, 43–44
  North Atlantic Deep Water and
    alternating dominance, 393–394
  water properties, 143F
Antarctic Circumpolar Current (ACC),
    29, 30F, 32, **141–153**
  Antarctic Bottom Water and, 10, 29
  Antarctic ice sheet formation and,
    384
  circulation, 145–146
    jets, 146
  current speeds, 142–143
  dynamics, 148–151
    bottom form stress, 150
    eddies, 150
  eddies, overturning circulation and,
    151–152
  eddy activity, 150
  effects on global circulation, 150
  flow
    route, 28, 29
  freshwater fluxes, 11F
  fronts, 141, 144
    biological populations and, 145
    seafloor topography and,
    145–146
    transport, 147
  isopycnals, 30, 31
  jets, 150F
  modeling, 148–150
  North Atlantic Deep Water flowing into,
    10

overturning circulation and, 151–152
  Polar Frontal Zone, 144–145
  speed, simulated, 151F
  structure, 141–145
  temperature and salinity, 28
  transport, 146–148
    longitude and, 147–148
    total, 141
    variability, 148
    volume, 146–147
  variability, 150–151
  water properties, 141–142, 143F
    latitude and, 142
  see also Antarctic Circumpolar Current;
    Antarctic Zone; Atlantic Ocean
    current systems; Polar Front; Polar
    Frontal Zone; Southern Antarctic
    Circumpolar Current front;
    Subantarctic Front; Subantarctic
    Zone
Antarctic continent, 30
  see also Antarctica
Antarctic continental shelf, sub ice-shelf
    circulation, 172–173, 174F
Antarctic Ice Sheet
  formation
    Oi1 event, 384
    oxygen isotope ratio and, 513–514,
    514F
  lithosphere depression effects,
    169–170
  sea level variations and, 510–511,
    511, 512
  surface melting projections, 511
Antarctic Intermediate Water (AAIW), 9F,
    141–142
  formation, 151, 152F
  water properties, 143F
Antarctic Ocean
  ice sheet formation, Oi1 event, 384
  ice-shelves, 169, 171F
Antarctic Peninsula, 286, 287–288
  ice shelf stability, 201, 203
Antarctic sea ice
  Bellingshausen and Amundsen Seas,
    286, 287F
  drift patterns, 224
  extents, 285, 286
  geographical extent, 239–243, 242F
  interannual variability, 243–244,
    243F
  Ross Sea, 286, 287F
  Southern Hemisphere, 286, 287–288,
    287F
  thickness, 247F, 248–250
    model evaluation, 197F
  Weddell Sea, 286, 287F
Antarctic Slope Front, water properties,
    143F
Antarctic Zone, 144–145
Anthropogenic carbon, 467
  definition, 475
Anthropogenic impacts
  climate change, 550
  noise see Marine mammals and ocean
    noise

of noise *see* Marine mammals and ocean noise
nutrient increases, 401–402
AO *see* Arctic Oscillation
AOP *see* Apparent optical properties (AOPs)
Apectodinium, 381F
Aphelion, 345–346
Aqua AMSR-E, 283–284, 288–289
Aquarius satellite, 130, 131F
Aqua satellite, 282, 288–289
Arabian Sea
    carbon isotope ratio, 79F, 82F
    monsoons, 76
        historical variability, 80
        indicators, 78–79
        long-term evolution, 82
    oxygen content, 79F
    salinity, 79F
    water temperature, 79F
Aragonite, 334–335
    saturation state, 542
    coastal waters, 542F
Arctic (region)
    climate, 286
    eastern, mean ice draft, 245F
    ecosystems, 286
    warming, 285, 286
Arctic Basin, 27–28
    circulation, 11
    sea ice
        cover, 234
        cover in future, 237
        thickness, 244
    see also Arctic Ocean; Arctic sea ice
Arctic Ice Dynamics Joint Experiment (ADJEX), 223, 224–226
Arctic Mediterranean Sea, 154, 155F
    intermediate circulation, 162F
    upper layer circulation, 159F
    water column profiles, 157F
    see also Arctic Ocean
Arctic Ocean, 27
    Atlantic water inflow, 156, 161F
        see also Atlantic water
    benthic foraminifera, 428T
    bottom water formation, 161
        geothermal heat flux and, 163
    circulation, 154–161, **154–168**
        drivers, 161–164
        mixing, 163–164
        surface water, 155
        transports, 166–167
        variability, 164–166
        see also Atlantic water
    deep water
        mixing, 163
        salinity, 163
    drifting stations, 226F
    eddies, 163–164
    effect on global climate, 167
    freshwater flux, 135
    halocline, 155–156
    heat flux, 167
    ice cover, 165
    ice export, 167

internal wave energy, 188
overview, 154
Pacific water, 161F
salinity profiles, 156F
sea ice, 224, 224F
    see also Arctic sea ice
shelf inflow, 163
temperature profiles, 156F
vertical structure, 154
water masses, 154, 158T
see also Arctic (region); Arctic Mediterranean Sea; Arctic Oscillation (AO); Arctic sea ice; Polar ecosystems
Arctic Oscillation (AO), 34–35, 131–132, 288–289
    Arctic Ocean Circulation and, 165
    North Atlantic Oscillation and, 33, 35
    sea ice cover and, 239
Arctic pressure fields, sea ice extent and, 240F
Arctic sea ice, 224F
    Arctic Basin *see* Arctic Basin
    Arctic Ocean, 285–286, 285F
    decrease, 285
    drift patterns, 224
    extents, 285, 288–289
    geographical extent, 234–236, 235F, 236–239, 237F
    Kara and Barents Seas, 285–286, 285F
    Northern Hemisphere, 285–286, 285F, 287–288
    oscillatory pattern, 285, 286
    Seas of Okhotsk and Japan, 285–286, 285F
    thickness, 244–248, 245F
    see also Arctic Ocean; North Atlantic Oscillation (NAO); Polar ecosystems
Argo, 275
Argon
    atmospheric abundance, 210T
    phase partitioning, 211T
    saturation responses, 212–213, 213F
    seawater concentration, 210T
    tracer applications, 211–212
ARGO project, 129–130, 137
Arrhenius, Svante, 349
ASCAT (Advanced Scatterometer), 111
Asia
    land-sea fluxes, human impact, 543–545
    monsoons
        evolution, 81–82
        global climate and, 83
        long-term evolution of, 83
    river water, composition, 537T
Astrochronology *see* Astronomical polarity timescale (APTS)
Atlantic *see* Atlantic Ocean
Atlantic cod (*Gadus morhua*)
    Baltic Sea populations, 555
    range extension, 554–555
Atlantic herring (*Clupea harengus*), 551
Atlantic meridional overturning circulation (MOC), 27, 30, 167
    cooling phase, 27

depth range, 27
mean temperature, 27
northward flow, 30
salt increase, 27
warming phase, 30
see also Meridional overturning circulation (MOC)
Atlantic Ocean
    benthic foraminifera, 428T
    circulation modes, 43–44, 44F
    climate effects on fisheries, 554–555
    geohistorical studies, 436
    heat transport, 132–134
        proportion of global total, 109
    North *see* North Atlantic
    North-eastern *see* North-eastern Atlantic
    North-west *see* North-west Atlantic
    oxygen isotope ratios, 314, 314F
    radiocarbon, 481F, 482
    radiocarbon circulation model, 473, 473F
    sea–air flux of carbon dioxide, 465, 465T
Atlantic sublayer, Arctic ocean, warming, 247–248
Atlantic thermohaline circulation, 43, 44
    see also Thermohaline circulation
Atlantic Water (AW)
    Arctic Ocean, 156
    Barents Sea, 158
    Barents Sea branch, 158
        variability, 165
    Canada basin, 159
    current interleaving, 160F
    Fram Strait branch, 159
    heat transport, 167
    Severnaya Zemla, 160F
    St. Anna Trough, 158
    variability, 164, 165
Atmosphere
    carbon dioxide flux, coastal zone, 542F
    circulation at low latitudes, 71
    convective zones, 71
    El Niño Southern Oscillation model and, 71
    heat exchange with ocean, 105
    heat transport, 103–104, 104F, 108
    measure/unit of pressure, 423
    oceanic forcing
        North Atlantic Oscillation (NAO), 39–40
    ocean surface interactions, 291
    as one-layer fluid on rotating sphere model, 71
Atmosphere-ocean general circulation models (AOGCMs)
    future climate, 509
    in paleoceanography, 364
    see also Paleoceanography
Atmospheric aerosols, infrared atmospheric correction algorithms, SST measurement, 293
Atmospheric carbon dioxide *see* Carbon dioxide (CO_2)

Atmospheric general circulation models
    (AGCMs)
  application to paleoclimate problems,
    362, 364, 366
  North Atlantic Oscillation, 39
Atmospheric pressure, 423
  Black Sea, 157F
  El Niño Southern Oscillation and, 57
  North Atlantic Oscillation (NAO) and,
    34F
Atmospheric windows, 291
Atomic absorption spectrophotometry
    (AAS), 334
Atomic emission spectrometry (AES),
    333–334
Austaush coefficient, Ae, 120–121
Australia
  El Niño events and, 57
  Northwest Shelf, shoreline
    reconstruction, 531
Autonomous underwater vehicles
    (AUVs)
  CTD profilers in, 129–130
AUVs see Autonomous underwater
    vehicles (AUVs)

B

Backscatter
  microwave, 110–111
  see also Scatterometry
Backstripping method (for sea level
    estimation), 517–519, 518F,
    521
Bacteria
  benefits to benthos, 419–420
  chemoautotrophic, 423
  symbiotic relationships
    deep-sea animals, 423
Baffin Bay
  sea ice cover, 235–236
  sea ice thickness, 244
Balaenids see Right whales (balaenids)
Balaenoptera musculus see Blue whale
Baltic Sea
  climate effects on fisheries, 555
  sea ice drift velocity, 227F
Barents Sea, 27–28, 154, 237–239
  Arctic sea ice, 285–286, 285F
  Atlantic water, 158
  inflow, 158–159
  sea ice cover, 234–235
Baroclinic instability (BI)
  Antarctic Circumpolar Current,
    150
Barrow, Alaska, temperature profiles,
    sub-sea permafrost, 253, 254F
Barrow Strait, 166
Basal melting, 169
  ice shelf stability, 201
Basin (ocean)
  changing volume over time, sea level
    variations and, 515–517, 516F
  water volume, sea level variations and,
    513–515, 514F
Bathyal zone, 419T, 422–423, 424

Bathymetric map
  Black Sea, 155F
  Southern Ocean, 149F
Bathysiphon filiformis foraminifer, 429F
Bay of Bengal, 134–135
BBL see Benthic boundary layer (BBL)
Beach(es)
  sub-sea permafrost, 255
Beam transmissometry see
    Transmissometry
Beaufort Gyre, 155
Beaufort scale, 97
Beaufort Sea
  mean ice draft, 245F
  sea ice cover, interannual variation,
    241F
Beer's law, 128
Bellingshausen Sea, northern, seabird
    responses to climate change, 562,
    562F
Benguela Current
  seabird responses to climate change, 565
Benthal environment, definition, 424
Benthic, definition, 416
Benthic foraminifera, **425–436**
  ecology, 428–429
    abundance and diversity, 429–430
    hard-substrate habitats, 429
    soft sediment habitats, 428
    well-oxygenated sites, 428, 428T
    environmental distribution controls,
      433–434
    CaCO₃ dissolution, 431
    currents, 433
    depth, 431–432
    lateral advection of water masses,
      431–432
    organic matter fluxes, 432
    species' optimal habitats, 432–433,
      433F
    low-oxygen environments, 434–436
    foraminifera tolerances, 433–434,
      434
    related to organic flux, 433
    microhabitats and temporal
      variability, 431–433
    deep-sea diversity, 429F
    factors influencing distribution, 430,
      430–431
    food and oxygen variability, 431,
      432F
    species distributions, 430, 430F, 431F
    role in benthic communities, 430–431
    biostabilization, 429–430
    bioturbation, 429–430
    organic carbon cycling, 429
    place in food webs, 429
  examples, 426F, 427F
  general characteristics, 425
    cell body, 425
    test, 425
  morphological/taxonomic diversity,
    425–428
    range of morphologies, 425
    sizes, 425
    taxonomic test characteristics, 425

δ¹⁸O records, 319–320, 320–321, 321F
  long-term patterns, 321–322, 321F,
    322F
  Mg/Ca ratios and, 323
  planktonic foraminifers vs., 321,
    321F, 322F
research history, 425
  multidisciplinary research, 425
research methodology, 428–429
  collection methods, 426
  distinguishing live and dead
    individuals, 427–428
  influence of mesh size, 428
use in geological research, 425
use in paleo-oceanography, 434T, 436
  example, 436
  factors making foraminifera useful,
    434–435
  limitations of accuracy, 435–436
  paleoenvironmental attributes
    studied, 435
  see also Cenozoic
  see also Planktonic foraminifera
Benthic infauna
  classification, 417–418
    alternative groupings, 419
    by size groups, 418, 418F
  communities, 418
  percentage of benthos, 418
Benthic organisms, **416–424**
'benthic' defined, 416
  boundary layer see Benthic boundary
    layer (BBL)
  classification, 417–419
    epifauna, 417–418
    percentage of benthos, 418
    infauna, 417–418
    alternative groupings, 419
    communities, 418, 419T, 420T
    percentage of benthos, 418
    size groups, 418, 418F
  classification of zones, 419T
  deep-sea environment, 422–423
    abyssal gigantism, 423
    biodiversity theories, 423
    cyclical events, 423
    depth and food availability, 423
    dominant species, 423
    energy sources, 423
    zones, 422–423
    see also Deep-sea fauna; Deep-sea fish
  depth divisions, 416
  distribution
    sediment influence on, 420
  feeding habits, 419–420
    bacterial breakdown of food,
      419–420
    dependence on detritus, 419–420
    detrivores, 420
    grazing and browsing, 420
    predatory behavior, 420
    sediment influence on distribution,
      420, 420T
  foraminifera see Benthic foraminifera
  larvae, pelagic vs. nonpelagic, 421, 421F,
    422, 422F

physical conditions, 416–417, 417F
  exposure to air, 417F
  level-bottom sediment, 417
  light, 416, 417F
  salinity, 417, 417F
  substratum material, 417, 417F
  temperature, 416–417, 417F
  turbulence, 417F
  water level, 416
  water pressure, 416
reproduction, 421–422
  fecundity and mortality, 422
  oviparity, 421
  settlement process, 422
  sexual/asexual reproduction, 421–422
  strategies for dispersal/nondispersal, 421
  viviparity, 421
spatial distribution, 420–421
  competition for space, 420
  horizontal, 420–421
  vertical, 421
see also Benthic boundary layer (BBL);
  Benthic foraminifera; Demersal
  fish(es); Demersal fisheries;
  Macrobenthos; Meiobenthos;
  Microphytobenthos; Phytobenthos
Benthos, 424
  carbon isotype profile, Arabian Sea, 82F
  oxygen isotype profile, Arabian Sea, 81F
  see also Benthic organisms
Bering Sea
  1998 regime shifts, fisheries, 554
  kittiwake, 564
  seabird responses to climate change, 564
  sea ice cover, 237–239
  sub-sea permafrost, 259
Bering Strait
  throughflow, 135
    Arctic Ocean halocline and, 155–156
  transport, 166
Beta effect, 9, 9F
  see also Coriolis force
BHSZ (base hydrate stability zone), 586
  see also Hydrate stability zone (HSZ)
Bicarbonate (HCO$_3^-$)
  in oceanic carbon cycle, 450–451
  river water concentration, 537T
Biogenic flocculent material see Floc
Biogeochemical cycles
  Southern Ocean overturning and, 152
Biogeochemical models
  carbon cycle, 473–474
  ocean circulation and, 469–471
Biogeochemistry
  process rates, 468
Biological models see Lagrangian
  biological models; Models/
  modeling
Biological pump
  marine organism calcification, 457F, 458
Bioturbation
  benthic foraminifera ecology, 429–430
'Bipolar seesaw', 41–42, 44
  in Holocene millennial-scale climate
    fluctuation, 393–394, 393F, 394F

Bjerkenes compensation mechanism, heat
  transport, 109
Black-body
  calibration, satellite remote sensing,
    291, 294
  radiance/radiation, 103, 103F
Black guillemot (Cepphus grylle),
  559–560, 560F
  see also Alcidae (auks)
  atmospheric pressure, 157F
  bathymetric map, 155F
  circulation
    geostrophic currents, 162F
  drainage area, 154
  history, 154
  islands, 154
  salinity, transect profile, 160F
  shelf, 154
  temperature, transect profile, 160F
  vertical profiles, 159F
Boltzmann's constant, 291
Bonarelli event, 379
Bond cycles, 310–311
Bosphorus Strait, circulation, 154
Bottom-simulating seismic reflector (BSR),
  586
Bottom water formation
  general circulation models (GCM),
    24
  sea ice, 280
Boundary
  surface water masses, satellite remote
    sensing of SST, 299, 300F
Boundary layer
  planetary, 179, 179–180
Boussinesq approximation
  density, 14–15
Box models
  carbon cycle, 470–471, 471F, 499–500,
    500F, 501–502
    refinements, 471
  chemical tracers and, 470–471
  concept of, 499
  fluxes, 499
    differential equations, 499
  reservoirs, 499
  turnover time, 499
Brazil
  El Niño events and, 57
Brightness temperature
  SST measurement by satellite, 291
  temperature deficit relationship,
    292
Brine drainage, 256–257
Brine rejection, 28
Brittle stars (Ophiuroidea), 423
Brown pelican
  climate change responses, 560
Brunhes-Matuyama event, 347, 349F
Brunt Ice Shelf, 205
Brunt-Vaisala frequency, 188
BSR see Bottom-simulating seismic
  reflector (BSR)
BT see Bathythermograph
Bulgaria, Black Sea coast, 154
Bulk formula, evaporation, 122, 125

Buoyancy flux(es), 10–11
  definition, 10–11
  thermohaline circulation, 10
Buoys
  data from, 288
  Tropical Atmosphere-Ocean array,
    60–63, 63F

C

Calanus, 398, 399F
Calcification
  definition and equation, 411
  marine organisms, biological pump
    and, 457F, 458
Calcite
  magnesium in, 333
  saturation state, 542
    coastal waters, 542F
Calcite compensation depth (CCD)
  Cenozoic, 497, 497F
Calcium (Ca, and Ca^{2+})
  river water concentration, 537T
Calcium carbonate (CaCO$_3$), 411, 414
  chemical weathering, 495
  climate proxy, 309–310
  content, Paleocene-Eocene Thermal
    Maximum, 380–382
  foraminiferal distributions, 431
  impact of CO$_2$ levels, 402
  oxygen isotopic ratio determination,
    316, 316–317
    see also Oxygen isotope ratio
  saturation state, coastal waters,
    542F
  in sedimentary sequences, 78
  see also Aragonite; Calcite; Carbonate
Calibration
  black-body, satellite remote sensing,
    291, 294
  radiocarbon (carbon-14, ^{14}C) ocean
    models, 487–490
  spacecraft instruments, 294
California, Gulf of, sedimentary records
  of Holocene climate variability,
  390
California Current
  seabird responses to climate change
    direct, 560–562, 561F, 562F
    indirect, 563
Caloric half-years, 348, 349F
Calving
  icebergs, 203
Calving
  ice shelf stability, 203, 205–206
Calypso sediment coring system,
  305–307
Canada Basin, 154
  boundary current, 161
  eddies, 163–164
  mean ice draft, 245F
  temperature and salinity profiles, 156F,
    157F
Canadian Arctic Archipelago
  sea ice cover, 234, 234–235
    interannual trend, 239

Canadian Basin Deep Water (CBDW), 162
  eddies, 164F
Carbon (C)
  anthropogenic, 467
    definition, 475
  coastal fluxes, 541F
  $\delta^{13}C$, 493
    see also Carbon isotope ratios ($\delta^{13}C$)
  $\delta^{13}C$, coral-based paleoclimate records, 438T, 444, 445
    decadal variability, 442
    upwelling, 439–440
    see also Radiocarbon (carbon-14, ^{14}C)
  effect on pH, 566–567, 567F
  exospheric pool, 495, 495–496
  importance of, 449
  isotopes
    variations in ocean, **493**
  isotopic fractionation, Cenozoic, 497–498, 498F
  isotopic ratio, planktonic-benthic foraminifera differences, 82F
  in marine biomass, 467–468
  radiocarbon see Radiocarbon (carbon-14, ^{14}C)
  river fluxes, 539
Carbon-12 (^{12}C), 493
Carbon-13 (^{13}C), 493
  $\delta^{13}C$, 493
Carbonate (CO_3^{2-})
  accumulation flux during Cenozoic, 497, 497F
    model, 502F
  in oceanic carbon cycle, 450–451
  pump, 453F, 457F, 458
  rocks, chemical weathering, 495
  sediments
    land-sea carbon flux and, 542
  sediments, Cenozoic, 497–498, 498F
    organic carbon and, 498F
Carbon cycle, 397, 468F, **449–458**
  biological pump, 453F, 456, 456F
  carbonate, 453F, 457F, 458
  contribution of POC vs. DOC, 456–457
  factors affecting efficiency, 456–457
    community structure, 458
    nutrient supply, 457–458
    iron fertilization and, 573, 581–582, 581F, 582, 582F
  box models, 470–471, 471F, 499–500, 500F, 501–502
    refinements, 471
  Cenozoic changes and associated processes, 494–496, 496–497
    carbonate accumulation, 497, 497F
    Himalayan uplift, 496–497, 498
    long-term regulation of atmospheric CO_2, 495–496
    lysocline, 497, 497F, 502
    organic carbon subcycle, 497–499, 498F
    strontium isotopic ratio, 496–497, 496F, 502
    weathering see Chemical weathering

global, 449–450
  gas hydrates and, 450F
  research programs, 450
  intermediate complexity models, 471–472
  inverse models, 472
  models, 474–475
    Cenozoic oceans, **494–503**
    inverse (isotopic data use), 498F, 501–502
    role of, 502
  oceanic, 451–452
  solubility/exchange of CO_2 between ocean/atmosphere, 452
    carbon redistribution and, 453–456
    oceanic circulation and, 453–456
    ocean structure and, 452–453
    solubility pump and, 452–453, 453F
Carbon dioxide (CO_2)
  anthropogenic, 449–450, 451F
  anthropogenic emission perturbations, 538, 540F, 541–542
  atmospheric, 32
    dissolved inorganic carbon and, 467
    global warming and, 467
    nutrient utilization and, 470–471
    paleoceanographic research, 359
  atmospheric-coastal water flux, 542F
  Cenozoic
    indicators of, 494, 494F
    long-term regulation, 495–496
  dissolved
    oceanic carbon cycle, 450–451
    transfer to atmosphere, 452, 454F
  effect of $CaCO_3$, 402
  effect on pH, 402, 414
  effects of coccolithophores, 412
  gas analyzer, 460
  greenhouse effect and, 349
  physical state at various pressures/temperatures, 569, 570F
  prehistoric atmospheric concentration, 378–379, 589
    Cretaceous, 379, 379–380
    Oi1 event, 384
    Phanerozoic, 379F
  radiocarbon, measurement of, 480
  solubility in sea water, 452
  sources, 461–463
    primary, 566
  transfer across air-water interface, 452, 452F
  solubility pump, 453–456, 453F, 454F
Carbon dioxide (CO_2) cycle, **459–466**
  atmospheric change, implications of, 459
  biological utilization, 461
  history, 460
  industrial emission rate, 459
  methods, 460–461
  sinks, 461–463
  sources, 461–463
  units, 459–460
  see also Carbon cycle; Carbon dioxide (CO_2); Primary production processes

Carbonic acid (H_2CO_3), in oceanic carbon cycle, 451
Carbon isotope excursions (CIE), 589
  Paleocene-Eocene Thermal Maximum, 380
  see also Carbon dioxide (CO_2); Methane hydrate
Carbon isotope ratios ($\delta^{13}C$)
  ancient soil carbonate measurements, 494, 497–498, 498F
  coral-based paleoclimate records, 438–439, 438T, 440
    cloud cover and upwelling, 439–440
Carbon sequestration by direct injection, **566–572**
  carbon's effect on pH, 566–567, 567F
  CO_2 from power plants/factories, 567
  effectiveness, 568–569
    atmospheric CO_2, 568, 568T
    computer modeling studies, 568
    net retention over time, 568, 569F
    outgassing timescale, 568
    use of tracers, 568
  factors favoring oceanic methods, 566
  injection methods, 569–571, 570F
    bypassing equilibration timescale, 570–571
    CO_2 lake, 569–570
    hydrate reactor, 569
    introduction to outflows, 569
    rising droplet plume, 569
    sinking plume, 570
  interest in technology, 566
  local impacts and public perception, 571
    effects of pH and CO_2 changes, 571
    focus of studies, 571
    inclusion in research/debate, 571
    significance of impacts, 571
  ocean capacity, 567–568
    carbonate system, 567
    effects of long travel times, 567
    effects of short travel times, 567
    impacts on deep-ocean environments, 567–568
    storage capacity, 567
  target stabilization of CO_2, 566
  terrestrial methods, 566
Carbon system, modeling, **467–476**
  3D biogeochemical simulations, 473–474
  intermediate complexity/inverse models, 471–472
  ocean circulation and biogeochemical models, 469–471
  ocean tracers and dynamics, 469
  as research tools, 474–475
  thermocline models, 472–473, 472F
  see also Carbon cycle
Cardigan Strait, 166
Cargo ships see World fleet
Caribbean Sea
  diurnal heating, 134F
Cartesian coordinate system, current flow, 3, 4F
Cassin's auklet (Ptychoramphus aleuticus), 400

CBDW *see* Canadian Basin Deep Water (CBDW)
CCD *see* Calcite compensation depth (CCD)
CDOM *see* Gelbstoff
CDW *see* Circumpolar Deep Water (CDW)
Cenozoic
  carbon releases, 585
  climate, **316–327**
    oxygen isotope records *see* Oxygen isotope ratio ($\delta^{18}O$)
  climate change, 316, 494
    mechanisms, 325–326
  cooling, paleoceanographic research, 359
  interocean gateways, opening/closing of, 362, 363F
  paleo-ocean modeling, 362
Central American Passage *see* Isthmus of Panama
Central Pacific, seabird responses to climate change, 563
Cepphus grylle (Black guillemot), 559–560, 560F
CFCs *see* Chlorofluorocarbon(s) (CFCs)
Challenger Expedition, 355
  sediment core collection, 355, 356F
Chemical tracers, 467
  model constraints, 473–474
  see also Tracer budgets
Chemical weathering
  Cenozoic
    carbonate minerals, 495
    François', 497
    Raymo's, 496, 497
    Walker's, 495–496, 497
    silicate minerals, 495
    air temperature dependency, 495–496
    flux model, 501, 502F
    Himalayan uplift and, 496–497
    rates, 496
China Sea, monsoons
  indicators, 79–80
  long-term evolution of, 82–83
Chinstrap penguin (*Pygoscelis antarctica*)
  response to climate change, 562, 562F
  prehistoric, 559
Chlorofluorocarbon(s) (CFCs)
  circulation tracers, 468–469, 469
Chlorofluorocarbon-11 (CFC-11)
  oceanic distribution/profile
    Southern Ocean, 143F
Chukchi Cap, mean ice draft, 245F
Chukchi Sea, 154, 161–162
  seabird responses to climate change
    direct, 559–560, 560F
    indirect, 564–565
  sub-sea permafrost, 259
Circulation *see* Ocean circulation
Circulation proxies, 386T
Circulation tracers, 468–469
Circumpolar Deep Water (CDW), 28–29, 141–142, 170
  radiocarbon, 483–484
  water properties, 143F

see also Lower Circumpolar Deep Water (LCDW); Upper Circumpolar Deep Water (UCDW)
'Classical limit,' theories, 23
Clathrates, 585–586
  see also Methane hydrate
Clausius-Clapeyron equation, 122–123
Clay minerals
  in sedimentary sequences, 78
CLIMAP project, 328, 329F, 338, 356
Climate
  ocean modeling, 20
  warming, 286, 288–289
    see also Global warming; Millennial-scale climate variability
  weather differences, 71
Climate: Long Range Investigation Mapping and Prediction (CLIMAP) project, 328, 329F, 338, 356
Climate change
  abrupt, **41–46**
    future risk, 45–46
    mechanisms, 43–45
    paleoclimatic data, 41–43
    thresholds, 43
  Cenozoic, 316, **316–327**
    mechanisms, 325–326
  climate variability, 403
  coral-based paleoclimate research, 440–442, 442–444, 442F, 443F
  cyclic global, glacial cycles and sea level change, 523–524, 524F
  definition, 546
  economics, 532
    see also Economics of sea level rise
  effect on marine mammals, **546–549**
    analysis of the research, 547–548
    animal/environment interactions, 546
    classifying effects, 547
    direct and indirect effects, 547
    evidence, 547
    lack/feebleness, 547
    future work, 548
    mammal resilience, 548
    metapopulation structures, 548
    risk-assessment processes, 548
    habitat change factors, 547
    normal environmental variation, 546–547
    harmful algal blooms, 546–547
    non-oscillatory climate change, 546
    oscillatory climate change, 546
    range retractions/population reductions, 546
  'global thermometer,' SST, 299
  ice shelf stability, 203, 206–207
  methane hydrate reserves and, 585
  ocean-driven, 389, 392–393
  paleoceanography, 354, 358, 359–360
  planktonic indicators, 397–398, 405F
  satellite oceanography, 276–277
  seabird responses, **558–565**
    direct, 558, 559–560
    California Current, 560–562, 561F, 562F

    Chukchi Sea, 559–560, 560F
    northern Bellinghausen Sea, 562, 562F
    Ross Sea, 562, 563F
    indirect, 558, 562–563
    Benguela Current, 565
    Bering Sea, 564
    California Current, 563
    central Pacific, 563
    Chukchi Sea, 564–565
    Gulf of Alaska, 564
    North Atlantic, 565
    Peru Current, 563–564, 564F
    prehistoric, 558–559
    north-west Atlantic/Gulf of Mexico, 559
    south-east Atlantic, 559
    Southern Ocean, 558–559
  upper ocean temperature and salinity and, 136–137
Climate change forecast models, 299–301
Climate models
  climate change forecast, 299–301
  fundamental methods, **14–21**
    subgrid-scale parametrization, 14–15
  greenhouse climate, 385–387
Climate systems, 403, 403–404
Climatic warming *see* Global warming
Clio pyrimidata (Pteropoda), 402
Cloud(s)
  contamination, infrared atmospheric corrected algorithms, SST measures, 293
  cover, coral-based paleoclimate research, 438T, 439–440
  movement, $\delta^{18}O$ values and, 317, 317F, 318F
  satellite radiometry, 113
Cloud condensation nuclei (CCN), 543–545
Clouds and Earth's Radiant Energy System (CERES) experiment, 113–114
Coastal zone(s)
  carbon dioxide flux, atmospheric, 542F
  land-sea fluxes, 538–541
    bioessential elements, 539–541
    global warming and, 541–543
  nutrient element fluxes, 538–539
Coccolithophores, 402, 403F, **407–415**
  biogeochemical impacts, 412–413
    atmospheric CO_2 levels, 412–413
    carbon assimilation, 412
    coccolith fluxes, 413
    dissolution of coccoliths, 412
    light scattering, 413
    production of dimethylsulfoniopropionate (DMSP), 413
  blooms, 408, 409F
  Calcidiscus quadriperforatus, 407F, 408F
  calcification, 411
    calcite, 411
    carbon uptake, 411

Coccolithophores (*continued*)
  physiological mechanisms, 411
  coccoliths, 410
    embedded coccoliths, 412F
    function, 410
    numbers, 410
    scale, 410
    types, 410
  Coccolithus pelagicus, 407F
  combination coccosphere, 408F
  definition and description, 407
  Discosphaera tubifera, 407F
  early studies, 407
  ecological niches, 411–412
    variability between species, 411–412
  electron microscope images, 407F
  Emiliania huxleyi, 407F
    blooms, 408, 409F, 412
    coccolith numbers, 410
    genome, 411
    research, 412
  Florisphaera profunda, 407F, 410
  fossil record, 413–414
    calcareous nanoplankton, 413
    chalk deposits, 413
    Cretaceous peak, 413
    K/T boundary data, 413
    survival/non-survival during past
      catastrophes, 414
    use in studying past environments,
      413–414
  future research, 414
    acidification, 414
    linking current and ancient records,
      414
    study of calcification rates, 414
  Gephyrocapsa oceanica, 407F
  life cycle, 410–411
    diploid and haploid stages, 410–411
    gametes, 410–411
    genome, 411
    life spans, 411
    sexual and asexual reproduction,
      410–411
  species and distribution, 407–410
    concentrations, 410
    difficulties in determining species
      numbers, 408
    distribution, 408
    species numbers, 408
  see also Phytoplankton
Coccolithus pelagicus (coccolithophore),
  407F
Cod (*Gadus*), 398
Cold dense polar water, 27, 27–28, 31,
  31–32
Colder equatorward flow, 27
Cold fresh water, 29, 30F, 31
Colombia
  El Niño events and, 57
Computer modeling, 286
  see also Models/modeling
Computers
  growth in power, general circulation
    model and, 23
  power, ocean modeling and, 14, 15–16

Conductivity-temperature-depth (CTD)
    profilers, 129–130
  AUVs and, 129–130
Congelation ice, 175
Continental breakup, sea level rise and,
    515, 516F, 517
Continental hyposography, estimation of
    long-term sea level changes, 517
Continental landmass
  elevation, and monsoon generation, 77
  monsoon indicators, 80
  topography/vegetation, and monsoon
    generation, 77
Continental shelf, 424
  mineral resources *see* Mineral resources
  width, 539
  see also Continental slope
Continental slope, 28, 28–29
  see also Coastal zone; Continental
    margins; Continental shelf
Contour-following bottom currents *see*
    Bottom currents, contour-following
Control systems, fishery management *see*
    Fishery management
Convection
  general circulation models (GCM), 24
Convective feedback, 43
Cooling phase – deep water formation,
    27–29
Copepod(s), 397
  phenology changes, 400
Coral(s)
  biology, 437
    animal-zooxanthellae symbiosis,
      437–438
    general, 437, 437F
    polyp, 437, 437F
    reproduction, 437
    skeleton, 437F, 438
      deposition, 438
      high-density bands, 438
      low-density bands, 438
  bleaching
    satellite remote sensing of sea surface
      temperatures, 299–301
  as climate proxy recorders, 437
    see also Coral-based paleoclimate
      research
  deep-sea, 444
  fossil, 444
  heterotrophy, 437–438, 439–440
  photosynthesis, 437–438, 439–440
Coral-based paleoclimate research, 440,
    437–446
  deep-sea corals, 444
  fossil corals, 444
  future directions, 445
  reconstruction of environmental
      variables, 438–439, 438T
    cloud cover, 438T, 439–440
    from fluorescence patterns, 438T, 440
    from growth records, 438, 438T
    from isotopes, 438–439, 438T
    methods, 438–439, 439F
    improved, 445
    pH, 438T, 440

river outflow, 438T, 440
    salinity, 438T, 439
    temperature, 438T, 439
    from trace and minor elements,
      438–439, 438T
    upwelling, 438T, 439–440
  records, 440
    Galapagos coral, 440–441, 442F
    interannual-to-decadal variation in
      climate and ENSO, 440–442,
      442F, 443F
    long-term trends, 442–444, 443F
    seasonal variation in, 440
  sea level rise data, 513
  sites of, 437F, 440
  see also El Niño Southern Oscillation
    (ENSO)
Coral reef(s)
  radiocarbon, 478–479, 478F, 479F
  raised, tectonics and sea level change,
    522, 523F
  threatened, time-series SST predicting,
    299
Coriolis force, 180
  beta effect, 9, 9F
  fluid parcels and, 18
  geostrophic balance, 7
  thermohaline circulation, 10
  wind driven circulation, 7–8, 9, 9F
  wind forcing and, 60
Coupled models
  climate models, 32
Coupled sea ice-ocean models,
    190–200
  basic principles, 191
  boundary conditions, 195–196
  coordinate systems, 196
  coupling, 195–196
    time resolution, 196
  dynamics, 193–195
    basic equation, 191
  evaluation, 196–197
    ice variables, 191, 196–197
    ocean variables, 197–198
    parameter choices, 198
  ice
    concentration, 192, 197
    drift, 193
    energy balances, 194F
    formation, 193
    internal forces, 194–195
    modeling approach, 191
    rheology, 193–194
    salinity, 193
    thickness, 192, 197
    volume, 193
  numerical aspects, 191
  open water, 193
  processes, 190, 190F, 195F
  regions relevant, 190–191
  snow layers, 191, 192
  subgrid scale parametrization, 191
    ice thickness, 191
    ice volume, 191–192
  thermodynamics, 192–193
    basic equation, 191

Cretaceous, 378
  climate, 379
    modeling, 385
    ocean anoxic events (OAE), 379
    proxy data, 381F
    thermal maximum, 379
  Late see Late Cretaceous
Cretaceous/Tertiary (K/T) boundary,
    374–375
  orbital tuning and, 374–375
  sedimentation rates across, 376F
Croll, James, 344
Crustacean(s)
  deep-sea communities, 423
Crustal dust, land-sea transfers, 544F
Crustal rebound phenomena, anatomy of
    sea level function, 525F, 527,
    528F
Current meters
  acoustic, 4, 4F
    definition, 4
  moorings, 4, 4F
  rotary, 4, 4F
    definition, 4
Cyttopsis rosea (rosy Dory), 550, 550F

D

Dam building
  in sea level rise reduction, 511
Dansgaard-Oeschger (D/O) cycles
  Heinrich events and, 394F
  Holocene, 390F, 392, 394–395, 394F
Dansgaard-Oeschger (D/O) events,
    41–42
  cyclicity, 44
  drivers, 43
  periodicity, 310–311
  warming distribution, 44
Danube, Black Sea input, 154
Darwin (Australia), atmospheric pressure
    anomaly time series, 60F
Dating techniques, orbital tuning,
    370–377
Deep boundary current, 28
Deep ocean
  temperatures, 12–13, 12F
Deep Sea Drilling Project (DSDP), 356
Deep-sea environment, 422–423
Deep water
  formation, 453, 454F
Deep-water species see Demersal fish
Deep Western Boundary Current
    (DWBC), 110
  see also Abyssal currents; Antarctic
    Bottom Water (AABW); North
    Atlantic Deep Water (NADW)
Defense Meteorological Space Program
    (DMSP) satellites, 114, 115
Deforestation, carbon emission and,
    467
Degassing flux (Fvol), 495
'Delayed oscillator' modes, 73–74
Denmark Strait, 27–28, 31
  NADW formation, 365
  paleoceanography

climate models in, 366
global ocean circulation and, 365–366
opening, 363F
see also Straits
Dense water formation, 210
Density
  distribution measurement, ocean
    circulation, 7
  surface water, 127
Density-driven circulation, 27, 31
Density profile
  Black Sea, 159F
  Southern Ocean, 143F
Deposition, 124
Detritus, 424
Detrivores, 420, 420–421, 424
Deuterium levels, 351F
Dew point temperature, definition, 121T
Diatom(s)
  paleothermometric transfer functions
    and, 341
DIC see Dissolved inorganic carbon
Diet
  benthic organisms, 419–420
Diffusion
  ocean climate models and, 18
Dimethylsulfide (DMS), 543–545
Dimethylsulfoniopropionate (DMSP), 413
Dinitrogen (N₂)
  fixation, 574
  iron availability and, 582
Dinoflagellates
  paleothermometric transfer functions
    and, 341–342
Diomedea immutabilis (Laysan albatross),
    expansion of geographical range,
    561
Direct deposit feeders, 420, 424
Direct numerical simulation (DNS), 15
Discosphaera tubifera coccolithophore,
    407F
Dissolved inorganic carbon (DIC)
  ariation of δ-carbon-13, 493, 493F
  atmospheric carbon and, 467
  carbon dioxide solubility and, 451
  components, 411
  oceanic carbon cycle, 451, 452, 455,
    455F, 456F
    vertical gradient, 452–453, 453F
  radiocarbon and, 473–474
  river fluxes, 539
Dissolved inorganic phosphorus (DIP)
  total, river water, 537T
Dissolved organic carbon (DOC)
  biological pump contribution, 453–455,
    455F, 456F
    particulate organic carbon vs.,
    456–457
  river fluxes, 539
Distribution
  coccolithophores, 408
DMSP (United States Defense
    Meteorological Satellite Program),
    280, 281–282, 284F, 285F,
    288–289, 288F
DNS see Direct numerical simulation

D/O see Dansgaard-Oeschger (D/O)
    events
DOC see Dissolved organic carbon
    (DOC)
DOP see Dissolved organic phosphorus
    (DOP)
Double diffusion, 183
Downwelling, 30F
Dpm, definition, 491
Drag coefficient, 97T, 181–182
  wind on sea surface, 97, 97T, 110–111
  see also Wind stress
Drake Passage, 28, 28F, 141
  paleoceanography
    climate and, 363
    climate models in, 364
    closure, opening of Central American
      Passage and, 364–365, 365F
    opening, 325–326, 363, 363F
Drake Passage effect, 384
Drifters, 5F
  position, determining, 4–5
  see also Float(s)
Drift ice, 223
  conservation law, 226–227
  equation of motion, 228–229
    drift in presence of internal friction,
      230–231, 230F
    drift problem, 229–230, 230–231
    free drift, 229–230, 230–231, 230F
    stationary, 229
  landscape features, 223
  modeling, 231
  rheology, 227–228
Drizzle, 129
Droughts, El Niño Southern Oscillation
    (ENSO) and, 67–68
DSDP see Deep Sea Drilling Project
    (DSDP)

E

Early Eocene Climatic Optimum (EECO),
    382
Earth
  elastic deformation, 526
  orbit, glacial cycles and, 345–347, 345F
    see also Milankovitch variability
  radiation budget, 103, 103F
  viscosity, sea level changes and, 522
Earth models of intermediate complexity
    (EMIC), 352
Earth Observing System (EOS), 282, 297
Earth Radiation Budget Experiment, 108
Earth System Models (ESM), 14
East Antarctic Ice Sheet
  disintegration, 512
East Australian Current (EAC)
  flow, 49F
Eastern Arctic, mean ice draft, 245F
Eastern oyster (Crassostrea virginica), 556
East Greenland Current (EGC), 10
  see also Atlantic Ocean current systems
East Siberian Coastal Current, 163
East Siberian Sea, 154
  sea ice cover, 234–235

EBDW see Eurasian Basin Deep Water
Eccentricity (orbital), 345–346, 370–371
    glacial cycles and, 348
ECMWF see European Center for
        Medium Range Weather Forecasts
        (ECMWF)
Economics of sea level rise, **532–535**
    adaptation costs, 533–534
        economy-wide implications, 534
        failure to adapt, 533
        methods of adaptation, 533
        protection and accommodation,
            533–534
        retreat, 534
        scale, 534
    adaptation methods/problems, 534
        coast protection as public good,
            534
        decision analysis of coastal protection,
            534
        'false security' problems, 534
        ignoring non-immediate threats, 534
        uniqueness of each case, 534
    climate change concerns and, 532
        adaptations costs, 532
        adaptations methods, 532
        costs of sea level rise, 532
    costs, 532–533
        episodic flooding, 533
        erosion and permanent flooding,
            532
        impacts, 532
        indirect costs, 533
        salt water intrusion, 533
        wetlands loss, 533
    distribution of impacts, 534–535
    greenhouse gas emissions, 532
    thermal expansion, 532
    uncertainties, 535
    see also Coastal topography impacted by
        humans; Coastal zone management
Eddies
    Antarctic Circumpolar Current, 150
    Arctic Ocean, 163–164, 164F
    diffusion, 120–121
    overturning circulation and, 151–152
Eddy correlation, 96
Eddy diffusion, 120–121
Eddy kinetic energy (EKE), 6F
    definition, 5–6
EECO see Early Eocene Climatic
        Optimum
EEZ see Exclusive economic zone (EEZ)
EIC see Equatorial Intermediate Current
        (EIC)
EKE see Eddy kinetic energy (EKE)
Ekman layer
    definition, 7–8
Ekman pumping, 9F
    definition, 7–8
    heat transfer, 12
Ekman transport, 8–9, 9F
    see also Ekman layer; Ekman pumping
Electrical conductivity see Conductivity
Electrically scanning microwave
        radiometer (ESMR), 281, 282, 284

El Niño, 70, 72F
    changes during late 1970s, 74
    as consequence of change in winds,
        72–73
    intense, in 1997, 74
    La Niña change to, 71, 73
    limited predictability, 74
    satellite remote sensing of SST, 299
        coupled ocean-atmosphere system
            perturbations, 297
        normal Pacific SST distribution,
            297–298
        seasonal predictions of disturbed
            patterns, 298–299, 298F
        Warm Pool movement, 297–298
    sea surface temperatures, 70, 70F
    thermocline and, 71, 72F
    tools for predicting, 70–71
        see also El Niño Southern Oscillation
            (ENSO) models
    warming of eastern tropical Pacific,
        71
    see also El Niño Southern Oscillation
        (ENSO)
El Niño forecasting
    benefits, 68
El Niño periods
    sea–air partial pressure of CO$_2$,
        461–463
El Niño Southern Oscillation (ENSO),
        100, 297, **57–69**
    1997–1998 event
        human impact, 68
        jet stream flow patterns, 65F
        precipitation and, 62F
        prediction, 68
        sea surface temperatures, 58F
        sea temperature depth profiles, 63F,
            64–66
        surface winds, 61F
    coral-based paleoclimate records,
        440–442, 442F, 443F, 444, 445
    cycle, 57
    ecological impact, 67
    effect on primary productivity, 403
    heat flux, 64
    human impact, 67–68
    impact on seabirds, 558
    interannual variations, 64–67
    mechanisms, 64
    ocean-atmosphere interactions, 57–64
    origin of term, 57
    oscillation, 64
    Pacific Ocean equatorial current system
        variability and, 54–55, 55–56,
        55F
    prediction, 68
    satellite remote sensing, 116–117, 117F,
        118F
    sea level variation and, 509
    sea surface temperatures, interannual
        variation, 66, 66F
    temperature variability and,
        131–132
    thermocline depth, 60–63
    time scale, 64

    upwelling and, 64, 67
    see also Climate change
El Niño Southern Oscillation (ENSO)
        models, **70–75**
    atmosphere circulation and, 71
    general circulation models, 71
    oceans and atmosphere interactions,
        73–74
        see also Ocean-atmosphere
            interactions
    oceans and thermoclines, 71–73
    purpose, 72–73
EMIC (earth models of intermediate
        complexity), 352
Emiliani, Cesare, 331–333
Emiliani, Cesarç, 319–320
Emiliania huxleyi coccolithophore,
        335–336, 335F, 407F
    blooms, 408, 409F, 412
    coccolith numbers, 410
    genome, 411
    research, 412
Energy
    conservation, ocean, 104
Energy Balance Models (EBM), 349
    annual mean atmospheric, 349–352
    climate/ice sheet, 352
    Northern Hemisphere ice sheet, 352
    seasonal atmospheric/mixed-layer
        occan, 352
Engraulis japonicus (Japanese sardine),
        553–554
    response to changes in production,
        553–554
Enhydra lutris see Sea otter
ENSO see El Niño Southern Oscillation
        (ENSO)
'Enstrophy', 24
Entrainment, 29
Environmental protection, from pollution
        see Global marine pollution;
        Pollution
Eocene, 378
    Early Climatic Optimum, 382
    ocean circulation, 366
    $\delta^{18}O$ records, 322F, 324, 324F
    paleo-ocean modeling, 362
    see also Cenozoic; Paleoceanography
Eocene/Oligocene boundary, oxygen
        isotope evidence, 324F
Eolian dust, 78
Epifauna, 417–418, 419F, 424
    see also Benthic boundary layer (BBL)
Epistominella exigua foraminifera, 435F,
        436
Epstein, Sam, 316–317
'Equation of state' (of sea water)
    general circulation models, 22
Equatorial Intermediate Current (EIC)
    flow, 52F, 53F
Equatorial Undercurrent (EUC)
    flow, 49F, 51, 52F, 53F
Equinoxes, precession, 344, 348
ERS-1 satellite, 111, 112T
    see also European Remote Sensing (ERS)
        satellites

ERS-2 satellite, 111, 112T
  see also European Remote Sensing
    (ERS) satellites
ESA (European Space Agency),
  274
ESM (Earth System Models), 14
ESMR (electrically scanning
    microwave radiometer), 281,
    282, 284
Eubalaena glacialis see North Atlantic
    right whale
Eulerian flow
  ocean circulation, 3, 4, 4F
Euphausiacea (krill) *see* Krill
    (Euphausiacea)
Euphausia superba (Antarctic krill),
    402
  see also Krill (Euphausiacea)
Eurasian Basin, 154
  mean ice drafts, 246F
Eurasian Basin Deep Water, 162
Europe *see* Europe/European Union
    (EU)
European Center for Medium Range
    Weather Forecasts (ECMWF),
    98, 124–125, 198
European Remote Sensing (ERS)
    satellites, 111
  see also ERS-1 satellite; ERS-2
    satellite
European Space Agency (ESA), 274
Europe/European Union (EU)
  paleo shorelines, migration and
    sea level change, 529–531,
    530F
  river water, composition, 537T
Eutrophication
  harmful algal blooms (red tides),
    401–402
Evaporation, 27, 32, **120–126**
  bulk formula, 122, 125
  definitions, 120–122
  eddy correlation equation, 121
  effect on $\delta^{18}O$ values, 317, 317F
  estimation by satellite data, 125
  global distribution, 135F
  global maxima, 134–135
  history, 120–122
  inertial dissipation *vs.* eddy correlation,
    121–122
  latitudinal variations, 124
  measuring, methods of, 121
  nomenclature, 120–122
  North Atlantic Oscillation and, 37
  regional variations, 124
  research directions, 125–126
  satellite remote sensing, 115
  sources of data, 124–125
Evolution
  coccolithophores, 413–414
Evolutive spectral analysis, orbital tuning
    and, 372–373
Excess helium-3, 475
Exploitation *see* Human exploitation
Export production
  definition, 475

F

Falsifiability, general circulation models/
    theory, 23
FAO *see* Food and Agriculture
    Organization (FAO)
Faroe Bank Channel, 27–28, 31
  see also Straits
Ferrel Cell, 9F
Fertilizers
  application levels, 540F
Filchner ice shelf, 169, 207
  climate change, effects of, 178
Filchner-Ronne Ice Shelf, 176–177
  see also Ronne Ice Shelf
Fisheries, and climate, **550–557**
  anthropogenic climate change, 550
  changes attributed to climate, 550
  effects of climate change, 550, 551–553
    bioclimate envelope, 552–553
    effects on individuals, 551–552
    effects on populations, 553
    identifying processes, 552
    recruitment, 552, 553F
    relation to fishing stresses, 553
    temperature effects, 552, 552F
  history, 551
    distribution shifts, 551
    historical patterns *vs.* current change,
      551
    lessons to be learned, 551
    climate influence on stocks, 551
    risk of stock collapse, 551
    stock recovery, 551
    Norwegian spring spawning herring,
      551
    distribution shifts, 551, 552F
    expansion of summer migration, 551
    historical account, 551
  impacts on human societies, 556–557
    economic consequences, 556
    examples, 556
    Greenland, 556
    projections of impacts, 556
  indirect effects and interactions,
    555–556
    changing primary production, 556
    spread of pathogens, 556
  North Pacific regime shifts (1998), 554
    California Current System, 554
    central North Pacific, 554
    Gulf of Alaska and Bering Sea, 554
  North Pacific regime shifts, history, 554
  northward spread of species, 550,
    550F
  regional effects of climate, 553–554
    Baltic Sea, 555
    coral reef fisheries, 555
    Greenland cod stocks, 554–555
    North Atlantic, 554–555
    North Pacific, 554
    species favored by climate trends, 555,
      555F
    tropical Pacific, 553–554
    tuna example, 553–554
  timescales and terminology, 550–551

Float(s)
  position, determining, 4–5
Florisphaera profunda coccolithophore,
    407F, 410
Flow
  Eulerian, ocean circulation, 3, 4, 4F
Fluid(s)
  microscopic/macroscopic laws for
    interactions, 23
Fluid dynamics
  coordinate transformation, 17
  modeling and, 16–17, 17
Fluid parcels
  definition, 17
  mass, 17
  mass conservation, 17
  momentum, 18
    time-stepping, 19
  tracers, 17–18
  velocity, 19
Fluorescence
  coral-based paleoclimate research, 438T,
    440
Fluorometry
  iron fertilization experiments, 578
Flux estimation, bulk formulae, 97
Food webs
  plankton, 399–400, 401
Foraminifera, 78, 81F
  abundance, sea surface temperature and,
    339F
  carbon isotope profile, Arabian Sea, 82F
  effects of global warming, 401, 401F
  fossilized, 77
  oxygen isotope variability, glacial cycles
    and, 345
  oxygen isotope profiles, Arabian Sea,
    81F
  paleothermometric transfer functions
    and, 340–341
  planktonic *see* Planktonic foraminifera
  shells, 332F
Forecasting/forecasts
  climate change, 299–301
Fossil(s)
  plankton, 77, 78
Fossil assemblages, SST transfer
    functions, 338–340
Fourier analysis, current flow variability,
    6–7, 6F
Fracture, ice shelves, 205–206
Fram, 188, 223
Fram Strait, 135
  transport, 166–167
  water column profiles, 157F
French Research Institute for Exploitation
    of the Sea (IFREMER), 276
Frequency spectrum, current flow
    variability, 6–7, 6F
Fresh water
  added in polar regions, 31–32
  balance, ocean circulation, 10–12, 11F
  flow from rivers, 27
  glacial meltwater, 28–29
Freshwater flux, 134–135
  satellite remote sensing, 115–116, 118F

Freshwater flux (*continued*)
upper ocean, 129
see also Precipitation; Salinity; Upper
Ocean
Friction
ocean models, 18
Friis, Peder Clauss, 551
Fronts
positions, satellite remote sensing of SST,
299

G

Gakkel Ridge, 154
Galapagos Islands
coral records, 440–441, 442F, 443F
Galapagos penguin (*Spheniscus mendiculus*)
climate change responses, 564, 564F
Gas(es)
ice solubility, 211
Gas hydrates, 252
clathrates, 585–586
GCM *see* General circulation models
(GCM)
General circulation models (GCM),
22–26
aims of large-scale models, 22
carbon cycle, 473–474
computing power, 23
correction by data assimilation, 22
definition, 22
heat equation, 22
momentum equation, 22
elements of, 22
El Niño Southern Oscillation, 71
glacial cycle modeling, 349
limits/errors, 24–25
conservation properties, 24
convection, 24
dissipation, 24
numerical viscosity, 24
momentum (vorticity) equation, 24
reality link (causal), 24
robustness, 22
sensitivity, 23
suggestive not predictive nature of,
23
theory and practice, 23–24
backward compatibility, 23
classical modes/physical laws, 23
confinement of form, 23
expression in quantity/extent and
duration, 23
falsifiability, 23, 24
violation of principles, 24
see also Coastal circulation models
Geochemical Ocean Sections Study
(GEOSECS), 460
carbon isotopes, 493
oxygen isotope ratios, 314
radiocarbon, 480–481, 484, 485F
Geoid
definition, 7
Geomagnetic polarity timescale (GPTS),
373

see also Astronomical polarity timescale
(APTS); Paleoceanography
George VI Ice Shelf, 176
temperature and salinity trajectories,
173F
Georgia, Black Sea coast, 154
Geos-3 satellite, 268
Geosat, 267F, 273–274
GEOSECS *see* Geochemical Ocean
Sections Study (GEOSECS)
GEOSS (Global Earth Observation
System of Systems), 277
Geostrophic balance, definition, 7
Geostrophic turbulence, 15
Geostrophy, 142
Geosynchronous orbit measurements of
SST, GOES Imager, 297
Gephyrocapsa oceanica coccolithophore,
407F
Gerard barrel, radiocarbon, 480
Gigatons, 459
Gimbals, 95–96
Glacial crustal rebound, **522–531**
Australia, Northwest Shelf, 531
Europe, 529–531, 530F
Scandinavia, 529–531, 530F
South East Asia, 530F, 531
Glacial cycles
ice extent, 348F, 350F
Milankovich variability and, **344–353**
history, 344–345
orbital parameters and insolation,
344
Plio-Pleistocene, 347–349
modeling, 349–352
monsoon abundance relative to
interglacials, 80
periodicity, 347F, 349F, 352
sea level changes, 523–524
from maximum levels, 524–526
Glacial maxima, 178
Glaciation period, sub-sea permafrost,
253, 253F
Glaciations
climate change within, 44
see also Dansgaard-Oeschger (D/O)
events; Heinrich events
cyclicity, 41
sea level variations and, 507, 513, 514,
514F
Glaciers
nonpolar, sea level variations and, 510,
510F, 511
Glacio-hydro-isostatic models, 526–527
elastic deformation of Earth, 526
eustatic sea level, 527
formulation of planetary deformation,
526–527
gravitational pull of ice-sheet, 526
ice-volume equivalent sea level, 524F,
526
isostatic contribution to relative sea
level, 526
isostatic contribution to sea level
change, 527
response parameters of Earth, 526–527

Global-average temperature, predicted
rise, 509
Global climate models (GCMs), 30
advantages, 475
carbon cycle, 473–474
chemical tracers and, 473
sea ice extent and global warming, 239
subgrid-scale parametrization, 473
see also Climate models
Global climate system, Asian monsoon
and, 83
Global Earth Observation System of
Systems (GEOSS), 277
Global heat transport, 30
Global ocean
sea–air flux of carbon dioxide, 465T
Global Positioning System (GPS)
determining data measurement
locations, 3, 4
Global Precipitation Measurement (GPM)
mission, 116
'Global thermometer,' satellite remote
sensing of SST, 301
climate change forecast models,
299–301
global climate change problems, 299
thermal inertia of the ocean, 299–301
Global warming, 27, 29, 31–32, 467
anthropogenic, 41, 45
sea ice coverage and, 239
see also Carbon dioxide (CO_2)
economics, 532
effects on phenology, 398–399, 400F
effects on plankton, 397–398, 399F,
405F
effects on stratification, 401
land-sea fluxes and, coastal zones,
541–543
see also Carbon sequestration by direct
injection; Climate change;
Economics of sea level rise;
Greenhouse climates
Globigerina bulloides foraminifera, 79,
436
Globigerinoroides bulloides, 334
Globigerinoroides ruber, 334, 339F
Globigerinoroides sacculifer, 334
Glycerol dialkyl glycerol tetraethers
(GDGT), 336–337, 337F
see also TEX86
GOES Imager, 297
GPS *see* Global Positioning System (GPS)
GPTS *see* Geomagnetic polarity timescale
(GPTS)
Grand Banks earthquake, 586
Gravitational tides *see* Tide(s)
Grazing
benthic organisms, 420
Great Barrier Reef Undercurrent
(GBRUC)
flow, 49F, 52
'Great Conveyor'
definition, 362
Great Salinity Anomaly, 38, 40
Greenhouse climates, **378–388**
Cretaceous, 379–380

key features, 378
modeling problems, 383F, 385–387,
  386T
Paleogene, 380–384
temperature gradient, 378
transition to current climate, 384–385
see also Climate change; Global
  warming
Greenhouse gas, 449
emissions, 532
  see also Carbon sequestration by
    direct injection
release during early Cenozoic, 325
SST modulates exchanges of heat and,
  301
see also Climate change
'Greenhouse world'
oxygen isotope evidence, 323
sea level variations, 515
Greenland
climate effects on fisheries, 556
Holocene climatic variability
  ice core records, 390–391, 392
  Little Ice Age, 391–392, 392F
paleoclimatic data, 41
sea ice cover, 234
Greenland Ice Sheet
formation, oxygen isotope ratio and,
  513–514, 514F
sea level variations and, 510–511,
  511
surface melting, 512
projections, 511
Greenland Sea, 27–28
sea ice
  cover, interannual trend, 239
  North Atlantic Oscillation and, 38
  thickness, 244
water column profiles, 157F
Greenland Sea gyre, sea ice cover, 239
Grey, Harold, 331
GSA (Great Salinity Anomaly), 38, 40
Guillemot(s)
black, 559–560, 560F
Gulf of Alaska
seabird responses to climate change, 564
Gulf of California, sedimentary records of
  Holocene climate variability, 390
Gulf of Mexico (GOM)
seabird responses to prehistoric climate
  change, 559
Gulf of St. Lawrence, 234–235
sea ice cover, 234
Gulf Stream, 9, 27, 72–73
flow measurement, telephone cable,
  105–106
heat transport, 106–107
identification, satellite remote sensing of
  SST, 299
see also Atlantic Ocean current systems;
  Florida Current, Gulf Stream and
  Labrador Currents; Gulf Stream
  System
Gyre ecosystems see Ocean gyre
  ecosystems
Gyre models, 472, 472F

H

H-0 event, dates, 307
H-1 event, dates, 307
H-2 event, dates, 307
H-3 event, dates, 307
H-4 event, dates, 307
H-5 event, dates, 307
H-6 event, dates, 307
Hadal zone, 419T, 422–423, 424
Hadley Cell, 9F, 12
Halmahera Eddy, 49F, 52–53
Halocline, 130
Arctic Ocean, 155–156
Haplosporidium nelsoni pathogen, 556
Haptophyte algae
long-chain alkenones, 335
Hawaii
El Niño events and, 57
Heat
fluid packets, 17
Heat dissipation
ocean, 16
Heat distribution, upper ocean, 130
Heat equation, general circulation
  models, 22
Heat flux
El Niño Southern Oscillation, 64
sea surface, **95–102**
  accuracy of estimates, 100–102
  data, sources of, 97–98
  measuring, 95–97
  North Atlantic sites, 100, 102F
  regional variation, 100
  seasonal variation, 100, 101F
  transfer coefficients, typical values,
    97, 97T
Heat transfer coefficient, 198
Heat transfer processes, sea surface, 110
Heat transport, 27, 30, 132–134, **103–
  109**
air–sea heat exchange, 105
atmospheric, 103–104, 104F, 108
Bjerkenes compensation mechanism,
  109
convergence, 104
definition, 104
direction, 103
eddies, 107–108
future prospects, 108–109
global distribution, 105–107, 105F,
  106F
global heat budget, 103–105
Indo-Pacific Ocean, 107–108
latitude and, 103F
models, 108–109
North Atlantic, 107
North Pacific, 107
ocean warming and, 108
power, 106–107
residual method of calculation, 108–109
Heinrich (H) events, 41–42, 307–309
correlating with other climate records,
  309
Dansgaard-Oeschger cycle and, 394F
drivers, 43

periodicity, 310–311
see also H-0 event; H-1 event
Helium, 172
atmospheric abundance, 210T
ice solubility, 211
mantle see Mantle helium
phase partitioning, 211T
saturation responses, 212–213, 213F
seawater concentration, 210T
see also Helium-3
Hell Gate, 166
Henry's Law, 451
Hermatypic corals
sea surface temperature
  paleothermometry, 330–331, 331T,
  334–335
strontium/calcium ratio, 334–335
Heterotrophy, corals, 437–438, 439–440
H events see Heinrich (H) events
HEXOS (Humidity Exchange Over the
  Sea) experiment, 122, 122F
Hibler model, 228
High Himalayan Crystalline Series,
  silicate minerals, strontium
  isotopic ratios, 497
High-nitrate, high-chlorophyll (HNHC)
  regions, 574–575, 574T
High-nitrate, low-chlorophyll (HNLC)
  regions, 574–575, 574T, 575–576,
  575F, 576F
iron deficiency, 573F, 575–576, 576
see also Iron fertilization
High Salinity Shelf Water (HSSW), 170,
  207
Antarctic continental shelf, 172–173
Himalayan uplift, carbon cycle changes
  and associated processes, 496–497,
  497, 498
Himalayas (mountains), 77
Holocene
climate variability, 390–391, 391F, **389–
  396**
  build up to, 389
  current understanding, 394–395,
    395F
  decadal-scale, 389
  Greenland see Greenland
  knowledge gaps, 394–395, 395F
  millennial-scale, 389, 390F, 394–395
    causes, 392–394, 393F, 394–395,
      394F
  Northwest Africa vs. North Atlantic,
    390–391, 391F
  sedimentary records, 390
  vegetation effects, 391
Dansgaard–Oeschger cycles, 390F, 392,
  394–395, 394F
land–sea carbon flux, 541
Little Ice Age, 390F, 391–392, 391F,
  392F, 394–395
millennial-scale climate variability, 310
paleoceanography, importance of,
  389–390
Holothuroidea (sea cucumbers), 423
Horned puffins, 559–560, 560F
see also Alcidae (auks)

HOZ (hydrate occurrence zone), 587
HSSW *see* High Salinity Shelf Water (HSSW)
HSZ *see* Hydrate stability zone
Hudson Bay, sea ice cover, 234
Hudson Strait
  Heinrich events, 41–42, 307–309
  ice core data, 308F
  sea ice cover, 234
Human activities, adverse effects
  increased atmospheric carbon dioxide concentrations, 449–450, 451F
Human health
  impact of phytoplankton, 397
Human population, land-sea fluxes and, 543
Humic substances *see* Colored dissolved organic matter (CDOM)
Humidity, **120–126**
  definitions, 120–122
  history, 120–122
  latitudinal variations, 124
  measurement *see* Humidity measurement
  nomenclature, 120–122
  regional variations, 124
  satellite remote sensing, 114–115
  sources of data, 124–125
  tropical conditions of, 123–124
  vertical structure of, 124
  see also Evaporation
Humidity Exchange Over the Sea (HEXOS) experiment, 122, 122F
Hungarian Academy of Sciences, 344–345
Hurricane(s), modeling *see* hurricane models
  1999 season (Atlantic), 299
  development, surface temperature, 135–136
  intensification
    prediction of development of storms, SST, 299
    satellite remote sensing of SST application, 299
  temperature and, 135–136
Hydrate occurrence zone (HOZ), 587
Hydrate stability zone (HSZ), 585F, 586
  base (BHSZ), 586
    depth and, 589
  hydrate distribution, 587
  Norwegian margin, 589
Hydrogen sulfide
  profile, Black Sea, 159F
Hydro-isostatic effect, anatomy of sea level function, 527–529, 528F
Hydrologic cycle, 31–32
  sublimation-deposition, 124
Hydrostatic approximations, 14–15, 18
Hydrothermal vent(s)
  energy in benthic environment, 423

I

IBM *see* Lagrangian biological models
Ice
  categories, 191–192, 194F

classes, 191, 194F
formation
  dense water and, 210
  noble gases and, **210–213**
  physical properties relevant, 210–211
  tracer applications, 211–213
  noble gas saturation and, 212, 212–213
  phase partitioning of dissolved gases, 211
marine, 173–174, 175
Ice, Cloud and land Elevation Satellite (ICE-*SAT*), 284
Ice ages
  astronomical theory (Milankovitch, Milutin), 344–345
  climate transitions within, 41–42
  end, 41
Ice-bearing permafrost, 253, 253F
Iceberg(s)
  calving, 203
Iceberg(s)
  ice shelf stability, 201
Iceberg-rafted detrital (IRD) peaks, 307
Ice-bonded permafrost
  conduction model, 253
Ice-bonded permafrost
  definition, 252–253
Ice-bonded permafrost
  ice-bearing permafrost, separation effects, 255
Ice caps, sea level variations and, 510
Ice cores, paleoclimatic data, 41
Ice cover
  internal wave field affects, 188
Ice floes, 223
  see also Drift ice
'Ice house world,' oxygen isotope evidence, 323–325
Iceland, 27–28, 28F, 31
  North Atlantic Oscillation, variability, 35
Ice–ocean interaction, **179–189**
  horizontal inhomogeneity
    summertime buoyancy flux, 186–188
    wintertime buoyancy flux, 184–186
  internal waves, ice cover, interaction effects, 188–189
  outstanding issues, 189
  under-ice boundary layer *see* Under-ice boundary layer
  wintertime convection
    heat balance, 182–184
    mass balance, 182–184
Ice-rafted detritus (IRD)
  deposition, 323, 325
  distribution, 323–324, 325
Ice-rich permafrost
  definition, 252–253
Ice-rich permafrost
  location, 255
Ice sheets, 312F
  decay over geological time, 508F, 513
  formation, 323–324
  glaciological modeling, 522

gravitational pull, glacio-hydro-isostatic models, 526
maximum glaciation, glacial cycles and sea level change, 523–524, 524F
ocean $\delta^{18}O$ values and, 319, 319–320, 320F
sea levels and, 510–511
see also Antarctic Ice Sheet; Greenland Ice Sheet
Ice shelf water (ISW), 172–173
Ice shelves, 201–203
  bay illustration with ice streams, 202F
  definition, 201
  features, 169
  locations, 169
  map, 170F
  stability, **201–209**
    break-up, reasons for, 205–206
    case studies, 204–205, 205
    disintegration, 204–205
    geographical setting, 201–203
    grounding lines, 207–209
    historical alterations, 203–204
    importance, 203
    physical setting, 201–203
    vulnerability of, 207–209
Ice states, 223
Ice-volume equivalent sea level function, 529, 529F
ICP *see* International Conference of Paleoceanography (ICP)
ICP-MS (inductively coupled plasma mass spectrometry), 333–334
Idealized ocean general circulation model (OGCM), 363
Igneous provinces, 379–380
  see also Large igneous provinces (LIPs)
IGOS (Integrated Global Observing Strategy), 277
Illite, 78
IMAGES (International Marine Global Change Study), 360
Indian Ocean
  benthic foraminifera, 428T
  coral records, 443F
  heat transport, 107
  monsoons
    nanofossil species diversity, 82F
    seasonal variability, 76
    sedimentary record, 76
  radiocarbon, 481F, 483F
  sea–air flux of carbon dioxide, 465T
  sea ice cover, 239–240
Indicator organisms
  climate change
    planktonic indicators, 397–398, 405F
Individual water parcels, 27
Indonesia
  El Niño events and, 57, 68
Indonesian-Malaysian Passages
  NADW formation, 365
  paleoceanography
    climate models in, 362
    gateway and global ocean circulation, 326, 365
Indo-Pacific Ocean, heat transport, 107

Inductively coupled plasma mass
    spectrometry (ICP-MS), 333–334
Inertial dissipation method, 96
    evaporation, 121–122
Infauna, 417–418, 424
    communities, 418
        latitudinal differences, 418–419,
        419F
        Petersen's, 418, 419T
        Thorson's, 420T
    size groups, 418F
Infrared atmospheric algorithms, satellite
    remote sensing of sea surface
    temperatures, 291–293
Infrared budget, 128
Infrared (IR) radiation
    seawater absorption depth, 128
Infrared (IR) radiometers
    application
        cooled detectors, 294
        microwave radiometers vs, satellite
        remote sensing of SST, 293–294
        new generation/future improvements,
        301
        satellite-borne, 292F, 294, 294T
In situ measurements, 284, 285
Insolation, 346F
    caloric summer half-year, 349F
    glacial cycles and, 344
    methods of calculation, 348
    orbital parameters affecting, 345–347,
    347F, 348, 370–371
    range, 128
    surface layer stability and, 130
Instabilities
    of MOC, 31–32
    in Pacific Ocean, 299, 301F
Integrated Global Observing Strategy
    (IGOS), 277
Integrated Ocean Drilling Program
    (IODP), 356
    IODP Site 302, 381F
Integrated water vapor (IWV), 114–115
Interglacials, monsoon abundance relative
    to glacials, 80
Intermediate Atlantic Water, 311
Intermediate complexity models, carbon
    cycle, 471–472
Internal wave(s)
    field, ice cover, 188
International Conference of
    Paleoceanography (ICP), 354
    meetings, 355T
International Decade of Ocean
    Exploration, 460
International Marine Global Change
    Study (IMAGES), 360
International Satellite Cloud Climatology
    Project, 113
Intertropical Convergence Zone (ITCZ),
    9F, 71, 98–99, 129
    El Niño and, 71
    El Niño Southern Oscillation and, 59
    salinity, 134
Introductions, of species see Exotic
    species introductions

Intrusions
    Arctic Ocean, 164
Inverse models/modeling
    carbon cycle, 472
IODP see Integrated Ocean Drilling
    Program (IODP)
IODP Site 302, 381F
IOP see Inherent optical properties (IOPs)
IRD see Ice-rafted detritus (IRD)
Iron (Fe)
    availability, dinitrogen (N_2) fixation and,
    582
    concentration
        seawater, 576
        vertical distribution in seawater, 576,
        577F
    in limitation of phytoplankton growth,
    575–576, 575F, 576, 577
    precipitation, 576
    role, 573, 576–577
IRONEX experiments, 577, 578F,
    579–580, 581, 582
Iron fertilization, 573, 577, 573–584
    carbon cycle and, 573, 581–582, 581F,
    582, 582F
    experimental measurements, 578–579
    experimental strategy, 577
        fluorometry, 578
        form of iron, 577, 578F
        inert tracer, 577–578, 578F
        Lagrangian point of reference, 578
        remote sensing, 578
        shipboard iron analysis, 578
    findings to date, 579
        biophysical response, 579
        carbon flux, 581–582, 581F, 582F
        growth response, 579–580, 580F
        heterotrophic community, 573–574
        nitrate uptake, 579
        nutrient uptake ratios, 580, 580F
        organic ligands, 580–581
    questions remaining, 582
    societal challenge, 582–583
    see also Nitrogen cycle; Phosphorus
    cycle
'Iron theory', 575–576, 575F
ISCCP (International Satellite Cloud
    Climatology Project), 113
Isopycnal coordinate systems, 20
Isopycnals (density isolines), 29, 30, 31F
    definition, 8–9
Isostatic basal melt rate, 182–183
Isotopic budget equation, 501
Isotopic ratios
    in carbon cycle models, 498F,
    501–502
    in coral-based paleoclimate research,
    438–439, 438T
Isotropic points, ice shelf stability, 207
Isthmus of Panama, paleoceanography
    climate models in, 364
    passage closure, 326, 363F, 364
        climate and, 363–364
    passage opening, closed Drake Passage
    and, 364–365, 365F
ISW (ice shelf water), 172–173

ITCZ see Intertropical convergence zone
    (ITCZ)
IWV (integrated water vapor), 114–115

J

Jan Mayen Current, 239
    sea ice cover and, 234
Japanese Marine Observation satellites,
    281–282
Japanese National Space Development
    Agency, 282
Japanese sardine see Engraulis japonicus
    (Japanese sardine)
Japan Sea
    monsoons
        historical variability, 81
        indicators, 80
        long-term evolution of, 83
    sea ice cover, interannual trend, 239
Jason-1 satellite, 275
JCOMM (Joint Technical Commission for
    Oceanography and Marine
    Meteorology), 274
Jet stream flow patterns
    El Niño event 1996-1997, 65F
    El Niño Southern Oscillation and,
    64
JOIDES Resolution, R/V, 360
Joint Oceanographic Institutions
    Incorporated (JOI), 270
Joint Technical Commission for
    Oceanography and Marine
    Meteorology (JCOMM), 274

K

Kanon der Erdbestrahlung, 345, 350F
'Kansas' experiment, 120
Kara Sea, 154
    sea ice cover, interannual trend, 237–239
    sub-sea permafrost, 259
Kaula report, 266–267
KEM see Mean kinetic energy (KEM)
Kennett, James, 320–321
Kilopascal, 424
Kinetic energy/unit volume (current flow),
    definition, 5
Kittiwake (Rissa tridactyla)
    Bering Sea, 564
Knipovich spectacles, 161F
Kolmogorov length, 14
Krill (Euphausiacea), 402
Kuroshio Current
    flow, 49F
    frontal position, satellite remote sensing
    of SST, 299
    western boundary current, heat
    transport, 107

L

Labrador Current (LC), 10
    sea ice cover and, 234
Labrador Sea, 28
    circulation, global warming and, 45

Labrador Sea (*continued*)
  North Atlantic Oscillation, 38
  sea ice
    North Atlantic Oscillation and, 38
Lagrangian biological models
  fluid dynamics, 17
Lagrangian flow, ocean circulation, 3,
    4–5, 4F
λ (decay constant), definition, 491
Lamb wave, 19
Land ice, 223
  melting, 137
  see also Glaciers
Land-ocean interface, 536
  see also Land-sea global transfers
Land-sea global transfers, **536–545**
  coastal zones, 538–541
    bioessential elements, 539–541
    global warming and, 541–543
  direction, 536
  human influence, 536, 540F
    Asia, 543–545
  mechanisms, 536–538
  trace elements, 538
Land surface(s)
  changes, relative to sea level changes,
    522
  instability, sea level changes, 522
  temperatures, North Atlantic
    Oscillation, 36F
La Niña, 57, 70, 71, 72F
  change to El Niño, 71, 73
  sea surface temperatures, 70, 70F
  see also El Niño Southern Oscillation
    (ENSO)
Laptev Sea, 154
  sea ice cover, 234–235
  sub-sea permafrost, 259
Large eddy simulation (LES), 15
Large igneous provinces (LIPs),
    379–380
  sea level rise and, 514F, 517
  see also Geomorphology; Mid-ocean
    ridge geochemistry and petrology;
    Mid-ocean ridge seismic structure;
    Mid-ocean ridge tectonics; Seismic
    structure; Volcanism
Larsen Ice Shelf, 201–203, 205
  breakup, 206F
  Larsen A, 205
  Larsen B, 205
  modeled strain-rate trajectories, 208F
Larvae
  lecithotrophic, 421, 424
  pelagic, 424
    benthic organisms, 421
    settlement process, 422
  planktotrophic, 424
    benthic organisms, 421, 422
    settlement process, 422
Last glacial maximum (LGM), 306F
  $\delta^{18}O$ values, 319, 320F
Late Cretaceous
  ocean circulation, 366–367
    heat transport, 366–367
    models, 367, 367F

surface, 367
  thermohaline, 367
Late Mesozoic
  paleo-ocean modeling, 362
Latent heat flux, 120, 125
  global distribution, 125, 126F
  satellite remote sensing, 114–115
Latent sea–air heat flux, 128–129
  global distribution, 133F
  see also Turbulent heat flux
Latitude
  heat radiation from Earth and, 103,
    103F
  heat transport and, 104–105
  sea ice depth and, 247T
Laurentide Ice Sheet, H events,
    41–42
Laysan albatross (*Diomedea immutabilis*)
  expansion of geographical range,
    561
LC *see* Loop Current (LC)
LCDW *see* Lower Circumpolar Deep
    Water (LCDW)
Lead convection
  modes of, 185F
Lead convection
  summertime studies, Surface Heat
    Budget, 188
Lead Experiment (LeadEx; 1992),
    185–186
  salinity profile, 185–186, 186F
  salt flux, 186, 187F
Leads (long linear cracks, ice), 184
Leucocarboninae *see* Shag(s)
Levitus atlas, 22
LGM *see* Last glacial maximum (LGM)
Life histories (and reproduction)
  coccolithophores, 410–411
Light
  effects of water depth, 416
LIP *see* Large igneous provinces (LIPs)
Lithosphere
  carbon dioxide cycle, 459
Little Ice Age (LIA), 390F, 391–392,
    391F, 392F, 394–395
  coral records, 442–444, 443F
Littoral zone, 419T
LNHS (low-nitrate, high-chlorophyll)
  regions, 574–575, 574T, 575F
LNLC (low-nitrate, low-chlorophyll)
  regions, 574–575, 574T
Log-layer solution, 179
Lomonosov Ridge, 154
Lonely, Alaska, sub-sea permafrost,
    257
Longitude of perihelion, 346
Long-wave heat flux, global distribution,
    133F
Lower Circumpolar Deep Water (LCDW)
  upwelling, 152F
  water properties, 143F
Low-nitrate, high-chlorophyll (LNHC)
  regions, 574–575, 574T, 575F
Low-nitrate, low-chlorophyll (LNLC)
  regions, 574–575, 574T
Lvitsa, 154

M

Maastrichtian, climate proxy records,
    381F
Mackenzie River Delta region
  sub-sea permafrost, 255, 257F, 260
Mackenzie River Delta region
  geological model, 260, 261F
  salt transport, 258
Macrofauna, 418, 418F, 424
  see also Macrobenthos
Magma lens *see* Axial magma chamber
    (AMC)
Magnesium (Mg^{2+}), concentrations
  in river water, 537T
Magnesium/calcium ratio, 382F
  calcification temperature and, 333F
  measurement in benthic foraminifers,
    323
  sea surface temperature
    paleothermometry and, 331T,
    333–334
  standard error, 334
Magnesium carbonate
  chemical weathering, 495
Magnetic field, Earth
  reversal, 347
Magnetic susceptibility, sediments, and
    origin of material, 78
Makarov Basin, deep water, 163
Makharov Basin, temperature and salinity
    profiles, 156F, 157F
Manganary, E.P., 154
Manganese (Mn)
  Black Sea profile, 159F
MAR *see* Mid-Atlantic Ridge (MAR)
Marchetti, Cesare, 569
Marginal ice zone (MIZ), 223
  sea ice deformation rate, 224–226
Marginal Ice Zone Experiments
    (MIZEX), 183
Marine aggregates *see* Aggregates, marine
Marine biodiversity
  late Eocene, 384
Marine finfish mariculture *see*
    Mariculture
Marine ice, 173–174, 175
  see also Ice; Sea ice
Marine otter *see* Sea otter
Marine Oxygen Isotope Stage, 391
Marine sediments *see* Sediment(s);
    Seafloor sediments
Marion Dufresne, 360
Mass
  fluid packets, 17
Mass spectrometry (MS), 305–307
  inductively-coupled, 333–334
MAT (modern analog technique),
    339–340
MC *see* Mindanao Current (MC)
MCP (Medieval Cold Period), 391–392
Mean ice draft
  Beaufort Sea, 245F
  Canada Basin, 245F
  Chukchi Cap, 245F
  Eastern Arctic, 245F

Eurasian Basin, 246*F*
Nansen Basin, 245*F*
North Pole, 245*F*
Mean kinetic energy (KEM), 6*F*
definition, 5–6
Mean life λ^{-1} (decay), 491
Medieval Cold Period (MCP), 391–392
see also Holocene
Medieval Warm Period (MWP), 391–392,
392*F*
see also Holocene
Mediterranean outflow
salty, 27, 29
Mediterranean Ridge, sedimentary
records of Holocene climate
variability, 390
Mediterranean Sea
carbon sequestration, 569
circulation, 11
salinity, 13
salty outflow, 27, 29
sediment sequences, orbital tuning, 374
Meiofauna, 418, 418*F*, 424
Melt ponds, 288
Meltwater events, bipolar seesaw model
and, 393–394, 394*F*
Mendeleyev Ridge, 154
Meridional overturning circulation
(MOC), 27
general theory for, 30–31
instability of, 31–32
Mesoscale eddies
fluid packet tracer mixing, 18
heat transport, 107–108, 108
modeling, 15–16
Mesozoic, 378
carbon releases, 585
Late, paleo-ocean modeling, 362
temperature gradients, 378
see also Paleoceanography
METEOR Expedition, 460
Meteor Expedition, 355
sediment core collection, 355, 356*F*
Meteoric ice, 250
Meteorite impacts, abrupt climate change
and, 45
Meteorological satellites, 275
Methane (CH_4)
atmospheric concentration, historical,
589
atmospheric sources and sinks, 100
historical emissions, 589–590
drivers, 590
Methane hydrate(s)
deposit stability, 585–586, 585*F*, 586
dissociation (methane release)
climatic effects, 325, 498
from methane hydrate reservoirs, 589
distribution in reservoirs, 587
quantification, 587
paleoceanographic research, 358
Paleocene-Eocene Thermal Maximum
and, 382
shallow-water reserves, 587
slide headwall and excess pore pressure,
589–590

slope instability regions, 588–589
stability
deposits, 585–586, 585*F*, 586
submarine slides and, **585–590**
evidence for causal relationship,
588–589
extent on continental margins
(largest), 588*F*
triggering mechanism, 588–589
METOP satellites, 111
MF *see* Midfrequency band (MF)
Mg/Ca *see* Magnesium/calcium ratio
Microfauna, 418, 418*F*, 424
Micropaleontology, 357
Microwave backscattering, 110–111
see also Scatterometry
Microwave data, atmospheric
interference, 281, 283
Microwave humidity sounder (MHS),
115
Microwave measurements, satellite
remote sensing of SST, 293–294
Microwave radiometers, 293
infrared radiometers vs, satellite remote
sensing of SST, 293–294
Microwave radiometry, 270
Microwave scanning radiometer, 114, 282
Microwave scattering, sea surface,
110–111
Microwave sensing, freshwater flux,
115–116
Middle East, North Atlantic Oscillation,
temperature, 35
Middle Ice, 234–235
Mid-ocean ridge(s) (MOR)
degassing, 495, 500*F*
carbon dioxide, 499–500
sea level fall and, 515, 516*F*, 517
Milankovich cycles, 41
Milankovitch, Milutin, 344–345
ice ages theory, 344–345
Milankovitch scale sea level variations,
515
see also Sea level changes/variations
Milankovitch variability, 310–311,
344–353
summer solstice insolation, 349, 351*F*
Millennial-scale climate variability, 305*F*,
305–313
Heinrich events, 307–309
history, 305–307
Holocene, 310
mechanisms, 310–312
other proxies, 309–310
proxy agreement, 311
proxy record examples, 307
Million year scale sea level variations,
507, 520*F*, 521
estimation, 520, 520*F*
research, 521
see also Sea level changes/variations
Mindanao Current (MC)
flow, 49*F*, 51–52
Mindanao Eddy, 49*F*
Mindanao Undercurrent (MUC), flow,
49*F*, 52–53

Miocene
$\delta^{18}O$ records, 322*F*, 324*F*, 325, 326
Mixed layer
upper, 328
Mixed layer temperature
depth, 128
global distribution, 132*F*
range, 127
Mixing ratio, definition, 121*T*
MIZ *see* Marginal ice zone
MIZEX (Marginal Ice Zone
Experiments), 183
Models/modeling
chemical tracers as calibrants, 468–469
complexity, 475
Moderate Resolution Imaging
Spectroradiometer (MODIS),
113–114, 297
36-band imaging radiometer, 297
narrower swath width than AVHRR,
297
Modern analog technique (MAT),
339–340
Modified warm deep water (MWDW),
170–171
MODIS *see* Moderate Resolution Imaging
Spectroradiometer (MODIS)
Mohovi^{1}i^o seismic boundary *see* Moho
Moles of carbon, 459
Mollusks
shells, radiocarbon, 478–479
Momentum
conservation/dissipation features in
equations, 24
upper ocean, general circulation model,
22
Momentum equation, general circulation
models, 22
Momentum fluxes
at sea surface, **95–102**
accuracy of estimates, 100–102
data, sources of, 97–98
measuring, 95–97
regional variation, 98–100
seasonal variation, 98–100
transfer coefficients, typical values,
97*T*
Monachus monachus (monk seal), 546
Monk seal (*Monachus monachus*), 546
Monsoon(s)
causes and generation, 76
effect of continental topography, 77
cyclicity, 83–84
equatorial upwelling, 83
geologic records, 76–77
historical variability, 80
history of, **76–84**
indicators
China Sea, 80
present, 78–79
in sediments, 77–78
salinity and, 134
seasonal variability
Northern Hemisphere, 76*F*
and sedimentary record, 76
teleconnections, 83

Moorings, 4, 4F
  sea ice thickness determination, 247
  see also Current meters
MOR *see* Mid-ocean ridge(s) (MOR)
MORB *see* Mid-ocean ridge basalt
  (MORB)
'Multisymplectic geometry', 24
Multiyear (MY) ice, 234
  Arctic, 236
Murre(s)
  common, 561–562
    climate change responses, 561–562,
    562F
MY ice *see* Multiyear (MY) ice
Mytilus *see* Mussels (*Mytilus*)

N

NADW *see* North Atlantic Deep Water
  (NADW)
Nanofossil species diversity, Indian
  Ocean, 82F
Nansen, Fridtjof, 188
Nansen Basin, 154
  deep water, 162–163
  mean ice draft, 245F
  temperature and salinity profiles, 156F,
  157F, 160F
NAO *see* North Atlantic Oscillation
  (NAO)
NASA (United States National
  Aeronautics and Space
  Administration), 281
  early data, 265–266, 265F, 266F
  Earth Observing System (EOS), 297
  missions *see* Satellite oceanography
  partnership with field centers, 269
  satellites *see* Satellite oceanography
NASA scatterometer (NSCAT), 111,
  112T, 271
NASA Team algorithm, 283, 288F
National Center for Atmospheric
  Research, 198
National Center for Environmental
  Prediction (NCEP), 98, 124–125,
  198
National Oceanographic and Atmospheric
  Administration (NOAA)
  satellites, 294
National Ocean Sciences AMS facility,
  482
Navier-Stokes equation, 24
Navy Remote Ocean Observing Satellite
  (NROSS), 267F, 271
NCC *see* Norwegian Coastal Current
  (NCC)
NCEP (National Center for
  Environmental Prediction), 98,
  124–125, 198
NECC *see* North Equatorial
  Countercurrent (NECC)
Neocalanus plumchrus copepods, 400
Neogloboquadrina pachyderma
  foraminifer, 436
Neon
  atmospheric abundance, 210T

ice solubility, 211
  phase partitioning, 211T
  seawater concentration, 210T
Neural networks
  paleothermometric transfer functions,
  341F
Newfoundland, 28
  sea ice cover and, 234
New Guinea Coastal Current (NGCC)
  flow, 49F, 52
New Guinea Coastal Undercurrent
  (NGCUC)
  flow, 49F, 52, 52–53
Newly formed water masses, 28
NGCUC *see* New Guinea Coastal
  Undercurrent (NGCUC)
Nimbus 5 satellite, 280, 281
Nimbus 6 satellite, 281–282
Nimbus 7 satellite, 234, 267F, 268, 280,
  281–282, 285F
  Arctic ice cover, 235F
Nitrate (NO_3^-)
  high-nitrate, high-chlorophyll (HNHC)
  regimes, 574–575, 574T
  low-nitrate, high-chlorophyll (LNHC)
  regimes, 574–575, 574T, 575F
  low-nitrate, low-chlorophyll (LNLC)
  regimes, 574–575, 574T
  profiles, 159F
  sea water concentrations
    phosphate *vs.*, 574, 574F
    deeper water *vs.*, 577F
Nitrogen (N)
  anthropogenic emission perturbations,
  540F
  coastal fluxes, 541F
  eutrophication, 401–402
  fertilizers, application levels, 540F
  limitation, phosphorus *vs.*, 574, 574F
  river fluxes, 539
NOAA *see* National Oceanographic and
  Atmospheric Administration
  (NOAA)
NOAA satellites, 294
NOAA/TIROS, 267F
Noble gases
  atmospheric abundance, 210T
  ice formation and, **210–213**
    phase partitioning, 210T, 211
    physical properties relevant, 211–213
  ice solubility, 211
  phase partitioning, 211T
  seawater concentration, 210T
  tracer applications, 210, 211–213
'Noether invariants', 24
Noise (acoustic)
  microwave scatterometry, 111
Nondimensional eddy coefficient, 181
Nondimensional surface layer thickness,
  181
Nonlinear SST (NLSST), 292–293
Nonpolar glaciers, sea level variations
  and, 510, 510F, 511
North Africa
  North Atlantic Oscillation, temperature,
  35

North America
  river water, composition, 537T
North Atlantic
  deep water formation, 27–28, 31, 32
  ecosystems, North Atlantic Oscillation
  and, 38
  freshwater sources, 312F
  heat fluxes (mean), 100, 102F
  heat transport, 107
  Holocene climate, 390–391, 391–392,
  391F, 392
  radiocarbon, meridional sections,
  483F
  seabird responses to climate change,
  565
  sills, 27–28, 31, 31F
  temperature changes, 12F
  thermohaline circulation, 10
  variability, 35
  warm water, 30, 31
North Atlantic Current (NAC), 10
  see also Atlantic Ocean current systems
North Atlantic Deep Water (NADW), 28,
  28F, 29, 32, 362, 392–393
  Antarctic Bottom Water and
  alternating dominance, 393–394
  areas of production, 312F
  Denmark Strait, 365
  flow routes, 9F, 10, 28, 28F, 29
  formation, 9F, 10
  global warming and, 45
  Heinrich events and, 41–42
  Indonesian-Malaysian Passages, 365
  paleoceanographic data, 362
  radiocarbon, 483
  reductions in intensity
    causes, 392–393, 393F
    Holocene, 391F, 392–393
  salinity, 28, 32
  Straits of Gibraltar, 362
  upwelling, Southern Ocean, 29, 152F
North Atlantic Intermediate Water, areas
  of production, 312F
North Atlantic Oscillation (NAO), 100,
  288–289, **33–40**
  anthropogenic warming and, 35–36,
  39
  Arctic Ocean, 234
  Arctic Ocean Circulation and, 165
  Arctic Oscillation and, 33
  atmospheric pressure anomaly, 34F
  centers of action, 33
  governing mechanisms, 38–39
    atmospheric, 39
    oceanic, 39–40
  influence, 35–36
    ecological impact, 38
    ocean variability and, 37–38
    precipitation, 36–37, 37F
    sea ice and, 38
    temperature, 35–36, 36F
  interdecadal variability, 39
  positive phase, 33, 36
  principal features, 33–35
  sea ice cover and, 239
North Atlantic Oscillation Index, 33, 34F

North Equatorial Countercurrent (NECC)
flow, 49F, 50, 50F, 51, 53F
North Equatorial Current (NEC)
flow, 49F, 50, 50F, 52F, 53F
Northern hemisphere
atmospheric meridional section, 9F
carbon dioxide sinks, 463
sea ice cover, interannual trend,
238F
subtropical gyre, 9
summer wind stress, 99–100, 99F
Northern Sea Route
navigability, 234–235
Northern Subsurface Countercurrent
(NSCC)
flow, 49F, 51, 52F
North Pacific
heat transport, 107, 132–134
oxygen and radiocarbon profiles, 469,
470F
North Pole, mean ice draft, 245F
North-west Africa, Holocene climate,
390–391, 391, 391F
North-west Atlantic
seabird responses to prehistoric climate
change, 559
Northwestern Europe, global warming
and, 45
Northwest Passage, sea ice cover,
234–235
Norway
herring, 551, 552F
North Atlantic Oscillation, ecological
effects, 38
Norwegian Atlantic Current, 156
Norwegian Current, 27–28
Norwegian margin, slides, methane
hydrate and, 589
Norwegian Sea, 27
NOSAMS (National Ocean Sciences AMS
facility), 482
NOSS (National Oceanographic Satellite
system), 269
Novorossiysk, 155
NROSS (Navy Remote Ocean Observing
Satellite), 267F, 271
NSCAT (NASA scatterometer), 271
Numerical models
limits/limitations, 24–25
Numerical weather prediction (NWP)
models, 98
Nutrient(s)
biological pump and, 457–458
land-sea global flux, 539–541
oceanic distribution, 574–576, 575F
high-nitrate, high-chlorophyll
(HNHC) regions, 574–575,
574T
low-nitrate, low-chlorophyll (LNLC)
regions, 574–575, 574T
Nutrient cycling
benthic foraminifera, 429, 436
Nutrient-rich waters
deep water, 27
NWP see Numerical weather prediction
(NWP) models

O

OAE (ocean anoxic events), 379, 380F
Objectively Analyzed air–sea Fluxes
(OAFlux) project, 115
Obliquity (orbital), 370–371
Ocean(s)
features
satellite remote sensing of SST
application, 297
thermal expansion, sea level variation
and, 509–510, 510F, 511
volume changes, sea level changes, 522
Ocean-atmosphere interactions
advection, 73
circulation system, 27, 30
coupled climate models, 32
El Niño, 73–74
exchanges, 282
long time-scales and slow adjustment to
winds, 73–74
stability properties, 74
Ocean basin
volume and sea level variations,
513–515, 514F
changes over time, 515–517, 516F
Ocean bottom
modeling, coordinate choice, 20
Ocean Carbon-Cycle Model
Intercomparison Project (OCMIP),
474F
Ocean Carbon Model Intercomparison
Project (OCMIP) -2 program,
488–490
participants, 489F, 490T
Ocean circulation, 27, 3–13
biogeochemical models and, 469–471
climate system, effect on, 12–13
deep ocean temperatures, 12–13, 12F
heat capacity of the ocean, 12
poleward heat transfer, 11F, 12
salinity measurements, 13
sea surface temperatures, 13
thermohaline circulation stability, 13
timescale variations, 12
see also Climate change; Ocean
climate models
definition, 3
determination of, 3–7
chemical tracers, 7
coordinates of measurement position,
3, 4
density distribution measurement, 7
dynamic method, 7
Eulerian flow, 3, 4, 4F
measurement techniques see Current
meters
flow variability, 5, 6F
time series analysis, 6–7, 6F
frequency spectrum, 6–7, 6F
geostrophic balance, definition, 7
geostrophic current, 7
Lagrangian flow, 3, 4–5, 4F
mariners' observations, 5
mean flow, 5
sea surface slope measurement, 7

vertical circulation, 7
see also World Ocean Circulation
Experiment (WOCE)
fresh water, role of, 10–12
Arctic Basin circulation, 11
buoyancy fluxes, 10–11
evaporation, 11
freshwater transfer, 11F
Mediterranean Sea, 11
polar oceans, 11
precipitation, 11
sea ice, 11
subpolar oceans, 11–12
general circulation model see Ocean
general circulation model
(OGCM)
mathematical methods see Elemental
distribution
meridional overturning, **27–32**
thermohaline see Thermohaline
circulation
wind driven see Wind-driven circulation
World Ocean Circulation Experiment
(WOCE), 7, 12–13
Ocean climate models
acoustic waves, 19
boundary fluxes, 20
climate, 20
eddies, 15–16
finite volume method, 19
fundamental budgets/methods, 17
gravity waves, 19
grid resolution, 14, 15
linear momentum budget, 18
mass conservation, 17
mesoscale eddies, modeling, 15–16
methods and budgets, 17
parameters, 16–17
problems, posing, 16–17
rounding errors, 20
scales of magnitude, 14
subgrid-scale parametrization, 14–15
time-stepping momentum, 18–19
tracer budgets, 17–18
truncation methods, 14–15
use/process, 20
vertical coordinates, 19–20
Ocean currents see Current(s)
Ocean Drilling Program (ODP), 356
sedimentary records of Holocene climate
variability and, 390
Ocean-driven climate change, 389,
392–393
see also Climate change
Ocean general circulation model
(OGCM), 7, 29
definition, 7
forcing, 7, 8F
model boxes, 7, 8F
momentum equation, 7, 8F
in paleoceanography, 363, 364
Denmark Strait, 366
Drake Passage, 363
Gibraltar, 365
Indonesian-Malaysian Passages, 365
Isthmus of Panama, 364

Ocean general circulation model (OGCM)
(*continued*)
    mid-Cretaceous, 367
    variable results, 368
  salinity equation, 7, 8F
  temperature equation, 7, 8F
  see also Atmosphere-ocean general
    circulation models (AOGCMs);
    General circulation models (GCM);
    Paleoceanography
Ocean global climate models (OGCM),
  30
  see also Global climate models (GCMs)
Oceania
  El Niño Southern Oscillation,
    precipitation, 59–60
  river water, composition, 537T
Oceanic anoxic events (OAE), 379, 380F
Oceanic carbon cycle, 451–452
  see also Carbon cycle
Ocean interior, modeling, vertical
    coordinate choice, 20
Ocean mixing *see* Mixing
Oceanography from Space (report), 266,
  270
Ocean surface, 291
  atmosphere interactions, 291
  high-resolution global coverage of
    satellites, 291
    see also Satellite oceanography
  ocean-atmosphere interactions, 291
  see also Sea surface
Ocean tides *see* Tide(s)
Ocean warming, methane hydrate
    reservoirs and, 587, 588F
OCMIP (Ocean Carbon-Cycle Model
    Intercomparison Project), 474F
OCMIP-2 (Ocean Carbon Model
    Intercomparison Project -2)
    program, 488–490
  participants, 489F, 490T
Odden, 234, 234–235, 239
ODP *see* Ocean Drilling Program (ODP)
Offshore permafrost *see* Sub-sea
    permafrost
OGCM *see* Ocean general circulation
    model (OGCM)
Oi1 event, 384
Okhotsk Sea
  sea ice cover, 234, 235–236
    interannual trend, 239
OML *see* Mixed layer
OMZ *see* Oxygen minimum zone (OMZ)
ON *see* Organic nitrogen (ON)
One-dimensional models
  oxygen and radiocarbon profiles, 469,
    470F
Ontong Java Plateau, 514F, 517
Open ocean convection, 27–28
  mixed layer *see* Surface mixed layer
  plumes *see* Ocean convection plumes
  see also Air–sea gas exchange; Breaking
    waves; Deep convection; Non-
    rotating gravity currents; Rotating
    gravity currents; Three-dimensional
    (3D) turbulence

Ophiuroidea (brittle stars), 421, 423
Optical oceanography *see* Ocean optics
Orbit (Earth)
  aphelion, 345–346
  Earth, glacial cycles and, 345–347
  long-term changes, uncertainty,
    371
  perihelion, 345–346
Orbital tuning, **370–377**
  accuracy limits, 371
  error sources, 372
  examples, 371F, 374–375
  future prospects, 375–376
  limitations, 373
  mapping function, 370
  methods, 372–373
    visual mapping, 372
  motivations, 375–376
  multiple parameters, 375–376, 375F
  necessary conditions, 373–374
  older sediments, 372
  proxy response lags and, 375–376
  verification, 373–374, 374
Organic carbon (OC)
  subcycle, Cenozoic, 497–499, 498F
  total, river water, 537T
Organic phosphorus
  Black Sea profile, 159F
OS3 event, 310
Östlund, Göte, 485
Overturning circulation, 151
  Antarctic Circumpolar Current and,
    151–152
  Southern Ocean, 152F
Oviparity, 424
Oxygen (O$_2$)
  concentration, 28, 29, 30F
  isotopes, **314–315**
  isotope stage (OS) events, 309–310
  profile
    Black Sea, 159F
    Southern Ocean, 143F
  tracer applications, 211–212
Oxygen isotope ratio ($\delta^{18}O$), 314
  Cenozoic records, 320–323, 324F,
    513–514
    caveats, 326
    foraminiferal, 320–321, 321F,
      322F
    'greenhouse world', 323
    'ice house world', 323–325
  clouds (values), 317, 317F, 318F
  coral-based paleoclimate records,
    438–439, 438T, 440
    climate trends, 442, 442–444,
      443F
    El Niño Southern Oscillation,
      440–441
    seasonal variation, 440
    temperature and salinity
      reconstructions, 439
  corals, 334
  Cretaceous, 381F
  determination, 316
  effects of evaporation, 317, 317F
  effects of precipitation, 317, 317F

estimation of sea level variations, 513
  glacial cycles and, 345, 347
  ice cores, 41, 42F, 307
  in marine sediment deposits, and
    monsoon activity, 77
  marine sediments, 42F
  ocean values, 318–319, 319F
    deep, 316
    ice sheet effects, 319–320
    surface, 316
  paleoclimate and, 378–379
  paleothermometry, 316–317
    equation, 316–317
  planktonic-benthic foraminifera
    differences, 81F
  as present monsoon indicator, 79
  rain (values), 317, 318F
  reporting standards, 316
  salinity and, 318–319, 319F
  sea surface temperature
    paleothermometry and, 330,
    331T
  snow (values), 317, 318F
  systematics, 316
  variations
    glacial-interglacial, 319–320, 321F
    interspecific, 326
    in natural environment, 317–318,
      317F, 318F
    Pleistocene, 319–320, 321F
    spatial, in modern sea water,
      318–319, 319F
    between species, 326
    temporal, 319, 320F
Oxygen isotope stage (OS) events,
  309–310
Oxygen minimum zone (OMZ)
  formation, 77–78
  and sediment bioturbation, 77, 77–78
Ozone (O$_3$)
  depletion, 29
  North Atlantic Oscillation and, 39

P

Pacific Decadal Oscillation (PDO),
  403–404, 404F
Pacific Deep Water (PDW)
  radiocarbon, 483–484
Pacific equatorial belt, carbon dioxide,
  463
Pacific-Indian throughflow, freshwater
  fluxes, 11F
Pacific Ocean
  benthic foraminifera, 428T
  climate effects on fisheries, 553–554
  coral records, 442, 442F, 443F
  coral reefs, radiocarbon record,
    478–479, 478F, 479F
  eastern equatorial, thermocline, 71
  eastern tropical
    atmosphere, 71
    sea surface temperature, 70, 70F
    warming, El Niño, 71
  El Niño Southern Oscillation (ENSO),
    403

frontal position instabilities, satellite remote sensing of SST, 299, 301F
Pacific Decadal Oscillation (PDO), 403–404, 404F
radiocarbon, 481F, 482
GEOSECS and WOCE comparisons, 485F
meridional sections, 483F, 488F
regime shifts, 554
sea–air flux of carbon dioxide, 465, 465T
sea ice cover, 234–235
sea surface temperatures, ENSO and, 59
tropical, SST distribution, satellite remote sensing, 297–298
western equatorial, thermocline, 71
wind change, El Niño and, 72–73
Pacific Ocean equatorial currents, 49F, 49–56
mean flow, 50–51
boundary flows, 49F, 51–52
subsurface, 52–53, 53F
surface, 51–52, 53F
interior flows, 49F, 50–51
subsurface, 51, 52F, 53F
surface, 50–51, 50F, 51F
variability, 53–55
annual cycle, 53–55, 54F
associated with El Niño events, 54–55, 55–56, 55F
high-frequency current, 56
meridional current, 53–54, 54F, 56, 56F
zonal current, 53–54, 54F, 55F, 56
Paleoceanography, 354–361
climate models, 362–369
applications, 362
history, 362
OGCM see Ocean general circulation model (OGCM)
seaways, 367–368
verification, 368
by incorporation of isotopic fractionation, 368
definition, 354–356
history, 354–356
Holocene, importance of, 389–390
interocean gateways and global climate systems, 362–363
Denmark Strait, 365–366
Drake Passage, 363, 364–365
implications, 366, 368
Indonesian-Malaysian Passages, 365
Isthmus of Panama, 363–364, 364–365
Straits of Gibraltar, 365
interocean gateways opened/closed during Cenozoic, 362, 363F
numerical dating techniques, 357–358
orbitally tuned timescales see Orbital tuning
oxygen isotope analysis, 356
proxy data, 354
measurement of, 356–357, 357F
relevance, 354
research directions, 360

sea surface temperature see Sea surface temperature paleothermography
sediment core recovery, 355, 356F
sediment records, correlation techniques, 357–358
studies, contributions to, 358–360
techniques, 356–358
Paleoceanography, 354
Paleocene
climate, 380
Paleocene-Eocene Thermal Maximum (PETM), 380, 381F
Paleoclimatology, 41–43
Paleogene
climate, modeling, 385
temperature gradients, 378
Paleotemperature
Mg/Ca and, 382F
oxygen isotope ratio and, 378–379
Paleothermometers, sea surface temperature paleothermography, 331–333
PALK method, natural radiocarbon, 486, 487F
Panamanian Isthmus see Isthmus of Panama
Partial pressure of atmospheric carbon dioxide (PCO$_2$), 451
Cenozoic
decreased, 494, 495–496, 496
Raymo's hypothesis, 496
reconstructions, 494, 494F
Henry's law, 451–452
response to changes in sea water properties, 451–452, 452F
sea–air, 461–463, 462F
see also Carbon dioxide (CO$_2$)
Partial pressure of CO$_2$ of ocean water (pCO$_2$)sw, 460
seasonal effects, 463
'skin' temperature, 461
Partial Test Ban Treaty, 478
Particulate organic carbon (POC)
biological pump contribution, 453–455
dissolved organic carbon vs., 456
Passive sensors, 98
Pathfinder SST (PSST), 292–293
P/E (salinity effect), 332
Pechora, sub-sea permafrost, 259
Peedee formation, 493
Pelecanus occidentalis see Brown pelican
Penguins see Sphenisciformes (penguins)
Perihelion, 345–346
Perkinsus marinus pathogen, 556
Permafrost, 252–253
high sea levels, 252
low sea levels, 252
onshore, submergence effects, 255
sea level variation and, 511
Permanent Service for Mean Sea Level (PSMSL), 507
see also Sea level changes/variations
Peru-Chile Undercurrent (PCUC)
flow, 49F, 53
Peru Current
flow, 49F, 52

seabird responses to climate change, 563–564, 564F
Peru Margin, monsoon evolution, 83
Peruvian anchoveta see *Engraulis ringens* (Peruvian anchoveta)
Petagrams, 459
Petersen, C.G.J.
classification of the benthos, 417–418, 418, 419T
PETM see Paleocene-Eocene Thermal Maximum
Petrels see Procellariiformes (petrels)
PFS *Polarstern*, 248
pH
effect of carbon, 566–567, 567F
effects on dissolved inorganic carbon, 414
impacts on phytoplankton, 402, 403F
impacts on zooplankton, 402
of sea water, coral-based paleoclimate research, 438T, 440
Phalaropus lobatus see Red-necked phalarope
Phase partitioning, noble gases, 211T
Phenology
defined, 398–399
effect of global warming, 398–399
Philippines
El Niño events and, 57
Phosphate (PO$_4{}^{3-}$)
profile
Black Sea, 159F
in sea water
nitrate vs., 574, 574F
Phosphorus (P)
accumulation in sediments through Cenozoic, 498–499, 499F
coastal fluxes, 541F
eutrophication
cause of, 401–402
fertilizers, application levels, 540F
limitation
nitrogen vs., 574, 574F
Photosynthesis
corals, 437–438, 439–440
equation, 453–455
Phytoplankton
harmful algal blooms (red tides), 397
high-nitrate, high-chlorophyll (HNHC) regimes, 574–575, 574T
low-nitrate, high-chlorophyll (LNHC) regimes, 574–575, 574T, 575F
low-nitrate, low-chlorophyll (LNLC) regimes, 574–575, 574T
nutrient(s)
requirements, 573–574
photosynthesis, 397
importance to marine life, 397
Phytoplankton blooms
global warming, 401–402
mammalian responses, 546–547
Pine Island Glacier, 176
Piston corers
historical use in paleoceanography, 355
Piston velocity, 461
Planck's constant, 291

Planck's equation, 291
Planck's function, linearized, 292
Planetary boundary layer, 179, 179–180
    under sea ice *see* Under-ice boundary
        layer
Planetary waves *see* Rossby waves
Plankton
    'bottom up'-controlled food web, 401
    carbon isotope profile, Arabian Sea, 82F
    fossilised, 77
    oxygen isotope profile, Arabian Sea, 81F
Plankton and climate, **397–406**
    beacons of climate change, 397–398
    carbon cycle, 397
    changes in abundance, 400–402
        effects on fisheries, 401
    changes in distribution, 398
        responses to global warming, 398
    changes in phenology, 398–400
        effects on food web, 399–400
    climate variability, 403–404
    consequences for the future, 404–406
        nutrient-phytoplankton-zooplankton
            (NPZ) models, 405, 405F
        plankton as models, 404–405
        results, 405–406
    global importance of plankton, 397
    impact of acidification, 402–403
        aragonite and calcite structures, 402
        effects on calcification, 402
    impacts of climate change, 398
    reflection of solar radiation, 397
    see also Upwelling ecosystems
Planktonic foraminifera
    $\delta^{18}O$ records, 319–320, 320–321, 323
        benthic foraminifers *vs.*, 321, 321F,
            322F
        long-term patterns, 321F, 322–323,
            322F
Plasticity, sea ice, 227–228
Plastic yield curves, sea ice, 228F
Plate tectonics
    long-term sea level change and, 515,
        516F, 517
Pleistocene
    glacial cycles, 347
    ice shelves, 203
    oxygen isotope variations, 319–320,
        321F
Pliocene
    glacial cycles, 347
    $\delta^{18}O$ records, 321F, 324F, 325
Plio-Pleistocene Ice Ages,
        paleoceanographic research, 359
PML (polar mixed layer), 154, 155–156,
        158T
Polar Front, 142F
    transport, 147
    water properties, 143F
Polar Frontal Zone, 144–145
Polar glaciation, paleoceanographic
        research, 359
Polarization, electric *see* Electrical
        properties of sea water
Polar mixed layer (PML), 154, 155–156,
        158T

Polar oceans
    ocean circulation, 11
Polychaetes/polychaete worms
    deep-sea communities, 423
Polynyas, 161, 240–241
Pore pressure, sediments, slides and,
        587–588
Potassium (K^+)
    concentrations
        river water, 537T
Potential temperature, 104
Prandtl (Schmidt) numbers *see* Schmidt
        (Prandtl) numbers
Prawns *see* Shrimps/prawns
Precession (orbital), 370–371
    equinoxes, 344, 348
Precipitation, 27, 28–29
    effect on $\delta^{18}O$ values, 317F,
        317
    El Niño Southern Oscillation
        and, 59–60, 62F
    freshwater flux and, 129
    global distributions, 134–135, 135F
    monsoonal
        oxygen isotope ratio and, 77
        seawater density and, 76
    North Atlantic Oscillation and, 36–37,
        37F
    tropics, sea surface temperature and,
        59–60
    see also Evaporation; Rain; Snow
Precipitation radar (PR), 115–116
Proximal continental shelf, 539
Prudhoe Bay, Alaska
    sub-sea permafrost, 255, 256F, 260,
        260F
Pteropods, 402
Ptychoramphus aleuticus (Cassin's
        auklet), 400
Puerto Rico Trench anomaly, 268
Puffins
    horned, 559–560, 560F
Pycnocline, 127, 188
Pyranometer, 96F
Pyrgeometer, 95–96

Q

Quartz
    sediment cores, 310
Quasigeostrophy, 23
Queensland (Australia), El Niño Southern
        Oscillation (ENSO), economic
        impact, 67–68
QuikSCAT satellite, 111, 112T, 275

R

Radiance
    black-body, 103, 103F
Radiation
    black-body, 103, 103F
    Earth's budget, 103, 103F
Radiative fluxes, 113–114
    global budget, 113T
    see also Air–sea heat flux

Radioactivity
    fluid packets and, 17
Radiocarbon (carbon-14, ^{14}C),
        477–492
    activity ratio relative, 478
    air–sea gas exchange, 490–491
    anthropogenic, 478
    application, 480
    atmospheric, 477
        historical recordings, 479, 479F
        history, 477, 477F
        production rate, 477
        testing, 478
    bomb produced, 478
    circulation tracer, 468–469, 469
    date tuning, 370
    decay rate, 477, 478
    deep ocean mixing, 486–487
    definition, 475
    $\delta^{14}C$, coral-based paleoclimate records,
        438T, 444, 445
        decadal variability, 442
        upwelling, 439–440
    dissolved inorganic carbon and,
        473–474, 473F
    distribution for large-scale circulation,
        482–485
    features, 477
    fractionation, 478
    Global Climate Models and, 473–474,
        473F
    implications for large-scale circulation,
        482–485
    measurements, 27
    measurement techniques, 479–480
    natural, 477
        and bomb components, separation of,
            485–486, 486F
    ocean models, calibration, 487–490
    oceanographic applications, 486
    oxygen utilization rate, 487, 488F
    primary application, 477
    residence time, 486–487
    sampling
        history, 480–482
        techniques, 479–480
    thermocline ventilation rate, 490–491,
        491F
    ventilation rate, 486–487
Radiocarbon carbon dioxide ($^{14}CO_2$),
        477
Radiolarians
    paleothermometric transfer functions
        and, 341
Radiometers, 291
    infrared *see* Infrared (IR) radiometers
    microwave *see* Microwave radiometers
    satellite, 113
Rain
    $\delta^{18}O$ values, 317, 318F
Rainfall
    global, 129
    heavy, eastern tropical Pacific Ocean and
        El Niño, 297–298
    tropics, 135
        patterns, El Niño and, 70

Rayleigh Distillation Model, 317, 317F, 318F
Raymo, M E, 496
Redfield ratio, 468, 573
    see also Carbon cycle; Nutrient(s); Phosphorus cycle
Red Sea
    salty outflow, 29
REEs see Rare earth elements (REEs)
Regime shifts
    climate change and, 585
Relative humidity (RH), definition, 121T
Relative sea level, isostatic contribution, 526
Relative variability of flow
    Agulhas Current, 6F
    definition, 5–6
Remote sensing
    iron fertilization experiments, 578
Reproduction
    coccolithophores, 410–411
    corals, 437
Restoration programs see Fishery stock manipulation
Revised analog method (RAM), 340
Reynolds number, 15, 183
    ocean models, 18
    see also Shear instability; Turbulence
Rheology
    sea ice, 227–228, 228F, 230
Richardson number, 180–181
River(s)
    anthropogenic loading, 540F
    Black Sea, 154
    carbon transport, 541F
    chemical flux to oceans, 536
    nitrogen transport, 541F
    outflow, coral-based paleoclimate research, 438T, 440
    phosphorus transport, 541F
    water composition, 537T
River outflow, coral-based paleoclimate research, 438T, 440
River water
    composition, 537T
Romania, Black Sea coast, 154
Ronne Ice Front, 175–176
Ronne Ice Shelf, 175, 207
    climate change, effects of, 178
    temperature and salinity trajectories, 173F
Rooth, Claes, 485
Rossby number, 180
Rossby radius, 15
Rossby similarity drag law, 182
Rossby waves, 72–73
    vorticity see Vorticity
    see also Coastal trapped waves; Ekman pumping; Ekman transport; Internal wave(s); Mesoscale eddies; Wind-driven circulation
Ross Ice Shelf, 169, 175
Ross Sea
    seabird responses to climate change, 562, 563F
    sea ice cover, 239–240, 241

Rosy Dory (*Cyttopsis rosea*), 550, 550F
Russia
    Black Sea coast, 154
Russian Cosmos satellite, 281
Rynchopidae (skimmers)
    climate change responses, 560

S

SAF see SubAntarctic Front (SAF)
Sahara, desertification, 45
St Anna Trough, Atlantic water, 158
St Lawrence River, 42–43
Salinity, 417
    Antarctic Circumpolar Current (ACC), 28
    Arctic Ocean, mixed layer, 154
    coupled sea ice-ocean models, 193
    definition, 127
    fluid packets and, 17
    Intertropical Convergence Zone (ITCZ), 134
    intrusions see Intrusions
    Labrador Sea see Labrador Sea
    measurement(s), 13
        Mediterranean Sea, 13
    monsoons and, 134
    North Atlantic Deep Water (NADW), 28, 32
    $\delta^{18}O$ values and, 318–319, 319F
    profiles
        Amundsen Basin, 156F, 157F, 160F
        Arctic Ocean, 156F
        Black Sea, 159F, 160F
        Nansen Basin, 156F, 157F, 160F
        Southern Ocean, 143F
    ranges, 127
    sea level effects, 509
    upper ocean
        global distribution, 134–135
        water-column stability and, 127
    see also Freshwater flux
Salinity effect (P/E), 332
Salmo salar (Atlantic salmon) fisheries see Atlantic salmon (*Salmo salar*) fisheries
Salt concentrations see Salinity
Salt fingers, 22
Salt transport, 132–134
SAMW see SubAntarctic Mode Water (SAMW)
Sapropels, 374
Satellite(s)
    infrared data, 288
    laser altimetry, 284
    passive-microwave instruments, 281–282
        see also Satellite passive-microwave measurements of sea ice; Satellite remote sensing SAR
Satellite-borne sensors, 98
Satellite oceanography, **265–279**
    challenges ahead, 274
        data policy, 274
        in situ observations, 274

        integrated observations, 274–275
        international integration, 277
        oceanographic institutional issue, 275–276
        transition from research to operations, 275
    first generation, 267–268
    missions, 272T
    origins, 265–267
    second generation, 268–269
    third generation, 269
        implementation, 271
        space policy, 271–273
        studies, 270
        understanding and consensus, 271
    joint missions, 273
    mission launch dates, 271
    new starts, 271
    partnerships in implementation, 269–270
    promotion and advocacy, 270
    research strategy for the decade, 270
Satellite passive-microwave measurements of sea ice, 280, 284, 285F, 287F, 288, 288F, **280–290**
    background, 280
    rationale, 280
    theory, 280–281
    see also Satellite oceanography
Satellite radiometers, 113
Satellite remote sensing, 278
    active sensors, 278, 278F
    air temperature, 114–115, 115
    applications, 116–118
    electromagnetic signals showing atmospheric transmittance, 276F
    fluxes, 116F, **110–119**
    history, 265
    humidity, 114–115
    oceanic observation techniques, 277F
    partnership with oceanography, 269
    passive sensors, 278, 278F
    turbulent heat fluxes, 114–115
    wind stress, 110–113
    see also Satellite radiometers; Satellite scatterometers
Satellite remote sensing of sea surface temperatures, 114, 118F, **291–302**
    accuracy, 294
    advantages, 291
    applications, 297–299
        air–sea exchanges, 301–302
        coral bleaching, 299–301
        coral reefs, threatened, time-series SST predicting, 299
        frontal positions, 299, 300F
        'global thermometer', 301
        hurricane intensification, 299
        instabilities in Pacific Ocean, 299, 301F
        oceanographic features resolution, 297

Satellite remote sensing of sea surface
    temperatures (*continued*)
    storm prediction, 299
    black-body calibration, 291, 294
    characteristics, 294–295
    coverage, 294
        global, 294
        limited to sea surface, 293–294
    future developments, 301–302
        Advanced ATSR, 302
        atmospheric correction algorithms
            improvement, 301
        infrared radiometers, 301
        SST retrieval algorithm validation,
            301
    infrared atmospheric corrected
        algorithms, 291–293, 301
        atmospheric aerosols, 293
        AVHRR *see* Advanced Very High
            Resolution Radiometer
            (AVHRR)
        constant temperature deficit, 291–292
        contamination by clouds, 293
        longer wavelength window, 291–292,
            292F
        Pathfinder SST (PSST), 292–293
        radiative transfer linearized, 292
        water vapor correction algorithm,
            292, 292F
    measurement principle, 291–293
        brightness temperature, 291
        infrared atmospheric windows, 291,
            292F
        microwave measurements, 293–294
        radiation attenuation, 291
        temperature deficit, 291
        thermal emission from sea, 291
        see also Electrical properties of sea
            water; Infrared (IR) radiometers;
            Penetrating shortwave radiation;
            Radiative transfer (oceanic)
    microwave measurements, 293–294
        infrared radiometers vs, 293, 293T
    penetration depth, 293–294
    temperature characteristic of 'skin of the
        ocean', 293–294
    time-series of global SST, 302
    see also Sea surface temperature (SST)
Satellite scatterometers, 112T
    deployment, 111
Saturation (of a dissolved gas)
    definition, 212
    noble gases, 212
Saturation humidity, definition, 121T
SAVE (South Atlantic Ventilation
    Experiment), 481–482
Scanning multichannel microwave
    radiometer (SMMR), 125, 234,
    280, 281–282, 283, 284, 285F
Scatterometers, 98, 125, 270
    Advanced Scatterometer (ASCAT), 111,
        112T
    NSCAT (NASA scatterometer), 271
    satellite *see* Satellite scatterometers
Scatterometry, 98, 110–111, 125, 270
    microwave, 111

rain and, 111
    tropical cyclones, 117F
Schmidt (Prandtl) numbers, 183
SCICEX *see* Scientific Ice Expeditions
Science Working Groups (SWGs), 269
Scientific Ice Expeditions (SCICEX), 244
Sea–air partial pressure of CO_2, 461–463,
    462F
Sea cucumbers (Holothuroidea), 423
Seafloor spreading
    sea level fall and, 515, 516F, 517
Sea ice, **234–251**
    area, 236
    bottom water formation, 280
    brightness temperatures, 280, 281F, 282,
        283, 288, 288F, 289
    concentration maps, 284, 284F
    concentrations, 280, 281, 282, 282–284,
        283, 288, 289
    conservation law, 226–227
    cover
        annual cycles, 280, 281
        extents, 284, 289
        extents and trends, 284–286, 288–289
        future monitoring, 288–289
        global, 281
        interannual differences, 281, 284
        monitoring, 280, 281F, 288–289
        seasonal cycle, 284, 285–286
        trends, 284, 288–289
    data, global, 280
    determinations from satellite passive-
        microwave data, 282–284, 284,
        285F, 287F, 288F
    drift problem, 229, 229F
    dynamics, **223–233**
        equation of motion, 228–229, 229F
        scaling, 229T
    formation, 28–29
    free drifting, 229–230, 230F
    geographical extent, 284, 289
        Antarctic seasonal variability,
            239–243
        Arctic interannual trends, 236–239,
            237F, 238F, 240F
        Arctic pressure fields, 240F
        Arctic seasonal variability, 234–236
        definition of 'extent', 236
        global interannual variability, 244F
        trends, 284–286, 288–289
    greenhouse climates, 385
    kinematics, 223–227
    melting, 226–227
    microwave emissions, 280
    modeling, 231
        short-term, 231
    multiyear, 234
        Arctic, 236
        Weddell Sea, 250
    North Atlantic Oscillation (NAO) and,
        38
    ocean circulation, 11
    pancake ice, 250
    plastic yield curves, 228F
    polar climate state, 280
    rheology, 227–228, 228F

satellite images, 224F
satellite passive-microwave
    measurements, **280–290**
    see also Satellite passive-microwave
        measurements of sea ice
sediment transport, 311
shear zone, 223
solar radiation reflection, 280, 282
stationary, 229
thermal growth, 226–227
thickness, 284, 285
    Antarctic, 248–250
    Arctic, 244–248, 245F, 247T
    data variability, 246, 247
    drilling from ships and, 250
    trends, 284–286, 288–289
types, 280, 282, 286–287
    first year ice, 286, 288F
    frazil ice, 286
    grease ice, 286
    multiyear ice, 286, 288F
    nilas, 286
    pancake ice, 286
    slush ice, 286
variables affecting, 287–288
    season length, 287–288
    summer melt, 287–288, 288
    temperature, 280, 281–282,
        287–288
    velocity, 287–288, 288
Weddell Sea, 224F
    multiyear, 250
see also Bottom Water Formation;
    Coupled sea ice-ocean models; Drift
    ice; Ice-Ocean Interaction
Sea level
    El Niño Southern Oscillation and, 64
    eustatic, glacio-hydro-isostatic models,
        527
    function, anatomy of, 527–529
        crustal rebound phenomena, 525F,
            527, 528F
        hydro-isostatic effect dominant,
            527–529, 528F
        local isostatic approximation, 527
        physical quantities estimated from
            observations, 528F, 529
        viscous response of the planet, 527,
            528F
    global, tectonics and, 522–523, 523F
    ice-volume equivalent, 524F, 526
    methane hydrate reservoirs and, 587,
        588F
    relative, isostatic contribution, 526
    Southern Ocean, 149F
Sea level changes/variations, **507–512,
    522–531**
    fall *see* Sea level fall
    geological timescales, **513–521**
        due to changing volume of ocean
            basin, 515–517, 516F, 521
        due to volume of water in ocean basin,
            513–515, 514F, 521
        Milankovitch scale, 515
        see also Million year scale sea level
            variations

geological timescales, estimations, 507
  backstripping method, 517–519,
    518F, 521
  continental hyposography, 517
  from location of strand line on stable
    continental craton, 517
  long-term change, 518F, 519–521
  million year scale, 520F, 521
  from observations on continents,
    517–519
  oxygen isotope ratio (δ¹⁸O), 513
  from sedimentary records, 517
  sequence stratigraphy, 520–521, 520F
  uncertainties, 517, 521
glacial cycles and, 523–524
  cyclic global climate change,
    523–524, 524F
  glacio- and hydro-isostatic effects, 524
isostatic contribution, 527
recent, observations, 507–508
  altimeter, 508–509
  tide-gauge, 507–508, 508F
recent, rate-determining processes,
    509–510
  Antarctic ice sheet, 510–511
  Greenland ice sheet, 510–511
  nonpolar glaciers and ice caps, 510,
    510F, 511
  ocean thermal expansion, 509–510,
    510F
  terrestrial storage changes, 511
relative change between land and, 522
rises see Sea level rise
shoreline migration see Shoreline
  migration
since last glacial maximum, 507,
    524–526
  areas of former major glaciation, 524,
    525F
  characteristic sea level curves, 525F
  positions of past shorelines, 523F,
    524, 525F
spectrum of variation, 526
tectonics and, 522–523
  see also Tectonics
  see also Climate change; Sea level fall;
    Sea level rise
Sea level fall
  mid-ocean ridges and, 515, 516F, 517
  seafloor spreading, 515, 516F, 517
  'snowball earth', 514
Sea level rise
  continental breakup and, 515, 516F, 517
  evidence from ancient corals, 513
  global-average rates
    estimates, 508, 509, 512
    recent acceleration, 508
  large igneous province (LIPs) and, 514F,
    517
  projections for twenty-first century, 511,
    511F, 512
    longer-term, 511–512
    regional, 511
  reduction, dam building and, 511
  volcanism and, 517
Sea lions see Otariinae (sea lions)

Sea of Okhotsk see Okhotsk Sea
Seasat-A satellite, 110–111, 112T
Seasat satellite, 267F, 268, 281–282
  global sea surface topography, 273F
Sea surface
  heat transfer processes, 110
  net carbon dioxide flux, 461, 463–465,
    464F, 465T
    sources of error, 465
Sea surface salinity (SSS)
  coral-based paleoclimate research, 438T,
    439
Sea surface temperature (SST), 13, 30
  characteristics, 328
  coral-based paleoclimate research, 438T,
    439
  depth profile, idealised, 329F
  eastern tropical Pacific Ocean, 70, 70F
  effect on meteorological parameters, 59
  El Niño events, 58F, 70, 70F
    time series 1950–1998, 66F
  El Niño Southern Oscillation
    east-west variations, 59
    interannual variation, 66
    prediction and, 68
  global distribution, 131F
  heat stored in the ocean indicated by,
    299
  hurricane development, 135–136
  ice core vs. marine sediment
    determination, 43F
  La Niña and, 70, 70F
  long-chain alkenone index and, 335–336
  mechanisms controlling, 73
  mixed layer temperature and, 130
  North Atlantic Oscillation and, 36F, 37
  ocean anoxic events, 379
  range, 127, 328
  tropical mean, 330F
  in tropics, thermoclines and, 71
  wind change effects, 73
  wind effect, 60
  winds causing/resulting in change, 70
Sea surface temperature
  paleothermography, **328–343**
  Last Glacial Maximum, 329F
  methods, 328–331
    alkenone unsaturation, 331T,
      335–336
    fossil assemblage transfer functions,
      338–340
    magnesium/calcium ratio, 333–334
    nature of SST signal in geology,
      330–331
    oxygen isotope ratio, 331–333
    paleothermometers, 331–333
    strontium/calcium ratio, 334–335
    tetraether index, 336–338
  proxy sources, 329F
Sea water
  constituents
    density, 127
    effect of monsoonal precipitation, 77
  freezing point, 192
Seaways, paleoceanographic models,
    367–368

SeaWinds scatterometer, 112T
  see also ADEOS II; QuikSCAT
SEC see South Equatorial Current
    (SEC)
SECC see South Equatorial
    Countercurrent (SECC)
Sediment(s)
  deposition, slides and, 586
  ice rafting, 311
  phosphorus
    accumulation through Cenozoic,
      498–499, 499F
  pore pressure, slides and, 587–588
  type, slumps and, 585
Sedimentary records
  Holocene climate variability, 390
  sea level changes and, 514–515,
    517
Sedimentary sequences
  calcium carbonate content, 78
  clay minerals, 78
  deposition rate, 77
  magnetic susceptibility, 78
  marine, at continental margin
    bioturbation, 77
    monsoon activity and, 76
  organic carbon content, 77–78
Sedimentation rate
  cyclic signals and, 373F
  slides and, 586
  slumps and, 585
Sediment cores
  dating, 305
  depth and age, 370
  high-accumulation-rate sediments,
    305–307
  oxygen isotope ratios, glacial cycles and,
    345, 347
  sea surface temperature determination
    from, 330
Sediment stratigraphy see Sediment cores
Seepage forces, seabed, slides and,
    586–587
Selective deposit feeders, 420, 420–421,
    424
Selli event, 379
Semi-enclosed evaporative seas, 29
Sensible heat flux, 128–129
  global distribution, 133F
  satellite remote sensing, 114–115
Sensors
  active, 98
  passive, 98
Sequence stratigraphy
  estimation of million year scale sea level
    variation, 520–521, 520F
  global sea level curve and, 520–521
  limitations, 520–521
SGS see Subgrid-scale parameterization
    (SGS)
Shackleton, Nicholas, 319–320
Ship(s)
  temperature and salinity measurements,
    129
Shipping
  future developments, 301–302

Shoreline migration, 522–531
past positions, glacial maximum and,
523F, 524, 525F
sea level change and, 529–531
Australia's past shorelines, 531
Europe's past shorelines, 529–531,
530F
ice-volume equivalent sea level
function, 529, 529F
South East Asia's past shorelines,
529F, 530F, 531,
see also Beach(es)
Shoreline reconstruction
Australia, Northwest Shelf, 531
Europe, 529–531, 530F
South East Asia, 529F, 530F, 531,
Shortwave heat flux, global distribution,
133F
Siberian shelf, sea ice cover, 234
Siderophores, 576, 580–581
see also Iron
Silica cycle, **573–584**
see also Biogenic silica; Carbon cycle
Silicate (Si[OH]4)
natural radiocarbon, 486, 486F
profiles, 159F
rocks
strontium isotopic ratios, Himalayan
area, 497
sea water concentrations
vertical profile, 577F
Silvery John Dory (Zenopsis conchifera),
550, 550F
Sinking phase, 27, 27–28
see also Ocean Subduction
Skampton ratio, 585
Skin of the ocean, satellite remote sensing
of SST, 293–294
Skipjack tuna (Katsuwonus pelamis)
response to changes in production,
553–554
Skylab, 268
Slides
global distribution, 588F
headwalls, pore pressure and, 589,
589–590
methane hydrate and, 587–588, 585–
590
slope inclination and, 586–587
slumps and, 585
trigger mechanisms, 586–588
Slumps
slides and, 585
Slush ice, 286
SMMR see Scanning multichannel
microwave radiometer
SMOS satellite, 130
SMOW (Standard Mean Ocean Water),
316
Snow
$\delta^{18}O$ values, 317, 318F
'Snowball earth,' sea level fall and, 514
SO see Southern Oscillation (SO)
Sodium (Na)
concentrations
river water, 537T

SOI see Southern Oscillation Index (SOI)
SOIREE see Southern Ocean Iron
Enrichment Experiment (SOIREE)
Solar constant, 103
Solar heat, 30
Solar radiation
clouds and, 113
emission levels, glaciation and, 45
seasonal variations, Southern Oscillation
and, 73, 74
Solubility pump, 453–456, 453F, 454F
Sooty shearwater (Puffinus griseus)
climate change responses, 561–562,
561F, 563
South America
El Niño region, interannual variability,
66
El Niño Southern Oscillation,
precipitation, 59–60
river water, composition, 537T
South Atlantic (Ocean)
Agulhas rings, 108
South Atlantic heat transport, 30
South Atlantic Ventilation Experiment
(SAVE), 481–482
South China Sea
monsoons, historical variability, 80–81
Southeast Asia
El Niño events and, 57
paleo shorelines, migration and sea level
change, 529F, 530F, 531,
South-east Atlantic, seabird responses to
climate change, 565
prehistoric, 559
South Equatorial Countercurrent (SECC)
flow, 49F, 50, 50F
South Equatorial Current (SEC)
flow, 49F, 50, 50F, 52F, 53F
Southern Antarctic Circumpolar Current
front, 142F
water properties, 143F
Southern hemisphere
carbon dioxide sinks, 463
subtropical gyre, 9
Southern Ocean
bathymetric map, 149F
benthic foraminifera, 428T
overturning
biogeochemical cycles and, 152
overturning circulation, 29, 30F,
151
predicted sea level rise, 511
radiocarbon, 481F, 482
sea–air flux of carbon dioxide, 465T
seabird responses to prehistoric climate
change, 558–559
sea level, 149F
sea surface height, latitude and, 150F
stratification, 142–143
warming, 150–151
water properties, latitude and, 143F
Southern Ocean Iron Enrichment
Experiment (SOIREE), 577, 578
questions remaining, 582
results, 579, 581
see also Iron fertilization

Southern Oscillation (SO), 57, 70, 73,
288–289
atmospheric pressure and, 59F
damped oscillation effects, 70
definition, 71
global effects, 67
gradual modulation of, 73–74
long-term variability, 67
models, 74
see also El Niño Southern Oscillation
(ENSO) models
thermocline depth effects, 74
weather prediction and, 71
Southern Oscillation Index (SOI), 57, 60F
time series 1866-1998, 67F
Southern Subsurface Countercurrent
(SSCC)
flow, 49F, 51, 52F
Southern Weddell Sea
hydrographic section over continental
slope, 172F
tidal model, 177F
South Pacific
El Niño, rainfall, 65F
heat transport, 107
South Pacific Convergence Zone, El Niño
Southern Oscillation and, 59
South Pacific Ocean
radiocarbon, 483–484, 484F, 485F
Soviet Union North Pole Drifting Stations
program, 223
Spacecraft instruments, 294–295
black-body calibration, 294
cooled detectors for infrared
radiometers, 294
geosynchronous orbit measurements of
SST, 297
GOES Imager, 297
TRMM Microwave Imager see TRMM
Microwave Imager (TMI)
SPCZ (South Pacific Convergence Zone),
El Niño Southern Oscillation and,
59
Special Sensor Microwave/Imager (SSM/
I), 114, 115, 125, 280, 281–282,
284, 285F, 288–289, 288F
Special sensor microwave temperature
sounder (SSM/T), 115
Specific heat capacity, 198
Specific humidity, definition, 121T
Spectra
infrared, 291, 292F
Spectral interval, 291–292
Spectral scattering coefficient see
Scattering coefficient
Sr/Ca (strontium/calcium ratio), sea
surface temperature and, 334–335
SSM/I see Special Sensor Microwave/
Imager (SSM/I)
SSS see Sea surface salinity (SSS)
SST see Sea surface temperature (SST)
St. Anna Trough, Atlantic water, 158
St. Lawrence River, 42–43
Standard Mean Ocean Water (SMOW),
316
Stanton numbers, 182, 183

Steady state, definition, 491
Sternidae (terns)
    climate change responses, 560
Storegga slide, 585
Storfjorden, 162
Storis, 234
Storm(s)
    North Atlantic Oscillation and, 36
    severe, heat and, 135–136
Storm forecasting, 136
Storm surges
    regional impacts, 511
Strait of Gibraltar
    NADW formation, 365
    paleoceanography
        climate models in, 365
        gateway, climate and, 365
        gateway closure during Cenozoic,
            363F
Stress patterns, ice shelf stability, 207
Strontium (Sr2+)
    isotope ratios
        Cenozoic carbon cycle changes and,
            496–497, 496F, 502
        present-day sea water, 496
Strontium/calcium ratio, sea surface
        temperature and, 334–335
Structure I gas hydrates, 585–586,
        587
Structure II gas hydrates, 585–586
SubAntarctic Front (SAF), 141, 142F
    transport, 147
    water properties, 143F
SubAntarctic Mode Water (SAMW),
        141–142, 143F
    formation, 151, 152F
SubAntarctic Zone, 144–145
Subgrid-scale parameterization (SGS),
        15–16
    Global Climate Models, 473
    vertical advection, 18
Sub ice-shelf, circulation and processes,
        169–178
    Antarctic continental shelf, 172–173,
        174F
    climate change, 177–178
    factors controlling, 169
    geographical setting, 169–170
    melting, seawater effects of, 171–172
    modes, 172–175
    numerical models, 169
    oceanographic setting, 170–171
    thermohaline modes, 172–175
        cold regime external ventilation,
            172–175
        internal recirculation, 175–176
        warm regime external ventilation,
            176, 176F
    tidal forcing, 176–177
Sublimation, 124
Sublittoral zone, 419T
Submarine(s)
    sea ice thickness determination, 244
Submarine measurements, 284, 285
Sub-sea permafrost, **252–262**
    characteristics, 254–255

definition, 252
distribution, 252F, 258–259
formation, 253–254
geothermal heat flow, 254
investigation, methods of, 252
models, 260–262
nomenclature, 252–253
occurrence, 258–259
processes, 255–257
    heat transport, 257–258
    salt transport, 257–258
    submergence, potential regions,
        255–257, 258F
temperature profiles, 253, 254F
thawing, 253–254, 254
    of negative seabed temperatures,
        257–258
Subtropical gyres, 8–9, 9F
    response timescales, 12
Sulfate (SO42−)
    concentrations
        in river water, 537T
Sulfur (S)
    anthropogenic emission perturbations,
        540F, 543–545
    atmospheric deposition, projected,
        544F
    river fluxes, 539
    total dissolved, river water, 537T
Sulfur dioxide (SO2)
    river water concentration, 537T
Summer (boreal), monsoon activity,
        Northern Hemisphere, 76, 76F
Summer melt
    onset, 280, 288
    process, 286
Summer solstice insolation, 349, 351F
Sun
    radiation emission levels, glaciation and,
        45
Sunlight, near sea surface temperatures
        and, 293–294
Surface freshwater exchange, 129
Surface heat exchange, 128–129
Surface heat flux, global distribution,
        133F
Surface polar currents, thermohaline
        circulation, 10
Surface radiation budget (GEWEX-SRB)
        project, 113
Surface wind velocity, measurement of,
        98
Suspension feeders, 420, 420–421,
        424
Sverdrup modeling, Antarctic
        Circumpolar Current and,
        148–150
Symbiotic relationships
    bacteria-deep-sea animals, 423
'Symplectic refinement', 24
Synthetic aperture radar (SAR), 270
    satellite remote sensing see Satellite
        remote sensing SAR
    see also Bio-optical models; Inherent
        optical properties (IOPs);
        Irradiance

T

t1/2, definition, 491
Tahiti, atmospheric pressure anomaly
        time series, 60F
Talik, definition, 252–253
Tasman-Antarctic Passage, opening of,
        363, 363F
Tasman Rise, 325–326
Taxonomy see Classification/taxonomy
        (organisms)
Tectonics
    sea level change and, 522–523, 523F
        global sea levels, 522–523, 523F
        local tectonic causes, 523
        raised coral reefs, 522, 523F
        vertical crustal displacements,
            522–523, 523F
Teleconnections, 33
Telephone cables, flow measurement
        using, 105–106
Temperature
    changes, noble gas saturations and,
        212–213
    deficit
        brightness temperatures, 292
        definition, 291
        water vapor as contributor,
            291–292
    global-average, predicted rise, 509
    latitudinal differences, 416–417
    profile
        Black Sea, 159F
        Southern Ocean, 143F
    upper ocean, distribution, 130
    vapor pressure vs., 123, 123F
    vertical differences, 416–417
    water-column stability and, 127
Temperature gradient
    global, greenhouse climates, 378,
        382–384, 385, 386T
Temperature proxies
    bias, 386T
Terrestrial biosphere, carbon dioxide
        cycle, 459
Terrestrial storage, changes, sea level and,
        511
Tethys Ocean, 154
    closure, 326
Tethys-Paratethys connection,
        paleoceanographic model, 367–368
TEX86 index, 336–338
    crenarchaeotal membrane lipids,
        336–338
    marine biocycle and global geochemical
        cycle, 357F
    Paleocene-Eocene Thermal Maximum,
        381F
    sea surface temperature and, 337F
    sea surface temperature
        paleothermometry and, 331T
    Tanzania, 383F
Thawed layer, development, 253
Theories, basic building blocks, 23
Thermal expansion, oceanic, sea level
        variation and, 509–510, 510F, 511

Thermocline, 29, 30, 30F, 31, 31F, 71
  deep, Southern Oscillation not possible, 74
  deepening, sea surface temperature rise, 71
  depth
    El Niño Southern Oscillation and, 60–63
    El Niño conditions, 60–63, 71, 72F
    La Nina conditions, 71, 72F
    main
      ocean circulation, 8–9, 9, 9F, 10
      thermohaline circulation, 10
    models, carbon cycle, 472–473, 472F
    shallow, Southern Oscillation short, 74
Thermocline ventilation rate
  radiocarbon, 490–491, 491F
Thermohaline circulation, 7–10, 27, 29
  abrupt climate change and, 43, 44
  Atlantic Ocean, 43, 44
  buoyancy fluxes, 10
  Coriolis force, 10
  deep ocean currents, 10
    see also Abyssal currents
  definition, 10
  density, horizontal variations, 10
  horizontal pressure gradient, 10
  hyperthetical ocean model, 10
  Late Cretaceous, 367
  North Atlantic, 10
  ocean basin bathymetry, 10
  salinity, 10
  stability, 13
  surface polar currents, 10
  thermocline, 10
Thermohaline forcing, 172
  cold regime external ventilation, 172–175
  internal recirculation, 175–176
  modes, 172–175
  warm regime external ventilation, 176, 176F
Thermometers, 129–130
Third-order sea level change see Million year scale sea level variations
Thorium (Th)
  isotopes, 468–469
Thorson, Gunnar, 420T
  classification of the benthos, 418
  distribution of benthic larvae, 421F, 421
Three-component ultrasonic anemometer, 96, 96F
Thulean basalts, eruption, 325
Thunnus albacares (yellowfin tuna)
  response to changes in production, 553–554
Tidal friction, orbital parameters and, 371
Tidal motions, 29
Tide gauges, in estimation of recent sea level variations, 508, 508F
Time-series analysis
  orbital tuning and, 374

Time-stepping momentum, ocean climate models, 18–19
TMI see TRMM Microwave Imager (TMI)
TOC (total organic carbon), river water, 537T
Toggweiler model, radiocarbon, 488
TOPEX/Poseidon satellite, 266–267
Topex/Poseidon satellite altimeter system mission, 509
Total dissolved nitrogen (TDN)
  river water, 537T
Total dissolved phosphorus, river water, 537T
Total dissolved sulfur, river water, 537T
Total organic carbon (TOC), river water, 537T
Total precipitable water, 114–115
TPD (Transpolar Drift), 155
Trace element(s)
  coral-based paleoclimate records, 438–439, 438T, 445
    seasonal variation, 440
    temperature reconstructions, 438T, 439
    upwelling, 438T, 439–440
  land-sea global transfers, 538
Tracer(s)
  chemical, 7
  transport, linear momentum, 18
Tracers in the Ocean (TTO) North Atlantic Study (NAS), 481–482
Trade winds
  El Niño events and, 64, 70, 71
Traenadjupet slide, 585
Transfer functions, fossil assemblage abundances and sea surface temperature, 338–340, 340F
  fossil groups used, 340–342
  modern analog technique, 339–340
  revised analog technique, 339–340
  weighted average, 339
Transient tracers
  definition, 475
Transpolar Drift, 155
Transport profile, Southern Ocean, 143F
Trinity Test, 478
Tritium (T, ^{3}H)
  circulation tracer, 468–469, 469
Tritium-helium-3 age
  definition, 475
TRMM Microwave Imager (TMI), 114, 297–299
  measurements restricted to tropical regions, 297
  provides SST in the tropics, 297
    see also Tropical Rainfall Measurement Mission (TRMM)
Tropical Atlantic Study (TAS), 481–482
Tropical Atmosphere-Ocean array, 60–63, 63F
  sea temperature depth profiles, 1996–1997, 63F

Tropical cyclones, 123–124
  heat potential, 136F
Tropical Rainfall Measurement Mission (TRMM), 115–116
  evaporation, estimation of, 125
    see also TRMM Microwave Imager (TMI)
Tropics
  rainfall, 135
Turborotalita quinqueloba, 339F
Turbulence
  geostrophic, 15
  shear stress, 179
Turbulent eddies, 120, 179
Turbulent heat flux, 128–129
  see also Latent heat flux
Turbulent mixing, 29
  laboratory studies see Laboratory turbulence studies
  timescale, general circulation model, 22
Turbulent shear stress, 179
Turkey, Black Sea coast, 154

U

UCDW see Upper Circumpolar Deep Water (UCDW)
UK37, 78
Ukraine, Black Sea coast, 154
ULES (upward-looking echo sounders), 250
Under-ice boundary layer (UBL)
  characteristics, 179–182, 180F
Under-ice boundary layer (UBL)
  drag, 179–182, 189
Under-ice boundary layer (UBL)
  molecular sublayer, 179
Under-ice boundary layer (UBL)
  outer layer, 180
Under-ice boundary layer (UBL)
  roughness elements, 179–180
Under-ice boundary layer (UBL)
  stress vectors, 181
Under-ice boundary layer (UBL)
  supercooling, 183
Under-ice boundary layer (UBL)
  surface layer, 179
Under-ice boundary layer (UBL)
  velocity vectors, 181
United States Defense Meteorological Satellite Program (DMSP), 280, 281–282, 284F, 285F, 288–289, 288F
United States of America (USA)
  East Coast, slope stability, 588
  El Niño, rainfall, 65F
Upper Circumpolar Deep Water (UCDW)
  overturning circulation and, 152F
  water properties, 143F
Upper mixed layer, 328
Upper ocean
  advective fluxes, 129
  distributions (heat/water), 130–134
  diurnal heating, 134F
  freshwater budget, **127–138**

climate change and, 136–137
processes governing, 127–129
freshwater distributions, 134–135, 135F
heat budget, 127–129, **127–138**
processes governing, 127–129
heat distribution, 130, 130–134
heat flux, 128–129
diurnal, 130
global distribution, 133F
measurements, 129–130
mixed layer, modeling, coordinate choice, 19–20
surface freshwater exchanges, 129
surface heat exchanges, 128–129
surface mixed layer see Surface mixed layer
temperature, interannual variation, 131–132
Upward-looking echo sounders (ULES), 250
Upwelling
coral-based paleoclimate research, 438T, 439–440
El Niño Southern Oscillation and, 64
modeling, radiocarbon and, 469
radiocarbon, 487
tropical Pacific, El Niño Southern Oscillation (ENSO) and, 64, 67
Urea profile, Black Sea, 159F
Urey, Harold, 316
UUV see Unmanned underwater vehicles (UUV)
UV see Ultraviolet radiation

V

Vaccination, mariculture see Mariculture diseases
Vapor flux, measuring, direct method, 120–121
Vapor pressure
definition, 121T
temperature vs., 123, 123F
Vegetation
terrestrial, and monsoon generation, 77
Vema, 355
Ventilation
definition, 475
Viral disease, mariculture see Mariculture
Viscous shear stress, 179, 180F
Visible and infrared scanning radiometer (VISR), 268
Visible radiometry, 270
VISR (visible and infrared scanning radiometer), 268
Viviparity, 424
Volcanism
abrupt climate change and, 45
Cretaceous, ocean circulation and, 379–380
sea level rise and, 517
Voluntary observing ships (VOS), 97, 110, 124–125

von Karman constant, 179
Vorticity equation, conservation/ dissipation features in equations, 24
VOS see Voluntary observing ships (VOS)
VSF see Volume scattering function (VSF)

W

Walker circulation, 76
Wanninkhof, 460–461
Warm core rings (WCR), 299
Warming phase, upwelling and return flow, 29–30
Warm poleward flow, 27
'Warm Pool,' El Niño, satellite remote sensing of SST, 297–298
Water mass(es)
conversion, 31
individual water parcels, 27
interdecadal changes, 30F
intermediate-depth, 29
newly formed, 28
Water packets, chemical tracers and, 472–473
Water pressure
relationship to depth, 416
Water vapor
flux, measuring, direct method, 120–121
temperature deficit and, 291–292
variability requiring atmospheric correction algorithm, 292
Wave height
North Atlantic Oscillation and, 36–37
Wavelet analysis, 288
Weather
climate differences, 71
prediction
Southern Oscillation and, 71
Weather center data, model evaluation and, 198
Weddell-Enderby Basin, sea ice thickness, 248
Weddell Polynya, 240–241
Weddell Sea, 28–29
sea ice, 224F
cover, 241, 248F
thickness, model evaluation, 197F
see also Sea ice
sea ice models, 231, 232F
see also Weddell Sea circulation
West Antarctic Ice Sheet (WAIS), 207–208
disintegration, implications of, 512
see also Antarctic ice sheet
Westerly wind(s), 29, 30F, 32
Westerly wind bursts
satellite remote imaging, 117
Western Atlantic salinity, 28, 28F
see also Salinity
Western equatorial Pacific Ocean, thermocline, 71

Western Interior Seaway of North America (Cretaceous), paleoceanographic model, 367–368
Western Mediterranean Deep Water (WMDW), 13
West Greenland Current (WGC), 10
see also Atlantic Ocean current systems
West Spitsbergen Current (WSC), 156
sea ice and, 234
water column profiles, 157F
Wet-bulb temperature, definition, 121T
Whalers' Bay, 234
WHOI see Woods Hole Oceanographic Institution
Wind(s)
cause/consequence of SST change, 70, 73, 73–74
El Niño as consequence, 72–73
El Niño Southern Oscillation, SST and, 60
Wind-driven circulation, 7–10, 27, 32
density stratification, 8–9, 9
fresh water exchange, 9
surface heat transfer, 9, 9F
see also Surface, gravity and capillary waves
Wind-driven currents, 29, 30
Wind-driven gyres, 27
Wind-forced currents
currents driven by, 29, 30
Wind forcing
Antarctic Circumpolar Current, 148, 150
Wind speed
sea–air carbon dioxide flux, 465
Wind stress
satellite remote sensing, 110–113
accuracy, 111
empirical function, 110–111
Wind velocity
El Niño events and, 61F
sea ice velocity and, 227F
Winged snails (Pteropoda), 402
Winter (boreal), monsoon activity, Northern Hemisphere, 76, 76F
Winter Weddell Gyre Study (WWGS), 248
Winter Weddell Sea Project (WWSP), 240–241
WMO see World Meteorological Organization (WMO)
WOCE see World Ocean Circulation Experiment (WOCE)
Woods Hole Oceanographic Institution (WHOI), 266
Wordie Ice Shelf, 204–205, 204F, 205
World Meteorological Organization (WMO), 97
World Ocean Circulation Experiment (WOCE), 106F, 107, 130
Antarctic Circumpolar Current transport, 146–147
Hydrographic survey, 106F, 107, 130

World Ocean Circulation Experiment
    (WOCE) (*continued*)
    measurements provided by, 7,
        12–13
    radiocarbon, 482, 483–484,
        485*F*
World Weather Watch system, 97, 98
WWGS (Winter Weddell Gyre Study),
    248
WWSP (Winter Weddell Sea Project),
    240–241

X

XKT *see* Expendable optical irradiance
    probe (XKT)

Y

Yellow River (Huanghe)
    land-sea fluxes and, 543
Younger Dryas, 42–43, 390*F*, 391
    coral-based paleoclimate records, 444

Z

Zenopsis conchifera (silvery John Dory),
    550, 550*F*
Zonal average models, 471–472
Zooplankton
    importance in food web, 397
    timing of biomass peak, 400